Marine Proteins and Peptides

Special Issue Editor
Se-Kwon Kim

MDPI • Basel • Beijing • Wuhan • Barcelona • Belgrade

MDPI

Special Issue Editor
Se-Kwon Kim
Korea Maritime and Ocean University
South Korea

Editorial Office
MDPI AG
St. Alban-Anlage 66
Basel, Switzerland

This edition is a reprint of the Special Issue published online in the open access journal *Marine Drugs* (ISSN 1660-3397) from 2016–2017 (available at: http://www.mdpi.com/journal/marinedrugs/special_issues/marine_proteins_pepti des).

For citation purposes, cite each article independently as indicated on the article page online and as indicated below:

Author; Author. Article title. *Journal Name*. **Year**. Article number, page range.

First Edition 2018

ISBN 978-3-03842-646-2 (Pbk)
ISBN 978-3-03842-647-9 (PDF)

Table of Contents

About the Special Issue Editor

Se-Kwon Kim, Ph.D., is presently working as a Distinguished Professor in Korea Maritime and Ocean University and Research advisor of Kolmar Korea Company. He was worked as distinguished Professor at Department of Marine Bio Convergence Science and Technology and Director of Marine Bioprocess Research Center (MBPRC) at Pukyong National University, Busan, South Korea.

He received his M.Sc. and Ph.D. degrees from Pukyong National University and conducted his postdoctoral studies at the Laboratory of Biochemical Engineering, University of Illinois, Urbana-Champaign, Illinois, USA. Later, he became a visiting scientist at the Memorial University of Newfoundland and University of British Colombia in Canada.

Dr. Kim served as president of the 'Korean Society of Chitin and Chitosan' in 1986-1990, and the 'Korean Society of Marine Biotechnology' in 2006-2007. To the credit for his research, he won the best paper award from the American Oil Chemists' Society In 2002. Dr. Kim was also the chairman for '7th Asia-pacific Chitin and Chitosan Symposium', which was held in South Korea in 2006. He was the chief-editor in the 'Korean Society of Fisheries and Aquatic Science' during 2008-2009. In addition, he is the board member of International Society of Marine Biotechnology Associations (IMBA) and International Society of Nutraceuticals and Functional Food (ISNFF).

His major research interests are investigation and development of bioactive substances from marine resources. His immense experience of marine bio-processing and mass-production technologies for marine bio-industry is the key asset of holding majorly funded Marine Bio projects in Korea. Furthermore, he expended his research fields up to the development of bioactive materials from marine organisms for their applications in oriental medicine, cosmeceuticals and nutraceuticals. To this date, he has authored around 850 research papers, 70 books, and 120 patents.

Preface to "Marine Proteins and Peptides"

In recent years, proteins and peptides from the marine resources have gained much attention in the field of pharmaceutical, cosmeceutical and nutraceuticals product development owing to the excellent biological properties. Proteins and peptides from marine sources are considered to be safe and inexpensive. Protein- and peptide-based drugs have been increasing in recent days to cure various diseases by serving multiple roles, such as antioxidants, anticancer drugs, antimicrobials, and anticoagulants. There are different marine sources (macroalgae, fish, shellfish, and bivalves), which possibly contain specific protein and peptides.

Totally, 27 articles were published in this special issue of "Marine Proteins and Peptides" including research and review articles which essentially explains about the antioxidant, antithrombic, neuroprotection, antimicrobial, antitumor, antifatigue, anticancer, angiotensin-I-converting enzymes and calcium binding capacity activities. Excellent reviews have been given on production of enzyme assisted discovery of marine antioxidative peptides from marine vertebrates, marine antifreeze proteins, alkynyl-containing peptides and marine skeletal matrix proteins.

In the applications part of this special issue, marine proteins and peptides and their usage in the field of cosmeceutical applications, treatment of type-2 diabetes, non-communicable diseases, as well as preclinical and clinical studies on antioxidative, antihypertensive and cardio protective are presented.

To compiling this special issue as a book, we planned to bring the latest technology to produce bioactive protein and peptides from marine organisms and their detailed mechanisms in terms of biological activity which lead to produce the several commercial products. Strong understanding the protein structure and their mechanisms are the ultimate goal to produce highly valuable, scientific and industrial applicable products. This book cover the recent technology on production and applications in terms of marine protein and peptides.

Se-Kwon Kim
Special Issue Editor

marine drugs

MDPI

Article

Identification of Angiotensin I-Converting Enzyme Inhibitory Peptides Derived from Enzymatic Hydrolysates of Razor Clam *Sinonovacula constricta*

Yun Li [1,*]**, Faizan A. Sadiq** [2]**, Li Fu** [1]**, Hui Zhu** [1]**, Minghua Zhong** [3] **and Muhammad Sohail** [4]

[1] School of Life Sciences and Food Technology, Hanshan Normal University, Chaozhou 521041, China; fl1990@163.com (L.F.); gdzhuhui@126.com (H.Z.)
[2] College of Biosystems Engineering and Food Science, Zhejiang University, Hangzhou 310058, China; faizan_nri@yahoo.co.uk
[3] School of Chemistry and Environmental Engineering, Hanshan Normal University, Chaozhou 521041, China; zhongmh@hstc.edu.cn
[4] National Institute of Food Science & Technology, University of Agriculture, Faisalabad 38040, Pakistan; Sohail.nifsat@gmail.com
* Correspondence: fgtmyself@163.com; Tel.: +86-768-231-7422

Academic Editor: Se-Kwon Kim
Received: 11 April 2016; Accepted: 30 May 2016; Published: 3 June 2016

Abstract: Angiotensin I-converting enzyme (ACE) inhibitory activity of razor clam hydrolysates produced using five proteases, namely, pepsin, trypsin, alcalase, flavourzyme and proteases from *Actinomucor elegans* T3 was investigated. Flavourzyme hydrolysate showed the highest level of degree of hydrolysis (DH) (45.87%) followed by *A. elegans* T3 proteases hydrolysate (37.84%) and alcalase (30.55%). The *A. elegans* T3 proteases was observed to be more effective in generating small peptides with ACE-inhibitory activity. The 3 kDa membrane permeate of *A. elegans* T3 proteases hydrolysate showed the highest ACE-inhibitory activity with an IC_{50} of 0.79 mg/mL. After chromatographic separation by Sephadex G-15 gel filtration and reverse phase-high performance liquid chromatography, the potent fraction was subjected to MALDI/TOF-TOF MS/MS for identification. A novel ACE-inhibitory peptide (VQY) was identified exhibiting an IC_{50} of 9.8 µM. The inhibitory kinetics investigation by Lineweaver-Burk plots demonstrated that the peptide acts as a competitive ACE inhibitor. The razor clam hydrolysate obtained by *A. elegans* T3 proteases could serve as a source of functional peptides with ACE-inhibitory activity for physiological benefits.

Keywords: ACE-inhibitory peptides; razor clam; enzymatic hydrolysis; *Actinomucor elegans* proteases; identification; MALDI/TOF-TOF MS/MS

1. Introduction

Hypertension is one of the major global health issues, owing to its chronic nature, wide prevalence and linkage with increased mortality and morbidity which affects approximately 16%–37% of the global population [1]. Long term hypertension is one of the major risk factors and clinical manifestations of arteriosclerosis, cardiovascular diseases, strokes, heart failures, and chronic renal diseases [2,3]. Angiotensin-converting enzyme (ACE, EC 3.4.15.1) is a key enzyme of renin-angiotensin system (RAS) which is known as a cascade that controls the regulation of arterial blood pressure and cardiac output. Angiotensin I is a ten-amino acid peptide produced by the action of rennin on angiotensinogen. Once angiotensin I is formed, it is converted to angiotensin II through the removal of two *C*-terminal residues (His-Leu) by the action of ACE, thus resulting in vasoconstriction, ultimately leading to the increase in blood pressure [4]. In addition, ACE is also known to catalyze the degradation of the vasodilator bradykinin into inactive fragments, which leads to the decrease in vasodilation [5]. Thus, the inhibition of ACE is considered as an effective strategy in designing pharmaceutical drugs for the treatment of hypertension. Synthetic drugs targeting inhibition of ACE are normally used for the clinical treatment of hypertension such as captopril, enalapril, and alcacepril. However, therapies with these drugs are believed to cause side effects including dry cough, renal failure, skin rashes, and angioneurotic edema [6]. So, there is a dire need to find natural ACE inhibitors with lower or no side effect in order to development pharmaceuticals and nutraceuticals for the prevention and remedy of hypertension.

Food protein-derived bioactive peptides are naturally physiologically active peptide fragments encrypted within the sequence of food proteins, and can be released through enzymatic hydrolysis and microbial fermentation. Besides providing adequate nutrients, food protein-derived bioactive peptides possess beneficial pharmacological properties such as antihypertensive, antioxidant, antiproliferative, and immunomodulatory activities [7]. There is great interest among researchers to unreveal food based bioactive peptides which are encrypted within food proteins, with a view to develop functional foods and nutraceuticals. Compared with chemosynthetic drugs, bioactive peptides of food origin are usually considered safe, effective and economical and thus these are healthier and more natural alternative to synthetic drugs [8]. Since the discovery of first ACE-inhibitory peptides from snake venome [9], many ACE-inhibitory peptides have been reported from the protein hydrolysates of foods [10].

Marine fishes, due to phenomenal biodiversity of their habitat and broad spectra of bioactivities, are relatively untapped and rich sources of proteins of high biological value as compared to land animals [5]. Thus, fish and sea food are excellent sources of proteins and can be utilized as an ideal starting material for the production of novel ACE-inhibitory peptides. Enzymatic hydrolysis is a widely used method to release ACE-inhibitory peptides from marine fish proteins. The effectiveness of using this method to generate specific peptide fragments with inhibitory activity mainly depends on the proteolytic enzyme used, hydrolysis conditions and the degree of hydrolysis (DH) achieved. A variety of enzymes including commercial proteases and proteases of microbial origin have been reported for the production of ACE-inhibitory peptides from various marine fish proteins. In particular, a number of novel ACE-inhibitory peptides with good activity have been reported from the enzymatic hydrolysate of shellfish such as oyster [11,12], shrimp [13], hard clam [14] and cuttlefish muscle [15].

Razor clam (*Sinonovacula constricta*) is one of the four major economically cultivated shellfish in China, which has been cultured for hundreds of years [16]. Due to its high nutritional and economical values, razor clam is a popular shellfish food and has been widely cultivated along east coast of China. According to 2015 Fisheries Statistical Yearbook of China (2015), the cultured razor clam yield was more than 786,000 tons in 2014. To date, there is no study aiming to investigate the potential of razor clam to generate ACE-inhibitory peptides which could be exploited as antihypertensive agents in functional foods and nutraceuticals. Therefore, the objectives of this work are two folds: first, to evaluate the ACE-inhibitory activity of the hydrolysates produced with different proteases. Secondly, to purify and identify the potential ACE-inhibitory peptides from the hydrolysate. Furthermore, the inhibitory kinetics of the identified peptide based on Lineweaver-Burk plots were also studied.

2. Results and Discussion

2.1. Production of Enzymatic Hydrolysates

2.1.1. Proximate Composition of Razor Clam

The results of the proximate composition of razor clam are shown in Table 1. The average values for moisture, protein, fat, carbohydrate and ash are 80.32, 13.68, 1.89, 2.13 and 1.93 g/100 g (fresh weight), respectively. On a dry weight basis, protein was the predominant proximate composition, occupying 69.51% of the dry weight. The protein content of razor clam determined in the present study was higher than reported values for protein (9.09–12.75 g/100 g fresh weight) in Asian hard clam (*Meretrix lusoria*) [17], Veneridae clams (9.00–12.51 g/100 g fresh weight) [18] and surf clam (*Mactra violacea*) (11.9 g/100 g fresh weight) [19]. The value of fat content was consistent with previously reported values for fat content in surf clam (1 g/100 g fresh weight) and Veneridae clams (1.32–2.4 g/100 g fresh weight). Similarly, the reported carbohydrate content in the current study is in the range of carbohydrate value that was previously reported in Veneridae clams (1.72–3.61 g/100 g fresh weight). However, a comparatively higher value for fat content has previously been reported for Asian hard clam (1.58–6.58 g/100 g fresh weight). The results of proximate analysis indicate that razor clam is a rich source of nutrients, particularly protein content, and can be used to produce bioactive peptides.

Table 1. Proximate composition of razor clam.

Composition	Contents (g/100 g Fresh Weight)
Moisture	80.32 ± 0.53
Protein	13.68 ± 0.62
Fat	1.89 ± 0.13
Carbohydrate	2.13 ± 0.31
Ash	1.93 ± 0.08

2.1.2. Degree of Hydrolysis and ACE-Inhibitory Activity of Hydrolysates by Different Proteases

Enzymatic hydrolysis was performed using pepsin, trypsin, alcalase, flavourzyme and crude proteases from *A. elegans* T3. Hydrolysis efficiency was evaluated by measuring degree of hydrolysis (DH) in the hydrolysates that had been generated by using five different proteases (Figure 1a). Overall, the hydrolysis of the razor clam proteins was characterized by a high rate of hydrolysis during the initial 1–2 h; 1 h for pepsin and trypsin hydrolysis, and within 2 h for alcalase, flavourzyme and crude proteases from *A. elegans* T3. The rapid increase in DH indicates that a large amount of peptides were cleaved from proteins and released into hydrolysates at the initial stage. After that, the hydrolysis entered into stationary phase where no apparent increase in DH was observed (Figure 1a). These results represent similar hydrolysis curves that are previously reported for the protein hydrolysates of sardinelle (*Sardinella aurita*) by-products [20], sole and squid [21], yellow stripe trevally (*Selaroides leptolepis*) [22] and catfish (*Pangasius sutchi*) [23]. The rate of enzymatic cleavage of peptide bonds is an important factor determining the rate of DH [24]. During the initial phase of the reaction kinetics, the reaction speed is very fast and thus peptide bonds are easily cleaved resulting in a large number of soluble peptides in the reaction mixture. These peptides also act as effective substrate competitors to undigested or partially digested compact proteins in substrate [25]. Decreased hydrolysis reaction rate during the stationary phase can also be attributed to the limited availability of the substrate, as it is known that the substrate decreases by the reaction time. Also, decrease in enzymatic activity or partial enzymatic inactivation by the time is an important reason of slower degree of hydrolysis during the later stages of the reaction [26].

Among the proteases investigated, hydrolysis with flavourzyme showed higher level of DH during the whole process, reaching a maximum level of 45.87% after 3 h, followed by *A. elegans* T3 proteases (37.84%) and alcalase (30.55%), whereas the lower DH values were observed with pepsin (18.72%) and trypsin (15.67%). The efficiency of proteases in catalyzing the hydrolysis depends on the nature of the substrate proteins and the specificity of proteases towards these proteins. Lower DH value obtained upon tryptic hydrolysis is probably due to trypsin's specificity, as it is known that trypsin preferentially catalyzes polypeptides on the carboxyl side of basic amino acids (arginine or lysine). In case of pepsin, the enzyme exhibits preferential cleavage for hydrophobic residues, preferably cleaves aromatic residues. However, pepsin is unable to hydrolyse the proline peptide bond efficiently [27]. This may cause resistance to hydrolysis when using pepsin to digest protein substrate containing high content of proline. Similar inefficiency of pepsin has previously been reported when the lowest DH was observed in the pepsin hydrolysate among all the proteases used for barley hordein proteolysis [28].

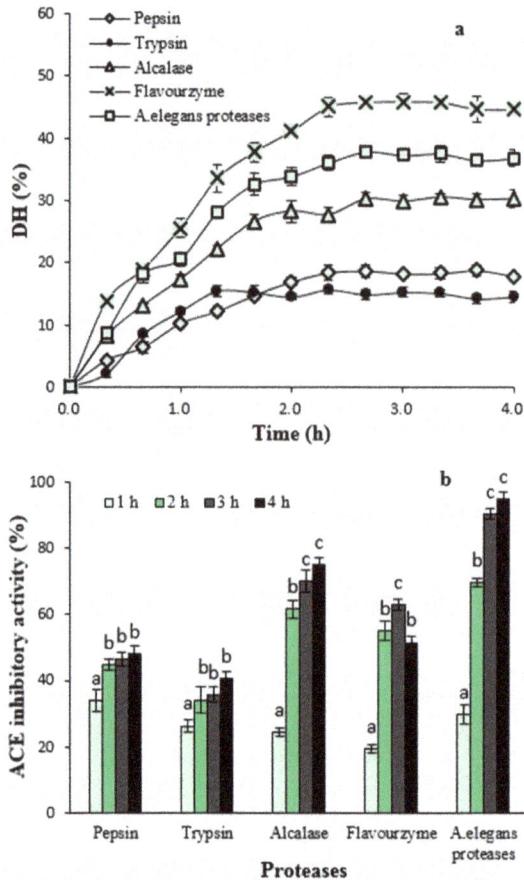

Figure 1. Degree of hydrolysis with proteases during hydrolysis (**a**) and effect of hydrolysis time on angiotensin I-converting enzyme (ACE)-inhibitory activity of hydrolysates (**b**). Different letters indicate significant differences in the same group ($p < 0.05$).

To investigate the effect of hydrolysis time on ACE-inhibitory activity, samples were taken from the hydrolysates at different time intervals and subjected to ACE-inhibitory activity assay at a concentration of 2 mg peptide/mL (Figure 1b). Among all hydrolysates, the ACE-inhibitory activity increased with increasing hydrolysis time except for flavourzyme-generated hydrolysates. The highest ACE inhibition at a level of 94.79% was observed for the hydrolysates of *A. elegans* T3 proteases after 4 h of hydrolysis. In particular, ACE-inhibitory activity significantly increased during the first stage of hydrolysis which depicts a fast increase in DH at the beginning and its positive influence on the generation of ACE-inhibitory peptides ($p < 0.05$). DH was defined as the percent ration between the fraction of peptide bonds cleaved to the total number of peptide bonds [29], and it has been widely used to evaluate hydrolytic progress. The positive correlation between DH value and ACE-inhibitory activity has been reported in studies on the proteolysis of canola meal [30], cuttlefish muscle [15], palm kernel cake [31] and bovine collagen [32] proteins. It has been suggested that reaching a certain level of DH was contributive to release more active peptides from protein precursors [30]. In the present study, the results of hydrolysis using pepsin, trypsin, alcalase and *A. elegans* T3 proteases were in agreement with these studies. The hydrolysate as a result of *A. elegans* T3 proteases, having higher

DH values, showed better ACE-inhibitory activity as well. However, the similar observation was not found in the case of treatment with flavourzyme hydrolysate, which, despite having the highest DH value, showed lower inhibitory activity. Flavourzyme is a complex of protease and peptidases having endoprotease as well as exopeptidase activities. It has been applied to prepare short chain peptides [28] and lower bitter taste of hydrolysates [33]. Action of peptidases can promote the production of peptides of small molecular weight. On the other hand, using this enzyme may also cause degradation of active peptides into shorter inactive peptides or amino acids. Similar inefficiency of using flavourzyme in the production of ACE-inhibitory peptides was reported for red scorpion fish proteins [34].

2.1.3. Peptide Content and ACE-Inhibitory Activity of Ultra-Filtration Fractions

After 4 h of hydrolysis, the hydrolysates obtained with different proteases were further separated by ultra-filtration into three molecular weight fractions, <3 kDa, 3–10 kDa and >10 kDa. The peptide contents and the molecular weight distributions are shown in Figure 2a. The peptide contents of *A. elegans* T3 proteases hydrolysate was significantly higher than that of other hydrolysates ($p < 0.05$), indicating that more peptides were released from protein precursors. Furthermore, *A. elegans* proteases hydrolysate contained larger proportion of the peptides with size below 3 kDa (45.0%) as compared to the other hydrolysates. These results suggest that *A. elegans* T3 proteases is more effective in generating peptides of low molecular weight from razor clam proteins. For flavourzyme hydrolysis, the higher DH did not lead to the higher content of peptides. This can be explained by the fact that flavourzyme contain exopeptidases which release more free amino acids. So the DH value for this enzyme hydrolysate correlates with the content of free amino acids and not with the content of peptides.

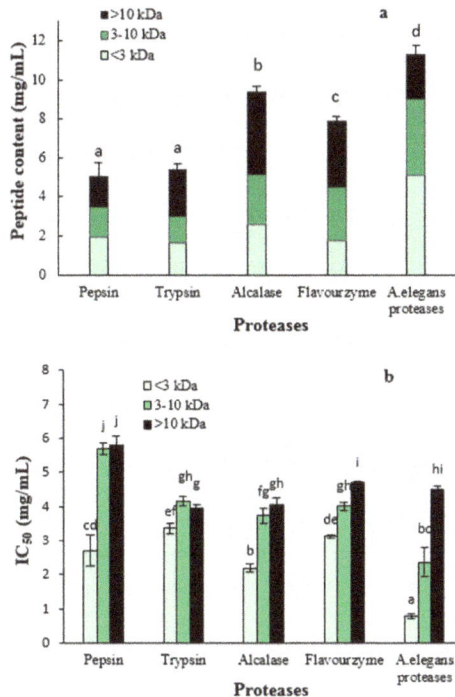

Figure 2. Peptide content (**a**) and IC$_{50}$ value (**b**) of fractions from hydrolysates separated by ultra-filtration. Different letters indicate the mean values are significantly different ($p < 0.05$).

The ACE-inhibitory activity was found to be significantly dependent on peptide fraction molecular weight (Figure 2b). The <3 kDa peptide fraction showed significantly higher ACE-inhibitory activity than those of higher molecular weight fractions (3–10 kDa and >10 kDa) for each protease hydrolysate ($p < 0.05$). Specifically, the 3–10 kDa fractions from flavourzyme and *A. elegans* proteases hydrolysates had significantly ($p < 0.05$) higher ACE-inhibitory properties in comparison with >10 kDa fractions. Pepsin, trypsin and alcalase hydrolysates, on the contrary, showed no significant difference in the activity of fractions (3–10 kDa and >10 kDa). The highest ACE-inhibitory activity (lowest IC_{50} value) was found in the <3 kDa fraction of *A. elegans* T3 proteases hydrolysate, with an IC_{50} value of 0.79 mg/mL. Molecular weight is an important determinant for the ACE-inhibitory activity of peptides. It was reported that food protein derived ACE-inhibitory peptides are in the molecular weight range of below 3 kDa [35]. The weak inhibitory activity of high MW peptides are primarily due to the inability of the ACE-catalytic site to bind large molecules [36]. Therefore, based on these result, the <3 kDa fraction of *A. elegans* T3 proteases hydrolysate was used for further purification and identification of active peptides.

2.2. Identification of ACE-Inhibitory Peptides

2.2.1. Isolation and Purification of ACE-Inhibitory Peptides

The <3 kDa fraction of *A. elegans* T3 proteases hydrolysate was separated by Sephadex G-15 gel filtration chromatography into five major absorbance peaks at 220 nm (Figure 3). Fractions (G1–G5) associated with the peaks were pooled and lyophilized for ACE-inhibitory activity assay. The fraction G5 exhibited the highest ACE-inhibitory activity among the collected fractions, with IC_{50} value of 0.17 mg/mL. Therefore, the fraction G5 was subjected to RP-HPLC for further purification. Eight peaks (F1–F8) were obtained separately according to the chromatogram (60 min) (Figure 4a). The highest inhibitory activity was observed in fraction F7, with an IC_{50} value of 29.3 µg/mL. Fraction F7 was further purified by the second step of RP-HPLC and fractionated into six major sub-fractions (F7.1–F7.6, Figure 4b). Most of the ACE-inhibitory activity occurred in fraction F7.5, which inhibited 96.2% of the ACE activity at the concentration of 30 µg/mL, whereas the inhibitory activities of the other sub-fractions were below 35%. Thereafter, fraction F7.5 was selected to identify its sequence by MALDI/TOF-TOF MS/MS.

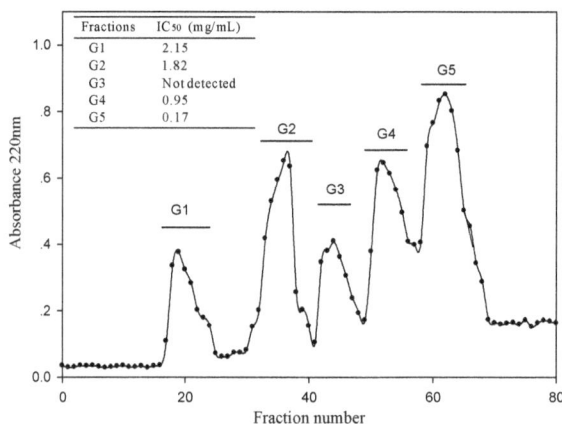

Figure 3. Gel filtration chromatography profile of <3 kDa fraction of *A. elegans* T3 proteases hydrolysate on Sephadex G-15 column.

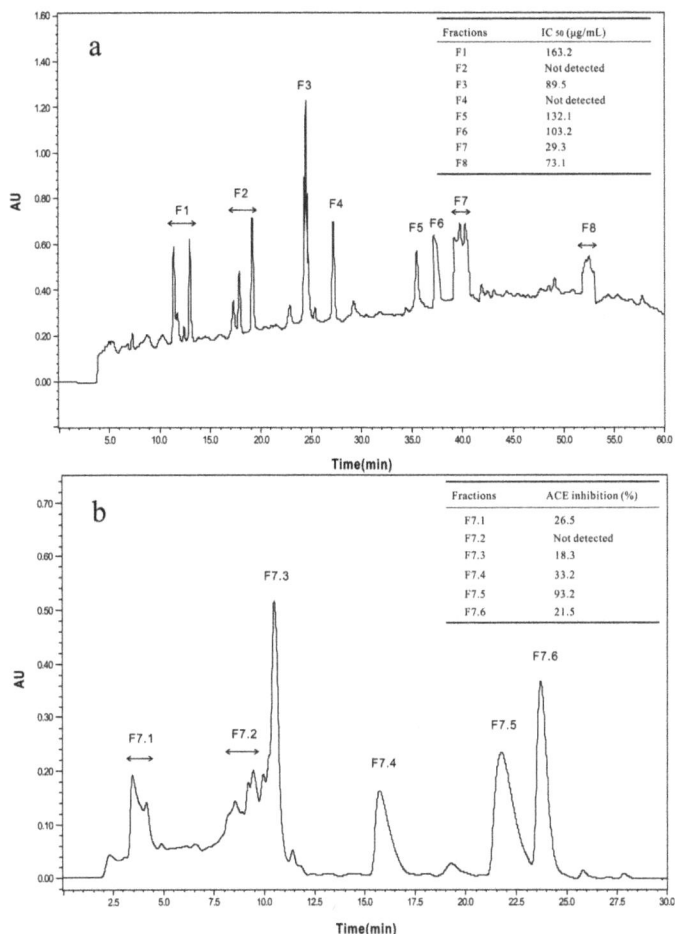

Figure 4. Chromatograms of RP-HPLC for the two-step method used to purify and assay the ACE-inhibitory peptides. (**a**) First step of RP-HPLC for fraction G5 from the Sephadex G-15 gel filtration; (**b**) Second step of RP-HPLC for fraction F7, the ACE-inhibitory activities of factions (F7.1–F7.6) were determined at a concentration of 30 µg/mL.

2.2.2. Determination of Amino Acids Sequence

The mass spectrum of fraction F7.5 revealed one most intensive signal, indicating a single positively charged ion ([M + H]$^+$) at 409.2 (Figure 5a). Several other signals with moderate intensity were seen on the spectrum. Tandem mass spectra confirmed that they are not peptides. The molecular mass of fraction F7.5 was determined to be 408.2 Da, and ion at m/z 409.2 was selected as precursor ion for TOF-TOF tandem MS analysis. The amino acid sequence was obtained by de novo sequencing using software from the MS/MS spectrum (Figure 5b). Also, the masses of the singly charged ions were matched to the single peptide fragment by manual validation. Therefore, the amino sequence of fraction F7.5 was identified as Val-Gln-Tyr.

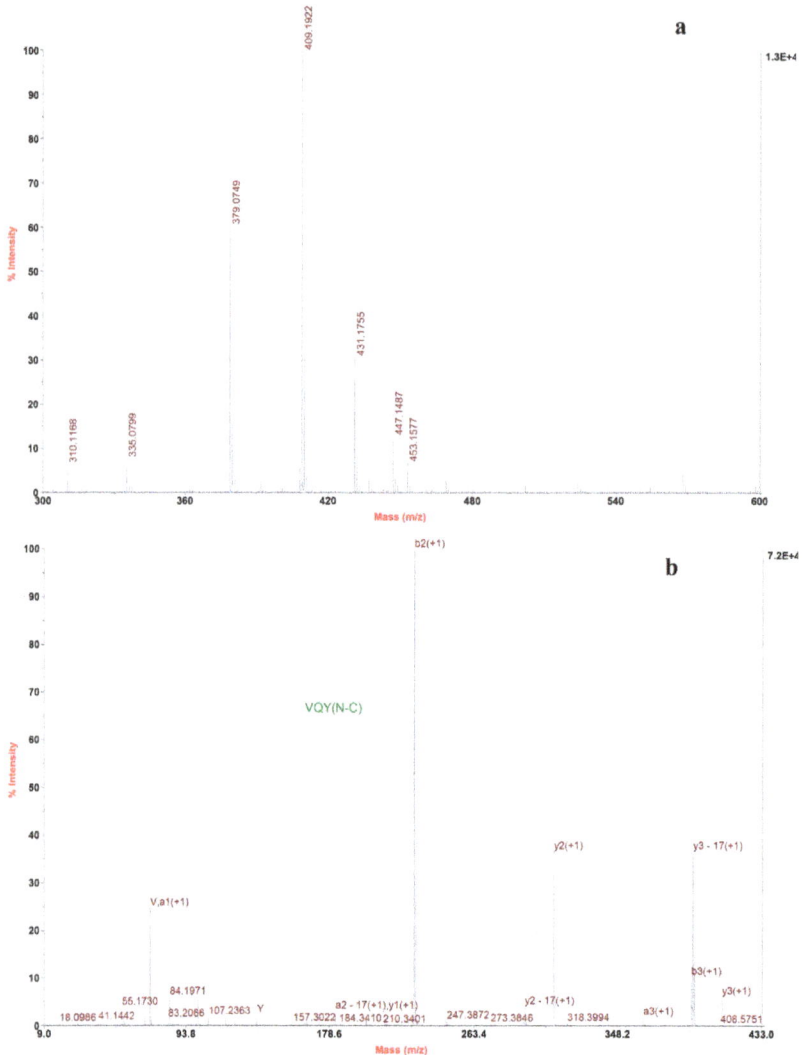

Figure 5. *De novo* sequencing of purified ACE-inhibitory peptide from RP-HPLC. (**a**) MALDI/TOF-TOF MS spectrum of the purified peptide; (**b**) MALDI/TOF-TOF MS/MS spectrum of the ion 409.2 *m/z*.

2.2.3. IC$_{50}$ Value and Inhibition Pattern of Val-Gln-Tyr

To determine the IC$_{50}$ value and ACE inhibition pattern, Val-Gln-Tyr (VQY) was chemically synthesized with a purity of greater than 98% by solid-phase technique (Chinese peptide Co., Ltd., Hangzhou, China). The IC$_{50}$ value of VQY was estimated by non-linear regression by fitting the results of ACE-inhibitory activity (assayed at different concentrations of inhibitor, 0.25–100 µM) to a four-parameter logistic equation (Figure 6). The nonlinear regression coefficient of the equation ($R = 0.977$) demonstrates that the actual value of the experimental data corresponds well with the value predicted by the equation. The IC$_{50}$ value of VQY was determined as 9.8 µM by solving the equation. Many potent ACE-inhibitory peptides have been isolated and identified from various food proteins. Among them, IPP and VPP are well characterized ACE-inhibitory peptides from fermented milk with IC_{50} values of 5 µM and 9 µM, respectively. The IC_{50} value of VQY reported in this study is close to these two peptides and another peptide VLP isolated from freshwater clam (*Corbicula fluminea*) with an IC_{50} value of 3.7 µM [37]. However, the IC_{50} value of VQY peptide reported in this study is much lower than YN peptide (51 µM) isolated from the hard clam *Meretrix lusoria* [14]. To the best of our knowledge, this peptide (VQY) is a novel peptide derived from razor clam proteins exhibiting a strong ACE-inhibitory activity. Structure-activity correlation among ACE-inhibitory peptides shows that their activity is strongly influenced by amino acid residues of peptide sequence [38,39]. Many studies have shown that potential ACE-inhibitory peptides exhibit hydrophobic amino acid residues (tryptophan, phenylalanine, tyrosine, or proline) at their C-terminus while contain branched aliphatic amino acid residues (Val, Ile, Leu) at the N-terminus [40,41]. The peptide VQY is in accordance with this rule, containing valine at the N-terminal and tyrosine at the C-terminal. Lineweaver-Burk plots of VQY for ACE inhibition showed three lines, representing ACE reaction performed in the absence and presence of the peptide. The lines intersected at one point on the vertical axis, which indicates a competitive inhibition pattern (Figure 7). This result suggests that the peptide (VQY) acts as a competitive inhibitor and razor clam hydrolysate is a potential candidate of antihypertensive nutraceuticals.

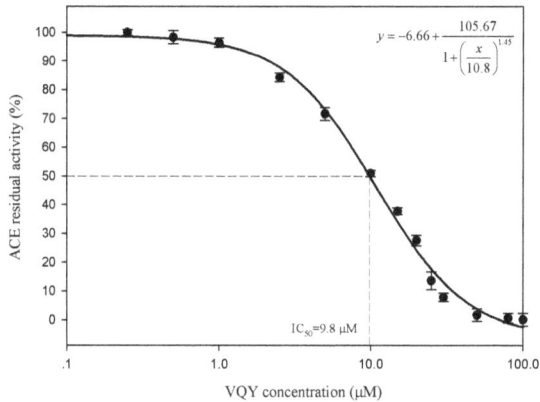

Figure 6. Determination of IC_{50} value of VQY.

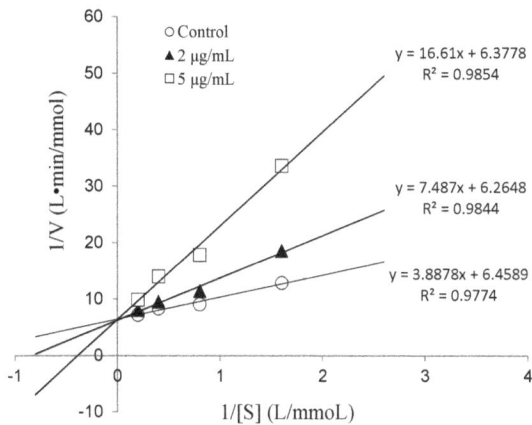

Figure 7. Lineweaver-Burk plots of VQY inhibition on ACE.

3. Materials and Methods

3.1. Materials

Samples of razor clams (*Sinonovacula constricta*) were obtained from local market. *Actinomucor elegans* T3 with strong proteolytic activity was isolated from a traditional fermented soybean product. ACE (EC 3.4.15.1, from rabbit lung), Hippurl-1-histidyl-l-leucine (HHL), Pepsin (P6887) and Trypsin (T1426) were purchased from Sigma-Aldrich (St. Louis, MO, USA). Alcalase 2.4 L and Flavourzyme 500 MG were purchased from Novozyme (Bagasvaerd, Denmark). All other chemicals were also of analytical grade.

3.2. Preparation of Crude Proteases from Actinomucor elegans T3

Production of crude enzyme from *Actinomucor elegans* T3 was obtained according to the following method. *A. elegans* T3 was grown on Potato Dextrose Agar (PDA) at 28 °C for 72 h. Firstly, the inoculum was prepared by transferring three round blocks (6 mm in diameter), cut from the plate culture, into 100 mL PDB (Potato Dextrose Broth). The culture was allowed to grow at 28 °C for 2 days on a shaking incubator at 150 rpm. Twenty milliliters of the inoculum was then transferred into 500 mL flasks containing 180 mL of medium for proteases production. The composition of the medium was as given (L^{-1}): 15 g glucose, 10 g soy protein isolate, 2.5 g yeast extract, 2 g KH_2PO_4, 2 g $MgSO_4$ with a final pH of 6.0. After inoculation, the medium containing the culture was incubated at 28 °C on a shaking incubator upheld at 150 rpm for 60 h. The supernatants were collected by centrifugation ($10,000 \times g$, 15 min) at 4 °C and then passed through 0.45 µm filters. The filtrates were lyophilized and used as crude proteases. The lyophilized filtrates were stored at −20 °C until use. One unit of proteases was defined as the amount of enzyme required to bring an increase of 0.01 OD units at 280 nm per minute at assay conditions and measured as 0.4 M Trichloroacetic acid (TCA) soluble products using hemoglobin as substrate.

3.3. Enzymatic Hydrolysis

Meat of razor clams was stripped from the shell completely and washed carefully with distilled water to remove sand. Clean tissues were homogenized with distilled water (two times the volume of the tissues). The homogenate was heated at 85 °C for 10 min to inactivate endogenous proteases and then lyophilized. The resulting razor clam powder was kept at −20 °C until hydrolysis. Proximate composition of razor clam was determined according to the method of the Association of Official Analytical Chemists [42].

For hydrolysis with each protease, twenty grams of razor clam powder was mixed with 200 mL of distilled water in a blender for 2 min. Protease was added to the mixture at the enzyme/substrate ratio of 3000 U/g. The hydrolysis reactions were conducted under optimal conditions of different proteases (Table 2). During the hydrolysis, the pH value was kept at the optimal level by adding 1 M HCL or 1 M NaOH. The reaction was stopped by heating the mixture at 90 °C for 10 min followed by centrifugation at $8000 \times g$ for 20 min at 4 °C. Samples from the supernatants were subjected to peptide content assay. The other collected supernatants were ultra-filtrated sequentially through 3 and 10 kDa molecular weight cutoff membranes (MWCO) (Millipore). The supernatants were first passed through the membranes with MWCO of 10 kDa. The retentate from 10 kDa membrane was collected and designated as >10 kDa fraction. The permeate solution collected from 10 kDa membrane was then filtered through the membrane with MWCO of 3 kDa. Retentate and permeate samples collected from 3 kDa membrane were designated as 3–10 kDa and <3 kDa fractions, respectively. All these collected fractions were then lyophilized and stored at −20 °C until further analysis.

Table 2. Hydrolysis conditions of proteases.

Protease	Source	Temperature (°C)	pH
Pepsin	porcine gastric mucosa	37	2.0
Trypsin	bovine pancreas	37	8.0
Alcalase	*Bacillus licheniformis*	40	8.0
Flavourzyme	*Aspergillus oryzae*	50	6.0
Crude proteases	*Actinomucor elegans*	55	6.0

3.4. Analytical Methods

3.4.1. Angiotensin-Converting Enzyme Inhibition Assay

The ACE-inhibitory activity was measured by HPLC according to the method described by Cushman and Cheung [43] using HHL as a substrate. The total volume of ACE reaction system was 100 uL consisting of the following components: 50 µL substrate solution (5 mM HHL in 50 mM HEPES with 300 mM NaCl, pH 8.3), 40 µL test sample and 10 µL ACE (0.1 U/mL). The substrate solution and sample were mixed and incubated at 37 °C for 5 min in a water bath. Then ACE was added and incubated at 37 °C for 30 min. The reaction was terminated by adding 250 µL of 1 M HCl. Hippuric acid (HA) released from ACE reaction was measured by RP-HPLC (Agilent Inc., Santa Clara, CA, USA) equipped with C18 column (4.6 × 150 mm, 5 µm, Thermo Scientific, Waltham, MA, USA) and absorbance detector set at 228 nm. The HHL and HA were eluted using a gradient of 21% (v/v) acetonitrile containing 0.5% (v/v) trifluoroacetic acid at a flow rate of 1 mL/min. The inhibitory activity was calculated using the following formula:

$$I\,(\%) = \frac{A - B}{A} \times 100 \tag{1}$$

where I is the percentage of ACE inhibition by sample, A is the concentration of HA of blank test by using distilled water instead of sample and B is the concentration of HA with sample added. The IC_{50} value was defined as the concentration of peptide inhibiting 50% of the ACE activity under the assayed conditions, which was estimated by non-linear regression by fitting data to a four-parameter logistic curve using SigmaPlot software (version 10.0, SPSS Inc., Chicago, IL, USA).

3.4.2. Degree of Hydrolysis Evaluation

Degree of hydrolysis (DH) was estimated by measuring the content of α-amino groups released by hydrolysis according to the *o*-phthaldialdehyde (OPA) method [44]. The content of α-amino groups was expressed as the concentration of serine corresponding to standard curve. The DH was calculated using the following equation.

$$DH\,(\%) = \frac{B - A}{C - A} \times 100 \tag{2}$$

A is the content of α-amino group at the beginning of protease hydrolysis, and B is the content of α-amino group in the supernatant after hydrolysis. C is the content of α-amino group from the razor clam powder hydrolyzed with 6 M HCl (containing 1% (v/v) phenol) at 110 °C for 12 h in tubes sealed under nitrogen.

3.4.3. Determination of Peptide Content

The peptide content was determined by the Folin phenol method [45] using synthetic peptide Tyr-Gly-Gly-Phe-Leu-Arg-Lys-Tyr (with molecular weight of 1003.17 g/mol, Chinese peptide Co. Ltd., Hangzhou, China) as standard.

3.5. Purification and Identification of ACE-Inhibitory Peptides

3.5.1. Gel Filtration Chromatography

The lyophilized powder of ultra-filtration permeate was dissolved in distilled water at a concentration of 100 mg/mL. Two milliliter of the solution was loaded onto a Sephadex G-15 column (1.8 × 60 cm) eluted with distilled water at a flow rate of 0.5 mL/min. Fractions were collected at 5 min intervals and the absorbance was measured at 220 nm. The active fractions were pooled and lyophilized for further purification.

3.5.2. Reversed-Phase High-Performance Liquid Chromatography

The selected fraction obtained from gel filtration was re-dissolved in ultrapure water at a concentration of 10 mg/mL. Five hundred microliters was injected into Waters 600 HPLC system (semi-preparative RP-HPLC, Waters, Milford, MA, USA) equipped with Kromasil C18 column (10 × 250 mm, 10 μm). Solvent A was 0.1% (*v/v*) trifluoroacetic acid (TFA) in ultrapure water and solvent B was 0.1% (*v/v*) TFA in 80% (*v/v*) acetonitrile. The elution was 100% solvent A for 5 min, followed by a linear gradient from 0% to 55% of solvent B in 60 min at a flow rate of 2 mL/min. The absorbance of eluent was detected with a UV detector at 220 nm. Fractions were collected separately through repeated chromatography using RP-HPLC and concentrated for ACE-inhibitory activity assay. The fraction with the highest inhibitory activity was lyophilized and dissolved at 5 mg/mL concentration for the second step RP-HPLC separation under the similar conditions. Two hundred microliters of the samples was injected and further separated at a flow rate of 1 mL/min with a linear gradient elution of 25%–40% solvent B for 30 min. The peak with the most of the inhibitory activity was collected and lyophilized.

3.5.3. Identification of the Amino Acid Sequence by MALDI/TOF-TOF MS/MS

The amino acid sequence of the purified peptide was identified by MALDI–TOF–MS/MS. Peptide sample (0.5 μL) was mixed with 0.5 μL of a saturated solution of α-cyano-4-hydroxycinnamic acid in 50% (*v/v*) acetonitrile containing 0.1% (*v/v*) TFA. The mixture was spotted on the target plate and analyzed in ABI 5700 MALDI-TOF/TOF MS/MS (AB Sciex, Framingham, MA, USA) in positive reflector mode with a mass range from 300 to 1000 *m/z*. The amino acid sequence of peptide fragments was determined by de novo sequencing using the software DeNovo Explorer (version4.5, AB Sciex, Framingham, MA, USA) and confirmed by manual validation.

3.5.4. Determination of ACE-Inhibition Pattern

The inhibition kinetics of the peptide on ACE was investigated using HHL as a substrate. Lineweaver-Burk plot was used to determine the type of inhibition of the peptide. The ACE reactions were carried out at various substrate concentrations (0.625, 1.25, 2.5 and 5 mM) in the absence and presence of two different concentrations of the peptide (2 and 5 μg/mL). Linear interpolation was plotted with the reciprocal of HHL concentration ($1/[S]$) as the independent variable and with the reciprocal of HA production ($1/[V]$) as the dependent variable [46].

3.6. Statistical Analysis

The results were expressed as mean ±SD (standard deviation). The statistics analysis was carried out using SPSS 20.0 (version 20, SPSS Inc., Chicago, IL, USA). Differences among treatments were determined by one way ANOVA. The *p* value less than 0.05 was considered as statistically significant.

4. Conclusions

The present study revealed that enzyme hydrolysates of razor clam have good potential for the production of ACE-inhibitory peptides. Among the proteases tested in this trial, *A. elegans* T3 proteases was found to be the most efficient in producing small peptides with the best ACE-inhibitory activity.

A novel potent ACE-inhibitory peptide, VQY, with the IC_{50} value of 9.8 µM, was purified from the hydrolysate by a series of chromatographic separations and identified by MALDI/TOF-TOF MS/MS. Lineweaver-Burk plots revealed that the peptide exhibits strong competitive inhibition activity against ACE. This is the first report of ACE-inhibitory peptides derived from enzymatic hydrolysates of razor clam. It is highly recommended that the ACE-inhibitory peptides from razor clam hydrolysates be employed in the development of nutraceuticals and pharmaceuticals for the treatment of hypertension.

Acknowledgments: This work was supported by the Technology Support Project of Hanshan Normal University, Chaozhou, China (QD20150910).

Author Contributions: Yun Li developed the design and ideas of this work; Yun Li, Hui Zhu, Li Fu and Minghua Zhong carried out the experiments; Li Fu performed the data analysis; Faizan A. Sadiq and Yun Li interpreted the results and contributed towards the manuscript preparation and writing; Muhammad Sohail reviewed the manuscript and provided useful suggestion to improve the manuscript.

Conflicts of Interest: The authors declare no conflict of interest.

References

1. Poulter, N.R.; Prabhakaran, D.; Caulfield, M. Hypertension. *Lancet* **2015**, *386*, 730–731. [CrossRef]
2. Lackland, D.T.; Weber, M.A. Global burden of cardiovascular disease and stroke: Hypertension at the core. *Can. J. Cardiol.* **2015**, *31*, 569–571. [CrossRef] [PubMed]
3. George, B.; Pantelis, S.; Rajiv, A.; Luis, R. Review of blood pressure control rates and outcomes. *J. Am. Soc. Hypertens.* **2014**, *8*, 127–141.
4. FerrãO, F.M.; Lara, L.S.; Lowe, J. Renin-angiotensin system in the kidney: What is new? *World J. Nephrol.* **2014**, *3*, 64–76. [CrossRef] [PubMed]
5. Cheung, R.C.; Ng, T.B.; Wong, J.H. Marine peptides: Bioactivities and applications. *Mar. Drugs* **2015**, *13*, 4006–4043. [CrossRef] [PubMed]
6. Wu, R.; Wu, C.; Liu, D.; Yang, X.; Huang, J.; Zhang, J.; Liao, B.; He, H.; Li, H. Overview of antioxidant peptides derived from marine resources: The sources, characteristic, purification, and evaluation methods. *Appl. Biochem. Biotechnol.* **2015**, *176*, 1815–1833. [CrossRef] [PubMed]
7. Fan, X.; Bai, L.; Zhu, L.; Yang, L.; Zhang, X. Marine algae-derived bioactive peptides for human nutrition and health. *J. Agric. Food Chem.* **2014**, *62*, 9211–9222. [CrossRef] [PubMed]
8. Bartneck, M.; Heffels, K.H.; Pan, Y.; Bovi, M.; Zwadlo-Klarwasser, G.; Groll, J. Inducing healing-like human primary macrophage phenotypes by 3D hydrogel coated nanofibres. *Biomaterials* **2012**, *33*, 4136–4146. [CrossRef] [PubMed]
9. Ferreira, S.H.; Bartelt, D.C.; Greene, L.J. Isolation of bradykinin-potentiating peptides from *Bothrops jararaca* venom. *Biochemistry* **1970**, *9*, 2583–2593. [CrossRef] [PubMed]
10. Saadi, S.; Saari, N.; Anwar, F.; Abdul Hamid, A.; Ghazali, H.M. Recent advances in food biopeptides: Production, biological functionalities and therapeutic applications. *Biotechnol. Adv.* **2015**, *33*, 80–116. [CrossRef] [PubMed]
11. Shiozaki, K.; Shiozaki, M.; Masuda, J.; Yamauchi, A.; Ohwada, S.; Nakano, T.; Yamaguchi, T.; Saito, T.; Muramoto, K.; Sato, M. Identification of oyster-derived hypotensive peptide acting as angiotensin-I-converting enzyme inhibitor. *Fish. Sci.* **2010**, *76*, 865–872. [CrossRef]
12. Wang, J.; Hu, J.; Cui, J.; Bai, X.; Du, Y.; Miyaguchi, Y.; Lin, B. Purification and identification of a ACE inhibitory peptide from oyster proteins hydrolysate and the antihypertensive effect of hydrolysate in spontaneously hypertensive rats. *Food Chem.* **2008**, *111*, 302–308. [CrossRef] [PubMed]
13. He, H.L.; Chen, X.L.; Sun, C.Y.; Zhang, Y.Z.; Zhou, B.C. Analysis of novel angiotensin-I-converting enzyme inhibitory peptides from protease-hydrolyzed marine shrimp *Acetes chinensis*. *J. Pept. Sci.* **2006**, *12*, 726–733.
14. Tsai, J.S.; Pan, C.B.S. ACE-inhibitory peptides identified from the muscle protein hydrolysate of hard clam (*Meretrix lusoria*). *Process Biochem.* **2008**, *43*, 743–747. [CrossRef]
15. Balti, R.; Nedjar-Arroume, N.; Adje, E.Y.; Guillochon, D.; Nasri, M. Analysis of novel angiotensin I-converting enzyme inhibitory peptides from enzymatic hydrolysates of cuttlefish (*Sepia officinalis*) muscle proteins. *J. Agric. Food Chem.* **2010**, *58*, 3840–3846. [CrossRef] [PubMed]

16. Feng, B.; Dong, L.; Niu, D.; Meng, S.; Zhang, B.; Liu, D.; Hu, S.; Li, J. Identification of immune genes of the *Agamaki clam* (*Sinonovacula constricta*) by sequencing and bioinformatic analysis of ests. *Mar. Biotechnol.* **2010**, *12*, 282–291. [CrossRef] [PubMed]

17. Karnjanapratum, S.; Benjakul, S.; Kishimura, H.; Tsai, Y.H. Chemical compositions and nutritional value of Asian hard clam (*Meretrix lusoria*) from the coast of Andaman Sea. *Food Chem.* **2013**, *141*, 4138–4145. [CrossRef] [PubMed]

18. Yoon, H.; An, Y.K.; Choi, S.D.; Kim, J. Proximate composition in the muscle and viscera of five Veneridae clams (bivalvia) from southern coast of Korea. *Korean J. Matacol.* **2008**, *24*, 67–72.

19. Laxmilatha, P. Proximate composition of the surf clam *Mactra violacea* (Gmelin 1791). *Indian J. Fish.* **2009**, *56*, 147–150.

20. Bougatef, A.; Nedjar-Arroume, N.; Ravallec-Plé, R.; Leroy, Y.; Guillochon, D.; Barkia, A.; Nasri, M. Angiotensin I-converting enzyme (ACE) inhibitory activities of sardinelle (*Sardinella aurita*) by-products protein hydrolysates obtained by treatment with microbial and visceral fish serine proteases. *Food Chem.* **2008**, *111*, 350–356. [CrossRef] [PubMed]

21. Giménez, B.; Alemán, A.; Montero, P.; Gómez-Guillén, M. Antioxidant and functional properties of gelatin hydrolysates obtained from skin of sole and squid. *Food Chem.* **2009**, *114*, 976–983. [CrossRef]

22. Klompong, V.; Benjakul, S.; Kantachote, D.; Shahidi, F. Antioxidative activity and functional properties of protein hydrolysate of yellow stripe trevally (*Selaroides leptolepis*) as influenced by the degree of hydrolysis and enzyme type. *Food Chem.* **2007**, *102*, 1317–1327. [CrossRef]

23. Mahmoodani, F.; Ghassem, M.; Babji, A.S.; Yusop, S.M.; Khosrokhavar, R. ACE inhibitory activity of pangasius catfish (*Pangasius sutchi*) skin and bone gelatin hydrolysate. *J. Food Sci. Technol.* **2014**, *51*, 1847–1856. [CrossRef] [PubMed]

24. Benjakul, S.; Morrissey, M.T. Protein hydrolysates from pacific whiting solid wastes. *J. Agric. Food Chem.* **1997**, *45*, 3423–3430. [CrossRef]

25. Nguyen, H.T.M.; Sylla, K.S.B.; Randriamahatody, Z.; Donnay-Moreno, C.; Moreau, J.; Tran, L.T.; Bergé, J.P. Enzymatic hydrolysis of yellowfin tuna (*Thunnus albacares*) by-products using protamex protease. *Food Technol. Biotechnol.* **2011**, *49*, 48–55.

26. Guerard, F.; Guimas, L.; Binet, A. Production of tuna waste hydrolysates by a commercial neutral protease preparation. *J. Mol. Catal. B Enzym.* **2002**, *19*, 489–498. [CrossRef]

27. Hausch, F.; Shan, L.; Santiago, N.A.; Gray, G.M.; Khosla, C. Intestinal digestive resistance of immunodominant gliadin peptides. *Am. J. Physiol. Gastrointest. Liver Physiol.* **2002**, *283*, 996–1003. [CrossRef] [PubMed]

28. Bamdad, F.; Wu, J.P.; Chen, L.Y. Effects of enzymatic hydrolysis on molecular structure and antioxidant activity of barley hordein. *J. Cereal Sci.* **2011**, *54*, 20–28. [CrossRef]

29. Adler-Nissen, J. Limited enzymic degradation of proteins: A new approach in the industrial application of hydrolases. *J. Chem. Technol. Biotechnol.* **1982**, *32*, 138–156. [CrossRef]

30. Wu, J.P.; Aluko, R.E.; Muir, A.D. Production of angiotensin I-converting enzyme inhibitory peptides from defatted canola meal. *Bioresour. Technol.* **2009**, *100*, 5283–5287. [CrossRef] [PubMed]

31. Zarei, M.; Forghani, B.; Ebrahimpour, A.; Abdul-Hamid, A.; Anwar, F.; Saari, N. *In vitro* and *in vivo* antihypertensive activity of palm kernel cake protein hydrolysates: Sequencing and characterization of potent bioactive peptides. *Ind. Crop. Prod.* **2015**, *76*, 112–120. [CrossRef]

32. Zhang, Y.; Olsen, K.; Grossi, A.; Otte, J. Effect of pretreatment on enzymatic hydrolysis of bovine collagen and formation of ACE-inhibitory peptides. *Food Chem.* **2013**, *141*, 2343–2354. [CrossRef] [PubMed]

33. Cheung, I.W.Y.; Li-Chan, E.C.Y. Application of taste sensing system for characterisation of enzymatic hydrolysates from shrimp processing by-products. *Food Chem.* **2014**, *145*, 1076–1085. [CrossRef] [PubMed]

34. Aissaoui, N.; Abidi, F.; Marzouki, M.N. ACE inhibitory and antioxidant activities of red scorpionfish (*Scorpaena notata*) protein hydrolysates. *J. Food Sci. Technol. Mysore* **2015**, *52*, 7092–7102. [CrossRef]

35. Hernandez-Ledesma, B.; Contreras, M.D.; Recio, I. Antihypertensive peptides: Production, bioavailability and incorporation into foods. *Adv. Colloid Interface Sci.* **2011**, *165*, 23–35. [CrossRef] [PubMed]

36. Ramanathan, N.; Schwager, S.L.U.; Sturrock, E.D.; Acharya, K.R. Crystal structure of the human angiotensin-converting enzyme-lisinopril complex. *Nature* **2003**, *421*, 551–554.

37. Tsai, J.S.; Lin, T.C.; Chen, J.L.; Pan, B.S. The inhibitory effects of freshwater clam (*Corbicula fluminea*, Muller) muscle protein hydrolysates on angiotensin I converting enzyme. *Process Biochem.* **2006**, *41*, 2276–2281. [CrossRef]

38. Cristina, M.; Maria, D.M.Y.; Justo, P.; Hassan, L.; Julio, G.C.; Manuel, A.; Francisco, M.; Javier, V. Purification of an ACE inhibitory peptide after hydrolysis of sunflower (*Helianthus annuus* L.) protein isolates. *J. Agric. Food Chem.* **2004**, *52*, 1928–1932.

39. Wu, J.; Ding, X. Hypotensive and physiological effect of angiotensin converting enzyme inhibitory peptides derived from soy protein on spontaneously hypertensive rats. *J. Agric. Food Chem.* **2001**, *49*, 501–506. [CrossRef] [PubMed]

40. Wu, J.; Aluko, R.E.; Shuryo, N. Structural requirements of angiotensin I-converting enzyme inhibitory peptides: Quantitative structure-activity relationship study of di- and tripeptides. *J. Agric. Food Chem.* **2006**, *54*, 732–738. [CrossRef] [PubMed]

41. Rohrbach, M.S.; Williams, E.B.; Rolstad, R.A. Purification and substrate specificity of bovine angiotensin-converting enzyme. *J. Biol. Chem.* **1981**, *256*, 225–230. [PubMed]

42. Helrich, K. *Official Methods of Analysis of the AOAC*, 14th ed.; Association of Official Analytical Chemists: Washington, DC, USA, 1990.

43. Cushman, D.W.; Cheung, H.S. Spectrophotometric assay and properties of the angiotensin I-converting enzyme of rabbit lung. *Biochem. Pharmacol.* **1971**, *20*, 1637–1648. [CrossRef]

44. Nielsen, P.M.; Petersen, D.; Dambmann, C. Improved method for determining food protein degree of hydrolysis. *J. Food Sci.* **2001**, *66*, 642–646. [CrossRef]

45. Lowry, O.H.; Rosebrough, N.J.; Farr, A.L.; Randall, R.J. Protein measurement with the Folin phenol reagent. *J. Biol. Chem.* **1951**, *193*, 265–275. [PubMed]

46. Rao, S.Q.; Ju, T.; Sun, J.; Su, Y.J.; Xu, R.R.; Yang, Y.J. Purification and characterization of angiotensin I-converting enzyme inhibitory peptides from enzymatic hydrolysate of hen egg white lysozyme. *Food Res. Int.* **2012**, *46*, 127–134. [CrossRef]

marine drugs

MDPI

Article

Coral Carbonic Anhydrases: Regulation by Ocean Acidification

Didier Zoccola [1,2], Alessio Innocenti [3], Anthony Bertucci [1,†], Eric Tambutté [1,2], Claudiu T. Supuran [3,*] and Sylvie Tambutté [1,2,*]

[1] Marine Biology Department, Centre Scientifique de Monaco, 8 Quai Antoine 1°, 98 000 Monaco, Monaco; zoccola@centrescientifique.mc (D.Z.); anthony.bertucci@u-bordeaux.fr (A.B.); etambutté@centrescientifique.mc (E.T.)

[2] Laboratoire International Associé 647 BIOSENSIB, Centre Scientifique de Monaco-Centre National de la Recherche Scientifique, 8 Quai Antoine 1°, 98 000 Monaco, Monaco

[3] Neurofarba Department, University of Florence, Via Ugo Schiff 6, Polo Scientifico, Sesto Fiorentino, 50019 Firenze, Italy; alessio.innocenti@unifi.it

* Correspondence: claudiu.supuran@unifi.it (C.T.S.); stambutte@centrescientifique.mc (S.T.); Tel.: +39-055-4573729 (C.T.S); +377-97-77-44-70 (S.T.)

† Present address: University of Bordeaux, UMR EPOC CNRS 5805, 33400 Talence, France

Academic Editor: Se-Kwon Kim
Received: 30 March 2016; Accepted: 30 May 2016; Published: 3 June 2016

Abstract: Global change is a major threat to the oceans, as it implies temperature increase and acidification. Ocean acidification (OA) involving decreasing pH and changes in seawater carbonate chemistry challenges the capacity of corals to form their skeletons. Despite the large number of studies that have investigated how rates of calcification respond to ocean acidification scenarios, comparatively few studies tackle how ocean acidification impacts the physiological mechanisms that drive calcification itself. The aim of our paper was to determine how the carbonic anhydrases, which play a major role in calcification, are potentially regulated by ocean acidification. For this we measured the effect of pH on enzyme activity of two carbonic anhydrase isoforms that have been previously characterized in the scleractinian coral *Stylophora pistillata*. In addition we looked at gene expression of these enzymes *in vivo*. For both isoforms, our results show (1) a change in gene expression under OA (2) an effect of OA and temperature on carbonic anhydrase activity. We suggest that temperature increase could counterbalance the effect of OA on enzyme activity. Finally we point out that caution must, thus, be taken when interpreting transcriptomic data on carbonic anhydrases in ocean acidification and temperature stress experiments, as the effect of these stressors on the physiological function of CA will depend both on gene expression and enzyme activity.

Keywords: coral; calcification; ocean acidification; carbonic anhydrase; gene expression; enzyme activity; temperature; pH

1. Introduction

Anthropogenic greenhouse gas emissions have increased since the pre-industrial era, which has led to an increase in atmospheric concentrations of carbon dioxide (CO_2), methane, and nitrous oxide. Their effects are extremely likely to have been the dominant cause of the observed warming since the mid-20th century (IPCC, 2014) [1]. In addition to atmospheric and oceanic warming, the subsequent uptake of additional CO_2 by the oceans causes ocean acidification (OA), which results in pH decrease and changes in seawater carbonate chemistry. Earth system models project a global increase in ocean acidification for all representative concentration pathway (RCP) scenarios by the end of the 21st century, with a slow recovery after mid-century under RCP2.6 (IPCC, 2014) [1]. The decrease in surface ocean pH is in the range of 0.06 to 0.07 (15% to 17% increase in acidity) for RCP2.6, 0.14 to 0.15 (38% to 41%) for RCP4.5, 0.20 to 0.21 (58% to 62%) for RCP6.0, and 0.30 to 0.32 (100% to 109%) for RCP8.5 (IPCC, 2014) [1]. Ocean acidification by decreasing pH and changing carbonate chemistry challenges marine organisms, especially those that form calcareous shells and skeletons, such as scleractinian corals, the major contributors to the structural foundation of coral-reef ecosystems. Meta-analysis of data obtained from laboratory and field-based studies indicate declines in coral calcification of 15%–22% at levels of OA predicted to occur under a business-as-usual scenario of CO_2 emissions by the end of the century [2] (note that this scenario predicts a pCO_2 of 800 ppm by the end of the century which corresponds to the prediction of scenario RCP6.0 in the report of IPCC 2014). Despite the high number of studies that have investigated how rates of calcification are affected by ocean acidification scenarios, comparatively few studies tackle how ocean acidification impacts the physiological mechanisms that drive calcification itself. Carbonic anhydrases (CAs, EC 4.2.1.1) play a major role in the physiology of coral calcification [3]. These enzymes catalyze the interconversion of CO_2 to bicarbonate ions and protons according to the following reaction: $CO_2 + H_2O \leftrightarrow HCO_3^- + H^+$. Even if the reaction of CO_2 hydration/HCO_3^- dehydration occurs spontaneously at reasonable rates in the absence of catalysts, their presence can speed up the reaction up to 10^7 times (hydration reaction occurs at a rate of 0.15 s^{-1} in water, whereas the rate for the most active human CA, hCAII is about 1.4×10^6 s^{-1}). In corals, several CAs have been identified at the molecular level in different coral species and the phylogenetic tree reveals three main clusters, the cytosolic and mitochondrial proteins, the membrane-bound or secreted proteins, and the carbonic anhydrase-related proteins [3]. These enzymes play major roles in two essential processes of coral physiology: they are involved in carbon supply for calcification as well as in carbon concentrating mechanisms for symbiont photosynthesis. However, the full molecular sequence together with the tissular localization have only been obtained for two isoforms of the coral *Stylophora pistillata* [3,4]. Both of these isoforms, STPCA and STPCA2, have been localized in the coral-calcifying cells, named calicoblastic cells. These cells also transport ions (calcium and bicarbonate) [5–7], regulate pH at the site of calcification [8], and synthesize organic matrix molecules which are then incorporated in the skeleton [9]. In the process of calcification two roles have been attributed to the two CA isoforms in *S. pistillata*: (1) STPCA catalyzes the interconversion between the different inorganic forms of dissolved inorganic carbon at the site of calcification [3,9,10]; (2) STPCA2 is an intracellular enzyme which is then found as an organic matrix protein incorporated in the skeleton [11–13]. As is the case for other enzymes, carbonic anhydrases are sensitive to environmental conditions and the pH dependency of the activity of bovine CA is well described [14,15]. Contrarily to mammals, to our knowledge, there are no data in corals concerning the dependency of the activity of carbonic anhydrase isoforms as a function of pH. The aim of our paper was, thus, to determine how the carbonic anhydrases characterized in corals are regulated by ocean acidification. For this we measured, *in vitro*, the kinetic constant (kcat) and the catalytic efficiency (kcat/Km where Km is the Michaelis-Menten constant) which both reflect the enzyme activity of STPCA and STPCA2 under a range of pH from 6 up to 9.5. In addition, we looked at gene expression of these enzymes *in vivo*, in corals maintained under conditions of CO_2-driven seawater acidification from pH 8 down to values of pH 7.2. This range of seawater pH has proved informative in several previous investigations that sought to identify clear patterns of physiological responses in corals under seawater acidification.

2. Results and Discussion

2.1. pH Dependency of Coral Carbonic Anhydrases

The pH dependency of CAs is primarily due to the protonation state of Zn-bound water at the active site. The curve describing the pH dependency of mammalian CAs activity for hydration of CO_2 is typically sigmoidal with a plateau obtained for alkaline values, a decrease in enzyme activity with decreasing pH, and a plateau for the most acidic values [14,15]. As can be seen on Figure 1, the catalytic efficiency (kcat/Km) of human CAII (hCAII) and two coral CAs, STPCA and STPCA2, shows a sigmoidal curve with a similar IC_{50} around 7.9. The linear part of the curve of enzyme activity *vs.* pH is obtained in the same range of pH for the three enzymes (between 7 and 8.9). STPCA is the membrane bound/secreted isoform localized in the calcifying cells and this enzyme is supposed to play a key role by modifying the kinetics of CO_2/HCO_3^- hydration reactions at the site of calcification [3,10,16]. The linear part of the curve for STPCA fits within the physiological range of this enzyme as pH at the site of calcification varies during diurnal cycles [17,18]. For STPCA2, which has been localized in the cytosol of calcifying cells, the linear part of the curve fits within the physiological pH value which remains almost constant at a pH of 7.4 during the diurnal cycle [18].

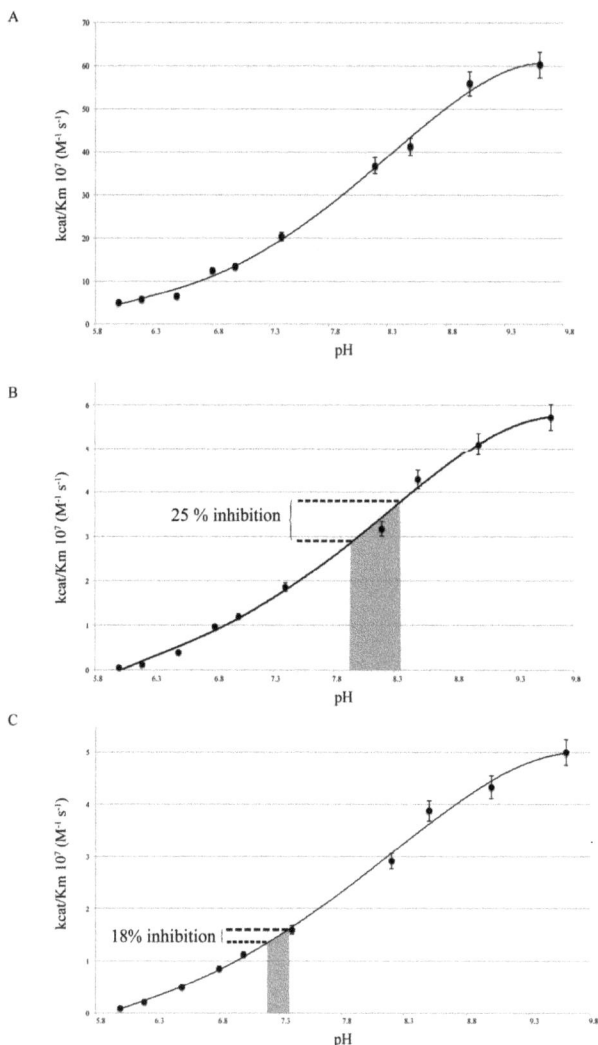

Figure 1. Catalytic efficiency (rate constant kcat/Km) for CO_2 hydration of carbonic anhydrase isoforms as a function of pH for (**A**) human CA II (hCAII); (**B**) coral STPCA; and (**C**) coral STPCA2. Decrease in catalytic efficiency for the coral CAs due to an increase in acidification between pH 8.36 to 7.93 for STPCA and between pH 7.38 to 7.19 for STPCA2 is highlighted in grey.

2.2. pH Dependency of Coral Carbonic Anhydrases: Effect of Ocean Acidification

Mechanistic studies on the response of corals to ocean acidification rely on physiological [8,17,18] and transcriptomic data [19–24]. It has been shown that the expression of several proteins changes under ocean acidification, some of them being upregulated, whereas others are downregulated. Moya *et al.* [21] have observed that in coral larvae, the expression of an *Acropora millepora* membrane/bound CA orthologous to STPCA is decreased under short-term exposure to moderate acidification (pH 7.96 and 7.86). Vidal-Dupiol *et al.* [24] observed that genes coding for CAs (with significant similarities with proteins that were previously shown to be involved in *Stylophora pistillata* calcification) were upregulated at moderate pH values of 7.8 and 7.4, but downregulated at the extreme level of pH 7.2 for the adult coral *P. damicornis* during a three-week exposure. Rocker *et al.* [23] showed that there was no change in genes coding for CAs for the adult coral *A. millepora* after 14 days of exposure to a pH of 7.57 (these CAs are not orthologous neither to STPCA nor to STPCA2). Hoadley *et al.* [25] have reported that there is no effect on gene expression of extra- and intra-cellular CAs (respectively, orthologous to STPCA and STPCA2) for two adult corals *P. damicornis* and *A. millepora* after 24 days of exposure to pH 7.90 and 7.83. Such discrepancies in the results have been attributed to species differences and/or stage-specific responses and/or experimental conditions. In the present study we focused on the coral *Stylophora pistillata* for which many physiological and molecular data related to calcification are available [10]. We measured the expression of genes coding for two isoforms of carbonic anhydrases, STPCA and STPCA2 (Figure 2) after one-year exposure of adult colonies to a pH of 7.2. These samples were part of a larger experiment in which we measured calcification rates and other physiological parameters linked to calcification. We have shown that calcification decreases under acidification, whereas photosynthesis and symbiont density were not affected [17]. Our present results clearly show that the effect of OA on the expression of genes coding for CAs is different when considering STPCA or STPCA2 with 3.85-fold and only 1.64-fold under-expression, respectively.

The range of physiological values that enzymes face within the coral when external seawater pH decreases from 8.0 to 7.2 is different for these two enzymes. At the site of calcification, STPCA, the membrane bound/secreted isoform, faces a decrease of 0.43 pH units (from 8.36 to 7.93, [17]). Within this range of pH, the activity of STPCA (kcat/Km) decreases of 25% (Figure 1). In the cells, STPCA2 faces a change in pH of only 0.19 (from 7.38 to 7.19, [8]) while its activity (kcat/Km) decreases of 18% (Figure 1). Thus, under acidification there is, at the same time, both an under-expression of the two isoforms of CAs and an inhibition of their activity (see schematic representation Figure 3). The results that we obtained during this experiment show that calcification is affected (rates of calcification measured by the buoyant weight technique decreased by about 20% at pH 7.2 compared to pH 8) with more porous skeletons under acidification [17]. We have shown that the decrease in pH at the site of calcification and inside the cells, together with a decrease in organic matrix proteins content, can explain such a pattern [17]. The results of the present study clearly show that CAs are affected by acidification. This enzymatic response could, thus, be another parameter which explains that calcification is affected under acidification, as suggested in Venn *et al.* [8].

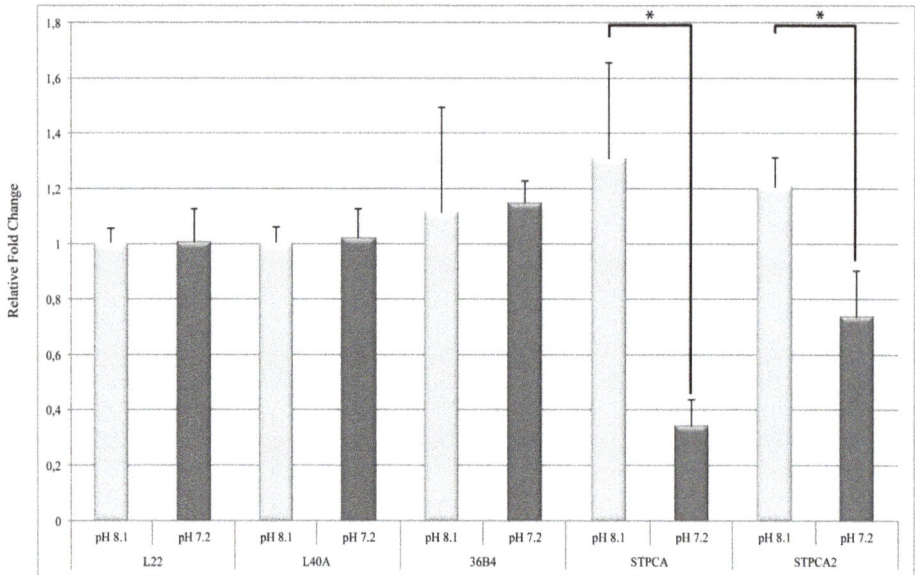

Figure 2. Relative gene expression of STPCA and STPCA2 by qPCR in *Stylophora pistillata*. Gene expression is relative to RPL22 expression, as well as RPL40A or RPLP0 (36B4) expression. Gene expression was measured in control sea water (pH 8.1 light grey) or after one-year exposure to a pH of 7.2 (dark grey). Errors bars represent standard error of the mean. * One-way ANOVA with $p < 0.05$.

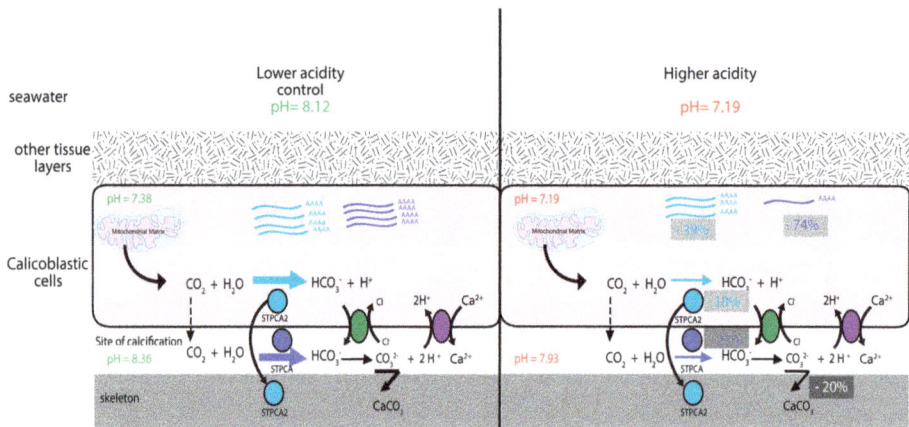

Figure 3. Schematic representation of the impact of ocean acidification on STPCA and STPCA2. Under seawater acidification, the intracellular pH decreases together with the pH at the site of calcification [8,16]. In the present study we have shown that under these conditions the expression of the transcripts coding for the intracellular CA isoform, STPCA2, and the membrane-bound/secreted isoform STPCA, decreases by, respectively, 39% and 74%, and their activity decreases, respectively, by 18% and 25%. This decrease of both gene expression and enzyme activity will affect the CO_2/HCO_3^- hydration and can explain that there will be less bicarbonate (and ultimately carbonate) available for the calcification process (calcification is decreased by 20% under these conditions).

2.3. pH and Temperature Dependency of Coral Carbonic Anhydrases

In this study we looked at ocean acidification, one of the side effects of the increase in atmospheric CO_2. Another one is global warming of the oceans [26]. As for pH, different scenarios of temperature increase have been proposed (IPCC, 2014) [1], depending on greenhouse gas emissions, with RCP2.6 being representative of a scenario that aims to keep global warming likely below 2 °C above pre-industrial temperatures. Since in the future ocean corals will face the combined effect of temperature increase and pH decrease, we have, thus, looked at the activity of STPCA and STPCA2 when these two stressors are combined. As can be seen on Figure 4, for a given pH, CA activity (kcat/Km) increases with increasing temperature which is usually observed for enzymes when they work in their physiological temperature range. However, what is noteworthy is that for a combined increase in temperature and decrease in pH, there is an opposite effect on CA activity (kcat/Km) suggesting that the effect of one of these stressors can counterbalance the effect of the other. For example, the catalytic constant (kcat) of STPCA at the site of calcification is similar at control pH and control temperature (25 °C and pH 8.36) as at increased acidification and increased temperature (28 °C and pH 7.93, Table 1) since the decrease in CA activity when pH decreases is counterbalanced by the increase in CA activity when temperature increases. The same effect is observed for STPCA2 where the catalytic constant is even slightly higher under acidification, combined with increased temperature than in control conditions (Table 1). There are only four studies that have looked at the combined effect of temperature increase and pH decrease on gene expression of CAs. Two carbonic anhydrase transcripts were down regulated in the coral *A. aspera* after a 14 day exposure at pH 7.9 and 35.2 °C (compared to control at pH 8.1 and 31 °C; [22]), two CAs transcripts were upregulated in the coral *A. millepora* after a 21 day exposure at pH 7.98 and 30.83 °C compared to control at pH 8.15 and 28.07 °C [23], six CA transcripts were downregulated in *A. millepora* after a five week exposure to pH of 7.85 and 7.68, with respective temperatures of 26 °C and 28 °C compared to control conditions at pH 8.02 and 24 °C [20]. Finally, another study, on *A. millepora* and *P. damicornis* CAs orthologous to *S. pistillata* STPCA and STPCA2, was performed during a 24-day exposure to pH 7.83, 7.9, and 8.07 (control) at two different temperatures (control 26.5 °C and 31.5°). It was observed that gene expression was only affected for the intracellular isoform of *A. millepora* under a temperature increase [25]. The different trends in gene expression in these four studies can be explained, for example, by a difference in the experimental protocols (different pH/temperature values, different time of exposure), or by a difference in the CA isoforms that were measured (however, molecular data on CAs are not available for all these studies). Regardless of the trend in gene expression, our results show that changes in CA activity with increasing temperature/decreasing pH can modulate the effect of the stressors on gene expression. Studies dealing only with the effect of temperature show that CA gene expression is downregulated when temperature increases [27–30], but in light of our results, we suggest that this could be, at least in part, counterbalanced by an increase in enzyme activity. However, it is not possible to determine quantitatively how respectively gene expression and enzymatic activity affect the physiological function of the enzyme.

Table 1. Catalytic activity (kcat) of coral carbonic anhydrase isoforms at different temperatures and pH. The values of kcat for a decrease in pH observed at the site of calcification and inside the calcifying cells (when seawater pH is decreased from control to 7.19) is highlighted in green boxes for STPCA and in red boxes for STPCA2.

	STPCA		STPCA2	
pH	25 °C	28 °C	25 °C	28 °C
8.36	3.943×10^6 s^{-1}	4.929×10^6 s^{-1}	3.200×10^6 s^{-1}	4.309×10^6 s^{-1}
7.93	2.965×10^6 s^{-1}	3.766×10^6 s^{-1}	2.410×10^6 s^{-1}1	3.286×10^6 s^{-1}
7.38	1.856×10^6 s^{-1}	2.401×10^6 s^{-1}	1.494×10^6 s^{-1}	2.098×10^6 s^{-1}
7.19	1.530×10^6 s^{-1}	1.981×10^6 s^{-1}	1.221×10^6 s^{-1}	1.737×10^6 s^{-1}

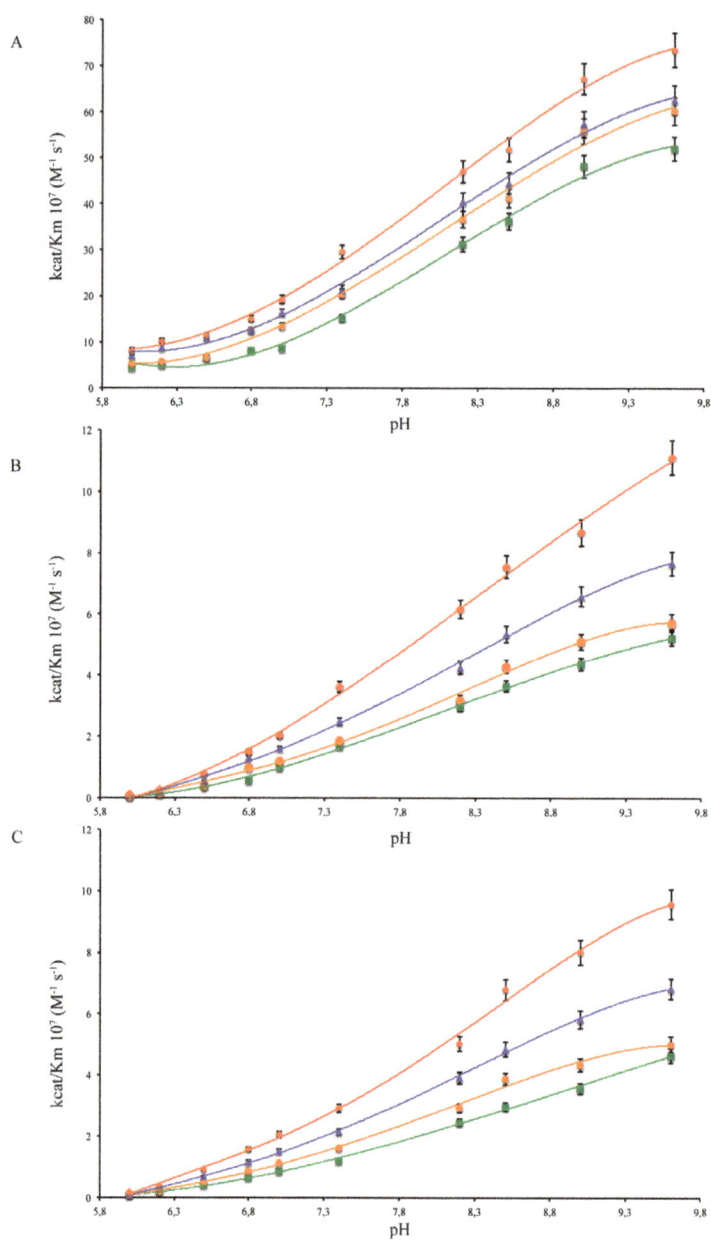

Figure 4. Catalytic efficiency (rate constant, kcat/Km) for CO_2 hydration activity of carbonic anhydrase isoforms as a function of pH at different temperatures (**A**) human CA II (hCAII) (**B**) coral STPCA, and (**C**) coral STPCA2. pH variation is measured at 23 °C (■ green), 25 °C (● orange), 28 °C (▲ blue), and 31 °C (◆ red).

3. Material and Methods

Biological material and treatments—Colonies of the tropical coral *Stylophora pistillata* were exposed to one-year seawater acidification as described previously [8,17]. Briefly corals were kept in aquaria supplied with Mediterranean seawater (exchange rate 70%/h) at a salinity of 38, temperature 25 C and irradiance of 170 µmol photons m^{-2}·s^{-1} on a 12 h/12 h photoperiod provided by HQI-10,000K metal halide lamps (BLV Nepturion, Steinhöring, Germany). Carbonate chemistry was manipulated by bubbling with CO_2 to reduce pH to the target values of pH 7.2. Control treatment was pH 8.1. Values of carbonate chemistry parameters are those measured in Tambutté *et al.* [17].

CA activity—An Applied Photophysics stopped-flow instrument has been used for assaying the CA-catalyzed CO_2 hydration activity [14]. Assay was performed on recombinant human and coral CAs (hCAII, STPCA, STPCA2, [4,31–33]). Phenol red (at a concentration of 0.2 mM) was used as indicator, working at the maximum absorbance of 557 nm, with 10 mM TRIS at ten different pH levels (6.0; 6.2; 6.5; 6.8; 7.0; 7.4; 8.2; 8.5; 9.0; 9.6), and 20 mM Na_2SO_4 or 20 mM NaCl (for maintaining constant the ionic strength), following the CA-catalyzed CO_2 hydration reaction for a period of 10–100 s. The CO_2 concentrations ranged from 1.7 to 17 mM for the determination of the kinetic parameters and inhibition constants. For each inhibitor at least six traces of the initial 5%–10% of the reaction have been used for determining the initial velocity. The uncatalyzed rates were determined in the same manner and subtracted from the total observed rates. Stock solutions of inhibitor (1 mM) were prepared in distilled-deionized water with 10%–20% (*v*/*v*) DMSO (which is not inhibitory at these concentrations) and dilutions up to 0.01 nM were done thereafter with distilled-deionized water. Inhibitor and enzyme solutions were preincubated together for 15 min at room temperature prior to assay, in order to allow for the formation of the E–I complex. The inhibition constants were obtained by non-linear least-squares methods using PRISM 3, from Lineweaver-Burk plots, as reported earlier, and represent the mean from at least three different determinations.

The temperature was controlled by an automatic thermostat, with a precision of ±0.2 °C. The solution of substrate and enzyme were thermostated at the required temperatures for 30 min before assay, and the same temperatures have been applied to the spectrophotometric cell where the reaction occurred.

Real-Time PCR experiments—Total RNAs extraction and cDNA synthesis were performed as described previously [34]. Briefly, cDNAs were synthesized using the Superscript®III kit (Invitrogen, Courtaboeuf, France). The experiment was repeated three times on clonal individuals. For each biological replicate, real-time PCR was then performed in technical triplicate with cDNAs diluted at a final concentration of 2 ng/µL and using the Express SYBR® greenER™ SuperMix with premixed ROX (Invitrogen, Courtaboeuf, France) in ABI 7300 Real-Time PCR System (Applied Biosystems, Courtaboeuf, France). Primers used (STPCA, STPCA2,) are from [35] and control gene 36B4 from [34]. We used two other control genes, ribosomal protein L22 (L22 Forward: 5′-TGATGTGTCCATTGATCGTC-3′ and L22 Reverse 5′-CATAGGTAGCTTGTGCAGATG-3′) and L40A genes (L40A Forward: 5′-CGACTGAGG GGAGGAGCCAA-3′ and L40A Reverse 5′-CTCATTTGGACACTCCCTT-3′). Relative expressions were calculated using Biogazelle qbase + 2.6™ (Gent, Belgium). Results are presented as mean ± SEM. Data were checked for normality using a Kolmogorov–Smirnov test with Lilliefors correction and log-transformed, if required. One-way ANOVA was used to test the effect of pH on STPCA and STPCA2. Differences were considered significant for *p*-values < 0.05. Statistics were performed using Statistica 10 (Statsoft, Tulsa, OK, USA).

4. Conclusions

Our results on the response of carbonic anhydrases to ocean acidification in the coral *Stylophora pistillata* show that these enzymes are affected by ocean acidification via an effect on both gene expression and enzyme activity. Our results also clearly show that temperature increase affects CA activity and we suggest that this could counterbalance the effect of acidification. Finally, we point out that caution must, thus, be taken when interpreting transcriptomic data on CAs in ocean acidification and temperature stress experiments as the effect of these stressors on the physiological function of CAs will depend both on gene expression and enzyme activity.

Acknowledgments: We thank Dominique Desgré for assistance with coral culture, Natacha Segonds and Nathalie Techer for assistance with maintenance of OA experiments. We thank Alexander Venn and Philippe Ganot for fruitful discussions. We thank two anonymous reviewers for their constructive comments. This work was funded by the Government of the Principality of Monaco.

Author Contributions: C.T.S., S.T. and D.Z. conceived and designed the experiments; A.B., A.I., E.T. and D.Z. performed the experiments; A.B., C.T.S., S.T. and D.Z. analyzed the data; A.B., C.T.S., S.T. and D.Z. wrote the paper.

Conflicts of Interest: The authors declare no conflict of interest.

Abbreviations

CA	Carbonic Anhydrase
STPCA	*Stylophora pistillata* carbonic anhydrase (ACA53457)
STPCA2	*Stylophora pistillata* carbonic anhydrase (ACE95141)
OA	Ocean Acidification

References

1. *IPCC, 2014: Climate Change 2014: Synthesis Report. Contribution of Working Groups i, ii and iii to the Fifth Assessment Report of the Intergovernmental Panel on Climate Change*; Intergovernmental Panel on Climate Change (IPCC): Geneva, Switzerland, 2014; p. 151.
2. Chan, N.C.S.; Connolly, S.R. Sensitivity of coral calcification to ocean acidification: A meta-analysis. *Glob. Chang. Biol.* **2013**, *19*, 282–290. [CrossRef] [PubMed]
3. Bertucci, A.; Moya, A.; Tambutté, S.; Allemand, D.; Supuran, C.T.; Zoccola, D. Carbonic anhydrases in anthozoan corals—A review. *Bioorg. Med. Chem.* **2013**, *21*, 1437–1450. [CrossRef] [PubMed]
4. Moya, A.; Tambutté, S.; Bertucci, A.; Tambutté, E.; Lotto, S.; Vullo, D.; Supuran, C.T.; Allemand, D.; Zoccola, D. Carbonic anhydrase in the scleractinian coral *Stylophora pistillata*: Characterization, localization, and role in biomineralization. *J. Biol. Chem.* **2008**, *283*, 25475–25484. [CrossRef] [PubMed]
5. Zoccola, D.; Ganot, P.; Bertucci, A.; Caminiti-Segonds, N.; Techer, N.; Voolstra, C.R.; Aranda, M.; Tambutté, E.; Allemand, D.; Casey, J.R.; *et al.* Bicarbonate transporters in corals point towards a key step in the evolution of cnidarian calcification. *Sci. Rep.* **2015**, *5*, 9983. [CrossRef] [PubMed]
6. Zoccola, D.; Tambutté, É.; Kulhanek, E.; Puverel, S.; Scimeca, J.-C.; Allemand, D.; Tambutté, S. Molecular cloning and localization of a PMCA P-type calcium ATPase from the coral *Stylophora pistillata*. *Biochim. Biophys. Acta* **2004**, *1663*, 117–126. [CrossRef] [PubMed]
7. Zoccola, D.; Tambutté, É.; Sénegas-Balas, F.; Michiels, J.-F.; Failla, J.-P.; Jaubert, J.; Allemand, D. Cloning of a calcium channel A1 subunit from the reef-building coral, *Stylophora pistillata*. *Gene* **1999**, *227*, 157–167. [CrossRef]
8. Venn, A.A.; Tambutté, E.; Holcomb, M.; Laurent, J.; Allemand, D.; Tambutté, S. Impact of seawater acidification on pH at the tissue-skeleton interface and calcification in reef corals. *PNAS* **2013**, *110*, 1634–1639. [CrossRef] [PubMed]
9. Allemand, D.; Tambutté, É.; Zoccola, D.; Tambutté, S. Coral calcification, cells to reefs. In *Coral Reefs: An Ecosystem in Transition*; Dubinsky, Z., Stambler, N., Eds.; Springer: Dordrecht, The Netherlands, 2011; Volume 3, pp. 119–150.
10. Tambutté, S.; Holcomb, M.; Ferrier-Pagès, C.; Reynaud, S.; Tambutté, E.; Zoccola, D.; Allemand, D. Coral biomineralization: From the gene to the environment. *J. Exp. Mar. Biol. Ecol.* **2011**, *408*, 58–78. [CrossRef]
11. Drake, J.L.; Mass, T.; Haramaty, L.; Zelzion, E.; Bhattacharya, D.; Falkowski, P.G. Proteomic analysis of skeletal organic matrix from the stony coral *Stylophora pistillata*. *Proc. Natl. Acad. Sci. USA* **2013**, *110*, 3788–3793. [CrossRef] [PubMed]
12. Mass, T.; Drake, J.L.; Peters, E.C.; Jiang, W.; Falkowski, P.G. Immunolocalization of skeletal matrix proteins in tissue and mineral of the coral *Stylophora pistillata*. *Proc. Natl. Acad. Sci. USA* **2014**, *111*, 12728–12733. [CrossRef] [PubMed]

13. Ramos-Silva, P.; Kaandorp, J.; Herbst, F.; Plasseraud, L.; Alcaraz, G.; Stern, C.; Corneillat, M.; Guichard, N.; Durlet, C.; Luquet, G.; *et al.* The skeleton of the staghorn coral *Acropora millepora*: Molecular and structural characterization. *PLoS ONE* **2014**, *9*, e97454.

14. Khalifah, R.G.; Edsall, J.T. Carbon dioxide hydration activity of carbonic anhydrase: Kinetics of alkylated anhydrases B and C from humans (metalloenzymes-isoenzymes-active sites-mechanism). *Proc. Natl. Acad. Sci. USA* **1972**, *69*, 172–176. [CrossRef] [PubMed]

15. Lindskog, S.; Coleman, J.E. The catalytic mechanism of carbonic anhydrase. *Proc. Natl. Acad. Sci. USA* **1973**, *70*, 2505–2508. [CrossRef] [PubMed]

16. Allemand, D.; Ferrier-Pages, C.; Furla, P.; Houlbreque, F.; Puverel, S.; Reynaud, S.; Tambutté, E.; Tambutté, S.; Zoccola, D. Biomineralisation in reef-building corals: From molecular mechanisms to environmental control. *C. R. Palevol* **2004**, *3*, 453–467. [CrossRef]

17. Tambutté, E.; Venn, A.A.; Holcomb, M.; Segonds, N.; Techer, N.; Zoccola, D.; Allemand, D.; Tambutté, S. Morphological plasticity of the coral skeleton under CO_2-driven seawater acidification. *Nat. Commun.* **2015**, *6*, 7368. [CrossRef] [PubMed]

18. Venn, A.A.; Tambutté, E.; Holcomb, M.; Allemand, D.; Tambutté, S. Live tissue imaging shows reef corals elevate pH under their calcifying tissue relative to seawater. *PLoS ONE* **2011**, *6*, e20013. [CrossRef] [PubMed]

19. Kaniewska, P.; Campbell, P.R.; Kline, D.I.; Rodriguez-Lanetty, M.; Miller, D.J.; Dove, S.; Hoegh-Guldberg, O. Major cellular and physiological impacts of ocean acidification on a reef building coral. *PLoS ONE* **2012**, *7*, e34659. [CrossRef] [PubMed]

20. Kaniewska, P.; Chan, C.K.; Kline, D.; Ling, E.Y.; Rosic, N.; Edwards, D.; Hoegh-Guldberg, O.; Dove, S. Transcriptomic changes in coral holobionts provide insights into physiological challenges of future climate and ocean change. *PLoS ONE* **2015**, *10*, e0139223. [CrossRef] [PubMed]

21. Moya, A.; Huisman, L.; Ball, E.E.; Hayward, D.C.; Grasso, L.C.; Chua, C.M.; Woo, H.N.; Gattuso, J.P.; ForêT, S.; Miller, D.J. Whole transcriptome analysis of the coral *Acropora millepora* reveals complex responses to CO_2-driven acidification during the initiation of calcification. *Mol. Ecol.* **2012**, *21*, 2440–2454. [CrossRef] [PubMed]

22. Ogawa, D.; Bobeszko, T.; Ainsworth, T.; Leggat, W. The combined effects of temperature and CO_2 lead to altered gene expression in *Acropora aspera*. *Coral Reefs* **2013**, *32*, 895–907. [CrossRef]

23. Rocker, M.M.; Noonan, S.; Humphrey, C.; Moya, A.; Willis, B.L.; Bay, L.K. Expression of calcification and metabolism-related genes in response to elevated pCO_2 and temperature in the reef-building coral *Acropora millepora*. *Mar. Genom.* **2015**, *24*, 313–318. [CrossRef] [PubMed]

24. Vidal-Dupiol, J.; Zoccola, D.; Tambutté, E.; Grunau, C.; Cosseau, C.; Smith, K.M.; Freitag, M.; Dheilly, N.M.; Allemand, D.; Tambutté, S. Genes related to ion-transport and energy production are upregulated in response to CO_2-driven pH decrease in corals: New insights from transcriptome analysis. *PLoS ONE* **2013**, *8*, e58652.

25. Hoadley, K.D.; Pettay, D.T.; Grottoli, A.G.; Cai, W.J.; Melman, T.F.; Schoepf, V.; Hu, X.; Li, Q.; Xu, H.; Wang, Y.; *et al.* Physiological response to elevated temperature and pCO_2 varies across four pacific coral species: Understanding the unique host + symbiont response. *Sci. Rep.* **2015**, *5*, 18371. [CrossRef] [PubMed]

26. Hoegh-Guldberg, O.; Mumby, P.-J.; Hooten, A.J.; Steneck, R.S.; Greenfield, P.; Gomez, E.; Harvell, C.D.; Sale, P.F.; Edwards, A.J.; Caldeira, K.; *et al.* Coral reefs under rapid climate change and ocean acidification. *Science* **2007**, *318*, 1737–1742. [CrossRef] [PubMed]

27. Barshis, D.J.; Ladner, J.T.; Oliver, T.A.; Seneca, F.O.; Traylor-Knowles, N.; Palumbi, S.R. Genomic basis for coral resilience to climate change. *Proc. Natl. Acad. Sci. USA* **2013**, *110*, 1387–1392. [CrossRef] [PubMed]

28. Edge, S.E.; Morgan, M.B.; Gleason, D.F.; Snell, T.W. Development of a coral cDNA array to examine gene expression profiles in Montastraea faveolata exposed to environmental stress. *Mar. Pollut. Bull.* **2005**, *51*, 507–523. [CrossRef] [PubMed]

29. Kenkel, C.D.; Meyer, E.; Matz, M.V. Gene expression under chronic heat stress in populations of the mustard hill coral (*Porites astreoides*) from different thermal environments. *Mol. Ecol.* **2013**, *22*, 4322–4334. [CrossRef] [PubMed]

30. Maor-Landaw, K.; Karako-Lampert, S.; Waldman Ben-Asher, H.; Goffredo, S.; Falini, G.; Dubinsky, Z.; Levy, O. Gene expression profiles during short-term heat stress in the red sea coral *Stylophora pistillata*. *Glob. Chang. Biol.* **2014**, *20*, 3026–3035. [CrossRef] [PubMed]

31. Bertucci, A.; Innocenti, A.; Scozzafava, A.; Tambutté, S.; Zoccola, D.; Supuran, C.T. Carbonic anhydrase inhibitors. Inhibition studies with anions and sulfonamides of a new cytosolic enzyme from the scleractinian coral *Stylophora pistillata*. *Bioorg. Med. Chem. Lett.* **2011**, *21*, 710–714. [PubMed]
32. Bertucci, A.; Innocenti, A.; Zoccola, D.; Scozzafava, A.; Allemand, D.; Tambutté, S.; Supuran, C.T. Carbonic anhydrase inhibitors: Inhibition studies of a coral secretory isoform with inorganic anions. *Bioorg. Med. Chem. Lett.* **2009**, *19*, 650–653. [CrossRef] [PubMed]
33. Bertucci, A.; Innocenti, A.; Zoccola, D.; Scozzafava, A.; Tambutté, S.; Supuran, C.T. Carbonic anhydrase inhibitors. Inhibition studies of a coral secretory isoform by sulfonamides. *Bioorg. Med. Chem.* **2009**, *17*, 5054–5058. [PubMed]
34. Moya, A.; Tambutté, S.; Beranger, G.; Gaume, B.; Scimeca, J.C.; Allemand, D.; Zoccola, D. Cloning and use of a coral 36B4 gene to study the differential expression of coral genes between light and dark conditions. *Mar. Biotechnol. (N.Y.)* **2008**, *10*, 653–663. [CrossRef] [PubMed]
35. Bertucci, A.; Tambutté, S.; Supuran, C.T.; Allemand, D.; Zoccola, D. A new coral carbonic anhydrase in *Stylophora pistillata*. *Mar. Biotechnol. (N.Y.)* **2011**, *13*, 992–1002. [CrossRef] [PubMed]

marine drugs

MDPI

Article

Recombinant Expression of a Modified Shrimp Anti-Lipopolysaccharide Factor Gene in *Pichia pastoris* GS115 and Its Characteristic Analysis

Hui Yang [1,2], Shihao Li [1,3], Fuhua Li [1,3,*], Kuijie Yu [1], Fusheng Yang [4] and Jianhai Xiang [1]

[1] Key Laboratory of Experimental Marine Biology, Institute of Oceanology, Chinese Academy of Sciences, Qingdao 266071, China; victor1900@163.com (H.Y.); lishihao@qdio.ac.cn (S.L.); shihao235@163.com (K.Y.); jhxiang@qdio.ac.cn (J.X.)
[2] University of Chinese Academy of Sciences, Beijing 100049, China
[3] Laboratory for Marine Biology and Biotechnology, Qingdao National Laboratory for Marine Science and Technology, Qingdao 266071, China
[4] Hangzhou Xiaoshan Donghai Aquaculture Company Limited, Hangzhou 311200, China; xiaoshanji2005@163.com
* Correspondence: fhli@qdio.ac.cn; Tel.: +86-532-8289-8836; Fax: +86-532-8289-8578

Academic Editor: Se-Kwon Kim
Received: 28 April 2016; Accepted: 25 July 2016; Published: 9 August 2016

Abstract: Anti-lipopolysaccharide factors (ALFs) with a LPS-binding domain (LBD) are considered to have broad spectrum antimicrobial activities and certain antiviral properties in crustaceans. FcALF2 was one isoform of ALFs isolated from the Chinese shrimp *Fenneropenaeus chinensis*. Our previous study showed that a modified LBD domain (named LBDv) of FcALF2 exhibited a highly enhanced antimicrobial activity. In the present study, a modified FcALF2 gene (*mFcALF2*), in which the LBD was substituted by LBDv, was designed and synthesized. This gene was successfully expressed in yeast *Pichia pastoris* GS115 eukaryotic expression system, and the characteristics of the recombinant protein mFcALF2 were analyzed. mFcALF2 exhibited apparent antibacterial activities against Gram-negative bacteria, including *Escherichia coli*, *Vibrio alginolyticus*, *Vibrio harveyi*, and *Vibrio parahaemolyticus*, and Gram-positive bacteria, including *Bacillus licheniformis* and *Staphylococcus epidermidis*. In addition, mFcALF2 could reduce the propagation of white spot syndrome virus (WSSV) in vivo by pre-incubation with virus. The present study paves the way for developing antimicrobial drugs in aquaculture.

Keywords: anti-lipopolysaccharide factors; recombinant protein; antibacterial activity; antiviral activity

1. Introduction

Antimicrobial peptides (AMPs), isolated from a variety of different living organisms, have received more and more attention for their contribution to host defense [1,2]. They are considered to be an essential part of the innate immune system since they possess a broad spectrum of antimicrobial activities against bacteria, fungi, some virus, and provide protection against microbial invasion [3,4]. Extensive researches have demonstrated that these AMPs could act not only as direct antimicrobial agents, but also as important regulators of the innate immune system [5–7]. AMPs exhibit microbicidal activity mostly by targeting the membrane of microorganisms to destroy their cell membrane [8–10]. AMPs could also eliminate bacteria by stimulating the non-inflammatory host immune responses, and inhibiting the cellular process, such as DNA replication, protein biosynthesis and folding or impairment of protein functions [11]. Therefore, AMPs are regarded as potential alternatives to conventional antibiotics since AMPs could hardly lead to bacterial resistance.

Anti-lipopolysaccharide factors (ALFs) isolated from crustaceans are regarded as important components of the innate immune system [12]. Multiple isoforms of ALFs exhibited different antimicrobial activities against Gram-positive or Gram-negative bacteria, and antiviral activity [13–15]. The LPS-binding domain (LBD) of ALFs was regarded as the functional domain for their antibacterial and antiviral activities [16,17]. The synthetic LBD peptides exhibited antibacterial and antiviral activity with high-efficiency [18,19]. Hence, ALFs could be a potential option to replace the conventional antibiotics in aquaculture.

In our previous studies, seven isoforms of ALF were identified from the Chinese shrimp *Fenneropenaeus chinensis* [20,21]. The transcriptional level of one isoform of ALF named *FcALF2* showed about 35-fold up-regulation when shrimp was at the acute infection stage of white spot syndrome virus (WSSV) compared with that at the latent infection stage [20]. The expression of *FcALF2* was significantly up-regulated when the shrimp was injected with *Micrococcus lysodeikticus* or *Vibrio anguillarum*, and the synthesized peptide of LBD from *FcALF2* possessed strong antibacterial activity and significant inhibition activity against WSSV [22]. Nowadays, more and more researches have focused on the rational design of AMPs [23–25]. In our previous study, we modified the LBD of FcALF2 by using lysine to substitute some non-ionized polar amino acids. The modified LBD peptide (LBDv) exhibited stronger antibacterial activities and broader antimicrobial spectrum than the original LBD peptide [22,26]. Since the cost for chemical synthesis of peptides is too expensive to be used in aquaculture, recombinant expressions should be a more practical way to obtain the proteins with bioactivity at large scale.

Yeast *Pichia pastoris* expression system has become a highly successful system for the large expression of heterologous genes [27]. In the present study, we synthesized the nucleotide sequence of a modified *FcALF2* (*mFcALF2*) gene, in which the original LBD sequence of FcALF2 was substituted by LBDv, and expressed mFcALF2 in the yeast *P. pastoris* GS115 expression system successfully. The recombinant mFcALF2 protein showed certain antimicrobial and antiviral activities. These data showed that a modified gene of AMPs could be expressed in *P. pastoris*, which will pave the way for developing antimicrobial drugs in aquaculture.

2. Results

2.1. Expression, Purification and Detection of mFcALF2 Protein

We designed the amino acid sequence of mFcALF2 (shown in Figure 1) in which the original LBD of FcALF2 was replaced by LBDv. Then we reversely translated the amino acid sequence into nucleotide sequence, and optimized the codon usage according to the codon bias for the yeast, and synthesized the nucleotide sequences of *mFcALF2*.

The protein expression vector pPIC9K containing a signal peptide of α-Factor with 85 amino acids was utilized in the present study (Figure 1A). The *mFcALF2* gene was comprised of 342 bp, with the restriction enzyme sites *EcoRI* (GAATTC) and *Not I* (GCGGCCGC) at the opposite ends of the sequence respectively. The mFcALF2 protein contained a 6× His-tag (112–117 aa) (Figure 1B). The deduced molecular mass of mFcALF2 was 13.79 kDa and its theoretical isoelectric point was 8.61. Multiple sequences alignment (Figure 1C) among mFcALF2, FcALF2 and LBDv revealed that only the LBD of FcALF2 was replaced, and the *mFcALF2* gene was successfully synthesized.

The recombinant plasmid was constructed using the *EcoRI* and *Not I* restriction enzyme. The recombinant plasmid was linearized and transformed into *P. pastoris* GS115 competent cell by electroporation. After transformation, the transformants were grown on MD plates. Some colonies were selected randomly and identified by PCR reaction with 5'AOX1 and 3'AOX1. Four positive colonies were picked and cultured for small-scale expression trials. Then we selected a positive transformant for large-scale production. The culture supernatant was analyzed by 15% SDS-PAGE and one major protein band with the molecular weight of about 15 kDa was detected (Figure 2). After Ni^{2+}-chelating chromatography purification, the recombinant mFcALF2 protein was detected by HRP-conjugated anti His-Tag mouse monoclonal antibody, which showed that the recombinant protein was the target protein (Figure 2). Using the constructed recombination system, about 1.2 mg recombinant mFcALF2 protein could be obtained from 1000 mL crude extract. The molecular mass of purified mFcALF2 protein was determined using matrix-assisted laser desorption ionization mode (MALDI/TOF) mass spectrometry, and the molecular weight of the purified mFcALF2 protein was about 13781.8320 Da (Figure 3). All these data indicated that the purified recombinant protein was mFcALF2 protein.

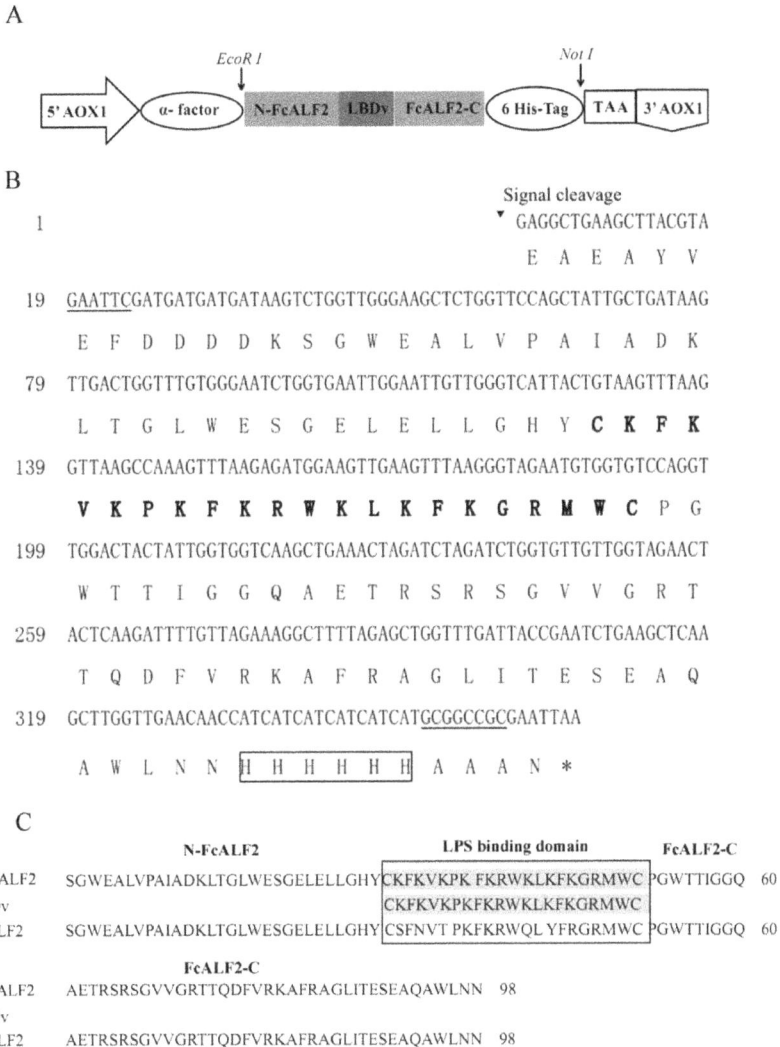

Figure 1. The nucleotide sequence and its deduced amino acid sequence of the modified anti-lipopolysaccharide factor isoform 2 from *Fenneropenaeus chinensis* (*FcALF2*) gene (*mFcALF2*). (**A**) Schematic representation of the vector pPIC9K-mFcALF2; (**B**) The LBD region of mFcALF2 is shown in bold and the stop codon is indicated by an asterisk. The restriction enzyme sites are underlined. The 6× His-tag is shown in box; (**C**) Multiple sequence alignment among mFcALF2, FcALF2 and LBDv.

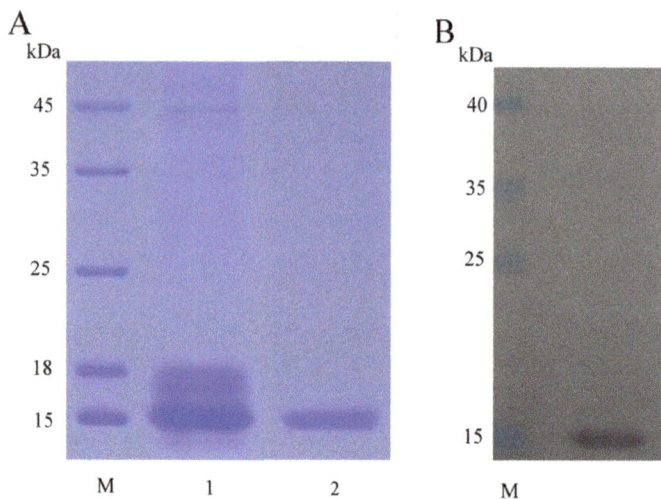

Figure 2. Detection of the recombinant mFcALF2 protein. (**A**) SDS-PAGE analyses of the recombinant mFcALF2 protein. Lane M in A and B represent molecular mass standards. Lane 1 shows the concentrated protein in supernatant secreted in GS115. Lane 2 shows the purified mFcALF2 protein; (**B**) Western blot analysis of the recombinant protein by anti-His tag antibody.

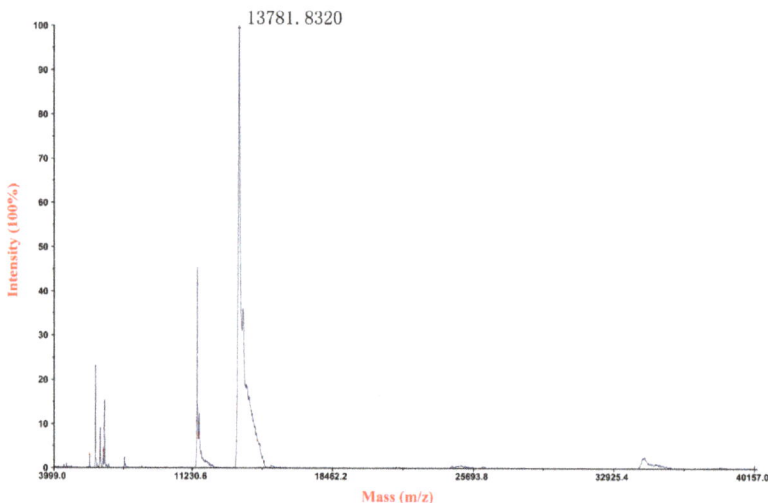

Figure 3. Molecular weight analysis of the recombinant mFcALF2 protein by MALDI/TOF (matrix-assisted laser desorption ionization mode) mass spectrometry.

2.2. Binding Assay of mFcALF2 to Bacteria

To detect the characteristic of recombinant mFcALF2 protein, we tested its binding activities to different Gram-negative and Gram-positive bacteria according to the method described previously [28]. The detected bacteria included *Escherichia coli*, *Vibrio alginolyticus*, *Bacillus licheniformis* and *Staphylococcus epidermidis*. The data revealed that the recombinant mFcALF2 protein could bind to the tested bacteria including *E. coli*, *V. alginolyticus*, *B. licheniformis* and *S. epidermidis* (Figure 4).

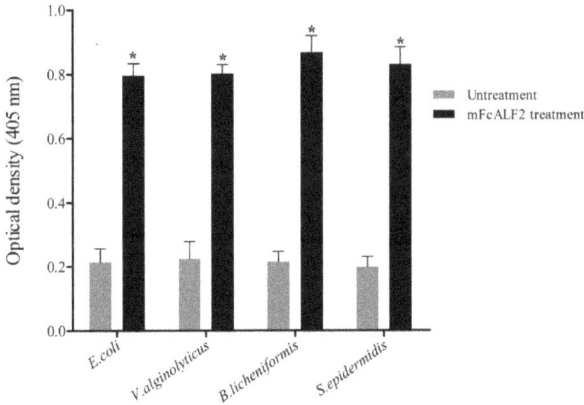

Figure 4. Binding activity analysis of recombinant mFcALF2 to bacteria. Star (*) indicates significant differences ($p < 0.05$) between the treated and untreated groups of different bacteria. The data are analyzed based on ANOVA with post hoc.

2.3. Observation on the Morphology of Bacterial Cells after Incubation with mFcALF2

The morphology of different bacteria including *E. coli*, *V. alginolyticus* and *S. epidermidis* after incubation with mFcALF2 were observed under scanning electron microscopy (SEM). The bacteria without any treatment displayed a smooth surface, with no apparent cellular debris. After incubation with mFcALF2 for 1 h, *E. coli* and *V. alginolyticus* exhibited remarkable changes on their surface, and *S. epidermidis* showed some leakage of the cytoplasm on their surface (Figure 5).

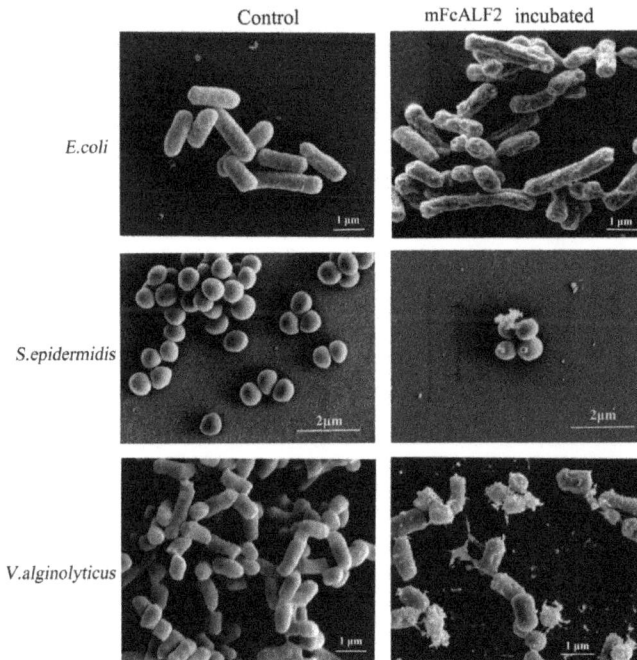

Figure 5. Morphology of bacteria after treatments by recombinant mFcALF2. The 10^8 cfu/mL different bacteria are incubated with 32 μM LBDv peptide for 2 h. The bacteria treated with same concentration pGFP peptide are used as negative control. Bar scale is 1 μM.

2.4. The Antibacterial Activity of Recombinant mFcALF2 Protein

The minimal growth inhibition concentration (MIC) assay and inhibition zone test were used to measure the antimicrobial activity of the purified mFcALF2 protein. The MICs to *V. alginolyticus*, *Vibrio harveyi*, *Vibrio parahaemolyticus*, *B. licheniformis* and *S. epidermidis* were 8–16 μM, while that to *E. coli* was 4–8 μM (Table 1). Obvious inhibition zone of recombinant mFcALF2 to *E. coli*, *V. alginolyticus*, *B. licheniformis* and *S. epidermidis* was detected (Figure 6).

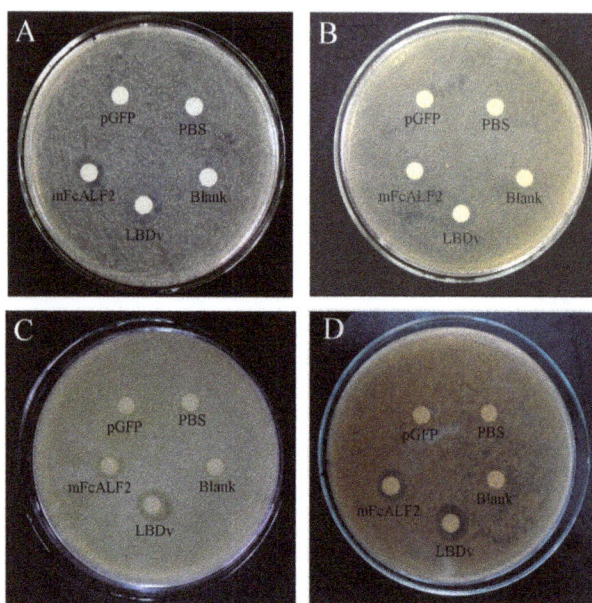

Figure 6. Inhibition zones of recombinant mFcALF2 to different bacteria: (**A**) *E. coli*; (**B**) *V. anguillarum*; (**C**) *B. licheniformis*; and (**D**) *S. epidermidis*. "Blank" represents blank group with nothing added. "PBS" represents control group with only PBS. "pGFP" represents negative control with synthetic pGFP peptide. "LBDv" represents positive control with synthetic LBDv peptide. mFcALF2 represents the recombinant protein. Twenty microliters of 32 μM protein/peptide solution is added to the center of filter paper.

Table 1. Minimal inhibitory concentration (MIC) of mFcALF2 to different bacteria.

Microorganisms	mFcALF2 MIC [a] (μM)
Gram negative bacteria:	-
Vibrio alginolyticus	8–16
Escherichia coli	4–8
Vibrio harveyi	8–16
Vibrio parahaemolyticus	8–16
Gram positive bacteria:	-
Bacillus licheniformis	8–16
Staphylococcus epidermidis	8–16

[a] MIC, minimal inhibitory concentration.

2.5. The Hemolytic Activities of mFcALF2

The hemolytic activity of mFcALF2 was checked on sheep blood agar plates. No obvious hemolytic activity was observed for mFcALF2 (Figure 7).

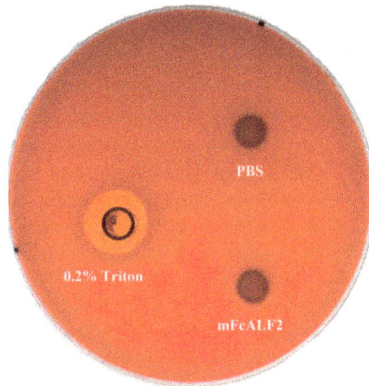

Figure 7. Hemolytic phenotypes of mFcALF2 on sheep blood agar. The same amount of PBS buffer (pH 7.4) and 0.2% Triton X-100 were used as negative and positive controls, respectively. The 60 µL purified mFcALF2 protein (32 µM) was added into the Oxford cup.

2.6. Inhibition of WSSV Replication by mFcALF2 in Litopenaeus vannamei

Litopenaeus vannamei were used as the experimental animals for WSSV infection. The antiviral activity of recombinant mFcALF2 protein was detected according to the method described previously [18,22,29]. Four groups including "Blank", "PBS + WSSV", "pGFP + WSSV", and "mFcALF2 + WSSV" were set. The WSSV copy numbers in the pleopods of shrimp from different groups at 24 h and 36 h after injection were shown in Figure 8. The WSSV copies per ng pleopods DNA in "mFcALF2 + WSSV" group was markedly lower than those in group "PBS+WSSV" and "pGFP + WSSV" at 24 h and 36 h after WSSV injection.

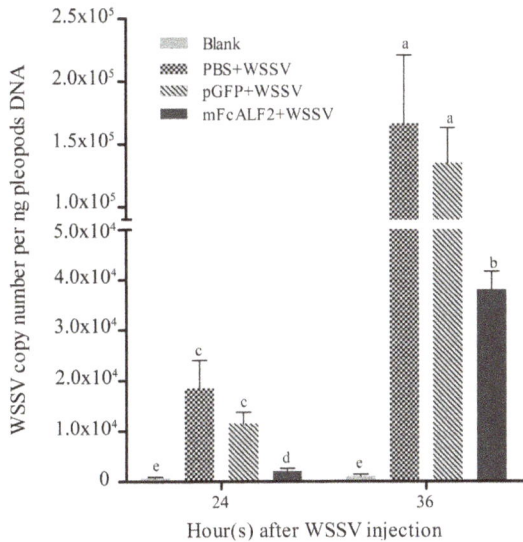

Figure 8. Detection of viral loads in *L. vannamei* after injection of WSSV incubated with recombinant mFcALF2. Data represent the means ± S.E. Lowercase letters (a, b, c, d and e) represent significant difference among treatments at $p < 0.05$. Three replicate experiments are performed. The data were analyzed based on ANOVA with post hoc.

3. Discussion

Currently, more than 1000 AMPs have been isolated or predicted by computational programs and divided into different subgroups [9,11,30]. Although some synthetic AMPs show certain activities, the high cost of synthetic peptides have driven the exploration of mass production by microbial expression systems, including prokaryotic and eukaryotic expression, through biotechnological approach [31]. The development of different heterologous expression systems exhibits many advantages, and one advantage is the mass production at low cost [32]. The prokaryotic expression system, such as *E. coli* system, is not usually used for the production of AMPs, especially for those with high inhibition activity to bacteria [28,33]. For yeast *P. pastoris* expression system, the recombinant proteins without toxicity to yeast can be effectively expressed and secreted into the medium under the direction of a signal peptide that is fused to the exogenous protein at the N-terminus [34]. With the development of synthetic biology approach, it has become reality to produce the recombinant proteins of the synthetic genes with high biological activity [35]. In the present study, the protein expression vector pPIC9K contained a strong and inducible promoter, and the α-Factor signal peptide for processing the fusion proteins was used to drive the expression of the synthetic gene encoding the mFcALF2 protein in *P. pastoris* GS115.

The recombinant mFcALF2 protein exhibited apparent antimicrobial activity to the detected Gram-positive and Gram-negative bacteria by binding to the bacteria. Strongly cationic peptides can potentially bind to negatively charged lipids on the outer leaflets of the bacterial membranes [36]. The cationic AMPs could bind to lipopolysaccharides (LPS) of Gram-negative bacteria and lipoteichoic acids (LTA) of Gram-positive bacteria [36,37]. Thus, we speculated the mFcALF2 protein with a highly cationic region could bind to both Gram-positive and Gram-positive bacteria, mostly the same as other cationic AMPs. Though some ALF isoforms have high affinities to LPS or LTA [38], whether the binding mechanisms are the same as cationic AMPs needs further investigation. In the present study, mFcALF2 has been proven to destroy the bacterial cell membrane, and lead to the leakage of the cytoplasm from bacteria. This is very similar to that of the reported ALF isoforms without any modification [39] and other AMPs [40]. Different from the traditional antibiotics, which have specific molecular targets, mFcALF2 might function by binding to the cell membrane of the bacteria through physical process, which is similar to that for other AMPs [41]. mFcALF2 showed some typical characteristic of AMPs. In our previous study, we found that 80% of Sf9 cells and *Cherax quadricarinatus* hemocytes could survive from the treatment with up to 16 μM synthetic peptide of LBDv [26]. Because the highly cationic region of mFcALF2 is responsible for cytotoxicity, it is reasonable to speculate the recombinant mFcALF2 protein would show little cytotoxicity at a concentration below 16 μM. Absence of hemolytic activity of mFcALF2 protein indicated that mFcALF2 might have a potential application in aquaculture in the future.

The purified mFcALF2 protein exhibited inhibition activity to both Gram-positive and Gram-negative bacteria, but the specific antibacterial activities to different bacteria were different. Compared with the recombinant protein FcALF5 from *Fenneropenaeus chinensis* and the recombinant protein of ALF4 from *Portunus trituberculatus* which were expressed in *E. coli* system [42,43], mFcALF2 showed a higher inhibition activity against *E. coli*. Although the recombinant protein of an ALF isoform from *Macrobrachium rosenbergii* expressed in the *Saccharomyces cerevisiae* showed an inhibition activity to *E. coli* and other bacteria, the MIC value was higher than that of mFcALF2 [28]. Therefore, we suggested that the GS115/pPIC9K-mFcALF2 vector and the *P. pastoris* expression system are suitable for a large-scale production of mFcALF2 with high activity.

WSSV was the most dangerous virus to shrimp aquaculture throughout the world [44]. Different ALFs isoforms exhibited certain inhibition activity against WSSV [19,45]. In our previous studies, the designed LBD analogous peptide showed strong antiviral activity when incubating with WSSV [18,22]. In the present study, the recombinant mFcALF2 protein also showed high inhibition activity to WSSV. This may provide a new strategy for the control of WSSV disease in aquaculture.

4. Materials and Methods

4.1. Synthesis of the Modified Sequence of FcLAF2 (mFcALF2)

We designed the *mFcALF2* gene in which the nucleotide sequence encoding the original LBD of FcALF2 was replaced by the nucleotide sequence encoding LBDv. During designing the new gene, the codon adaptation index (CAI) was used to measure the codon bias patterns by comparing those codons used in the translated sequence with the patterns of codon usage of yeast, using the Rare Codon Analysis Tool (http://www.genscript.com/cgi-bin/tools/rare_codon_analysis). A 6× His-tag and two restriction enzyme sites (*EcoR I* and *Not I*) were added. Then the optimized modified gene sequence named *mFcALF2* was synthesized by Sangon Biotech Company (Shanghai, China).

4.2. Construction of the Expression Plasmid, Transformation and Selection of Recombinant Clones

The *mFcALF2* gene was cloned into pUC57 vector. Then the plasmid was digested with the restriction enzymes and cloned into *EcoR I/Not I* sites of the *P. pastoris* expression vector pPIC9K (Invitrogen, Waltham, MA, USA), downstream of the α-factor secretion sequence and the Glu-Ala-Glu-Ala repeat sequence. The recombinant plasmid was transferred into *Escherichia coli* DH5α for its massive production. The sequence of the recombinant plasmid was confirmed by nucleotide sequencing.

P. pastoris GS115 was grown at 30 °C overnight, 280 rpm in YPD medium (1% yeast extract, 2% tryptone, 2% glucose). Then the yeast cells were harvested, washed twice with ice-cold sterile water and resuspended in 1 M sorbitol. The purified pPIC9K-*mFcALF2* was linearized by *Sac I* and 10 μg of plasmid was transformed into *P. pastoris* competent cell by electroporation following the manufacturer's instructions (Gene PulserXcell, Bio-Rad, Hercules, CA, USA). One milliliter of 1 M sorbitol precooled on ice was added into the cuvette immediately. The cells were then spread on MD plates containing 0.5 mg/mL G418 (1.34% YNB, 4×10^{-5}% biotin, 2% dextrose, and 2% agar). The plates were incubated at 30 °C and checked daily until positive colonies were observed. The positive colonies were identified by PCR reaction with the specific primer 5'AOX1 (5'-GACTGGTTCCAATTGACAAGC-3') and 3'AOX1 (5'-GCAAATGGCATTCTGACATCC-3').

4.3. Production and Purification of the Recombinant Protein

Single clone were grown overnight in 9 mL YPD medium at 30 °C for 24 h and then used to inoculate 35 mL of BMGY medium including 1% yeast extract, 2% tryptone, 100 mM potassium phosphate (pH 6.0), 1.34% YNB (yeast nitrogen base with ammonium sulfate without amino acid), 4×10^{-5} biotin, and 1% glycerol, for 48 h. Then the cells were harvested by centrifugation at 10,000 rpm for 5 min at room temperature and resuspended in 35 mL BMMY medium including 1% yeast extract, 2% tryptone, 100 mM potassium phosphate (pH 6.0), 1.34% YNB, 4×10^{-5} biotin, and 0.5% methanol with a concentration of 1.2×10^{9} cfu/mL. To induce the expression of mFcALF2, 100% methanol was added every 24 h to a final concentration of 0.5%. After 72 h, the supernatant was collected and analyzed by Dot Blot using the mouse anti-His tag monoclonal antibody to detect the expression of mFcALF2. The clone that expressed the highest amount of recombinant protein was selected for further large-scale production.

The culture medium system was amplified to 1 L, and the condition of the culture was the same as above. After cultured for 72 h, the entire medium was harvested by centrifugation at 10,000 rpm for 5 min and the supernatant was concentrated by PEG20,000. Then the concentrated product was purified by affinity chromatography using Ni-IDA-Sepharose CL-6B column (GE Healthcare, Uppsala, Sweden). The samples were loaded slowly at the rate of 0.5 mL/min and then the column was washed with washing buffer (20 mM Tris-HCl, 20 mM imidazole, 0.15 M NaCl) at the rate of 1.0 mL/min until the absorbance at 280 nm reached 0. Then the column was eluted with an elution buffer (20 mM Tris-HCl, 250 mM imidazole, 0.15 M NaCl). The purified protein was dialyzed in PBS (137 mmol/L NaCl, 2.7 mmol/L KCl, 10 mmol/L Na_2HPO_4, 1.8 mmol/L KH_2PO_4, pH 7.4) for 12 h.

Concentration of the mFcALF2 protein was tested by the Bradford method using Bradford Assay kit (TianGen, Beijing, China).

4.4. Western Blot Detection and Mass Spectrometry Analysis

The purified protein mFcALF2 was separated by 15% sodium dodecyl sulfate-polyacrylamide gel electrophoresis (SDS-PAGE) and visualized with Coomassie brilliant blue R250. Western-blot analysis was also used to detect the expression of mFcALF2 protein. After SDS-PAGE, mFcALF2 protein was transferred onto polyvinylidenefluoride (PVDF) membrane (Millipore, Temecula, CA, USA) and blocked with 5% nonfat milk in Tris-buffered saline (TBS) (10 mM Tris-HCl, 150 mM NaCl, pH 7.4) with 0.05% Tween-20 for 2 h at room temperature. Then it was incubated with HRP-conjugated anti His-Tag mouse monoclonal antibody overnight (1/1000 diluted in TBS). After the membrane was washed with TBST (TBS buffer with 0.05% Tween-20), the signal was detected using enhanced chemiluminescence detection assay kit (Tiangen, Beijing, China). The molecular mass of the purified mFcALF2 protein was determined using matrix-assisted laser desorption ionization mode (MALDI/TOF) mass spectrometry. The MALDI-TOF mass spectrometry was acquired in linear mode using a AB SCIEX MALDI-TOF/TOF 5800 System (ABSciex, Framingham, MA, USA) in positive reflector mode (10 kV) with a matrix of CHCA (Sigma, St. Louis, MO, USA). Two thousand laser shots were accumulated for each spectrum. MS data were calibrated by external calibration using the 5800 Mass Standards. Mass accuracy of MALDI/TOF mass spectra, after external calibration, resulted in approximately 100 ppm. Data were aquired and analyzed with 4000 Series Explorer Software V3.5 (Applied Biosystems, Waltham, MA, USA).

4.5. Bacteria Binding Assay

The binding of mFcALF2 to four species of bacteria, including *E. coli*, *V. alginolyticus*, *B. licheniformis* and *S. epidermidis*, was examined by indirected ELISA according to the method described previously [28]. The freshly cultured bacteria were collected and washed with PBS three times. Then the bacteria were resuspended by coating buffer (Na_2CO_3 1.59 g/L, $NaHCO_3$ 2.93 g/L, pH 9.6) to 10^8 cfu/mL. A 96-well plate was coated with 100 μL of bacteria suspension at 4 °C overnight. Then the wells were washed and blocked with 5% nonfat milk in Tris-buffered saline (TBS) buffer at 37 °C for 2 h. After three washes with TBS, 100 μL of mFcALF2 (32 μM) were added and incubated at 37 °C for 2 h. The wells were washed three times, and 100 μL HRP-conjugated anti-His Tag mouse monoclonal antibody (1/2000 diluted in TBS) was added. After incubation at 37 °C for 2 h and washing as described above, the reactivity was measured using 100 μL soluble TMB substrate solution (TianGen, Beijing, China). The absorbance was measured at 405 nm. The assay was performed in triplicates in three independent experiments.

4.6. Scanning Electron Microscopy (SEM) Detection

The morphology of *E. coli*, *V. alginolyticus* and *S. epidermidis* after incubation with 32 μM mFcALF2 was observed under scanning electron microscopy (SEM). Firstly, mid-logarithmic phase cultures of bacteria were harvested by centrifugation at $1000 \times g$ for 10 min and resuspended in PBS at 10^8 cfu/mL. Cells were incubated with 32 μM mFcALF2 for 1 h. The bacteria cells treated with the same amount of pGFP peptide were used as control. The collected cells were subsequently fixed in 2.5% (v/v) glutaraldehyde in 0.1 M phosphate buffer (pH 7.4) for 1 h and dehydrated with a graded ethanol series. After critical-point drying and gold coating, the samples were visualized by Hitachi S-3400N Scanning Electron Microscope (Hitachi High-Technologies, Tokyo, Japan).

4.7. Antimicrobial Activity Assays

The antimicrobial activity of the purified mFcALF2 protein was exhibited by inhibition zone test and MIC assay against Gram-positive and Gram-negative bacteria. The MIC and inhibition zone test were performed according to the method described previously [18]. Briefly, the bacterial strains were grown in medium up to 1×10^8 cfu/mL. Then 2 µL of the bacterial cultures, 15 µL of 1/2-fold serially diluted mFcALF2 (320 µM–10 µM) in PBS (pH 7.4) and 133 µL of fresh medium were added into each well of the sterile 96-well plate, so that the bacterial cultures were diluted to 1×10^6 cfu/mL and the recombinant mFcALF2 were diluted 1/2-fold serially to the concentrations of 32 µM–1 µM in a final volume of 150 µL. Then the 96-well plates were incubated at the corresponding temperature for another 6 to 8 h. Absorbance at 600 nm for Gram-positive bacteria or 560 nm for Gram-negative bacteria was determined using a precision micro-plate reader (TECAN infinite M200 PRO, Salzburg, Austria). The assay was performed in triplicates in three independent experiments. The MICs were defined as the lowest concentration of the compounds to inhibit the growth of microorganisms based on the spectroscopic absorbance readings. The bacteria strains used in this study included four Gram-negative bacteria, including *E. coli*, *V. harveyi*, *V. parahaemolyticus* and *V. alginolyticus*, and two Gram-positive bacteria including *B. licheniformis* and *S. epidermidis*.

Two Gram-negative bacteria *E. coli* and *V. alginolyticus*, and two Gram-positive bacteria *B. licheniformis* and *S. epidermidis* were used for inhibition zone test. The overnight culture of bacteria was diluted 100 times and 200 µL of culture were spread on the solid LB medium uniformly. Sterile filter paper with a diameter of 5 mm was put on the surface of the solid medium. Twenty mircoliters of 32 µM recombinant protein solution was added to the center of filter paper. Moreover, 20 µL PBS and 20 µL 32 µM pGFP peptide solution were used as negative control, and 20 µL of 32 µM LBDv peptide solution was used as a positive control. The plates were cultured at 37 °C or 28 °C for 24 h.

4.8. The Hemolytic Activity Test of mFcALF2

The hemolytic activities of mFcALF2 with a concentration of 32 µM, were evaluated as previously described [46,47]. The hemolytic activity tests were checked on sheep blood agar plates (Qingdao Hope Bio. Technology Co., Ltd., Qingdao, China). The 60 µL purified mFcALF2 protein (32 µM) was added into the Oxford cup (a stainless cylinder, outer diameter 7.1 ± 0.1 mm, inner diameter 6.0 ± 0.1 mm and height 10 ± 0.1 mm), which was placed on the surface of the agar. The same volume of PBS buffer (pH 7.4) and 0.2% Triton X-100 were used as a negative and a positive control separately. Then the plates were incubated at 30 °C for 6 to 8 h, and the hemolytic halos were measured.

4.9. Detection on the Antiviral Activity of mFcALF2

In order to test the antiviral activity of mFcALF2, WSSV particles pre-incubated with the mFcALF2 protein were injected into *L. vannamei* and the WSSV copy number in the pleopods was tested by realtime PCR. The specific procedure for WSSV extraction from pathologically infected shrimp were the same as described previously [48]. *L. vannamei* with body weight of 1.2 ± 0.3 g were used as the experimental animals for WSSV infection. The experiment was divided into four groups and named as Blank, PBS + WSSV, pGFP + WSSV, and mFcALF2 + WSSV. WSSV was incubated with 32 µM pGFP or mFcALF2 peptides solutions for 2 h at room temperature, respectively. For Blank group, shrimp were only injected with PBS. For PBS + WSSV group, each shrimp was injected with 10 µL (5000 copies) WSSV after incubation with PBS for 2 h at room temperature. For the other groups (pGFP + WSSV and mFcALF2 + WSSV), each shrimp was injected with 10 µL (5000 copies) WSSV solutions after incubation with the corresponding peptides. At 24 h and 36h after WSSV injection, 12 shrimp were collected from each group and three individuals were put together as one sample to extract DNA using the Plant Genomic DNA Kit (Tiangen, Beijing, China) for quantifying the copy numbers of WSSV. The method for quantify WSSV copy number was described as previous research [49].

4.10. Statistical Analyses

The statistical analyses were carried out with SPSS 17.0 software (SPSS Inc., Chicago, IL, USA). Data were analyzed with analyses of variance (ANOVA) and Duncan's Multiple Comparisons. Differences between treatments and controls were considered significant at $p < 0.05$.

5. Conclusions

In conclusion, we have successfully obtained the recombinant protein of a synthesized gene *mFcALF2* through the yeast *P. pastoris* expression system. The mFcALF2 protein exhibited high antimicrobial and antiviral activity, which could be potentially used in aquaculture in the future. These data will pave the way for developing antimicrobial drugs in aquaculture.

Acknowledgments: This work was financially supported by China Agriculture Research system-47 (CARS-47), General Program of National Natural Science Foundation of China (31272683) to Fuhua Li, the Major State Basic Research Development Program of China (973 program) (2012CB114403), and the Scientific and Technological Innovation Project Financially Supported by Qingdao National Laboratory for Marine Science and Technology (No.2015ASKJ02). This work was also supported by the Hangzhou Qianjiang Distinguished Expert Program. We would like to thank Wei Liu from Institute of Oceanology, CAS for his help in preparing the samples for scanning electronic microscopy detection.

Author Contributions: F.L. and J.X. initiated the project. F.L. and S.L. designed and supervised the experimental work, which was performed by H.Y. and S.L. K.Y. and F.Y. cultured the experimental animals. H.Y. and S.L. performed data analysis. H.Y., F.L. and S.L. were in charge of writing and checking the manuscript. All the authors read and approved the final manuscript.

Conflicts of Interest: The authors declare no conflict of interest.

References

1. Merrifield, R.; Juvvadi, P.; Andreu, D.; Ubach, J.; Boman, A.; Boman, H.G. Retro and retroenantio analogs of cecropin-melittin hybrids. *Proc. Natl. Acad. Sci. USA* **1995**, *92*, 3449–3453. [CrossRef] [PubMed]
2. Zasloff, M. Antimicrobial peptides of multicellular organisms. *Nature* **2002**, *415*, 389–395. [CrossRef] [PubMed]
3. Boman, H.G. Peptide antibiotics and their role in innate immunity. *Annu. Rev. Immunol.* **1995**, *13*, 61–92. [CrossRef] [PubMed]
4. Nicolas, P. Multifunctional host defense peptides: Intracellular-targeting antimicrobial peptides. *FEBS J.* **2009**, *276*, 6483–6496. [CrossRef] [PubMed]
5. Boman, H.G. Antibacterial peptides: Basic facts and emerging concepts. *J. Int. Med.* **2003**, *254*, 197–215. [CrossRef]
6. Hancock, R.E.; Sahl, H.G. Antimicrobial and host-defense peptides as new anti-infective therapeutic strategies. *Nat. Biotechnol.* **2006**, *24*, 1551–1557. [CrossRef] [PubMed]
7. Zasloff, M. Antibiotic peptides as mediators of innate immunity. *Curr. Opin. Immunol.* **1992**, *4*, 3–7. [CrossRef]
8. Ahmad, A.; Yadav, S.P.; Asthana, N.; Mitra, K.; Srivastava, S.P.; Ghosh, J.K. Utilization of an amphipathic leucine zipper sequence to design antibacterial peptides with simultaneous modulation of toxic activity against human red blood cells. *J. Biol. Chem.* **2006**, *281*, 22029–22038. [CrossRef] [PubMed]
9. Brogden, K.A. Antimicrobial peptides: Pore formers or metabolic inhibitors in bacteria? *Nat. Rev. Microbiol.* **2005**, *3*, 238–250. [CrossRef] [PubMed]
10. Anne Pereira, H. Novel therapies based on cationic antimicrobial peptides. *Curr. Pharm. Biotechnol.* **2006**, *7*, 229–234. [CrossRef] [PubMed]
11. Strempel, N.; Strehmel, J.; Overhage, J. Potential Application of Antimicrobial Peptides in the Treatment of Bacterial Biofilm Infections. *Curr. Pharm. Des.* **2015**, *21*, 67–84. [CrossRef] [PubMed]
12. Tanaka, S.; Nakamura, T.; Morita, T.; Iwanaga, S. Limulus Anti-Lps Factor—An Anticoagulant Which Inhibits the Endotoxin-Mediated Activation of *Limulus* Coagulation System. *Biochem. Biophys. Res. Commun.* **1982**, *105*, 717–723. [CrossRef]
13. Chia, T.J.; Wu, Y.C.; Chen, J.Y.; Chi, S.C. Antimicrobial peptides (AMP) with antiviral activity against fish nodavirus. *Fish Shellfish Immunol.* **2010**, *28*, 434–439. [CrossRef] [PubMed]
14. Krepstakies, M.; Lucifora, J.; Nagel, C.H.; Zeisel, M.B.; Holstermann, B.; Hohenberg, H.; Kowalski, I.; Gutsmann, T.; Baumert, T.F.; Brandenburg, K.; et al. A new class of synthetic peptide inhibitors blocks attachment and entry of human pathogenic viruses. *J. Infect. Dis.* **2012**, *205*, 1654–1664. [CrossRef] [PubMed]
15. Nagoshi, H.; Inagawa, H.; Morii, K.; Harada, H.; Kohchi, C.; Nishizawa, T.; Taniguchi, Y.; Uenobe, M.; Honda, T.; Kondoh, M.; et al. Cloning and characterization of a LPS-regulatory gene having an LPS binding domain in kuruma prawn *Marsupenaeus japonicus*. *Mol. Immunol.* **2006**, *43*, 2061–2069. [CrossRef] [PubMed]

16. Hoess, A.; Watson, S.; Siber, G.R.; Liddington, R. Crystal structure of an endotoxin-neutralizing protein from the horseshoe crab, Limulus anti-LPS factor, at 1.5 A resolution. *EMBO J.* **1993**, *12*, 3351–3356. [PubMed]

17. Supungul, P.; Klinbunga, S.; Pichyangkura, R.; Hirono, I.; Aoki, T.; Tassanakajon, A. Antimicrobial peptides discovered in the black tiger shrimp *Penaeus monodon* using the EST approach. *Dis. Aquat. Organ.* **2004**, *61*, 123–135. [CrossRef] [PubMed]

18. Guo, S.Y.; Li, S.H.; Li, F.H.; Zhang, X.J.; Xiang, J.H. Modification of a synthetic LPS-binding domain of anti-lipopolysaccharide factor from shrimp reveals strong structure-activity relationship in their antimicrobial characteristics. *Dev. Comp. Immunol.* **2014**, *45*, 227–232. [CrossRef] [PubMed]

19. Tharntada, S.; Ponprateep, S.; Somboonwiwat, K.; Liu, H.P.; Soderhall, I.; Soderhall, K.; Tassanakajon, A. Role of anti-lipopolysaccharide factor from the black tiger shrimp, *Penaeus monodon*, in protection from white spot syndrome virus infection. *J. Gen. Virol.* **2009**, *90*, 1491–1498. [CrossRef] [PubMed]

20. Li, S.H.; Zhang, X.J.; Sun, Z.; Li, F.H.; Xiang, J.H. Transcriptome Analysis on Chinese Shrimp *Fenneropenaeus chinensis* during WSSV Acute Infection. *PLoS ONE* **2013**, *8*, e58627. [CrossRef] [PubMed]

21. Liu, F.S.; Liu, Y.C.; Li, F.H.; Dong, B.; Xiang, J.H. Molecular cloning and expression profile of putative antilipopolysaccharide factor in Chinese shrimp (*Fenneropenaeus chinensis*). *Mar. Biotechnol.* **2005**, *7*, 600–608. [CrossRef] [PubMed]

22. Li, S.H.; Guo, S.Y.; Li, F.H.; Xiang, J.H. Characterization and function analysis of an anti-lipopolysaccharide factor (ALF) from the Chinese shrimp *Fenneropenaeus chinensis*. *Dev. Comp. Immunol.* **2014**, *46*, 349–355. [CrossRef] [PubMed]

23. Javadpour, M.M.; Juban, M.M.; Lo, W.-C.J.; Bishop, S.M.; Alberty, J.B.; Cowell, S.M.; Becker, C.L.; McLaughlin, M.L. De novo antimicrobial peptides with low mammalian cell toxicity. *J. Biol. Chem.* **1996**, *39*, 3107–3113. [CrossRef] [PubMed]

24. Maloy, W.L.; Kari, U.P. Structure-activity studies on magainins and other host defense peptides. *Biopolymers* **1995**, *37*, 105–122. [CrossRef] [PubMed]

25. Mangoni, M.L.; Shai, Y. Short native antimicrobial peptides and engineered ultrashort lipopeptides: Similarities and differences in cell specificities and modes of action. *Cell Mol. Life Sci.* **2011**, *68*, 2267–2280. [CrossRef] [PubMed]

26. Yang, H.; Li, S.H.; Li, F.H.; Xiang, J.H. Structure and bioactivity of a modified peptide derived from the LPS-binding domain of an anti-lipopolysaccharide factor (ALF) of Shrimp. *Mar. Drugs* **2016**, *14*, 96. [CrossRef] [PubMed]

27. Cregg, J.M.; Cereghino, J.L.; Shi, J.; Higgins, D.R. Recombinant protein expression in *Pichia pastoris*. *Mol. Biotechnol.* **2000**, *16*, 23–52. [CrossRef]

28. Liu, C.C.; Chung, C.P.; Lin, C.Y.; Sung, H.H. Function of an anti-lipopolysaccharide factor (ALF) isoform isolated from the hemocytes of the giant freshwater prawn *Macrobrachium rosenbergii* in protecting against bacterial infection. *J. Invertebr. Pathol.* **2014**, *116*, 1–7. [CrossRef] [PubMed]

29. Li, S.H.; Guo, S.Y.; Li, F.H.; Xiang, J.H. Functional diversity of anti-lipopolysaccharide factor isoforms in shrimp and their characters related to antiviral activity. *Mar. Drugs* **2015**, *13*, 2602–2616. [CrossRef] [PubMed]

30. Hancock, R.E. Peptide antibiotics. *Lancet* **1997**, *349*, 418–422. [CrossRef]

31. Otero-González, A.J.; Magalhães, B.S.; Garcia-Villarino, M.; López-Abarrategui, C.; Sousa, D.A.; Dias, S.C.; Franco, O.L. Antimicrobial peptides from marine invertebrates as a new frontier for microbial infection control. *FASEB. J.* **2010**, *24*, 1320–1334. [CrossRef] [PubMed]

32. Arbulu, S.; Jiménez, J.J.; Gútiez, L.; Cintas, L.M.; Herranz, C.; Hernández, P.E. Cloning and expression of synthetic genes encoding the broad antimicrobial spectrum bacteriocins SRCAM 602, OR-7, E-760, and L-1077, by recombinant *Pichia pastoris*. *Biomed. Res. Int.* **2015**, *2015*, e767183. [CrossRef] [PubMed]

33. Ponprateep, S.; Somboonwiwat, K.; Tassanakajon, A. Recombinant anti-lipopolysaccharide factor isoform 3 and the prevention of vibriosis in the black tiger shrimp, *Penaeus monodon*. *Aquaculture* **2009**, *289*, 219–224. [CrossRef]

34. Cereghino, J.L.; Cregg, J.M. Heterologous protein expression in the methylotrophic yeast *Pichia pastoris*. *Microbiol. Lett.* **2000**, *24*, 45–66.

35. Jiménez, J.J.; Borrero, J.; Gútiez, L.; Arbulu, S.; Herranz, C.; Cintas, L.M.; Hernández, P.E. Use of synthetic genes for cloning, production and functional expression of the bacteriocins enterocin A and bacteriocin E 50-52 by *Pichia pastoris* and *Kluyveromyces lactis*. *Mol. Biotechnol.* **2014**, *56*, 571–583. [CrossRef] [PubMed]

36. Malanovic, N.; Lohner, K. Gram-positive bacterial cell envelopes: The impact on the activity of antimicrobial peptides. *Biochim. Biophys. Acta (BBA) Biomembr.* **2015**, *1858*, 936–946. [CrossRef] [PubMed]
37. Devine, D.A.; Hancock, R.E. Cationic peptides: Distribution and mechanisms of resistance. *Curr. Pharm. Des.* **2002**, *8*, 703–714. [CrossRef] [PubMed]
38. Somboonwiwat, K.; Bachère, E.; Rimphanitchayakit, V.; Tassanakajon, A. Localization of anti-lipopolysaccharide factor (ALFPm3) in tissues of the black tiger shrimp, *Penaeus monodon*, and characterization of its binding properties. *Dev. Comp. Immunol.* **2008**, *32*, 1170–1176. [CrossRef] [PubMed]
39. Jaree, P.; Tassanakajon, A.; Somboonwiwat, K. Effect of the anti-lipopolysaccharide factor isoform 3 (ALFPm3) from *Penaeus monodon* on *Vibrio harveyi* cells. *Dev. Comp. Immunol.* **2012**, *38*, 554–560. [CrossRef] [PubMed]
40. Zhang, L.; Yang, D.; Wang, Q.; Yuan, Z.; Wu, H.; Pei, D.; Cong, M.; Li, F.; Ji, C.; Zhao, J. A defensin from clam *Venerupis philippinarum*: Molecular characterization, localization, antibacterial activity, and mechanism of action. *Dev. Comp. Immunol.* **2015**, *51*, 29–38. [CrossRef] [PubMed]
41. Wang, Y.; Chen, J.; Zheng, X.; Yang, X.; Ma, P.; Cai, Y.; Zhang, B.; Chen, Y. Design of novel analogues of short antimicrobial peptide anoplin with improved antimicrobial activity. *J. Pept. Sci.* **2014**, *20*, 945–951. [CrossRef] [PubMed]
42. Liu, Y.; Cui, Z.; Li, X.; Song, C.; Li, Q.; Wang, S. A new anti-lipopolysaccharide factor isoform (PtALF4) from the swimming crab *Portunus trituberculatus* exhibited structural and functional diversity of ALFs. *Fish Shellfish Immunol.* **2012**, *32*, 724–731. [CrossRef] [PubMed]
43. Yang, H.; Li, S.H.; Li, F.H.; Lv, X.J.; Xiang, J.H. Recombinant expression and functional analysis of an isoform of anti-lipopolysaccharide factors (FcALF5) from Chinese shrimp *Fenneropenaeus chinensis*. *Dev. Comp. Immunol.* **2015**, *53*, 47–54. [CrossRef] [PubMed]
44. Lo, C.F.; Ho, C.H.; Chen, C.H.; Liu, K.F.; Chiu, Y.L.; Yeh, P.Y.; Peng, S.E.; Hsu, H.C.; Liu, H.C.; Chang, C.F.; et al. Detection and tissue tropism of white spot syndrome baculovirus (WSBV) in captured brooders of *Penaeus monodon* with a special emphasis on reproductive organs. *Dis. Aquat. Organ.* **1997**, *30*, 53–72. [CrossRef]
45. Liu, H.P.; Jiravanichpaisal, P.; Soderhall, I.; Cerenius, L.; Soderhall, K. Antilipopolysaccharide factor interferes with white spot syndrome virus replication in vitro and in vivo in the crayfish *Pacifastacus leniusculus*. *J. Virol.* **2006**, *80*, 10365–10371. [CrossRef] [PubMed]
46. Liu, F.; Liu, G.X.; Li, F.H. Characterization of two pathogenic Photobacterium strains isolated from *Exopalaemon carinicauda* causing mortality of shrimp. *Aquaculture* **2016**, *464*, 129–135. [CrossRef]
47. Bicca, F.C.; Fleck, L.C.; Ayub, M.Z. Production of biosurfactant by hydrocarbon degrading *Rhodococcus ruber* and *Rhodococcus erythropolis*. *Rev. Microbiol.* **1999**, *30*, 231–236. [CrossRef]
48. Sun, Y.M.; Li, F.H.; Chi, Y.H.; Xiang, J.H. Enhanced resistance of marine shrimp *Exopalamon carincauda* Holthuis to WSSV by injecting live VP28-recombinant bacteria. *Acta Oceanol. Sin.* **2013**, *32*, 52–58. [CrossRef]
49. Sun, Y.M.; Li, F.H.; Xiang, J.H. Analysis on the dynamic changes of the amount of WSSV in Chinese shrimp *Fenneropenaeus chinensis* during infection. *Aquaculture* **2013**, *376*, 124–132. [CrossRef]

marine drugs

MDPI

Article

Anticancer Activity of a Hexapeptide from Skate (*Raja porosa*) Cartilage Protein Hydrolysate in HeLa Cells

Xin Pan, Yu-Qin Zhao *, Fa-Yuan Hu, Chang-Feng Chi and Bin Wang *

School of Food and Pharmacy, Zhejiang Ocean University, 1st Haidanan Road, Changzhi Island, Lincheng, Zhoushan 316022, China; Uniquepan2015@163.com (X.P.); moonriveryue@163.com (F.-Y.H.); chichangfeng@hotmail.com (C.-F.C.)
* Correspondence: zhaoy@hotmail.com (Y.-Q.Z.); wangbin4159@hotmail.com (B.W.);
   Tel.: +86-580-255-5085 (Y.-Q.Z. & B.W.); Fax: +86-580-255-4781 (Y.-Q.Z. & B.W.)

Academic Editor: Se-Kwon Kim
Received: 30 June 2016; Accepted: 8 August 2016; Published: 16 August 2016

Abstract: In this study, the hexapeptide Phe-Ile-Met-Gly-Pro-Tyr (FIMGPY), which has a molecular weight of 726.9 Da, was separated from skate (*Raja porosa*) cartilage protein hydrolysate using ultrafiltration and chromatographic methods, and its anticancer activity was evaluated in HeLa cells. Methylthiazolyldiphenyl-tetrazolium bromide (MTT) assay indicated that FIMGPY exhibited high, dose-dependent anti-proliferation activities in HeLa cells with an IC_{50} of 4.81 mg/mL. Acridine orange/ethidium bromide (AO/EB) fluorescence staining and flow cytometry methods confirmed that FIMGPY could inhibit HeLa cell proliferation by inducing apoptosis. Western blot assay revealed that the Bax/Bcl-2 ratio and relative intensity of caspase-3 in HeLa cells treated with 7-mg/mL FIMGPY were 2.63 and 1.83, respectively, significantly higher than those of the blank control ($p < 0.01$). Thus, FIMGPY could induce apoptosis by upregulating the Bax/Bcl-2 ratio and caspase-3 activation. Using a DNA ladder method further confirmed that the anti-proliferation activity of FIMGPY was attributable to its role in inducing apoptosis. These results suggest that FIMGPY from skate cartilage protein hydrolysate may have applications as functional foods and nutraceuticals for the treatment and prevention of cancer.

Keywords: skate (*Raja porosa*); cartilage; peptide; anticancer activity; apoptosis

---

1. Introduction

Cancer is one of the single most important causes of death in humans, inducing approximately 8.2 million deaths or 14.6% of all human deaths in 2012 [1]. Currently, chemotherapy is the most common method used to eliminate cancer cells, prevent cancer recurrence, control cancer by slowing cell growth, and reduce symptoms [2]. However, healthy cells may be damaged by the many side effects of anticancer drugs, and resistance to anticancer drugs has been observed [3]. Therefore, substantial attention is being paid to identifying anticancer drugs with high efficiency and low toxicity from natural sources [4].

Bioactive peptides, which consist of 2–20 amino acid residues, are inactive in the sequence of their parent proteins and can be released by enzymatic hydrolysis either during gastrointestinal digestion in the body or during food processing. To date, some peptides with anticancer and antioxidant activities have been purified from various protein hydrolysates [5]. (Leu-Ala-Asn-Ala-Lys) LANAK, which has a MW of 515.29 Da and is from oyster protein hydrolysates, was shown to initiate cancer cell death by inhibiting cancer cell growth, increasing DNA damage and apoptosis in the HT-29 colon cancer cell line, and displaying strong antioxidant potential as a 2,2-diphenyl-1-picrylhydrazyl radical (DPPH•) scavenger [1]. (Gln-Pro-Lys) QPK, which was isolated from a sepia ink protein hydrolysate, could significantly inhibit the proliferation of DU-145, PC-3, and LNCaP cells in a time- and dose-dependent manner. This peptide induced apoptosis by decreasing the expression of the anti-apoptotic protein Bcl-2 and increasing the expression of the apoptotic protein Bax [6]. Tyr-Ala-Leu-Arg-Ala-His (YALRAH), which has a MW of 670.77 Da and is from half-fin anchovy (*Setipinna taty*) hydrolysates, exhibited strong anti-proliferation effects on human prostate cancer PC-3 cells, with an IC_{50} of 11.1 µM [7]. Arg-Gln-Ser-His-Phe-Ala-Asn-Ala-Gln-Pro (RQSHFANAQP), which has a MW of 1155 Da and is from chickpea protein hydrolysates, showed significant dose-dependent activities in hydroxyl radical (HO•)-(EC_{50} 2.03 µM), DPPH•-(EC_{50} 3.15 µM) and 2,2′-azino-bis-3-ethylbenzothiazoline-6-sulfonic acid radical (ABTS$^+$•)-(EC_{50} 2.31 µM) scavenging assays. Additionally, cell viability assays showed high anti-proliferative activities on the breast cancer cells MCF-7 and MDA-MB-231, with IC_{50} values of 2.38 and 1.50 µmol/mL, respectively. Furthermore, the key tumor suppressor protein (p53) level was shown to increase with increasing RQSHFANAQP concentrations by enzyme-linked immunosorbent assay (ELISA) [8]. WPP, which has a MW of 398.44 Da and is isolated from blood clam (*Tegillarca granosa*), showed significant antioxidant activities against DPPH•, HO•, O_2^-•, and ABTS$^+$• with EC_{50} values of 1.388, 0.406, 0.536, and 2.75 mg/mL, respectively. Furthermore, this peptide exhibited strong, dose-dependent cytotoxicity toward PC-3, DU-145, H-1299, and HeLa cell lines and significantly changed the morphologies of PC-3 cells [2]. Previous research indicated that food-derived peptides could have the potential to prevent and treat diseases associated with reactive oxygen species (ROS), specifically cancers [9]. Therefore, consuming antioxidant peptides could dramatically reduce organismal ROS levels and contribute substantially to maintaining health and preventing ROS-associated diseases, especially cancers [10–12].

Cartilaginous fishes (Chondrichthyes) are a commercially important species. During processing, large quantities of cartilage are discarded as waste because of their low economic value. In our previous studies, three antioxidant hexapeptides—Phe-Ile-Met-Gly-Pro-Tyr (FIMGPY), Gly-Pro-Ala-Gly-Asp-Tyr (GPAGDY) and Ile-Val-Ala-Gly-Pro-Gln (IVAGPQ)—were isolated from skate (*Raja porosa*) cartilage protein hydrolysates [13], and FIMGPY exhibited good scavenging activities on (DPPH•)-(EC_{50} 3.5768 M), (HO•)-(EC_{50} 4.1821 M), (O_2^-•)-(EC_{50} 3.1181 M) and (ABTS$^+$•)-(EC_{50} 1.4307 M), respectively. In addition, FIMGPY showed the higher anti-proliferation activity in HeLa cells than those of GPAGDY and IVAGPQ. Therefore, the objective of the present study was to investigate the anticancer activities and molecular mechanisms of FIMGPY in HeLa cells.

2. Results and Discussion

2.1. Proliferation Inhibition of HeLa Cell Lines

Cell proliferation is a physiological process that occurs in almost all tissues and under many circumstances. Under normal conditions, the balance between proliferation and programmed cell death, which usually occurs via apoptosis, is maintained by regulating both processes to ensure the integrity of tissues and organs. However, uncontrolled cell division can induce tissue proliferation and even cancer [14]. Therefore, the inhibition of cell proliferation is thought to be an effective method for tumor therapy. In this study, the HeLa cell line was used to measure the proliferation inhibition rate of FIMGPY. The peptide was evaluated in mouse embryo fibroblast NIH3T3 cells under the same experimental conditions to determine its cytotoxic effect on normal cells. As shown in Figure 1, FIMGPY showed strong, dose-dependent cytotoxicity against HeLa cell lines, with an IC_{50} of 4.81 mg/mL for 24 h. The IC_{50} value of FIMGPY in HeLa cells was lower than those of GPAGDY (4.86 mg/mL) and IVAGPQ (6.26 mg/mL), which are also from skate cartilage protein hydrolysates. The results indicate that FIMGPY exerted higher cytotoxic activity against HeLa cells under identical conditions than the other two peptides. The proliferation-inhibition rate of FIMGPY in NIH3T3 cells (IC_{50} 4.81 mg/mL for 24 h) was also far below than that in HeLa cells (data not shown), suggesting that FIMGPY has almost no cytotoxic effects on normal cells. Therefore, that FIMGPY is cell selective, destroying tumor cells rather than normal cells.

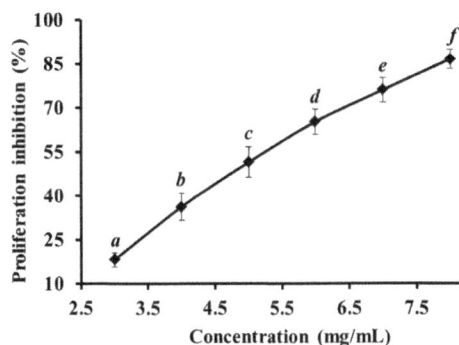

Figure 1. Proliferation inhibition of HeLa cell lines treated by FIMGPY for 24 h. All data are presented as the mean ± standard deviation (SD) of three experiments. $^{a-f}$ Values with same letters indicate no significant difference for each group of samples at the same concentration ($p > 0.05$).

The composition of cell membrane bilayers and the distribution of phospholipids determine cell selectivity and cell susceptibility to lysis. The amount of phosphatidylserine (PS) located in the outer leaflets of cancer cell membranes is 3–7 times that found in the inner leaflets of normal cell membranes [15]. FIMGPY is composed of the hydrophobic amino acids Phe (F), Ile (I), Met (M), and Pro (P), which could lead to increased interactions between FIMGPY and the outer leaflets of tumor cell membrane bilayers, which have high phospholipid contents. These increased interactions may explain FIMGPY's cell selectivity.

2.2. Morphological Observations by Acridine Orange/Ethidium Bromide (AO/EB) Staining

Apoptosis is a process of programmed cell death characterized by biochemical and morphological processes and plays a crucial role in developing and maintaining the health of the body by eliminating old, unnecessary and unhealthy cells [16]. During different stages of apoptosis, some characteristic cell morphologies include blebbing, shrinkage, nuclear fragmentation, chromatin condensation, poly-nucleosomal DNA fragmentation, global mRNA decay, and the fragmentation of cells into apoptotic bodies [17]. Therefore, fluorescence microscopy and AO/EB staining methods were employed to observe the cell changes to distinguish between apoptotic and normal cells, and determine the effects of external factors on cancer cells [6].

As shown in Figure 2, HeLa cells showed significant, morphological, apoptotic changes after treatment with 0-, 3-, 5-, and 7-mg/mL FIMGPY for 24 h. Green, yellow/green and reddish/orange staining of the cells indicate viable, early apoptotic, and late apoptotic cells, respectively. The yellow/green staining in Figure 2(A-2,A-3) shows HeLa cells that was at an early stage of apoptosis. Typical apoptotic changes, such as condensed chromatin, cytoplasmic blebs, and fragmented nuclei, were also observed in the HeLa cells after exposure to 3- and 5-mg/mL FIMGPY for 24 h. In Figure 2(A-4), additional features—i.e., orange necrotic cell apoptotic bodies—were observed, indicating that the HeLa cells were in the final stages of apoptosis after exposure to 7-mg/mL FIMGPY for 24 h. The AO/EB staining results revealed that the morphological features of the apoptotic HeLa cells were dose dependent, similar to previous AO/EB staining results obtained for DU-145 and PC-3 cells treated with QPK from cuttlefish ink [6], PC-3 cells treated with Arg-Ala-Ala-Leu-Ala-Val-Val-Leu-Gly-Arg-Gly-Gly-Pro-Pro (RAALAVVLGRGGPR) and Arg-Asp-Gly-Asp-Ser-Cys-Arg-Gly-Gly-Gly-Pro-Val (RDGDSCRGGGPV) from *Bullacta exarata* [18], and PC-3 cells treated with Trp-Pro-Pro (WPP) from blood clam [2].

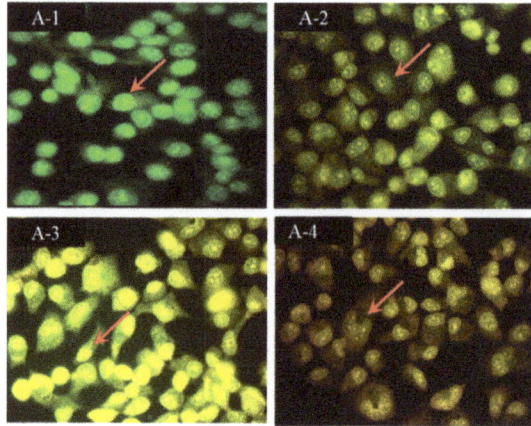

Figure 2. Morphological observation with AO/EB staining at 400× actual magnification. HeLa cells were treated with FIMGPY at (**A-1**) 0, (**A-2**) 3, (**A-3**) 5, and (**A-4**) 7 mg/mL for 24 h. (**A-1**) Cell indicated by the arrow indicates viable cell; (**A-2** and **A-3**) Cells indicated by the arrow indicates early apoptotic cells; (**A-4**) Cell indicated by the arrow indicates late apoptotic cell. Each experiment was performed in triplicate and generated similar morphological features.

2.3. Cell Apoptotic Rate Detected by Flow Cytometry

In normal cells, PS distributes only on the inner side of the cytomembrane and transfers to the outer side of the cytomembrane during early cell apoptosis. Therefore, Annexin V can bind to PS that is expressed on the outer layer of the cytomembrane and is used to identify cells entering apoptosis [19]. Propidium iodide (PI) is used as a DNA stain for flow cytometry to evaluate cell viability or DNA content via cell cycle analysis and to differentiate necrotic, apoptotic, and normal cells [20]. Thus, Annexin V-fluorescein isothiocyanate (FITC)/PI can identify distinct cell stages and quantitatively illustrate the apoptotic process [21].

The percentages of Annexin V-stained HeLa cells treated with FIMGPY at concentrations ranging from 3 to 7 mg/mL are depicted in Figure 3. The percentage of Annexin V-stained HeLa cells was 4.54% for the control. After 24 h of exposure to FIMGPY, the apoptosis percentages increased to 8.64 ± 0.31, 11.72 ± 0.57 and $19.25 \pm 0.76\%$ for concentrations of 3, 5, and 7 mg/mL, respectively. Compared with the control, the apoptotic effect on the HeLa cells markedly increased as the FIMGPY concentration increased. Therefore, FIMGPY displayed a high capacity to induce apoptosis in HeLa cells.

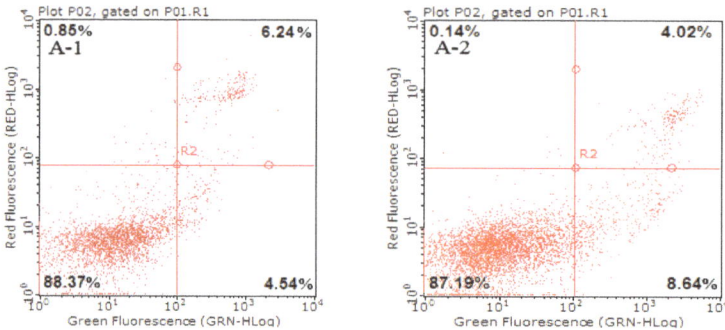

Figure 3. *Cont.*

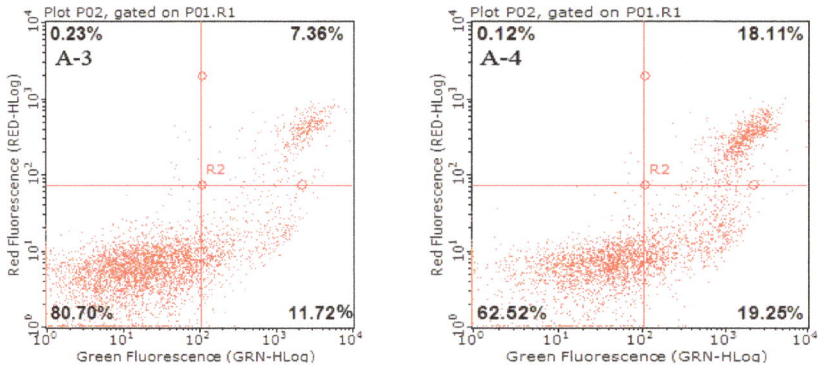

Figure 3. Flow cytometry analysis of HeLa cells by double-labeling with Annexin-V and PI. Quadrants: lower left-live, cells; upper left, necrotic cells; lower right, early apoptotic cells; upper right, late apoptotic cells. The percentages of early apoptotic cells were (**A-1**) 4.54% in the blank control cells; (**A-2**) 8.64% in the 3-mg/mL FIMGPY-treated cells; (**A-3**) 11.72% in the 5-mg/mL FIMGPY-treated cells; and (**A-4**) 19.25% in the 7-mg/mL FIMGPY-treated cells. All data are presented as the mean ± standard deviation (SD) of three experiments.

2.4. Western Blotting Results for Bcl-2, Bax, and Caspase-3 in FIMGPY-Treated HeLa Cells

The flow cytometry assay indicated that the apoptosis rate increased in HeLa cells as the FIMGPY concentration increased. Apoptosis is a highly regulated and controlled process that confers advantages during an organism's life cycle. Therefore, the initiation of apoptosis is precisely regulated by activation mechanisms involving specific factors; for example, caspases and Fas receptors promote apoptosis, whereas some members of the Bcl-2 family of proteins inhibit apoptosis [22]. To further confirm the effects of FIMGPY in HeLa cells and explain the reasons for the observed apoptosis, western blot assay was performed to investigate anti- and pro-apoptosis protein expression levels in treated HeLa cells.

Two distinct pathways (intrinsic and extrinsic) can lead to the activation of apoptosis. The intrinsic or mitochondrial apoptosis is crucially regulated by the interplay/balance between the pro- and anti-apoptotic Bcl-2 family members. Consequently, the Bcl-2 family proteins play a pivotal role in determining whether a cell will live or die [23,24]. Members of the Bcl-2 family, such as Bax, Bak, Bad, and Bcl-Xs, possess pro-apoptotic characteristics, whereas other members, such as Bcl-2, Bcl-XL, Bcl-W, Bfl-1, and Mcl-1, act as anti-apoptotic regulators. The apoptosis-inducing effect is more dependent on the balance between Bcl-2 and Bax than on Bcl-2 alone. Typically, the ratio of Bcl-2 and Bax protein expression is used as an index for apoptosis [6]. In this experiment, the levels of pro-apoptotic Bax and anti-apoptotic Bcl-2 proteins were measured by Western blot analysis in the presence of different doses of FIMGPY (0, 3, 5, and 7 mg/mL). As shown in Figure 4, a remarkable upregulation of Bax protein levels and a decrease in the Bcl-2 protein levels were observed as the FIMGPY concentration increased (Figure 4A), eventually leading to an increase in the Bax/Bcl-2 ratio in FIMGPY-treated HeLa cells (Figure 4B). The Bax/Bcl-2 ratio in HeLa cells treated with 7-mg/mL FIMGPY was 2.63, which was significantly higher than that of the blank control ($p < 0.01$). The result indicated that FIMGPY could promote apoptosis in HeLa cells by upregulating the Bax/Bcl-2 ratio.

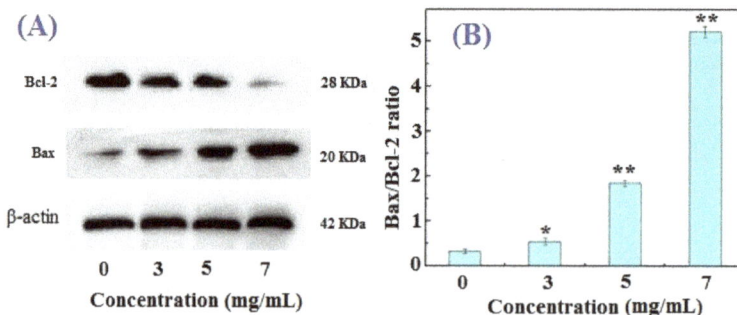

Figure 4. Expression of the apoptosis-associated proteins Bax and Bcl-2 in HeLa cells treated with FIMGPY for 24 h. (**A**) Sodium dodecyl sulfate polyacrylamide gel electrophoresis (SDS-PAGE) patterns for Bax and Bcl-2 and (**B**) the Bax/Bcl-2 ratio. * $p < 0.05$ and ** $p < 0.01$ vs. control.

Caspases are the executioners of apoptosis and are divided into the following two types according to their functions in apoptosis: (1) initiator (apical) caspases and (2) effector (executioner) caspases [25,26]. Initiator caspases (e.g., caspase-2, 8, 9, and 10) cleave inactive pro-forms of effector caspases, thereby activating them; effector caspases (e.g., caspase-3, 6, 7), in turn, cleave other protein substrates within the cell to trigger apoptosis. Among them, caspase-3 interacts with caspase-8 and caspase-9 in apoptosis, and plays a central role in the execution phase of apoptosis [22,27]. Figure 5A shows that FIMGPY noticeably upregulated caspase-3 levels in HeLa cells and that its relative intensity increased from 0.72 to 1.83 when the peptide concentration ranged from 0 to 7 mg/mL. The relative intensity of caspase-3 at 8 mg/mL was significantly higher than that of the blank control ($p < 0.01$) (Figure 5B). Slee, Adrain, and Martin reported that caspase-3 is the primary executioner caspase in apoptotic death and is necessary for the cytochrome c/dATP-inducible cleavage of fodrin, gelsolin, and U1 small nuclear ribonucleoprotein and DNA fragmentation factor 45/inhibitor of caspase-activated DNase [28]. Caspase-3 is also essential for apoptosis-associated chromatin margination, DNA fragmentation, and nuclear collapse in this system. Therefore, based on the activation of caspase-3, the FIMGPY-induced apoptosis of HeLa cells seemed to be related to the mitochondria-mediated pathway. Therefore, the apoptotic signal will be amplified step by step and the apoptotic process promoted as the caspase-3 level increases.

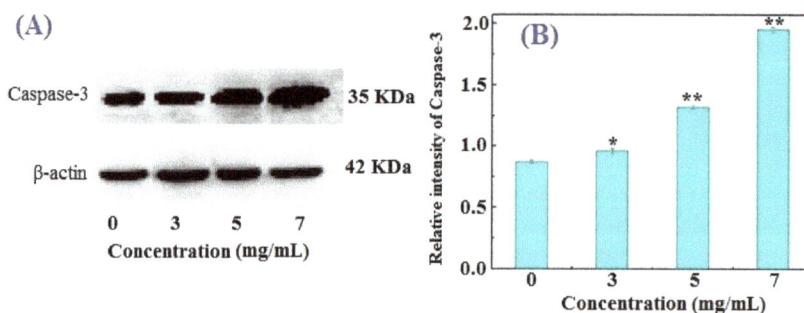

Figure 5. Caspase-3 expression in HeLa cells treated with FIMGPY for 24 h. (**A**) SDS-PAGE pattern of caspase-3 and (**B**) the relative intensity of caspase-3. * $p < 0.05$ and ** $p < 0.01$ vs. control.

2.5. DNA Ladder Analysis

The degradation of nuclear DNA into nucleosomal units is one of the hallmarks of apoptotic cell death. During this process, chromatin DNA is cleaved into inter-nucleosomal fragments, which will show a ladder pattern in agarose gel electrophoresis; thus, apoptosis can be detected via a DNA laddering assay [29,30]. As shown in Figure 6, the DNA bands from the control group of HeLa cells remained intact, whereas DNA ladder patterns were observed for the HeLa cells treated with different concentrations of FIMGPY for 24 h. DNA fragmentation also increased as the FIMGPY concentration increased. These results indicated that FIMGPY could induce apoptosis in HeLa cells and that the number of apoptotic cells increases as the FIMGPY concentration increases. This finding is in good agreement with the AO/EB staining, flow cytometry, and Western blotting analysis results.

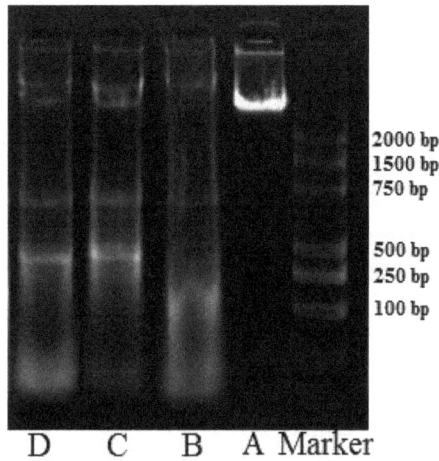

Figure 6. DNA fragmentation assay of HeLa cells treated with different concentrations of FIMGPY for 24 h. (**A**) Blank control; (**B**) 3 mg/mL; (**C**) 5 mg/mL; (**D**) 7 mg/mL, and the MV 2000 marker.

2.6. Discussion

The structural properties can provide effective guides for evaluating food proteins as potential precursors of bioactive peptides, and design the rational enzymolysis technology to prepare the bioactive peptides from various food-resources proteins [13]. At present, there is still a shortage of solid evidence to clarify the relationship between structural properties of peptides and their anticancer property. However, hydrophobicity, molecular size, amino acid composition, and sequence are deemed to play an essential role in bioactivity of peptides [31,32]. Molecular size ranged from 0.5 to 3 kDa has been supposed to be a key factor affecting the bioactivity of oxidant activity of protein hydrolysates and peptides [33]. CPe-III (RQSHFANAQP) with a MW of 1155 Da showed high inhibition activity on MCF-7 and MDA-MB-231 cells with EC_{50} of 2.38 and 1.50 µM. QPK with a MW of W 387.4 Da could significantly inhibit the proliferation of DU-145, PC-3, and LNCaP cells in a time- and dose-dependent manner [6]. Therefore, the anticancer activity of FIMGPY might be due to its small molecules (MW 726.9 Da).

In addition, hydrophobic properties could play an important role in their anticancer activities. For example, hydrophobic peptide fractions separated from anchovy sauce have been shown to exhibit cancer-chemopreventive effects in human lymphoma cells (U937) by inducing apoptosis in cancer cells; Ala (A) and Phe (F) were supposed to be the key factors underlying this activity [34,35]. Hydrophobic Ala (A) and Leu (L) residues in the peptide YALPAH were confirmed to be important for this peptide's anti-proliferative activities in PC-3 cells [7]. Chi et al. reported that the hydrophobic residues Trp (W) and Pro (P) in WPP play a vital role in its proliferation-inhibition ability in PC-3 cell lines. Therefore, Phe (F), Ile (I), Met (M), Pro (P), and Tyr (Y) in the sequence of FIMGPY should contribute to its high anticancer activities [2].

3. Experimental Section

3.1. Chemicals and Reagents

Skates (*R. porosa*) were purchased from Nanzhen market in Zhoushan City, China. NIH3T3 and HeLa cell lines were purchased from the China Cell Bank of the Institute of Biochemistry and Cell Biology in Shanghai, China. Methylthiazolyldiphenyl-tetrazolium bromide (MTT) and Annexin V-FITC Apoptosis Detection Kits were purchased from Sigma-Aldrich Trading Co., Ltd. (Shanghai, China). All other chemicals and reagents were of analytical grade and were obtained from Sinopharm Chemical Reagent Co., Ltd. (Shanghai, China).

3.2. Preparation of Hexapeptide FIMGPY

The hexapeptide FIMGPY, which has a molecular weight of 726.9 Da, was separated from skate (*Raja porosa*) cartilage protein hydrolysate according to the method of Pan et al. [13].

3.3. Anti-Tumor Activity

3.3.1. Anti-Proliferative Activity

Anti-proliferative activity was evaluated in vitro by MTT assay using the method of Chi et al. and expressed as IC_{50} values (defined as the concentration of peptide that caused 50% cell death) [2]. Briefly, cells were seeded at a density of 1×10^4 cells per well in a 96-well plate for 24 h at 37 °C in a 5% CO_2 incubator. Then, the cells were treated with FIMGPY at final concentrations of 3, 4, 5, 6, 7, and 8 mg/mL. Untreated cells were used as a negative control. The cell proliferation-inhibition rate (%) was calculated as follows:

$$\text{Inhibition rate (\%)} = [(A_{control} - A_{treated})/(A_{control} - A_{blank})] \times 100\%$$

3.3.2. Morphological Study with Fluorescence Microscopy

Apoptosis morphology was evaluated using AO/EB fluorescence staining [6]. Briefly, HeLa cells were seeded in a six-well plate (1×10^5 cells/well) and incubated overnight before treatment. Then, the cells were exposed to FIMGPY at concentrations of 3, 5, and 7 mg/mL for 24 h. Untreated cells served as the negative control. After the designated time, 25 µL of 100-µg/mL AO/EB dye mixture in PBS (pH 7.4) was added to the FIMGPY-treated cells. After staining, the cells were immediately visualized and imaged under a fluorescence microscope (Leica DM 3000, Leica Microsystems, Wetzlar, Germany). Each image was collected with excitation at 488 nm and emission at 520 nm.

3.3.3. Flow Cytometry Analysis

The apoptosis rate was quantitatively detected with an Annexin V-FITC/PI double-staining assay using a FACS Calibur flow cytometer (Becton Dickinson, New York, NY, USA) [2]. Annexin-V binding was performed using an Annexin-V-FITC kit as described by Sigma-Aldrich Trading Co., Ltd. (Shanghai, China). Briefly, HeLa cells were seeded at a density of 1×10^5 cells/well in six-well plates for 24 h and then treated with FIMGPY at concentrations of 3, 5, and 7 mg/mL for 24 h. Then, 1×10^5 cells were collected by centrifugation at 9000× *g* for 5 min at 4 °C, rinsed twice with cold PBS (pH 7.0), gently re-suspended in 400 µL of binding buffer, and incubated with 5 µL of Annexin V-FITC for 15 min and 5 µL of PI (100 µg/mL) for 5 min in the dark. Finally, the cells were analyzed with a flow cytometer. Data analysis was performed with BD FACStation Software (Becton Dickinson,

New York, NY, USA). Apoptosis was quantitatively confirmed by analyzing the percentage of early apoptotic cells using Annexin-V-FITC/PI double staining.

3.3.4. Western Blot Analysis

Western blot analysis was conducted according to the method of Huang et al. [6]. HeLa cells were seeded at a density of 1×10^5 cells/well in six-well plates for 24 h and then treated with FIMGPY at concentrations of 3, 5, and 7 mg/mL for 24 h; cell culture medium was used as the negative control. Then, 1×10^5 cells were collected by centrifugation at $9000 \times g$ for 5 min at 4 °C and rinsed twice with cold PBS (pH 7.2). The cells from the six-well plates were treated with 200 µL of lysis buffer containing phenylmethanesulfonyl fluoride for 30 min. Subsequently, the treated cells were centrifuged for 5 min at $12,000 \times g$, and the protein in the supernatant was measured by bicinchoninic acid (BCA) assay and separated by SDS-PAGE. After SDS-PAGE, proteins were transferred to a polyvinylidene difluoride (PVDF) membrane, and the membrane was blocked with 10% non-immune serum for 2 h and then incubated with primary antibody (Cell Signaling, rabbit monoclonal antibody, 1:1000) overnight at 4 °C. After washing three times with Tris-buffered saline with 0.1% Tween-20 (TBST) buffer, the membrane was incubated with the secondary antibody (goat-anti-rabbit horseradish peroxidase [HRP]-conjugated 1:3000) at room temperature for 2 h and subsequently washed with TBST. The intensity of the specific immunoreactive bands was detected by enhanced chemiluminescence (ECL), quantified by densitometry and expressed as a ratio to β-actin.

3.3.5. DNA Ladder Analysis

HeLa cells were seeded at a density of 1×10^5 cells/well in six-well plates for 24 h and then treated with FIMGPY at concentrations of 3, 5, and 7 mg/mL for 24 h; cell culture medium was used as the negative control. The cells were collected by centrifugation at $9000 \times g$ for 5 min at 4 °C, rinsed twice with cold PBS (pH 7.2), and treated with 500 µL of lysis buffer at 50 °C. After 12 h, an equal volume of phenol-chloroform-isoamyl alcohol was added, and the solution was mixed gently and centrifuged at $12,000 \times g$ for 10 min. Then, the aqueous phase was transferred to new Eppendorf tubes. An equal volume of cold chloroform-isoamyl alcohol was added to the tubes and mixed gently by inversion. The aqueous phase was once again transferred to new Eppendorf tubes, treated with 60 µL of ammonium acetate (10 M) and 600 µL of absolute ethanol, and then stored at −20 °C. After 12 h, the solution was centrifuged at $12,000 \times g$ for 10 min, and the precipitate was collected and air-dried for 30 min. Then, the dried DNA was dissolved in 40 µL of Tris-ethylenediaminetetraacetic acid (EDTA) (TE) buffer (pH 7.4) and electrophoresed on a 1.0% agarose gel. The gel was examined and photographed with an ultraviolet gel documentation system (iNTAS, Goettingen, Germany).

3.4. Statistical Analysis

The results are presented as the mean ± SD ($n = 3$). An ANOVA test using SPSS 19.0 (Statistical Program for Social Sciences, SPSS Corporation, Chicago, IL, USA) was used to analyze the experimental data. Significant differences were determined using Duncan's multiple-range test ($p < 0.05$ and 0.01).

4. Conclusions

In this study, the anticancer activities of the skate (*R. porosa*) cartilage protein hydrolysate peptide FIMGPY were evaluated in HeLa cells. FIMGPY displayed high anti-proliferation activities in HeLa cells, inducing apoptosis by upregulating the Bax/Bcl-2 ratio and caspase-3 activation. Thus, FIMGPY has great potential as an anti-carcinogen in the food and pharmaceutical industries. However, studies on the structure-activity relationship of bioactive peptides and in vivo studies on this peptide's anticancer activities remain to be performed.

Acknowledgments: This work was funded by the Public Projects of Zhejiang Province (2014C33034) and the International S & T Cooperation Program of China (2012DFA30600).

Author Contributions: Bin Wang and Yu-Qin Zhao conceived and designed the experiments. Fa-Yuan Hu, Xin Pan and Bin Wang performed the experiments. Xin Pan, Chang-Feng Chi and Bin Wang analyzed the data. Bin Wang contributed the reagents/materials/analytical tools and wrote the paper.

Conflicts of Interest: The authors declare no conflict of interest.

References

1. Umayaparvathi, S.; Meenakshi, S.; Vimalraj, V.; Arumugam, M.; Sivagami, G.; Balasubramanian, T. Antioxidant activity and anticancer effect of bioactive peptide from enzymatic hydrolysate of oyster (*Saccostrea cucullata*). *Biomed. Prev. Nutr.* **2014**, *4*, 343–353. [CrossRef]

2. Chi, C.F.; Hu, F.Y.; Wang, B.; Li, T.; Ding, G.F. Antioxidant and anticancer peptides from protein hydrolysate of blood clam (*Tegillarca granosa*) muscle. *J. Funct. Foods* **2015**, *15*, 301–313. [CrossRef]

3. Taddia, L.; D'Arca, D.; Ferrari, S.; Marraccini, C.; Severi, L.; Ponterini, G.; Assaraf, Y.G.; Marverti, G.; Costi, M.P. Inside the biochemical pathways of thymidylate synthase perturbed by anticancer drugs: Novel strategies to overcome cancer chemoresistance. *Drug Resist. Update* **2015**, *23*, 20–54. [CrossRef] [PubMed]

4. Zhu, C.Z.; Zhang, W.G.; Zhou, G.H.; Xu, X.L.; Kang, Z.L.; Yin, Y. Isolation and identification of antioxidant peptides from Jinhua ham. *J. Agric. Food Chem.* **2013**, *61*, 1265–1271. [CrossRef] [PubMed]

5. De Castro, R.J.S.; Sato, H.H. Biologically active peptides: Processes for their generation, purification and identification and applications as natural additives in the food and pharmaceutical industries. *Food Res. Int.* **2015**, *74*, 185–198. [CrossRef]

6. Huang, F.; Yang, Z.; Yu, D.; Wang, J.; Li, R.; Ding, G. Sepia ink oligopeptide induces apoptosis in prostate cancer cell lines via caspase-3 activation and elevation of Bax/Bcl-2 ratio. *Mar. Drugs* **2012**, *10*, 2153–2165. [CrossRef] [PubMed]

7. Song, R.; Wei, R.; Luo, H.; Yang, Z. Isolation and identification of an antiproliferative peptide derived from heated products of peptic hydrolysates of half-fin anchovy (*Setipinna taty*). *J. Funct. Foods* **2014**, *10*, 104–111. [CrossRef]

8. Xue, Z.; Wen, H.; Zhai, L.; Yu, Y.; Li, Y.; Yu, W.; Cheng, A.; Wang, C.; Kou, X. Antioxidant activity and anti-proliferative effect of a bioactive peptide from chickpea (*Cicer arietinum* L.). *Food Res. Int.* **2015**, *77*, 75–81. [CrossRef]

9. Wattanasiritham, L.; Theerakulkait, C.; Wickramasekara, S.; Maier, C.S.; Stevens, J.F. Isolation and identification of antioxidant peptides from enzymatically hydrolyzed rice bran protein. *Food Chem.* **2016**, *192*, 156–162. [CrossRef] [PubMed]

10. Memarpoor-Yazdi, M.; Asoodeh, A.; Chamani, J. A novel antioxidant and antimicrobial peptide from hen egg white lysozyme hydrolysates. *J. Funct. Foods* **2012**, *4*, 278–286. [CrossRef]

11. Wang, B.; Gong, Y.; Li, Z.; Yu, D.; Chi, C.; Ma, J. Isolation and characterisation of five novel antioxidant peptides from ethanol-soluble proteins hydrolysate of spotless smoothhound (*Mustelus griseus*) muscle. *J. Funct. Foods* **2014**, *6*, 176–185. [CrossRef]

12. Wang, S.; Mateos, R.; Goya, L.; Amigo-Benavent, M.; Sarriá, B.; Bravo, L. A phenolic extract from grape by-products and its main hydroxybenzoic acids protect Caco-2 cells against pro-oxidant induced toxicity. *Food Chem. Toxicol.* **2016**, *88*, 65–74. [CrossRef] [PubMed]

13. Pan, X.; Zhao, Y.; Hu, F.; Wang, B. Preparation and identification of antioxidant peptides from protein hydrolysate of skate (*Raja porosa*) cartilage. *J. Funct. Foods* **2016**, *25*, 220–230. [CrossRef]

14. Ibrahim, B.; Sowemimo, A.; Spies, L.; Koekomoer, T.; van de Venter, M.; Odukoya, O.A. Antiproliferative and apoptosis inducing activity of *Markhamia tomentosa* leaf extract on HeLa cells. *J. Ethnopharmacol.* **2013**, *149*, 745–749. [CrossRef] [PubMed]

15. Leuschner, C.; Hansel, W. Membrane disrupting lytic peptides for cancer treatments. *Curr. Pharm. Des.* **2004**, *10*, 2299–2310. [CrossRef] [PubMed]

16. Lee, S.; Ryu, B.; Je, J.; Kim, S. Diethylaminoethyl chitosan induces apoptosis in HeLa cells via activation of caspase-3 and p53 expression. *Carbohydr. Polym.* **2011**, *84*, 571–578. [CrossRef]

17. Degterev, A.; Yuan, J. Expansion and evolution of cell death programmes. *Nat. Rev. Mol. Cell Biol.* **2008**, *9*, 378–390. [CrossRef] [PubMed]

18. Ma, J.; Huang, F.; Lin, H.; Wang, X. Isolation and purification of a peptide from *Bullacta exarata* and its impaction of apoptosis on prostate cancer cell. *Mar. Drugs* **2013**, *11*, 266–273. [CrossRef]

19. Vermes, I.; Haanen, C.; Steffens-Nakken, H.; Reutelingsperger, C. A novel assay for apoptosis. Flow cytometric detection of phosphatidylserine expression on early apoptotic cells using fluorescein labelled Annexin V. *J. Immunol. Methods* **1995**, *184*, 39–51. [CrossRef]

20. Lecoeur, H. Nuclear apoptosis detection by flow cytometry: Influence of endogenous endonucleases. *Exp. Cell Res.* **2002**, *277*, 1–14. [CrossRef] [PubMed]

21. Chen, J.; Zhao, Y.; Tao, X.; Zhang, M.; Sun, A. Protective effect of blueberry anthocyanins in a CCL4-induced liver cell model. *LWT Food Sci. Technol.* **2015**, *60*, 1105–1112. [CrossRef]

22. Morales-Cano, D.; Calviño, E.; Rubio, V.; Herráez, A.; Sancho, P.; Tejedor, M.C.; Diez, J.C. Apoptosis induced by paclitaxel via Bcl-2, Bax and caspases 3 and 9 activation in NB4 human leukaemia cells is not modulated by ERK inhibition. *Exp. Toxicol. Pathol.* **2013**, *65*, 1101–1108. [CrossRef] [PubMed]

23. Czabotar, P.E.; Lessene, G.; Strasser, A.; Adams, J.M. Control of apoptosis by the BCL-2 protein family: Implications for physiology and therapy. *Nat. Rev. Mol. Cell Biol.* **2014**, *15*, 49–63. [CrossRef] [PubMed]

24. Nys, K.; Agostinis, P. Bcl-2 family members: Essential players in skin cancer. *Cancer Lett.* **2012**, *320*, 1–13. [CrossRef] [PubMed]

25. Kumar, S. Caspase function in programmed cell death. *Cell Death Differ.* **2007**, *14*, 32–43. [CrossRef] [PubMed]

26. Brentnall, M.; Rodriguez-Menocal, L.; De Guevara, R.L.; Cepero, E.; Boise, L.H. Caspase-9, caspase-3 and caspase-7 have distinct roles during intrinsic apoptosis. *BMC Cell Biol.* **2013**, *14*, 32. [CrossRef] [PubMed]

27. Porter, A.G.; Jänicke, R.U. Emerging roles of caspase-3 in apoptosis. *Cell Death Differ.* **1999**, *6*, 99–104. [CrossRef] [PubMed]

28. Slee, E.A.; Adrain, C.; Martin, S.J. Executioner caspase-3, -6, and -7 perform distinct, non-redundant roles during the demolition phase of apoptosis. *J. Biol. Chem.* **2001**, *276*, 7320–7326. [CrossRef] [PubMed]

29. Wyllie, A.H. Glucocorticoid-induced thymocyte apoptosis is associated with endogenous endonuclease activation. *Nature* **1980**, *284*, 555–556. [CrossRef] [PubMed]

30. Vethakanraj, H.S.; Babu, T.A.; Sudarsanan, G.B.; Duraisamy, P.K.; Kumar, S.A. Targeting ceramide metabolic pathway induces apoptosis in human breast cancer cell lines. *Biochem. Biophys. Res. Commun.* **2015**, *464*, 833–839. [CrossRef] [PubMed]

31. Chi, C.F.; Hu, F.Y.; Wang, B.; Li, Z.R.; Luo, H.Y. Influence of amino acid compositions and peptide profiles on antioxidant capacities of two protein hydrolysates from skipjack tuna (*Katsuwonus pelamis*) dark muscle. *Mar. Drugs* **2015**, *13*, 2580–2601. [CrossRef] [PubMed]

32. Harnedy, P.A.; FitzGerald, R.J. Bioactive peptides from marine processing waste and shellfish: A review. *J. Funct. Foods* **2012**, *4*, 6–24. [CrossRef]

33. Sila, A.; Bougatef, A. Antioxidant peptides from marine by-products: Isolation, identification and application in food systems. A review. *J. Funct. Foods* **2016**, *21*, 10–26. [CrossRef]

34. Lee, Y.G.; Kim, J.Y.; Lee, K.W.; Kim, K.H.; Lee, H.J. Peptides from anchovy sauce induce apoptosis in a human lymphoma cell (U937) through the increase of caspase-3 and -8 activities. *Ann. N. Y. Acad. Sci.* **2003**, *1010*, 399–404. [CrossRef] [PubMed]

35. Lee, Y.G.; Lee, K.W.; Kim, J.Y.; Kim, K.H.; Lee, H.J. Induction of apoptosis in a human lymphoma cell line by hydrophobic peptide fraction separated from anchovy sauce. *Biofactors* **2004**, *21*, 63–67. [CrossRef] [PubMed]

marine drugs

MDPI

Review

An Overview of the Medical Applications of Marine Skeletal Matrix Proteins

M. Azizur Rahman [1,2]

1   Department of Chemical & Physical Sciences, University of Toronto Mississauga,
    Mississauga, ON L5L 1C6, Canada; mazizur.rahman@utoronto.ca or azizurr142@gmail.com;
    Tel.: +1-647-892-4221
2   HiGarden Inc., Markham, ON L3R 3W4, Canada

Academic Editor: Se-Kwon Kim
Received: 9 May 2016; Accepted: 7 September 2016; Published: 12 September 2016

Abstract: In recent years, the medicinal potential of marine organisms has attracted increasing attention. This is due to their immense diversity and adaptation to unique ecological niches that has led to vast physiological and biochemical diversification. Among these organisms, marine calcifiers are an abundant source of novel proteins and chemical entities that can be used for drug discovery. Studies of the skeletal organic matrix proteins of marine calcifiers have focused on biomedical applications such as the identification of growth inducing proteins that can be used for bone regeneration, for example, 2/4 bone morphogenic proteins (BMP). Although a few reports on the functions of proteins derived from marine calcifiers can be found in the literature, marine calcifiers themselves remain an untapped source of proteins for the development of innovative pharmaceuticals. Following an overview of the current knowledge of skeletal organic matrix proteins from marine calcifiers, this review will focus on various aspects of marine skeletal protein research including sources, biosynthesis, structures, and possible strategies for chemical or physical modification. Special attention will be given to potential medical applications and recent discoveries of skeletal proteins and polysaccharides with biologically appealing characteristics. In addition, I will introduce an effective protocol for sample preparation and protein purification that includes isolation technology for biopolymers (of both soluble and insoluble organic matrices) from coralline algae. These algae are a widespread but poorly studied group of shallow marine calcifiers that have great potential for marine drug discovery.

Keywords: biomineralization; coralline algae; chitin; collagen; marine calcifiers; marine skeletal proteins; proteomics

1. Introduction

Skeletal proteins and polysaccharides in marine organisms are present as complex mixtures within organic matrices. The organic matrices of marine calcifiers, for example, are a potentially untapped source of skeletal proteins [1–6]. Organic matrices have the advantage of being naturally produced, retaining the native, functional conformation of the original proteins. Moreover, a significant number of calcifying marine invertebrates produce polysaccharides within their extracellular matrices and connective tissues [7,8] that have molecular structures and functions similar to human versions [1,6]. Polysaccharides derived from marine invertebrate extracellular matrices encompass an enormous variety of structures and should be considered as an extraordinary source of biochemical diversity. However, they remain largely under-exploited with respect their potential in medical applications [9,10]. Macromolecules derived from marine calcifiers that hold promise for biomedical applications include a broad range of protein and sugar (carbohydrates and lectins) molecules that participate in signaling, development, regeneration, and metabolism.

It is has been hypothesized that marine skeletal proteins that function in biomineral growth, maintenance, and repair could facilitate tissue engineering. For example, some of these proteins with human physiological activity can help accelerate lab-based bone morphogenesis and increase bone volumes with efficacies equivalent to currently used recombinant proteins [1]. Proteins with potential for bone repair and drug discovery, extracted either from naturally occurring skeletal organic matrices or derived from cultivated tissues, can be identified and isolated using chromatography, cell assays and proteomic methods [1,9,11]. Proteomics is a high-throughput analytical method for rapidly identifying known or unknown proteins in complex mixtures [5]. If purification methods can be established for skeletal proteins derived from calcifying marine organisms, researchers in the emerging fields of proteomics and medicinal chemistry could utilize these methods for subsequent drug discovery and, as a more specific example, bone repair. Currently, primary sequences of different skeletal proteins from marine organisms are available in public databases, and this information can be used to infer the biological function and origin of individual proteins and provide clues related to the mechanisms of formation of any skeleton.

Pharmaceutical industries now accept the world's oceans as a major frontier for medical research. The emergence of this relatively new area of scientific exploration has been of enormous interest to the popular and scientific press, and several review publications have appeared on the topic [1,9,11]. In the review presented here, we focus on recent progress in the discovery and production of new marine skeletal proteins and polysaccharides of pharmaceutical interest. We also introduce a new technique for purifying compounds derived from the skeletal organic matrices of coralline algae that will be useful for proteomic analysis and purifying biopolymers such as chitin and collagen. Overall, this review demonstrates the existence of unique biomineralization-related skeletal proteins in marine calcifiers that hold promise for drug development, and moreover, provides the first description of proteinaceous components in coralline red algae.

2. Applications and Modification Strategies of Marine Skeletal Proteins for Drug Discovery

Marine calcifiers (shallow, mid-shelf, and deep sea) are widespread in oceans globally. However, due to the lack of effective extraction/analytical methods, the applications of these potential resources for drugs are comparatively fewer than for other marine organisms. Recently, we perceived protein induced crystallization [2,7,8,12,13], which showed potential crystal design and growth that could help medicinal chemistry in drug design. Our primary chemical proteomic results from soft coral revealed a number of molecules with high concentrations [5,14]. In addition, some proteins extracted from soft corals are homologous with many human proteins, making them useful due to their similarity [15,16]. The information with respect to the close homology of soft coral and human proteins provides us functional and evolutionary clues on the structure and functions of their sequences. These homologous proteins could lead to possible drug discovery and form a potential resource for biotechnological research. It is our hope that further sequence studies of these materials will contribute to a better understanding of structural proteins in soft corals. Bioassay-directed fractionation of octocoral *Cespitularia hypotentaculata*, which has a novel endoskeleton, yielded the diterpene cespitularin A–D, the norditerpene cespitularin E and three other diterpenes, cespitularin F–H [17]. Two new dolabellane-type diterpenoids and the known diterpene clavenone [18] were isolated from a octocoral *Clavularia* species [19]. A saponin compound was isolated from the octocoral *Lobophytum* spp., which was collected from Hainan Island, China.

Among the marine calcifiers, very few scleractinian corals were investigated. In a recent review, the authors discussed the potential of scleractinian coral, which has therapeutic characteristics, including anti-inflammatory properties, anticancer properties, bone repair, and neurological benefits [6]. Research on the scleractinian coral *Montipora* spp. from the republic of Korea (South Korea) found three diacetylenes (1, 4, 6). One of these was a potent cytotoxin with respect to a range of tumor cell lines [20]. The authors tested the extracted compounds against a panel of human cancer cell lines and the structures have been interpreted on the basis of spectroscopic evidence.

These three compounds showed a structural activity profile to similar to those previously reported [21]. The results showed that the compound 6 with b-hydroxy ketone functionality has strong cytotoxic properties and Methyl montiporate C (1) was active only against a skin cancer cell line, while compound 4 was moderately active. Extracts from the calcifying octocorals *Pseudopterogorgia elizabethae* (which contains pseudopterosins) and *Eunicea fusca* (which contains fucoside-A) can be used in the cosmetic industry [22]. Similarly, coral (endoskeletons and exoskeletons) and coralline algal skeletons could be used for cosmetics as both contain a high concentration of organic matrix components [7,13,23].

In recent years, numerous applications have been proposed for chitosan-based delivery devices [24–26], however, most of these were unrelated to marine calcifiers. Chitosan is a copolymer of β-(1-4)-linked 2-acetamido-2-deoxy-D-glucopyranose and 2-amino-2-deoxy-D-glucopyranose, obtained by deacethylation of the naturally occurring chitin. Chitin was firstly extracted from the exoskeleton of marine organisms, mainly crabs and shrimps, as described by Burrows [27]. This polymer has also recently been extracted from coralline algae [7], which opened the doors for possible applications of these biomaterials using a group of marine calcifers which are found in shallow water and are easy-to-collect, abundant and widespread. The major applications of chitosan are for biomaterials, pharmaceuticals, foodstuff treatment (e.g., flocculation, clarification, etc., due to its efficient interaction with other polyelectrolytes), cosmetics, metal ion sequestration, and agriculture [28–31]. Development of chitosan chemistry has relevant biomedical applications, particularly in the field of drug delivery [32]. While chitin is insoluble in most common solvents, chitosan can be readily turned into fibers as well as films, or triggered in a variety of micromorphologies from its acidic aqueous solutions. Protein-polysaccharides play an important role in biomedical and pharmaceutical applications. However, at times the properties of such biomaterials do not meet the needs for exact applications. As a result, approaches that chemically or physically modify their structure and, thus, physical-chemical properties are increasingly gaining interest [33,34]. With respect to the polysaccharides' chitin and chitosan, it is possible to target the reaction using sulfur trioxide-pyridine at two sites or at only one specific site, following different pathways of synthesis [35]. Great efforts have thus focused on the progress of efficient modification reactions in well-controlled conditions under tolerable temperatures [35]. For example, modification reactions of water-soluble chitin can be conducted in aqueous solutions or in organic solvents in an engorged state under mild conditions, and selective *N*-acetylation [35]. Some significant chemical reactions of acylation, alkylation, Schiff base formation and reductive *N*-alkylation, carboxyalkylation, *N*-phthaloylation are well described [35].

3. A Promising Future for Marine Calcifiers in Drug Discovery

Marine resources such as coral, mollusk and coralline algae could be a major source of medicines over the next decades. It is estimated that marine ecosystems, such as those found in coral reefs or at a deep sea level have greater biological diversity than those of tropical rain forests. However, as with tropical rain forests, coral reefs represent considerable untouched potential in the science of medicine.

At present, marine calcifier collection and drug appraisal occurs successfully. However, there is no question that these resources are inadequate and it is possible that collectable marine organisms will be almost completely explored within the next 20 years. There is still a doubt as to where scientists will turn in order to ensure a continuing flow of new medicines. The solution is difficult, however drugs can now be developed using many methods such as computer-aided design, combinatorial synthesis and proteomics. The chemical multiplicity of marine ecosystems, from simple to complex peptide and protein extraction, draws us in the direction of the discovery of new marine natural products in various therapeutic areas such as cancer, inflammation, microbial infections, and various other deadly diseases [36]. Cancer is the biggest challenge of the current century, and marine calcifying organisms show new promise in fighting against this and other dangerous diseases.

4. A Novel Approach to Isolation, Purification and Characterization of Marine Skeletal Proteins

Isolation and purification of skeletal proteins from marine calcifiers are complex because of the potential for contamination of the soft tissues and the high sensitivity of organic matrices to handling. However, successfully purified skeletal proteins from several groups of marine calcifiers have recently emerged [4,5,14,22,23,37]. The overview concerning marine skeletal proteins presented above allows us to understand some newly developed techniques [5,12,14–16,23,38–40] as well as useful methods for isolating and purifying skeletal proteins and proteomic analysis. Among marine calcifiers, we recently investigated coralline red algae, which has specific biological characteristics [7] and contains high concentrations of soluble organic matrix (SOM) and insoluble organic matrix (IOM) fractions. High concentrations of both chitin and collagen biopolymers are present in SOM and IOM (Figure 1). Coralline algal concentrations of SOM (0.9%) and IOM (4.5%) are significantly higher than those of other skeletal marine calcifiers such as octocorals, with SOM and IOM concentrations of 0.03% and 0.05%, respectively [5,13,15]. The highly concentrated biopolymers present in skeletal organic matrices open up the possibility for future drug development, because these two polymers are frequently utilized in drug design [24,29–31,41–46].

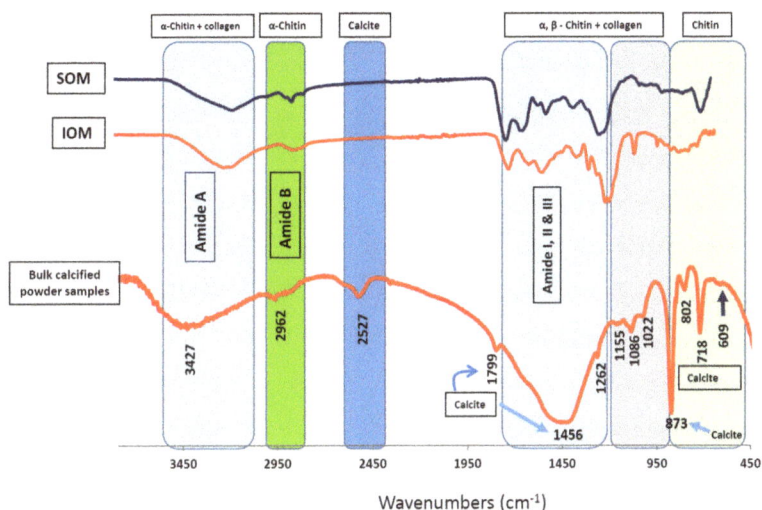

Figure 1. Identification of chitin and collagen in algal skeletal protein-polysaccharides complexes. Structural comparison of FTIR spectra between organic matrix fractions (soluble organic matrix (SOM) and insoluble organic matrix (IOM)) and bulk skeletal powder. Graphs for SOM, IOM fractions and bulk skeletal powder are indicated. Different colored boxes in the spectra indicate involvement of molecules in SOM and IOM fractions in forming skeletal structure in coralline algal calcification system. (Reproduced from Scientific Reports, Rahman and Halfar 2014 [7]).

Detailed geochemical studies of coralline algae [47–51] provide a broad spectrum of environmental and structural background information. However, there is a lack of information on the protein-polysaccharide complex in the coralline algal skeleton, which plays a key role in the regulation of biocalcification [7] and may contain prospective biomaterials for drug development. Hence, we have developed a useful technique from sample preparation to protein isolation for the Sub-Arctic coralline alga *Clathromorphum compactum* (Figure 2, see Ref. [7] for details) using recently developed analytical approaches for other marine calcifiers (Figure 2, References [5,12,23]). We characterized the SOM-polysaccharide complex from its $CaCO_3$ skeleton, which is involved in the biocalcification process. Sodium dodecyl sulfate-polyacrylamide gel electrophoresis (SDS-PAGE) analysis [52] of the preparations [5,14,23] showed two bands of proteins with molecular masses of 250-kDa and 30-kDa (Figure 3A, lane 1 and 2). The protein with molecular masses of 30-kDa was by far the most abundant protein, whereas the 250 kDa protein band was weak and somewhat faint (Figure 3A, lane 2). Periodic acid-Schiff (PAS) staining was used to identify chitin associated glycoproteins. Interestingly, the 250-kDa protein was identified with high abundance as the only glycoprotein contained in the skeleton (Figure 3B, lane 1 and 2), even though it only appeared as a weak band in Coomassie Brilliant Blue (CBB) staining solution (Figure 3A). Chitin is the main component of the protein-polysaccharide complex of cell walls [7,53,54]), which is also composed of glycoprotein [55]. Protein-polysaccharide complexes are also present in coralline algal cell structures [7]. Therefore, detection of a strong glycosylation protein in coralline algal skeletons reveals the presence of highly abundant chitin. The chitin found in coralline alga has been recognized to be involved in the calcification process [7] and this polymer is considered highly useful for drug design [24,29–31,41–46]. Our observations therefore strongly suggest that the skeletal matrix proteins in coralline alga are not only a structural protein but also have potential for drug development.

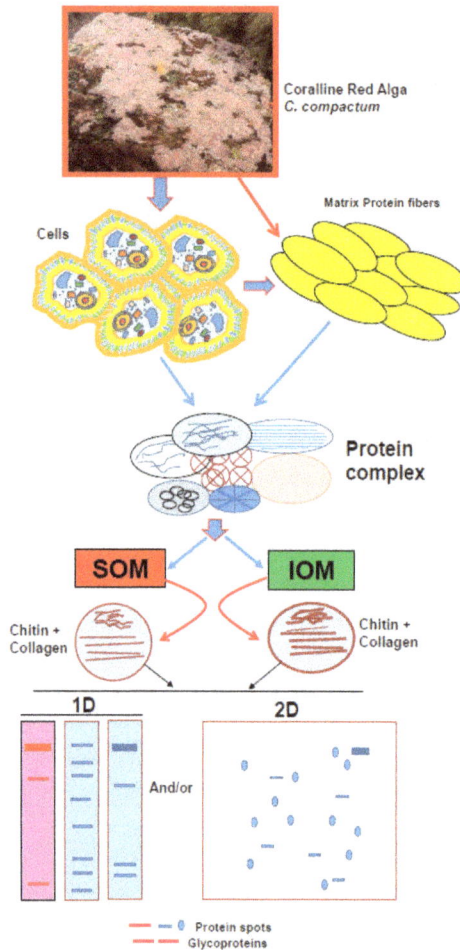

Figure 2. Model of general strategy for analyzing protein-polysaccharides complex from skeletal organic matrix of marine calcifiers.

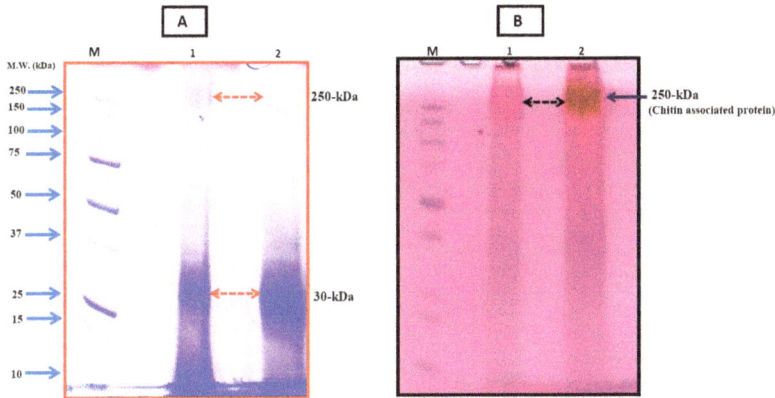

Figure 3. Electrophoretic analysis of skeletal matrix proteins extracted from the coralline red alga *C. compactum*. (**A**) SDS-PAGE fractionation with Coomassie Brilliant Blue (CBB) staining after purification of the skeletal proteins. Lane 1 and 2 indicate purified skeletal proteins. Arrows indicate protein bands; (**B**) SDS-PAGE gel with Periodic Acid-Schiff (PAS) staining to identify glycoprotein in skeletal matrix proteins of *C. compactum*. Lane 1 and 2, a strong abundant chitin associated glycoprotein was identified (indicated by arrow) by periodic Acid-Schiff staining. An eluate (derived from 5 g of algal skeleton) was run on 12% polyacrylamide gel M, protein marker. The Precision Plus SDS-PAGE standard (Bio-Rad) was used as protein marker for electrophoresis.

5. Conclusions

In this brief review, recent advances in applications of protein-polysaccharides of marine calcifiers in the medical and pharmaceutical fields have been discussed. The results demonstrate the potential for marine calcifiers to generate new drugs. Understanding the proteinaceous components of marine calcifiers is an important step toward advancing the science of marine medicinal chemistry. Among the different sources of polysaccharides, algal polysaccharides such as chitin and collagen could play an important role in future development of tissue engineering, bone regeneration, and much more. In light of these emerging findings, in the near future established techniques may also be potentially useful for isolating skeletal proteins from similar marine calcifiers for drug discovery. As a discovery-driven science, the techniques discussed here allow researcher to identify candidate proteins for drug discovery and identify unknowns without missing unanticipated interactions. These techniques can be employed to dramatically improve the range of applications within the field of marine drug discovery. Since the marine realm consists of diverse ecosystems and matrices in which these proteins reside, the development of effective methods for accessing proteins will be a continuing challenge in future years.

Acknowledgments: The author thanks Jochen Halfar for providing coralline samples. The author also thanks Steven Short for providing electrophoresis apparatus.

Conflicts of Interest: The author declares no conflict of interest.

References

1. Green, D.W.; Padula, M.P.; Santos, J.; Chou, J.; Milthorpe, B.; Ben-Nissan, B. A therapeutic potential for marine skeletal proteins in bone regeneration. *Mar. Drugs* **2013**, *11*, 1203–1220. [CrossRef] [PubMed]
2. Rahman, M.A.; Fujimura, H.; Shinjo, R.; Oomori, T. Extracellular matrix protein in calcified endoskeleton: A potential additive for crystal growth and design. *J. Cryst. Growth* **2011**, *324*, 177–183. [CrossRef]

3. Rahman, M.A.; Isa, Y.; Uehara, T. Proteins of calcified endoskeleton: Ii partial amino acid sequences of endoskeletal proteins and the characterization of proteinaceous organic matrix of spicules from the alcyonarian, synularia polydactyla. *Proteomics* **2005**, *5*, 885–893. [CrossRef] [PubMed]

4. Drake, J.L.; Mass, T.; Haramaty, L.; Zelzion, E.; Bhattacharya, D.; Falkowski, P.G. Proteomic analysis of skeletal organic matrix from the stony coral stylophora pistillata. *Proc. Natl. Acad. Sci. USA* **2013**, *110*, 3788–3793. [CrossRef] [PubMed]

5. Rahman, M.A.; Karl, K.; Nonaka, M.; Fujimura, H.; Shinjo, R.; Oomori, T.; Worheide, G. Characterization of the proteinaceous skeletal organic matrix from the precious coral *Corallium konojoi*. *Proteomics* **2014**, *14*, 2600–2606. [CrossRef] [PubMed]

6. Cooper, E.L.; Hirabayashi, K.; Strychar, K.B.; Sammarco, P.W. Corals and their potential applications to integrative medicine. *Evid. Based Complement. Altern. Med.* **2014**, *2014*, 184959. [CrossRef] [PubMed]

7. Rahman, M.A.; Halfar, J. First evidence of chitin in calcified coralline algae: New insights into the calcification process of clathromorphum compactum. *Sci. Rep.* **2014**, *4*, 6162. [CrossRef] [PubMed]

8. Rahman, M.A.; Oomori, T. In vitro regulation of $CaCO_3$ crystal growth by the highly acidic proteins of calcitic sclerites in soft coral, *Sinularia polydactyla*. *Connect. Tissue Res.* **2009**, *50*, 285–293. [CrossRef] [PubMed]

9. Laurienzo, P. Marine polysaccharides in pharmaceutical applications: An overview. *Mar. Drugs* **2010**, *8*, 2435–2465. [CrossRef] [PubMed]

10. Senni, K.; Pereira, J.; Gueniche, F.; Delbarre-Ladrat, C.; Sinquin, C.; Ratiskol, J.; Godeau, G.; Fischer, A.M.; Helley, D.; Colliec-Jouault, S. Marine polysaccharides: A source of bioactive molecules for cell therapy and tissue engineering. *Mar. Drugs* **2011**, *9*, 1664–1681. [CrossRef] [PubMed]

11. Rahman, M.A. The medicinal potential of promising marine organisms: A review. *Blue Biotechnol. J.* **2012**, *1*, 318–333.

12. Rahman, M.A.; Oomori, T. Analysis of protein-induced calcium carbonate crystals in soft coral by near-field IR microspectroscopy. *Anal. Sci.* **2009**, *25*, 153–155. [CrossRef] [PubMed]

13. Rahman, M.A.; Oomori, T. Structure, crystallization and mineral composition of sclerites in the alcyonarian coral. *J. Cryst. Growth* **2008**, *310*, 3528–3534. [CrossRef]

14. Rahman, M.A.; Shinjo, R.; Oomori, T.; Worheide, G. Analysis of the proteinaceous components of the organic matrix of calcitic sclerites from the soft coral *Sinularia* sp. *PLoS ONE* **2013**, *8*, e58781. [CrossRef]

15. Rahman, M.A.; Isa, Y. Characterization of proteins from the matrix of spicules from the alcyonarian, *Lobophytum crassum*. *J. Exp. Mar. Biol. Ecol.* **2005**, *321*, 71–82. [CrossRef]

16. Rahman, M.A.; Isa, Y.; Uehara, T. Studies on two closely related species of octocorallians: Biochemical and molecular characteristics of the organic matrices of endoskeletal sclerites. *Mar. Biotechnol.* **2006**, *8*, 415–424. [CrossRef] [PubMed]

17. Duh, C.Y.; Chien, S.C.; Song, P.Y.; Wang, S.K.; El-Gamal, A.A.H.; Dai, C.F. New cadinene sesquiterpenoids from the formosan soft coral xenia puerto-galerae. *J. Nat. Prod.* **2002**, *65*, 1853–1856. [CrossRef] [PubMed]

18. Mori, K.; Iguchi, K.; Yamada, N.; Yamada, Y.; Inouye, Y. Bioactive marine diterpenoids from Japanese soft coral of *Clavularia* sp. *Chem. Pharm. Bull.* **1988**, *36*, 2840–2852. [CrossRef] [PubMed]

19. Iguchi, K.; Sawai, H.; Nishimura, H.; Fujita, M.; Yamori, T. New dolabellane-type diterpenoids from the Okinawan soft coral of the genus Clavularia. *Bull. Chem. Soc. Jpn.* **2002**, *75*, 131–136. [CrossRef]

20. Alam, N.; Hong, J.K.; Lee, C.O.; Choi, J.S.; Im, K.S.; Jung, J.H. Additional cytotoxic diacetylenes from the stony coral *Montipora* sp. *Chem. Pharm. Bull.* **2002**, *50*, 661–662. [CrossRef] [PubMed]

21. Alam, N.; Bae, B.H.; Hong, J.; Lee, C.O.; Im, K.S.; Jung, J.H. Cytotoxic diacetylenes from the stony coral Montipora species. *J. Nat. Prod.* **2001**, *64*, 1059–1063. [CrossRef] [PubMed]

22. Roussis, V.; Wu, Z.D.; Fenical, W.; Strobel, S.A.; Vanduyne, G.D.; Clardy, J. New antiinflammatory pseudopterosins from the marine octocoral pseudopterogorgia-elisabethae. *J. Org. Chem.* **1990**, *55*, 4916–4922. [CrossRef]

23. Rahman, M.A.; Oomori, T.; Worheide, G. Calcite formation in soft coral sclerites is determined by a single reactive extracellular protein. *J. Biol. Chem.* **2011**, *286*, 31638–31649. [CrossRef] [PubMed]

24. Zhang, J.; Xia, W.; Liu, P.; Cheng, Q.; Tahirou, T.; Gu, W.; Li, B. Chitosan modification and pharmaceutical/biomedical applications. *Mar. Drugs* **2010**, *8*, 1962–1987. [CrossRef] [PubMed]

25. Hejazi, R.; Amiji, M. Chitosan-based gastrointestinal delivery systems. *J. Control. Release* **2003**, *89*, 151–165. [CrossRef]

26. Illum, L.; Jabbal-Gill, I.; Hinchcliffe, M.; Fisher, A.N.; Davis, S.S. Chitosan as a novel nasal delivery system for vaccines. *Adv. Drug. Deliv. Rev.* **2001**, *51*, 81–96. [CrossRef]

27. Burrows, F.; Louime, C.; Abazinge, M.; Onokpise, O. Extraction and evaluation of chitin from crub exoskeleton as a seed fungicide and plant growth enhancer. *Am. Eurasian J. Agric. Environ. Sci.* **2007**, *2*, 103–111.

28. Rinaudo, M. Chitin and chitosan: Properties and application. *Prog. Polym. Sci.* **2006**, *31*, 603–632. [CrossRef]

29. Da Sacco, L.; Masotti, A. Chitin and chitosan as multipurpose natural polymers for groundwater arsenic removal and as 203 delivery in tumor therapy. *Mar. Drugs* **2010**, *8*, 1518–1525. [CrossRef] [PubMed]

30. Aam, B.B.; Heggset, E.B.; Norberg, A.L.; Sorlie, M.; Varum, K.M.; Eijsink, V.G. Production of chitooligosaccharides and their potential applications in medicine. *Mar. Drugs* **2010**, *8*, 1482–1517. [CrossRef] [PubMed]

31. Khoushab, F.; Yamabhai, M. Chitin research revisited. *Mar. Drugs* **2010**, *8*, 1988–2012. [CrossRef] [PubMed]

32. Kumar, M.N.; Muzzarelli, R.A.; Muzzarelli, C.; Sashiwa, H.; Domb, A.J. Chitosan chemistry and pharmaceutical perspectives. *Chem. Rev.* **2004**, *104*, 6017–6084. [CrossRef] [PubMed]

33. Holte, O.; Onsoyen, E.; Myrvold, R.; Karlsen, J. Sustained release of water-soluble drug from directly compressed alginate tablets. *Eur. J. Pharm. Sci.* **2003**, *20*, 403–407. [CrossRef] [PubMed]

34. Tonnesen, H.H.; Karlsen, J. Alginate in drug delivery systems. *Drug Dev. Ind. Pharm.* **2002**, *28*, 621–630. [CrossRef] [PubMed]

35. Kurita, K. Chitin and chitosan: Functional biopolymers from marine crustaceans. *Mar. Biotechnol.* **2006**, *8*, 203–226. [CrossRef] [PubMed]

36. Rawat, D.S.; Joshi, M.C.; Joshi, P.; Atheaya, H. Marine peptides and related compounds in clinical trial. *Anticancer Agents Med. Chem.* **2006**, *6*, 33–40. [CrossRef] [PubMed]

37. Mann, K.; Poustka, A.J.; Mann, M. In-depth, high-accuracy proteomics of sea urchin tooth organic matrix. *Proteome Sci.* **2008**, *6*, 33. [CrossRef] [PubMed]

38. Cusack, M.; Freer, A. Biomineralization: Elemental and organic influence in carbonate systems. *Chem. Rev.* **2008**, *108*, 4433–4454. [CrossRef] [PubMed]

39. Debreuil, J.; Tambutte, E.; Zoccola, D.; Deleury, E.; Guigonis, J.M.; Samson, M.; Allemand, D.; Tambutte, S. Molecular cloning and characterization of first organic matrix protein from sclerites of red coral, *Corallium rubrum*. *J. Biol. Chem.* **2012**, *287*, 19367–19376. [CrossRef] [PubMed]

40. Rahman, M.A.; Oomori, T.; Uehara, T. Carbonic anhydrase in calcified endoskeleton: Novel activity in biocalcification in alcyonarian. *Mar. Biotechnol.* **2008**, *10*, 31–38. [CrossRef] [PubMed]

41. Ruiz-Herrera, J.; San-Blas, G. Chitin synthesis as target for antifungal drugs. *Curr. Drug Targets Infect. Disord.* **2003**, *3*, 77–91. [CrossRef] [PubMed]

42. Chaudhary, P.M.; Tupe, S.G.; Deshpande, M.V. Chitin synthase inhibitors as antifungal agents. *Mini Rev. Med. Chem.* **2013**, *13*, 222–236. [CrossRef] [PubMed]

43. Reese, T.A.; Liang, H.E.; Tager, A.M.; Luster, A.D.; van Rooijen, N.; Voehringer, D.; Locksley, R.M. Chitin induces accumulation in tissue of innate immune cells associated with allergy. *Nature* **2007**, *447*, 92–96. [CrossRef] [PubMed]

44. Natali, I.; Tempesti, P.; Carretti, E.; Potenza, M.; Sansoni, S.; Baglioni, P.; Dei, L. Aragonite crystals grown on bones by reaction of CO_2 with nanostructured $Ca(OH)_2$ in the presence of collagen. Implications in archaeology and paleontology. *Langmuir* **2014**, *30*, 660–668. [CrossRef] [PubMed]

45. Addad, S.; Exposito, J.Y.; Faye, C.; Ricard-Blum, S.; Lethias, C. Isolation, characterization and biological evaluation of jellyfish collagen for use in biomedical applications. *Mar. Drugs* **2011**, *9*, 967–983. [CrossRef] [PubMed]

46. Yang, T.L. Chitin-based materials in tissue engineering: Applications in soft tissue and epithelial organ. *Int. J. Mol. Sci.* **2011**, *12*, 1936–1963. [CrossRef] [PubMed]

47. Halfar, J.; Adey, W.H.; Kronz, A.; Hetzinger, S.; Edinger, E.; Fitzhugh, W.W. Arctic sea-ice decline archived by multicentury annual-resolution record from crustose coralline algal proxy. *Proc. Natl. Acad. Sci. USA* **2013**, *110*, 19737–19741. [CrossRef] [PubMed]

48. Adey, W.H.; Halfar, J.; Williams, B. The coralline genus *clathromorphum* foslie emend. Adey: Biological, physiological, and ecological factors controlling carbonate production in an arctic-subarctic climate archive. *Smithson. Contrib. Mar. Sci.* **2013**, *40*, 1–41. [CrossRef]

49. Ries, J.B.; Anderson, M.A.; Hill, R.T. Seawater Mg/Ca controls polymorph mineralogy of microbial caco3: A potential proxy for calcite-aragonite seas in precambrian time. *Geobiology* **2008**, *6*, 106–119. [CrossRef] [PubMed]

50. Ries, J.B.; Cohen, A.L.; McCorkle, D.C. Marine calcifiers exhibit mixed responses to CO_2-induced ocean acidification. *Geology* **2009**, *37*, 1131–1134. [CrossRef]

51. Halfar, J.; Zack, T.; Kronz, A.; Zachos, J.C. Growth and high-resolution paleoenvironmental signals of rhodoliths (coralline red algae): A new biogenic archive. *J. Geophys. Res.* **2000**, *105*, 22107–22116. [CrossRef]

52. Laemmli, U.K. Cleavage of structural proteins during the assembly of the head of bacteriophage T4. *Nature* **1970**, *227*, 680–685. [CrossRef] [PubMed]

53. Banks, I.R.; Specht, C.A.; Donlin, M.J.; Gerik, K.J.; Levitz, S.M.; Lodge, J.K. A chitin synthase and its regulator protein are critical for chitosan production and growth of the fungal pathogen cryptococcus neoformans. *Eukaryot. Cell* **2005**, *4*, 1902–1912. [CrossRef] [PubMed]

54. Sendbusch, P.V. Cell Walls of Algae. Available online: https://s10.lite.msu.edu/res/msu/botonl/b_online/e26/26d.htm (accessed on 8 September 2016).

55. Lipke, P.N.; Ovalle, R. Cell wall architecture in yeast: New structure and new challenges. *J. Bacteriol.* **1998**, *180*, 3735–3740. [PubMed]

marine drugs

MDPI

Review

Marine Microbiological Enzymes: Studies with Multiple Strategies and Prospects

Yan Wang, Qinghao Song and Xiao-Hua Zhang *

College of Marine Life Sciences, Ocean University of China, Qingdao 266003, China;
wangy12@ouc.edu.cn (Y.W.); 016080910050@sjtu.edu.cn (Q.S.)
* Correspondence: xhzhang@ouc.edu.cn; Tel./Fax: +86-532-82032767

Academic Editor: Se-Kwon Kim
Received: 12 July 2016; Accepted: 14 September 2016; Published: 22 September 2016

Abstract: Marine microorganisms produce a series of promising enzymes that have been widely used or are potentially valuable for our daily life. Both classic and newly developed biochemistry technologies have been broadly used to study marine and terrestrial microbiological enzymes. In this brief review, we provide a research update and prospects regarding regulatory mechanisms and related strategies of acyl-homoserine lactones (AHL) lactonase, which is an important but largely unexplored enzyme. We also detail the status and catalytic mechanism of the main types of polysaccharide-degrading enzymes that broadly exist among marine microorganisms but have been poorly explored. In order to facilitate understanding, the regulatory and synthetic biology strategies of terrestrial microorganisms are also mentioned in comparison. We anticipate that this review will provide an outline of multiple strategies for promising marine microbial enzymes and open new avenues for the exploration, engineering and application of various enzymes.

Keywords: AHL lactonase; polysaccharide-degrading enzymes; marine microorganism

1. Introduction

Microorganisms produce series of enzymes [1–3]. Given the complicated diversity and ease of large-scale fermentation, microorganisms are widely used in the exploration of enzyme resources. In recent decades, bacteria and fungi from the terrestrial environment have served as the most important and best-studied sources for promising industrial enzymes and secondary metabolites. Recent technological developments have made it easier to utilize marine resources, especially from the deep sea [4,5]. The ocean occupies greater than 70% of total surface of the earth, thus serving as a habitat for numerous microorganisms with vast diversity. The special environmental conditions, involving low temperature, low light, high pressure and high salinity, give marine residents multiple novel characteristic features, which have been attracting increasing attention from marine biologists. Correspondingly, these organisms also produce various novel enzymes and secondary metabolites, some of which have already been used as food additives and potential drugs [6–8]. For example, various PKS (polyketide synthase) and NRPS (non-ribosomal peptide synthetase) enzymes responsible for producing secondary metabolites have been identified in marine bacteria, particularly *Streptomyces*, and fungi in recent years [9–12]. However, compared to terrestrial resources, marine microbial resources, e.g., amylase and alginate lyase, are largely unexplored, although this pool of marine resources is huge. Moreover, the regulatory mechanisms of promising genes and signaling pathway cascades of marine microorganisms are also largely unknown.

In this review, we provide a brief description of two types of promising marine enzymes: acyl-homoserine lactones (AHL) lactonase and polysaccharide-degrading enzymes. We will present the recent research progress regarding these enzymes and discuss potential strategies for further studies. Using well-studied terrestrial microorganisms as references, we hope to open new avenues of exploration, engineering and regulatory mechanisms of marine enzymes.

2. AHL Lactonase

2.1. Introduction

Quorum sensing is a population-dependent reaction for microorganisms that occurs via the up/down-regulation of downstream gene expression [13]. The process is also essential for biofilm formation and the secretion of virulence factors, especially in pathogenic bacteria, and causes a series of bacterial diseases [14]. Correspondingly, quorum-quenching technology is an environmentally friendly strategy for disease control [15,16]. AHL lactonase, which degrades molecular *N*-acyl homoserine lactone (AHL) signals, is one of the two types of enzymes involved in quorum quenching (Figure 1). AHL lactonase can open the lactonic ring, and the ring-open molecule is ineffective and cannot reorganize downstream receptor proteins. Several AHL lactonases have been isolated and well-studied in recent years (Table 1). The first reported and well-studied AHL lactonase is AiiA, which inhibits the pathogenic bacterium *Erwinia carotovora* and other plant-related pathogenic bacteria with considerably high activities [17,18]. AiiA is one type of metalloprotein that contains the highly conserved amino acid sequence HXDH-H-D, which serves as a zinc-binding site [17,18]. The conserved sequence is also present in a series of AHL lactonase family members and is essential for normal protein activity [19]. Previous studies revealed that the combination of a yeast strain overexpressing AiiA and the pathogenic bacterium *Aeromonas hydrophila* significantly decreased the death rate of cultivated carp [20,21]. In addition, AiiA, AiiB, AttM, AhlD, QsdA, AiiM and AidH were cloned and characterized from soil *Bacillus*, *Agrobacterium*, *Arthrobacter*, *Rhodococcus*, *Microbacterium* and *Ochrobactrum*, respectively [22–28]. All of these AHL lactonases were from terrestrial bacteria (Figure 2).

Table 1. Properties of well-studied acyl-homoserine lactones (AHL) lactonase.

QQ Enzyme	Length (aa)	Predictable Domains	Signal Peptide	Host Organisms	Origin	Structure	References
AiiA	231aa	Beta-lactamase family (15–216)	No signal	*Bacillus*	terrestrial	3DHB	[18]
AiiB	276aa	Beta-lactamase family (42–259)	No signal	*Agrobacterium*	terrestrial	unknown	[22]
AttM	295aa	Beta-lactamase family (78–282)	1–17	*Agrobacterium*	terrestrial	unknown	[25]
QsdA	323aa	Phosphotriesterase family (11–322)	No signal	*Rhodococcus*	terrestrial	unknown	[24]
AidH	279aa	Alpha/beta hydrolase (25–147)	No signal	*Ochrobactrum*	terrestrial	unknown	[27,28]
GKL	330aa	Phosphotriesterase family (16–329)	No signal	*Geobacillus*	terrestrial	unknown	[26]
MomL	293aa	Beta-lactamase family (72–277)	1–21	*Muricauda*	oceanic	unknown	[29]
QsdH	968aa	AcrB/AcrD/AcrF family (182–964)	1–23	*Pseudoalteromonas*	oceanic	unknown	[30]

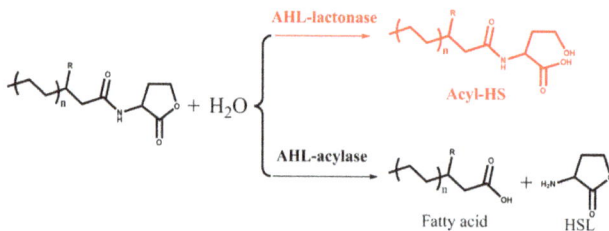

Figure 1. Catalytic mechanism of quorum quenching enzymes.

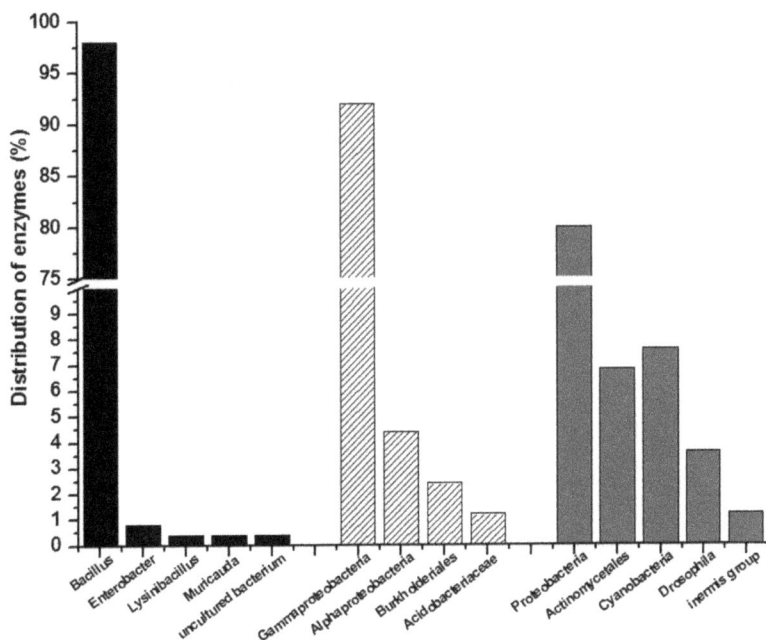

Figure 2. Distribution of AHL lactonase (**left**), amylase (**middle**) and alginate lyase (**right**). Calculation method of the percentage: The reported enzymes were bioinformatic analyzed and classified based on different species source.

2.2. Marine Resources of AHL Lactonase and Research Methods

Only a few types of AHL lactonases have been discovered from marine environments. QsdH, which was discovered from marine *Pseudoalteromonas*, is not a single protein but combines with a special transporter of small molecules that belongs to the resistance-nodulation-cell division (RND) superfamily [30]. It is promising to exploit the quorum-quenching enzymes from relatively unknown marine environments. Several new methods have been studied for quorum quenching enzymes exploration. Recently, our group developed a novel high-throughput strategy for identifying bacterial strains with quorum-quenching activity [31]. This method (A136 liquid X-gal (5-bromo-4-chloro-3-indolyl-β-D-galactopyranoside) assay) is based on the measurement of residual AHL molecules after the reaction (Figure 3). Compared with previous strategies, the main improvement of the A136 liquid X-gal assay involves the detection of β-galactosidase activity in a rapid and quantitative manner. Although the ONPG assay (2-nitrophenyl β-D-galactopyranoside) is a broadly approved strategy to measure β-galactosidase activity, the complex process of the ONPG assay causes low throughput and relatively poor efficiency [31]. However, the A136 liquid X-gal assay can be performed in a 96-well plate layout with one-step detection of enzymatic activity in a high-throughput manner. With the A136 liquid X-gal assay, 25 quorum-quenching bacterial strains belonging to different species were identified from hundreds of candidate marine bacterial strains. Additionally, with this method, several genera with quorum quenching activity, such as *Flaviramulus*, *Muricauda* and *Rhodobacter*, were identified [31].

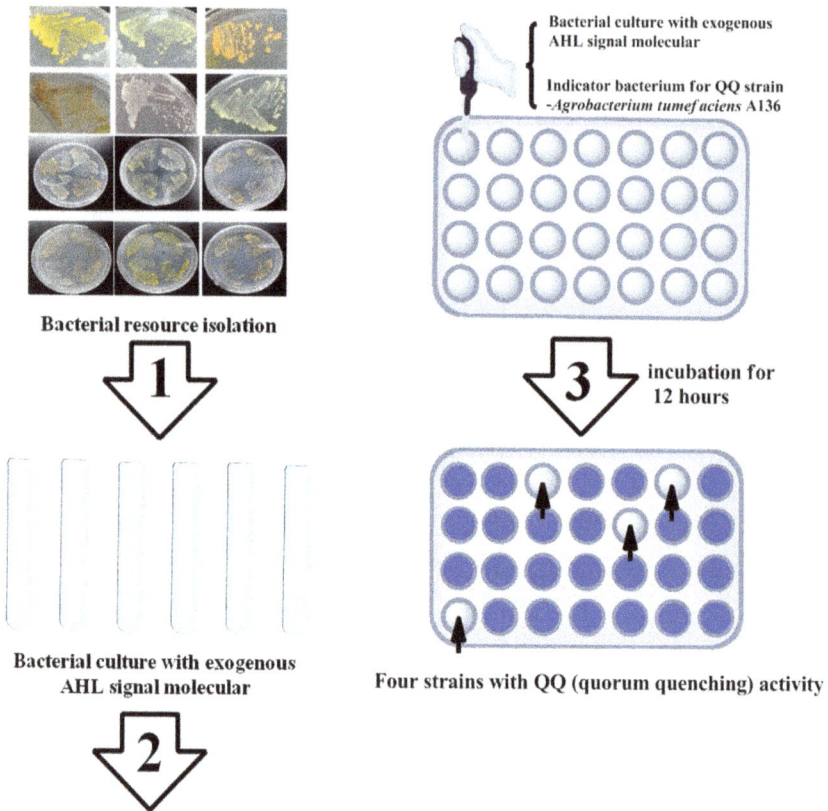

Figure 3. Schematic diagram of high-throughput method for identifying quorum quenching bacteria.

The Gram-negative strain *Muricauda olearia* Th120, which was isolated from *Paralichthys olivaceus*, exhibits high-level quorum-quenching activity. Sequencing and bioinformatic results demonstrate that the genome contains a gene encoding AHL lactonase named MomL. MomL consists of 294 amino acids with a molecular weight of approximately 38.4 KDa [29]. MomL belongs to the metallo-β-lactamase family, and the homolog exhibits a 24.5% similarity with AiiA. In addition to containing the conserved ion-binding site HXDH-H-D, MomL possesses two additional novel characteristic features: (1) the protein possesses one special signal peptide consisting of 21 amino acids at the *N*-terminus of the sequence that presumably facilitate the protein secretory ability; (2) The highest amino acid identity compared with other terrestrial homologs is only approximately 20%. Using the pTWIN1 vector, MomL protein was heterologously-expressed and purified, and the expected approximately 30 KDa band was observed in SDS-PAGE (SDS-polyacrylamide gel electrophoresis). LC-MS (liquid chromatography-mass spectrometry) results demonstrated that MomL degrades C6 or C12-HSL to linear products by hydrolyzing lactonic rings. In vitro enzyme assays indicated that MomL possesses high activity and broad substrate selectivity. Kinetic results indicated MomL had 10-fold increased C6-HSL (C6-homoserine lactones) degrading activity compared with AiiA protein [29].

Moreover, using several developed and promising technologies, we performed the following assays using MomL. First, we performed directed evolution of MomL (Figure 4). Although this strategy is not broadly used to engineer marine microbial enzymes, the strategy has been widely used in terrestrial microorganisms. Using directed evolution, Yi Tang's group at the University of California Los Angeles enhanced the activity of LovD up to 11-fold, which is an acyltransferase that converts the inactive secondary metabolite monacolin J acid into the cholesterol-lowering lovastatin [32]. Regarding the directed mutation strategy, the first step involves constructing a random mutation gene pool of the target protein. To maintain the activity of the mutated protein, the mutation rate was controlled by adjusting the error-prone PCR and sequential error-prone PCR protocols. A mutation rate of 1 to 3 mutated points in every 100 amino acids was suitable for an ideal gene pool [32]. Several screening strategies could be used for positive mutation selection of different proteins with improved activity. These strategies can be easily used for MomL engineering when combined with the high-throughput selection method described above.

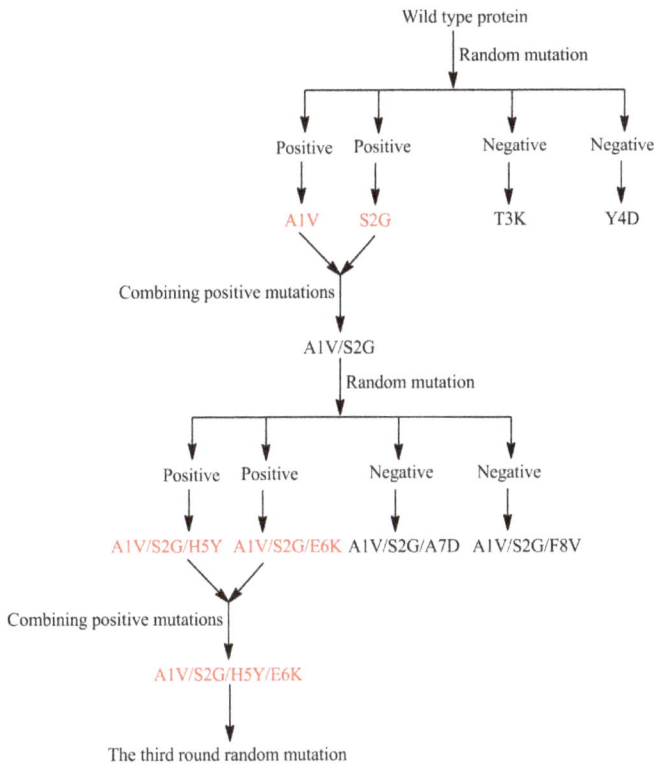

Figure 4. Schematic diagram of directed evolution assay. (A1V means alanine was replaced by valine).

Another novel research area for marine microbial enzymes involves gene regulation and signaling cascade pathways. Regulatory studies may aid in the exploration of silencing functional genes and identifying positive regulatory elements of target genes. After identifying positive regulatory genes, another important strategy involves synthetic biology. Through the construction of vectors overexpressing positive regulatory genes, the transcriptional level of the target gene is upregulated, and enzymatic expression is increased. The study of synthetic biology and related research in marine microorganisms is also a relatively new area compared with terrestrial microorganisms. Instead of reviewing marine bacteria, we use soil bacteria as references. *Lysobacter*, a genus of Gram-negative gliding bacteria, has emerged as a novel group of biocontrol agents [33]. Additionally, these species are a new bacterial source of bioactive natural products [34,35]. Liangcheng Du's laboratory at the University of Nebraska-Lincoln developed a simple method to identify target transformants based on yellow to black color change as a selection marker [36]. Using this special overexpression vector, these researchers constructed various vectors that overexpressed a positive regulator gene identified in the WAP-8294A biosynthetic gene cluster that acted as a potent anti-MRSA (Methicillin-resistant *Staphylococcus aureus*) antibiotic [37], and another regulator gene from the HSAF (Heat Stable Antifungal Factor) biosynthetic gene cluster that acted as an antifungal compound with a novel mode of action [38,39]. The enzymes that produced both WAP-8294A2 and HSAF were upregulated in the strain with overexpressed TonB-dependent receptor, and the production of WAP-8294A and HSAF increased by 2-fold and 7-fold, respectively, compared with the wild type. This work represents a successful metabolic engineering technique in a terrestrial microorganism that can also be used to manipulate unexplored marine bacteria enzyme sources.

Signal molecules regulate the transcriptional level of multiple genes, which serves as an additional strategy for exploring functional genes, especially for discovery of silenced gene clusters. This strategy is also relatively newly developed for marine microorganisms compared with soil studies. Recently, with *Lysobacter*, researchers identified a small molecule metabolite (*Le*DSF3) that regulates the biosynthesis of HSAF [40]. The addition of *Le*DSF3 in *L. enzymogenes* cultures increases HSAF biosynthetic gene transcription and HSAF yield. Additionally, the researchers identified the signaling cascade pathway. *Le*DSF3-regulated HSAF transcription and production are dependent on the two-component regulatory system, RpfC/RpfG (histidine kinase sensor/response regulator). Moreover, the global regulator cAMP receptor-like protein, which is a product of the *clp* gene, is another essential element in this signaling cascade pathway [40]. In addition to *Le*DSF3, AHL, indole, diffusible signal factor (DSF), yellow pigments and several other small molecules regulate the transcription of enzyme-encoding genes (Figure 5) [41–46]. We have used these strategies to identify the signal cascade of MomL, and the experiment is ongoing with promising progress.

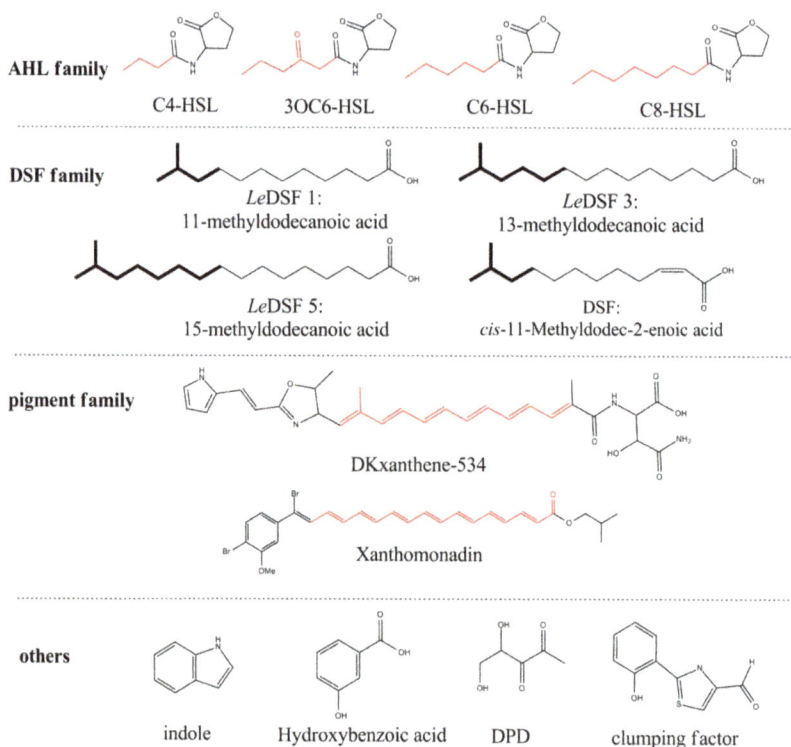

Figure 5. The list of well-known signal molecules.

2.3. Prospects for AHL Lactonase

Quorum quenching strategy is attracting increased attentions. AHL lactonase, as one kind of quorum quenching enzyme, can be applied in a range of industries. For instance, it can be used in biological control and aquaculture fields to inhibit the toxicities of pathogenic bacterium. It can be used as antistaling agents of fruits and vegetables during long-distance transport. Also, this novel quorum quenching enzyme will be widely applied in antifouling fields to inhibit the formation of bacterial biofilm. Marine-derived AHL lactonase is under highly undeveloped state. Through the updated techniques and methods mentioned above, it is believed that more enzyme resources will be exploited for various applications.

3. Amylase

3.1. Introduction

Starch, an important component of the human diet, is one of the main energy storage forms of commercial crops, such as wheat, rice, corn, potatoes and cassava. Based on structure, starches are categorized as amylose and amylopectin. Amylose is linked by a α-1,4 glucosidic bond, whereas amylopectin is linked by a α-1,6 branch bond as well as a α-1,4 linked bond [47]. Currently, acid-catalyzed hydrolysis and enzyme-catalyzed hydrolysis are the main methods of amylo-degradation. Compared with acid-catalyzed hydrolysis, enzyme-catalyzed hydrolysis exhibits considerable advantages, such as substrate specificity and low-energy consumption. Here we listed the properties of well-studied and representative alpha-amylases (Table 2).

Table 2. Properties of well-studied alpha-amylases.

Stain	UniProtKB	Molecular Mass (kDa)	Signal Peptide (aa)	Temperature Optimum (°C)	Thermostabiliy	pH Optimum	pH Stability	Specific Activity with Soluble Starch (U/mg)	Reference
Luteimonas abyssi	NM	49	35	50	34%, 50 °C, 20 min	9	>50%, 6–11, 50 °C, 1 h	8881 [a]	[48]
Bacillus licheniformis	Q208A7	55	29	90	Clear halos, 100 °C, 120 min	NM	NM	NM	[49]
Bacillus amyloliquefaciens	P00692	54.8	31	60	NM	NM	NM	NM	[50]
Alteromonas haloplanctis A23	P29957	50	24	25	6%, 25 °C	7	NM	NM	[51]
Bacillus sp. strain KSM-K38	Q93I48	55	21	55–60	20%, 50 °C, 30 min	8.0–9.5	>80%, 6–11, 40 °C, 30 min	4221 [a]	[52]
alkaliphilic bacterium N10	Q6WUB6	61	31	50	71%, 50 °C, 30 min	9.5	>80%, 8.5–11, 50 °C, 10 min	7826 [a]	[53]
Bacillus sp. XAL601	Q45643	225	31	70	NM	9.0	NM	57.3 [a]	[54]
Bacillus sp.	O82839	53	31	55	25%, 80 °C, 10 min	8.0–8.5	>50%, 6–9, 40 °C, 30 min	5009 [a]	[55]
Nocardiopsis sp. 7326	NM	55	NM	35	18%, 55 °C, 30 min	8.0	>60%, 7–9, 4 °C, 24 h	548 [a]	[56]
Bacillus sp. strain GM8901	NM	97	NM	60	37%, 60 °C, 2 h (−Ca), 78%, 60 °C, 2 h (+Ca)	11–12	>85%, 6–13, 50 °C, 1 h	157.5 [a]	[57]
Bacillus sp. NRRL B-3881	NM	NM	NM	50	50%, 55 °C	9.2	>50%, 7.0–10.5	3485 [a]	[58]
Bacillus acidicola	J9PQD2	62	no signal	60	50%, 90 °C, 10 min	4	100%, 4, 12 h, 100%, 3, 1 h	1166 [a]	[59]
Lipomyces kononenkoae	Q01117	76	28	70	0, 70 °C, 10 min	4.5–5.0	>70%, 3–8, 1 h	258 [a]	[60]
Alicyclobacillus acidocaldarius	C8WUR2	160	23	75	NM	3	NM	16.9 [b]	[61]
Bacillus sp. Ferdowsicous	P86331	53	NM	70	75%, 75 °C, 45 min	4.5	>75%, 3.5–6, 60 min	267 [a]	[62]
Bacillus acidocaldarius	NM	68	NM	75	50%, 60 °C, 5 days	3.5	Stable below 4.5	257 [b]	[63]
Aspergillus penicillioides	NM	42	NM	80	60%, 100 °C	9	>80%, 7–10	118.42 [a]	[64]
Talaromyces pinophilus 1-95	NM	58	NM	55	<45 °C, 1 h	4–5	5–9.5, 24 h	673.08 [a]	[65]
Thermococcus sp. HJ21	B4X9V8	51.4	NM	95	50%, 90 °C, 5 h, 40%, 30%; 100 °C; 2 h, 3 h	5	5–9	8.3 [a]	[66]
Malbranchea cinnamomea	K9L8F3	60.3	21	65	50%, 60 °C, 41.1 min	6.5	>90%, 5–10, 30 min	514.6 [a]	[67]

Table 2. Cont.

Stain	UniProtKB	Molecular Mass (kDa)	Signal Peptide (aa)	Temperature Optimum (°C)	Thermostability	pH Optimum	pH Stability	Specific Activity with Soluble Starch (U/mg)	Reference
Aspergillus niveus	NM	60	NM	65	50%, 70 °C, 20 m	6	4–7, 24 h	168 [a]	[68]
Thermoactinomyces vulgaris	G8ZE61	40.6	NM	50	50%, 50 °C, 2 h	6–7	4–9	127,100.33 [b]	[69]
Bacillus sp. AAH-31	S6BGD1	91	28	70	<60 °C	8.5	6.4–10.3	16.7 [a]	[70]
Paecilomyces variotii	NM	75	NM	60	50%, 60 °C, 53 min	4	>70%, 5–8, 1 h	612.5 [a]	[71]
Pseudoalteromonas arctica GS230	NM	55	24	30	49%, 30 °C, 150 min	7.5	>60%, 7–8.5, 1 h	25.5 [a]	[72]
Bacillus sp. YX-1	A9YDD9	56	31	40–50	60%, 60 °C, 1 h	5	>80%, 4.5–11, 1 h	607 [b]	[73]
Geobacillus thermoleovorans	NM	26	NM	100	50%, 100 °C, 3.6 h	8	50%, 6, 4.5 h, 50%, 7, 7.5 h	450 [a]	[74]
Fusicoccum sp. BCC4124	Q0Z8K1	50	no signal	70	95%, 50 °C, 1 h	7	NM	90 [a]	[75]
Bacillus subtilis KCC103	A8VWC5	53	33	65–70	50%, 70 °C, 7 min	6–7	>98%, 5–9.5	483 [a]	[76]
Bacillus subtilis AX20	NM	149	NM	55	50 °C, 30 min	6	5–9, 24 h	4133 [a]	[77]
Halothermothrix orenii	Q8GPL8	60	23	65	37–75 °C	7.5	6–9.5	22.32 [a]	[78]
Bacillus stearothermophilus	NM	64	NM	50	92%, 100 °C, 1 h	7	23%, 3, 1 h, 26%, 10, 1 h	77.2 [b]	[79]
Thermus filiformis Ork A2	NM	60	NM	95	50%, 95 °C, 19 min	5.5–6	>80%, 4.5–8, 1 h	6352 [b]	[80]
Bacillus subtilis	NM	48	NM	50	70%, 60 °C, 1 h	6.5	5–6.5	772.7 [a]	[81]
Clostridium perfringens NCTC 8679	NM	76	NM	30	NM	6.5	NM	NM	[82]
Escherichia coli (strain K12)	P25718	75.7	17	NM	NM	8	NM	NM	[83]
Bacillus subtilis	P00691	67	27	NM	NM	8.5	NM	NM	[84]

[a] Enzyme activity are measured by DNS method; [b] Enzyme activity are measured by the colored starch-I_2 complex method; NM means not mention in the essays.

3.2. Marine Resources of Amylase and Related Catalytic Mechanisms

Currently, heat-stable α-amylases have been well studied and applied in a series of industrial fields. As the development of industry, the demand for cold-active amylase is attracting increasingly attention. Different from terrestrial resources of amylase, many of the marine resources of amylase has high activity in low temperature as over 75% of the ocean is in 0–6 °C. Therefore, the marine environment is an ideal area to find cold-active amylase. In Table 2, we can clearly find that the optimum temperature of α-amylase from *Alteromonas haloplanctis* A23, isolated from Antarctica, is 25 °C and it has high activity in low temperature. In deep sea sediment of Prydz Bay, Antarctic, a cold-adapted α-amylase from *Nocardiopsis* sp. 7326 was identified. It can retain 38% of its highest activity at 0 °C. Recently, our group found a novel alkalophilic α-amylase LaaA which has the highest specific activity reported. The specific activity reached highly to 8881 U/mg. it was cloned from deep-sea bacterium *Luteimonas abyssi* XH031^T which isolated from the sediment of the South Pacific Gyre with low temperature. It even maintained 38% residual activity at 10 °C [48].

Given the convenience of genetic and molecular biological manipulation, prokaryotes are the most important resource for α-amylase exploration (Figure 2). Previous studies demonstrated that α-amylase consists of three domains: A, B and C [85]. Domain A, which includes active site residues and is directly related to catalytic reaction, forms the core portion of α-amylase. Domains B and C are related to substrate specificity and active site stability [86]. Thermostability is one of the most valuable characteristics features of amylase. Studies have demonstrated that protein thermostability is mainly influenced by hydrogen bonding, hydrophobic interaction, electrostatic interactions and packing [87]. Through studies of α-amylases from *Bacillus*, two stages of thermo-inactivation have been revealed: the partial unfolded state, which is a reversible step, and the fully unfolded state, which is an irreversible step. With increasing temperatures, the enzyme reaches the partial unfolded state first. When the temperature reaches a certain level, the enzyme is fully unfolded and totally inactive [88]. It is hypothesized that the irreversible inactivation is due to covalent modifications of polypeptide chains or that a higher energy barrier is required during the folding process [86].

To date, most of the well-studied α-amylases contain a conserved calcium site, which is potentially related to enzyme stability and activity [89]. The conserved calcium site is located far away from the active site; thus, this site does not participate in the catalytic reaction but plays an important structural role in enzyme stability and activity [86]. The conserved calcium ion interacts with four amino acid residues, and three of them are strictly conserved in both structure and sequence [90]. Asn104, Asp200 and His235 are three conserved sites in *Bacillus licheniformis* α-amylase (BLA) [91]. Mischa et al. [90] first equivocally elucidated the mechanism of calcium-activating α-amylase, and this group identified a large region that contains 21 disordered residues. The disordered to ordered transition that occurs in this region is mediated by calcium, which leads to the formation of one wall of the cleft containing the extended substrate binding site [90]. If this region is disrupted by two extra residues (Glu-Gly), the conformation of calcium binding is also modified, and thermal stability is subsequently decreased [92].

Great efforts have been made during recent decades through site-directed mutagenesis and directed evolution to improve the properties of promising α-amylases, such as catalytic activity, oxidation resistance, pH tolerance, and temperature tolerance. Directed evolution mainly applies to the enzymes with biochemical and structural properties that remain poorly understood. Site-directed mutagenesis is based on thorough research of enzymatic properties. Using homologous sequence alignment or information regarding tertiary structure, enzymes can be purposefully modified. For example, to improve the thermal stability of α-amylase from *Bacillus megaterium* WHO (BMW), researchers compared BMW-amylase with the most similar protein (*Halothermothrix orenii* α-amylase, 67%) through bioinformatic methods and modified the protein using site-specific mutagenesis. The thermal stability was dramatically improved by H58I mutation, which corresponds to Ile50 in *H. orenii* α-amylase [93].

The deletion of residues is an effective method to improve thermostability. Studies demonstrated that BLA and BAA (*Bacillus amyloliquefaciens* α-amylase) are highly similar in structure, but a significant difference in thermostability was noted. The sequence alignment of BAA and BLA demonstrated the absence of two amino acids, 209E and 210G, in BLA compared with BAA. Then, the mutant strain BAA-△EG (209E and 210G were mutated) was constructed. The results demonstrated that the maximal thermostability was increased by ten degrees compared with the wild type BAA [92]. Similar studies by Mamdouh et al. [94] reported that the amylase of *Bacillus stearothermophilus* US100 has an additional loop compared with the model of BLA. The deletion of two residues (Ile214 and Gly215) increased thermostability and reduced calcium requirements. Further studies revealed that the stability of the loop affects the thermostability. The additional loop (containing residues Gly213, Ile214 and Gly215) along with the neighboring residues Arg212 and Lys216 play a critical role in stabilizing the structure of α-amylase and the calcium-binding site of calcium I. The structure was stabilized via the interaction between Lys216 and Phe194 and Asp238. Moreover, the stability of Lys216 is directly related to the stability of the GIG loop [95]. Hydrogen bonding and salt bridges have the important function of maintaining the stability of α-amylase at low pH [96–98]. The stability of mutants at low pH was particularly increased by H275D, H293D and H310D mutations. Histidine (His) is a basic amino acid with positive charges, whereas aspartic acid (Asp) is an acidic amino acid with negative charges. The survey demonstrates that Asp can stabilize the structure of α-amylase by interacting with hydrogen bonding and salt bridges under acidic conditions [99]. Salt bridges are also related by the high thermal stability of BLA [97]. The lysine residues (Lys88, Lys253 and Lys385) interact with each other to form a stable salt bridge [97]. It is proposed that the interaction between two residues with similar charges enhances protein stability. Karimeh et al. [100] found that P407H mutations improved the thermal stability of BAA. Regarding mutated BAA, His407 is located in calcium III, which forms a His-His pair with the neighboring amino acid His406.

3.3. Prospects for Amylase

α-amylase is widely applied in food, fermentation and detergent industries. Furthermore, in the medical field, α-amylase can be used as drug targets for treating diabetes, obesity and high cholesterol, etc. As the increasing demand of different special properties of α-amylase in the industry and research fields, the complex marine environments provide the possibility of finding various of α-amylases. The study of marine resources of amylase is becoming increasingly popular. Besides, protein engineering is an efficient method to improve the properties of α-amylase and plays an increasingly important role in the research of α-amylase.

4. Alginate Lyase

4.1. Introduction

Algin is a linear complex copolymer composed of mannuronate acid (M) and guluronic acid (G) that was originally extracted from the mesenchyme of kelp, gulfweed, and seaweed (Figure 6). Alginate lyase degradation products include oligosaccharides, which exhibit various bioactivities, such as antibiosis, anti-cancer, anti-tumor and promoting plant growth. Theses characteristic features allow these proteins to have broad application prospects in medical and agricultural fields. The methods of algin degradation include acid hydrolysis, chemical oxidation and enzymolysis, among which enzymolysis has great potential usage given its high specificity and reaction efficiency.

Alginate lyase is a type of enzyme that specifically degrades algin and is primarily found in microorganisms, animals and plants. Most bacterial alginate lyases are from *Gamma-proteobacteria* (Figure 2). Alginate lyases derived from microorganisms are more stable and exhibit increased activity compared with those obtained from plants and animals. Various types of classification have been reported, including those based on the specificity of the degradation substrate. In addition, algin enzymes can be classified as a mannuronate acid lyase, which exclusively degrades the M section, or a guluronic acid lyase, which exclusively degrades the G section (Figure 6). At present, most of the well-studied alginate lyases exhibit M section-degrading activity, whereas a small proportion degrades both M and G sections (Figure 6) [101].

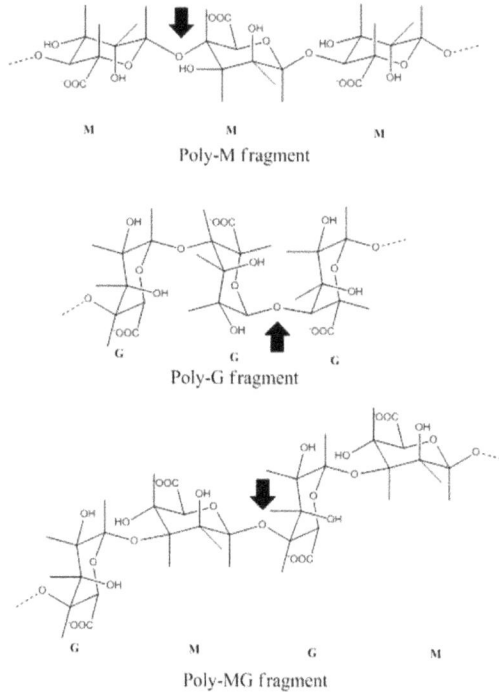

Figure 6. Structure of algin and the mode of action of alginate lyase.

4.2. Marine Resources of Alginate Lyase and Catalytic Mechanisms

Given the special environment and promising potential, marine alginate lyase studies have become common in the last several decades. Tseng and colleagues isolated two types of alginate lyases from *Vibrio* sp. AL-9 in 1992. One type breaks the α-1,4 glycosidic bond of guluronic acid, whereas the other type breaks the β-1,4 glycosidic bond of mannuronate acid [102]. Moreover, this group also identified an alginate lyase that degrades guluronic acid in *Vibrio* sp. AL-128 [103]. The alginate lyase isolated from *Vibrio* sp. QY101 by Song et al. [104] exhibited obvious activity for both mannuronate acid and guluronic acid. For the first time, this group reported that the 9-amino acid region (YXRESLREM) appears not only in guluronic acid lyase but also in mannuronate acid lyase [105]. Takeshita et al. identified one novel alginate lyase from a *Vibrio* species isolated from the intestinal tract of red snapper [106]. The special alginate lyase retains 45% of catalytic activity under heat shock conditions of 100 °C. Moreover, activity is also retained after treatment with 3% SDS in 25 °C for 30 min [106].

In 2009, Liu et al. [107] constructed the recombinant plasmid pINA1317-Y1CWP110 with alginate lyase (AlyVI) isolated from *Vibrio* sp. QY101 as the target gene. This enzyme has considerable catalytic degradation activity if expressed in *Yarrowia lipolytica*. Moreover, this enzyme degrades mannuronate acid, guluronic acid, and alginate and produces a series of oligos with different lengths [107]. Alginate lyase was expressed in *Saccharomycetes* for the first time, and different oligosaccharide lengths were produced. In the bacterial strain *Pseudoalteromonas atlantica* AR06, some research groups used homologous recombination technology to fuse green fluorescent protein (GFP) to the C-terminus of alginate lyase AlyA, demonstrating that the bacterial strain has normal degradation activity and is able to release fluorescence. As a visible gene expression tool, GFP is convenient for biochemical and catalytic studies of alginate lyase. Recently, several additional alginate lyases have been well studied. Liu et al. [108] reported for the first time that the extracellular alginate lyase-like protein from *Pseudomonas fluorescens* exhibits high degradation activity for alginate. Dong and other scientists performed a study on bacteria isolated from the Arctic marine seaweed that produces alginate lyase [109]. In total, 65 bacterial strains were isolated from the kelp specimen, 21 of which exhibited alginate lyase activity. Among these isolates, 11 bacterial strains exhibited an optimum temperature between 20 °C and 30 °C, which indicates that the extracellular lyases are cryophilic enzymes [109]. Moreover, the bacterial strains *Psychrobacter, Winogradskyella, Psychromonas* and *Polaribacter* produce alginate lyase. We have recently focused our studies on *Luteimonas abyssi* sp. nov., which was isolated from the abyssal sediment under the circulating area of the South Pacific Ocean. Given that the strain lives in low-temperature environments and possesses alginate lyase activity, we hypothesize that alginate lyase is a cold-adaptive enzyme.

Previous studies identified different degradation products of alginate lyase under different pH conditions. For example, vAL-1 from PL-14 exhibits glucuronic acid degradation activity. In addition, this enzyme also degrades algin under alkaline conditions [110]. The purified alginate lyase Smlt1473 from *S. maltophilia* inducing exhibits hyaluronic acid degradation activity at a pH of 5, guluronic acid degradation activity at a pH of 7 and alginate degradation activity at a pH of 9 [111].

Secondly, according to the similarity of amino acid sequence, alginate lyases may be classified into the PL-5, PL-7 and PL-15 families. In the PL-5 family [112], 41 alginate lyases have been identified to date, all of which are from microorganisms. In addition, 86 alginate lyases have been discovered in the PL-7 family, including 84 from bacteria and only 2 from eukaryotic cells. The PL-15 family currently has 11 alginate lyases, and all of them are from bacteria. Thus, most of the identified alginate lyases have been identified in bacteria.

Thirdly, alginate lyases can be divided into three categories based on molecular weight [113]: 20 to 35 kD, approximately 40 kD and approximately 60 kD. The first type of enzyme (20 to 35 kD) exhibits a variety of substrate specificity. Given that the alginate lyases with a molecular weight of approximately 40 kD have degradation activity specific to the M section [114] and most of them contain the common conserved sequence NNHSYW, it is inferred that this sequence is related to the M section-degradation activity. The homologous sequence YFKAGXYXQ of the C-terminus is another characteristic feature of small alginate lyases [101]. However, this same sequence was recently identified in AlyPI lyase, which is a member of the 60-kD family [115]. By comparing the homology of abalone alginate lyases (HdAly) with members of the PL-14 family, such as turban shell SP2 and chlorella virus CL2, Sayo Yamamoto et al. [116] reported that Arg92, Lys95, Arg110, Arg119, and Lys19 are particularly highly conservative amino acid residues. Utilizing site-specific mutagenesis technology, it was discovered that Lys95 mutations cause complete enzymatic inactivation, whereas Arg92, Arg110 and Arg119 mutations lead to reductions in activity greater than 65%, thus suggesting that the Arg92 to Arg119 region is closely related to HdAly activity [116].

To date, all of the alginate degradation enzymes have been classified as lyases, and no hydrolytic enzymes have been reported [117]. Alginate lyases catalyze the degradation of alginate via a β-elimination mechanism by breaking α-1,4 glycosidic bonds and yielding oligosaccharides with an unsaturated double bond at the non-reducing end. A portion of the reaction in which alginate lyase is cleaved can be induced by a specific substrate. For example, Matsushima et al. [118,119] reported that alyA gene transcription in *Pseudoalteromonasatlantica* AR06 is induced by alginate in basic medium.

4.3. Prospects for Alginate Lyase

Alginate lyases can be utilized in multiple industrial and medical areas. Alginate lyase can be used for developing new methods of tissue regeneration. For example, stem cell cultivation needs alginate hydrogel as a biological support. However, mammals cannot produce alginate lyase, and thus, alginate hydrogel cannot be easily degraded. Theoretically, alginate lyase provides a viable way to solve this problem [120]. Alginate lyase can also improve the drug susceptibility of biofilms, working as quorum sensing inhibitors [121]. Moreover, it has an inhibitory effect on the pathogens of pulmonary cystic fibrosis. *Pediococcus* sp. Ab1, which has alginate lyase activity, can be used as a probiotic to improve the intestinal microflora and nutritional status of abalone during their cultivation [122].

5. Chitinase

5.1. Introduction

Chitin is an *N*-acetyl-glucosamine homopolymer linked by a β-1,4-glucosidic bond [123] that is structurally identical to cellulose, except that the hydroxyl group in cellulose at C2 is replaced by an acetamide group [124]. Chitin is a particularly rich and important nutrient and energy source for maintaining the ecosystems of marine environments. The majority of chitin originates from marine ecosystems, and chitinolytic marine bacteria play a critical role in recycling chitinous materials, such as exoskeletons of crustaceans and insects. Moreover, chitinolytic marine bacteria are also important to maintain the balance of marine ecosystem. The biofunction of chitin mainly involves its degradation product oligochitosan, which is capable of resisting fungus, tumors, plant diseases, and related pests. Oligochitosan also alters the body's immunological function. Compared with traditional chemical approach, oligochitosan production by biological methods is milder and more environmentally friendly. Thus, chintinase studies have drawn increased attention during recent decades.

5.2. Marine Resources of Chitinase and Catalytic Mechanisms

Currently, chitinases have been identified in a series of marine organisms, such as *Alteromonas* sp. O7, *V. parahaemolyticus* [125], *Salinivibrio costicola* [126] and *Microbulbifer degradans* [127]. Chitinase is responsible for hydrolyzing the β-1,4-glycosidic bond of different types of chitin. Based on catalytic domain similarities, chitinase can be divided into two families: 18 and 19. Family 18 includes chitinases from bacteria, fungi, viruses, animals and some plants, whereas family 19 includes chitinases from most plants and special bacteria, such as chitinase C (ChiC) identified from *Streptomyces griseus* [128]. The chitinases of the two families, which potentially evolved from different ancestors, possess different catalytic mechanisms [129]. The catalytic domain of the GH18 family has a typical triosephosphate isomerase (TIM) structure that forms an inner barrel with 8 α-helixes surrounded by an outer barrel of 8 β-sheets. Whereas the catalytic domain of the GH19 family has a high proportion of α-helixes, and its structure is similar to chitosanase and lysozyme. GH18 contains 3 groups (ChiA/ChiB/ChiC), which are classified according to amino acid sequence differences in the catalytic domain. ChiA and ChiB hydrolyze chitin chains towards opposite direction, whereas ChiC is an endochitinase [130]. Most chitinases from marine microorganisms belong to family 18 [131,132]. Comparing with chitinases isolated from terrestrial bacteria, marine chitinase exhibit better pH and salinity tolerance, which may represent promise for some special applications. Wang et al. [133] isolated marine *Bacillus cereus* that express two chitinases with optimum pH, optimum temperature, pH stability, and thermal stability values of 9, 50 °C, 3 to 11, 50 °C and 5, 40 °C, 3 to 11, 60 °C, respectively. These enzymes retain 61%, 60%, 73%, and 100% as well as 60%, 60%, 71%, and 96% of their original activity in the presence of 2% Tween 20, 2% Tween 40, 2% Triton X-100, and 1 mM SDS, respectively [133]. The marine psychrophilic bacterium isolated by Stefanidi et al. from a sample raised from a depth of 1200 meters in the northern Pacific Ocean, secretes several chitinases in response to chitin induction. These chitinase genes encode a protein of 550 amino acids. The optimum pH and temperature of this chitinase are 5.0 and 28 °C, respectively. There are two crucial residues, Trp275 and Trp397, in the catalytic domain of the chitinase [134]. A chitinase directed mutagenesis study by Suginta et al. [134] revealed that Gly and Phe instead of Trp275 and Trp397, respectively, heavily altered the selection of β substrate isomers. The Trp275 mutation alters the chitinase's kinetics characteristic features by increasing the catalytic constants (k_{cat}) and the specificity (k_{cat}/K_M) of all substrates 5- to 10-fold. In contrast, the Trp397 mutation decreases the strength of binding between chitinase and substrate and the rate of soluble substrate degradation.

5.3. Prospects for Chitinase

The degradation of chitin into oligosaccharides has promise that may be useful in numerous biological functions, such as antimicrobial activity and antitumor activity [135]. At present, chitooligosaccharides are mainly produced through chemical reactions in industry. This process has many drawbacks, such as the production of a series of unexpected short strain oligosaccharides. Moreover, chemical reactions can easily cause serious environmental pollution. Comparatively, biological degradation using chitinase has many advantages, such as being environmentally friendly, inexpensive and repeatable. Because of these advantages, the use of chitinase to hydrolyze chitin has drawn increasing attention during recent years.

6. Cellulase

6.1. Introduction

Cellulases are multi-component enzymes that can be divided into three components based on catalytic function, including endoglucanase (endo-1,4-β-D-glucanase, EC3.2.1.4), exoglucanase (exo-1,4-β-D-glucanase, EC3.2.1.91), and cellobiase (β-1,4-glucosidase, EC3.2.1.21). Cellulases exhibit great potential in various applications, including papermaking, detergents, bioenergy and effluent treatment. Previous studies were limited to terrestrial-derived cellulases. As research on marine microbes and their enzymes advance, it was discovered that enzymes secreted by marine microbes possess several characteristic features, such as pressure tolerance, alkali resistance, cold resistance and heat resistance. As a result, marine cellulase resources have drawn increasing attention [136,137].

6.2. Marine Resources of Cellulase and Catalytic Mechanisms

Thermophiles and cryophiles are two main types of marine microbes in the research of cellulose. For example, *Rhodothermus marinus* and *Thermotoga neapolitana* are well characterized. Extreme thermophile cellulose is composed of several domains. The cellulose and hemi-cellulase domains are catalytic domains that are linked by several cellulose binding domains (CBDs) [138]. *Rhodothermus marinus* express a type of cellulose with an optimal temperature that is greater than 90 °C. Bioinformatics and three-dimensional structure comparisons suggest that the aromatic amino acid cluster exposed on the surface of the protein is responsible for its thermophile activity [139]. Hakamada and coworkers [140] analyzed the structures of thermophile cellulose and alkali cellulose and revealed that three lysines located at 137, 179 and 194 are responsible for heat resistance. Moreover, they also demonstrated that increases in arginine, histidine and glutamine residues and decreases in aspartate and lysine residues are also related with alkali cellulose stability. *Pseudoalteromonas* sp. is another species gaining considerable attention. The optimal temperature of its cellulose, which ranges from 45 to 60 °C, is lower than that of other microbial celluloses. Lee and his partner [141] isolated a type of marine bacteria *Bacillus subtilis*. The optimal temperature of its cellulose is 50 °C, and the optimum reaction pH is 6.5. Alfredsson reported on the marine bacteria *Rhodothermus marinus* isolated from alkaline underwater hot springs in Iceland. The optimum growth conditions of the enzyme are 65 °C, pH 7.0 and 2% NaCl [142]. Trivedi [143] reported on *Bacillus flexus* from alga that produces alkaline cellulose. The molecular mass of the cellulose is 97 kDa, and good stability is noted at pH ranges of 9.0 to 12.0. Moreover, approximately 70% of activity is maintained in 15% NaCl. Fellerand and his co-workers [51] isolated the psychrophilic filamentous bacterium near the Zhongshan Station and the Great Wall Station. This species produces cellulase and decomposes cellulase at 0 or 5 °C, thus maintaining proliferation at low temperatures [51]. Recently, Fang [144] identified a β-glucosidase gene named *bgl1A* from a marine microbial metagenomic library by functional screening. This gene permitted tolerance of high glucose concentrations. The protein BgllA was identified as a member of the glycoside hydrolase 1 family. The recombinant β-glucosidase Bgl1A exhibited a high level of stability in the presence of various cations and high concentrations of NaCl. The protein was activated by glucose with low concentrations. The enzymatic activity of Bgl1A was gradually inhibited by increasing concentrations of glucose, but 50% of the original value remained even in up to 1000 mM glucose.

6.3. Prospects for Cellulase

Cellulases from marine microorganisms exhibit activities under extreme conditions, such as high salt, high pressure, pH and high/low temperature. Thus, microorganisms abundant in the unique marine environment provide an important material base for exploiting new source of cellulases. For example, alkaline cellulose is used in the detergent industry. Moreover, alkaline celluloses exhibit more advantages in the disposal of sewage from spinning, papermaking, pickling and sauce production. Moreover, laundry processes require cellulases with the properties of alkali resistance, heat resistance and insensitivity to surfactants to simultaneously cut losses in detergent processing, storing and transiting. With the rapid development of the seaweed industry, there is an enormous demand for cellulases used in algae wall solutions and degradation of algae processing wastes. With the development of biotechnologies, especially bioinformatics and metagenomics, we believe that more cellulases with important functions that were previously unable to be discovered will be exploited in the future.

7. Conclusions

Marine microorganisms contain a series of novel and studied enzymes. However, due to limitations of exploration, a large proportion of these organisms have not been identified. This review presents several classic methods for enzyme transcriptional regulation and engineering, which can be used in marine microbial enzyme exploration. Additionally, this review describes the mechanism and current status of several polysaccharide-degrading enzymes. Hopefully, these novel strategies and well-studied catalytic mechanisms can serve as a reference for identifying novel enzymes from marine environments.

Acknowledgments: This work was supported by projects from the National Natural Science Foundation of China (No. 41276141, 41506160 and 31571970).

Conflicts of Interest: The authors declare no conflict of interest.

References

1. Antranikian, G.; Vorgias, C.E.; Bertoldo, C. Extreme environments as a resource for microorganisms and novel biocatalysts. *Adv. Biochem. Eng. Biotechnol.* **2005**, *96*, 219–262. [PubMed]
2. Bhattacharya, A.; Pletschke, B.I. Review of the enzymatic machinery of *Halothermothrix orenii* with special reference to industrial applications. *Enzyme Microb. Technol.* **2014**, *55*, 159–169. [CrossRef] [PubMed]
3. Liu, X.; Huang, Z.; Zhang, X.; Shao, Z.; Liu, Z. Cloning, expression and characterization of a novel cold-active and halophilic xylanase from *Zunongwangia profunda*. *Extremophiles* **2014**, *18*, 441–450. [CrossRef] [PubMed]
4. Jaiganesh, R.; Sampath Kumar, N.S. Marine bacterial sources of bioactive compounds. *Adv. Food Nutr. Res.* **2012**, *65*, 389–408. [PubMed]
5. Zotchev, S.B. Marine actinomycetes as an emerging resource for the drug development pipelines. *J. Biotechnol.* **2012**, *158*, 168–175. [CrossRef] [PubMed]
6. Lee, J.S.; Kim, Y.S.; Park, S.; Kim, J.; Kang, S.J.; Lee, M.H.; Ryu, S.; Choi, J.M.; Oh, T.K.; Yoon, J.H. Exceptional production of both prodigiosin and cycloprodigiosin as major metabolic constituents by a novel marine bacterium, *Zooshikella rubidus* S1-1. *Appl. Environ. Microbiol.* **2011**, *77*, 4967–4973. [CrossRef] [PubMed]
7. Martins, A.; Vieira, H.; Gaspar, H.; Santos, S. Marketed marine natural products in the pharmaceutical and cosmeceutical industries: Tips for success. *Mar. Drugs* **2014**, *12*, 1066–1101. [CrossRef] [PubMed]
8. Rahman, H.; Austin, B.; Mitchell, W.J.; Morris, P.C.; Jamieson, D.J.; Adams, D.R.; Spragg, A.M.; Schweizer, M. Novel anti-infective compounds from marine bacteria. *Mar. Drugs* **2010**, *8*, 498–518. [CrossRef] [PubMed]

9. Abdelmohsen, U.R.; Yang, C.; Horn, H.; Hajjar, D.; Ravasi, T.; Hentschel, U. Actinomycetes from Red Sea sponges: Sources for chemical and phylogenetic diversity. *Mar. Drugs* **2014**, *12*, 2771–2789. [CrossRef] [PubMed]

10. Komaki, H.; Ichikawa, N.; Hosoyama, A.; Fujita, N.; Igarashi, Y. Draft genome sequence of marine-derived *streptomyces* sp. TP-A0873, a producer of a pyrrolizidine alkaloid bohemamine. *Genome Announc.* **2015**, *3*. [CrossRef] [PubMed]

11. Zhou, K.; Zhang, X.; Zhang, F.; Li, Z. Phylogenetically diverse cultivable fungal community and polyketide synthase (PKS), non-ribosomal peptide synthase (NRPS) genes associated with the South China Sea sponges. *Microb. Ecol.* **2011**, *62*, 644–654. [CrossRef] [PubMed]

12. Zhu, P.; Zheng, Y.; You, Y.; Yan, X.; Shao, J. Molecular phylogeny and modular structure of hybrid NRPS/PKS gene fragment of *Pseudoalteromonas* sp. NJ6-3-2 isolated from marine sponge Hymeniacidon perleve. *J. Microbiol. Biotechnol.* **2009**, *19*, 229–237. [PubMed]

13. Bassler, B.L. Cell-to-cell communication in bacteria: A chemical discourse. *Harvey Lect.* **2004**, *100*, 123–142. [PubMed]

14. Miyamoto, C.M.; Meighen, E.A. Involvement of LuxR, a quorum sensing regulator in *Vibrio harveyi*, in the promotion of metabolic genes: *argA*, *purM*, *lysE* and *rluA*. *Biochim. Biophys. Acta* **2006**, *1759*, 296–307. [CrossRef] [PubMed]

15. Choudhary, S.; Schmidt-Dannert, C. Applications of quorum sensing in biotechnology. *Appl. Microbiol. Biotechnol.* **2010**, *86*, 1267–1279. [CrossRef] [PubMed]

16. Mangwani, N.; Dash, H.R.; Chauhan, A.; Das, S. Bacterial quorum sensing: Functional features and potential applications in biotechnology. *J. Mol. Microbiol. Biotechnol.* **2012**, *22*, 215–227. [CrossRef] [PubMed]

17. Liu, D.; Momb, J.; Thomas, P.W.; Moulin, A.; Petsko, G.A.; Fast, W.; Ringe, D. Mechanism of the quorum-quenching lactonase (AiiA) from *Bacillus thuringiensis*. 1. Product-bound structures. *Biochemistry* **2008**, *47*, 7706–7714. [CrossRef] [PubMed]

18. Momb, J.; Wang, C.; Liu, D.; Thomas, P.W.; Petsko, G.A.; Guo, H.; Ringe, D.; Fast, W. Mechanism of the quorum-quenching lactonase (AiiA) from *Bacillus thuringiensis*. 2. Substrate modeling and active site mutations. *Biochemistry* **2008**, *47*, 7715–7725. [CrossRef] [PubMed]

19. Thomas, P.W.; Stone, E.M.; Costello, A.L.; Tierney, D.L.; Fast, W. The quorum-quenching lactonase from *Bacillus thuringiensis* is a metalloprotein. *Biochemistry* **2005**, *44*, 7559–7569. [CrossRef] [PubMed]

20. Cao, Y.; He, S.; Zhou, Z.; Zhang, M.; Mao, W.; Zhang, H.; Yao, B. Orally administered thermostable *N*-acyl homoserine lactonase from *Bacillus* sp. strain AI96 attenuates *Aeromonas hydrophila* infection in zebrafish. *Appl. Environ. Microbiol.* **2012**, *78*, 1899–1908. [CrossRef] [PubMed]

21. Chen, R.; Zhou, Z.; Cao, Y.; Bai, Y.; Yao, B. High yield expression of an AHL-lactonase from *Bacillus* sp. B546 in *Pichia pastoris* and its application to reduce *Aeromonas hydrophila* mortality in aquaculture. *Microb. Cell Fact.* **2010**, *9*, 39. [CrossRef] [PubMed]

22. Carlier, A.; Uroz, S.; Smadja, B.; Fray, R.; Latour, X.; Dessaux, Y.; Faure, D. The Ti plasmid of *Agrobacterium tumefaciens* harbors an *attM*-paralogous gene, *aiiB*, also encoding *N*-Acyl homoserine lactonase activity. *Appl. Environ. Microbiol.* **2003**, *69*, 4989–4893. [CrossRef] [PubMed]

23. Park, S.Y.; Lee, S.J.; Oh, T.K.; Oh, J.W.; Koo, B.T.; Yum, D.Y.; Lee, J.K. AhlD, an *N*-acylhomoserine lactonase in *Arthrobacter* sp., and predicted homologues in other bacteria. *Microbiology* **2003**, *149*, 1541–1550. [CrossRef] [PubMed]

24. Uroz, S.; Oger, P.M.; Chapelle, E.; Adeline, M.T.; Faure, D.; Dessaux, Y. A *Rhodococcus qsdA*-encoded enzyme defines a novel class of large-spectrum quorum-quenching lactonases. *Appl. Environ. Microbiol.* **2008**, *74*, 1357–1366. [CrossRef] [PubMed]

25. Wang, W.Z.; Morohoshi, T.; Ikenoya, M.; Someya, N.; Ikeda, T. AiiM, a novel class of *N*-acylhomoserine lactonase from the leaf-associated bacterium *Microbacterium testaceum*. *Appl. Environ. Microbiol.* **2010**, *76*, 2524–2530. [CrossRef] [PubMed]

26. Zhang, H.B.; Wang, L.H.; Zhang, L.H. Genetic control of quorum-sensing signal turnover in *Agrobacterium tumefaciens*. *Proc. Natl. Acad. Sci. USA* **2002**, *99*, 4638–4643. [CrossRef] [PubMed]

27. Gao, A.; Mei, G.Y.; Liu, S.; Wang, P.; Tang, Q.; Liu, Y.P.; Wen, H.; An, X.M.; Zhang, L.Q.; Yan, X.X.; et al. High-resolution structures of AidH complexes provide insights into a novel catalytic mechanism for *N*-acyl homoserine lactonase. *Acta Crystallogr. D Biol. Crystallogr.* **2013**, *69*, 82–91. [CrossRef] [PubMed]

28. Mei, G.Y.; Yan, X.X.; Turak, A.; Luo, Z.Q.; Zhang, L.Q. AidH, an alpha/beta-hydrolase fold family member from an *Ochrobactrum* sp. strain, is a novel *N*-acylhomoserine lactonase. *Appl. Environ. Microbiol.* **2010**, *76*, 4933–4942. [CrossRef] [PubMed]

29. Tang, K.; Su, Y.; Brackman, G.; Cui, F.; Zhang, Y.; Shi, X.; Coenye, T.; Zhang, X.H. MomL, a novel marine-derived *N*-acyl homoserine lactonase from *Muricauda olearia*. *Appl. Environ. Microbiol.* **2015**, *81*, 774–782. [CrossRef] [PubMed]

30. Huang, W.; Lin, Y.; Yi, S.; Liu, P.; Shen, J.; Shao, Z.; Liu, Z. QsdH, a novel AHL lactonase in the RND-type inner membrane of marine *Pseudoalteromonas byunsanensis* strain 1A01261. *PLoS ONE* **2012**, *7*, e46587. [CrossRef] [PubMed]

31. Tang, K.; Zhang, Y.; Yu, M.; Shi, X.; Coenye, T.; Bossier, P.; Zhang, X.H. Evaluation of a new high-throughput method for identifying quorum quenching bacteria. *Sci. Rep.* **2013**, *3*, 2935. [CrossRef] [PubMed]

32. Gao, X.; Xie, X.; Pashkov, I.; Sawaya, M.R.; Laidman, J.; Zhang, W.; Cacho, R.; Yeates, T.O.; Tang, Y. Directed evolution and structural characterization of a simvastatin synthase. *Chem. Biol.* **2009**, *16*, 1064–1074. [CrossRef] [PubMed]

33. Sullivan, R.F.; Holtman, M.A.; Zylstra, G.J.; White, J.F.; Kobayashi, D.Y. Taxonomic positioning of two biological control agents for plant diseases as *Lysobacter enzymogenes* based on phylogenetic analysis of 16S rDNA, fatty acid composition and phenotypic characteristics. *J. Appl. Microbiol.* **2003**, *94*, 1079–1086. [CrossRef] [PubMed]

34. Wang, Y.; Qian, G.; Li, Y.; Wright, S.; Shen, Y.; Liu, F.; Du, L. Biosynthetic mechanism for sunscreens of the biocontrol agent *Lysobacter enzymogenes*. *PLoS ONE* **2013**, *8*, e66633. [CrossRef] [PubMed]

35. Xie, Y.; Wright, S.; Shen, Y.; Du, L. Bioactive natural products from *Lysobacter*. *Nat. Prod. Rep.* **2012**, *29*, 1277–1287. [CrossRef] [PubMed]

36. Wang, Y.; Qian, G.; Liu, F.; Li, Y.Z.; Shen, Y.; Du, L. Facile method for site-specific gene integration in *Lysobacter enzymogenes* for yield improvement of the anti-MRSA antibiotics WAP-8294A and the antifungal antibiotic HSAF. *ACS Synth. Biol.* **2013**, *2*, 670–678. [CrossRef] [PubMed]

37. Zhang, W.; Li, Y.; Qian, G.; Wang, Y.; Chen, H.; Li, Y.Z.; Liu, F.; Shen, Y.; Du, L. Identification and characterization of the anti-methicillin-resistant *Staphylococcus aureus* WAP-8294A2 biosynthetic gene cluster from *Lysobacter enzymogenes* OH11. *Antimicrob. Agents Chemother.* **2011**, *55*, 5581–5589. [CrossRef] [PubMed]

38. Lou, L.; Qian, G.; Xie, Y.; Hang, J.; Chen, H.; Zaleta-Rivera, K.; Li, Y.; Shen, Y.; Dussault, P.H.; Liu, F.; et al. Biosynthesis of HSAF, a tetramic acid-containing macrolactam from *Lysobacter enzymogenes*. *J. Am. Chem. Soc.* **2011**, *133*, 643–645. [CrossRef] [PubMed]

39. Yu, F.; Zaleta-Rivera, K.; Zhu, X.; Huffman, J.; Millet, J.C.; Harris, S.D.; Yuen, G.; Li, X.C.; Du, L. Structure and biosynthesis of heat-stable antifungal factor (HSAF), a broad-spectrum antimycotic with a novel mode of action. *Antimicrob. Agents Chemother.* **2007**, *51*, 64–72. [CrossRef] [PubMed]

40. Han, Y.; Wang, Y.; Tombosa, S.; Wright, S.; Huffman, J.; Yuen, G.; Qian, G.; Liu, F.; Shen, Y.; Du, L. Identification of a small molecule signaling factor that regulates the biosynthesis of the antifungal polycyclic tetramate macrolactam HSAF in *Lysobacter enzymogenes*. *Appl. Microbiol. Biotechnol.* **2015**, *99*, 801–811. [CrossRef] [PubMed]

41. Di Cagno, R.; De Angelis, M.; Calasso, M.; Gobbetti, M. Proteomics of the bacterial cross-talk by quorum sensing. *J. Proteom.* **2011**, *74*, 19–34. [CrossRef] [PubMed]

42. Dong, Y.H.; Zhang, L.H. Quorum sensing and quorum-quenching enzymes. *J. Microbiol.* **2005**, *43*, 101–109. [PubMed]

43. Martino, P.D.; Fursy, R.; Bret, L.; Sundararaju, B.; Phillips, R.S. Indole can act as an extracellular signal to regulate biofilm formation of *Escherichia coli* and other indole-producing bacteria. *Can. J. Microbiol.* **2003**, *49*, 443–449. [CrossRef] [PubMed]

44. Meiser, P.; Bode, H.B.; Muller, R. The unique DKxanthene secondary metabolite family from the myxobacterium *Myxococcus xanthus* is required for developmental sporulation. *Proc. Natl. Acad. Sci. USA* **2006**, *103*, 19128–19133. [CrossRef] [PubMed]

45. Poplawsky, A.R.; Walters, D.M.; Rouviere, P.E.; Chun, W. A gene for a dioxygenase-like protein determines the production of the DF signal in *Xanthomonas campestris* pv. *campestris*. *Mol. Plant. Pathol.* **2005**, *6*, 653–657. [CrossRef] [PubMed]

46. Stevens, A.M.; Queneau, Y.; Soulere, L.; von Bodman, S.; Doutheau, A. Mechanisms and synthetic modulators of AHL-dependent gene regulation. *Chem. Rev.* **2011**, *111*, 4–27. [CrossRef] [PubMed]

47. Guzman-Maldonado, H.; Paredes-Lopez, O. Amylolytic enzymes and products derived from starch: A review. *Crit. Rev. Food Sci. Nutr.* **1995**, *35*, 373–403. [CrossRef] [PubMed]
48. Song, Q.; Wang, Y.; Yin, C.; Zhang, X.H. LaaA, a novel high-active alkalophilic alpha-amylase from deep-sea bacterium *Luteimonas abyssi* XH031(T). *Enzyme Microb. Technol.* **2016**, *90*, 83–92. [CrossRef] [PubMed]
49. Sibakov, M.; Palva, I. Isolation and the 5'-end nucleotide sequence of *Bacillus licheniformis* alpha-amylase gene. *Eur. J. Biochem.* **1984**, *145*, 567–572. [CrossRef] [PubMed]
50. Takkinen, K.; Pettersson, R.F.; Kalkkinen, N.; Palva, I.; Soderlund, H.; Kaariainen, L. Amino acid sequence of alpha-amylase from *Bacillus amyloliquefaciens* deduced from the nucleotide sequence of the cloned gene. *J. Biol. Chem.* **1983**, *258*, 1007–1013. [PubMed]
51. Feller, G.; Payan, F.; Theys, F.; Qian, M.; Haser, R.; Gerday, C. Stability and structural analysis of alpha-amylase from the antarctic psychrophile *Alteromonas haloplanctis* A23. *Eur. J. Biochem.* **1994**, *222*, 441–447. [CrossRef] [PubMed]
52. Hagihara, H.; Igarashi, K.; Hayashi, Y.; Endo, K.; Ikawa-Kitayama, K.; Ozaki, K.; Kawai, S.; Ito, S. Novel alpha-amylase that is highly resistant to chelating reagents and chemical oxidants from the alkaliphilic *Bacillus* isolate KSM-K38. *Appl. Environ. Microb.* **2001**, *67*, 1744–1750. [CrossRef] [PubMed]
53. Wang, N.; Zhang, Y.; Wang, Q.; Liu, J.; Wang, H.; Xue, Y.; Ma, Y. Gene cloning and characterization of a novel alpha-amylase from alkaliphilic *Alkalimonas amylolytica*. *Biotechnol. J.* **2006**, *1*, 1258–1265. [CrossRef] [PubMed]
54. Lee, S.P.; Morikawa, M.; Takagi, M.; Imanaka, T. Cloning of the aapT gene and characterization of its product, alpha-amylase-pullulanase (AapT), from thermophilic and alkaliphilic *Bacillus* sp. strain XAL601. *Appl. Environ. Microbiol.* **1994**, *60*, 3764–3773. [PubMed]
55. Igarashi, K.; Hatada, Y.; Hagihara, H.; Saeki, K.; Takaiwa, M.; Uemura, T.; Ara, K.; Ozaki, K.; Kawai, S.; Kobayashi, T.; et al. Enzymatic properties of a novel liquefying alpha-amylase from an alkaliphilic *Bacillus* isolate and entire nucleotide and amino acid sequences. *Appl. Environ. Microbiol.* **1998**, *64*, 3282–3289. [PubMed]
56. Zhang, J.W.; Zeng, R.Y. Purification and characterization of a cold-adapted alpha-amylase produced by *Nocardiopsis* sp. 7326 isolated from Prydz Bay, Antarctic. *Mar. Biotechnol.* **2008**, *10*, 75–82. [CrossRef] [PubMed]
57. Kim, T.U.; Gu, B.G.; Jeong, J.Y.; Byun, S.M.; Shin, Y.C. Purification and characterization of a maltotetraose- forming alkaline (alpha)-amylase from an alkalophilic *Bacillus* Strain, GM8901. *Appl. Environ. Microbiol.* **1995**, *61*, 3105–3112. [PubMed]
58. Boyer, E.W.; Ingle, M.B. Extracellular alkaline amylase from a *Bacillus* species. *J. Bacteriol.* **1972**, *110*, 992–1000. [PubMed]
59. Sharma, A.; Satyanarayana, T. Cloning and expression of acidstable, high maltose-forming, Ca^{2+}-independent alpha-amylase from an acidophile *Bacillus acidicola* and its applicability in starch hydrolysis. *Extremophiles* **2012**, *16*, 515–522. [CrossRef] [PubMed]
60. Prieto, J.A.; Bort, B.R.; Martinez, J.; Randezgil, F.; Buesa, C.; Sanz, P. Purification and characterization of a new alpha-amylase of intermediate thermal-stability from the yeast *Lipomyces kononenkoae*. *Biochem. Cell Biol.* **1995**, *73*, 41–49. [CrossRef] [PubMed]
61. Schwermann, B.; Pfau, K.; Liliensiek, B.; Schleyer, M.; Fischer, T.; Bakker, E.P. Purification, properties and structural aspects of a thermoacidophilic alpha-amylase from *Alicyclobacillus acidocaldarius* atcc 27009. Insight into acidostability of proteins. *Eur. J. Biochem.* **1994**, *226*, 981–991. [CrossRef] [PubMed]
62. Asoodeh, A.; Chamani, J.; Lagzian, M. A novel thermostable, acidophilic alpha-amylase from a new thermophilic "*Bacillus* sp. Ferdowsicous" isolated from Ferdows hot mineral spring in Iran: Purification and biochemical characterization. *Int. J. Biol. Macromol.* **2010**, *46*, 289–297. [PubMed]
63. Buonocore, V.; Caporale, C.; De Rosa, M.; Gambacorta, A. Stable, inducible thermoacidophilic alpha-amylase from *Bacillus acidocaldarius*. *J. Bacteriol.* **1976**, *128*, 515–521. [PubMed]
64. Ali, I.; Akbar, A.; Anwar, M.; Prasongsuk, S.; Lotrakul, P.; Punnapayak, H. Purification and characterization of a polyextremophilic alpha-amylase from an obligate halophilic *Aspergillus penicillioides* isolate and its potential for souse with detergents. *Biomed. Res. Int.* **2015**. [CrossRef] [PubMed]
65. Xian, L.; Wang, F.; Luo, X.; Feng, Y.L.; Feng, J.X. Purification and characterization of a highly efficient calcium-independent alpha-amylase from *Talaromyces pinophilus* 1-95. *PLoS ONE* **2015**, *10*, e0121531. [CrossRef] [PubMed]

66. Wang, S.; Lu, Z.; Lu, M.; Qin, S.; Liu, H.; Deng, X.; Lin, Q.; Chen, J. Identification of archaeon-producing hyperthermophilic alpha-amylase and characterization of the alpha-amylase. *Appl. Microbiol. Biotechnol.* **2008**, *80*, 605–614. [CrossRef] [PubMed]

67. Han, P.; Zhou, P.; Hu, S.Q.; Yang, S.Q.; Yan, Q.J.; Jiang, Z.Q. A novel multifunctional alpha-amylase from the thermophilic fungus *Malbranchea cinnamomea*: Biochemical characterization and three-dimensional structure. *Appl. Biochem. Biotechnol.* **2013**, *170*, 420–435. [CrossRef] [PubMed]

68. Silva, T.M.; Damasio, A.R.; Maller, A.; Michelin, M.; Squina, F.M.; Jorge, J.A.; Polizeli Mde, L. Purification, partial characterization, and covalent immobilization-stabilization of an extracellular alpha-amylase from *Aspergillus niveus*. *Folia Microbiol.* **2013**, *58*, 495–502. [CrossRef] [PubMed]

69. El-Sayed, A.K.; Abou Dobara, M.I.; El-Fallal, A.A.; Omar, N.F. Purification, sequencing, and biochemical characterization of a novel calcium-independent alpha-amylase AmyTVE from *Thermoactinomyces vulgaris*. *Appl. Biochem. Biotechnol.* **2013**, *170*, 483–497. [CrossRef] [PubMed]

70. Kim, D.H.; Morimoto, N.; Saburi, W.; Mukai, A.; Imoto, K.; Takehana, T.; Koike, S.; Mori, H.; Matsui, H. Purification and characterization of a liquefying alpha-amylase from alkalophilic thermophilic *Bacillus* sp. AAH-31. *Biosci. Biotechnol. Biochem.* **2012**, *76*, 1378–1383. [CrossRef] [PubMed]

71. Michelin, M.; Silva, T.M.; Benassi, V.M.; Peixoto-Nogueira, S.C.; Moraes, L.A.; Leao, J.M.; Jorge, J.A.; Terenzi, H.F.; Polizeli Mde, L. Purification and characterization of a thermostable alpha-amylase produced by the fungus *Paecilomyces variotii*. *Carbohydr. Res.* **2010**, *345*, 2348–2353. [CrossRef] [PubMed]

72. Lu, M.; Wang, S.; Fang, Y.; Li, H.; Liu, S.; Liu, H. Cloning, expression, purification, and characterization of cold-adapted alpha-amylase from *Pseudoalteromonas arctica* GS230. *Protein J.* **2010**, *29*, 591–597. [CrossRef] [PubMed]

73. Liu, X.D.; Xu, Y. A novel raw starch digesting alpha-amylase from a newly isolated *Bacillus* sp. YX-1: Purification and characterization. *Bioresour. Technol.* **2008**, *99*, 4315–4320. [CrossRef] [PubMed]

74. Uma Maheswar Rao, J.L.; Satyanarayana, T. Purification and characterization of a hyperthermostable and high maltogenic alpha-amylase of an extreme thermophile *Geobacillus thermoleovorans*. *Appl. Biochem. Biotechnol.* **2007**, *142*, 179–193. [CrossRef] [PubMed]

75. Champreda, V.; Kanokratana, P.; Sriprang, R.; Tanapongpipat, S.; Eurwilaichitr, L. Purification, biochemical characterization, and gene cloning of a new extracellular thermotolerant and glucose tolerant maltooligosaccharide-forming alpha-amylase from an endophytic ascomycete *Fusicoccum* sp. BCC4124. *Biosci. Biotechnol. Biochem.* **2007**, *71*, 2010–2020. [CrossRef] [PubMed]

76. Nagarajan, D.R.; Rajagopalan, G.; Krishnan, C. Purification and characterization of a maltooligosaccharide-forming alpha-amylase from a new *Bacillus subtilis* KCC103. *Appl. Microbiol. Biotechnol.* **2006**, *73*, 591–597. [CrossRef] [PubMed]

77. Najafi, M.F.; Deobagkar, D.; Deobagkar, D. Purification and characterization of an extracellular alpha-amylase from *Bacillus subtilis* AX20. *Protein Expr. Purif.* **2005**, *41*, 349–354. [CrossRef] [PubMed]

78. Mijts, B.N.; Patel, B.K. Cloning, sequencing and expression of an alpha-amylase gene, *amyA*, from the thermophilic halophile *Halothermothrix orenii* and purification and biochemical characterization of the recombinant enzyme. *Microbiology* **2002**, *148*, 2343–2349. [CrossRef] [PubMed]

79. Chakraborty, K.; Bhattacharyya, B.K.; Sen, S.K. Purification and characterization of a thermostable alpha-amylase from *Bacillus stearothermophilus*. *Folia Microbiol.* **2000**, *45*, 207–210. [CrossRef]

80. Egas, M.C.; da Costa, M.S.; Cowan, D.A.; Pires, E.M. Extracellular alpha-amylase from *Thermus filiformis Ork A2*: Purification and biochemical characterization. *Extremophiles* **1998**, *2*, 23–32. [CrossRef] [PubMed]

81. Marco, J.L.; Bataus, L.A.; Valencia, F.F.; Ulhoa, C.J.; Astolfi-Filho, S.; Felix, C.R. Purification and characterization of a truncated *Bacillus* subtilis alpha-amylase produced by *Escherichia coli*. *Appl. Microbiol. Biotechnol.* **1996**, *44*, 746–752. [PubMed]

82. Shih, N.J.; Labbe, R.G. Purification and characterization of an extracellular alpha-amylase from *Clostridium perfringens* type A. *Appl. Environ. Microbiol.* **1995**, *61*, 1776–1779. [PubMed]

83. Spiess, C.; Happersberger, H.P.; Glocker, M.O.; Spiess, E.; Rippe, K.; Ehrmann, M. Biochemical characterization and mass spectrometric disulfide bond mapping of periplasmic alpha-amylase MalS of *Escherichia coli*. *J. Biol. Chem.* **1997**, *272*, 22125–22133. [CrossRef] [PubMed]

84. Mantsala, P.; Zalkin, H. Membrane-bound and soluble extracellular alpha-amylase from *Bacillus subtilis*. *J. Biol. Chem.* **1979**, *254*, 8540–8547. [PubMed]

85. Buisson, G.; Duee, E.; Haser, R.; Payan, F. Three dimensional structure of porcine pancreatic alpha-amylase at 2.9 Å resolution. Role of calcium in structure and activity. *EMBO J.* **1987**, *6*, 3909–3916. [PubMed]

86. Nielsen, J.E.; Borchert, T.V. Protein engineering of bacterial α-amylases. *Biochim. Biophys. Acta* **2000**, *1543*, 253–274. [CrossRef]

87. Tang, S.Y.; Le, Q.T.; Shim, J.H.; Yang, S.J.; Auh, J.H.; Park, C.; Park, K.H. Enhancing thermostability of maltogenic amylase from *Bacillus thermoalkalophilus* ET2 by DNA shuffling. *FEBS J.* **2006**, *273*, 3335–3345. [CrossRef] [PubMed]

88. Tomazic, S.J.; Klibanov, A.M. Mechanisms of irreversible thermal inactivation of *Bacillus* alpha-amylases. *J. Biol. Chem.* **1988**, *263*, 3086–3091. [PubMed]

89. Vallee, B.L.; Stein, E.A.; Sumerwell, W.N.; Fischer, E.H. Metal content of alpha-amylases of various origins. *J. Biol. Chem.* **1959**, *234*, 2901–2905. [PubMed]

90. Machius, M.; Declerck, N.; Huber, R.; Wiegand, G. Activation of *Bacillus licheniformis* alpha-amylase through a disorder-order transition of the substrate-binding site mediated by a calcium–sodium–calcium metal triad. *Structure* **1998**, *6*, 281–292. [CrossRef]

91. Machius, M.; Wiegand, G.; Huber, R. Crystal structure of calcium-depleted *Bacillus licheniformis* alpha-amylase at 2.2 Å resolution. *J. Mol. Biol.* **1995**, *246*, 545–559. [CrossRef] [PubMed]

92. Li, L.; Yang, J.; Li, J.; Long, L.; Xiao, Y.; Tian, X.; Wang, F.; Zhang, S. Role of two amino acid residues' insertion on thermal stability of thermophilic alpha-amylase AMY121 from a deep sea bacterium *Bacillus* sp. SCSIO 15121. *Bioprocess Biosyst. Eng.* **2014**, *38*, 871–879. [CrossRef] [PubMed]

93. Ghollasi, M.; Khajeh, K.; Naderi-Manesh, H.; Ghasemi, A. Engineering of a *Bacillus* alpha-amylase with improved thermostability and calcium independency. *Appl. Biochem. Biotechnol.* **2010**, *162*, 444–459. [CrossRef] [PubMed]

94. Ben Ali, M.; Khemakhem, B.; Robert, X.; Haser, R.; Bejar, S. Thermostability enhancement and change in starch hydrolysis profile of the maltohexaose-forming amylase of *Bacillus stearothermophilus* US100 strain. *Biochem. J.* **2006**, *394*, 51–56. [CrossRef] [PubMed]

95. Khemakhem, B.; Ben Ali, M.; Aghajari, N.; Juy, M.; Haser, R.; Bejar, S. The importance of an extra loop in the B-domain of an alpha-amylase from *B. stearothermophilus* US100. *Biochem. Biophys. Res. Commun.* **2009**, *385*, 78–83. [CrossRef] [PubMed]

96. Binter, A.; Staunig, N.; Jelesarov, I.; Lohner, K.; Palfey, B.A.; Deller, S.; Gruber, K.; Macheroux, P. A single intersubunit salt bridge affects oligomerization and catalytic activity in a bacterial quinone reductase. *FEBS J.* **2009**, *276*, 5263–5274. [CrossRef] [PubMed]

97. Liu, Y.H.; Lu, F.P.; Li, Y.; Wang, J.L.; Gao, C. Acid stabilization of *Bacillus licheniformis* alpha amylase through introduction of mutations. *Appl. Microbiol. Biotechnol.* **2008**, *80*, 795–803. [CrossRef] [PubMed]

98. Yang, G.; Bai, A.; Gao, L.; Zhang, Z.; Zheng, B.; Feng, Y. Glu88 in the non-catalytic domain of acylpeptide hydrolase plays dual roles: Charge neutralization for enzymatic activity and formation of salt bridge for thermodynamic stability. *Biochim. Biophys. Acta* **2009**, *1794*, 94–102. [CrossRef] [PubMed]

99. Yang, H.; Liu, L.; Shin, H.D.; Chen, R.R.; Li, J.; Du, G.; Chen, J. Structure-based engineering of histidine residues in the catalytic domain of alpha-amylase from *Bacillus subtilis* for improved protein stability and catalytic efficiency under acidic conditions. *J. Biotechnol.* **2013**, *164*, 59–66. [CrossRef] [PubMed]

100. Haghani, K.; Khajeh, K.; Naderi-Manesh, H.; Ranjbar, B. Evidence regarding the hypothesis that the histidine-histidine contact pairs may affect protein stability. *Int. J. Biol. Macromol.* **2012**, *50*, 1040–1047. [CrossRef] [PubMed]

101. Wong, T.Y.; Preston, L.A.; Schiller, N.L. ALGINATE LYASE: Review of major sources and enzyme characteristics, structure-function analysis, biological roles, and applications. *Annu. Rev. Microbiol.* **2000**, *54*, 289–340. [CrossRef] [PubMed]

102. Tseng, C.-H.; Yamaguchi, K.; Kitamikado, M. Two types of alginate lyase from a marine bacterium *Vibrio* sp. Al-9. *Nippon Suisan Gakkaishi* **1992**, *58*, 743–749. [CrossRef]

103. Tseng, C.-H.; Yamaguchi, K.; Kitamikado, M. Isolation and some properties of alginate lyase from a marine bacterium *Vibrio* sp. Al-128. *Nippon Suisan Gakkaishi* **1992**, *58*, 533–538. [CrossRef]

104. Song, Y.; Yu, W.G.; Han, F.; Han, W.J.; Li, J.B. Purification and characterization of aginate lyase from marine bacterium *Vibrio* sp. QY101. *Acta Biochim. Biophys. Sin.* **2003**, *35*, 473–477. [PubMed]

105. Han, F.; Gong, Q.H.; Song, K.; Li, J.B.; Yu, W.G. Cloning, sequence analysis and expression of gene alyVI encoding alginate lyase from marine bacterium *Vibrio* sp. QY101. *DNA Seq.* **2004**, *15*, 344–350. [CrossRef] [PubMed]

106. Takeshita, S.; Sato, N.; Igarashi, M.; Muramatsu, T. A highly denaturant-durable alginate Lyase from a marine bacterium: Purification and properties. *Biosci. Biotechnol. Biochem.* **1993**, *57*, 1125–1128. [CrossRef] [PubMed]

107. Liu, G.; Yue, L.; Chi, Z.; Yu, W.; Madzak, C. The surface display of the alginate lyase on the cells of *Yarrowia lipolytica* for hydrolysis of alginate. *Mar. Biotechnol.* **2009**, *11*, 619–626. [CrossRef] [PubMed]

108. Li, L.; Jiang, X.; Guan, H.; Wang, P.; Guo, H. Three alginate lyases from marine bacterium *Pseudomonas fluorescens* HZJ216: Purification and characterization. *Appl. Biochem. Biotechnol.* **2011**, *164*, 305–317. [CrossRef] [PubMed]

109. Dong, S.; Yang, J.; Zhang, X.Y.; Shi, M.; Song, X.Y.; Chen, X.L.; Zhang, Y.Z. Cultivable alginate lyase-excreting bacteria associated with the Arctic brown alga *Laminaria*. *Mar. Drugs* **2012**, *10*, 2481–2491. [CrossRef] [PubMed]

110. Rahman, M.M.; Inoue, A.; Tanaka, H.; Ojima, T. cDNA cloning of an alginate lyase from a marine gastropod *Aplysia kurodai* and assessment of catalytically important residues of this enzyme. *Biochimie* **2011**, *93*, 1720–1730. [CrossRef] [PubMed]

111. MacDonald, L.C.; Berger, B.W. A polysaccharide lyase from *Stenotrophomonas maltophilia* with a unique, pH-regulated substrate specificity. *J. Biol. Chem.* **2014**, *289*, 312–325. [CrossRef] [PubMed]

112. Kobayashi, T.; Uchimura, K.; Miyazaki, M.; Nogi, Y.; Horikoshi, K. A new high-alkaline alginate lyase from a deep-sea bacterium *Agarivorans* sp. *Extremophiles* **2009**, *13*, 121–129. [CrossRef] [PubMed]

113. Miyake, O.; Ochiai, A.; Hashimoto, W.; Murata, K. Origin and diversity of alginate lyases of families PL-5 and -7 in *Sphingomonas* sp. strain A1. *J. Bacteriol.* **2004**, *186*, 2891–2896. [CrossRef] [PubMed]

114. Osawa, T.; Matsubara, Y.; Muramatsu, T.; Kimura, M.; Kakuta, Y. Crystal structure of the alginate (poly alpha-L-guluronate) lyase from *Corynebacterium* sp. at 1.2 Å resolution. *J. Mol. Biol.* **2005**, *345*, 1111–1118. [CrossRef] [PubMed]

115. Duan, G.; Han, F.; Yu, W. Cloning, sequence analysis, and expression of gene *alyPI* encoding an alginate lyase from marine bacterium *Pseudoalteromonas* sp. CY24. *Can. J. Microbiol.* **2009**, *55*, 1113–1118. [CrossRef] [PubMed]

116. Yamamoto, S.; Sahara, T.; Sato, D.; Kawasaki, K.; Ohgiya, S.; Inoue, A.; Ojima, T. Catalytically important amino-acid residues of abalone alginate lyase HdAly assessed by site-directed mutagenesis. *Enzyme Microb. Technol.* **2008**, *43*, 396–402. [CrossRef]

117. Zhu, B.; Yin, H. Alginate lyase: Review of major sources and classification, properties, structure-function analysis and applications. *Bioengineered* **2015**, *6*, 125–131. [CrossRef] [PubMed]

118. Matsushima, R.; Danno, H.; Uchida, M.; Ishihara, K.; Suzuki, T.; Kaneniwa, M.; Ohtsubo, Y.; Nagata, Y.; Tsuda, M. Analysis of extracellular alginate lyase and its gene from a marine bacterial strain, *Pseudoalteromonas atlantica* AR06. *Appl. Microbiol. Biotechnol.* **2010**, *86*, 567–576. [CrossRef] [PubMed]

119. Matsushima, R.; Watanabe, R.; Tsuda, M.; Suzuki, T. Analysis of extracellular alginate lyase (*alyA*) expression and its regulatory region in a marine bacterial strain, *Pseudoalteromonas atlantica* AR06, using a *gfp* gene reporter system. *Mar. Biotechnol.* **2013**, *15*, 349–356. [CrossRef] [PubMed]

120. Ashton, R.S.; Banerjee, A.; Punyani, S.; Schaffer, D.V.; Kane, R.S. Scaffolds based on degradable alginate hydrogels and poly(lactide-co-glycolide) microspheres for stem cell culture. *Biomaterials* **2007**, *28*, 5518–5525. [CrossRef] [PubMed]

121. Hoiby, N.; Bjarnsholt, T.; Givskov, M.; Molin, S.; Ciofu, O. Antibiotic resistance of bacterial biofilms. *Int. J. Antimicrob. Agents* **2010**, *35*, 322–332. [CrossRef] [PubMed]

122. Iehata, S.; Inagaki, T.; Okunishi, S.; Nakano, M.; Tanaka, R.; Maeda, H. Improved gut environment of abalone *Haliotis gigantea* through *Pediococcus* sp Ab1 treatment. *Aquaculture* **2010**, *305*, 59–65. [CrossRef]

123. Younes, I.; Rinaudo, M. Chitin and chitosan preparation from marine sources. Structure, properties and applications. *Mar. Drugs* **2015**, *13*, 1133–1174. [CrossRef] [PubMed]

124. Wang, S.; Shao, B.; Fu, H.; Rao, P. Isolation of a thermostable legume chitinase and study on the antifungal activity. *Appl. Microbiol. Biotechnol.* **2009**, *85*, 313–321. [CrossRef] [PubMed]

125. Hirono, I.; Yamashita, M.; Aoki, T. Note: Molecular cloning of chitinase genes from *Vibrio anguillarum* and *V. parahaemolyticus*. *J. Appl. Microbiol.* **1998**, *84*, 1175–1178. [CrossRef] [PubMed]

126. Aunpad, R.; Panbangred, W. Cloning and characterization of the constitutively expressed chitinase C gene from a marine bacterium, *Salinivibrio costicola* strain 5SM-1. *J. Biosci. Bioeng.* **2003**, *96*, 529–536. [CrossRef]

127. Howard, M.B.; Ekborg, N.A.; Taylor, L.E.; Weiner, R.M.; Hutcheson, S.W. Genomic analysis and initial characterization of the chitinolytic system of *Microbulbifer degradans* strain 2-40. *J. Bacteriol.* **2003**, *185*, 3352–3360. [CrossRef] [PubMed]

128. Tsujibo, H.; Kubota, T.; Yamamoto, M.; Miyamoto, K.; Inamori, Y. Characterization of chitinase genes from an alkaliphilic actinomycete, *Nocardiopsis prasina* OPC-131. *Appl. Environ. Microbiol.* **2003**, *69*, 894–900. [CrossRef] [PubMed]

129. Fukamizo, T. Chitinolytic enzymes: Catalysis, substrate binding, and their application. *Curr. Protein Pept. Sci.* **2000**, *1*, 105–124. [CrossRef] [PubMed]

130. Horn, S.J.; Sorbotten, A.; Synstad, B.; Sikorski, P.; Sorlie, M.; Varum, K.M.; Eijsink, V.G. Endo/exo mechanism and processivity of family 18 chitinases produced by *Serratia marcescens*. *FEBS J.* **2006**, *273*, 491–503. [CrossRef] [PubMed]

131. Han, Y.; Yang, B.; Zhang, F.; Miao, X.; Li, Z. Characterization of antifungal chitinase from marine *Streptomyces* sp. DA11 associated with South China Sea sponge *Craniella australiensis*. *Mar. Biotechnol.* **2009**, *11*, 132–140. [CrossRef] [PubMed]

132. Stefanidi, E.; Vorgias, C.E. Molecular analysis of the gene encoding a new chitinase from the marine psychrophilic bacterium *Moritella marina* and biochemical characterization of the recombinant enzyme. *Extremophiles* **2008**, *12*, 541–552. [CrossRef] [PubMed]

133. Wang, S.L.; Liang, T.W.; Lin, B.S.; Wang, C.L.; Wu, P.C.; Liu, J.R. Purification and characterization of chitinase from a new species strain *Pseudomonas* sp. TKU008. *J. Microbiol. Biotechnol.* **2010**, *20*, 1001–1005. [PubMed]

134. Suginta, W.; Songsiriritthigul, C.; Kobdaj, A.; Opassiri, R.; Svasti, J. Mutations of Trp275 and Trp397 altered the binding selectivity of *Vibrio carchariae* chitinase A. *Biochim. Biophys. Acta* **2007**, *1770*, 1151–1160. [CrossRef] [PubMed]

135. De Assis, C.F.; Costa, L.S.; Melo-Silveira, R.F.; Oliveira, R.M.; Pagnoncelli, M.G.; Rocha, H.A.; De Macedo, G.R.; Santos, E.S. Chitooligosaccharides antagonize the cytotoxic effect of glucosamine. *World J. Microbiol. Biotechnol.* **2012**, *28*, 1097–1105. [CrossRef] [PubMed]

136. Chi, Z.; Zhang, T.; Liu, G.; Li, J.; Wang, X. Production, characterization and gene cloning of the extracellular enzymes from the marine-derived yeasts and their potential applications. *Biotechnol. Adv.* **2009**, *27*, 236–255. [CrossRef] [PubMed]

137. Mba Medie, F.; Davies, G.J.; Drancourt, M.; Henrissat, B. Genome analyses highlight the different biological roles of cellulases. *Nat. Rev. Microbiol.* **2012**, *10*, 227–234. [CrossRef] [PubMed]

138. Gibbs, M.D.; Reeves, R.A.; Farrington, G.K.; Anderson, P.; Williams, D.P.; Bergquist, P.L. Multidomain and multifunctional glycosyl hydrolases from the extreme thermophile *Caldicellulosiruptor* isolate Tok7B.1. *Curr. Microbiol.* **2000**, *40*, 333–340. [CrossRef] [PubMed]

139. Bronnenmeier, K.; Kern, A.; Liebl, W.; Staudenbauer, W.L. Purification of *Thermotoga maritima* enzymes for the degradation of cellulosic materials. *Appl. Environ. Microbiol.* **1995**, *61*, 1399–1407. [PubMed]

140. Hakamada, Y.; Koike, K.; Yoshimatsu, T.; Mori, H.; Kobayashi, T.; Ito, S. Thermostable alkaline cellulase from an alkaliphilic isolate, *Bacillus* sp. KSM-S237. *Extremophiles* **1997**, *1*, 151–156. [CrossRef] [PubMed]

141. Lee, B.-H.; Kim, B.-K.; Lee, Y.-J.; Chung, C.-H.; Lee, J.-W. Industrial scale of optimization for the production of carboxymethyl cellulase from rice bran by a marine bacterium, *Bacillus subtilis* subsp. *subtilis* A-53. *Enzyme Microb. Technol.* **2010**, *46*, 38–42. [CrossRef]

142. Alfredsson, G.A.; Kristjansson, J.K.; Hjorleifsdottir, S.; Stetter, K.O. *Rhodothermus marinus*, gen. nov., sp. nov., a thermophilic, halophilic bacterium from submarine hot springs in iceland. *J. Gen. Microbiol.* **1988**, *134*, 299–306. [CrossRef]

143. Trivedi, N.; Gupta, V.; Kumar, M.; Kumari, P.; Reddy, C.R.K.; Jha, B. An alkali-halotolerant cellulase from *Bacillus flexus* isolated from green seaweed Ulva lactuca. *Carbohydr. Polym.* **2011**, *83*, 891–897. [CrossRef]

144. Fang, Z.; Fang, W.; Liu, J.; Hong, Y.; Peng, H.; Zhang, X.; Sun, B.; Xiao, Y. Cloning and characterization of a beta-glucosidase from marine microbial metagenome with excellent glucose tolerance. *J. Microbiol. Biotechnol.* **2010**, *20*, 1351–1358. [CrossRef] [PubMed]

marine drugs

Article

Purification of Antioxidant Peptides by High Resolution Mass Spectrometry from Simulated Gastrointestinal Digestion Hydrolysates of Alaska Pollock (*Theragra chalcogramma*) Skin Collagen

Liping Sun, Weidan Chang, Qingyu Ma and Yongliang Zhuang *

Yunnan Institute of Food Safety, Kunming University of Science and Technology, No. 727 South Jingming Road, Kunming 650500, Yunnan, China; kmlpsun@163.com (L.S.); changweidan08023@163.com (W.C.); maqingyu0323@163.com (Q.M.)
* Correspondence: ylzhuang@kmust.edu.cn; Tel./Fax: +86-871-6592-0216

Academic Editor: Se-Kwon Kim
Received: 14 July 2016; Accepted: 26 September 2016; Published: 17 October 2016

Abstract: In this study, the stable collagen hydrolysate was prepared by alcalase hydrolysis and twice simulated gastrointestinal digestion from Alaska pollock skin. The characteristics of hydrolysates and antioxidant activities in vitro, including 2,2′-azino-bis (3-ethylbenzothiazoline-6-sulfonic acid) radical (ABTS$^{\bullet+}$) scavenging activity, ferric-reducing antioxidant power (FRAP) and hydroxyl radical (OH·) scavenging activity, were determined. After twice simulated gastrointestinal digestion of skin collagen (SGI-2), the degree of hydrolysis (DH) reached 26.17%. The main molecular weight fractions of SGI-2 were 1026.26 and 640.53 Da, accounting for 59.49% and 18.34%, respectively. Amino acid composition analysis showed that SGI-2 had high content of total hydrophobic amino acid (307.98/1000). With the simulated gastrointestinal digestion progressing, the antioxidant activities increased significantly ($p < 0.05$). SGI-2 was further purified by gel filtration chromatography, ion exchange chromatography and high performance liquid chromatography, and the A_{1a3c-p} fraction with high hydroxyl radical scavenging activity (IC50 = 7.63 µg/mL) was obtained. The molecular weights and amino acid sequences of key peptides of A_{1a3c-p} were analyzed using high resolution mass spectrometry (LC-ESI-LTQ-Orbitrap-MS) combined with de novo software and UniProt of MaxQuant software. Four peptides were identified from A_{1a3c-p}, including YGCC (444.1137 Da) and DSSCSG (554.1642 Da) identified by de novo software and NNAEYYK (900.3978 Da) and PAGNVR (612.3344 Da) identified by UniProt of MaxQuant software. The molecular weights and amino acid sequences of four peptides were in accordance with the features of antioxidant peptides. The results indicated that different peptides were identified by different data analysis software according to spectrometry mass data. Considering the complexity of LC-ESI-LTQ-Orbitrap-MS, it was necessary to use the different methods to identify the key peptides from protein hydrolysates.

Keywords: Alaska pollock skin collagen; simulated gastrointestinal digestions; antioxidant peptide; peptide purification; de novo software; UniProt of MaxQuant

1. Introduction

Reactive oxygen species (ROS) and free radicals are very unstable and react rapidly with lipids and proteins in the body, generating damage in the biological system, such as DNA, protein and membrane lipid, if the human body cannot control their formation or eliminate them [1]. Therefore, it is important to inhibit oxidation reactions and the formation of free radicals in the living body. In recent years, some peptides have been found to possess antioxidant activity [2,3]. Collagen is rich in hydrophobic amino acids, and the abundance of these amino acids favors higher affinity to oil and better emulsifying ability. Therefore, collagen is expected to provide natural antioxidant peptides and exert higher antioxidant effects than other proteins. In addition, collagen peptides have good biocompatibility, good penetrability and cause no irritation to the body [4]. Researchers have reported the characteristic and antioxidant activity of different collagen resources, especially from aquatic animals, such as sea jumbo squid skin [5], jellyfish umbrella [2], smooth hound protein [6], sea cucumber [7], pacific hake [8] and tilapia skin [3].

The structure of bioactive peptides may be changed when they are digested, absorbed and transferred in gastrointestinal tract [9]. It is expected that bioactive peptides would not be further digested in the gastrointestinal tract and thereby ensure its stability after digestion. The stable peptides should ideally be isolated and identified in vivo. However, in vivo studies are costly and time-consuming, are rather complicated to perform. As an alternative to in vivo studies, a simple, rapid, and inexpensive in vitro simulated gastric and intestinal digestive method has been established to isolate bioactive peptides. In this method, the prior hydrolysis with the endopeptidases would increase the stability and bioavailability of the bioactive peptide in vivo. In the previous studies, it has been shown that gastrointestinal enzyme digestion results in more potent peptides compared with other enzymatic digestions [10].

Alaska pollock (*Theragra chalcogramma*) is one of the commercially important fish species in China, but a large number of scraps containing fish skin, head and bones are left in the processing of fillet production. It has been reported that 70% of the dry matter of fish skin is collagen. Some studies reported the characteristics of collagen and bioactivities of collagen peptides from Alaska pollock skin. Yan et al. [11] reported the characterization of collagen from the skin of Alaska pollock. Hou et al. [12] studied the immunomodulatory activity of Alaska pollock hydrolysates and the active peptide was identified. Byun and Kim [13] purified the key peptide from Alaska Pollock skin with angiotensin I converting enzyme inhibitory activity. Jia et al. [14] studied the enzymatic hydrolysates of Alaska Pollock skin with antioxidant activity.

In this study, twice simulated gastrointestinal (GI) digestion was used to prepare hydrolysates from the collagen of Alaska Pollock skin, in order to obtain the stable antioxidant peptides. The antioxidant activity of the hydrolysates was evaluated in vitro. The bioactive fraction from hydrolysates with the highest antioxidant activity was separated by gel filtration chromatography, ion exchange chromatography and high performance liquid chromatography. Furthermore, two methods, including de novo analysis software and UniProt of MaxQuant software, were used to identify the key peptides from protein hydrolysates.

2. Results and Discussion

2.1. Analyses, DH, Molecular Weight (MW) and Amino Acid Composition of Hydrolysates

In this study, Alaska Pollock skin collagen was hydrolysed by alcalase to obtain its hydrolysates (ASCH). Two successive simulated GI digestions of ASCH were conducted in order to get the stable collagen hydrolysates. The DHs of collagen hydrolysates were studied at different stages, including the first simulated gastric digestion (SG-1), the first simulated intestinal digestion (SGI-1), the second simulated gastric digestion (SG-2) and the second simulated intestinal digestion (SGI-2). As shown in Figure 1, the DH of ASCH was 13.17%. It increased from 16.92% (SG-1) to 22.65% (SGI-1) at the first simulated GI digestion. It was similar to the study of You et al. which indicated that more peptide bonds were broken using pancreatin digestion than using pepsin digestion [15]. In the second simulated GI digestion, the increase of DH was significant ($p < 0.05$) at the stage of SG-2 (25.47%), but it was not significant between the SG-2 and SGI-2 stage (26.17%) ($p > 0.05$). It might be because the DH became basically stable as the second simulated intestinal digestion was processed.

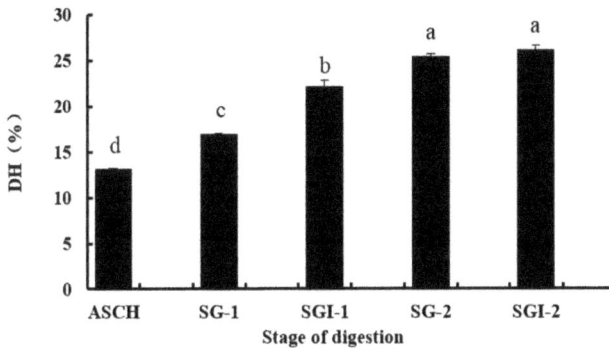

Figure 1. The hydrolysis degree changes of different stages of digestion from skin collagen of Alaska pollock ASCH: alcalase hydrolysates; SG-1: the first simulated gastric digestion; SGI-1: the first simulated intestinal digestion; SG-2: the second simulated gastric digestion; SGI-2: the second simulated intestinal digestion. Different letters indicated significant differences ($p < 0.05$).

Furthermore, the molecular weight distributions of the different stages of digestion from skin collagen of Alaska pollock were shown in Table 1. The main molecular weight fractions of ASCH, SGI-1 and SGI-2 were 3198.76 (66.82%), 1552.34 (74.66%) and 1026.26 Da (59.49%), respectively. With the increasing of DH, the molecular weight of the hydrolysates significantly decreased. As shown in Table 1, the MW distribution of SGI-2 was 1026.26, 640.53, 284.97 and 96.58 Da. Based on the peak area, they accounted for 59.49, 18.34, 16.60 and 4.56%, respectively.

Table 1. Molecular weight distributions of three stages of hydrolysates.

Hydrolysates	ASCH		SGI-1		SGI-2	
Num	MW (Da)	Content (%) *	MW (Da)	Content (%)	MW (Da)	Content (%)
1	3198.76	66.82	1552.34	74.66	1026.26	59.49
2	2245.49	16.45	976.83	18.04	640.53	18.34
3	647.39	8.09	505.59	5.08	284.97	16.60
4	199.63	8.08	180.09	2.15	96.58	4.56

* the percentage of the peak area.

The amino acid compositions of the different stages of digestion were showed in Table 2. The compositions of three stages were similar and they were rich in Gly, Glu, Pro, Asp, and Arg. The result was similar to that of other fish skin hydrolysates [4,5] and in accordance with the characteristics of collagen hydrolysates. In this study, the total hydrophobic amino acid (THAA) contents of three hydrolysates were high, showing 315.11, 302.22 and 307.98 per 1000 residues, respectively. High content of hydrophobic amino acids could increase the solubility in lipids and therefore enhance the antioxidative activity of hydrolysates [4].

Table 2. Amino acid compositions of three stages of hydrolysates (No. of residues per 1000 residues).

Amino Acids	ASCH	SGI-1	SGI-2
Asp	54.19	56.37	56.81
Thr	25.56	26.43	27.37
Ser	64.78	64.81	62.17
Glu	75.24	76.53	76.38
Gly	314.42	325.46	319.66
Ala	109.31	107.95	105.06

* THAA: total hydrophobic amino acid.

Table 2. *Cont.*

Amino Acids	ASCH	SGI-1	SGI-2
Cys	31.80	19.17	23.18
Val	17.90	19.44	20.56
Met	15.01	14.53	14.39
Ile	11.82	12.49	13.42
Leu	22.25	22.71	23.58
Tyr	3.40	3.77	4.90
Phe	13.04	12.81	13.89
Lys	27.95	28.70	30.59
NH_3	57.51	54.29	69.71
His	8.83	8.87	9.17
Arg	53.01	52.55	35.25
Pro	93.99	93.13	93.90
THAA*	315.11	302.22	307.98
Total	1000	1000	1000

* THAA: total hydrophobic amino acid.

2.2. Analyses of Antioxidant Activities

Since there is no single antioxidant standard method to test for antioxidant capacity, it is recommended to use different methods for investigating the different mechanisms of antioxidant capacity [16]. In order to evaluate the antioxidant activity of collagen hydrolysates, the pHs was adjusted to about 7.0 in this study. As shown in Figure 2, the $ABTS^{\bullet+}$ scavenging activity, FRAP and OH· scavenging activity were measured.

The ABTS assay is often used to evaluate the ability of antioxidants to scavenge free radicals. After simulated GI digestion, the $ABTS^{\bullet+}$ scavenging activity obviously increased (Figure 2a). The $ABTS^{\bullet+}$ scavenging activity was 39.79% in the SGI-1 stage and 59.17% in the SGI-2 stage at the concentrations of 0.5 mg/mL. The FRAP assay is based on the ability of antioxidants to reduce Fe^{3+} to Fe^{2+} in the presence of 1,3,5-tri(2-pyridyl)-2,4,6-triazine (TPTZ). As shown in Figure 2b, with the simulated gastrointestinal digestion progressing, FRAP of the hydrolysates increased significantly ($p < 0.05$). At the dose of 10 mg/mL, the activities of SGI-1 and SGI-2 were 129.42 and 209.27 µmol/L $FeSO_4$, respectively. Hydroxyl radical is the most reactive radical, which has been demonstrated to be a highly damaging species in free radical pathology, attacking almost every molecule in living cells. As shown in Figure 2c, the simulated GI digestion significantly increased the OH· scavenging activity ($p < 0.05$). Indeed, the OH· scavenging activity of ASCH was 15.53% at the concentration of 2 mg/mL, and the activities of SGI-1 and SGI-2 were 34.77 and 59.11%, respectively. The increase of scavenging activity of the pancreatin digestion process was higher than the pepsin digestion.

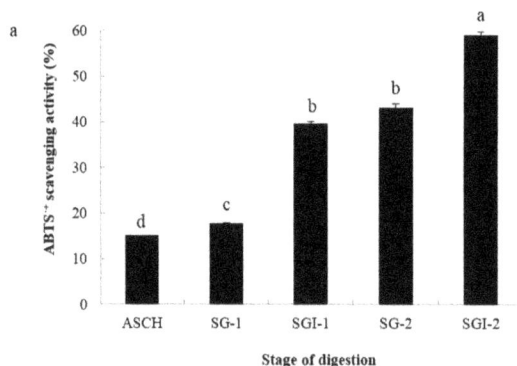

Figure 2. *Cont.*

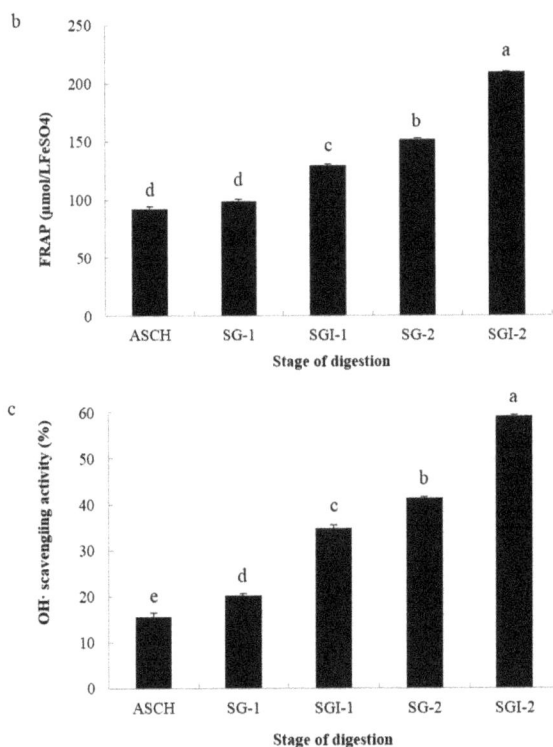

Figure 2. The antioxidant acitivties of different stages of digestion of skin collagen of Alaska pollock. (**a**): ABTS$^{\bullet+}$ scavenging activity (at 0.5 mg/mL); (**b**): FRAP (at 10 mg/mL); (**c**): OH· scavenging activity (at 2 mg/mL). Different letters indicated significant differences ($p < 0.05$).

2.3. Purification of the Antioxidant Peptides from SGI-2

The removal of OH· is probably one of the most effective defenses of a living body against various diseases [17]. Based on this reason, the OH· scavenging activity was selected as the indicator of purification of antioxidant peptides in the study.

The SGI-2 solution was purified by a Sephadex G-25 gel filtration column, and four fractions were obtained, noting A–D respectively (Figure 3a). Four fractions were collected, concentrated and the OH· scavenging activities were determined. Results showed that fraction A had the highest OH· scavenging activity among the four fractions, with the IC$_{50}$ value being 0.26 mg/mL (Figure 3b).

Ion-exchange chromatography was a method of separation according to the substance with a different acid-base property and polarity. SP Sephadex C-25 was one of the strong cation exchangers with a main functional group of sulfopropyl and it was widely used in separating bioactive peptides [18]. The fraction A collected from Sephadex G-25 was further separated by the SP Sephadex C-25 column and five fractions were obtained, noting A$_1$, A$_2$, A$_3$, A$_4$ and A$_5$, respectively (Figure 4a). The OH· scavenging activities of these five fractions were shown in Figure 4b; fraction A$_1$ had the highest OH· scavenging activity with the IC$_{50}$ value being 81.15 μg/mL. Thus, the fraction A$_1$ was selected for next separation.

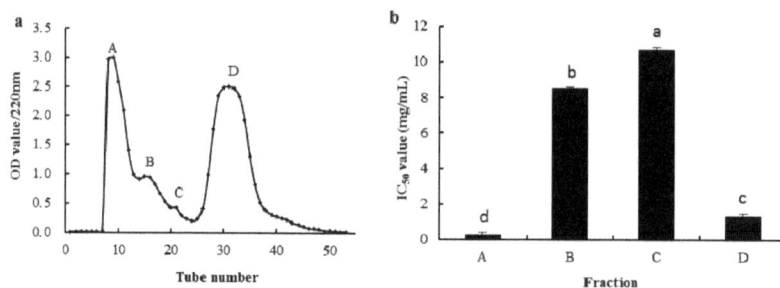

Figure 3. Sephadex G-25 gel chromatography (**a**) and the IC_{50} value (mg/mL) of each fraction was measured by OH· scavenging activities (**b**). Different letters indicate significant differences ($p < 0.05$).

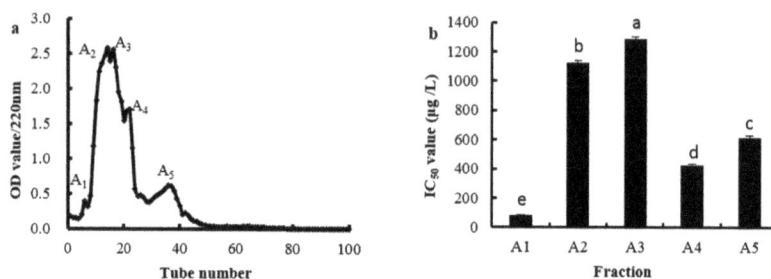

Figure 4. Elution profile of fraction A separated by SP Sephadex C-25 chromatography (**a**) and the IC_{50} value (μg/mL) of the OH· scavenging activities of each fraction (**b**). Different letters indicate significant differences ($p < 0.05$).

Sephadex G-15 was used to remove NaCl from the eluate of SP Sephadex C-25 in the fraction A_1. As shown in Figure 5, four fractions (A_{1a}, A_{1b}, A_{1c}, and A_{1d}) were obtained and their OH· scavenging activities were measured. The OH· scavenging activity of the fraction A_{1a} was the highest compared with the other three fractions, and the IC_{50} value was 73.52 μg/mL.

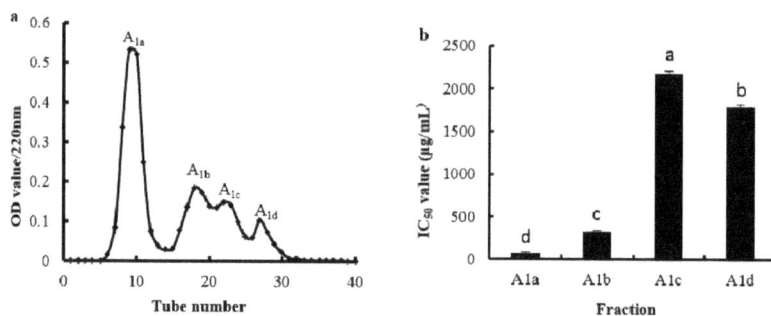

Figure 5. Elution profile of fraction A_1 separated by Sephadex G-15 chromatography (**a**) and the IC_{50} value (μg/mL) of the OH· scavenging activities of each fraction (**b**). Different letters indicate significant differences ($p < 0.05$).

The fraction A_{1a} was further isolated by a Shim-pack GIS C18 preparative column with a liner gradient of acetonitrile containing 0.1% trifluoroacetic acid (TFA) from 5% to 30% in 30 min. The elution profile was shown in Figure 6. A total of 12 peaks were obtained and named as A_{1a1}–A_{1a12}, respectively.

Each peak was collected, concentrated and it's OH· scavenging activities were measured. The result showed that A_{1a3} had the highest antioxidant activity and the IC_{50} value of A_{1a3} was 31.72 μg/mL.

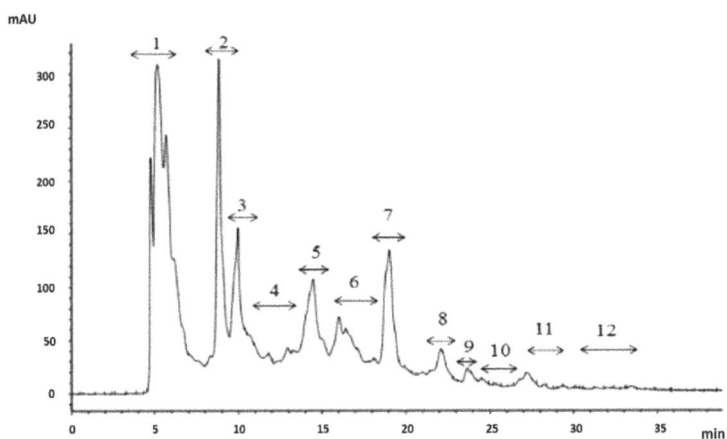

Figure 6. Chromatography of A_{1a} separated by a Shim-pack GIS C18. Liner gradient was 5%–30% acetonitrile containing 0.1% TFA from 0 to 30 min.

The fraction A_{1a3} was isolated by HPLC on the semi-preparative C18 column using a liner gradient of acetonitrile containing 0.1% TFA from 5% to 25% in 30 min. Seven fractions were collected and designated as A_{1a3a}–A_{1a3g} in turn respectively (Figure 7). After the OH· scavenging activities of seven fractions were determined, we found that A_{1a3c} had the highest OH· scavenging activity, and the IC_{50} value of A_{1a3c} was 14.36 μg/mL. The most active A_{1a3c} faction was isolated again by the semi-preparative C18 column using a different liner gradient of acetonitrile containing 0.1% TFA from 5% to 20% in 30 min. The main peak A_{1a3c-p} with high antioxidant activity was collected and concentrated (Figure 8). The IC_{50} value of OH· scavenging activity of the A_{1a3c-p} fraction was 7.63 μg/mL.

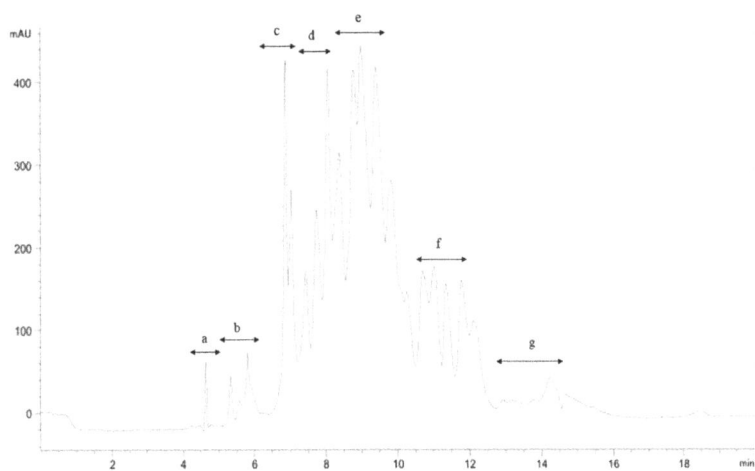

Figure 7. Chromatography of A_{1a3} separated by C18 semi-preparing HPLC. The liner gradient was 5%–25% acetonitrile containing 0.1% TFA from 0 to 30 min.

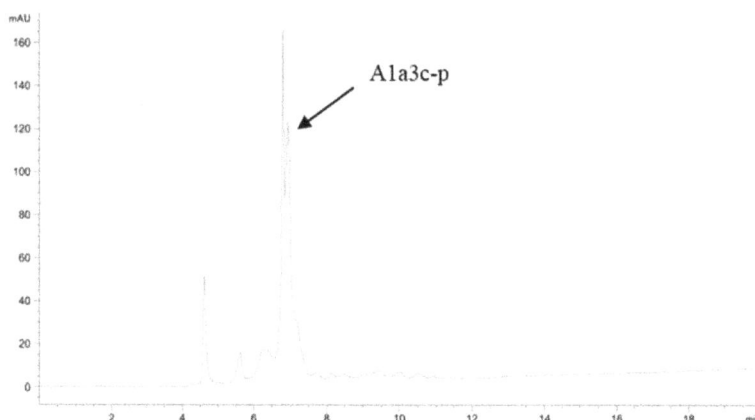

Figure 8. Chromatography of A$_{1a3c}$ separated by C18 semi-preparing HPLC. The liner gradient was 5%–20% acetonitrile containing 0.1% TFA from 0 to 30 min.

2.4. Identification of Purified Peptide

The antioxidant activity of peptides is connected with their molecular weights, amino acid compositions, amino acid sequences and so on [19]. In this study, the fraction A$_{1a3c-p}$ was analyzed by high resolution mass spectrometry combined with two methods, including de novo software and MaxQuant software. As shown in Figure 9, two peptides were obtained by de novo software. The peptide sequences were Try–Gly–Cys–Cys (YGCC) and Asp–Ser–Ser–Cys–Ser–Gly (DSSCSG), and their molecular weight was 444.1137 and 554.1642 Da, respectively. The data was scanned in the "fish collagen" database by UniProt of MaxQuant software. Two peptides, Asn-Asn-Ala-Gln-Tyr-Tyr-Lys (NNAEYYK) and Pro-Ala-Gly-Asn-Val-Arg (PAGNVR), were identified, and their molecular weight was 900.3978 and 612.3344 Da, respectively. The amino acids at the C-terminus of two peptides were K and R, which conformed to the fracture characters of simulated gastric and intestinal digestions.

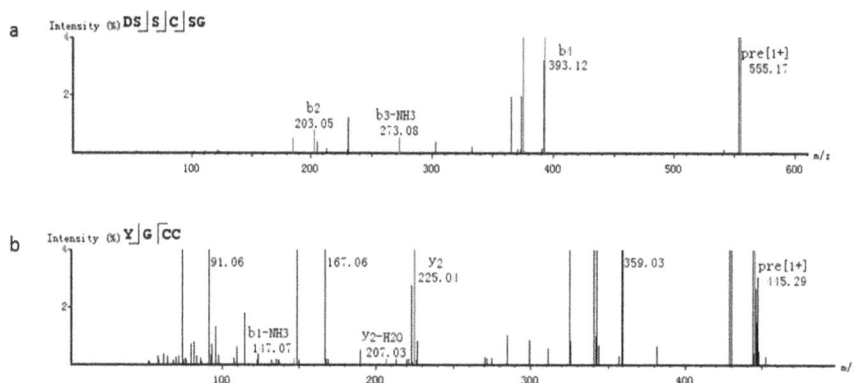

Figure 9. MS/MS spectrum analysis of the active peptides. (**a**): DSSCSG; (**b**): YGCC.

Generally, there is no direct relationship between antioxidant activity and molecular weight. However, the previous study indicated that the peptides with smaller molecular weights have stronger antioxidant activities, more resistant to the gastrointestinal digestion and easier to cross the intestinal barrier to exert biological activities than larger peptides [20]. Antioxidative peptides usually contain 2–20 amino acids with molecular weights below 3000 Da [5]. In this study, the amino acid numbers of four peptides identified from Alaska pollock skin collagen were 4, 6, 7, and 6 and the molecular weight was lower 1000 Da. The amino acid number and molecular weight was in accordance with the feature of antioxidant peptide. Our study was similar to the peptides purified from walnut [21] and loach protein hydrolysates [15].

Moreover, compositions and the specific position of amino acids in the peptide may play an important role in its antioxidant activities. High content of hydrophobic amino acids, especially at the *N*- or *C*-terminus of peptides, could enhance the activities of antioxidative peptides by interacting with lipid molecules and donating protons into radicals to scavenge radicals [22]. Moreover, polar/charged amino acids such as Arg at the *C*-terminus position also contribute to the antioxidant activity [23]. Our results were similar to these previous reports, and hydrophobic amino acids or arginine existed in the terminus of four peptides.

Some studies have reported that peptide sequences containing Tyr show strong antioxidant activity, especially when the presence of Tyr was at terminals of the peptide sequence. The antioxidant activity of Tyr may be explained by the special capability of phenolic groups to serve as hydrogen donors, which is one mechanism of inhibiting the radical-mediated peroxidizing chain reaction [24]. YGCC obtained from this study had a Tyr*N*-terminus, and this might be one of the reasons why YGCC showed higher radical scavenging activity. In addition, NNAEYYK had two Tyr, which could effectively increase its antioxidant activity. Previous studies show that Cys is hydrophobic in nature and can interact directly with free radicals by donating the sulfur hydrogen, so the presence of Cys is one of the reasons for the good antioxidant activity of the isolated peptide [25]. Li et al. considered that Cys residue at the *C*-terminus or next to the *C*-terminus plays an important role in antioxidative activities [26]. It was similar to our study and YGCC and DSSCSG contained Cys, YGCC in particular had two Cys at the *C*-terminus, which might improve its antioxidant activity. Acidic amino acids, such as Asn and Gln, play important roles in the chelation of metal ions by their side chains, which may inhibit the formation of the hydroxyl radical [27]. Rajapakse et al. reported that the presence of Asp seemed to play a vital role, irrespective of its position, as observed in several antioxidative peptide sequences [28]. It was similar to our results. DSSCSG and NNAEYYK had Asp and Asn at the *N*-terminus, respectively.

The fraction A_{1a3c-p} was analyzed using high resolution mass spectrometry, and the key peptides obtained by de novo and MaxQuant software were different. Therefore, to adequately identify the key peptides from protein hydrolysates fractions, it was necessary to use different methods to analyze mass data. A further study about the quantitative analysis of key peptides will be carried out.

3. Materials and Methods

3.1. Materials

Collagen of Alaska pollock skin was prepared by the previous methods. Alcalase was purchase from Genencor International Co. (Wuxi, China); Pepsin, pancreatin, 2,2'-azino-bis (3-ethylbenzothiazoline-6-sulfonic acid) (ABTS) and 2,4,6-Tris(2-pyridyl)-*S*-triazine (TPTZ) were purchased from Sigma Chemical Co., (St. Louis, MO, USA). Sephadex G-25, Sephadex G-15 and SP Sephadex C-25 were purchased from GE Healthcare (Fairfield, CT, USA). Acetonitrile (HPLC grade) was purchased from Merck KGaA (Darmstadt, Germany). All other reagents used in this study were analytical grade.

3.2. Preparation of Skin Collagen Hydrolysates of Alaska Pollock

The Alaska pollock skin collagen was mixed with distilled water at a concentration of 1% (*w/v*). The mixture was adjusted to pH 9.0 by 1 M NaOH solution and then hydrolyzed using Alcalase (E/S: 5/100, *w/w*) at 55 °C for 2 h. Alcalase were inactivated in boiling water for 10 min and centrifuged at 5000 rpm for 20 min. The supernatants (ASCH) were collected and then lyophilized. ASCH was hydrolyzed by the simulated GI digestion [15,29]. A total of 2 g ASCH was dissolved in 150 mL distilled water and adjusted to pH 2.5 with 6 M HCl. Then, pepsin was added at a ratio of enzyme to substrate of 1:35 (w/w). After the mixture was incubated at 37 °C for 1 h with shaking (SG-1), sodium cholate (0.2 mM) and SG-1 (1:1, v/v) were mixed. The pH was adjusted to 7.5 using 2 M NaOH. Then, pancreatin was added at a ratio of enzyme to substrate of 1:25 (w/w). The mixture was incubated at 37 °C for 2 h with shaking and then inactivated in boiling water for 15 min and centrifuged at 5000 rpm for 20 min. The supernatants (SGI-1) were desalinized and then lyophilized. The second simulated gastric and intestinal digestions were conducted using the same method with the above processes and SG-2 and SGI-2 were obtained.

3.3. Determination of Characteristics of Hydrolysates

3.3.1. Determination of the Degree of Hydrolysis (DH)

The contents of free amino ($-NH_2$) and protein (N) were evaluated according to the ninhydrin colorimetric method and kjeldah method, respectively. DH was calculated as follows [30]:

$$DH = \frac{h(mmol/g)}{h_{tot}(mmol/g)} \times 100\% = [\frac{M_1(\mu mol/mL)}{N(mg/mL)} - M_0(mmol/g)] \div h_{tot}(mmol/g) \times 100\% \quad (1)$$

where h is the number of broken peptide bonds per gram protein; h_{tot} is the total number of peptide bonds per gram original protein (the h_{tot} of collagen was 8.41 mmol per gram protein); M_1 is the content of $-NH_2$ in hydrolysate; M_0 is the content of $-NH_2$ in original protein; N is the content of protein in hydrolysate.

3.3.2. Molecular Weight (MW) Distribution

The molecular weight distribution of the different hydrolysates was measured using a high-performance liquid chromatography (HPLC) system (1260 series, Agilent Scientific, Santa Clara, CA, USA) with a TSK gel 3000 PWXL column (30 mm i.d. × 7.8 mm, Tosoh, Tokyo, Japan) [17]. The mobile phases were acetonitrile-water (1:1, v/v) in the presence of 0.1% (v/v) trifluoroacetic acid, and the flow rate was 0.6 mL/min. The process was monitored at 220 nm at 30 °C. A calibration curve of molecular weight was prepared according to the following standards: cytochrome C (12,500 Da), insulin (5734 Da), vitamin B_{12} (1355 Da), hippuryl-histydilleucine (429.5 Da), and glutathione (309.5 Da). The logarithm of molecular weight (MW) and the retention time (tR) were in a linear relationship and the formula was calculated as lg MW = $-0.284tR + 7.310$ ($R^2 = 0.9922$, $p < 0.01$).

3.3.3. Amino Acid Composition

The different hydrolysates were hydrolyzed under reduced pressure with 6 mol/L HCl at 110 °C for 22 h and the amino acid compositions were analyzed on a Hitachi amino acid analyzer 835-50 (Hitachi, Tokyo, Japan).

3.4. Antioxidative Activity Assay

3.4.1. ABTS$^{\bullet+}$ Scavenging Activity Assay

ABTS$^{\bullet+}$ scavenging activities were determined as described by previous method with a slight modification [31]. A total of 5 mL of 7 mM ABTS and 88 μL of 40 mM potassium persulfate was mixed to prepare ABTS$^{\bullet+}$ stock solution. The mixture was left in the dark at room temperature for 12 h. The ABTS$^{\bullet+}$ stock solution was diluted with PBS (2 mM, pH 7.4) to an absorbance of 0.70 ± 0.02 at 734 nm. Then, 0.5 mL of samples were mixed with 4 mL ABTS$^{\bullet+}$ stock solution. The mixture was shaken for 10 s and left in the 30 °C water bath for 6 min. The absorbance was measured at 734 nm. The capability of ABTS$^{\bullet+}$ scavenging was calculated according to the following equation.

$$\text{Radical scavenging activity (\%)} = \frac{A_c - (A_s - A_{cs})}{A_c} \times 100 \tag{2}$$

where A_c was 0.5 mL ethanol + 4.0 mL ABTS$^{\bullet+}$ solution; A_s was 0.5 mL sample + 4.0 mL ABTS$^{\bullet+}$ solution; A_{cs} was 0.5 mL sample + 4.0 mL ethanol.

3.4.2. FRAP Assay

FRAP was determined according to the method of Alemania et al. [32] with a slight modification. A total of 300 mM acetic acid buffer solution (pH 3.6) was mixed with 10 mM TPTZ and 20 mM FeCl$_3\cdot$6H$_2$O according to the rate of 10:1:1. Then the mixture was left in a 37 °C water bath to prepare the FRAP solution. The mixture of 150 μL samples and 4.5 mL FRAP solution was reacted at 37 °C for 10 min and then determined the absorbance at 593 nm. A total of 150 μL of distilled water was used instead of samples solution as a control. The absorbance of different concentrations of FeSO$_4$ solution (0–500 μmol/mL) were determined at 593 nm. The FRAP of samples were expressed as equal to μmol/mL FeSO$_4$.

3.4.3. OH· Scavenging Activity Assay

Hydroxyl radical (OH·) scavenging activity was determined by the previous method with slight modification [33]. Briefly, after 1 mL of samples mixed with 0.3 mL of FeSO$_4$ (8 mM), 1 mL of salicylic acid (3 mM) and 0.25 mL of H$_2$O$_2$ (20 mM), the mixture was incubated at 37 °C for 30 min. The reaction mixture was cooled by flowing water to room temperature. Then, 0.45 mL distilled water was added into the mixture to make the end volume 3.0 mL. The mixture was centrifuged at 3000 rpm for 10 min. The absorbance of supernatant was measured at 510 nm, and 1 mL of the solvent solution was used instead of the sample solution as a control. The capability of scavenging the hydroxyl radical was calculated according to following equation:

$$\text{Radical scavenging activity (\%)} = \frac{A_0 - (A_1 - A_2)}{A_0} \times 100 \tag{3}$$

where A_0 was the absorbance of the control without a sample, A_1 was the absorbance with a sample, and A_2 was the absorbance of the reagent blank. The IC$_{50}$ value was defined as an effective concentration that is required to scavenge 50% of radical activity.

3.5. Purification of Antioxidant Peptides

The SGI-2 was dissolved in distilled water and preliminarily separated by a Sephadex G-25 gel filtration column (Φ 2.6 cm × 30 cm). The SGI-2 was eluted at a flow rate of 0.5 mL/min and collected every 6 min. Then, the eluted solution was monitored at 220 nm. The peptide fraction showing the highest OH· scavenging activity was collected and concentrated.

The peptide fraction with the highest OH· scavenging activity was loaded onto a SP Sephadex C-25 of cationic exchange column (Φ 1.6 cm × 80 cm), which was previously equilibrated with a 0.02 M sodium acetate buffer (pH 4.0). The peptide fraction was eluted with a linear gradient of NaCl concentration from 0 to 1.0 M in the same buffer at a flow rate of 0.8 mL/min and monitored at 220 nm. The elution solution was collected at 6 min intervals and then concentrated. The OH· scavenging activity of isolated fractions was determined.

The peptide fraction with the highest OH· scavenging activity obtained from SP Sephadex C-25 was dissolved and further separated and desalinated by the Sephadex G-15 gel filtration column (Φ 2.6 cm × 30 cm). The peptide fraction was eluted at a flow rate of 0.5 mL/min and collected every 6 min. The solution was monitored at 220 nm. The OH· scavenging activity of isolated fractions was determined.

The highest active faction after Sephadex G-15 was further purified by preparative high performance liquid on a Shim-pack GIS C18 preparative column (Φ 20 mm × 250 mm, Shimadzu, Kyoto, Japan). The mobile phase A was water, and mobile phase B was acetonitrile containing 0.1% TFA. The column was eluted by a linear gradient of 5% B to 30% B in 30 min. The flow rate was 10.0 mL/min, and detection wavelength was 220 nm. The above steps were repeated several times until the different eluted fractions were able to measure the OH· scavenging activity and purify further. The same fractions were pooled and concentrated to remove acetonitrile and TFA.

The fraction with the highest OH· scavenging activity was passed through a Zorbax semi-preparative SB-C18 column (Φ 9.4 mm × 250 mm, Agilent Scientific, Santa Clara, CA, USA) by Agilent HPLC 1260 system (Agilent Scientific, Santa Clara, CA, USA). The fraction was eluted using a linear gradient of 5% to 25% acetonitrile containing 0.1% TFA (0 to 30 min) at a flow rate of 2.0 mL/min [3]. The column temperature was controlled at 35 °C and the detection wavelength was 220 nm. The fraction showing the high antioxidant activities was concentrated to remove acetonitrile and TFA and lyophilized.

3.6. Analysis and Identification of Purified Peptide

3.6.1. Assay of High Resolution Mass Spectrometry (LC-ESI-LTQ-Orbitrap-MS)

Purified peptides were eluted from Q Exactive Focus (Thermo Fisher, Tewksbury, MA, USA) with a Hypersil Gold C18 chromatographic column (1.9 μm, Φ 2.1 mm × 100 mm) at a flow rate of 0.2 mL/min. The mobile phase A was acetonitrile containing 0.1% formic acid, and mobile phase B was water with 0.1% formic acid. The column was equilibrated for 1 min at 5% A and eluted as the following flow gradient: 1–2.5 min, 5.0%–10.0% A; 2.5–12.5 min, 10.0%–25.0% A; 12.5–20 min, 25.0%–52.5% A; 20–22 min, 52.5%–95.0% A; 22–24 min, 95.0%–5.0% A; 24–30 min, 5.0% A. The mass spectrogram was scanned in the positive ion mode. The instrument was set up as follows: scanning mode, Full MS-ddMS2; resolution, Full MS 35000, ddMS2 17500; scan range: 120~1800 m/z; stepped CE: 10 eV, 20 eV, 30 eV; AGC target: 1×10^5.

3.6.2. Identification of the Key Peptides

The molecular weights and amino acid sequences of purified peptides were identified by two software methods: (1) De novo analysis software. The peptide was automatically selected for fragmentation. The molecular weight and amino acid sequence of the MS date was processed using de novo software. Peptide identifications were accepted if they could be established at greater than 85% probability; (2) UniProt of MaxQuant software. Peptides identification was achieved by comparing mass data against the UniProt data using MaxQuant Server (version 1.5.3.28) [34]. The "fishcollagen" database was downloaded from http://www.uniprot.org/. The parameters of database searches were as follows: variable oxidation of methionins, and tolerance of the ions at 5 ppm for parents and 0.5 Da for fragments [35]. No enzyme or static modification was set for database searching. No missed cleavage was allowed.

3.7. Statistical Analysis

All results obtained were expressed as means ± standard deviation and analyzed by the SPSS 19.0 statistical software (Armonk, NY, USA). Data were analyzed using one-way analysis of variance (ANOVA). $p < 0.05$ indicated statistical significance.

4. Conclusions

In this study, the stable collagen hydrolysate of Alaska pollock skin was prepared by successive simulated gastrointestinal digestion. The DHs, molecular weight distributions, amino acid compositions and antioxidant activities in vitro were evaluated. With the simulated gastrointestinal digestions, The DHs and antioxidant activities increased obviously. An antioxidant fraction (A$_{1a3c-p}$) was purified by gel filtration chromatography, ion exchange chromatography and high performance liquid chromatography, and the IC$_{50}$ value of hydroxyl radical scavenging activity was 7.63 μg/mL.

Furthermore, four key peptides of A_{1a3c-p}, including YGCC, DSSCSG, NNAEYYK and PAGNVR, were analyzed by high resolution mass spectrometry combined with de novo software and UniProt of MaxQuant software. This paper could provide some help for the application of fish skin collagen and the identification of key peptides from protein hydrolysates.

Acknowledgments: This work was financially supported by National Grants of China (31360381).

Author Contributions: The authors contributions were as follows: Liping Sun, Weidan Chuang: performing of the experiments, analysis and interpretation of data, drafting of manuscript; Qingyu Ma: collection of reagents/materials/analysis tools; Yongliang Zhuang: study concept and design, providing guidance on revising the manuscript.

Conflicts of Interest: The authors declare that there are no conflict of interest.

References

1. Wattanasiritham, L.; Theerakulkait, C.; Wickramasekara, S.; Maier, C.; Stevens, J.F. Isolation and identification of antioxidant peptides from enzymatically hydrolyzed rice bran protein. *Food Chem.* **2016**, *192*, 156–162. [CrossRef] [PubMed]

2. Zhuang, Y.; Sun, L.; Zhao, X.; Wang, J.; Hou, H.; Li, B. Antioxidant and melanogenesis inhibitory activities of collagen peptide from jellyfish (*Rhopilemaesculentum*). *J. Sci. Food Agric.* **2009**, *89*, 1722–1727. [CrossRef]

3. Zhang, Y.F.; Duan, X.; Zhuang, Y.L. Purification and characterization of novel antioxidant peptides from enzymatic hydrolysates of tilapia (*Oreochromis niloticus*) skin gelatin. *Peptides* **2012**, *38*, 13–22. [CrossRef] [PubMed]

4. Zhuang, Y.L.; Sun, L.P.; Zhao, X.; Hou, H.; Li, B.F. Investigation of gelatin polypeptides of jellyfish (*Rhopilema esculentum*) for their antioxidant activity in vitro. *Food Technol. Biotechnol.* **2010**, *48*, 222–228.

5. Mendis, E.; Rajapakse, N.; Byun, H.G.; Kim, S.K. Investigation of jumbo squid (*Dosidicus gigas*) skin gelatin peptides for their in vitro antioxidant effects. *Life Sci.* **2005**, *17*, 2166–2178. [CrossRef] [PubMed]

6. Bougatef, A.; Hajji, M.; Balti, R.; Lassoued, I.; Triki-Ellouz, Y.; Nasri, M. Antioxidant and free radical-scavenging activities of smooth hound (*Mustelus mustelus*) muscle protein hydrolysates obtained by gastrointestinal proteases. *Food Chem.* **2009**, *114*, 1198–1205. [CrossRef]

7. Zhou, X.; Wang, C.; Jiang, A. Antioxiant peptides isolated from sea cucumber *Stichopus Japonicus*. *Eur. Food Res. Technol.* **2012**, *234*, 441–447. [CrossRef]

8. Cheung, I.W.Y.; Cheung, L.K.Y.; Tan, N.Y.; Li-Chan, E.C.Y. The role of molecular size in antioxidant activity of peptide fractions from Pacific hake (*Merluccius productus*) hydrolysates. *Food Chem.* **2012**, *134*, 1297–1306. [CrossRef] [PubMed]

9. Minekus, M.; Alminger, M.; Alvito, P.; Balance, S.; Bohn, T.; Bourlieu, C.; Carrière, F.; Boutrou, R.; Corredig, M.; Dupont, D.; et al. A standardised static in vitro digestion method suitable for food—An international consensus. *Food Funct.* **2014**, *5*, 1113–1124. [CrossRef] [PubMed]

10. Samaranayaka, A.G.; Kitts, D.D.; Li-Chan, E.C. Antioxidative and angiotensin-I-converting enzyme inhibitory potential of a Pacific Hake (*Merluccius productus*) fish protein hydrolysate subjected to simulated gastrointestinal digestion and Caco-2 cell permeation. *J. Agric. Food Chem.* **2010**, *58*, 1535–1542. [CrossRef] [PubMed]

11. Yan, M.; Li, B.; Zhao, X.; Yi, J. Physicochemical properties of gelatin gels from walleye pollock (*Theragra chalcogramma*) skin cross-linked by gallic acid and rutin. *Food Hydrocol.* **2011**, *25*, 907–914. [CrossRef]

12. Hou, H.; Fan, Y.; Wang, S.; Si, L.; Li, B. Immunomodulatory activity of Alaska Pollock hydrolysates obtained by glutamic acid biosensor-Artificial neural network and the identification of its active central fragmen. *J. Funct. Foods* **2016**, *24*, 37–47. [CrossRef]

13. Byun, H.G.; Kim, S.K. Purification and characterization of angiotensin I converting enzyme (ACE) inhibitory peptides from Alaska pollock (*Theragra chalcogramma*) skin. *Process Biochem.* **2001**, *36*, 1155–1162. [CrossRef]

14. Jia, J.; Zhou, Y.; Lu, J.; Chen, A.; Li, Y.; Zheng, G. Enzymatic hydrolysis of Alaska Pollock (*Theragra chalcogramma*) skin and antioxidant activity of the resulting hydrolysate. *J. Sci. Food Agric.* **2010**, *90*, 635–640. [PubMed]

15. You, L.; Zhao, M.; Regenstein, J.M. Changes in the antioxidant activity of loach (*Misgurnus anguillicaudatus*) protein hydrolysates during a simulated gastrointestinal digestion. *Food Chem.* **2010**, *120*, 810–816. [CrossRef]

16. Sun, L.; Bai, X.; Zhuang, Y. Effect of different cooking methods on total phenolic contents and antioxidant activities of four Boletus mushrooms. *J. Food Sci. Technol.* **2014**, *51*, 3362–3368. [CrossRef] [PubMed]

17. Zhuang, Y.; Zhao, X.; Li, B. Optimization of antioxidant activity by response surface methodology in hydrolysates of jellyfish (*Rhopilema esculentum*) umbrella collagen. *J. Zhejiang Univ. Sci. B* **2009**, *10*, 572–579. [CrossRef] [PubMed]

18. Hou, H.; Fan, Y.; Li, B. Purification and identification of immunomodulating peptides from enzymatic hydrolysates of Alaska pollock frame. *Food Chem.* **2012**, *134*, 821–828. [CrossRef] [PubMed]

19. Song, R.; Wei, R.B.; Ruan, G.Q. Isolation and identification of antioxidative peptides from peptic hydrolysates of half-fin anchovy (*Setipinna taty*). *LWT Food Sci. Technol.* **2015**, *60*, 221–229. [CrossRef]

20. Nimalaratne, C.; Bandara, N.; Wu, J. Purification and characterization of antioxidant peptides from enzymatically hydrolyzed chicken egg white. *Food Chem.* **2015**, *188*, 467–472. [CrossRef] [PubMed]

21. Chen, N.; Yang, H.; Sun, Y. Purification and identification of antioxidant peptides from walnut (*Juglans regia* L.) protein hydrolysates. *Peptides* **2012**, *38*, 344–349. [CrossRef] [PubMed]

22. Hong, J.; Chen, T.T.; Hu, P. Purification and characterization of an antioxidant peptide (GSQ) from Chinese leek (*Allium tuberosum Rottler*) seeds. *J. Funct. Foods* **2014**, *10*, 144–153. [CrossRef]

23. Li, Y.W.; Li, B. Characterization of structure-antioxidant activity relationship of peptides in free radical systems using QSAR models: Key sequence positions and their amino acid properties. *J. Theor. Biol.* **2013**, *318*, 29–43. [CrossRef] [PubMed]

24. Fan, J.; He, J.; Zhuang, Y.; Sun, L. Purification and Identification of Antioxidant Peptides from Enzymatic Hydrolysates of Tilapia (*Oreochromis niloticus*) Frame Protein. *Molecules* **2012**, *17*, 12836–12850. [CrossRef] [PubMed]

25. Kumar, N.S.S.; Nazeer, R.A.; Jaiganesh, R. Purification and biochemical characterization of antioxidant peptide from horse mackerel (*Magalaspis cordyla*) viscera protein. *Peptides* **2011**, *32*, 1496–1501. [CrossRef] [PubMed]

26. Li, Y.W.; Li, B.; He, J. Structure–activity relationship study of antioxidative peptides by QSAR modeling: The amino acid next to C-terminus affects the activity. *J. Pept. Sci.* **2011**, *17*, 454–462. [CrossRef] [PubMed]

27. Zhang, J.H.; Zhang, H.; Wang, L.; Guo, X.N.; Wang, X.G.; Yao, H.Y. Isolation and identification of antioxidative peptides from rice endosperm protein enzymatic hydrolysate by consecutive chromatography and MALDI-TOF/TOF MS/MS. *Food Chem.* **2010**, *119*, 226–234. [CrossRef]

28. Rajapakse, N.; Mendis, E.; Byun, H.G.; Kim, S.K. Purification and in vitro antioxidative effects of giant squid muscle peptides on free radical-mediated oxidative systems. *J. Nutr. Biochem.* **2005**, *16*, 562–569. [CrossRef] [PubMed]

29. Liang, Q.; Wang, L.; He, Y.; Wang, Z.; Xu, J.; Ma, H. Hydrolysis kinetics and antioxidant activity of collagen under simulated gastrointestinal digestion. *J. Funct. Food* **2014**, *11*, 493–499. [CrossRef]

30. Hou, H.; Li, B.; Zhao, X.; Zhang, Z.; Li, P. Optimization of enzymatic hydrolysis of Alaska pollock frame for preparing protein hydrolysates with low-bitterness. *LWT Food Sci. Technol.* **2011**, *44*, 421–428. [CrossRef]

31. Ma, Y.; Xiong, Y.L.; Zhai, J. Fractionation and evaluation of radical scavenging peptides from in vitro digests of buckwheat protein. *Food Chem.* **2010**, *118*, 582–588. [CrossRef] [PubMed]

32. Aleman, A.; Gimenez, B.; Perez-Santin, E. Contribution of Leu and Hyp residues to antioxidant and ACE-inhibitory activities of peptide sequences isolated from squid gelatin hydrolysate. *Food Chem.* **2011**, *125*, 334–341. [CrossRef]

33. Guo, Z.; Liu, H.X.; Ji, X. Hydroxyl radicals scavenging activity of *N*-substituted chitosan and quaternized chitosan. *Bioorg. Med. Chem. Lett.* **2006**, *16*, 6348–6350. [CrossRef] [PubMed]

34. MaxQuant Server (version 1.5.3.28). Available online: http://www.coxdocs.org/doku.php?id=maxquant:common:download_and_installation (accessed on 12 October 2016).

35. Orsini Delgado, M.C.; Nardo, A.; Pavlovic, M.; Rogniaux, H.; Añón, M.; Tironi, V.A. Identification and characterization of antioxidant peptides obtained by gastrointestinal digestion of amaranth proteins. *Food Chem.* **2016**, *197*, 1160–1167. [CrossRef] [PubMed]

marine drugs

MDPI

Article

Enzymatic Pre-Treatment Increases the Protein Bioaccessibility and Extractability in Dulse (*Palmaria palmata*)

Hanne K. Mæhre *, Ida-Johanne Jensen and Karl-Erik Eilertsen

Faculty of Biosciences, Fisheries and Economics, Norwegian College of Fishery Science,
UIT The Arctic University of Norway, N-9037 Tromsø, Norway; ida-johanne.jensen@uit.no (I.-J.J.);
karl-erik.eilertsen@uit.no (K.-E.E.)
* Correspondence: hanne.maehre@uit.no; Tel: +47-776-46793

Academic Editor: Se-Kwon Kim
Received: 21 September 2016; Accepted: 21 October 2016; Published: 26 October 2016

Abstract: Several common protein extraction protocols have been applied on seaweeds, but extraction yields have been limited. The aims of this study were to further develop and optimize existing extraction protocols and to examine the effect of enzymatic pre-treatment on bioaccessibility and extractability of seaweed proteins. Enzymatic pre-treatment of seaweed samples resulted in a three-fold increase in amino acids available for extraction. Combining enzymatic pre-treatment with alkaline extraction resulted in a 1.6-fold increase in the protein extraction yield compared to a standard alkaline extraction protocol. A simulated in vitro gastrointestinal digestion model showed that enzymatic pre-treatment of seaweed increased the amount of amino acids available for intestinal absorption 3.2-fold. In conclusion, enzymatic pre-treatment of seaweeds is effective for increasing the amount of amino acids available for utilization and may thus be an effective means for increasing the utilization potential of seaweed proteins. However, both the enzymatic pre-treatment protocol and the protein extraction protocol need further optimization in order to obtain optimal cost-benefit and results from the in vitro gastrointestinal digestion model need to be confirmed in clinical models.

Keywords: *Palmaria palmata*; amino acids; protein; extraction; bioaccessibility; enzymatic treatment; gastrointestinal digestion

1. Introduction

Along with the expected world population growth in the coming decades, there will be a general increased demand for food, and particularly for proteins. Around 70% of the Earth is covered by water, but despite this, only 6.5% of the current global food protein consumption origins from the ocean, the main sources being fish and shellfish [1]. Besides fish and shellfish, there are many other marine species that could serve as valuable protein sources and among these are seaweeds. Seaweeds have long been a part of the diet in East Asia, but are not frequently used in other regions. The global production of seaweeds was around 25 million tons in 2012, of which 95% came from aquaculture with China and Indonesia as the main contributors [2]. In order to ensure a stable delivery of raw materials for industrial or nutritional purposes, cultivation is considered necessary.

Being plants, seaweeds are primary producers of macronutrients, such as carbohydrates, lipids and proteins. Elements like carbon, nitrogen and phosphorus are efficiently taken up from the environment into the cells and enzymatically converted to macronutrients, which are further used for growth or maintenance or stored intracellularly [3]. A large part of seaweed protein is thus situated intracellularly, in forms of newly formed amino acids or proteins, along with a wide range of enzymes. In order to optimize the commercial utilization of seaweed proteins, degrading the cell wall and liberating the intracellular proteins is of great importance. Seaweed cells are, like other plant cells and unlike animal cells, surrounded by a rigid cell wall functioning mainly as structural support and protection. The main constituents of the cell walls are complex polysaccharides, but also some proteins are embedded in it [4]. The cell wall polysaccharides are considered indigestible for humans, as the human gastrointestinal system does not contain the enzymes necessary for hydrolyzing the (1-4)-β-D-glycosidic bonds within them. In addition, they make ionic interactions with the attached proteins, hindering efficient extraction of these [5]. The protein bioaccessibility and extractability of seaweeds are thus lower than that of proteins of animal origin.

In a previous study [6], it was shown that several seaweed species are rich in proteins of good quality and thus that they could be utilized as protein sources in food and feed, or as ingredients in such [6]. Among the species in the mentioned paper, *Palmaria palmata* was found to be the best candidate for utilization in food and feed [6], and was thus chosen as a model species for this study. However, cultivation of this species has been shown to face some challenges when economic viability is concerned, and prior to commercial utilization of this species, these challenges have to be solved.

For protein extraction, several protocols have been developed, exploring the effects of osmotic shock, mechanical grinding, ultrasonic and enzymatic degradation of the cell walls [7,8]. The extraction yields have generally been low and also varied between studies, indicating that there is potential for optimization of these protocols. Concerning bioaccessibility, evaluated as the amount of amino acids available for absorption after gastrointestinal digestion, the literature is scarcer.

The aims of this study were thus to further develop and optimize existing extraction protocols and to examine the effect of enzymatic pre-treatment on bioaccessibility and extractability of *P. palmata* proteins.

2. Results

The amino acid composition and protein content in untreated, homogenized and enzymatically treated *Palmaria palmata* is shown in Table 1. In the untreated samples, both total amino acid (TAA) content and the relative proportion of essential amino acids (EAA) were within the same ranges as previously described [6,9,10]. The amount of available TAA increased significantly both after homogenization and after enzymatic treatment with the polysaccharidases xylanase and cellulose, 1.7-fold and three-fold, respectively. Enzymatic treatment also increased the amount of available amino acids significantly compared to homogenization alone. However, there were no significant differences between the different enzyme concentrations.

Table 1. Amino acid composition in raw (A), homogenized (B) and enzymatically treated *Palmaria palmata* (C–E). The enzymes used were xylanase and cellulose in concentrations of 10 (C), 50 (D) and 100 (E) $U \cdot g^{-1} \cdot$alga. Values are given as mean $\pm$ SD ($n = 5$) and in mg$\cdot$AA$\cdot$g$^{-1}\cdot$DW. Different letters indicate significant differences ($p < 0.05$) between treatments.

	A.	B.	C.	D.	E.
	Raw Material	After Homogenization	After Enzyme Pre-Treatment (10 U)	After Enzyme Pre-Treatment (50 U)	After Enzyme Pre-Treatment (100 U)
Essential Amino Acids (EAA)					
Threonine	8.9 ± 0.7 [a]	16.8 ± 3.3 [b]	25.7 ± 2.9 [c]	30.4 ± 2.6 [c]	27.1 ± 2.2 [c]
Valine	12.4 ± 0.9 [a]	22.6 ± 3.6 [b]	36.9 ± 5.0 [c]	42.9 ± 4.0 [c]	38.5 ± 2.5 [c]
Methionine	4.0 ± 0.4 [a]	9.0 ± 1.7 [b]	13.1 ± 2.2 [b,c]	14.7 ± 1.3 [c]	13.2 ± 0.8 [c]
Isoleucine	8.2 ± 0.7 [a]	15.8 ± 3.0 [b]	26.2 ± 4.1 [c]	30.0 ± 3.2 [c]	26.8 ± 2.7 [c]
Leucine	14.3 ± 1.8 [a]	27.7 ± 4.7 [b]	46.1 ± 6.3 [c]	53.2 ± 6.0 [c]	47.1 ± 4.8 [c]
Phenylalanine	8.7 ± 0.6 [a]	16.6 ± 2.9 [b]	26.5 ± 3.0 [c]	31.4 ± 2.4 [c]	28.4 ± 1.6 [c]
Lysine	11.0 ± 2.0 [a]	19.5 ± 3.6 [b]	28.8 ± 4.3 [c]	33.7 ± 2.1 [c]	29.5 ± 3.4 [c]
Histidine	2.6 ± 0.5 [a]	4.7 ± 0.7 [b]	7.0 ± 0.8 [c]	8.7 ± 0.5 [c]	7.9 ± 0.8 [c]

Table 1. *Cont.*

	A.	B.	C.	D.	E.
	Raw Material	After Homogenization	After Enzyme Pre-Treatment (10 U)	After Enzyme Pre-Treatment (50 U)	After Enzyme Pre-Treatment (100 U)
Non-Essential Amino Acids (NEAA)					
Aspartic acid *	21.9 ± 1.2 [a]	32.1 ± 5.5 [a]	50.9 ± 6.2 [b]	59.2 ± 5.7 [b]	52.8 ± 4.4 [b]
Serine	10.5 ± 0.8 [a]	20.5 ± 3.9 [b]	31.8 ± 4.3 [c]	37.7 ± 3.9 [c]	33.2 ± 3.0 [c]
Glutamic acid *	20.4 ± 1.8 [a]	27.7 ± 5.6 [a]	43.1 ± 5.3 [b]	50.3 ± 5.2 [b]	44.1 ± 3.1 [b]
Proline	9.1 ± 0.4 [a]	14.0 ± 3.1 [a]	23.8 ± 2.2 [b]	27.7 ± 3.8 [b]	25.3 ± 2.4 [b]
Glycine	12.1 ± 0.8 [a]	20.7 ± 3.9 [b]	32.4 ± 3.3 [c]	37.6 ± 3.5 [c]	34.6 ± 2.0 [c]
Alanine	16.4 ± 1.4 [a]	28.7 ± 5.2 [b]	44.7 ± 7.0 [c]	50.5 ± 5.1 [c]	43.7 ± 3.1 [c]
Cysteine	1.4 ± 0.4 [a]	3.0 ± 0.8 [a]	4.4 ± 1.7 [b]	7.1 ± 1.2 [b]	7.1 ± 1.4 [b]
Tyrosine	6.9 ± 0.9 [a]	13.3 ± 2.8 [a]	23.6 ± 3.2 [a,b]	29.2 ± 2.7 [b]	26.2 ± 2.7 [b]
Arginine	11.5 ± 1.1 [a]	22.5 ± 4.4 [b]	34.7 ± 4.9 [c]	41.6 ± 2.3 [c]	35.8 ± 3.6 [c]
Sum	180.5 ± 12.3 [a]	312.0 ± 54.2 [b]	495.2 ± 59.5 [c]	586.1 ± 53.5 [c]	521.2 ± 40.7 [c]
Relative amount EAA (%)	38.9 ± 0.6 [a]	42.6 ± 0.9 [b]	42.5 ± 1.2 [b]	41.8 ± 0.3 [b]	41.9 ± 0.5 [b]

* Aspartic acid and glutamic acid represent the sums of aspartic acid + asparagine and glutamic acid + glutamine, respectively, as asparagine and glutamine are present in their acidic forms after acidic hydrolysis. Tryptophan is lacking due to destruction during acidic hydrolysi.

In Figure 1, it is shown that the amount of each essential amino acid in raw and enzymatically treated (50 U xylanase and cellulose $g^{-1} \cdot$alga) *P. palmata* proteins, is equal to or higher than the corresponding amount in the reference protein defined by FAO/WHO/UNU [11].

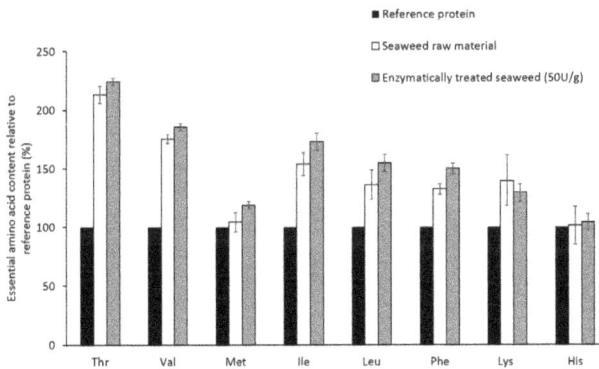

Figure 1. Essential amino acids composition (mg·EAA·g^{-1} protein) in raw and enzymatically treated *Palmaria palmata* relative to the reference protein set by WHO/FAO/UNU. The values are given as mean ± SD (n = 5) and in percent of the reference protein. Tryptophan is lacking due to destruction during acidic hydrolysis.

The results of the protein extraction experiment are shown in Table 2. Here, it is seen that alkaline extraction, either alone or in combination with 3.5% saline, was more efficient than 3.5% saline alone and ethanol in extracting alga proteins. Extraction at 60 °C seemed to increase the extraction efficiency compared to extraction at 23 °C, however, this effect was significant only for 0.05 M NaOH, 0.1 M NaOH and 0.1 M NaOH in 3.5% saline. Alkaline extraction following enzymatic pre-treatment increased the protein extraction yield significantly compared to all other extraction solvents and on both temperatures.

Following a simulated in vitro gastrointestinal digestion of raw and enzymatically treated *P. palmata*, the amount of amino acids available for absorption were significantly ($p < 0.05$) higher in all of the treated samples than in the raw sample (Figure 2). There were no significant differences between the different enzyme concentrations.

Table 2. Total amino acids and extraction yield in extracts of *Palmaria palmata* using solutions as described in Table 3, along with alkaline extraction following enzymatic pre-treatment (50 U·g⁻¹·alga). Values are reported as mean ± SD ($n = 5$) and in mg·AA·g⁻¹·DW for total amino acids and in percent of raw material DW for extraction yields. Different small letters indicate significant differences ($p < 0.05$) between extractions at 23 °C, while different capital letters indicate significant differences ($p < 0.05$) between extractions at 60 °C. * indicate significant differences ($p < 0.05$) between 23 °C and 60 °C using the same extraction solvent.

| | | Extraction Temperature | | | |
| | | 23 °C | | 60 °C | |
	Solvent	Amount Extracted Amino Acids (mg·g⁻¹·DW)	Extraction Yield (%)	Amount Extracted Amino Acids (mg·g⁻¹·DW)	Extraction Yield (%)
A	0.01 M NaOH	55.8 ± 10.2 [b]	17.9	59.9 ± 7.2 [B]	19.2
B	0.05 M NaOH	80.6 ± 9.5 [b,c]	25.8	118.1 ± 25.2 [B,C,*]	37.9
C	0.1 M NaOH	90.1 ± 7.9 [c]	28.9	122.0 ± 10.5 [C,*]	39.1
D	3.5% NaCl	18.3 ± 4.7 [a]	5.9	26.6 ± 7.0 [A]	8.5
E	70% Ethanol	23.5 ± 4.6 [a]	7.5	27.3 ± 4.6 [A]	8.8
F	0.1 M NaOH in 3.5% NaCl	58.8 ± 13.3 [b]	18.8	114.6 ± 19.2 [C,*]	36.7
G	0.1 M NaOH following enzymatic pre-treatment	409.2 ± 46.0 [d]	69.8	442.8 ± 86.5 [D]	75.6

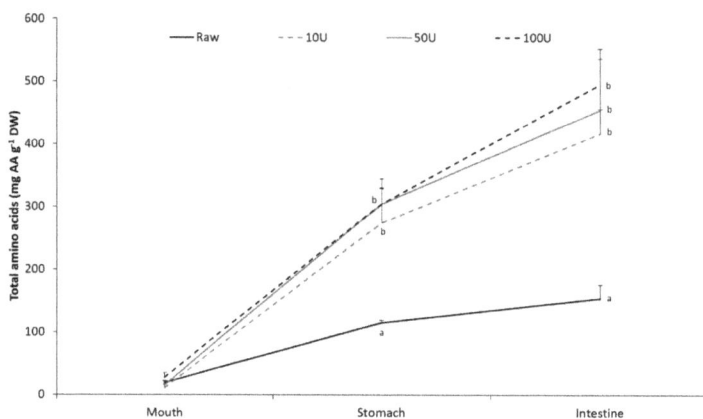

Figure 2. Total amino acids (AA) liberated in the mouth, stomach and intestinal fluids during gastrointestinal (GI) digestion of raw and enzymatically treated *Palmaria palmata*. Values are reported as mean ± SD ($n = 5$) and in mg·AA·g⁻¹·DW of the source material. Different letters indicate significant differences ($p < 0.05$) between treatments, within each GI phase.

3. Discussion

Plant cells, including seaweeds, are surrounded by a rigid cell wall comprised of complex polysaccharides, with small amounts of proteins embedded in it [4]. Being primary producers of macronutrients, the algal cells contain large amounts of various enzymes involved in the conversion of absorbed elements to macronutrients. In addition, newly formed amino acids and proteins are stored intracellularly [3]. Extraction and subsequent utilization of these depend on disruption of the cell wall. In this study, cell wall disruption was performed using mechanical force, namely by

Ultra Turrax homogenization, and enzymatic degradation. Cellulose is present in most plant cell walls. However, in red algae, the class in which *P. palmata* belong, xylans has been shown to make up a large proportion of the extracellular matrix, along with cellulose [12]. Thus, it was decided to use a combination of enzymes directed towards both of these polysaccharides for the experiments in this study.

Ensuring an adequate intake of EAA is necessary and when examining efficient protocols for increasing the amino acids available for hydrolysis it is important that the composition of EAA is not negatively altered. Previously [10], it has been shown that *P. palmata* proteins fulfill the demands of a complete protein, as defined by FAO/WHO/UNU [11]. This study confirmed the previous findings regarding protein quality (Figure 1) and the enzymatic treatment did not alter the EAA composition. Most common protein extraction protocols are based on the principle that cells burst due to osmotic shock when exposed to hypotonic conditions, and involve exposing the tissue to water or weak buffer solutions. This is a valid principle and an efficient procedure when extracting proteins of animal origin. In plants, however, the cell wall complicates protein extraction procedures. Plant cells hold a defense mechanism against osmotic variations, a mechanism in which intracellular vacuoles containing fluid of high ionic strength are central. When exposed to hypotonic solutions, water or buffer will flow into the vacuole, increasing its size and pushing the other cell organelles towards the cell wall. The intracellular pressure will thus increase, but the cell wall will prevent the cell from bursting [13]. Previous studies have shown that protein extraction protocols based solely on the osmotic shock principle are not very efficient for the extraction of seaweed proteins [5,14]. Several extraction protocols aiming at destruction of the cell wall, either by applying mechanical force or by enzymatic treatment, has been developed in order to overcome this problem [5,7,8]. The extraction yields have, however, been limited in most protocols.

In this study, several common extraction protocols were examined and modified in order to increase the protein extraction yield. It is well-known that the solubility of different proteins depends on the solvent used and in a previous paper it was shown that heat treatment increased the bioaccessibility of dulse proteins [10]. The extraction variables chosen were thus two extraction temperatures (23 °C and 60 °C), along with different types and concentrations of extraction solvents based on the solubility properties of different proteins. The solvents used were water, sodium hydroxide, sodium chloride and ethanol, along with combinations of these. Alkaline extraction following enzymatic pre-treatment was also included.

The extraction yields ranged from around 5% using 3.5% saline as extraction solvent at room temperature to 40% using 0.1 M NaOH as solvent at 60 °C (Table 2). Applying polysaccharidases for enzymatic destruction of the cell wall was shown to be more efficient than mechanical degradation, as extraction of the pre-treated alga resulted in an extraction yield of 75% at 60 °C, a 1.63-fold increase compared to the water-alkaline protocol (Table 2). This yield is markedly higher than reported in other studies using enzymatic degradation of the cell wall [5,8]. It is, however, difficult to compare results from different studies directly due to differences in the methods of protein determination, along with type and concentration of polysaccharidases used.

After enzymatic treatment, the algae samples were subjected to a simulated in vitro gastrointestinal model (Figure 2) in order to investigate the effect of enzymatic treatment on the bioaccessibility of seaweed proteins. The liberation of amino acids into the digesta increased during the digestion process both for the raw samples and the samples exposed for enzymatic treatment. At the end of the process, simulating the end of the small intestine, the liberation of TAA was around 2.5–3.2 times higher in the enzymatically treated samples than in the raw samples. This increase corresponds well with the increased amount of amino acids available for digestion as a result of the enzymatic treatment as seen in Table 1 and indicates that GI digestion did not contribute to a further increase. The fact that the GI digestion was not more efficient in liberating amino acids from the enzymatically treated algae compared to the raw samples may indicate that the increased amount of amino acids released during the enzymatic treatment was not released as intact proteins, but rather as smaller peptides or free amino acids. As some of the intracellular proteins are non-specific hydrolytic enzymes normally participating in the cellular protein turnover, it is likely to believe that these may have contributed to a partial degradation of the intracellular proteins prior to the GI digestion. Around 4%–17% of the intracellular amino acids have also been shown not to be protein bound [3].

To sum up the findings in this study, it was shown that enzymatic pre-treatment of *P. palmata* increased the protein bioaccessibility and extractability, mainly by increasing the amount of amino acids available for hydrolysis. These results indicate that enzymatic pre-treatment of algae may increase the utilization potential of seaweed proteins. However, both the enzymatic pre-treatment protocol and the protein extraction protocol need to be optimized further in order to obtain optimal cost–benefit and results from the in vitro gastrointestinal digestion model need to be confirmed in clinical studies.

4. Experimental Section

4.1. Raw Material

Dehydrated *Palmaria palmata* was purchased from "Fremtidens mat" (Oslo, Norway). The seaweed was harvested at the south coast of Iceland. Following harvest, the seaweed was flushed with seawater and dehydrated at 40 °C for 24 h using electrical fans driven by geothermal energy. The dried seaweed was thereafter packed in airtight bags, before transport to Norway. Seaweed samples were cut into pieces of 0.5 cm × 0.5 cm prior to treatments. All chemical used in these experiments were of analytical grade and purchased from Sigma Chemical Co. (St. Louis, MO, USA) unless otherwise stated.

4.2. Water Content

Water content was determined using a modified version of AOAC method 950.46B [15]. Approximately 0.5 g of seaweed sample ($n = 5$), was dried at 105 °C until constant weight. Water content was determined gravimetrically. The water contents were only used for calculation of dry matter in the different fractions and results are thus presented as supplementary material (Table S1).

4.3. Protein Extraction

Protein extraction was performed according to Barbarino and Lourenço [7], with some modifications (Figure 3). In short, approximately 100 mg of milled seaweed samples were dissolved in 8 mL distilled water, homogenized using an Ultra Turrax T8 basic homogenizer (IKA Werke GmbH, Staufen, Germany) and incubated for 24 h at either 23 °C or 60 °C. The samples were centrifuged at $4000 \times g$ at 4 °C for 15 min. The supernatant was removed and the pellet was re-dissolved in 8 mL of the different solvents described in Table 3 and incubated for 24 h at 23 °C or 60 °C. The samples were exposed to constant shaking during both incubations. Samples were then centrifuged at $4000 \times g$ at 4 °C for 15 min. The two supernatants were combined and the final extracts were subjected to amino acid analysis.

Table 3. Overview of the types and concentrations of the different extraction solvents used in this study, along with the types of extracted protein relevant for each solvent.

Extract	Extraction Solvent	Type of Extracted Protein
All	Water	Albumins
A	0.01 M NaOH	Glutelins
B	0.05 M NaOH	Glutelins
C	0.1 M NaOH	Glutelins
D	3.5% NaCl	Globulins
E	70% Ethanol	Prolamines
F	0.1 M NaOH in 3.5% NaCl	Combination
G	Enzymes + 0.1 M NaOH	Combination

Alga raw material (Sample A, Table 1)

+

H_2O

↓

Homogenization (Ultra Turrax, 20,000 rpm, 30 s)

↓

Incubation (24 h, 23 or 60 °C, constant shaking)

↓

Centrifugation (20 min, 4 °C, 4000× *g*)

Pellet   Supernatant

AAA   Extraction
(Sample B, Table 1)   solutions (A–F, Table 3)

↓

Incubation (24 h, 23 or 60 °C, constant shaking)

↓

Centrifugation (20 min, 4 °C, 4 000× *g*)

Pellet   Supernatant

AAA (Samples A–F, Table 2)

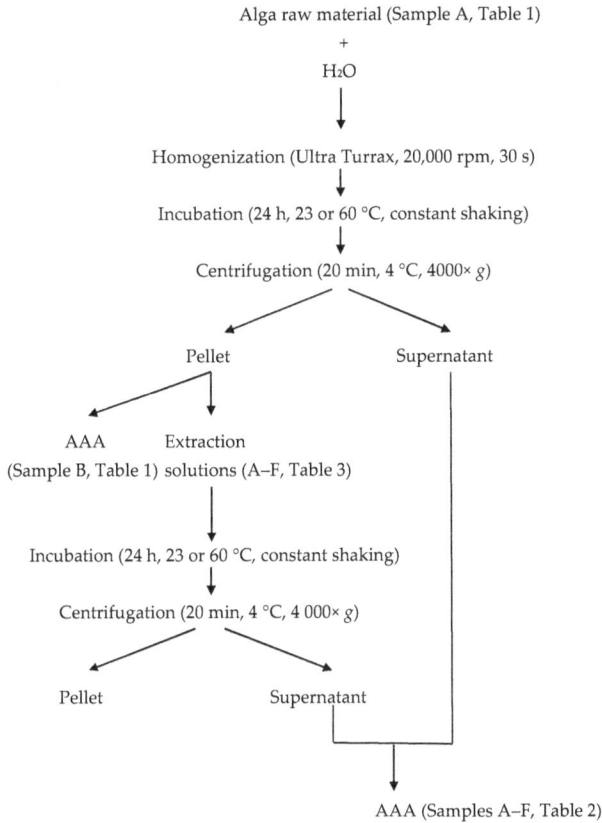

Figure 3. Flowchart of protein extraction and sample collections. AAA: amino acid analysis.

4.4. Enzymatic Pre-Treatment

Enzymatic pre-treatment was performed according to Harnedy and FitzGerald [8], with some modifications (Figure 4). Approximately one gram of seaweed was dissolved in 28 mL of 0.05 M sodium acetate buffer (pH 5.0), homogenized for 30 s using an Ultra Turrax T25 and incubated for 30 min at 40 °C under constant shaking. Enzyme solutions containing 10, 50 or 100 U xylanase and cellulase (both from *Trichoderma longibrachiatum*) in 2 mL sodium acetate buffer was added and incubation continued for 18 h at 40° C under constant agitation. Thereafter, the samples were centrifuged at 4000× *g* at 4 °C for 15 min, before separating supernatants and pellets. Pellets were subjected to amino acid analysis, in vitro gastrointestinal digestion and alkaline protein extraction. Algae samples without enzymes and buffer samples with enzymes were used as controls.

For alkaline extraction following enzymatic pre-treatment, pellets were re-dissolved in 8 mL 0.1 M NaOH, incubated for 24 h either at 23 °C or 60 °C with constant shaking and centrifuged at 4000× *g* at 4 °C for 15 min. Supernatants and pellets were separated and supernatants were subjected to amino acid analysis.

Alga raw material
+
Na-acetate buffer, pH 5.0
+
Cellulase + Xylanase

↓

Incubation (24 h, 40 °C, constant shaking)

↓

Centrifugation (20 min, 4 °C, 4000× *g*)

Pellet   Supernatant

+

GI digestion   0.1 M NaOH   AAA (Samples C, D and E, Table 1)

↓

Incubation (24 h, 23 or 60 °C, constant shaking)

↓

Centrifugation (20 min, 4 °C, 4000× *g*)

Pellet   Supernatant

↓

AAA (Sample G, Table 2)

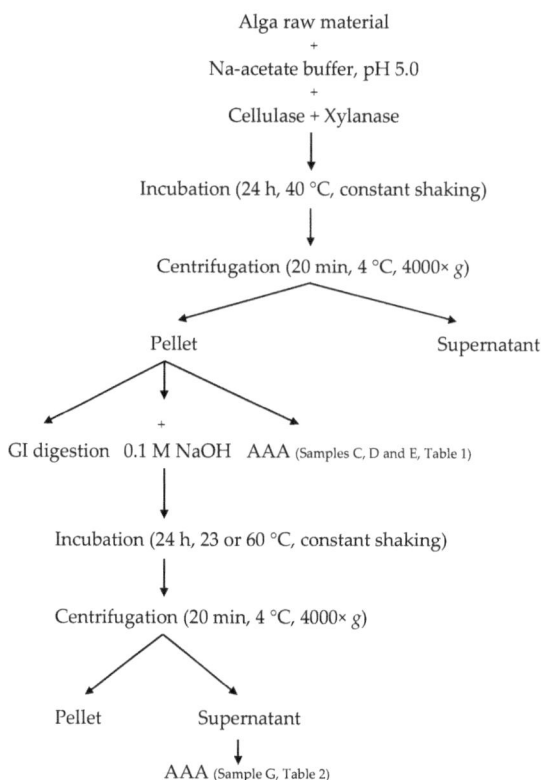

Figure 4. Flowchart of enzymatic pre-treatment and sample collections. GI: gastrointestinal, AAA: amino acid analysis.

4.5. In Vitro Gastrointestinal Digestion

Raw seaweed and seaweed after enzymatic pre-treatment were subjected to a simulated in vitro gastrointestinal digestion model as described by Versantvoort et al. [16], with the modifications described by Maehre et al. [10]. Approximately 0.5 g of the seaweed samples were mixed with 6 mL of a solution mimicking salivary fluid (pH 6.80 ± 0.02) and homogenized with an Ultra Turrax T25 for 30 s, followed by incubation at 37 °C for 5 min under constant rotation. After centrifugation at 2750× *g* for 3 min, a 2 mL sample from the supernatant was collected and to the rest of the digesta, 12 mL of a solution mimicking gastric fluid (pH 1.30 ± 0.01) was added. The mixture was incubated at 37 °C for 120 min under constant rotation and the sampling procedure was repeated. Then, 12 mL of a solution mimicking duodenal fluid (pH 8.10 ± 0.02), 6 mL of bile solution (pH 8.22 ± 0.02) and 2 mL of 1 M NaHCO$_3$ was added and another 120-min incubation at equal conditions was applied, followed by collection of a final 2 mL sample. For inactivation of the enzymes, all of the GI samples were kept at 90 °C for 5 min and then put on ice. Samples without seaweed were subjected to the same procedure and used for adjustment of amino acid contribution from the digestive enzymes.

4.6. Amino Acid Analysis

Raw and homogenized seaweed samples, along with pellets from enzymatic pre-treatment and supernatant samples from the different experiments were subjected to analysis of total amino acids (TAA). Sample preparations were similar to those described previously [10]. Approximately 200 mg of raw seaweed samples and pellets after enzymatic pre-treatment were dissolved in 0.7 mL distilled water, 0.5 mL 20 mM norleucine (internal standard) and 1.2 mL of 12 M hydrochloric acid (HCl). Samples were flushed in N_2-gas for 15 s and hydrolyzed at 110 °C for 24 h, according to Moore and Stein [17]. Aliquots of 0.1 mL of the hydrolyzed samples were evaporated under N_2 and re-dissolved in 1 mL lithium citrate buffer, pH 2.2. For the liquid samples (supernatants from all experiments and digesta from the GI model), 0.05 mL 20 mM norleucine and 0.55 mL 12 M HCl were added to 0.5 mL sample, before flushing with N_2 and hydrolysis as described above. After hydrolysis, 0.1 mL sample was evaporated and re-dissolved in 0.5 mL lithium citrate buffer, pH 2.2.

All amino acid samples were analyzed chromatographically and identified as described previously [18] using a Biochrom 30 amino acid analyzer (Biochrom Co., Cambridge, UK). Tryptophan is destroyed during acidic hydrolysis and is thus not included in the results.

4.7. Statistics

Statistical analysis was performed using SPSS 23 (SPSS Inc., Chicago, IL, USA). Tests of normality (Shapiro–Wilk's test) and homogeneity of variance (Levene's test) returned normal distribution with unequal variance for all chemical variables. Thus, one-way analysis of variance (ANOVA) was performed, followed by the Dunnett's T3 post hoc test for evaluation of statistics. Means were considered significantly different at $p < 0.05$.

Supplementary Materials: The following are available online at www.mdpi.com/1660-3397/14/11/196/s1. Table S1: Water content of raw, homogenized and enzyme-treated *Palmaria palmata*. Values are presented as mean ± SD (*n* = 5) and in g·kg^{-1}·alga.

Acknowledgments: This work was supported by the Publication Fund of UIT The Arctic University of Norway.

Author Contributions: H.K.M. has contributed to planning the experiments, conducted analytical work and been the main author of the manuscript; I.-J.J. has contributed to planning the experiments, conducted analytical work, and contributed to discussions and preparation of the manuscript; K.-E.E. has contributed to planning the experiments, and contributed to discussions and preparation of the manuscript.

Conflicts of Interest: None of the authors report any conflicts of interest.

References

1. Béné, C.; Barange, M.; Subasinghe, R.; Pinstrup-Andersen, P.; Merino, G.; Hemre, G.I.; Williams, M. Feeding 9 billion by 2050-Putting fish back on the menu. *Food Secur.* **2015**, *7*, 261–274. [CrossRef]
2. Food and Agricultural Organization (FAO). *The State of World Fisheries and Aquaculture 2014—Opportunities and Challenges*; Food and Agricultural Organization of the United Nations: Rome, Italy, 2014; p. 223.
3. Hurd, C.L.; Harrison, P.J.; Bischof, K.; Lobban, C.S. Nutrients. In *Seaweed Ecology and Physiology*; Hurd, C.L., Harrison, P.J., Bischof, K., Lobban, C.S., Eds.; Cambridge University Press: Cambridge, UK, 2014; pp. 238–293.
4. Kloareg, B.; Quatrano, R.S. Structure of the cell walls of marine algae and ecophysiological functions of the matrix polysaccharides. *Oceanogr. Mar. Biol.* **1988**, *26*, 259–315.
5. Joubert, Y.; Fleurence, J. Simultaneous extraction of proteins and DNA by an enzymatic treatment of the cell wall of *Palmaria palmata* (Rhodophyta). *J. Appl. Phycol.* **2008**, *20*, 55–61. [CrossRef]
6. Maehre, H.K.; Malde, M.K.; Eilertsen, K.E.; Elvevoll, E.O. Characterization of protein, lipid and mineral contents in common Norwegian seaweeds and evaluation of their potential as food and feed. *J. Sci. Food Agric.* **2014**, *94*, 3281–3290. [CrossRef] [PubMed]
7. Barbarino, E.; Lourenço, S.O. An evaluation of methods for extraction and quantification of protein from marine macro- and microalgae. *J. Appl. Phycol.* **2005**, *17*, 447–460. [CrossRef]
8. Harnedy, P.A.; FitzGerald, R.J. Extraction of protein from the macroalga *Palmaria palmata*. *LWT Food Sci. Technol.* **2013**, *51*, 375–382. [CrossRef]
9. Galland-Irmouli, A.V.; Fleurence, J.; Lamghari, R.; Lucon, M.; Rouxel, C.; Barbaroux, O.; Bronowicki, J.P.; Villaume, C.; Gueant, J.L. Nutritional value of proteins from edible seaweed *Palmaria palmata* (Dulse). *J. Nutr. Biochem.* **1999**, *10*, 353–359. [CrossRef]

10. Maehre, H.K.; Edvinsen, G.K.; Eilertsen, K.E.; Elvevoll, E.O. Heat treatment increases the protein bioaccessibility in the red seaweed dulse (*Palmaria palmata*), but not in the brown seaweed winged kelp (*Alaria esculenta*). *J. Appl. Phycol.* **2016**, *28*, 581–590. [CrossRef]

11. Food and Agriculture Organization (FAO); World Health Organization (WHO); United Nations University (UNU). *Protein and Amino Acid Requirements in Human Nutrition: Report of a Joint FAO/WHO/UNU Expert Consultation*; World Health Organization: Geneva, Switzerland, 2007; p. 265.

12. Popper, Z.A.; Michel, G.; Herve, C.; Domozych, D.S.; Willats, W.G.T.; Tuohy, M.G.; Kloareg, B.; Stengel, D.B. Evolution and diversity of plant cell walls: From algae to flowering plants. *Annu. Rev. Plant Biol.* **2011**, *62*, 567–588. [CrossRef] [PubMed]

13. Karsten, U. Seaweed acclimation to salinity and desiccation stress. In *Seaweed Biology: Novel Insights into Ecophysiology, Ecology and Utilization*; Wiencke, C., Bischof, K., Eds.; Springer: Heidelberg, Germany, 2012; pp. 87–107.

14. Fleurence, J.; LeCoeur, C.; Mabeau, S.; Maurice, M.; Landrein, A. Comparison of different extractive procedures for proteins from the edible seaweeds *Ulva rigida* and *Ulva rotundata*. *J. Appl. Phycol.* **1995**, *7*, 577–582. [CrossRef]

15. Horwitz, W. *Official Methods of Analysis of AOAC International*; AOAC International: Gaithersburg, MD, USA, 2004.

16. Versantvoort, C.H.M.; Oomen, A.G.; Van de Kamp, E.; Rompelberg, C.J.M. Sips AJAM Applicability of an in vitro digestion model in assessing the bioaccessibility of mycotoxins from food. *Food Chem. Toxicol.* **2005**, *43*, 31–40. [CrossRef] [PubMed]

17. Moore, S.; Stein, W.H. Chromatographic determination of amino acids by the use of automatic recording system. *Methods Enzymol.* **1963**, *6*, 819–831.

18. Maehre, H.K.; Hamre, K.; Elvevoll, E.O. Nutrient evaluation of rotifers and zooplankton: Feed for marine fish larvae. *Aquac. Nutr.* **2013**, *19*, 301–311. [CrossRef]

marine drugs

MDPI

Review

Natural Proline-Rich Cyclopolypeptides from Marine Organisms: Chemistry, Synthetic Methodologies and Biological Status

Wan-Yin Fang [1],[†], Rajiv Dahiya [2],*,[†], Hua-Li Qin [1],*, Rita Mourya [3] and Sandeep Maharaj [2]

[1] School of Chemistry, Chemical Engineering and Life Science, Wuhan University of Technology, Wuhan 430070, China; wanyinfang@whut.edu.cn

[2] Laboratory of Peptide Research and Development, School of Pharmacy, Faculty of Medical Sciences, The University of the West Indies, Saint Augustine, Trinidad and Tobago, West Indies; Sandeep.Maharaj@sta.uwi.edu

[3] School of Pharmacy, College of Medicine and Health Sciences, University of Gondar, Gondar 196, Ethiopia; ritz_pharma@yahoo.co.in

* Correspondence: Rajiv.Dahiya@sta.uwi.edu (R.D.); qinhuali@whut.edu.cn (H.-L.Q.);
  Tel.: +1868-493-5655 (R.D.); +86-27-8774-9379 (H.-L.Q.)

† These authors contribute equally to this work.

Academic Editor: Se-Kwon Kim

Received: 11 September 2016; Accepted: 15 October 2016; Published: 26 October 2016

Abstract: Peptides have gained increased interest as therapeutics during recent years. More than 60 peptide drugs have reached the market for the benefit of patients and several hundreds of novel therapeutic peptides are in preclinical and clinical development. The key contributor to this success is the potent and specific, yet safe, mode of action of peptides. Among the wide range of biologically-active peptides, naturally-occurring marine-derived cyclopolypeptides exhibit a broad range of unusual and potent pharmacological activities. Because of their size and complexity, proline-rich cyclic peptides (PRCPs) occupy a crucial chemical space in drug discovery that may provide useful scaffolds for modulating more challenging biological targets, such as protein-protein interactions and allosteric binding sites. Diverse pharmacological activities of natural cyclic peptides from marine sponges, tunicates and cyanobacteria have encouraged efforts to develop cyclic peptides with well-known synthetic methods, including solid-phase and solution-phase techniques of peptide synthesis. The present review highlights the natural resources, unique structural features and the most relevant biological properties of proline-rich peptides of marine-origin, focusing on the potential therapeutic role that the PRCPs may play as a promising source of new peptide-based novel drugs.

Keywords: proline-rich cyclic peptide; marine sponge; marine tunicate; peptide synthesis; stereochemistry; lipophilicity parameter; pharmacological activity

---

1. Introduction

An interesting class of marine cyclic peptides is represented by the proline-rich compounds usually containing more than six or seven amino acid residues. The role of proline in these molecules has been linked to the control of the conformation of the molecule in solution because of the restricted φ of proline. The proline-rich cyclic peptides (PRCPs) are formed by linking one end of the peptide and the other with an amide bond or other chemically-stable bonds. Some of them are used in the clinic, e.g., gramicidin S and tyrocidine with bactericidal activity, while others are in clinical trials, e.g., dehydrodidemnin B, and most of them originate from natural resources. Although the literature is enriched with reports concerned with marine-derived linear proline-rich bioactive peptides [1–5], e.g., dolastatin 15, kurahyne B, jahanyne, cemadotin, koshikamide A_1, etc., PRCPs from marine resources are becoming popular and attracting the attention of scientists nowadays, due to their unique structural features and a wide range of the biological properties, like cytotoxicity [6], antibacterial activity [7], antifungal activity [8], immunosuppressive activity [9], anti-inflammatory activity [10], anti-HIV activity [11], repellent (antifouling) activity [12], antitubercular activity [13] and antiviral activity [14], associated with them. PRCPs include a large and heterogeneous group of small to large-sized oligopeptides characterized by the presence of proline units often constituting peculiar sequences, which confers them a typical structure that determines the various biological functions endowed by these molecules. As several features make PRCPs attractive lead compounds for drug development, as well as nice tools for biochemical research, scientists are focusing and giving diverse efforts to develop biologically-active proline-rich cyclic peptide compounds.

1.1. Natural Resources

Various natural sources of PRCPs include marine sponges, ascidians, different genera of cyanobacteria and higher plants. One of the potent resources is sessile aquatic animals, i.e., sponges like Kenyan sponge *Callyspongia abnormis* [15], Dominican sponge *Eurypon laughlini* [16], Indonesian sponge *Callyspongia aerizusa* [17], sponge *Ircinia* sp. [18], Jamaican sponge *Stylissa caribica* [19], Yongxing Island sponge *Reniochalina stalagmitis* [20], Vanuatu sponge *Axinella carteri* [21], Korean sponge *Clathria gombawuiensis* [22], Fijian sponge *Stylotella aurantium* [23], Papua New Guinea sponge *Stylissa massa* [24], South China sponge *Phakella fusca* [25], Lithistid sponge *Scleritoderma nodosum* [26], Borneo sponge *Pseudaxinyssa* sp. [27], Philippines sponge *Myriastra clavosa* [28], Papua New Guinea sponge *Stylotella* sp. [29], Comoros sponge *Axinella* cf. *carteri* [30], Okinawan sponge *Hymeniacidon* sp. [31], Indo-Pacific sponges *Phakellia costata* and *Stylotella aurantium* [32], Indonesian sponge *Stylissa* sp. [33], Red sea sponge *Stylissa carteri* [34], Western Pacific Ocean sponge *Hymeniacidon* sp. [35], Puerto Rican sponge *Prosuberites laughlini* [36], Micronesian sponge *Cribrochalina olemda* [37], Indonesian sponge *Sidonops microspinosa* [38], Palau sponge *Axinella* sp. [39], etc. The structures of various proline-rich cyclopolypeptides from marine sponges are compiled in Figure 1.

Other sources of proline-rich cyclooligopeptides are marine tunicates, like compound ascidian *Didemnum molle* [40], Ishigaki Island sea slug *Pleurobranchus forskalii* [41], Fijian ascidian *Eudistoma* sp. [42], Caribbean tunicate *Trididemnum solidum* [43], unidentified Brazilian ascidian (family Didemnidae) [44], Mediterranean ascidian *Aplidium albicans* [45], cyanobacteria like Papua New Guinea cyanobacterium *Lyngbya semiplena* [46], Red Sea cyanobacterium *Moorea producens* [47], Florida Everglades cyanobacterium *Lyngbya* sp. [48], Northern Wisconsin cyanobacterium *Trichormus* sp. UIC 10339 [49], toxic cyanobacterium *Nostoc* sp. 152 [50], Kenyan cyanobacterium *Lyngbya majuscule* [51], mollusks like Papua New Guinea mollusk (sea hare) *Dolabella auricularia* [52] and alga like Indonesian red alga (Rhodophyta) *Ceratodictyon spongiosum* containing the symbiotic sponge *Sigmadocia symbiotica* [10]. Structures of diverse proline-rich cyclopeptides from marine tunicates and cyanobacteria are tabulated in Figure 2. Besides this, proline-containing cyclooligopeptides are also obtained from roots, stems, barks, seeds, fruit peels of higher plants, as well as from bacteria and fungi [53–66].

Purification procedures of PRCPs isolated from sea animals, like ascidians, sponges and mollusk, usually include initial extraction with methanol (MeOH), partitions of these extracts with organic solvents of increasing polarities to render diverse organic fractions and chromatographic steps on silica and Sephadex LH-20 columns, as well as the use of reversed phase C18 HPLC for the final purification [67].

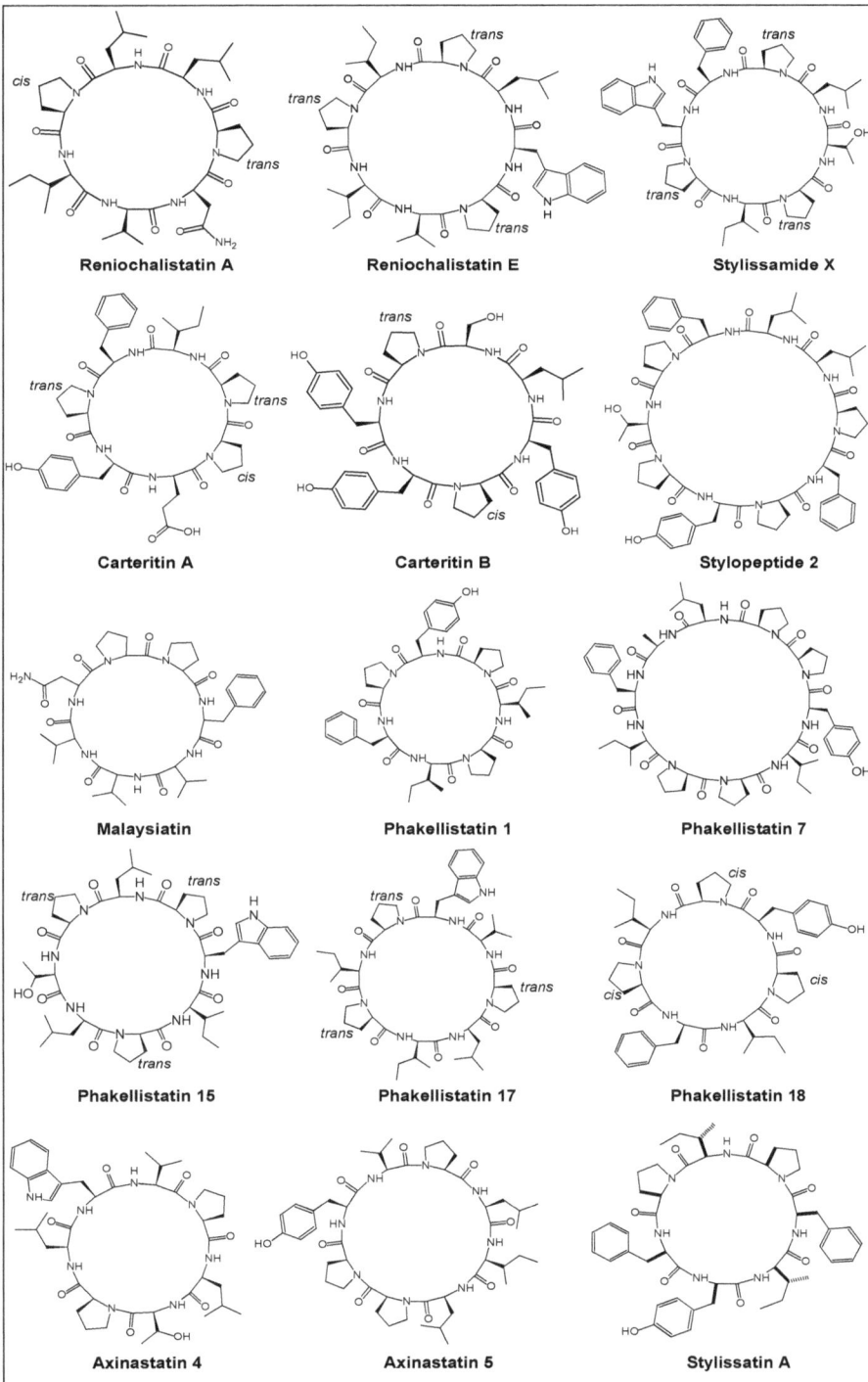

Figure 1. *Cont.*

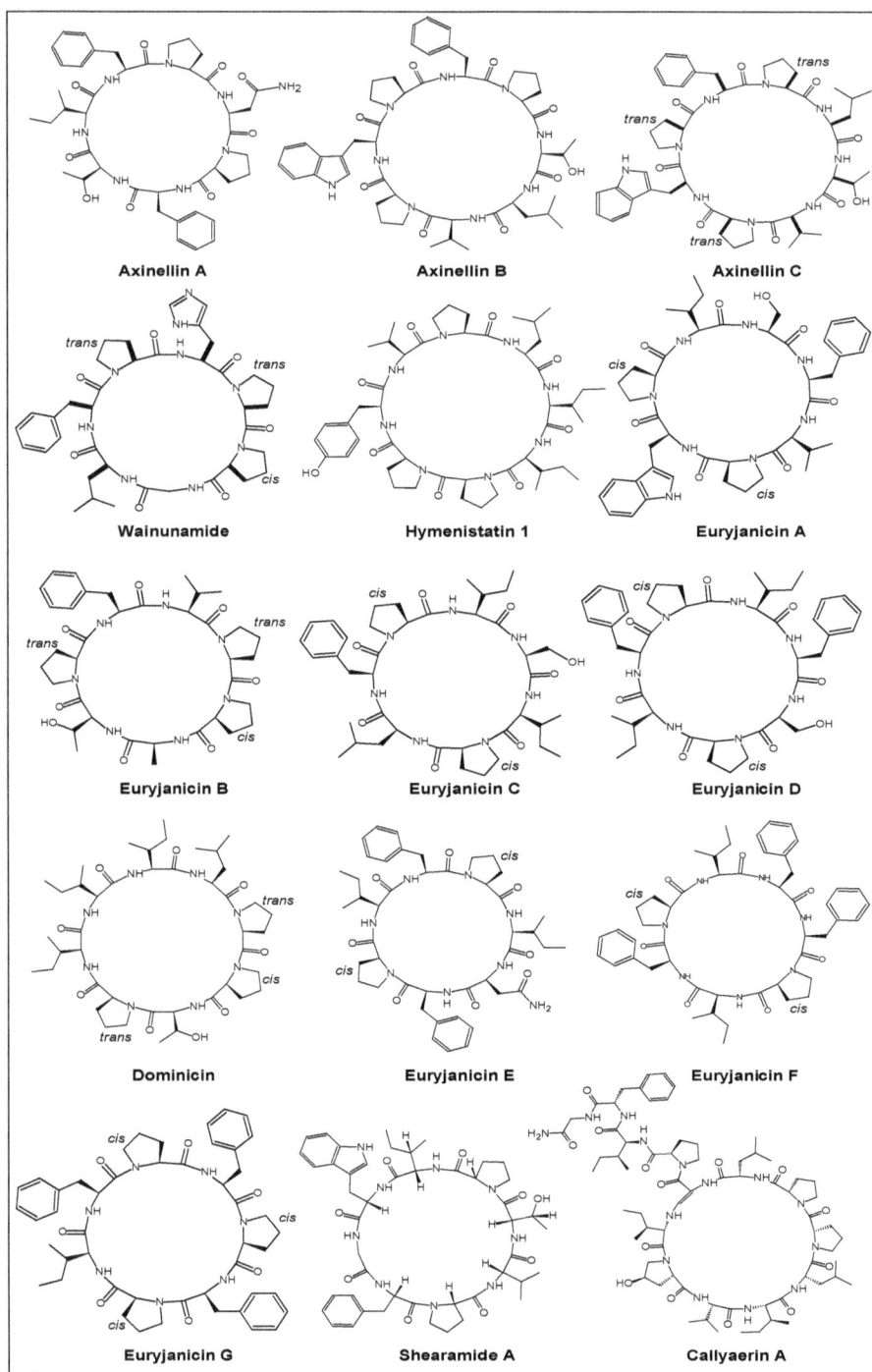

Figure 1. Proline-rich cyclic peptides (PRCPs) from marine sponges.

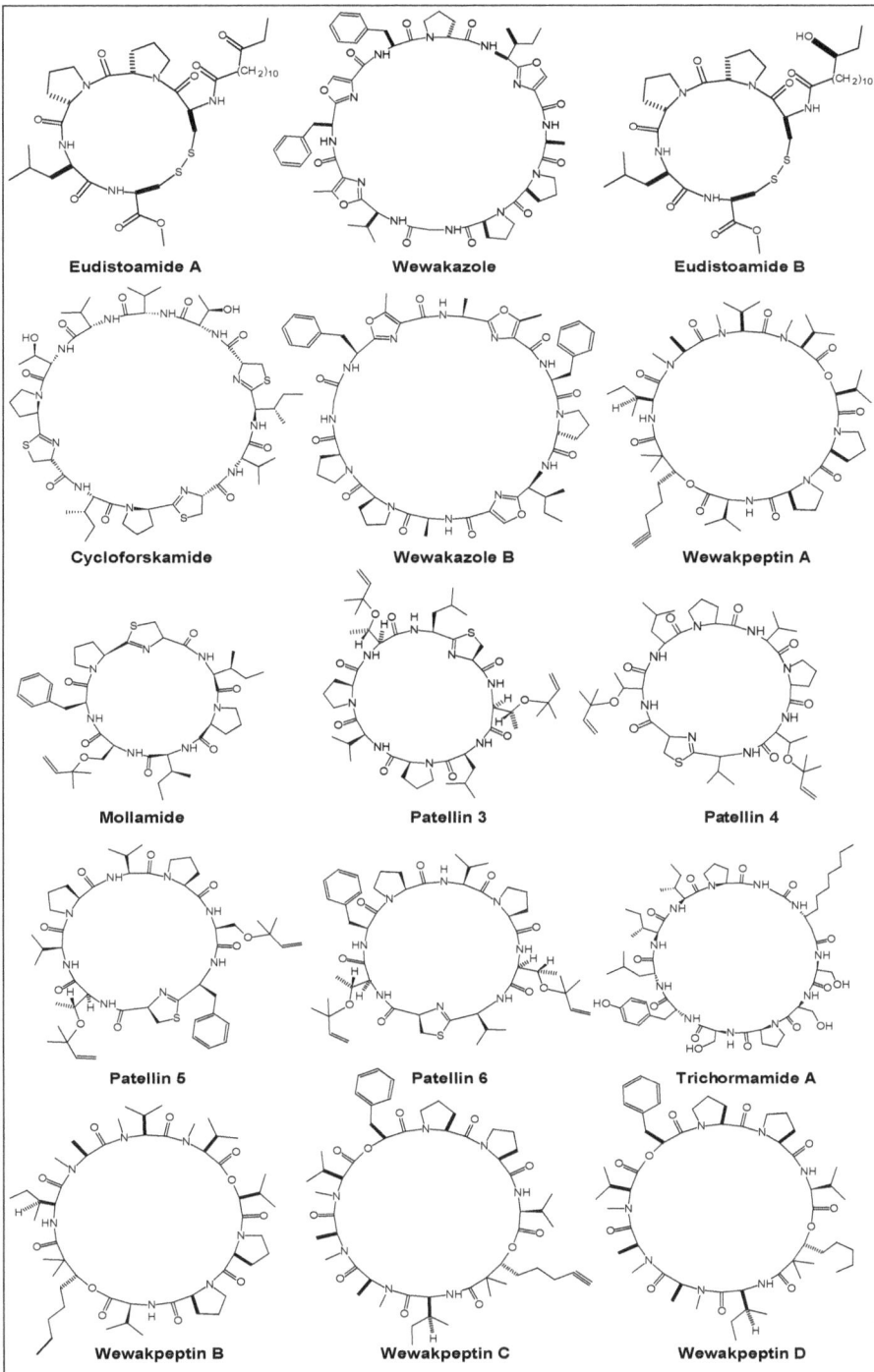

Figure 2. PRCPs from marine ascidians (tunicates) and cyanobacteria.

1.2. Stability and Comparison with Linear Peptides

Linear peptides that contain less than 10 amino acid residues are especially flexible in solution. Once the length of linear peptides extends to between 10 and 20 amino acid residues, random linear peptide sequences can begin to obtain secondary structures, including α-helices, turns and β-strands. These secondary structures impose constraints that reduce the free energy of linear peptides and limit their conformations to those that may be more biologically active. The constraints imposed by cyclization force cyclic peptides to adopt a limited number of molecular conformations in solution. Generally, if cyclization limits conformations to those required for optimum receptor binding, these cyclic peptides would be more useful compared with their linear counterparts that can adopt more conformations, which are not useful for receptor binding. Cyclization has been shown to increase the propensity for β-turn formation in peptides, which is of vital utility since β-turns are often found in native proteins. Although peptide cyclization generally induces structural constraints, the site of cyclization within the sequence can affect the binding affinity of cyclic peptides.

In the case of proline, which is a proteinogenic amino acid with a secondary amine that does not follow along with the typical Ramachandran plot, the ψ and φ angles about the peptide bond have fewer allowable degrees of rotation due to the ring formation connected to the beta carbon. As a result, it is often found in "turns" of peptides/proteins, as its free entropy (ΔS) is not as comparatively large as other amino acids, and thus, in a folded form vs. unfolded form, the change in entropy is less. Furthermore, proline is rarely found in α and β structures, as it would reduce the stability of such structures, because its side chain α-N can only form one hydrogen bond.

Further, the hydroxylation of proline by prolyl hydroxylase and other additions of electron-withdrawing substituents, such as fluorine, increases the conformational stability of collagen significantly. Hence, the hydroxylation of proline is a critical biochemical process for maintaining the connective tissue of higher organisms. Polypeptide chains containing proline lack the flexibility of other peptides, because the proline ring has only one available angle for backbone rotation. Rotation occurs around the angles φ, ψ and ω [68,69].

The cyclization of linear peptide sequences can create constrained geometries that can alter the specificity of cyclic peptides to different isoforms or subtypes of targeted receptors. Peptides can be cyclized in order to reduce the overall numbers of interchanging conformers in the hope of limiting them to those selective for the desired receptors while avoiding degradation by not forming conformers susceptible to interacting with proteolytic enzymes [70].

In general, cyclization often increases the stability of peptides [71,72], which can prolong their biological activity. This prolonged activity may even be the result of additional resistance to enzymatic degradation by exoproteases that preferentially cleave near the *N*- or *C*-termini of peptide sequences. In particular, cyclization can create peptides with the ability to penetrate tumors in order to enhance the potency of anticancer drugs [73]. Cyclic peptides can potentially obtain desirable constrained geometries that are responsible for increasing their binding affinity, specificity or stability compared with their linear counterparts. Cyclic peptides are of considerable interest as potential protein ligands and might be more cell permeable than their linear counterparts due to their reduced conformational flexibility. However, it is important to note that cyclization does not necessarily lead to improvements in all of these properties, e.g., linear peptides can contain sequences that can support rigid structures without the need for cyclization [74].

2. Chemistry

2.1. Structural Features

The distinctive cyclic structure of proline's side chain gives proline an exceptional conformational rigidity compared to other amino acids, which affects the rate of peptide bond formation between proline and other amino acids. The exceptional conformational rigidity of proline affects the secondary structure of proteins near a proline residue and may account for proline's higher prevalence in the proteins of thermophilic organisms. Proline acts as a structural disruptor in the middle of regular secondary structure elements, such as alpha helices and beta sheets; however, proline is commonly found as the first residue of an alpha helix and also in the edge strands of beta sheets. Multiple prolines and hydroxyprolines in a row can create a polyproline helix, the predominant secondary structure in collagen [75].

The number of proline units in a cyclic peptide structure varies from one to five (Table 1). In addition to normal hydrophobic amino acids, marine organism-derived cyclopolypeptides rich in proline units contain modified and unusual amino acid moieties and other rings, like hydroxyproline (Hyp), (Z)-2,3-diaminoacrylic acid (DAA), thiazoline (Tzn), thiazole (Tzl), oxazole, methyloxazoline, reverse prenylated ethers, i.e., serine and threonine carrying a dimethylallyl ether group, *para*-hydroxystyrylamide (*p*HSA), pyroglutamic acid (pyroGlu), 3a-hydroxypyrrolo[2,3-*b*]indoline (Hpi), the 12-hydroxy-tetradecanoyl moiety, 2-(1-amino-2-*p*-hydroxyphenylethane)-4-(4-carboxy-2,4-dimethyl-2Z,4E-propadiene)-thiazole (ACT), *O*-methyl-*N*-sulfo-D-serine, keto-*allo*-isoleucine, methyloxazoline, β-methoxyaspartic acid, β-aminodecanoic acid, 2,2-dimethyl-3-hydroxy-7-octynoic acid (Dhoya), β-amino acid 3-amino-2-methylbutanoic acid (Maba) and 2-Hydroxy-isovaleric acid (Hiva), *O*-prenyltyrosine (Ptyr) (2S,3R,5R)-3-amino-2,5-dihydroxy-8-phenyloctanoic acid (Ahoa), dolaphenvaline (Pval) and dolamethylleucine (Admpa), *N*-acetyl-*N*-methylleucine (Aml), E- and Z-dehydrobutyrines (Dhb), a homophenylalanine (homophe), (2S,3R)-β-hydroxy-*p*-bromophenylalanine and *N*,*O*-dimethyl tyrosine, hydroxyisovaleric acid (Hiv) (Figure 3).

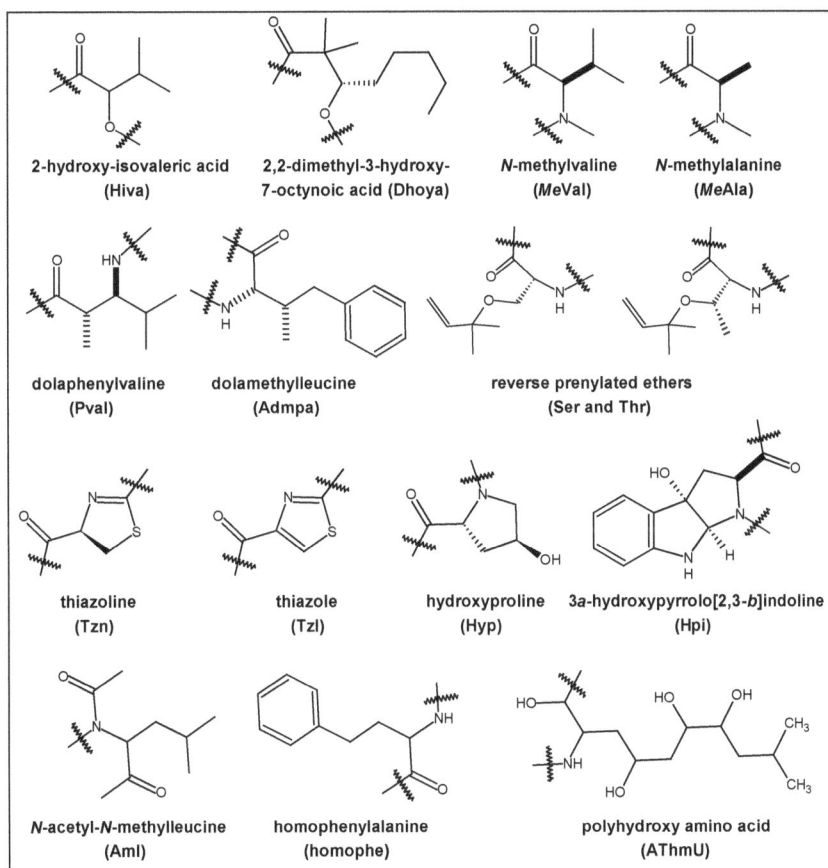

Figure 3. Modified amino acid moieties/heterocyclic rings present in marine-derived PRCPs.

Table 1. Proline-rich cyclopolypeptides from marine resources.

Year	Cyclic Peptide	Molecular Formula	No. of Proline Units	Composition
1981	Didemnin B [43]	$C_{57}H_{89}N_7O_{15}$		cyclodepsipeptide
1988	Aplidine [45]	$C_{57}H_{87}N_7O_{15}$		cyclodepsipeptide
1991	Axinastatin 1 [6]	$C_{38}H_{56}N_8O_8$		cycloheptapeptide
1992	Malaysiatin [27]	$C_{38}H_{56}N_8O_8$		cycloheptapeptide
1992	Polydiscamide A [7]	$C_{76}H_{109}BrN_{19}O_{20}SNa$		cyclodepsipeptide
1993	Axinastatin 4 [76]	$C_{42}H_{62}N_8O_8$		cycloheptapeptide
1993	Cyclooligopeptide [77]	$C_{24}H_{32}N_4O_5$		cyclotetrapeptide
1993	Hymenamide B [31]	$C_{43}H_{56}N_8O_{10}$		cycloheptapeptide
1993	Hymenamide C [8]	$C_{43}H_{54}N_8O_9$		cycloheptapeptide
1993	Hymenamide D [8]	$C_{38}H_{55}N_7O_{10}$		cycloheptapeptide
1993	Hymenamide E [8]	$C_{45}H_{55}N_7O_{10}$		cycloheptapeptide
1994	Mollamide [40]	$C_{42}H_{61}N_7O_7S$		cycloheptapeptide
1994	Schizotrin A [78]	$C_{72}H_{107}N_{13}O_{21}$		cycloundecapeptide
1994	Axinastatin 2 [39]	$C_{39}H_{58}N_8O_8$		cycloheptapeptide
1994	Axinastatin 3 [39]	$C_{40}H_{61}N_8O_8$		cycloheptapeptide
1995	Stylopeptide 1 [79]	$C_{40}H_{61}N_7O_8$		cycloheptapeptide
1996	Patellin 3 [80]	$C_{48}H_{78}N_8O_9S$		cyclooctapeptide
1996	Patellin 4 [80]	$C_{47}H_{76}N_8O_9S$		cyclooctapeptide
1996	Patellin 5 [80]	$C_{49}H_{72}N_8O_9S$		cyclooctapeptide
1996	Patellin 6 [80]	$C_{50}H_{74}N_8O_9S$		cyclooctapeptide
1996	Hymenamide F [81]	$C_{35}H_{60}N_{10}O_7S$		cycloheptapeptide
1996	Agardhipeptin B [82]	$C_{57}H_{69}N_{11}O_8$		cyclooctapeptide
1996	Kapakahine A [37]	$C_{58}H_{72}N_{10}O_9$		cyclooctapeptide
1996	Kapakahine C [37]	$C_{58}H_{72}N_{10}O_{10}$		cyclooctapeptide
1996	Kapakahine D [37]	$C_{58}H_{72}N_{10}O_{10}$		cyclooctapeptide
1998	Axinellin A [21]	$C_{42}H_{56}N_8O_9$		cycloheptapeptide
1998	Shearamide A [83]	$C_{47}H_{63}N_9O_9$		cyclooctapeptide
1999	Prenylagaramide B [84]	$C_{49}H_{68}N_8O_{10}$		cycloheptapeptide
1999	Nostophycin [50]	$C_{46}H_{64}N_8O_{10}$		cycloheptapeptide
2000	*trans,trans*-ceratospongamide [10]	$C_{41}H_{49}N_7O_6S$		cycloheptapeptide
2000	Tamandarine A [44]	$C_{54}H_{87}N_7O_{14}$	two	cyclodepsipeptide
2000	Tamandarine B [44]	$C_{53}H_{82}N_7O_{14}$		cyclodepsipeptide
2001	Microspinosamide [38]	$C_{75}H_{109}BrN_{18}O_{22}S$		cyclodepsipeptide
2003	Myriastramide C [28]	$C_{42}H_{53}N_9O_7S$		cyclooctapeptide
2004	Scleritodermin A [26]	$C_{42}H_{54}N_7O_{10}SNa$		cyclodepsipeptide
2004	Cyclonellin [85]	$C_{45}H_{62}N_{12}O_{12}$		cyclooctapeptide
2005	Wewakpeptin A [46]	$C_{52}H_{85}N_7O_{11}$		cyclodepsipeptide
2005	Wewakpeptin B [46]	$C_{52}H_{89}N_7O_{11}$		cyclodepsipeptide
2005	Wewakpeptin C [46]	$C_{54}H_{81}N_7O_{11}$		cyclodepsipeptide
2005	Wewakpeptin D [46]	$C_{54}H_{85}N_7O_{11}$		cyclodepsipeptide
2007	Pahayokolide A [48]	$C_{72}H_{105}N_{13}O_{20}$		cycloundecapeptide
2007	Pahayokolide B [48]	$C_{63}H_{90}N_{12}O_{18}$		cycloundecapeptide
2008	Polydiscamide B [18]	$C_{75}H_{110}BrN_{18}O_{21}S$		cyclodepsipeptide
2008	Polydiscamide C [18]	$C_{74}H_{107}BrN_{18}O_{21}S$		cyclodepsipeptide
2008	Polydiscamide D [18]	$C_{73}H_{105}BrN_{18}O_{21}S$		cyclodepsipeptide
2009	Euryjanicin A [36]	$C_{44}H_{58}N_8O_8$		cycloheptapeptide
2009	Euryjanicin C [14]	$C_{40}H_{61}N_7O_8$		cycloheptapeptide
2009	Euryjanicin D [14]	$C_{44}H_{59}N_7O_8$		cycloheptapeptide
2009	Eudistomide A [42]	$C_{37}H_{61}N_5O_8S_2$		cyclolipopeptide
2009	Eudistomide B [42]	$C_{37}H_{63}N_5O_8S_2$		cyclolipopeptide
2010	Anacyclamide A10 [86]	$C_{49}H_{72}N_{12}O_{14}$		cyclodecapeptide
2011	Duanbanhuain A [87]	$C_{43}H_{58}N_8O_{11}$		cyclooctapeptide
2011	Duanbanhuain B [87]	$C_{45}H_{57}N_9O_{10}$		cyclooctapeptide
2012	Mollamide F [12]	$C_{33}H_{46}N_6O_5S$		cyclohexapeptide
2013	Stylissatin A [24]	$C_{49}H_{63}N_7O_8$		cycloheptapeptide
2013	Euryjanicin E [88]	$C_{44}H_{60}N_8O_8$		cycloheptapeptide
2013	Euryjanicin F [88]	$C_{49}H_{63}N_7O_7$		cycloheptapeptide
2013	Gombamide A [22]	$C_{38}H_{45}N_7O_8S_2$		cyclothiohexapeptide
2013	Cycloforskamide [41]	$C_{54}H_{86}N_{12}O_{11}S_3$		cyclododecapeptide
2014	Trichormamide A [49]	$C_{58}H_{93}N_{11}O_{15}$		cycloundecapeptide
2014	Reniochalistatin A [20]	$C_{37}H_{62}N_8O_8$		cycloheptapeptide
2016	Carteritin B [34]	$C_{46}H_{57}N_7O_{11}$		cycloheptapeptide

Table 1. *Cont.*

Year	Cyclic Peptide	Molecular Formula	No. of Proline Units	Composition
1990	Hymenistatin 1 [35]	$C_{47}H_{72}N_8O_9$		cyclooctapeptide
1993	Phakellistatin 1 [32]	$C_{45}H_{61}N_7O_8$		cycloheptapeptide
1993	Hymenamide A [31]	$C_{46}H_{61}N_{11}O_7$		cycloheptapeptide
1993	Phakellistatin 2 [89]	$C_{45}H_{61}N_7O_8$		cycloheptapeptide
1994	Axinastatin 5 [30]	$C_{47}H_{72}N_8O_9$		cyclooctapeptide
1994	Hymenamide G [90]	$C_{47}H_{72}N_8O_9$		cyclooctapeptide
1994	Hymenamide H [90]	$C_{47}H_{69}N_9O_9$		cyclooctapeptide
1995	Phakellistatin 11 [91]	$C_{53}H_{67}N_9O_9$		cyclooctapeptide
1996	Waiakeamide [12]	$C_{37}H_{49}N_7O_8S_3$		cyclohexapeptide
1998	Axinellin B [21]	$C_{50}H_{67}N_9O_9$		cyclooctapeptide
2000	Haligramide A [92]	$C_{37}H_{49}N_7O_6S_3$		cyclohexapeptide
2000	Haligramide B [92]	$C_{37}H_{49}N_7O_7S_3$		cyclohexapeptide
2001	Haliclonamide A [93]	$C_{45}H_{60}N_8O_9$		cyclooctapeptide
2001	Haliclonamide B [93]	$C_{40}H_{52}N_8O_9$		cyclooctapeptide
2001	Wainunuamide [23]	$C_{38}H_{51}N_9O_7$		cycloheptapeptide
2002	Axinellin C [94]	$C_{50}H_{67}N_9O_9$		cyclooctapeptide
2002	Dolastatin 16 [52]	$C_{47}H_{70}N_6O_{10}$		cyclodepsipeptide
2002	Haliclonamide C [95]	$C_{45}H_{60}N_8O_{10}$		cyclooctapeptide
2002	Haliclonamide D [95]	$C_{40}H_{54}N_8O_{10}$		cyclooctapeptide
2002	Haliclonamide E [95]	$C_{45}H_{62}N_8O_{10}$	three	cyclooctapeptide
2003	Myriastramide A [28]	$C_{45}H_{58}N_8O_9$		cyclooctapeptide
2003	Myriastramide B [28]	$C_{45}H_{57}ClN_8O_9$		cyclooctapeptide
2003	Wewakazole [96]	$C_{59}H_{72}N_{12}O_{12}$		cyclododecapeptide
2005	Dominicin [16]	$C_{43}H_{72}N_8O_9$		cyclooctapeptide
2006	Stylisin 1 [19]	$C_{45}H_{61}N_7O_8$		cycloheptapeptide
2009	Euryjanicin B [14]	$C_{36}H_{51}N_7O_8$		cycloheptapeptide
2010	Phakellistatin 15 [25]	$C_{48}H_{71}N_9O_9$		cyclooctapeptide
2010	Phakellistatin 17 [25]	$C_{49}H_{73}N_9O_8$		cyclooctapeptide
2010	Phakellistatin 18 [25]	$C_{45}H_{61}N_7O_8$		cycloheptapeptide
2010	Callyaerin B [13]	$C_{65}H_{108}N_{12}O_{14}$		cyclooctapeptide [b]
2010	Callyaerin C [13]	$C_{70}H_{105}N_{13}O_{16}$		cycloheptapeptide [c]
2012	Stylissamide X [33]	$C_{51}H_{69}N_9O_9$		cyclooctapeptide
2013	Euryjanicin G [88]	$C_{48}H_{59}N_7O_7$		cyclooctapeptide
2014	Reniochalistatins E [20]	$C_{49}H_{73}N_9O_8$		cyclooctapeptide
2016	Carteritin A [34]	$C_{44}H_{57}N_7O_{10}$		cycloheptapeptide
2016	Stylissatin B [97]	$C_{38}H_{51}N_9O_7$		cycloheptapeptide
2016	Stylissatin C [97]	$C_{39}H_{55}N_7O_9$		cycloheptapeptide
2016	Stylissatin D [97]	$C_{40}H_{57}N_7O_9$		cycloheptapeptide
2016	Wewakazole B [47]	$C_{58}H_{70}N_{12}O_{12}$		cyclododecapeptide
1968	Antamanide [98]	$C_{64}H_{78}N_{10}O_{10}$		cyclodecapeptide
2004	Callynormine A [15]	$C_{61}H_{93}N_{11}O_{13}$		cycloheptapeptide [b]
2006	Stylisin 2 [19]	$C_{44}H_{57}N_7O_8$		cycloheptapeptide
2008	Stylopeptide 2 [29]	$C_{63}H_{84}N_{10}O_{12}$	four	cyclodecapeptide
2010	Callyaerin A [13]	$C_{69}H_{108}N_{14}O_{14}$		cyclooctapeptide [c]
2010	Callyaerin E [13]	$C_{66}H_{94}N_{12}O_{13}$		cycloheptapeptide [c]
2010	Callyaerin H [13]	$C_{54}H_{81}N_{11}O_{10}$		cycloheptapeptide [a]
2008	Callyaerin G [99]	$C_{69}H_{91}N_{13}O_{12}$	five	cycloheptapeptide [c]

With [a] dipeptide, [b] tripeptide and [c] tetrapeptide side chains.

Callynormine A represents a new class of heterodetic cyclic peptides possessing an α-amido-β-aminoacrylamide cyclization functionality. Hyp forms part of the composition of cyclic endiamino peptides like callynormine A [15] and callyaerin A–D. The unusual non-proteinogenic (Z)-DAA moiety is characteristic of the callyaerin series of peptides callyaerins A–M, which links the cyclic peptide part of the callyaerins with a linear peptide side chain [13]. Indo-Pacific ascidian *Didemnum molle* is found to be rich in thiazole-, oxazole- and thiazoline-containing peptides, like mollamide, which share the peculiar reverse prenylated ethers of serine and threonine amino acids [40].

Furthermore, unusual amino acid residues like *p*HSA and pyroGlu were found to be part of the structure of cyclothiopeptide gombamide A, which possess moderate inhibitory activity against

Na⁺/K⁺-ATPase [22]. Further, thiazoline-based proline containing doubly-prenylated cyclopeptides like trunkamide A contain reverse prenylated ethers of serine and threonine together in their composition. Heterocyclic amino acids like histidine and tryptophan also form part of the structures of proline-rich cyclic peptides, such as wainunuamide, phakellistatin 15, 17 and stylissatin B [23,25,97]. Moreover, cytotoxic phakellistatin 3 and isophakellistatin 3 represent a new class of proline-rich cycloheptapeptides containing an unusual amino acid unit "Hpi" that apparently derived from a photooxidation product of tryptophan [100].

Moreover, five-residue cystine-linked cyclic peptides like eudistomides A, B are flanked by a C-terminal methyl ester and a 12-oxo- or 12-hydroxy-tetradecanoyl moiety [42]. The structure of proline containing cytotoxic peptide scleritodermin A incorporates a novel conjugated thiazole moiety 2-(1-amino-2-*p*-hydroxyphenylethane)-4-(4-carboxy-2,4-dimethyl-2Z,4E-propadiene)-thiazole (ACT) and unusual amino acids O-methyl-N-sulfo-D-serine, keto-*allo*-isoleucine [26]. The proline unit may be part of a cyclic peptide and/or may be part of a side chain, e.g., scleritodermin A, didemnin B, C and plitidepsin [26,43,45], or may be part of a linear peptide, e.g., dolastatin 15 and koshikamide A₁ [1,5]. The methyloxazoline ring is the part of the composition of cyclohexapeptides ceratospongamides [10]. In addition, trichormamide A contains β-amino acid residue viz. β-aminodecanoic acid, in addition to two D-amino acid residues (D-Tyr and D-Leu) [49]. The wewakpeptins, proline-rich cyclic depsipeptides contain unusual moieties, like "Dhoya", "Maba" and "Hiva" [46], and prenylagaramides B and C contain a rare "Ptyr" unit. Moreover, nostophycin bears a novel β-amino acid moiety "Ahoa" in its structure [50]. Macrocyclic depsipeptides, homodolastatin 16 and dolastatin 16 contain the new and unusual amino acid units "Pval" and "Admpa" [51,52]. Besides this, structural features for pahayokolides A and B include a pendant N-acetyl-N-methylleucine, both E- and Z-dehydrobutyrines, a homophenylalanine and an unusual polyhydroxy amino acid [48]. Oxazole and methyloxazole rings were found to be part of the structures of cyclopolypeptides myriastramides A–C and haliclonamide A [28,93], whereas N,O-dimethyl tyrosine and "Hiv" moieties were found in the structures of cytotoxic depsipeptides, tamandarins A and B [44]. The presence of two dimethylallyl threonines (or one threonine and one serine) side chains and one thiazoline ring in the backbone of the patellins is the most important feature of these compounds termed as "cyanobactins", which have sparked attention due to their interesting bioactivities and for their potential to be prospective candidates in the development of drugs [101,102].

2.2. Stereochemical Aspects

Structurally, proline is the only unusual amino acid with a secondary amino group based on a pyrrolidine, which forms a ring structure with rigid conformation and a secondary amine compared to the other twenty natural amino acids. This significantly reduces the structural flexibility of the polypeptide chain, and the nitrogen in the pyrrolidine ring cannot participate in hydrogen bonding with other residues [103]. Many biologically-important cyclic peptide sequences and natural products contain multiple proline residues. As seen previously for peptide bonds, the proline amide bond can also exist in *trans* or *cis* conformations (Figure 4). Peptide bonds to proline, and to other N-substituted amino acids, are able to populate both the *cis* and *trans* isomers. Most peptide bonds overwhelmingly adopt the *trans* isomer (typically 99.9% under unstrained conditions), because the amide hydrogen (*trans* isomer) offers less steric repulsion to the preceding C_α atom than does the following C_α atom (*cis* isomer). By contrast, the *cis* and *trans* isomers of the X-Pro peptide bond (where X represents any amino acid) both experience steric clashes with the neighboring substitution and are nearly equal energetically. Hence, the fraction of X-Pro peptide bonds in the *cis* isomer under unstrained conditions ranges from 10% to 40%; the fraction depends slightly on the preceding amino acid, with aromatic residues favoring the *cis* isomer slightly. Proline *cis-trans* isomerization plays a key role in the rate-determining steps of protein folding [104]. Furthermore, proline *cis-trans* isomerization controls autoinhibition of a signaling protein [105].

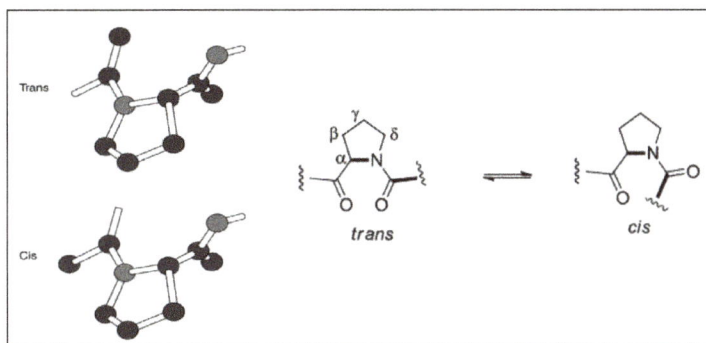

Figure 4. The two possible conformations for the proline peptide bond.

Although the *trans* amide bond is more common, the occurrence of *cis* geometry is more frequent for the proline peptide bond than for other amino acids. The frequency of the *cis* proline peptide bond is higher in cyclic peptides than in linear peptides. As per a statistical study performed on the Cambridge Structural Database, 57.4% of proline residues present in cyclic peptides were in the *cis* conformation as compared to only 5.6% in acyclic peptides [106]. The reason for this high proportion of *cis* proline in cyclopeptides is due to the conformational restrictions during the cyclisation step. The geometry of the proline amide can be determined on the basis of the difference in ^{13}C chemical shifts between Cβ and Cγ signals ($\Delta\delta\beta\gamma = \delta\beta - \delta\gamma$). A small ^{13}C chemical shift difference indicates that the proline peptide bond is *trans*, while a large ^{13}C chemical shift difference indicates a *cis* proline residue. The change in conformation of a cyclopolypeptide from "*trans*" to "*cis*" can result in loss of activity [10], e.g., the *trans, trans*-isomer of cyclic heptapeptide ceratospongamide showed potent inhibition of sPLA$_2$ expression in a cell-based model for anti-inflammation, whereas the *cis, cis*-isomer was inactive (Figure 5). The distribution of the peptide bond angle omega for peptidyl-prolyl bonds in proteins shows significant peaks at 180° (*trans* peptide bond) and 0° (*cis* peptide bond). Investigations on "peptidyl-prolyl bonds and secondary structure" showed that *trans* petidyl-prolyl bonds are distributed in all types of secondary structure, whereas *cis* peptidyl is found primarily in bends and turns, suggesting a specific structural role for this type of bonding.

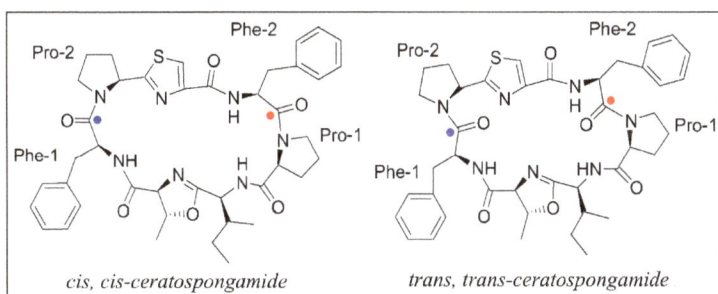

Figure 5. Different conformers of cyclopolypeptide ceratospongamide.

Most amino acids occur in two possible optical isomers, called D and L (Figure 6). The L-amino acids represent the vast majority of amino acids found in proteins. L-proline is a natural non-essential amino acid, and D-proline is an unnatural amino acid, with one basic and one acidic center each. In proline, only the L-stereoisomer is involved in the synthesis of mammalian peptides/proteins.

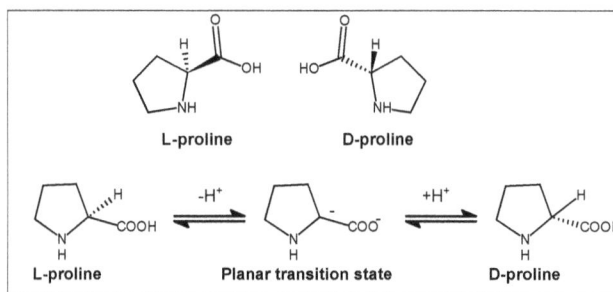

Figure 6. General structures of L- and D-proline and their isomerization via proline racemase.

The racemization of L-proline to D-proline proceeds through a planar transition state, where the tetrahedral α-carbon becomes trigonal as a proton leaves the L-proline. The transition-state analog for this step is pyrrolidin-2-ide-2-carboxylate (2^-). The absolute configuration of proline residue can be determined by Marfey's method using reagent 1-fluoro-2,4-dinitrophenyl-5-L-alanineamide (FDAA) [107]. The absolute configuration of amino proline was determined by comparing the retention time with the standard FDAA-derivatized amino acids, e.g., the structure of cyclooctapeptide reniochalistatin E contains three L-proline units with *trans* conformation [20] whereas the structure of cycloheptapeptide euryjanicin E contains three L-proline units with *cis* conformation [88]. Further, a novel cyclic tetrapeptide isolated from a *Pseudomonas* sp. (strain IM-1) associated with the marine sponge *Ircinia muscarum* was found to contain two proline units, one with L-configuration and the other with D-configuration [77].

2.3. Steric and Lipophilicity Parameters

In order to describe the intermolecular forces of drug receptor interaction, as well as the transport and distribution of drugs in a quantitative manner, various steric and lipophilicity parameters, like molar refractivity (MR^{20}), molar volume (MV^{20}), parachor (P_r), index of refraction (n^{20}), surface tension (γ^{20}), density (d^{20}), polarizability (α), etc., need to be calculated for natural cyclic peptides. Diverse parameters were calculated for proline-rich cyclopolypeptides of marine origin using ACD/ChemSketch software (Version 2.0, Toronto, ON, Canada) (Table S1, Supplementary Materials).

2.4. Synthetic Methodologies

Many proline-rich cyclic peptides were synthesized successfully by various research groups employing different techniques of peptide synthesis. The literature is enriched with reports explaining the synthesis of euryjanicin A [108], delavayin C [109], cherimolacyclopeptide G [110], psammosilenin A [111], hymenamide E [112], stylisin 1 [113], stylisin 2 [114], hymenistatin and yunnanin F [115], pseudostellarin B [116], segetalin E [117], rolloamide B [118] and pseudostellarin G [119] using the solution-phase method utilizing different carbodiimides as coupling agents, TEA/NMM as the base and the synthesis of euryjanicin B [120], mollamide [121], met-cherimolacyclopeptide B [122], axinellin A [123], phakellistatin 7 [124], phakellistatin 12 [125], petriellin A [126], hymenamide C [127], gombamide A [128] and scleritodermin A [129] by the solid-phase method of peptide synthesis. Solid-phase peptide synthesis (SPPS) results in high yields of pure products and works more quickly than classical synthesis, i.e., liquid-phase peptide synthesis (LPPS). Through the replacement of a complicated isolation procedure for each intermediate product with a simple washing procedure, much time is saved using SPPS. In addition, SPPS has proven possible to increase the yield in each individual step to 99.5% or better, which cannot be attained using conventional synthetic approaches. However, solution phase synthesis continues to be especially valuable for large-scale manufacturing and for specialized laboratory applications [130,131]. Moreover, in some cases, a mixed solid-phase/solution synthesis strategy is employed to accomplish total synthesis of the cyclopolypeptide [132], e.g., during the total synthesis of the naturally-occurring proline-rich cyclic octapeptide stylissamide X, the linear octapeptide was assembled first by standard Fmoc solid-phase peptide synthesis (SPPS), and cyclization was carried out subsequently by the solution method. Total synthesis can also be achieved via a convergent native chemical ligation-oxidation strategy [133], e.g., polydiscamides B–D, or utilizing diethyl phosphorocyanidate/BOP-Cl chemistry [134], e.g., axinastatins 2 and 3.

3. Biological Status

L-proline itself is an osmoprotectant and is used in many pharmaceutical and biotechnological applications, whereas the proline analogue *cis*-4-hydroxy-L-proline has been clinically evaluated as an anticancer drug. Although proline-rich cyclopolypeptides of marine origin are associated with a number of bioactivities, including anti-cancer, anti-tuberculosis, anti-inflammatory, anti-viral, immunosuppressive and anti-fungal activities, still the majority of them were found to exhibit cell growth inhibitory activity [135,136]. Various pharmacological activities associated marine-derived proline-rich cyclopeptides along with susceptible cell line/organism with minimum inhibitory concentration are compiled in Table 2.

Table 2. Marine-derived proline-rich cyclopeptides with diverse bioactivities.

PRCPs	Resource	Pharmacological Activity	
		Susceptibility	MIC Value
Axinastatin 1 [6]	marine sponge	Cytotoxicity against PS leukemia cell line	0.21 µg/mL
Polydiscamide A [7]	marine sponge	Antiproliferative activity against human lung cancer A549 cell line; antibacterial activity against *Bacillus subtilis*	0.7 µg/mL; 3.1 µg/mL
Hymenamide E [8]	marine sponge	Antifungal activity against pathogenic *Cryptococcus neoformans*	133 µg/mL
trans,trans-Ceratospongamide [10]	marine red alga	Inhibition of sPLA$_2$ expression in a cell-based model for anti-inflammation	0.0013 µg/mL
Mollamide F [12]	marine tunicate	Anti-HIV activity in cytoprotective cell-based assay and HIV integrase inhibition assay	0.0016 and 0.0031 µg/mL
Callyaerin A [13]	marine sponge	Anti-TB activity against *M. tuberculosis*, inhibitory activity toward *C. albicans*	7.37 µg/mL
Callyaerin B [13]	marine sponge	Anti-TB activity against *Mycobacterium tuberculosis*	7.8 µg/mL
Callyaerin E, H [13]	marine sponge	Cytotoxicity against L5178Y cell line	7.91 and 9.59 µg/mL
Euryjanicin C [14]	marine sponge	Inhibitory activity against human hepatitis B virus	49 µg/mL
Polydiscamides B–D [18]	marine sponge	Agonist activity against human sensory neuron-specific G protein couple receptor (SNSR) that is involved in the modulation of pain	-
Axinellin A, B [21]	marine sponge	Antitumor activity against human bronchopulmonary non-small-cell lung-carcinoma lines (NSCLC-N6)	3.0 and 7.3 µg/mL
Wainunuamide [23]	marine sponge	Cytotoxic activity against A2780 ovarian tumor and K562 leukemia cancer cells	19.15 and 18.36 µg/mL
Stylissatin A [24]	marine sponge	Inhibition of NO production in LPS-stimulated RAW264.7 cells	0.0011 µg/mL
Scleritodermin A [26]	marine sponge	Inhibition of tubulin polymerization and human tumor cell lines	-
Axinastatin 5 [30]	marine sponge	Cytotoxic activity against human and murine cancer cells	0.3–3.3 µg/mL
Phakellistatin 1 [32]	marine sponges	Cell growth inhibitory activity against P-388 murine leukemia	7.5 µg/mL
Stylissamide X [33]	marine sponge	Inhibitory activity against migration of HeLa cells	0.001–0.1 µg/mL
Carteritin A [34]	marine sponge	Cytotoxicity against HeLa, HCT116 and RAW264 cells	0.0012–0.0026 µg/mL
Hymenistatin 1 [35]	marine sponge	Cytotoxicity against P-388 leukemia cells	3.5 µg/mL
Kapakahine A, C [37]	marine sponge	Cytotoxicity against P-388 murine leukemia cells	5.4 and 5.0 µg/mL
Microspinosamide [38]	marine sponge	Anti-HIV activity in CEM-SS cells	0.2 µg/mL
Axinastatin 2 [39]	marine sponge	Cytotoxicity against murine leukemia P-388 cell line	0.02 µg/mL
Axinastatin 3 [39]	marine sponge	Cytotoxicity against PS leukemia cell line	0.4 µg/mL
Mollamide [40]	sea squirt	Cytotoxicity against P-388 (murine leukemia) and A549 (human lung carcinoma), HT29 (human colon carcinoma) cells	1.0–2.5 µg/mL
Cycloforskamide [41]	sea slug	Cytotoxicity against murine leukemia P-388 cells	8.51 µg/mL
Didemnin B [43]	marine tunicate	Cytotoxic activity against human L1210 lymphocytic leukemia cell lines; pancreatic carcinoma (BX-PC3) cell lines; prostatic cancer (DU-145) cell lines; head and neck carcinoma (UMSCC10b) cell lines	0.0025 µg/mL; 0.002 µg/mL; 0.0015 µg/mL; 0.0018 µg/mL

Table 2. *Cont.*

PRCPs	Resource	Pharmacological Activity	
		Susceptibility	MIC Value
Tamandarin A [44]	marine ascidian	Cytotoxic activity against human pancreatic carcinoma (BX-PC3) cell lines; prostatic cancer (DU-145) cell lines; head and neck carcinoma (UMSCC10b) cell lines	0.0018 µg/mL; 0.0014 µg/mL; 0.0009 µg/mL
Wewakpeptin A [46]	marine cyanobacterium	Cytotoxicity against NCI-H460 human lung tumor and the neuro-2a mouse neuroblastoma cell lines	0.001 µg/mL
Wewakazole B [47]	marine cyanobacterium	Cytotoxicity against human MCF7 breast/H460 lung cancer cells	8.87–15.29 µg/mL
Pahayokolide A [48]	marine cyanobacteria	Antibacterial activity against *Bacillus megaterium*, *Bacillus subtilis*	5 µg/mL
Trichormamide A [49]	marine cyanobacteria	Antiproliferative activities against the human melanoma cell line (MDA-MB-435) and the human colon cancer cell line (HT-29)	8.45 and 8.53 µg/mL
Axinastatin 4 [76]	marine sponge	Cytotoxic activity against P-388 lymphocytic leukemia cell line	0.057 µg/mL
Phakellistatin 2 [89]	marine sponge	Cell growth inhibitory activity against P-388 cell line	0.34 µg/mL
Phakellistatin 7–9 [137]	marine sponge	Cell growth inhibitory activity against P-388 murine leukemia	3.0, 2.9 and 4.1 µg/mL
Axinellin C [94]	marine sponge	Cytotoxic activity against A2780 ovarian tumor and K562 leukemia cancer cells	13.17 and 4.46 µg/mL
Callyaerin G [99]	marine sponge	Cytotoxic towards the mouse lymphoma cell line (L5178Y) and HeLa cells	0.53 and 5.4 µg/mL
Stylissatin B [97]	marine sponge	Inhibitory effects against human tumor cell lines including HCT-116, HepG2, BGC-823, NCI-H1650, A2780 and MCF7	0.0013 µg/mL
Phakellistatin 10, 11 [91]	marine sponge	Cell growth inhibitory activity against murine P-388 lymphocytic leukemia	2.1, 0.20 µg/mL
Stylopeptide 1 [79]	marine sponge	Cell growth inhibitory activity against murine P-388 lymphocytic leukemia	0.01 µg/mL
Phakellistatin 12 [138]	marine sponge	Cell growth inhibitory activity against murine P-388 lymphocytic leukemia	2.8 µg/mL

3.1. Mechanism of Action

In drug development, a good antimicrobial candidate should exhibit highly specific biological activity followed by a good pharmacokinetic profile and low immunogenicity. Studies have demonstrated that the members of the proline-rich peptide group and their derivatives act with a completely divergent mechanism than the lytic amphiphilic antimicrobial peptides. Retaining highly potent antimicrobial activities, proline-rich antimicrobial peptides subsequently act in a divergent way, including stereospecific interaction with the membrane translocation system followed by intracellular targeting, compared with the more general membrane disruption mode of action of traditional antimicrobial peptides. It has been further suggested that proline-rich antimicrobial peptides stereo-specifically bind to intracellular targets, such as the bacterial heat shock DnaK protein, and this binding can be correlated with the observed antimicrobial activity. Moreover, proline-rich peptides are characterized by good water solubility, high potency against bacteria killing and low cytotoxic effects at high concentrations, making them attractive lead candidates for the development of novel antimicrobial therapeutic agents [103].

Further, proline-rich antimicrobial peptides are actively transported inside the bacterial cell where they bind and inactivate specific targets like the bacterial ribosome and, thereby, inhibit protein synthesis. This implies that they can be used as molecular hooks to identify the intracellular or membrane proteins that are involved in their mechanism of action and that may be subsequently used as targets for the design of novel antibiotics with mechanisms different from those now in use. Didemnin B is a heterodetic non-polar cyclic peptide associated with antiviral, antitumor, immunomodulating properties, potently inhibits protein and DNA synthesis by binding to the eukaryotic translation elongation factor EF-1α in a GTP-dependent manner, and the formation of the didemnin B-GTP-EF-1α complex may be responsible for the observed inhibition of protein synthesis [139]. Inhibition of protein synthesis by didemnin B occurs by stabilization of aminoacyl-tRNA to the ribosomal A-site, preventing the translocation of phenylalanyl-tRNA from the A- to the P-site, but not preventing peptide bond formation. Tamandarin A may act by the same mechanism as didemnin B. Aplidine's (dehydrodidemnin B) mechanism of action involves several pathways, including cell cycle arrest and inhibition of protein synthesis. Aplidine induces early oxidative stress and results in a rapid and persistent activation of JNK and p38 MAPK phosphorylation with activation of both kinases occurring very rapid, long before the execution of apoptosis [140]. Didemnin B induces the death of a variety of transformed cells with apoptotic morphology, DNA fragmentation within the cytosol and the generation of DNA ladders. Scleritodermin A acts by tubulin polymerization inhibition [26].

The immunosuppressive activity of cyclolinopeptide A results from the formation of the complex with cyclophilin and inhibition of the phosphatase activity of calcineurin, a phosphatase that plays an important role in T lymphocyte signaling [141]. Cemadotin (LU103793) is a water-soluble synthetic analogue of linear peptide dolastatin 15, which is believed to act on microtubules involving binding to tubulin and strong suppression of microtubule dynamics.

3.2. Peptide Market and PRCPs in Clinical Trials

Currently, there are more than 60 U.S. Food and Drug Administration (FDA)-approved peptide medicines on the market, and this is expected to grow significantly, with approximately 140 peptide drugs currently in clinical trials and more than 500 therapeutic peptides in preclinical development. In terms of value, the global peptide drug market has been predicted to increase from US$14.1 billion in 2011 to an estimated US$25.4 billion in 2018, with an underlying increase in novel innovative peptide drugs from US$8.6 billion in 2011 (60%) to US$17.0 billion (66%) in 2018 [74]. Currently, most peptide drugs are administered by the parental route, and approximately 75% are given as injectables. However, alternative administration forms are gaining increasing traction, including oral, intranasal and transdermal delivery routes, according to the respective technology developments. The use of alternative administration forms could also enable greater usage of peptide therapeutics in other disease areas, such as inflammation, where topical administration of peptides could be the basis for highly efficacious novel treatments.

The cyclic depsipeptide didemnin B was the first marine-derived cyclopolypeptide to undergo clinical trials targeted at oncological patients. However, high toxicity, poor solubility and short life span led to the discontinuation of clinical trials of didemnin B and rendered it unsuitable for further drug development [142]. The linear depsipeptide kahalalide F is known for its antifungal and antitumor activities, and its phase II clinical trials are underway. Another cyclic depsipeptide plitidepsin (dehydrodidemnin B or aplidine) is in clinical development. In 2003, plitidepsin was granted orphan drug status by the European Medicines Agency for treating acute lymphoblastic leukemia. In 2007, it was undergoing multicenter phase II clinical trials, and in 2016, early results in a small phase I trial for multiple myeloma were announced. The two most promising peptides of antimitotic dolastatins group, dolastatin 10 and 15, were selected for development and are currently undergoing phase II clinical trials. Cemadotin, the synthetic analogue of dolastatin 15, is also in phase II clinical trials as a promising cancer chemotherapeutic agent [143,144].

4. Conclusions and Future Prospects

There is increased evidence of the emergence of resistance to conventional drugs illustrating the importance of research on natural peptide-based drug development. PRCPs have several structural features making them good drug leads, and there are several naturally-occurring cyclic peptides in clinical use and in clinical trials. In addition, biologically-active proline-rich cyclic peptides have been developed with synthetic approaches, and they are useful as therapeutics and biochemical tools. With the introduction of new high throughput screening methods, there will be more availability of marine-based PRCPs with interesting biological properties. PRCPs can work on their targets very selectively, as the interaction with the targets is very specific compared to small molecules. In addition to the merits of peptides, especially "proline-rich cyclic structures" as drug molecules, cyclopolypeptides could make even better peptide drugs for future use. Moreover, the future development of peptide drugs will continue to build upon the strengths of naturally-occurring proline-rich peptides, with the application of traditional rational design to improve their weaknesses, such as their chemical and physical properties. Further, emerging peptide technologies will help broaden the applicability of PRCPs as therapeutics. While still in the early stages of development, PRCPs drug leads have started gaining the attention of the pharmaceutical industry; however, their true potential is still very much unknown.

Supplementary Materials: The following are available online at www.mdpi.com/1660-3397/14/11/194/s1. Table S1: Various steric and lipophilic parameters for proline-rich cyclopolypeptides from diverse marine resources.

Acknowledgments: The authors wish to thank chief librarians of Central Drug Research Institute (CDRI), Lucknow, Uttar Pradesh, India, National Medical Library (NML), New Delhi, India, Faculty of Medical Sciences, The University of the West Indies, Trinidad and Tobago, West Indies and Wuhan University of Technology, Wuhan, China, for providing literature support.

Author Contributions: All authors were involved in all aspects of the work done for this paper.

Conflicts of Interest: The authors declare no conflict of interest.

References

1. Bai, R.; Friedman, S.J.; Pettit, G.R.; Hamel, E. Dolastatin 15, a potent antimitotic depsipeptide derived from *Dolabella auricularia*: Interaction with tubulin and effects on cellular microtubules. *Biochem. Pharmacol.* **1992**, *43*, 2637–2645. [CrossRef]

2. Okamoto, S.; Iwasaki, A.; Ohno, O.; Suenaga, K. Isolation and structure of kurahyne B and total synthesis of the kurahynes. *J. Nat. Prod.* **2015**, *78*, 2719–2725. [CrossRef] [PubMed]

3. Iwasaki, A.; Ohno, O.; Sumimoto, S.; Ogawa, H.; Nguyen, K.A.; Suenaga, K. Jahanyne, an apoptosis-inducing lipopeptide from the marine cyanobacterium *Lyngbya* sp. *Org. Lett.* **2015**, *17*, 652–655. [CrossRef] [PubMed]

4. Jordan, M.A.; Walker, D.; de Arruda, M.; Barlozzari, T.; Panda, D. Suppression of microtubule dynamics by binding of cemadotin to tubulin: Possible mechanism for its antitumor action. *Biochemistry* **1998**, *37*, 17571–17578. [CrossRef] [PubMed]

5. Fusetani, N.; Warabi, K.; Nogata, Y.; Nakao, Y.; Matsunaga, S.; van Soest, R.R.M. Koshikamide A_1, a new cytotoxic linear peptide isolated from a marine sponge *Theonella* sp. *Tetrahedron Lett.* **1999**, *40*, 4687–4690. [CrossRef]

6. Pettit, G.R.; Herald, C.L.; Boyd, M.R.; Leet, J.E.; Dufresne, C.; Doubek, D.L.; Schmidt, J.M.; Cerny, R.L.; Hooper, J.N.A.; Rutzler, K.C. Antineoplastic agents. 219. Isolation and structure of the cell growth inhibitory constituents from the western Pacific marine sponge *Axinella* sp. *J. Med. Chem.* **1991**, *34*, 3339–3340. [CrossRef] [PubMed]

7. Gulavita, N.K.; Gunasekela, S.P.; Pomponi, S.A.; Robinson, E.V. Polydiscamide A: A new bioactive depsipeptide from the marine sponge *Discodermia* sp. *J. Org. Chem.* **1992**, *57*, 1767–1772. [CrossRef]

8. Tsuda, M.; Shigemori, H.; Mikami, Y.; Kobayashi, J. Hymenamides C–E, new cyclic heptapeptides with two proline residues from the Okinawan marine sponge *Hymeniacidon* sp. *Tetrahedron* **1993**, *49*, 6785–6796. [CrossRef]

9. Cebrat, M.; Wieczorek, Z.; Siemion, I.Z. Immunosuppressive activity of hymenistatin 1. *Peptides* **1996**, *17*, 191–196. [CrossRef]

10. Tan, L.T.; Williamson, R.T.; Gerwick, W.H.; Watts, K.S.; McGough, K.; Jacobs, R. *cis,cis*- and *trans,trans*-Ceratospongamide, new bioactive cyclic heptapeptides from the indonesian red alga *Ceratodictyon spongiosum* and symbiotic sponge *Sigmadocia symbiotica*. *J. Org. Chem.* **2000**, *65*, 419–425. [CrossRef] [PubMed]

11. Lu, Z.; Harper, M.K.; Pond, C.D.; Barrows, L.R.; Ireland, C.M.; van Wagoner, R.M. Thiazoline peptides and a *tris*-phenethyl urea from *Didemnum molle* with anti-HIV activity. *J. Nat. Prod.* **2012**, *75*, 1436–1440. [CrossRef] [PubMed]

12. Sera, Y.; Adachi, K.; Fujii, K.; Shizuri, Y. A new antifouling hexapeptide from a palauan sponge, *Haliclona* sp. *J. Nat. Prod.* **2003**, *66*, 719–721. [CrossRef] [PubMed]

13. Ibrahim, S.R.; Min, C.C.; Teuscher, F.; Ebel, R.; Kakoschke, C.; Lin, W.; Wray, V.; Edrada-Ebel, R.; Proksch, P. Callyaerins A–F and H, new cytotoxic cyclic peptides from the Indonesian marine sponge *Callyspongia aerizusa*. *Bioorg. Med. Chem.* **2010**, *18*, 4947–4956. [CrossRef] [PubMed]

14. Vera, B.; Vicente, J.; Rodriguez, A.D. Isolation and structural elucidation of euryjanicins B–D, proline-containing cycloheptapeptides from the Caribbean marine sponge *Prosuberites laughlini*. *J. Nat. Prod.* **2009**, *72*, 1555–1562. [CrossRef] [PubMed]

15. Berer, N.; Rudi, A.; Goldberg, I.; Benayahu, Y.; Kashman, Y. Callynormine A, a new marine cyclic peptide of a novel class. *Org. Lett.* **2004**, *6*, 2543–2545. [CrossRef] [PubMed]

16. Williams, D.E.; Patrick, B.O.; Behrisch, H.W.; van soest, R.; Roberge, M.; Andersen, R.J. Dominicin, a cyclic octapeptide, and laughine, a bromopyrrole alkaloid, isolated from the Caribbean marine sponge *Eurypon laughlini*. *J. Nat. Prod.* **2005**, *68*, 327–330. [CrossRef] [PubMed]

17. Daletos, G.; Kalscheuer, R.; Koliwer-Brandl, H.; Hartmann, R.; de Voogd, N.J.; Wray, V.; Lin, W.; Proksch, P. Callyaerins from the marine sponge *Callyspongia aerizusa*: Cyclic peptides with antitubercular activity. *J. Nat. Prod.* **2015**, *78*, 1910–1925. [CrossRef] [PubMed]

18. Feng, Y.; Carroll, A.R.; Pass, D.M.; Archbold, J.K.; Avery, V.M.; Quinn, R.J. Polydiscamides B–D from a marine sponge *Ircinia* sp. as potent human sensory neuron-specific G protein coupled receptor agonists. *J. Nat. Prod.* **2008**, *71*, 8–11. [CrossRef] [PubMed]

19. Mohammed, R.; Peng, J.; Kelly, M.; Hamann, M.T. Cyclic heptapeptides from the jamaican sponge *Stylissa caribica*. *J. Nat. Prod.* **2006**, *69*, 1739–1744. [CrossRef] [PubMed]

20. Zhan, K.X.; Jiao, W.H.; Yang, F.; Li, J.; Wang, S.P.; Li, Y.S.; Han, B.N.; Lin, H.W. Reniochalistatins A–E, cyclic peptides from the marine sponge *Reniochalina stalagmitis*. *J. Nat. Prod.* **2014**, *77*, 2678–2684. [CrossRef] [PubMed]

21. Randazzo, A.; Piaz, F.D.; Orrù, S.; Debitus, C.; Roussakis, C.; Pucci, P.; Gomez-Paloma, L. Axinellins A and B: New proline-containing antiproliferative cyclopeptides from the Vanuatu sponge *Axinella carteri*. *Eur. J. Org. Chem.* **1998**, *11*, 2659–2665. [CrossRef]

22. Woo, J.K.; Jeon, J.E.; Kim, C.K.; Sim, C.J.; Oh, D.C.; Oh, K.B.; Shin, J. Gombamide A, a cyclic thiopeptide from the sponge *Clathria gombawuiensis*. *J. Nat. Prod.* **2013**, *76*, 1380–1383. [CrossRef] [PubMed]

23. Tabudravu, J.; Morris, L.A.; Kettenes-van den Bosch, J.J.; Jaspars, M. Wainunuamide, a histidine-containing proline-rich cyclic heptapeptide isolated from the Fijian marine sponge *Stylotella aurantium*. *Tetrahedron Lett.* **2001**, *42*, 9273–9276. [CrossRef]

24. Kita, M.; Gise, B.; Kawamura, A.; Kigoshi, H. Stylissatin A, a cyclic peptide that inhibits nitric oxide production from the marine sponge *Stylissa massa*. *Tetrahedron Lett.* **2013**, *54*, 6826–6828. [CrossRef]

25. Zhang, H.J.; Yi, Y.H.; Yang, G.J.; Hu, M.Y.; Cao, G.D.; Yang, F.; Lin, H.W. Proline-containing cyclopeptides from the marine sponge *Phakellia fusca*. *J. Nat. Prod.* **2010**, *73*, 650–655. [CrossRef] [PubMed]

26. Schmidt, E.W.; Raventos-Suarez, C.; Bifano, M.; Menendez, A.T.; Fairchild, C.R.; Faulkner, D.J. Scleritodermin A, a cytotoxic cyclic peptide from the lithistid sponge *Scleritoderma nodosum*. *J. Nat. Prod.* **2004**, *67*, 475–478. [CrossRef] [PubMed]

27. Fernandez, R.; Omar, S.; Feliz, M.; Quinoa, E.; Riguera, R. Malaysiatin, the first cyclic heptapeptide from a marine sponge. *Tetrahedron Lett.* **1992**, *33*, 6017–6020. [CrossRef]

28. Erickson, K.L.; Gustafson, K.R.; Milanowski, D.J.; Pannell, L.K.; Klose, J.R.; Boyd, M.R. Myriastramides A–C, new modified cyclic peptides from the Phillipines marine sponge *Myriastra clavosa*. *Tetrahedron* **2003**, *59*, 10231–10238. [CrossRef]

29. Brennan, M.R.; Costello, C.E.; Maleknia, S.D.; Pettit, G.R.; Erickson, K.L. Stylopeptide 2, a proline-rich cyclodecapeptide from the sponge *Stylotella* sp. *J. Nat. Prod.* **2008**, *71*, 453–436. [CrossRef] [PubMed]

30. Pettit, G.R.; Gao, F.; Schmidt, J.M.; Cerny, R. Isolation and structure of axinastatin 5 from a Republic of Comoros marine sponge. *Bioorg. Med. Chem. Lett.* **1994**, *4*, 2935–2940. [CrossRef]

31. Kobayashi, J.; Tsuda, M.; Nakamura, T.; Mikami, Y.; Shigemori, H. Hymenamides A and B, new proline-rich cyclic heptapeptides from the okinawan marine sponge *hymeniacidon* sp. *Tetrahedron* **1993**, *49*, 2391–2402. [CrossRef]

32. Pettit, G.R.; Cichacz, Z.; Barkoczy, J.; Dorsaz, A.C.; Herald, D.L.; Williams, M.D.; Doubek, D.L.; Schmidt, J.M.; Tackett, L.P.; Brune, D.C.; et al. Isolation and structure of the marine sponge cell growth inhibitory cyclic peptide phakellistatin 1. *J. Nat. Prod.* **1993**, *56*, 260–267. [CrossRef] [PubMed]

33. Arai, M.; Yamano, Y.; Fujita, M.; Setiawan, A.; Kobayashi, M. Stylissamide X, a new proline-rich cyclic octapeptide as an inhibitor of cell migration, from an Indonesian marine sponge of *Stylissa* sp. *Bioorg. Med. Chem. Lett.* **2012**, *22*, 1818–1821. [CrossRef] [PubMed]

34. Afifi, A.H.; El-Desoky, A.H.; Kato, H.; Mangindaan, R.E.P.; de Voogd, N.J.; Ammar, N.M.; Hifnawy, M.S.; Tsukamoto, S. Carteritins A and B, cyclic heptapeptides from the marine sponge *Stylissa carteri*. *Tetrahedron Lett.* **2016**, *57*, 1285–1288. [CrossRef]

35. Pettit, G.R.; Clewlow, P.J.; Dufrense, C.; Doubek, D.L.; Cerny, R.L.; Rutzler, K. Antineoplastic agents. 193. Isolation and structure of the cyclic peptide hymenistatin 1. *Can. J. Chem.* **1990**, *68*, 708–711. [CrossRef]

36. Vicente, J.; Vera, B.; Rodriguez, A.D.; Rodriguez-Escudero, I.; Raptis, R.G. Euryjanicin A: A new cycloheptapeptide from the Caribbean marine sponge *Prosuberites laughlini*. *Tetrahedron Lett.* **2009**, *50*, 4571–4574. [CrossRef] [PubMed]

37. Yeung, B.K.S.; Nakao, Y.; Kinnel, R.B.; Carney, J.R.; Yoshida, W.Y.; Scheuer, P.J.; Kelly-Borges, M. The Kapakahines, cyclic peptides from the marine sponge *Cribrochalina olemda*. *J. Org. Chem.* **1996**, *61*, 7168–7173. [CrossRef] [PubMed]

38. Rashid, M.A.; Gustafson, K.R.; Cartner, L.K.; Shigematsu, N.; Pannell, L.K.; Boyd, M.R. Microspinosamide, a new HIV-inhibitory cyclic depsipeptide from the marine sponge *Sidonops microspinosa*. *J. Nat. Prod.* **2001**, *64*, 117–121. [CrossRef] [PubMed]

39. Pettit, G.R.; Gao, F.; Cerny, R.L.; Doubek, D.L.; Tackett, L.P.; Schmidt, J.M.; Chapuis, J.C. Antineoplastic agents. 278. Isolation and structure of axinastatins 2 and 3 from a western Caroline Island marine sponge. *J. Med. Chem.* **1994**, *37*, 1165–1168. [CrossRef] [PubMed]

40. Carroll, A.R.; Bowden, B.F.; Coll, J.C.; Hockless, D.C.R.; Skelton, B.W.; White, A.H. Studies of australian ascidians. IV. Mollamide, a cytotoxic cyclic heptapeptide from the compound ascidian *Didemnum molle*. *Aust. J. Chem.* **1994**, *47*, 61–69. [CrossRef]

41. Tan, K.O.; Wakimoto, T.; Takada, K.; Ohtsuki, T.; Uchiyama, N.; Goda, Y.; Abe, I. Cycloforskamide, a cytotoxic macrocyclic peptide from the sea slug *Pleurobranchus forskalii*. *J. Nat. Prod.* **2013**, *76*, 1388–1391. [CrossRef] [PubMed]

42. Whitson, E.L.; Ratnayake, A.S.; Bugni, T.S.; Harper, M.K.; Treland, C.M. Isolation, structure elucidation and synthesis of eudistomides A and B, lipopeptides from a fijian ascidian *Eudistoma* sp. *J. Org. Chem.* **2009**, *74*, 1156–1162. [CrossRef] [PubMed]

43. Rinehart, K.L., Jr.; Gloer, J.B.; Cook, J.C., Jr.; Mizsak, S.A.; Scahill, T.A. Structures of the didemnins, antiviral and cytotoxic depsipeptides from a Caribbean tunicate. *J. Am. Chem. Soc.* **1981**, *103*, 1857–1859. [CrossRef]

44. Vervoort, H.; Fenical, W. Tamandarins A and B: New cytotoxic depsipeptides from a *Brazilian ascidian* of the family Didemnidae. *J. Org. Chem.* **2000**, *65*, 782–792. [CrossRef] [PubMed]

45. Mercader, A.G.; Duchowicz, P.R.; Sivakumar, P.M. *Chemometrics Applications and Research: QSAR in Medicinal Chemistry*; Apple Academic Press, Inc.: Oakville, ON, Canada, 2016; p. 278.

46. Han, B.; Goeger, D.; Maier, C.S.; Gerwick, W.H. The Wewakpeptins, cyclic depsipeptides from a papua new guinea collection of the marine cyanobacterium *Lyngbya semiplena*. *J. Org. Chem.* **2005**, *70*, 3133–3139. [CrossRef] [PubMed]

47. Lopez, J.A.V.; Al-Lihaibi, S.S.; Alarif, W.M.; Abdel-Lateff, A.; Nogata, Y.; Washio, K.; Morikawa, M.; Okino, T. Wewakazole B, a cytotoxic cyanobactin from the cyanobacterium *Moorea producens* collected in the red sea. *J. Nat. Prod.* **2016**, *79*, 1213–1218. [CrossRef] [PubMed]

48. An, T.; Kumar, T.K.; Wang, M.; Liu, L.; Lay, J.O., Jr.; Liyanage, R.; Berry, J.; Gantar, M.; Marks, V.; Gawley, R.E.; et al. Structures of pahayokolides A and B, cyclic peptides from a *Lyngbya* sp. *J. Nat. Prod.* **2007**, *70*, 730–735. [CrossRef] [PubMed]

49. Luo, S.; Krunic, A.; Kang, H.S.; Chen, W.L.; Woodard, J.L.; Fuchs, J.R.; Swanson, S.M.; Orjala, J. Trichormamides A and B with antiproliferative activity from the cultured freshwater cyanobacterium *Trichormus* sp. UIC 10339. *J. Nat. Prod.* **2014**, *77*, 1871–1880. [CrossRef] [PubMed]

50. Fujii, K.; Sivonen, K.; Kashiwagi, T.; Hirayama, K.; Harada, K.I. Nostophycin, a novel cyclic peptide from the toxic cyanobacterium *Nostoc* sp. 152. *J. Org. Chem.* **1999**, *64*, 5777–5782. [CrossRef]

51. Davies-Coleman, M.T.; Dzeha, T.M.; Gray, C.A.; Hess, S.; Pannell, L.K.; Hendricks, D.T.; Arendse, C.E. Isolation of homodolastatin 16, a new cyclic depsipeptide from a Kenyan collection of *Lyngbya majuscula*. *J. Nat. Prod.* **2003**, *66*, 712–715. [CrossRef] [PubMed]

52. Nogle, L.M.; Gerwick, W.H. Isolation of four new cyclic depsipeptides, antanapeptins A–D, and dolastatin 16 from a madagascan collection of *Lyngbya majuscula*. *J. Nat. Prod.* **2002**, *65*, 21–24. [CrossRef] [PubMed]
53. Dahiya, R. Cyclopolypeptides with antifungal interest. *Coll. Pharm. Commun.* **2013**, *1*, 1–15.
54. Dahiya, R.; Gautam, H. Synthesis, characterization and biological evaluation of cyclomontanin D. *Afr. J. Pharm. Pharmacol.* **2011**, *5*, 447–453. [CrossRef]
55. Dahiya, R.; Gautam, H. Synthetic and pharmacological studies on a natural cyclopeptide from *Gypsophila arabica*. *J. Med. Plant Res.* **2010**, *4*, 1960–1966.
56. Dahiya, R.; Singh, S. Synthesis, characterization and biological screening of diandrine A. *Acta Pol. Pharm.* **2016**, submitted.
57. Dahiya, R.; Gautam, H. Solution phase synthesis and bioevaluation of cordyheptapeptide B. *Bull. Pharm. Res.* **2011**, *1*, 1–10.
58. Dahiya, R. Synthesis of a phenylalanine-rich peptide as potential anthelmintic and cytotoxic agent. *Acta Pol. Pharm.* **2007**, *64*, 509–516. [PubMed]
59. Dahiya, R.; Gautam, H. Toward the first total synthesis of gypsin D: A natural cyclopolypeptide from Gypsophila arabica. *Am. J. Sci. Res.* **2010**, *11*, 150–158.
60. Dahiya, R.; Kaur, K. Synthesis and pharmacological investigation of segetalin C as a novel antifungal and cytotoxic agent. *Arzneimittelforschung* **2008**, *58*, 29–34. [CrossRef] [PubMed]
61. Dahiya, R. Synthetic and pharmacological studies on longicalycinin A. *Pak. J. Pharm. Sci.* **2007**, *20*, 317–323. [PubMed]
62. Dahiya, R.; Kumar, A. Synthetic and biological studies on a cyclopolypeptide of plant origin. *J. Zhejiang Univ. Sci. B* **2008**, *9*, 391–400. [CrossRef] [PubMed]
63. Dahiya, R.; Gautam, H. Synthesis and pharmacological studies on a cyclooligopeptide from marine bacteria. *Chin. J. Chem.* **2011**, *29*, 1911–1916.
64. Dahiya, R. Synthesis, characterization and biological evaluation of a glycine-rich peptide—Cherimolacyclopeptide E. *J. Chil. Chem. Soc.* **2007**, *52*, 1224–1229. [CrossRef]
65. Dahiya, R.; Gautam, H. Toward the synthesis and biological screening of a cyclotetrapeptide from marine bacteria. *Mar. Drugs* **2011**, *9*, 71–81. [CrossRef] [PubMed]
66. Dahiya, R.; Maheshwari, M.; Yadav, R. Synthetic and cytotoxic and antimicrobial activity studies on annomuricatin B. *Z. Naturforsch.* **2009**, *64*, 237–244. [CrossRef]
67. Aneiros, A.; Garateix, A. Bioactive peptides from marine sources: Pharmacological properties and isolation procedures. *J. Chromatogr. B Anal. Technol. Biomed. Life Sci.* **2004**, *803*, 41–53. [CrossRef] [PubMed]
68. Silver, F.H. *Mechanosensing and Mechanochemical Transduction in Extracellular Matrix. Biochemical, Chemical, Engineering, and Physiological Aspects. Macromolecular Structures in Tissues*; Springer: Berlin/Heidelberg, Germany, 2006; Volume XVI, p. 33.
69. Pandey, A.K.; Naduthambi, D.; Thomas, K.M.; Zondlo, N.J. Proline editing: A general and practical approach to the synthesis of functionally and structurally diverse peptides. Analysis of steric versus stereoelectronic effects of 4-substituted prolines on conformation within peptides. *J. Am. Chem. Soc.* **2013**, *135*, 4333–4363. [CrossRef] [PubMed]
70. Roxin, A.; Zheng, G. Flexible or fixed: A comparative review of linear and cyclic cancer-targeting peptides. *Future Med. Chem.* **2012**, *4*, 1601–1618. [CrossRef] [PubMed]
71. Goodwin, D.; Simerska, P.; Toth, I. Peptides as therapeutics with enhanced bioactivity. *Curr. Med. Chem.* **2012**, *19*, 4451–4461. [CrossRef] [PubMed]
72. Jensen, J.E.; Mobli, M.; Brust, A.; Alewood, P.F.; King, G.F.; Rash, L.D. Cyclisation increases the stability of the sea anemone peptide APETx2 but decreases its activity at acid-sensing ion channel 3. *Mar. Drugs* **2012**, *10*, 1511–1527. [CrossRef] [PubMed]
73. Roxin, A. Towards Targeted Photodynamic Therapy: Synthesis and Characterization of Aziridine Aldehyde-Cyclized Cancertargeting Peptides and Bacteriochlorin Photosensitizers. Ph.D. Thesis, Graduate Department of Pharmaceutical Sciences, University of Toronto, Toronto, ON, Canada, 2014.
74. Fosgerau, K.; Hoffmann, T. Peptide therapeutics: Current status and future directions. *Drug Discov. Today* **2015**, *20*, 122–128. [CrossRef] [PubMed]
75. Shanmugam, S.; Kumar, S.T.; Selvam, K.P. *Laboratory Handbook on Biochemistry*, 1st ed.; Prentice-Hall of India Private Limited: New Delhi, India, 2010.

76. Pettit, G.R.; Gao, F.; Cerny, R. Isolation and structure of axinastatin 4 from the western indian ocean marine sponge *Axinella* cf. carteri. *Heterocycles* **1993**, *35*, 711–718. [CrossRef]

77. Kawagishi, H.; Somoto, A.; Kuranari, J.; Kimura, A.; Chiba, S. A novel cyclotetrapeptide produced by *Lactobacillus helveticus* as a tyrosinase inhibitor. *Tetrahedron Lett.* **1993**, *34*, 3439–3440. [CrossRef]

78. Pergament, I.; Carmeli, S. Schizotrin A; a novel antimicrobial cyclic peptide from a cyanobacterium. *Tetrahedron Lett.* **1994**, *35*, 8473–8476. [CrossRef]

79. Pettit, G.R.; Srirangam, J.K.; Herald, D.L.; Xu, J.P.; Boyd, M.R.; Cichacz, Z.; Kamano, Y.; Schmidt, J.M.; Erickson, K.L. Isolation and crystal structure of stylopeptide 1, a new marine porifera cycloheptapeptide. *J. Org. Chem.* **1995**, *60*, 8257–8261. [CrossRef]

80. Carroll, A.R.; Coll, J.C.; Bourne, J.C.; MacLeod, J.K.; Zanriskie, T.M.; Ireland, C.M.; Bowden, B.F. Patellins 1-6 and Trunkamide A: Novel cyclic hexa-, hepta- and octa-peptides from colonial ascidians, *Lissoclinurn* sp. *Aust. J. Chem.* **1996**, *49*, 659–667.

81. Kobayashi, J.; Nakamura, T.; Tsuda, M. Hymenamide F, new cyclic heptapeptide from marine sponge *Hymeniacidon* sp. *Tetrahedron* **1996**, *52*, 6355–6360. [CrossRef]

82. Shin, H.J.; Matsuda, H.; Murakami, M.; Yamaguchi, K. Agardhipeptins A and B, two new cyclic hepta- and octapeptide, from the cyanobacterium *Oscillatoria agardhii* (NIES-204). *Tetrahedron* **1996**, *52*, 13129–13136. [CrossRef]

83. Belofsky, G.N.; Gloer, J.B.; Wicklow, D.T.; Dowd, P.F. Shearamide A: A new cyclic peptide from the ascostromata of *Eupenicillium shearii*. *Tetrahedron Lett.* **1998**, *39*, 5497–5500. [CrossRef]

84. Murakami, M.; Itou, Y.; Ishida, K.; Shin, H.J. Prenylagaramides A and B, new cyclic peptides from two strains of *Oscillatoria agardhii*. *J. Nat. Prod.* **1999**, *62*, 752–755. [CrossRef] [PubMed]

85. Milanowski, D.J.; Rashid, M.A.; Gustafson, K.R.; O'Keefe, B.R.; Nawrocki, J.P.; Pannell, L.K.; Boyd, M.R. Cyclonellin, a new cyclic octapeptide from the marine sponge *Axinella carteri*. *J. Nat. Prod.* **2004**, *67*, 441–444. [CrossRef] [PubMed]

86. Leikoski, N.; Fewer, D.P.; Jokela, J.; Wahlsten, M.; Rouhiainen, L.; Sivonen, K. Highly diverse cyanobactins in strains of the genus Anabaena. *Appl. Environ. Microbiol.* **2010**, *76*, 701–709. [CrossRef] [PubMed]

87. Cheng, Y.X.; Zhou, L.L.; Yan, Y.M.; Chen, K.X.; Hou, F.F. Diabetic nephropathy-related active cyclic peptides from the roots of *Brachystemma calycinum*. *Bioorg. Med. Chem. Lett.* **2011**, *21*, 7334–7439. [CrossRef] [PubMed]

88. Aviles, E.; Rodriguez, A.D. Euryjanicins E–G, poly-phenylalanine and poly-proline cyclic heptapeptides from the Caribbean sponge *Prosuberites laughlini*. *Tetrahedron* **2013**, *69*, 10797–10804. [CrossRef] [PubMed]

89. Pettil, G.R.; Tan, R.; Williams, M.D.; Tackett, L.; Schmidt, J.M.; Cerny, R.L.; Hooper, J.N.A. Isolation and structure of phakellistatin 2 from the eastern indian ocean marine sponge *phakellia carteri*. *Bioorg. Med. Chem. Lett.* **1993**, *3*, 2869–2874. [CrossRef]

90. Tsuda, M.; Sasaki, T.; Kobayashi, J. Hymenamides G, H, J, and K, four new cyclic octapeptides from the Okinawan marine sponge *Hymeniacidon* sp. *Tetrahedron* **1994**, *50*, 4667–4680. [CrossRef]

91. Pettit, G.R.; Tan, R.; Ichihara, Y.; Williams, M.D.; Doubek, D.L.; Tackett, L.P.; Schmidt, J.M.; Cerny, R.L.; Boyd, M.R.; Hooper, J.N. Antineoplastic agents, 325. Isolation and structure of the human cancer cell growth inhibitory cyclic octapeptides phakellistatin 10 and 11 from Phakellia sp. *J. Nat. Prod.* **1995**, *58*, 961–965. [CrossRef] [PubMed]

92. Rashid, M.A.; Gustafson, K.R.; Boswell, J.L.; Boyd, M.R. Haligramides A and B, two new cytotoxic hexapeptides from the marine sponge *Haliclona nigra*. *J. Nat. Prod.* **2000**, *63*, 956–959. [CrossRef] [PubMed]

93. Guan, L.L.; Sera, Y.; Adachi, K.; Nishida, F.; Shizuri, Y. Isolation and evaluation of nonsiderophore cyclic peptides from marine sponges. *Biochem. Biophy. Res. Commun.* **2001**, *283*, 976–981. [CrossRef] [PubMed]

94. Tabudravu, J.N.; Morris, L.A.; Kettenes-van den Bosch, J.J.; Jaspars, M. Axinellin C, a proline-rich cyclic octapeptide isolated from the Fijian marine sponge *Stylotella aurantium*. *Tetrahedron* **2002**, *58*, 7863–7868. [CrossRef]

95. Sera, Y.; Adachi, K.; Fujii, K.; Shizuri, Y. Isolation of haliclonamides: New peptides as antifouling substances from a marine sponge species, *Haliclona*. *Mar. Biotechnol.* **2002**, *4*, 441–446. [CrossRef] [PubMed]

96. Nogle, L.M.; Marquez, B.L.; Gerwick, W.H. Wewakazole, a novel cyclic dodecapeptide from a papua new guinea *Lyngbya majuscule*. *Org. Lett.* **2003**, *5*, 3–6. [CrossRef] [PubMed]

97. Sun, J.; Cheng, W.; de Voogd, N.J.; Proksch, P.; Lin, W. Stylissatins B–D, cycloheptapeptides from the marine sponge *Stylissa massa*. *Tetrahedron Lett.* **2016**, in press.

98. Wieland, T.; Luben, G.; Ottenheym, H.; Faesel, D.C.J.; de Vries, J.X.; Prox, A.; Schmid, D.C.J. The discovery, isolation, elucidation of structure, and synthesis of antamanide. *Angew. Chem. Int. Ed.* **1968**, *7*, 204–208. [CrossRef] [PubMed]

99. Ibrahim, S.R.M.; Edrada-Ebel, R.A.; Mohamed, G.A.; Youssef, D.T.A.; Wray, V.; Proksch, P. Callyaerin G, a new cytotoxic cyclic peptide from the marine sponge *Callyspongia aerizusa*. *ARKIVOC Arch. Org. Chem.* **2008**, *2008*, 164–171.

100. Pettit, G.R.; Tan, R.; Herald, D.L.; Cerny, R.L.; Williams, M.D. Antineoplastic agents. 277. Isolation and structure of phakellistatin 3 and isophakellistatin 3 from a republic of Comoros marine sponge. *J. Org. Chem.* **1994**, *59*, 1593–1595. [CrossRef]

101. Martins, J.; Vasconcelos, V. Cyanobactins from cyanobacteria: Current genetic and chemical state of knowledge. *Mar. Drugs* **2015**, *13*, 6910–6946. [CrossRef] [PubMed]

102. Donia, M.S.; Ravel, J.; Schmidt, E.W. A global assembly line to cyanobactins. *Nat. Chem. Biol.* **2008**, *4*, 341–343. [CrossRef] [PubMed]

103. Mojsoska, B.; Jenssen, H. Peptides and peptidomimetics for antimicrobial drug design. *Pharmaceuticals* **2015**, *8*, 366–415. [CrossRef] [PubMed]

104. Wedemeyer, W.J.; Welker, E.; Scheraga, H.A. Proline *cis-trans* isomerization and protein folding. *Biochemistry* **2002**, *41*, 14637–14644. [CrossRef] [PubMed]

105. Sarkar, P.; Reichman, C.; Saleh, T.; Birge, R.B.; Kalodimos, C.G. Proline *cis-trans* isomerization controls autoinhibition of a signaling protein. *Mol. Cell* **2007**, *25*, 413–426. [CrossRef] [PubMed]

106. Vitagliano, L.; Berisio, R.; Mastrangelo, A.; Mazzarella, L.; Zagari, A. Preferred proline puckerings in cis and trans peptide groups: Implications for collagen stability. *Protein Sci.* **2001**, *10*, 2627–2632. [CrossRef] [PubMed]

107. Bhushan, R.; Bruckner, H. Marfey's reagent for chiral amino acid analysis: A review. *Amino Acids* **2004**, *27*, 231–247. [CrossRef] [PubMed]

108. Anand, M.; Alagar, M.; Ranjitha, J.; Selvaraj, V. Total synthesis and anticancer activity of a cyclic heptapeptide from marine sponge using water soluble peptide coupling agent EDC. *Arab. J. Chem.* **2016**, in press.

109. Shinde, N.V.; Himaja, M.; Bhosale, S.K.; Ramana, M.V.; Sakarkar, D.M. Synthesis and biological evaluation of delavayin-C. *Indian J. Pharm. Sci.* **2008**, *70*, 827–831. [CrossRef] [PubMed]

110. Dahiya, R. Synthesis, spectroscopic and biological investigation of cyclic octapeptide: Cherimolacyclopeptide G. *Turk. J. Chem.* **2008**, *32*, 205–215.

111. Dahiya, R. Total synthesis and biological potential of psammosilenin A. *Arch. Pharm. Chem. Life Sci.* **2008**, *341*, 502–509. [CrossRef] [PubMed]

112. Dahiya, R.; Pathak, D.; Himaja, M.; Bhatt, S. First total synthesis and biological screening of hymenamide E. *Acta Pharm.* **2006**, *56*, 399–415. [PubMed]

113. Dahiya, R.; Kumar, A.; Gupta, R. Synthesis, cytotoxic and antimicrobial screening of a proline-rich cyclopolypeptide. *Chem. Pharm. Bull. (Tokyo)* **2009**, *57*, 214–217. [CrossRef] [PubMed]

114. Dahiya, R.; Gautam, H. Total synthesis and antimicrobial activity of a natural cycloheptapeptide of marine origin. *Mar. Drugs* **2010**, *8*, 2384–2394. [CrossRef] [PubMed]

115. Poojary, B.; Belagali, S.L. Synthetic studies on cyclic octapeptides: Yunnanin F and hymenistatin. *Eur. J. Med. Chem.* **2005**, *40*, 407–412. [CrossRef] [PubMed]

116. Poojary, B.; Kumar, K.H.; Belagali, S.L. Synthesis and biological evaluation of pseudostellarin B. *Pharmaco* **2001**, *56*, 331–334. [CrossRef]

117. Dahiya, R.; Kaur, K. Synthetic and biological studies on natural cyclic heptapeptide: Segetalin E. *Arch. Pharm. Res.* **2007**, *30*, 1380–1386. [CrossRef] [PubMed]

118. El Khatib, M.; Elagawany, M.; Caliskan, E.; Davis, E.F.; Faidallah, H.M.; El-Feky, S.A.; Katritzky, A.R. Total synthesis of cyclic heptapeptide rolloamide B. *Chem. Commun. (Camb.)* **2013**, *49*, 2631–2633. [CrossRef] [PubMed]

119. Poojary, B.; Kumar, K.H.; Belagali, S.L. Synthesis of a new cyclic peptide, pseudostellarin G. *Z. Naturforsch. B* **2004**, *59*, 817–820. [CrossRef]

120. Zhang, C.M.; Guo, J.X.; Wang, L.; Chai, X.Y.; Hu, H.G.; Wu, Q.Y. Total synthesis of cyclic heptapeptide euryjanicin B. *Chin. Chem. Lett.* **2011**, *22*, 631–634. [CrossRef]

121. McKeever, B.; Pattenden, G. Total synthesis of mollamide, a reverse prenyl substituted cytotoxic cyclic peptide from *Didemnum molle. Tetrahedron Lett.* **1999**, *40*, 9317–9320. [CrossRef]

122. Dellai, A.; Maricic, I.; Kumar, V.; Arutyunyan, S.; Bouraoui, A.; Nefzi, A. Parallel synthesis and anti-inflammatory activity of cyclic peptides cyclosquamosin D and Met-cherimolacyclopeptide B and their analogs. *Bioorg. Med. Chem. Lett.* **2010**, *20*, 5653–5657. [CrossRef] [PubMed]

123. Fairweather, K.A.; Sayyadi, N.; Roussakis, C.; Jolliffi, K.A. Synthesis of the cyclic heptapeptide axinellin A. *Tetrahedron* **2010**, *66*, 935–939. [CrossRef]

124. Napolitano, A.; Bruno, I.; Riccio, R.; Gomez-Paloma, L. Synthesis, structure, and biological aspects of cyclopeptides related to marine phakellistatins 7–9. *Tetrahedron* **2005**, *61*, 6808–6815. [CrossRef]

125. Ali, L.; Musharraf, S.G.; Shaheen, F. Solid-phase total synthesis of cyclic decapeptide phakellistatin 12. *J. Nat. Prod.* **2008**, *71*, 1059–1062. [CrossRef] [PubMed]

126. Sleebs, M.M.; Scanlon, D.; Karas, J.; Maharani, R.; Hughes, A.B. Total synthesis of the antifungal depsipeptide petriellin A. *J. Org. Chem.* **2011**, *76*, 6686–6693. [CrossRef] [PubMed]

127. Napolitano, A.; Bruno, I.; Rovero, P.; Lucas, R.; Peris, M.P.; Gomez-Paloma, L.; Riccio, R. Synthesis, structural aspects and bioactivity of the marine cyclopeptide hymenamide C. *Tetrahedron* **2001**, *57*, 6249–6255. [CrossRef]

128. Garcia-Barrantes, P.M.; Lindsley, C.W. Total synthesis of gombamide A. *Org. Lett.* **2016**, *18*, 3810–3813. [CrossRef] [PubMed]

129. Sellanes, D.; Manta, E.; Serra, G. Toward the total synthesis of scleritodermin A: Preparation of the C_1–N_{15} fragment. *Tetrahedron Lett.* **2007**, *48*, 1827–1830. [CrossRef] [PubMed]

130. Dahiya, R.; Pathak, D. First total synthesis and biological evaluation of halolitoralin A. *J. Serb. Chem. Soc.* **2007**, *72*, 101–107. [CrossRef]

131. Dahiya, R.; Maheshwari, M.; Kumar, A. Toward the synthesis and biological evaluation of hirsutide. *Monatsh. Chem.* **2009**, *140*, 121–127. [CrossRef]

132. Huang, T.; Zou, Y.; Wu, M.C.; Zhao, Q.J.; Hu, H.G. Total synthesis of proline-rich cyclic octapeptide stylissamide X. *Chem. Nat. Prod.* **2015**, *51*, 523–526. [CrossRef]

133. Santhakumar, G.; Payne, R.J. Total synthesis of polydiscamides B, C, and D via a convergent native chemical ligation-oxidation strategy. *Org. Lett.* **2014**, *16*, 4500–4503. [CrossRef] [PubMed]

134. Pettit, G.R.; Holman, J.W.; Boland, G.M. Synthesis of the cyclic heptapeptides axinastatin 2 and axinastatin 3. *J. Chem. Soc. Perkin Trans. 1* **1996**, 2411–2416. [CrossRef]

135. Dahiya, R.; Pathak, D. Cyclic peptides: New hope for antifungal therapy. *Egypt. Pharm. J. (NRC)* **2006**, *5*, 189–199.

136. Pathak, D.; Dahiya, R. Cyclic peptides as novel antineoplastic agents: A review. *J. Sci. Pharm.* **2003**, *4*, 125–131.

137. Pettit, G.R.; Xu, J.P.; Dorsaz, A.C.; Williams, M.D.; Boyd, M.R.; Cerny, R.L. Isolation and structure of the human cancer cell growth inhibitory cyclic decapeptides phakellistatins 7, 8 and 9. *Bioorg. Med. Chem. Lett.* **1995**, *5*, 1339–1344. [CrossRef]

138. Pettit, G.R.; Tan, R. Antineoplastic agents 390. Isolation and structure of phakellistatin 12 from a Chuuk Archipelago marine sponge. *Bioorg. Med. Chem. Lett.* **2003**, *13*, 685–688. [CrossRef]

139. Li, L.H.; Timmins, L.G.; Wallace, T.L.; Krueger, W.C.; Prairie, M.D.; Im, W.B. Mechanism of action of didemnin B, a depsipeptide from the sea. *Cancer Lett.* **1984**, *23*, 279–288. [CrossRef]

140. Zheng, L.H.; Wang, Y.J.; Sheng, J.; Wang, F.; Zheng, Y.; Lin, X.K.; Sun, M. Antitumor peptides from marine organisms. *Mar. Drugs* **2011**, *9*, 1840–1859. [CrossRef] [PubMed]

141. Siemion, I.Z.; Cebrat, M.; Wieczorek, Z. Cyclolinopeptides and their analogs—A new family of peptide immunosuppressants affecting the calcineurin system. *Arch. Immunol. Ther. Exp.* **1999**, *47*, 143–153.

142. Malaker, A.; Ahmad, S.A.I. Therapeutic potency of anticancer peptides derived from marine organism. *Int. J. Eng. Appl. Sci.* **2013**, *2*, 53–65.

143. Simmons, T.L.; Andrianasolo, E.; McPhail, K.; Flatt, P.; Gerwick, W.H. Marine natural products as anticancer drugs. *Mol. Cancer Ther.* **2005**, *4*, 333–342. [PubMed]

144. Proksch, P.; Ebel, R.; Edrada, R.A.; Wray, V.; Steube, K. Bioactive natural products from marine invertebrates and associated fungi. *Prog. Mol. Subcell. Biol.* **2003**, *37*, 117–142. [PubMed]

marine drugs

MDPI

Review

Preclinical and Clinical Studies on Antioxidative, Antihypertensive and Cardioprotective Effect of Marine Proteins and Peptides—A Review

Ida-Johanne Jensen * and Hanne K. Mæhre

Norwegian College of Fishery Science, Faculty of Biosciences, Fisheries and Economics,
UIT The Arctic University of Norway, N-9037 Tromsø, Norway; hanne.maehre@uit.no
* Correspondence: ida-johanne.jensen@uit.no; Tel.: +47-776-46-721

Academic Editor: Se-Kwon Kim
Received: 3 October 2016; Accepted: 11 November 2016; Published: 18 November 2016

Abstract: High seafood consumption has traditionally been linked to a reduced risk of cardiovascular diseases, mainly due to the lipid lowering effects of the long chained omega 3 fatty acids. However, fish and seafood are also excellent sources of good quality proteins and emerging documentation show that, upon digestion, these proteins are sources for bioactive peptides with documented favorable physiological effects such as antioxidative, antihypertensive and other cardioprotective effects. This documentation is mainly from in vitro studies, but also animal studies are arising. Evidence from human studies evaluating the positive health effects of marine proteins and peptides are scarce. In one study, a reduction in oxidative stress after intake of cod has been documented and a few human clinical trials have been performed evaluating the effect on blood pressure. The results are, however, inconclusive. The majority of the human clinical trials performed to investigate positive health effects of marine protein and lean fish intake, has focused on blood lipids. While some studies have documented a reduction in triglycerides after intake of lean fish, others have documented no effects.

Keywords: marine; proteins; peptides; bioactive; antioxidative; clinical trials; preclinical; animal studies

1. Introduction

Cardiovascular diseases (CVDs) are a group of diseases affecting the heart and blood vessels and they are the largest cause of morbidity and premature deaths worldwide [1] accounting for 31% of all global deaths in 2012 [2]. The development of CVDs is associated with several risk factors, both modifiable and non-modifiable, and the danger of developing CVD increases considerably with the number of risk factors present [3]. Gender, heredity and increasing age are risk factors that are non-modifiable. Modifiable risk factors are often life-style related and may be associated with oxidative stress. Tobacco smoking, physical inactivity, diabetes mellitus, obesity and overweight are, along with hypertension and dyslipidemia, examples of such preventable risk factors. Although the risk factors associated with the development of CVDs are many and affect several processes in the body, there are two major underlying causes for CVD, namely hypertension and atherosclerosis [1]. Nutritional intervention is well accepted as a safe and effective approach to health maintenance and it has been estimated that a change in nutritional pattern may reduce cardiovascular-related deaths by 60% [4]. Seafood consumption has been linked to a reduced risk of these illnesses, and traditionally the beneficial effects have been associated with the long-chained omega 3 polyunsaturated fatty acids eicosapentaenoic acid (EPA, 20:5n3) and docosahexaenoic acid (DHA, 22:6n3) [5–10]. Emerging evidence has now demonstrated that the proteins, as well as other bioactive compounds, may also be relevant for improving human health by different mechanisms. Fish and seafood are excellent sources of good quality proteins that upon digestion may be sources for bioactive peptides with documented favorable physiological effects such as antioxidative, antihypertensive and other cardioprotective effects. The documentation is mainly from in vitro studies, but the number of preclinical studies and human trials performed is arising. This review aims to summarize these preclinical and clinical studies.

Clinical Trials

In order to evaluate causal relationships between nutrients and chronic diseases, two main approaches are normally applied, namely epidemiological and experimental studies. There are advantages and disadvantages associated with both study types, and a combination of studies will probably return the most solid foundation for evidence. In brief, epidemiological studies range over a long period of time and include large population groups. Data material may be very large, there are few restrictions in diet and diseases can be included as endpoint. The main drawback is that they are poorly controlled and that the sources of error are many. Experimental studies are performed in a more controlled environment compared to epidemiological studies. Clinical trials involving human volunteers and preclinical trials involving animals fall into this category. In such studies, test subjects are enrolled into a controlled environment where their diets, together with other relevant measures, are regulated. Effects are registered through a range of different physiological parameters, depending on the aim of the study.

In our opinion, experimental clinical studies on humans are by far the most accurate way to evaluate the health effects of different diets or food components. However, such studies are also very expensive, time consuming and complex. Further, inclusion criteria for participants may vary according to the aim of the study and comparisons between studies may therefore be difficult.

2. Oxidative Stress and Antioxidative Status

Aerobic metabolism is accompanied by an inevitable production of reactive oxygen species, normally referred to as ROS. To reduce the production and counter the harm of these ROS, the human body is equipped with several antioxidant systems involving mechanisms that prevent free radicals from causing damage and mechanisms that repair or mitigate any occurred damage [11]. However, the balance between ROS and antioxidants may shift in favor of ROS, and a condition called oxidative stress arises. This condition has been related to several disorders, atherosclerosis [12] in particular. There is now a general acceptance that consumption of dietary antioxidants is an effective approach to increase the body's antioxidant load and mitigate the effects of ROS [13]. The mechanisms may be inactivation of ROS [14], scavenging of free radicals [15], chelating of pro-oxidative transition metals [16] and reduction of hydroperoxides [17,18]. Some amino acids, in particular histidine, glutamic acid, aspartic acid, along with phosphorylated serine and threonine, have the ability to chelate prooxidative transition metals [16]. Usually, peptides are considered more potent antioxidants due to the stability of the resultant peptide radical [13]. The antioxidant potential of a protein or peptide depends on the amino acids being exposed and accessible to prooxidants. Increased exposure of amino acids can be attained by food processing, fermentation or gastrointestinal digestion. The in vitro antioxidant activity of marine protein hydrolysates has been shown for several fish species, mollusks, crustaceans and microalgae. The link to a beneficial health outcome in humans is, however, still on a theoretical level. Despite evidence showing clear associations between oxidative stress and CVDs, epidemiological data on antioxidant intake and disease prevention are inconclusive. Natural antioxidant intake from foods has been proven beneficial [19], whereas analyses with antioxidant supplementation have been proven unfavorable or even resulting in adverse effects in preventing all-cause mortality [20,21].

2.1. Human Studies

One study has been published focusing on the effect of marine proteins on oxidative stress and antioxidative status (Table 1). During a randomized parallel intervention, 276 overweight subjects were following a diet designed for weight loss [22]. The subjects were randomized to four groups and followed a diet plan with either lean meat, lean meat supplemented with omega-3 fatty acids, cod or salmon during eight weeks. The oxidation product malondialdehyd and the antioxidative capacity were measured before and after the trial. After the intervention period, the amount of the oxidation product was significantly reduced in the group following the cod based diet (from 1.81 nM to 1.72 nM). At the same time, the antioxidative capacity in this group increased significantly (from 0.62 nM to 0.71 nM) and was significantly higher than that in the individuals following both the lean meat diet and the lean meat with omega-3 capsules. It was suggested by the authors that the specific protein characteristics of cod or the high amount of taurine may have contributed to this effect.

2.2. Animal Studies

Documentation from preclinical trials has been increasing. Two studies investigated the effect of fish protein compared to casein in the feed for male spontaneously hypertensive rats (SHR) over a two months period. In one study, lipid peroxidation (measured as TBARS) in heart and liver were significantly lower in the SHR receiving the fish protein diet compared to the SHR receiving the casein diet [23], whereas no difference was observed in muscle and adipose tissue and higher lipid peroxidation was observed in kidney. The antioxidant status in heart and liver increased with the fish protein diet, whereas it remained unchanged in plasma during the feeding trial. This suggests that fish protein plays an important role in the antioxidative defense system in heart and liver, but not in plasma. In the second study 50%, of the SHR were induced with diabetes after one month, which resulted in increased plasma antioxidative status in the fish protein fed SHR compared to the casein fed SHR [24]. In a recent study by Jensen et al. [25], apolipoprotein E-deficient (apoeE$-/-$) mice were used to evaluate the effect of dietary cod and scallop on atherosclerotic burden and related parameters, among them gene expressions of antioxidative proteins. Twenty-four 5-week-old female apoeE$-/-$ mice were fed Western type diets with chicken or cod and scallop as the protein sources for 13 weeks. It was shown that the hepatic endogenous antioxidant paraoxonase 2 (*Pon2* gene) was down regulated in mice fed the cod-scallop diet, suggesting lower oxidative stress in this group.

3. Atherosclerosis, Dyslipidemia and Inflammation

Atherosclerosis (originating from Greek: athero meaning gruel and sclerosis meaning hardness) is a complex, progressive and multifactorial inflammatory condition affecting the arteries. The arteries consist of three distinct layers: the outer layer, tunica adventitia, consists of flexible fibrous connective tissue, tunica media consists of smooth muscle cell tissue and elastic connective tissue, whereas the inner layer, tunica intima, consists of a membrane of collagen and glycoproteins lined by endothelial cells. The endothelial cells have a vast range of metabolic and regulatory functions, including transport of metabolic substances, regulation of vascular tone, defense against inflammation, angiogenesis and regulation of hemostasis and coagulation [26]. Disturbance of these regulatory processes, for instance by oxidative stress, is often the trigger for the onset of atherosclerosis. Under normal conditions, vasoactive substances are released from endothelial cells [27], but reduced bioavailability of these compounds, in combination with accumulated low density lipoprotein (LDL) could lead to activation of endothelial cells and subsequently a condition known as endothelial dysfunction [28]. Activation of endothelial cells leads to an inflammatory response involving the production of a cascade of chemokines, adhesion factors and integrins that are stimulated by transcription factors, such as nuclear factor kappa b (NFκB) [29]. These substances recruit monocytes to the endothelial surface, followed by adherence and transmigration into the intima. The influx of monocytes is often accompanied by influx of other inflammation cells, such as T-cells, dendritic cells and mast cells. Once placed in the intima, monocytes may differentiate into macrophages influenced by pro-inflammatory cytokines. Macrophages are phagocytic cells expressing scavenger receptors for uptake of modified LDL. The activated macrophages are programmed to protect our body against danger, and thus the normal processes for cholesterol handling and transport are impaired and accumulation of cholesteryl esters eventually leads to the formation of foam cells, and fatty streaks [30]. Continued inflammatory responses may further accelerate the atherosclerotic process. Stimulation of proliferation and migration of smooth muscle cells to the intima and release of intracellular contents (lipids, cholesterol) from both macrophages and smooth muscle cells, may build up a large plaque inside the intima. Protease secretion by macrophages degrade extracellular matrix, such as collagen, and a fibrous cap is formed around the excess lipids. Expression of collagen degrading enzymes can gradually weaken the fibrous cap leading to plaque rupture and release of intracellular content into the arteries, thrombus formation, and this may eventually result in myocardial infarction [31,32].

3.1. Inflammation

Very few studies documenting the effect of marine proteins on inflammation or parameters associated with inflammation are published (Table 1).

3.1.1. Human Studies

The effect of lean fish on inflammatory gene expression has been investigated in two published studies, one study evaluated the effect of lean fish in patients with coronary heart disease [33] and another study evaluated the effect in healthy subjects [34]. In the study with coronary heart disease patients, 27 subjects were divided into three groups eating either lean fish or fatty fish four portions a week for eight weeks. One group served as control and did not consume fish during the intervention period. No effect on the inflammatory gene expression was observed in this study [33]. In the clinical trial with healthy individuals, 71 subjects were divided into five groups eating 400 g cod per week for eight weeks or the same amount of smoked salmon or fresh salmon. One group maintained their regular diet, and another group maintained their regular diet only supplemented with 15 mL cod liver oil. No changes in the measured inflammatory parameters were observed. Ouellet et al. [35] investigated the effect of cod protein compared to other animal protein sources on C-reactive protein. For four weeks, 19 insulin resistant, overweight subjects participated in a crossover study and were given a diet with 60% of proteins as cod or other animal sources. After the four weeks, the subjects returned to their normal diet for two weeks, before they switched to the opposite diet. C-reactive protein was reduced by 24% in the cod group compared to an increase of 13% in the group eating other animal protein sources.

3.1.2. Animal Studies

In a study published by Jensen et al. [25], apolipoprotein E-deficient (apoeE−/−) mice were also used to evaluate the effect of dietary cod and scallop on atherosclerotic burden and related inflammatory parameters. Twenty-two five-week-old female apoeE−/− mice were fed Western type diets with chicken or cod and scallop as the protein sources for 13 weeks. After the study period the mice given cod-scallop as the protein source had a 24% lower atherosclerotic plaque compared to the mice eating the chicken feed. Additionally, the cod-scallop group had a 19% lower expression of the inflammatory gene vascular cell adhesion molecule 1. Dort et al. [36] investigated the effect of cod on the resolution of inflammation in 128 male Wistar rats. For three weeks, the rats had free access to feed with cod protein or casein protein. Thereafter the leg was injured with bupivacaine. The results showed that the inflammation due to the bupivacaine injection resoluted earlier in the cod group. At 14 and 24 days post damage, the amount of neutrophile granulocytes was significantly lower in the cod group compared to the casein group. In another study by the same group, it was confirmed that the anti-inflammatory effect of cod was due to the amino acids arginine, glycine and taurine [37].

3.2. Dyslipidemia

Lipids such as cholesterol and triglycerides are highly hydrophobic and have to be transported by lipoproteins in the blood stream. Both LDL and high density lipoproteins (HDL) are important parts of the regulation of the cholesterol homeostasis in the body; LDL delivers cholesterol from the liver to the various organs, whereas HDL is important for the reverse transport from the organs back to the liver. An imbalance between these two lipoproteins in favor of LDL will lead to accumulation of cholesterol in the vasculature and in tissues other than the liver [38] and a condition, named dyslipidemia occurs [39]. This condition is one of the most prominent risk factors for the development of atherosclerosis. Elevated plasma concentrations of triglycerides have in several prospective studies been shown to make up a considerable risk factor for atherosclerosis [40,41]. Lowering of LDL cholesterol, by medicinal treatment or by lifestyle/dietary changes, has been adapted as a means to reduce the risk of atherosclerosis. Increasing the level of HDL cholesterol is considered as a way of reducing the risk of atherosclerosis. In addition to its reverse cholesterol transport properties, HDL is also associated with vasodilation [42].

3.2.1. Human Studies

Several cross over studies have been conducted on healthy individuals comparing diets where lean seafood is the major protein source to diets with non-seafood, such as beef, chicken, eggs, and milk, as the major protein source (Table 1). A significant reduction in triglycerides was observed in some of the studies [43,44], whereas no significant difference between the diets was the conclusion in other studies [45,46]. In a study by Elvevoll et al. [47], 80 participants were given either a regular fish pate or a fish pate enriched with taurine. After the intervention period, subjects eating the taurine enriched fish pate experienced a reduction in cholesterol and LDL compared to the subjects eating fish pate without enrichment, suggesting an extensive beneficial effect of taurine. In other studies, the participants were selected based on being overweight, with a BMI over 27 (Table 1). In a double blind, randomized, placebo controlled study, 40 subjects were given supplements with fish protein or placebo over a period of eight weeks. No significant difference between the groups was observed for neither cholesterol, HDL nor triglycerides. LDL, however, was significantly reduced compared to baseline in the fish protein group [48]. Further the HDL/LDL ratio increased in the fish group during the intervention period. The effect of cod as the protein source in energy restricted diets for weight loss have been investigated in several studies [22,49,50]. In neither of the studies a significant effect on cholesterol was observed, whereas, in some studies, triglyceride levels were reduced compared to control diets [22,49]. In a crossover study by Ouellet et al. [35], cod was compared to other animal proteins in a four weeks crossover study with participants being overweight and diabetic (Table 1). In this study, the cholesterol and LDL was significantly reduced in the group eating the animal proteins compared to the group eating the cod protein. In addition, Erkkila et al. [51] conducted a clinical trial with patients with coronary heart disease (Table 1). For eight weeks the participants had either lean white fish or meat as protein source. No significant difference in blood lipids was found after the eight weeks. A similar study was repeated later with the same conclusion [52].

3.2.2. Animal Studies

The effects of marine proteins on blood lipids have been investigated in both mice and rat models. Cod and scallop were compared to chicken and casein as protein sources to assess their effect on blood lipids in a high fat diet and in a Western diet [25,53]. The level of triglycerides was reduced after seven weeks on the high fat diet with cod and scallop, whereas no effects on total cholesterol or HDL-cholesterol were observed. In the other study with the Western diet, the mice eating cod–scallop had lower LDL-cholesterol compared to those eating chicken feed. Liaset et al. [54] divided 15 rats into three groups and fed them saithe hydrolysate, soy or casein as protein source for almost four weeks. The plasma concentration of triglycerides was reduced in the saithe hydrolysate fed rats compared to soy and casein fed rats. In two studies, spontaneously hypertensive rats were given feed with 20% fish protein compared to 20% casein. The group eating fish protein had significantly reduced total cholesterol in both studies [24,55]. In the latter study, the triglyceride levels were reduced after the intervention period as well. The combined effect of cod protein and oil on triglyceride metabolism has been investigated in rats [56]. The rats were fed different protein sources and oils during four weeks. Cod protein alone did not affect the level of triglycerides, whereas together with menhaden oil, the cod protein reduced triglyceride levels by 50% compared to casein.

3.3. Coronary Heart Disease

Coronary heart disease is a collective term for heart attack and angina pectoris. An epidemiological study evaluated the association between increased seafood consumption and reduced risk for coronary heart disease.

Human Studies

Bernstein et al. [57] followed 84,136 30–55 year old women in the Nurses' health study (Table 1). The women in this study had no known cancer, diabetes mellitus, angina, myocardial infarction, stroke, or other vascular diseases, for 26 years. In a model which statistically controlled for energy intake, it was shown that one serving of fish per day was associated with a 30% reduction in risk for coronary heart disease compared with one serving of red meat.

4. Hypertension

The blood pressure is a measure of the heart's ability to pump blood and is presented as systolic above diastolic pressure. Systolic designates the pressure of the pumping heart and diastolic designates the pressure of the relaxed heart. A blood pressure of 120/80 mmHg is regarded normal and if one or both numbers are elevated, the heart's workload is increased and a condition called hypertension arises. This condition is one of the most important precursors for CVD, and is associated with heart failure, myocardial infarction and stroke [58], affecting almost one third of adults worldwide [59]. An increase of 20/10 mmHg above normal has been reported to double the risk of fatal CVD among people between 40 and 49 years [60].

The regulation of blood pressure is a complex process involving several mechanisms. Some of these, such as change of arteries diameter, regulation of blood volume in the blood stream and addition or removal of fluids in the blood stream, are purely mechanic, whereas others are more complex regulatory systems. One of these is the renin-angiotensin-aldosterone-system (RAAS). When blood flow or volume through the kidney decreases, the enzyme renin is excreted from the glomerulus. Renin cleaves angiotensinogen produced in the liver to form the decapeptide angiotensin I (Ang I). Angiotensin Converting Enzyme (ACE) produced mainly in the lungs further cleaves Ang I to the octapeptide Angiotensin II (Ang II) which constricts the arterial vessels and induces a rise of the blood pressure. In addition, it stimulates the adrenal cortex to produce aldosterone, which increases the reabsorption of sodium and water from the kidneys and further increases the blood pressure [61]. Another regulatory system is the kinin-kallicrein system (KKS). The ACE also participates in this system where it inactivates the vasodilator bradykinin [62]. Hence, inhibition of ACE will in both regulatory systems result in a prevention of blood pressure rising. In addition to being an independent risk factor for CVDs, high blood pressure is also recognized as a risk of atherosclerosis [63,64].

The effect of marine protein on blood pressure has been evaluated in several animal models, but limited data from epidemiological studies have suggested any association between fish intake and blood pressure.

4.1. Human Studies

The effect on blood pressure of lean fish as the protein source, has been evaluated and documented in two dietary intervention studies (Table 1). Erkkila et al. [51] randomized 33 medicated patients with coronary heart disease into three groups eating lean fish, fatty fish or lean meat as protein sources four times a week during eight weeks. After the intervention period, both systolic and diastolic blood pressure was reduced in the group eating lean fish. Ramel et al. [50] investigated the dose-response effect of number of cod meals per week. They randomized 126 healthy, overweight individuals into three groups, all following an energy restricted diet with either no cod, cod three times, or five times a week for eight weeks. The results from the blood pressure measurements were, however, inconsistent and therefore not reliable. Double-blind placebo controlled studies are generally regarded as a gold standard for evaluating the effect of different substances (Table 1). In one study, 34 overweight adults received supplementation of fish protein capsules or placebo tablets for eight weeks [48]. The intake of the supplement was 3 g per day the first four weeks and thereafter 6 g per day. No effect on blood pressure was observed. In another similar study, the effect of a salmon peptide on blood pressure was evaluated [65]. A number of 52 mild hypertensive individuals were divided into three groups drinking a beverage (50 mL/day) with 1 g, 0.3 g or no salmon peptide for four weeks. The systolic blood pressure was significantly reduced (140 to 135 mmHg) in the group receiving 1 g salmon peptide. Kawasaki et al. [66] evaluated the effect of a peptide administered to 29 individuals with high-normal blood pressure and mild essential hypertension. The subjects were randomized into two groups for a cross over placebo-controlled trial. The dipeptide drink significantly reduced the blood pressure in the dipeptide group, whereas no change was observed in the placebo group.

Results on blood pressure are not easily extrapolated between fish species, as taurine is known for its blood pressure reducing effect [67]. Taurine content varies greatly between fish species [68], however, compared to other foods, it is generally high in marine foods.

4.2. Animal Studies

While the documentation of blood pressure reducing effect in humans is scarce, several studies performed on animal are published. The, by far, commonest model for hypertension evaluation is using spontaneously hypertensive rats (SHR). These rats are bread to develop high blood pressure, and are well suited for monitoring through both acute and chronic studies. The majority of the studies published on the effect of marine protein on blood pressure, are acute studies. The SHR are given marine hydrolysates or peptides orally and the blood pressure has been measured before administration, just after administration up to several hours after administration. The first studies documenting the antihypertensive effect of bonito in SHR were published already in the 1990s [69–71]. Later, single oral doses of 10 mg/kg body weight of tuna hydrolysate [72,73] and yellow fin sole hydrolysate [74] have been shown to significantly reduce blood pressure. A blood pressure reducing effect has been documented in hydrolysates from shrimp [75], oyster [76], loach [77], sea cucumber [78], sardine [66], jellyfish [79], salmon [65,80], cobia [81] and skate [82]. In these studies, the test doses vary, making any comparison difficult. Nevertheless, the results may give an indication that marine protein in general is potential as a blood pressure reducing nutraceutical, food ingredient or food.

Some studies have also evaluated the chronic effect of marine hydrolysates on blood pressure. In such studies, the SHR are given the test items daily. Negative control is normally water or saline, while the positive control commonly is captopril. Jellyfish hydrolysate [79], sardine peptide [83] and sea bream hydrolysate [84] have all been tested in chronic studies. SHR have been administered daily over a period of four weeks with the hydrolysates in different dosages, resulting in significantly lowered blood pressure, even comparable to that of captopril. Hydrolysates of cod, haddock and salmon did not significantly reduce blood pressure in SHR during a four-week study [85], although the blood pressure in the group treated with cod hydrolysate did not increase after day 7. Fish has also been evaluated as part of the feed itself. Spontaneously hypertensive rats fed a standard chow supplemented with tuna hydrolysate, Katsuo-bushi, for seven weeks, experienced reduced blood pressure [71]. In three studies lasting for two months, SHR were fed standard animal chow where 20% of the feed was either fish protein or casein protein [24,55,86]. The blood pressure in the SHR eating fish protein was significantly reduced compared to that in those eating the casein protein. However, when this was investigated in rats with diabetes, no effect on blood pressure was observed [24].

Table 1. Clinical trials investigating cardio protective of marine proteins and peptides.

Parameter	Study	Subjects, Inclusion Criteria	Protein Source	Result	Year	References
Oxidative stress	8 weeks, randomized parallel intervention	276 (4 groups), overweight, healthy	Cod, salmon, fish oil, control	Oxidation product reduced, AOC increased in cod group	2007	[22]
Blood pressure	8 weeks, double blind, randomized, controlled intervention	34 (2 groups), overweight	Fish protein capsules, placebo	No significant effect	2013	[48]
	8 weeks controlled, parallel dietary intervention	126 (3 groups), overweight	150 g cod 1/week, 150 g cod 3/week, no cod	No results	2009	[50]
	8 weeks controlled, parallel intervention	31 (3 groups), myocardial infarction	Lean fish, fatty fish, no fish	Blood pressure reduced in lean fish group	2008	[51]
	4 weeks double blind, placebo-controlled	52 (3 groups), mild hypertension	Salmon peptide, placebo	Systolic blood pressure reduced in peptide group	2008	[65]
	4 weeks randomized, double blind, placebo-controlled	29 (2 groups), high-normal blood pressure and mild essential hypertension	Sardine peptide	Blood pressure reduced in peptide group	2000	[66]
Inflammation	8 weeks, randomized, parallel dietary intervention	27 (3 groups) coronary heart disease	Lean fish, fatty fish, no fish	No significant effect	2009	[33]
	2 × 4 weeks crossover design	19 overweight/obesity insulin-resistance	Cod, other animal protein sources	24% reduction in plasma CRP	2008	[35]
Blood lipids	2 × 4 weeks, randomize, crossover design	20 healthy	Lean seafood, non-seafood	Reduced TG in lean seafood-group	2015	[43]
	4 weeks prospective, randomized crossover design	10 healthy	Lean seafood, beef diet	Reduced TG, cholesterol and VLDL	2009	[44]
	2 × 4 weeks randomized crossover design	11 healthy men	Lean fish, non-fish	No significant effect	2000	[46]
	2 × 4 weeks crossover design	14 healthy premenopausal women	Fish, non-fish	No significant effect	1996	[45]
	7 weeks dietary intervention	80 (2 groups) healthy	Fish pate, fish pate with taurine	Reduced cholesterol and LDL with taurine	2008	[47]
	8 weeks, double blind, randomized, controlled intervention	34 (2 groups) overweight	Fish protein capsules, placebo	Reduced LDL in fish group compared to baseline	2013	[48]
	8 weeks, randomized parallel dietary intervention	276 (4 groups) overweight, healthy	Cod, salmon, fish oil, control	Reduced TG in cod-group	2007	[22]
	8 weeks, randomized, parallel dietary intervention	324 (4 groups), overweight	Lean fish, oily fish, control, fish oil	Reduced TG	2008	[49]
	8 weeks controlled, parallel dietary intervention	126 (3 groups) overweight	150 g cod 1/week, 150 g cod 3/week, no cod	No results		[50]
	2 × 4 weeks crossover design	19 overweight/obese insulin-resistant subjects	Cod, other animal protein sources	Reduced cholesterol and LDL	2008	[35]
	8 weeks controlled, parallel dietary intervention	31 (3 groups) subjects with myocardial infarction	Lean fish, fatty fish, no fish	No significant effect	2008, 2014	[51,52]
Coronary heart disease	Epidemiological study, 26 years	Healthy women aged 30-55	Fish	Reduced risk for coronary heart disease		[57]

AOC, antioxidative capacity; CRP, C-reactive protein; TG, triglycerides; VLDL, very low density lipoprotein; LDL, low density lipoprotein.

5. Conclusions

Focus on health benefits from marine resources has traditionally been on the long chain omega-3 fatty acids. However, emerging evidence points out that other nutrients, such as peptides and proteins also play a major role. The current review sums up preclinical and clinical trials on the cardioprotective effects of marine protein and peptides. Clinical studies on humans are the superior method for evaluation of health effects, but also the most expensive, time consuming and complex way. The number of studies is thus quite low, but there are indications that marine proteins may have a positive effect on oxidative stress. Studies on inflammation parameters, blood lipid and hypertension are inconclusive. Further, as inclusion criteria for participants in each study vary greatly, depending on weight, gender, age and health status, conclusions from the different studies are difficult to draw and the clinical relevance is therefore limited. The number of animal studies published is larger and, particularly, the effects of marine proteins on hypertension are well documented. However, documentation of the effect on atherosclerosis and inflammation is scarce and further research on this field is also acquired. It is therefore of utmost importance to include more research from both animal, and most importantly, human studies on cardiovascular health effects of marine proteins and peptides.

Acknowledgments: This work was supported by the Publication Fund of UIT The Arctic University of Norway.

Author Contributions: The authors contributed equally in the writing of this manuscript.

Conflicts of Interest: The authors declare no conflict of interest.

References

1. Mendis, S.; Puska, P.; Norrving, B. *Global Atlas on Cardiovascular Disease Prevention and Control*; WHO: Gevena, Switzerland, 2011.
2. Cardiovascular Diseases. Available online: http://www.who.int/mediacentre/factsheets/fs317/en/ (accessed on 28 September 2016).
3. Yusuf, H.R.; Giles, W.H.; Croft, J.B.; Anda, R.F.; Casper, M.L. Impact of multiple risk factor profiles on determining cardiovascular disease risk. *Prev. Med.* **1998**, *27*, 1–9. [CrossRef] [PubMed]
4. Kris-Etherton, P.M.; Harris, W.S.; Appel, L.J.; American Heart Association Nutrition Committee. Fish consumption, fish oil, omega-3 fatty acids, and cardiovascular disease. *Circulation* **2002**, *106*, 2747–2757. [CrossRef] [PubMed]
5. De Leiris, J.; de Lorgeril, M.; Boucher, F. Fish oil and heart health. *J. Cardiovasc. Pharmacol.* **2009**, *54*, 378–384. [CrossRef] [PubMed]
6. He, K.; Song, Y.; Daviglus, M.L.; Liu, K.; Horn, L.V.; Dyer, A.R.; Greenland, P. Accumulated evidence on fish consumption and coronary heart disease mortality: A meta-analysis of cohort studies. *Circulation* **2004**, *109*, 2705–2711. [CrossRef] [PubMed]
7. Marik, P.E.; Varon, J. Omega-3 dietary supplements and the risk of cardiovascular events: A systematic review. *Clin. Cardiol.* **2009**, *32*, 365–372. [CrossRef] [PubMed]
8. Mozaffarian, D.; Rimm, E.B. Fish intake, contaminants, and human health: Evaluating the risks and the benefits. *JAMA* **2006**, *296*, 1885–1899. [CrossRef] [PubMed]
9. Saremi, A.; Arora, R. The utility of omega-3 fatty acids in cardiovascular disease. *Am. J. Ther.* **2009**, *16*, 421–436. [CrossRef] [PubMed]
10. Zheng, J.; Huang, T.; Yu, Y.; Hu, X.; Yang, B.; Li, D. Fish consumption and CHD mortality: An updated meta-analysis of seventeen cohort studies. *Public Health Nutr.* **2012**, *15*, 725–737. [CrossRef] [PubMed]
11. Lakshmi, S.V.V.; Padmaja, G.; Kuppusamy, P.; Kutala, V.K. Oxidative stress in cardiovascular disease. *Indian J. Biochem. Biophys.* **2009**, *46*, 421–440. [PubMed]
12. Bonomini, F.; Tengattini, S.; Fabiano, A.; Bianchi, R.; Rezzani, R. Atherosclerosis and oxidative stress. *Histol. Histopatol.* **2008**, *23*, 381–390.
13. Elias, R.J.; Kellerby, S.S.; Decker, E.A. Antioxidant activity of proteins and peptides. *Crit. Rev. Food Sci. Nutr.* **2008**, *48*, 430–441. [CrossRef] [PubMed]
14. Fang, Y.Z.; Yang, S.; Wu, G.Y. Free radicals, antioxidants, and nutrition. *Nutrition* **2002**, *18*, 872–879. [CrossRef]

15. Guiotto, A.; Calderan, A.; Ruzza, P.; Borin, G. Carnosine and carnosine-related antioxidants: A review. *Curr. Med. Chem.* **2005**, *12*, 2293–2315. [CrossRef] [PubMed]
16. Seth, A.; Mahoney, R.R. Iron chelation by digests of insoluble chicken muscle protein: The role of histidine residues. *J. Sci. Food Agric.* **2001**, *81*, 183–187. [CrossRef]
17. Garner, B.; Witting, P.K.; Waldeck, A.R.; Christison, J.K.; Raftery, M.; Stocker, P. Oxidation of high density lipoproteins 1. Formation of methionine sulfoxide in apolipoproteins AI and AII is an early event that accompanies lipid peroxidation and can be enhanced by alpha-tocopherol. *J. Biol. Chem.* **1998**, *273*, 6080–6087. [CrossRef] [PubMed]
18. Pryor, W.A.; Jin, X.; Squadrito, G.L. One-electron and 2-electron oxidations of methionine by peroxynitrite. *Proc. Natl. Acad. Sci. USA* **1994**, *91*, 11173–11177. [CrossRef] [PubMed]
19. Qureshi, S.A.; Lund, A.C.; Veierød, M.B.; Carlsen, M.H.; Blomhoff, R.; Andersen, L.F.; Ursin, G. Food items contributing most to variation in antioxidant intake; A cross-sectional study among Norwegian women. *BMC Public Health* **2014**. [CrossRef] [PubMed]
20. Bjelakovic, G.; Nikolova, D.; Gluud, L.L.; Simonett, R.G.; Gludd, C. Antioxidant supplements for prevention of mortality in healthy participants and patients with various diseases. *Cochrane Database Syst. Rev.* **2012**. [CrossRef]
21. Bjelakovic, G.; Nikolova, D.; Gluud, C. Meta-regression analyses, meta-analyses, and trial sequential analyses of the effects of supplementation with beta-carotene, vitamin A, and vitamin E singly or in different combinations on all-cause mortality: Do we have evidence for lack of harm? *PLoS ONE* **2013**, *8*, e74558. [CrossRef] [PubMed]
22. Parra, D.; Bandarra, N.M.; Kiely, M.; Thorsdottir, I.; Martinez, J.A. Impact of fish intake on oxidative stress when included into a moderate energy-restricted program to treat obesity. *Eur. J. Nutr.* **2007**, *46*, 460–467. [CrossRef] [PubMed]
23. Yahia, D.A.; Madani, S.; Prost, E.; Prost, J.; Bouchenak, M.; Belleville, J. Tissue antioxidant status differs in spontaneously hypertensive rats fed fish protein or casein. *J. Nutr.* **2003**, *133*, 479–482. [PubMed]
24. Boukortt, F.O.; Girard, A.; Prost, J.L.; Ait-Yahia, D.; Bouchenak, M.; Belleville, J. Fish protein improves the total antioxidant status of streptozotocin-induced diabetes in spontaneously hypertensive rat. *Med. Sci. Monit.* **2004**, *10*, 397–404.
25. Jensen, I.J.; Walquist, M.; Liaset, B.; Elvevoll, E.O.; Eilertsen, K.E. Dietary intake of cod and scallop reduces atherosclerotic burden in female apolipoprotein E-deficient mice fed a Western-type high fat diet for 13 weeks. *Nutr. Metab.* **2016**, *13*. [CrossRef] [PubMed]
26. Galley, H.F.; Webster, N.R. Physiology of the endothelium. *Br. J. Anaesth.* **2004**, *93*, 105–113. [CrossRef] [PubMed]
27. Deanfield, J.E.; Halcox, J.P.; Rabelink, T.J. Endothelial function and dysfunction: testing and clinical relevance. *Circulation* **2007**, *115*, 1285–1295. [PubMed]
28. Bonetti, P.O.; Lerman, L.O.; Lerman, A. Endothelial dysfunction: A marker of atherosclerotic risk. *Arterioscler. Thromb. Vasc. Biol.* **2003**, *23*, 168–175. [CrossRef] [PubMed]
29. Sprague, A.H.; Khalil, R.A. Inflammatory cytokines in vascular dysfunction and vascular disease. *Biochem. Pharmacol.* **2009**, *78*, 539–552. [CrossRef] [PubMed]
30. McLaren, J.E.; Michael, D.R.; Ashlin, T.G.; Ramji, D.P. Cytokines, macrophage lipid metabolism and foam cells: implications for cardiovascular disease therapy. *J. Am. Heart Assoc.* **2011**, *50*, 331–347. [CrossRef] [PubMed]
31. Szmitko, P.E.; Wang, C.H.; Weisel, R.D.; de Almeida, J.R.; Anderson, T.J.; Verma, S. New markers of inflammation and endothelial cell activation: Part I. *Circulation* **2003**, *108*, 1917–1923. [CrossRef] [PubMed]
32. Szmitko, P.E.; Wang, C.H.; Weisel, R.D.; Jeffries, G.A.; Anderson, T.J.; Verma, S. Biomarkers of vascular disease linking inflammation to endothelial activation: Part II. *Circulation* **2003**, *108*, 2041–2048. [CrossRef] [PubMed]
33. De Mello, V.D.; Erkkila, A.T.; Schwab, U.S.; Pulkkinen, L.; Kolehmainen, M.; Atalay, M.; Mussalo, H.; Lankinen, M.; Oresic, M.; Lehto, S.; et al. The effect of fatty or lean fish intake on inflammatory gene expression in peripheral blood mononuclear cells of patients with coronary heart disease. *Eur. J. Nutr.* **2009**, *48*, 447–455. [CrossRef] [PubMed]

34. Elvevoll, E.O.; Barstad, H.; Breimo, E.S.; Brox, J.; Eilertsen, K.E.; Lund, T.; Olsen, J.O.; Osterud, B. Enhanced incorporation of n-3 fatty acids from fish compared with fish oils. *Lipids* **2006**, *41*, 1109–1114. [CrossRef] [PubMed]

35. Ouellet, V.; Weisnagel, S.J.; Marois, J.; Bergeron, J.; Julien, P.; Gougeon, R.; Tchernof, A.; Holub, B.J.; Jacques, H. Dietary cod protein reduces plasma C-reactive protein in insulin-resistant men and women. *J. Nutr.* **2008**, *138*, 2386–2391. [CrossRef] [PubMed]

36. Dort, J.; Sirois, A.; Leblanc, N.; Cote, C.H.; Jacques, H. Beneficial effects of cod protein on skeletal muscle repair following injury. *Appl. Physiol. Nutr. Metab.* **2012**, *37*, 489–498. [CrossRef] [PubMed]

37. Dort, J.; Leblanc, N.; Maltais-Giguere, J.; Liaset, B.; Cote, C.H.; Jacques, H. Beneficial effects of cod protein on inflammatory cell accumulation in rat skeletal muscle after injury are driven by its high levels of arginine, glycine, taurine and lysine. *PLoS ONE* **2013**, *8*, e77274. [CrossRef] [PubMed]

38. Babiak, J.; Rudel, L.L. Lipoproteins and atherosclerosis. *Baillieres Clin. Endocrinol. Metab.* **1987**, *1*, 515–550. [CrossRef]

39. Angelico, F.; Baratta, F.; Ben, M.D. Current ways of treating dyslipidemias to prevent atherosclerosis. *Ther. Apher. Dial.* **2013**, *17*, 125–129. [CrossRef] [PubMed]

40. Hokanson, J.E.; Austin, M.A. Plasma triglyceride level is a risk factor for cardiovascular disease independent of high-density lipoprotein cholesterol level: A meta-analysis of population-based prospective studies. *J. Cardiovasc. Risk* **1996**, *3*, 213–219. [CrossRef] [PubMed]

41. Nordestgaard, B.G.; Benn, M.; Schnohr, P.; Tybjaerg-Hansen, A. Nonfasting triglycerides and risk of myocardial infarction, ischemic heart disease, and death in men and women. *JAMA* **2007**, *298*, 299–308. [CrossRef] [PubMed]

42. Mineo, C.; Yuhanna, I.S.; Quon, M.J.; Shaul, P.W. High density lipoprotein-induced endothelial nitric-oxide synthase activation is mediated by Akt and MAP kinases. *J. Biol. Chem.* **2003**, *278*, 9142–9149. [CrossRef] [PubMed]

43. Aadland, E.K.; Lavigne, C.; Graff, I.E.; Eng, O.; Paquette, M.; Holthe, A.; Mellgren, G.; Jacques, H.; Liaset, B. Lean-seafood intake reduces cardiovascular lipid risk factors in healthy subjects: Results from a randomized controlled trial with a crossover design. *Am. J. Clin. Nutr.* **2015**, *102*, 582–592. [CrossRef] [PubMed]

44. Leaf, D.A.; Hatcher, L. The effect of lean fish consumption on triglyceride levels. *Phys. Sportsmed.* **2009**, *37*, 37–43. [CrossRef] [PubMed]

45. Gascon, A.; Jacques, H.; Moorjani, S.; Deshaies, Y.; Brun, L.D.; Julien, P. Plasma lipoprotein profile and lipolytic activities in response to the substitution of lean white fish for other animal protein sources in premenopausal women. *Am. J. Clin. Nutr.* **1996**, *63*, 315–321. [PubMed]

46. Lacaille, B.; Julien, P.; Deshaies, Y.; Lavigne, C.; Brun, L.D.; Jacques, H. Responses of plasma lipoproteins and sex hormones to the consumption of lean fish incorporated in a prudent-type diet in normolipidemic men. *J. Am. Coll. Nutr.* **2000**, *19*, 745–753. [CrossRef] [PubMed]

47. Elvevoll, E.O.; Eilertsen, K.E.; Brox, J.; Dragnes, B.T.; Falkenberg, P.; Olsen, J.O.; Kirkhus, B.; Lamglait, A.; Osterud, B. Seafood diets: Hypolipidemic and antiatherogenic effects of taurine and n-3 fatty acids. *Atherosclerosis* **2008**, *200*, 396–402. [CrossRef] [PubMed]

48. Vikoren, L.A.; Nygard, O.K.; Lied, E.; Rostrup, E.; Gudbrandsen, O.A. A randomised study on the effects of fish protein supplement on glucose tolerance, lipids and body composition in overweight adults. *Br. J. Nutr.* **2013**, *109*, 648–657. [CrossRef] [PubMed]

49. Gunnarsdottir, I.; Tomasson, H.; Kiely, M.; Martinez, J.A.; Bandarra, N.M.; Morais, M.G.; Thorsdottir, I. Inclusion of fish or fish oil in weight-loss diets for young adults: Effects on blood lipids. *Int. J. Obes.* **2008**, *32*, 1105–1112. [CrossRef] [PubMed]

50. Ramel, A.; Jonsdottir, M.T.; Thorsdottir, I. Consumption of cod and weight loss in young overweight and obese adults on an energy reduced diet for 8-weeks. *Nutr. Metab. Cardiovasc. Dis.* **2009**, *19*, 690–696. [CrossRef] [PubMed]

51. Erkkila, A.T.; Schwab, U.S.; de Mello, V.D.F.; Lappalainen, T.; Mussalo, H.; Lehto, S.; Kemi, V.; Lamberg-Allardt, C.; Uusitupa, M.I.J. Effects of fatty and lean fish intake on blood pressure in subjects with coronary heart disease using multiple medications. *Eur. J. Nutr.* **2008**, *47*, 319–328. [CrossRef] [PubMed]

52. Erkkila, A.T.; Schwab, U.S.; Lehto, S.; de Mello, V.D.; Kangas, A.J.; Soininen, P.; Ala-Korpela, M.; Uusitupa, M.I. Effect of fatty and lean fish intake on lipoprotein subclasses in subjects with coronary heart disease: A controlled trial. *J. Clin. Lipidol.* **2014**, *8*, 126–133. [CrossRef] [PubMed]

53. Tastesen, H.S.; Ronnevik, A.K.; Borkowski, K.; Madsen, L.; Kristiansen, K.; Liaset, B. A mixture of cod and scallop protein reduces adiposity and improves glucose tolerance in high-fat fed male C57BL/6 J mice. *PLoS ONE* **2014**, *9*, e112859. [CrossRef] [PubMed]

54. Liaset, B.; Madsen, L.; Hao, Q.; Criales, G.; Mellgren, G.; Marschall, H.U.; Hallenborg, P.; Espe, M.; Froyland, L.; Kristiansen, K. Fish protein hydrolysate elevates plasma bile acids and reduces visceral adipose tissue mass in rats. *Biochim. Biophys. Acta* **2009**, *1791*, 254–262. [CrossRef] [PubMed]

55. Yahia, D.A.; Madani, S.; Prost, J.; Bouchenak, M.; Belleville, J. Fish protein improves blood pressure but alters HDL2 and HDL3 composition and tissue lipoprotein lipase activities in spontaneously hypertensive rats. *Eur. J. Nutr.* **2005**, *44*, 10–17. [CrossRef] [PubMed]

56. Demonty, I.; Deshaies, Y.; Lamarche, B.; Jacques, H. Cod protein lowers the hepatic triglyceride secretion rate in rats. *J. Nutr.* **2003**, *133*, 1398–1402. [PubMed]

57. Bernstein, A.M.; Sun, Q.; Hu, F.B.; Stampfer, M.J.; Manson, J.E.; Willett, W.C. Major Dietary Protein Sources and Risk of Coronary Heart Disease in Women. *Circulation* **2010**, *122*, 876–883. [CrossRef] [PubMed]

58. Harris, T.; Cook, E.F.; Kannel, W.; Schatzkin, A.; Goldman, L. Blood pressure experience and risk of cardiovascular disease in the elderly. *Hypertension* **1985**, *7*, 118–124. [CrossRef] [PubMed]

59. Saleh, A.S.; Zhang, Q.; Shen, Q. Recent Research in Antihypertensive Activity of Food Protein-derived Hydrolyzates and Peptides. *Crit. Rev. Food Sci. Nutr.* **2016**, *56*, 760–787. [CrossRef] [PubMed]

60. Lewington, S.; Clarke, R.; Qizilbash, N.; Peto, R.; Collins, R. Age-specific relevance of usual blood pressure to vascular mortality: A meta-analysis of individual data for one million adults in 61 prospective studies. *Lancet* **2002**, *360*, 1903–1913. [CrossRef]

61. Goodfriend, T.L.; Elliott, M.E.; Catt, K.J. Angiotensin receptors and their antagonists. *N. Engl. J. Med.* **1996**, *334*, 1649–1654. [PubMed]

62. Witherow, F.N.; Helmy, A.; Webb, D.J.; Fox, K.A.; Newby, D.E. Bradykinin contributes to the vasodilator effects of chronic angiotensin-converting enzyme inhibition in patients with heart failure. *Circulation* **2001**, *104*, 2177–2181. [CrossRef] [PubMed]

63. Graninger, M.; Reiter, R.; Drucker, C.; Minar, E.; Jilma, B. Angiotensin receptor blockade decreases markers of vascular inflammation. *J. Cardiovasc. Pharmacol.* **2004**, *44*, 335–339. [CrossRef] [PubMed]

64. McGraw, A.P.; Bagley, J.; Chen, W.S.; Galayda, C.; Nickerson, H.; Armani, A.; Caprio, M.; Carmeliet, P.; Jaffe, I.Z. Aldosterone increases early atherosclerosis and promotes plaque inflammation through a placental growth factor-dependent mechanism. *J. Am. Heart Assoc.* **2013**. [CrossRef] [PubMed]

65. Enari, H.; Takahashi, Y.; Kawarasaki, M.; Tada, M.; Tatsuta, K. Identification of angiotensin I-converting enzyme inhibitory peptides derived from salmon muscle and their antihypertensive effect. *Fish. Sci.* **2008**, *74*, 911–920. [CrossRef]

66. Kawasaki, T.; Seki, E.; Osajima, K.; Yoshida, M.; Asada, K.; Matsui, T.; Osajima, Y. Antihypertensive effect of valyl-tyrosine, a short chain peptide derived from sardine muscle hydrolyzate, on mild hypertensive subjects. *J. Hum. Hypertens.* **2000**, *14*, 519–523. [CrossRef] [PubMed]

67. Sun, Q.; Wang, B.; Li, Y.; Sun, F.; Li, P.; Xia, W.; Zhou, X.; Li, Q.; Wang, X.; Chen, J.; et al. Taurine Supplementation Lowers Blood Pressure and Improves Vascular Function in Prehypertension: Randomized, Double-Blind, Placebo-Controlled Study. *Hypertension* **2016**, *67*, 541–549. [CrossRef] [PubMed]

68. Dragnes, B.T.; Larsen, R.; Ernstsen, M.H.; Mæhre, H.K.; Elvevoll, E.O. Impact of processing on the taurine content in processed seafood and their corresponding unprocessed raw materials. *Int. J. Food Sci. Nutr.* **2008**, *12*, 1–10. [CrossRef] [PubMed]

69. Fujii, M.; Matsumura, N.; Mito, K.; Shimizu, T.; Kuwahara, M.; Sugano, S.; Karaki, H. Antihypertensive effects of peptides in autolysate of bonito bowels on spontaneously hypertensive rats. *Biosci. Biotechnol. Biochem.* **1993**, *57*, 2186–2188. [CrossRef] [PubMed]

70. Karaki, H.; Kuwahara, M.; Sugano, S.; Doi, C.; Doi, K.; Matsumura, N.; Shimizu, T. Oral administration of peptides derived from bonito bowels decreases blood pressure in spontaneously hypertensive rats by inhibiting angiotensin converting enzyme. *Comp. Biochem. Physiol. C* **1993**, *104*, 351–353. [PubMed]

71. Fujita, H.; Yokoyama, K.; Yasumoto, R.; Yoshikawa, M. Antihypertensive effect of thermolysin digest of dried bonito in spontaneously hypertensive rat. *Clin. Exp. Pharmacol. Physiol. Suppl.* **1995**, *22*, 304–305. [CrossRef]

72. Lee, S.H.; Qian, Z.J.; Kim, S.W. A novel angiotensin I converting enzyme inhibitory peptide from tuna frame protein hydrolysate and its antihypertensive effect in spontaneously hypertensive rats. *Food Chem.* **2010**, *118*, 96–102. [CrossRef]

73. Qian, Z.J.; Je, J.Y.; Kim, S.K. Antihypertensive effect of angiotensin I converting enzyme-inhibitory peptide from hydrolysates of Bigeye tuna dark muscle, Thunnus obesus. *J. Agric. Food Chem.* **2007**, *55*, 8398–8403. [CrossRef] [PubMed]

74. Jung, W.K.; Mendis, E.; Je, J.Y.; Park, P.J.; Son, B.W.; Kim, H.C.; Choi, J.K.; Kim, S.K. Angiotensin I-converting enzyme inhibitory peptide from yellowfin sole (Limanda aspera) frame protein and its antihypertensive effect in spontaneously hypertensive rats. *Food Chem.* **2006**, *94*, 26–32. [CrossRef]

75. Cao, W.; Zhang, C.; Hong, P.; Ji, H.; Hao, J. Purification and identification of an ACE inhibitory peptide from the peptic hydrolysate of Acetes chinensis and its antihypertensive effects in spontaneously hypertensive rats. *Int. J. Food Sci. Technol.* **2010**, *45*, 959–965. [CrossRef]

76. Xie, C.L.; Kim, J.S.; Ha, J.M.; Choung, S.Y.; Choi, Y.J. Angiotensin I-converting enzyme inhibitor derived from cross-linked oyster protein. *Biomed. Res. Int.* **2014**. [CrossRef] [PubMed]

77. Li, Y.; Zhou, J.; Huang, K.; Sun, Y.; Zeng, X. Purification of a novel angiotensin I-converting enzyme (ACE) inhibitory peptide with an antihypertensive effect from loach (*Misgurnus anguillicaudatus*). *J. Agric. Food Chem.* **2012**, *60*, 1320–1325. [CrossRef] [PubMed]

78. Zhao, Y.; Li, B.; Dong, S.; Liu, Z.; Zhao, X.; Wang, J.; Zeng, M. A novel ACE inhibitory peptide isolated from *Acaudina molpadioidea* hydrolysate. *Peptides* **2009**, *30*, 1028–1033. [CrossRef] [PubMed]

79. Liu, X.; Zhang, M.; Zhang, C.; Liu, C. Angiotensin converting enzyme (ACE) inhibitory, antihypertensive and antihyperlipidaemic activities of protein hydrolysates from *Rholipema esculentum*. *Food Chem.* **2012**, *134*, 2134–2140. [CrossRef] [PubMed]

80. Ono, S.; Hosokawa, M.; Miyashita, K.; Takahashi, K. Isolation of Peptides with Angiotensin I-converting Enzyme Inhibitory Effect Derived from Hydrolysate of Upstream Chum Salmon Muscle. *J. Food Sci.* **2003**, *68*, 1611–1614. [CrossRef]

81. Yang, P.; Jiang, Y.; Hong, P.; Cao, W. Angiotensin I converting enzyme inhibitory activity and antihypertensive effect in spontaneously hypertensive rats of cobia (*Rachycentron canadum*) head papain hydrolysate. *Food Sci. Technol. Int.* **2013**, *19*, 209–215. [CrossRef] [PubMed]

82. Ngo, D.H.; Kang, K.H.; Ryu, B.; Vo, T.S.; Jung, W.K.; Byun, H.G.; Kim, S.K. Angiotensin-I converting enzyme inhibitory peptides from antihypertensive skate (*Okamejei kenojei*) skin gelatin hydrolysate in spontaneously hypertensive rats. *Food Chem.* **2015**, *174*, 37–43. [CrossRef] [PubMed]

83. Otani, L.; Ninomiya, T.; Murakami, M.; Osajima, K.; Kato, H.; Murakami, T. Sardine peptide with angiotensin I-converting enzyme inhibitory activity improves glucose tolerance in stroke-prone spontaneously hypertensive rats. *Biosci. Biotechnol. Biochem.* **2009**, *73*, 2203–2209. [CrossRef] [PubMed]

84. Fahmi, A.; Morimura, S.; Guo, H.C.; Shigematsu, T.; Kida, K.; Uemura, Y. Production of angiotensin I converting enzyme inhibitory peptides from sea bream scales. *Process. Biochem.* **2004**, *39*, 1195–1200. [CrossRef]

85. Jensen, I.J.; Eysturskareth, J.; Madetoja, M.; Eilertsen, K.E. The potential of cod hydrolyzate to inhibit blood pressure in spontaneously hypertensive rats. *Nutr. Res.* **2014**, *34*, 168–173. [CrossRef] [PubMed]

86. Ait-Yahia, D.; Madani, S.; Savelli, J.L.; Prost, J.; Bouchenak, M.; Belleville, J. Dietary fish protein lowers blood pressure and alters tissue polyunsaturated fatty acid composition in spontaneously hypertensive rats. *Nutrition* **2003**, *19*, 342–346. [CrossRef]

marine drugs

Review

Alkynyl-Containing Peptides of Marine Origin: A Review

Qiu-Ye Chai [1,2,†], Zhen Yang [3,†], Hou-Wen Lin [1,*] and Bing-Nan Han [1,*]

[1] Research Center for Marine Drugs, Department of Pharmacy, State Key Laboratory of Oncogenes and Related Genes, Renji Hospital, School of Medicine, Shanghai Jiao Tong University, Shanghai 200127, China; chaiqiuye123@sina.cn

[2] School of Pharmacy, Jiangxi University of Traditional Chinese Medicine, Nanchang 330000, China

[3] Department of Pharmacy, Graduate School, Hunan University of Chinese Medicine, Changsha 410208, China; zyyangzhen1991@163.com

* Correspondence: franklin67@126.com (H.-W.L.); hanbingnan@shsmu.edu.cn (B.-N.H.); Tel.: +86-21-6838-3346 (B.-N.H.); Fax: +86-21-5873-2594 (B.-N.H.)

† These authors contributed equally to this work.

Academic Editor: Se-Kwon Kim

Received: 19 September 2016; Accepted: 16 November 2016; Published: 23 November 2016

Abstract: Since the 1990s, a number of terminal alkynyl residue-containing cyclic/acyclic peptides have been identified from marine organisms, especially cyanobacteria and marine mollusks. This review has presented 66 peptides, which covers over 90% marine peptides with terminal alkynyl fatty acyl units. In fact, more than 90% of these peptides described in the literature are of cyanobacterial origin. Interestingly, all the linear peptides featured with terminal alkyne were solely discovered from marine cyanobacteria. The objective of this article is to provide an overview on the types, structural characterization of these unusual terminal alkynyl fatty acyl units, as well as the sources and biological functions of their composed peptides. Many of these peptides have a variety of biological activities, including antitumor, antibacterial, antimalarial, etc. Further, we have also discussed the evident biosynthetic origin responsible for formation of terminal alkynes of natural PKS (polyketide synthase)/NRPS (nonribosome peptide synthetase) hybrids.

Keywords: marine cyanobacteria; mollusk; alkynyl peptides; biological activity; absolute configuration

1. Introduction

As oceans comprise over 70% of the earth's surface and harbor a tremendous variety of flora and fauna, marine habitat represents a rich source of diverse chemical structures and biological activities of natural products [1], which include alkaloids, terpenoids, peptides, polyketides, steroids, etc. Peptides as an important bioactive natural product, present in many marine species, including sponges, ascidians, seaweeds, mollusks, and marine microorganisms, have been extensively studied [2,3]. Interestingly, diverse structural classes of peptides such as linear peptides, linear depsipeptides, linear lipopeptides, cyclic peptides, cyclic depsipeptides, and cyclic lipopeptides have been discovered from all of these marine species. The broad bioactivity spectrum of marine peptides has high medicinal potential which attracts the attention of the pharmaceutical industry. Since the discovery of the first marine-derived antitumorcyclic peptide, ulithiacyclamide, many marine anticancerpeptides have entered into clinical trials with good prospects for drug development [4–6], such as kahalalide F, hemiasterlin, dolastatins, cemadotin, soblidotin, didemnins, aplidine, etc. [7]. Cyclic peptides as a valuable lead for drug discovery with better resistance to enzymatic degradation and higher bioavailability in vivo have attracted considerable attention for further study in the areas of marine natural products [4,8]. Acyclic peptides with the prospect of pharmacological activity are also promising, such as the well-known anticancer lead dolastatin 10 isolated from both sea hare *Dollabella auricularia* [9] and its diet of marine cyanobacterium, the *Symploca* species [10], whose synthetic derivatives have been used in clinical phase III trials [7]. In recent years, a number of structurally intriguing peptides containing diverse fatty acyl units with a terminal alkyne functional group have been found in multiple marine organisms [11–14], especially marine cyanobacteria and mollusks. The structural characteristics of these peptides with various unusual amino acid residues have displayed their variety of biological functions as antitumor, antibacterial, antimalarial activities, etc., which seemed in some cases correlated to the presence of the terminal alkynyl moieties [14–16]. Cyanobacteria, also known as blue-green algae, are ancient photosynthetic prokaryotes living in a wide range of habitats including open oceans, tropical reefs, shallow water environments, and terrestrial substrates. The rich elaboration of biologically active natural products has assisted some of these organisms to survive in predator-rich ecosystems. A major part of cyanobacterial secondary metabolites arepeptides or possess peptidic substructures, which contribute to the more than 600 cyanobacterial peptides discovered thus far [17,18]. Mollusks are the largest marine phylum, comprising about 23% of all the named marine organisms. The gastropods (snails and slugs) are by far the most numerous mollusks in terms of classified species, and account for 80% of the total [19]. To date, over 100 mollusks peptides with diverse structures have been reported (Data based on reviewing the literatures, Marine Natrual Products in Natural Product Reports published during 1985–2015), some of which displayed a variety of bioactivities as antitumor, anti-HIV, ion blockers, etc. [20,21].

In this review, we have provided an overview of the types and structural characterization of these unusual terminal alkynyl fatty acyl units, as well as the sources and biological functions of their composed peptides from marine cyanobacteria and mollusks. Further, we have also discussed the evident biosynthetic origins responsible for formation of terminal alkynes of natural PKS (polyketide synthase)/NRPS (nonribosome peptide synthetase) hybrids, providing perspective insight for drug discovery research.

2. Cyclic Peptides Containing Terminal Alkyne

A number of terminal alkynylfatty acyl moieties are identified in the cyclic/acyclic marine peptides, which are different by structure and bioactivities (Table 1, Figure 1). Onchidin as the first terminal alkynyl-containing cyclic peptide, featured with 3-amino-2-methyl-7-octynoicacid (Amoya, **a**) moiety was isolated as a molluscan metabolite in 1994 [11]. Since then, Amoya as a component of cyclic peptides has been identified from many marine cyanobacterial metabolites including ulongapeptin, guineamide C, and companeramides A and B. It is likely that the 3-hydroxy-2-methyloct-7-ynoic acid (Hmoya, **b**) moiety was originally discovered in onchidin B from a marine mollusk, and subsequently identified in many cyanobacterial metabolites such as antanapeptin A and D, trungapeptin A, and hantupeptin A. Interestingly, abromine-containing 3-hydroxy-2-methyloct-7-ynoic acidmoiety (Br-Hmoya, **c**) was subsequently identified in several veraguamides isolated from marine cyanobacteria as well. The 2,2-dimethyl-3-hydroxy-7-octynoic acid (Dhoya, **d**) moiety was first discovered as a fatty acyl component in kulolide-1, from a cephalaspidean mollusk, *Philinopsis speciosa*, thereafter reported in many cyclic peptides with cyanobacteria origin as yanucamides A and B, pitipeptolide A, viequeamides A, and more. The 3-amino-6-octyneoic acid (Aoy, **e**) residue and the 5,7-dihydroxy-2,6-dimethyldodec-2-en-11-ynoic acid (Dddd, **f**) residue have been only identified in dolastatin 17 from a marine mollusk *Dolebella auricularia* and in Palau'amide from a marine cyanobacteria *Lyngbya* sp., respectively.

Table 1. Terminal alkynyl-containing cyclic/acyclic peptides from marine cycanobacteria and mollusks.

Moiety Unit	Compound	Organism	Bioactivities	Reference
Dhoya	Yanucamides A (1) and B (2)	Marine cyanobacterium *Lyngbya majuscule, Schizothrix* sp.	Strong brine shrimp toxicity	[12]
	Pitipeptolides A (3) Pitipeptolides D–F (4–6)	Marine cyanobacterium *Lyngbya majuscula*	Antitumor cytotoxicity	[22,23]
	Georgamide (7)	Marine cyanobacterium	anti-HIV cytotoxicity	[24]
	Mantillamide (10) Dudawalamide A (11)	Marine cyanobacterium *Lyngbya* sp.	Antitumor cytotoxicity Antimalaria parasites	[25]
	Guineamide G (12)	Marine cyanobacterium *Lyngbya majuscula*	Brine shrimp toxicity Antitumor cytotoxicity	[26]
	Cocosamides A–B (13–14)	Marine cyanobacterium *Lyngbya majuscula*	Antitumor cytotoxicity	[27]
	Viequeamides A–B (15–16) and E–F (17–18)	Marine cyanobacterium *Rivularia* sp.	Antitumor cytotoxicity	[28]
	Kulolide-1 (38)	Marine mollusk *Philinopsis speciosa* Pease	Antitumor cytotoxicity	[29]
	Kulokainalide-1 (39)	Marine cephalaspidean mollusk *Philinopsis speciosa*	Moderate antitumor cytotoxicity	[30]
Dhoaa	Wewakpeptins A and C (8a–9)	Marine cyanobacterium *Lyngbya semiplena*	Antitumor cytotoxicity	[31]
Amoya	Malevamide C (19)	Marine cyanobacterium *Symplocalaete-viridis*	No cytotoxicity	[32]
	Guineamide C (20)	Marine cyanobacterium *Lyngbya majuscula*	Antitumor cytotoxicity	[33]
	Ulongapeptin (21)	Marine cyanobacterium *Lyngbya* sp.	Antitumor cytotoxicity	[34]
	Companeramides A–B (22–23)	Marine cyanobacterium *Leptolyngbya* sp.	Antiplasmodial activity	[35]
	Onchidin (36)	Marine pulmonate mollusk *Onchidium* sp.	Strong antitumor cytotoxicity	[11,36]
Hmoya	Antanapeptin A and D (24–25)	Marine cyanobacterium *Lyngbya majuscula*	Na$^+$ channel modulation Antimicrobial activity	[37]
	Trungapeptins A (26)	Marine cyanobacterium *Lyngbya majuscula*	Brine shrimp toxicity and ichthyotoxicity	[30,38]
	Hantupeptin A (27)	Marine cyanobacterium *Lyngbya majuscula*	Brine shrimp toxicity Antitumor cytotoxicity	[39]
	Veraguamides B–F (29–33)	Marine cyanobacterium *Symploca* cf. *hydnoides*	Veraguamides A and C, antitumor cytotoxicity	[40]
	Veraguamides H (34)	Marine cyanobacterium *Oscillatoria margaritifera*	No cytotoxicity	[13]
	Onchidin B (37)	Marine pulmonate mollusk *Onchidium* sp.	Strong antitumor cytotoxicity	[11,36]
	Kulomo'opunalide-1 (40) and (41)	Marine cephalaspidean mollusk *Philinopsis speciosa*	Moderate antitumor cytotoxicity	[30]
Dddd	Palau'amide (35)	Marine cyanobacterium *Lyngbya* sp.	Strong antitumor cytotoxicity	[41]
Aoy	Dolastatin 17 (42)	Marine mollusk *Dolebella auricularia*	Antitumor cytotoxicity	[12,42]
Oya	Apramides B and G (44,47)	Marine cyanobacterium *Lyngbya majuscula*	Apramide A exhibited stimulating elastase activity	[43]
Moya	Apramides A,D and G (43,45–46)	Marine cyanobacterium *Lyngbya majuscula*	Apramide A exhibited stimulating elastase activity	[43]
	Dragonamides A–B (48–49)	Marine cyanobacterium *Lyngbya majuscule* Gomont	Antileishmaniasis	[44–47]
	Dragonamides C–E (50–52)	Marine cyanobacterium *Lyngbya polychroa*	Antileishmaniasis	[47]

Table 1. *Cont.*

Moiety Unit	Compound	Organism	Bioactivities	Reference
Moya	Dragomabin (**53**)	Marine cyanobacterium *Lyngbya majuscula*	Antiparasite toxicity	[45]
	Almiramide B (**54**)	Marine cyanobacterium *Lyngbya majuscule*	Antitumor cytotoxicity	[14]
	Almiramides D–H (**55–59**)	Marine cyanobacterium *Oscillatoria nigroviridis*	Antitumor cytotoxicity	[48]
Mdyna	Viridamides A–B (**61–62**)	Marine cyanobacterium *Oscillatoria nigro-Wiridis*	Antitrypanosomal activity Antileishmanial activity	[49]
Br-Hmoya	Veraguamides A (**28**)	Marine cyanobacterium *Symploca* cf. *hydnoides*	Veraguamides A and C, antitumor cytotoxicity	[40]
	Viridamides K–L (**63–64**)	Marine cyanobacteria, cf. *Oscillatoria margaritifera*	Antitumor cytotoxicity	[13]
2,4-dimethyl-9-decynoic acid	Carmabins A (**60**)	Marine cyanobacterium *Lyngbya majuscula*	Antimalaria against the W2 chloroquine-resistant malaria strain	[50]
9-(chloromethylene)-6-methyltetradec-4-en-13-ynoic acid	Jamaicamide A–B (**65–66**)	Marine Cyanobacterium *Lyngbya majuscula*	not mentioned	[51–53]

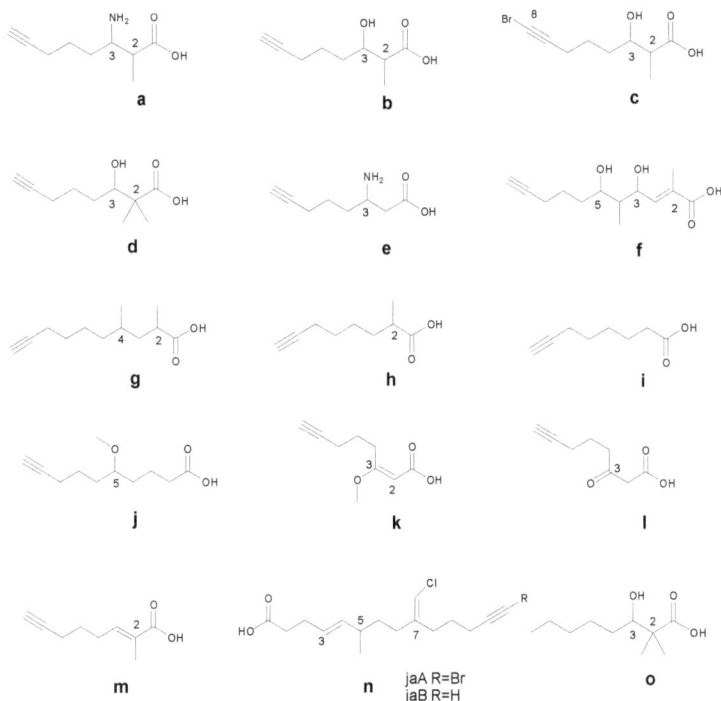

Figure 1. Structures of the terminal alkynyl fatty acyl moieties identified in cyclic/acyclic marine peptides. **a**. 3-amino-2-methyl-7-octynoicacid (Amoya); **b**. 3-hydroxy-2-methyloct-7-ynoic acid (Hmoya); **c**. bromine-containing 3-hydroxy-2-methyloct-7-ynoic acid (Br-Hmoya); **d**. 2,2-dimethyl-3-hydroxy-7-octynoic acid (Dhoya); **e**. 3-amino-6-octyneoic acid (Aoy); **f**. 5,7-dihydroxy-2,6-dimethyldodec-2-en-11-ynoic acid (Dddd); **g**. 2,4-dimethyl-9-decynoic acid; **h**. 2-methyl-7-octynoic acid (Moya); **i**. 7-octynoic acid (Oya); **j**. 5-methoxydec-9-ynoic acid (Mdyna); **k**. 3-methoxy-2-en-7-octynoic acid; **l**. 3-keto-7-octynoic acid; **m**. (*E*)-2-methyloct-2-en-7-ynoic acid; **n**. (4*E*,9*E*)-9- (chloromethylene)-6-methyltetradec-4-en-13-ynoic acid; **o**. 2,2-dimethyl-3-hydroxy-7-octanoic acid (Dhoaa).

2.1. Cyclic Peptides with Dhoya Unit from Marine Cyanobacteria

Cyclic peptides are representative secondary metabolites of cyanobacteria, and in recent years a number of structurally diverse terminal alkynyl-containing cyclic peptides have been found in marine cyanobacteria. The 2,2-dimethyl-3-hydroxy-7-octynoic acid (Dhoya) moiety appeared to be most frequently identified in the terminal alkynyl-containing cyclic peptides. The first two Dhoya unit-containing cyanobacterial cyclic depsipeptides, yanucamides A (**1**) and B (**2**, Table 1, Figure 2), were isolated from the lipid extract of a *Lyngbya majuscula* and *Schizothrix* sp. assemblage collected at Yanuca Island, Fiji, in 2000 [12]. Interestingly, the Dhoya unit had previously been found only in kulolide-1 (**38**) and kulokainalide-1 (**39**), metabolites isolated from the marine mollusk *Philinopsis speciosa*. Thus, the discovery of the yanucamides from a field-collected marine cyanobacterium substantiated the hypothesis that marine cyanobacteria are the probable source of the kulolides and their related metabolites. Both yanucamides A and B displayed strong brine shrimp toxicity (LD_{50}, 5 ppm). In 2001, Luesch et al. reported isolation and identification of two new cyclic depsipeptides, pitipeptolides A (**3**, Figure 2) and B, from a population of the marine cyanobacterium *Lyngbya majuscula* collected at Piti Bomb Holes, Guam [22]. Pitipeptolide A with a Dhoya unit and B with a reduced form of Dhoya unit, both showed potent in vitro cytotoxicity against LoVo cells with IC_{50} values of 2.25 and 1.95 µg/mL, respectively; and also exhibited certain growth inhibition for *Mycobacterium tuberculosis* strains ATCC 25177 and ATCC 35818 in the diffusion susceptibility assay. Both compounds were also observed to increase elastase activity (2.76-fold and 2.55-fold, respectively, at 50 µg/mL). Further, in 2011, Luesch et al. revisited larger collections of the same cyanobacterium and obtained additional analogs of pitipeptolides A and B, as well aspitipeptolides C (tetrahydro analog of **3**) and D–F (**4c**, **5–6**, Figure 2) [23]. Pitipeptolide A as the major metabolite in this series was reported to act as a feeding deterrent at natural concentrations against a range of marine grazers, suggesting that pitipeptolide A may play an important ecological role among these organisms [54]. Although pitipeptolides C–F were less potent than pitipeptolides A and B against HT-29 colon adenocarcinoma and MCF7 breast cancer cell lines, pitipeptolides C and E showed similar antimycobacterial activities comparable to pitipeptolides A and B. Among them, pitipeptolide F exhibited the highest potency, but pitipeptolide D did not show activities against both mammalian and bacterial cells. As a result, it indicates that the activities of pitipeptolides are not strongly impacted by the Dhoya unit in the structure. Georgamide (**7**, Figure 2), another analog of pitipeptolides featuring Dhoya residue, was obtained from an Australian cyanobacterium Q66C5927 at the head of the King George River, Northwestern Australia [24].

In 2005, an assay-based screening program for anticancer compounds from the marine cyanobacterium *Lyngbya semiplena* collected from Papua New Guinea led to the discovery of four new depsipeptides: wewakpeptins A–D featured with Dhoya or its fully reduced form (Dhoaa, **o**) residues [31]. Intriguingly, wewakpeptins A (**8a**, Figure 2) and B were approximately 10-fold more toxic than C (**9**) and D, with an LC_{50} of approximately 0.4 µM to NCI H-460 human lung tumor and mouse neuroblastoma cells. These cyclic depsipeptides most likely derive from a nonribosomal polypeptide synthetase (NRPS) pathway, and thus, the structural variation of wewakpeptins is intriguing and might suggest that adenylation domains with relaxed substrate specificity are involved in their biosynthesis [31]. Mantillamide (**10**), and dudawalamide A (**11**) featured with Dhoya residues were obtained from the marine cyanobacterium *Lyngbya* sp. because of their biological activity to cancer cells or malaria parasites, and they were able to be identified in a rapid manner using an annotation program developed from tandem mass spectra called MS-CPA available as a web tool (http://lol.ucsd.edu/ms-cpa_v1/Input.py) [25]. Isolation of a new cyclic depsipeptide, guineamide G (**12**) was reported in 2011 from the marine cyanobacterium *Lyngbya majuscula*, collected from Papua New Guinea. Guineamide G was the only cyclic depsipetide featuring Dhoya residue in the series of guineamides, which showed potent brine shrimp toxicity and moderate cytotoxicity to a mouse neuroblastoma cell line with LC_{50} value of 2.7 µM [26]. In 2011, Paul et al. reported isolation and identification of cocosamides A (**13**) and B (**14**) from the lipophilic extract of a collection of *Lyngbya majuscula* from Cocos Lagoon, Guam [27]. Cocosamide A consisting of Dhoea (a reduced form of Dhoya residue) was less potent than cocosamide B (featuring Dhoyaresidue) against HT-29 cells with IC_{50} values of 24 and 11 μM, respectively, indicating the presence of Dhoya moiety may have a slight effect on the cytotoxicity. In 2012, the family of viequeamides A–F was discovered from a shallow subtidal collection of a cyanobacterium (*Rivularia* sp.) near the island of Vieques, Puerto Rico, among which viequeamides A–B (**15–16**) and E–F (**17–18**, Figure 2) are 2,2-dimethyl-3-hydroxy-7-octynoic acid (Dhoya)-containing cyclic depsipeptides [28]. Intriguingly, viequeamide A was found to be the most active ($IC_{50}= 60 \pm 10$ nM) against H460 human lung cancer cell line, whereas the other viequeamides with quite similar structures were inactive.

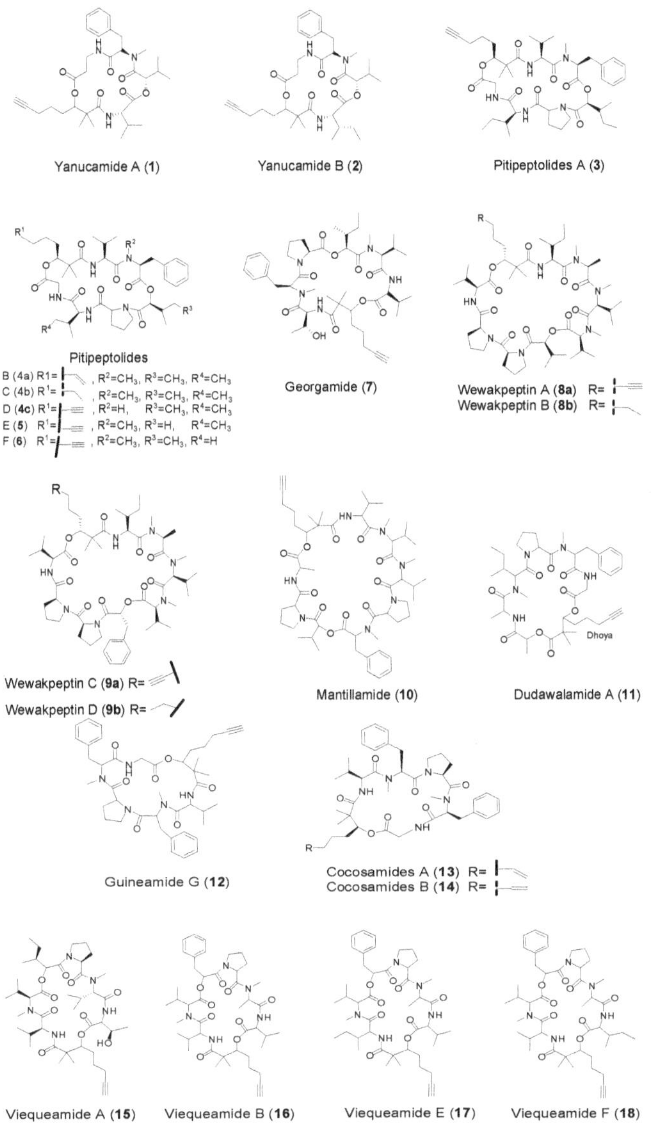

Figure 2. Structures of cyclic peptides with Dhoya residue from marine cyanobacteria.

2.2. Cyclic Peptides with Amoya Unit from Marine Cyanobacteria

Malevamide C (**19**, Table 1, Figure 3), as the first reported 3-amino-2-methyl-7-octynoic acid (Amoya)-containing cyanobactrial peptide, was obtained from a cyanobactrium *Symplocalaete-viridis* collected in waters adjacent to AlaMoana Beach Park, Hawaii in 2000. The unusual β-amino acid residue, Amoya, was only previously identified in onchidin, a cyclic depsipeptide isolated from a marine mollusk *Onchidium* spp. [32]. However, malevamide C did not display potent cytotoxicity against a variety of cancer cell lines. In 2003, another Amoya-containing cyclic depsipeptide, guineamide C (**20**, Figure 3) was discovered by William Gerwick's group from a Papua New Guinea collection of the marine cyanobacterium *Lyngbya majuscula*. As malevamide C, guineamide C, only exhibited moderate cytotoxicity against neuroblastoma cells with an IC_{50} value of 16 μM [33]. Meanwhile, Williams et al. reported discovery of ulongapeptin (**21**) featuring Amoya residue, isolated from a dark reddish-black clump of cyanobacterium, designated VP755 collected at Ulong Channel in Palau. Interestingly, ulongapeptin showed strong cytotoxicity against KB cells at an IC_{50} value of 0.63 μM [34]. Just recently, two new cyclic depsipeptides, companeramides A (**22**) and B (**23**) containing Amoya unit, were obtained from a marine cyanobacterial assemblage comprising a small filament *Leptolyngbya* species, from Coiba Island, Panama. It is interesting to note that companeramides A and B showed high nanomolar in vitro antiplasmodial activity, though not quite cytotoxic to human cancer cell lines [35].

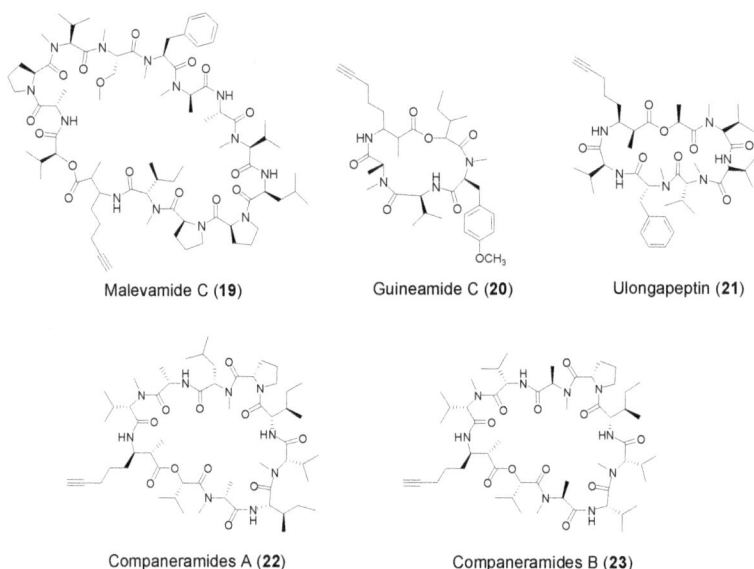

Malevamide C (**19**) Guineamide C (**20**) Ulongapeptin (**21**)

Companeramides A (**22**) Companeramides B (**23**)

Figure 3. Structures of cyclic peptides with Amoya residue from marine cyanobacteria.

2.3. Cyclic Peptides with Hmoya/Br-Hmoya/Dddd Units from Marine Cyanobacteria

While the 3-hydroxy-2-methyloctynoic acid (Hmoya) residue was initially identified in the molluscan metabolite onchidin B [11,36], antanapeptin A (**24**) and antanapeptin D (**25**, Figure 4) are the first two cyclic peptides containing Hmoya residue, obtained from a cyanobacterium *Lyngbya majuscule* collected from Antany Mora, Madagascar [37]. The antanapeptins were observed inactive in brine shrimp toxicity, sodium channel modulation, and antimicrobial bioassays. Subsequently, Sitachitta et al. in 2006, reported isolation and identification of three new cyclic peptides, trungapeptins A (**26**)–C, containing Hmoya residue, 3-hydroxy-2-methyl-7-octenoic acid (Hmoea), and 3-hydroxy-2-methyl-7-octanoic acid (Hmoaa) residues, respectively [38]. The relative stereochemistry of Hmoya residue of trungapeptin A was determined to be *syn* configuration between H-2 and H-3 by measurement of homonuclear coupling constant as well as comparison of the literature value. The absolute stereochemistry of the Hmoya unit was established as 2*S*, 3*R* by Mosher's analysis. Intriguingly, herein the stereochemistry of the Hmoya unit is identical to that of kulomo'opunalides [30], but is diastereomeric to that of onchidin B (2*R*, 3*R*). Unlike antanapeptins, trungapeptin A exhibited potent brine shrimp toxicity and ichthyotoxicity at 10 ppm and 6.25 ppm, respectively. However, it was inactive against KB and LoVo cells at 10 μg/mL. In 2009, a new Hmoya-containing analog of trungapeptin A, hantupeptin A (**27**, Figure 4) was discovered from the marine cyanobacterium *Lyngbya majuscula* from PulauHantuBesar, Singapore [39]. The absolute configuration at C-3 was determined to be *S* by Mosher's analysis following methanolysis of hantupeptin A and isolation of the Hmoya fragment. However, the stereochemistry at C-3 of the Hmoya unit in hantupeptin A is different from that of trungapeptin A. Further, hantupeptin A afforded both brine shrimp toxicity at 10 ppm and strong cytotoxicity against the leukemia cell line MOLT-4 with an IC_{50} value of 32 nM.

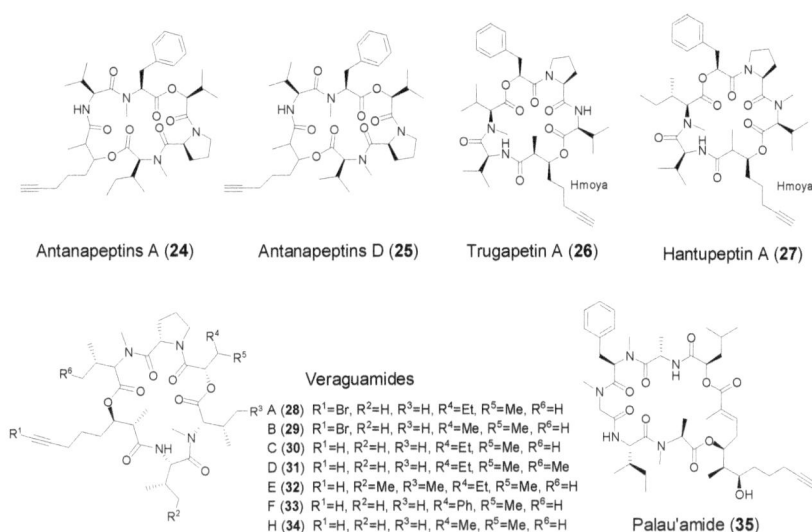

Antanapeptins A (**24**)   Antanapeptins D (**25**)   Trugapetin A (**26**)   Hantupeptin A (**27**)

Veraguamides
R^3 A (**28**) R^1=Br, R^2=H, R^3=H, R^4=Et, R^5=Me, R^6=H
B (**29**) R^1=Br, R^2=H, R^3=H, R^4=Me, R^5=Me, R^6=H
C (**30**) R^1=H, R^2=H, R^3=H, R^4=Et, R^5=Me, R^6=H
D (**31**) R^1=H, R^2=H, R^3=H, R^4=Et, R^5=Me, R^6=Me
E (**32**) R^1=H, R^2=Me, R^3=Me, R^4=Et, R^5=Me, R^6=H
F (**33**) R^1=H, R^2=H, R^3=H, R^4=Ph, R^5=Me, R^6=H
H (**34**) R^1=H, R^2=H, R^3=H, R^4=Me, R^5=Me, R^6=H

Palau'amide (**35**)

Figure 4. Structures of cyclic peptides with Hmoya/Br-Hmoya/Dddd residue from marine cyanobacteria.

In 2011, the Luesch group and Gerwick group coincidently reported isolation and identification of a series of peptides featured with Hmoya and its derived residues, veraguamides A–F (**28–33**), from a cyanobacterium *Symploca* cf. *hydnoides* at Cetti Bay, Guam [40], and veraguamides H (**34**), I–L from the marine cyanobacterium cf. *Oscillatoria margaritifera* at the Coiba National Park, Panama [13], respectively. Among them, veraguamides A and B are 8-bromo-3-hydroxy-2-methyl-7-octynoic acid (Br-Hmoya) moiety-containing cyclic peptides, while veraguamides K and L (**63–64**) are Br-Hmoya-containing linear peptides (more in Section 3). It is interesting to note that veraguamides D and E were five-fold more potent than their related congener veraguamide C against HT29 colorectal and HeLa cervical adenocarcinoma cells, while veraguamides A, B and F were inactive againstthese cancer cell lines. Surprisingly, veraguamide A exhibited strong potency in the H-460 cytotoxicity assay (LD_{50} = 141 nM), but veraguamides B, C, K and L were much less active.

Palau'amide (**35**, Figure 4) is a unique terminal alkynyl-containing cyclic depsipeptide, consisting of a novel polyketide unit, 5,7-dihydroxy-2,6-dimethyldodec-2-en-11-ynoic acid (Dddd), which was obtained from a *Lyngbya* sp. from Palau. Palau'amide showed strong cytotoxicity against KB cells with an IC_{50} value of 13 nM [41].

2.4. Cyclic Peptides from Marine Mollusks

Onchidin (**36**, Figure 5) as the first report of a dimeric depsipeptide from a mollusc, featured with two 3-amino-2-methyl-7-octynoicacid (Amoya, **a**) residues, was obtained from the pulmonate mollusk *Onchidium* sp. collected off New Caledonian 1994 [11]. Onchidin B (**37**) isolated and identified along with onchidin from the same extract, shares quite similar structural features with onchidin. Interestingly, onchidin B featured with two 3-hydroxy-2-methyloct-7-ynoic acid (Hmoya, **b**) does not have a C_2 axis of symmetry as does onchidin, due to the presence of the two enantiomers of proline that renders the two halves of the molecule different [36]. Onchidin and onchidin B exhibited identical cytotoxicity against P-388 murine leukemia cells (IC_{50} = 8 µg/mL) and Kb human epidermoid carcinoma cells (IC_{50} = 8 µg/mL), respectively.

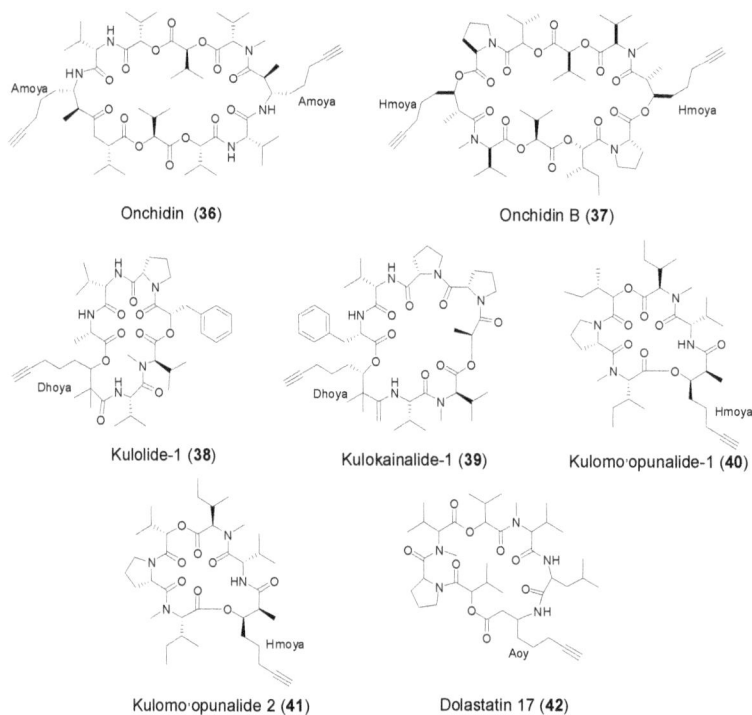

Onchidin (**36**) Onchidin B (**37**)

Kulolide-1 (**38**) Kulokainalide-1 (**39**) Kulomo·opunalide-1 (**40**)

Kulomo·opunalide 2 (**41**) Dolastatin 17 (**42**)

Figure 5. Structures of cyclic peptides with Amoya/Hmoya/Dhoya/Aoy residue from marine mollusks.

A cephalaspidean mollusk, *Philinopsis speciosa* Pease, 1860 collected off North Shore, Oahu's (Hawaiian Islands) Shark Bay, afforded the first 2,2-dimethyl-3-hydroxy-7-octynoic acid (Dhoya)-containing cyclic depsipeptide, kulolide-1 (**38**, Figure 5) [29]. Kulolide-1was active against L-1210 leukemia cells and P388 murine leukemia cells at IC_{50} values of 0.7 and 2.1 µg/mL, respectively. Along with kulolide-1, three other terminal alkynyl-containing cyclic depsipeptides, kulokainalide-1 (Dhoya, **39**), kulomo'opunalide-1 (Hmoya, **40**) and kulomo'opunalide-2 (Hmoya, **41**), were also discovered from the same sample of the cephalaspidean mollusk, *Philinopsis speciosa* [30].

3-amino-6-octyneoic acid (Aoy, **e**) as an unprecedented terminal alkynyl moiety, was only identified in a novel cyclic depsipeptide, dolastatin 17, isolated from a sea hare *Dolebella auricularia* [12]. Dolastatin 17 (**42**, Figure 5) displayed significant growth-inhibitory activity against OVCAR-3 (GI50 0.67 µg/mL), SF-295 (GI50 0.55 µg/mL), NCI-H460 (GI50 0.74 µg/mL), KM20L (GI50 0.45 µg/mL) human cancer cell lines [42].

3. Acyclic Lipopeptides Containing Terminal Alkyne from Marine Cyanobacteria

It is interesting to note that many linear peptides have also been found to possess the terminal alkynyl fatty acyl moieties, including 2,4-dimethyl-9-decynoic acid (**g**), 2-methyl-7-octynoic acid (Moya, **h**), 7-octynoic acid unit (Oya, **i**), 5-methoxydec-9-ynoic acid (Mdyna, **j**), 3-methoxy-2-en-7-octynoic acid (MeO-Oya-2-ene, **k**), 3-keto-7-octynoic acid (**l**), and (*E*)-2-methyloct-2-en-7-ynoic acid (**m**), which are different from that of cyclic peptides, except for Hmoya and Br-Hmoya residues present in both linear and cyclic veraguamides (Table 1). In addition, an acyclic amide-like secondary metabolite from the marine cyanobacteria *Lyngbya majuscula*, termed jamaiapcamides A, has provided an alkynyl bromide, vinyl chloride, β-methoxyeneone moiety (**n**) to the terminal alkynyl-containing peptides.

All the terminal alkynyl-containing linear peptides were solely discovered from marine cyanobacteria. In 2000, Luesch et al. reported the isolation and identification of six new linear peptides, apramides A–G (Figure 6), from the marine cyanobacterium *Lyngbya majuscule* collected at Apra Harbor, Guam [43]. Apramides A (**43**), D (**45**) and G (**46**) are Moya-containing acylic peptides, while apramides C and F consist of 2-methyl-7-octenoic acid moiety (Moea) in their structures. Apramides B (**44**) and E (**47**) possess a 7-octynoicacid unit (Oya) in lieu of the Moya moiety, and the rest of the structures are identical to apramides A and D, respectively. Apramides A–G was inactive in cytotoxic, antibacterial, antifungal assays, but apramide A exhibited stimulating elastase activity.

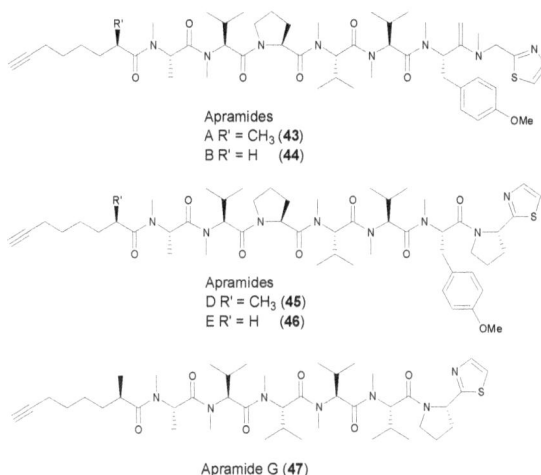

Apramides
A R' = CH₃ (**43**)
B R' = H   (**44**)

Apramides
D R' = CH₃ (**45**)
E R' = H   (**46**)

Apramide G (**47**)

Figure 6. Structures of linear peptides (apramides A–G) from marine cyanobacteria.

Dragonamides are a family of structurally close linear peptides composing of a variety of terminal alkynyl units (Figure 7). Several separate Panamanian collections of *Lyngbya majuscule* Gomont afforded dragonamides A, B (**48–49**) and E [44–46], while the collection of brown *Lyngbya polychroa* from Hollywood Beach, Fort Lauderdale, FL led to the discovery of dragonamides C and D [47]. Dragonamides A and B contain a terminal 2-methyl-7-octynoic acid unit (Moya), whereas dragonamides C, D and E (**50–52**) possess three different terminal acetylene units, 3-methoxy-2-en-7-octynoic acid (**k**), 3-keto-7-octynoic (**l**) (*E*)-2-methyloct-2-en-7-ynoic acid (**m**), respectively, which were not previously reported from marine peptides. Dragonamides did not exhibit strong activities against a variety of tumor cell lines, except dragonamides A and E which showed moderate in vitroactivity against leishmaniasis. Along with dragonamides A and B, another terminal Moya-containing linear peptide, dragomabin (**53**, Figure 7), was isolated and identified in 2007, from a Panamanian strain of the marine cyanobacterium *Lyngbya majuscula* [45]. Dragomabin possesses the best differential toxicity between parasite and mammalian cells, with IC$_{50}$ value of 6.0 μM against the W2 chloroquine-resistant malaria strain and IC_{50} value of 182.3 µM against Vero cells (kidney epithelial cells).

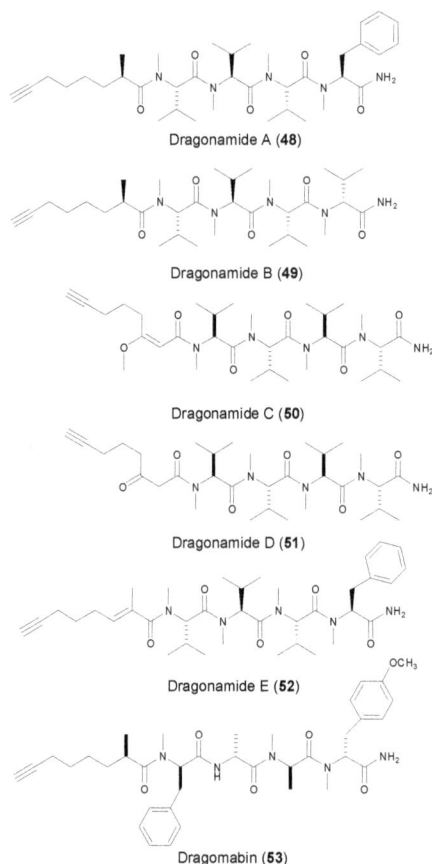

Dragonamide A (**48**)

Dragonamide B (**49**)

Dragonamide C (**50**)

Dragonamide D (**51**)

Dragonamide E (**52**)

Dragomabin (**53**)

Figure 7. Structures of linear peptides (dragonamides A–E, dragomabin) from marine cyanobacteria.

In 2010, Linington et al. reported the isolation and identification of a series of terminal fatty acyl units-containing linear peptides, almiramides A–C, from a Panamanian strain of the marine cyanobacterium *Lyngbya majuscule* [14]. Among them, almiramide B (**54**) is featured with a terminal Moya unit (Figure 8), whereas almiramide C contains a reduced form of Moya as a 2-methyloct-7-enoic acid residue. Biological evaluation of these three compounds showed that almiramides B and C possessed good selectivity between parasite and mammalian cells with strong in vitro antiparasitic activity against *leishmania* (IC_{50} = 2.4 and 1.9 µM, respectively), and weak activity against Vero cells (IC_{50} = 52.3 and 33.1 µM, respectively). Just recently, a series of new terminal Moya-containing linear peptides, almiramides D–H (**55–59**) along with known almiramide B (Figure 8), were isolated and identified from a cyanobacterium sample of *Oscillatoria nigroviridis* collected at the Colombian Caribbean Sea [48]. Intriguingly, two structurally representative almiramides B and D showed mild toxicity against five human tumor cell lines, but high toxicity against the gingival fibroblast cell line was used as reference to evaluate selectivity against tumor cell lines compared with primary cell line.

Two novel terminal fatty acyl-containing linear peptides, carmabins A (**60**) and B were discovered from a collection of the marine cyanobacterium *Lyngbya majuscule* at Barbara Beach (Spanish Waters),

Curacao, Netherlands Antilles in 1998 [50]. Carmabin A (Figure 9) is featured with a novel terminal 2,4-dimethyl-9-decynoic acid residue, but in carmabin B, the acetylene functional group is replaced with a methyl ketone. To the best of our knowledge, carmabin A is the only reported compound containing a 2,4-dimethyldec-9-ynoic acid moiety. Carmabin A exhibited moderate cytotoxicity to Vero cells (IC_{50} = 9.8 μM), and mild activity against the W2 chloroquine-resistant malaria strain (IC_{50} = 4.3μM).

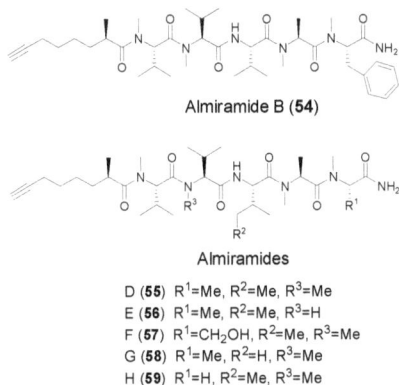

Almiramide B (**54**)

Almiramides

D (**55**) R^1=Me, R^2=Me, R^3=Me
E (**56**) R^1=Me, R^2=Me, R^3=H
F (**57**) R^1=CH$_2$OH, R^2=Me, R^3=Me
G (**58**) R^1=Me, R^2=H, R^3=Me
H (**59**) R^1=H, R^2=Me, R^3=Me

Figure 8. Structures of linear peptides (almiramide B, D–H) from marine cyanobacteria.

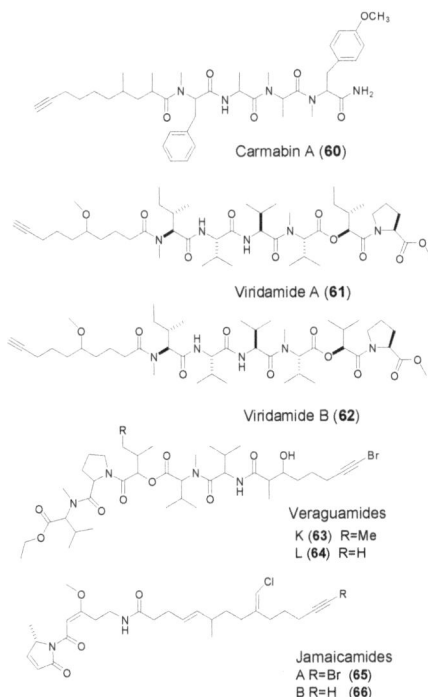

Carmabin A (**60**)

Viridamide A (**61**)

Viridamide B (**62**)

Veraguamides
K (**63**) R=Me
L (**64**) R=H

Jamaicamides
A R=Br (**65**)
B R=H   (**66**)

Figure 9. Structures of linear peptides (carmabin A, viridamide A–B, veraguamides K and L, and jamaicamides A–B) from marine cyanobacteria.

In 2008, Simmons et al. reported discovery of two new linear peptides, viridamides A and B (**61–62**, Figure 9) isolated from the marine cyanobacterium *Oscillatoria nogroviridis* [51] (Figure 9), whose structures contain a novel terminal 5-methoxydec-9-ynoic acid moiety (Mdyna). Viridamide A displayed antitrypanosomal activity (IC$_{50}$ 1.1 µM to *Trypanosoma cruzi*) and antileishmanial activity (IC$_{50}$ 1.5 µM to *Leishmania mexicana*).

4. Different Methods to Determine the Absolute Configuration of Different Alkynyl Fragments

4.1. Amoya (**a**)

Determination of stereochemistry of the 3-amino-2-methyl-7-octynoic acid (Amoya, **a**) residue in the cyclic depsipeptides was established using differential methods such as NMR or Marfey's analysis. The configuration of an Amoya unit in onchidin was found to be threo through analysis of the NOE data and their coupling constants for critical protons, which indicated the relative stereochemistry of the pentine side chain on the same side as the neighboring MeVal and Val isopropyl groups. As a result, the absolute configuration of an Amoya unit in onchidin was determined to be 7*S*, 9*S* [11]. The stereochemistry of the Amoya unit in ulongapeptin was determined using the synthetically saturated 3-amino-2-methyloctanoic acid C-2 diastereomers (2*R*, 3*R* and 2*S*, 3*R*) as standards for Marfey's analysis. Comparison with the derivatized hydrogenated hydrolysate of ulongapeptin established the absolute configuration of the Amoya as 2*S*, 3*S* [34]. Surprisingly, the absolute configuration of the Amoya unit in companeramides A (**22**) and B (**23**) was determined to be 2*S*, 3*R* using the method of Marfey's analysis in comparison with synthetically saturated 3-amino-2-methyloctanoic acid C-2 diastereomeric (2*R*, 3*R* and 2*S*, 3*R*) standards [35].

4.2. Hmoya (**b**)

Determination of stereochemistry of 3-hydroxy-2-methyloct-7-ynoic acid (Hmoya) was first accomplished in the work of identification of onchidin B [36]. As beginning of the work, all four possible stereoisomers of Hmoya were synthesized in a diastereo selective mode. However, direct comparative analysis of the methyl esters of the four synthetic standards with the methyl ester of the natural Hmoya hydrolyzed from onchidin B using chiral gas chromatography (GC) and HPLC was not successful due to a separation issue. Consequently, the problem was overcome by derivation of the four hydroxy esters with (−)-(*R*)-α-methoxy-α-(9-anthryl) acetic acid as well as the natural Hmoya component to obtain good resolution of the four synthetic stereoisomers in LC-MS analysis, which indicated that the absolute configuration of Hmoya moiety in onchidin B was 2*R*, 3*R*. The stereochemistry of the Hmoya unit in Kulomo'opunalide-1 (**40**) and kulomo'opunalide-2 (**41**) was initially worked on comparison of chemical shifts of the *p*-bromobenzoyl derivatized synthetic standards with the derivatized natural Hmoaa (hydrogenated form of Hmoya) in ¹H NMR spectra to provide the relative stereochemistry of 2*S**, 3*R**. Comparison of retention time and co-injection of the standards with hydrolyte of the hydrogenated (**40**) and (**41**) confirmed the absolute stereochemistry of the Hmoya unit as 2*S*, 3*R* [30], which is surprisingly different from 2*R*, 3*R* of the Hmoya unit in onchidin B. Interestingly, the absolute configuration of the Hmoya unit in trungapeptin A (**26**) was determined to be 2*S*, 3*R* by application of the *J*-based configuration analysis as well as Mosher's method [38]. Further, the stereochemistry of Hmoya in hantupeptin A (**27**) was determined to be *S* at C-3 using Mosher's analysis, but the configuration at C-2 was not established [39]. In addition, the absolute configuration of the Br-Hmoya unit in veraguamide A (**28**) was also determined to be 2*S*, 3*R* identical to that of trungapeptin A using the *J*-based configuration analysis as well as the Mosher's method subjected to the linear veraguamide A following methanolysis of 28 [40].

4.3. Dhoya (**d**)

Determination of absolute configuration of 2,2-dimethyl-3-hydroxy-7-octynoic acid (Dhoya) residue was initially achieved in the structure elucidation of kulolide-1 (**38**), which was treated with

NaOMe to release the free hydroxyl functional group in the Dhoya-containing fragment, followed by Mosher's analysis to reveal the *R*-configuration at C-3 of Dhoya [29]. Interestingly, the stereochemistry of the Dhoya unit in kulokainalide-1 was determined to be 3*S* by comparing the values of optical rotation of Dhoaa (saturated form of Dhoya) residues obtained from the acid hydrolysates of both hydrogenated kulolide-1 and kulokainalide-1 [30]. Further, Ye et al. achieved a total synthesis of yanucamide A to confirm the absolute configuration of Dhoya to be the same (3*S*) as in kulokainalide-1 [55]. The stereochemistry of the Dhoya unit in pitipeptolide A (**3**) was also revealed as 3*S* using the optical rotation data of the obtained Dhoaa unit [22]. Interestingly, the absolute configuration of the Dhoya unit in wewakpeptin A (**8**) was determined to be *R* by chiral GC-MS analysis of the hydrogenated Dhoya in **8** possessing the same retention time as synthetic *R*-Dhoaa [31]. The chiral center of Dhoya residue in cocosamide B (**14**), was suggested to possess the same 3*S* configuration as in pitipeptolide A, by comparison of the NOE correlations of specific protons observed for Dhoya as well as related protons in the structures of cocosamide B and pitipeptolide A [27]. The configuration of Dhoya residue in viequeamide A was revealed to be *S* by chiral GC-MS analysis of the synthetic standards and the obtained natural Dhoya unit [28].

4.4. Moya (**h**)

The 2-methyl-7-octynoic acid (Moya, **h**) unit is the most frequently identified terminal alkynyl residue in the linear peptides. The absolute configuration at C-2 in apramides was proposed to be *R* based on the negative contribution of the C-2 stereocenter to the molar optical rotation of the molecule [50], because it is known for a closely related model compound that the 2*S* epimer gives a more positive rotation in CHCl₃ than the corresponding epimer with *R* configuration in the lipid chain [56]. The stereochemistry of Moya residue in dragonamide A was initially determined to be *R*, which was inferred by comparison of optical rotation data of 2-methyloctanoic acid obtained from hydrolyte of hydrogenated dragonamide A with literature values of other 2-methylalkanoic acids [57,58]. Subsequently, the later total synthesis of dragonamide A has led to a reassignment of the configuration as *S* at the stereogenic center of the Moya unit of the molecule [16]. Further, dragonamide B and dragomabin were isolated with dragonamide A from a Panamanian collection of *Lyngbya majuscule* Gomont, while the NMR and optical rotation data for this dragonamide A closely match the 2*S* synthetic product, but differ significantly from the 2*R* synthetic product [45]. Therefore, it was concluded that dragonamide A, dragonamide B, and dragomabin all contain 2*S*-methyloct-7-ynoic acid. The stereochemistry at C-2 of Moya residue in almiramides B and C was investigated by comparison of commercial standards with obtained natural Moya derivatives using GC-MS, which was determined to be *R* configuration [46], surprisingly opposite to the absolute configuration of the Moya unit in dragonamides.

4.5. Other Special Fragments

Determination of stereochemistry of 5,7-dihydroxy-2,6-dimethyldodec-2-en-11-ynoic acid (Dddd, **f**) residue in Palau'amide was a bit complex, due to an inter-converting mixture of rotamers around these stereocenters of Dddd. With the secured NMR assignments for the two major conformers of Palau'amide in CDCl₃ (C-R1/-R2), subsequent NOE experiments recorded in CDCl₃ revealed a strong correlation between H-40 and H-46 that indicated the *erythro* configuration of C-38 and C-39. The Mosher's analysis of the absolute configuration of C-39 was carried on the α-methoxy phenyl acetic acid (MPA) derivatives of Palau'amide. Comparison of the $\Delta\delta^{RS}$ values for these derivatives established the *R* configuration of C-39 [41]. While the configuration of C-37 could not be rigorously established by chemical means, analysis of molecular models in conjunction with NOE data suggested an *S*-configuration for this chiral center. The double bond configuration of 3-methoxy-2-en-7-octynoic acid (**k**) in dragonamide C and that of 2-methyloct-2-en-7-ynoic acid (**m**) in dragonamide E, were both assigned as E-geometry by NOE analysis [47,48].

5. Conclusions

A number of structurally intriguing peptides containing diverse terminal alkynyl fatty acyl residues, such as Dhoya, Hmoya, Amoya, Aoy, Moya, etc., have been found in multiple marine organisms, especially marine mollusk and cyanobacteria. In 1998, a study about the biological origin of Dhoya-containing cyclic depsipeptide, kulolide-1, by Scheuer and coworkers showed that the marine mollusk *Philinopsis speciosa* preyed on the herbivorous sea hare *Stylocheilus longicaudus* that is well recognized to possess the predator-prey relationship with cyanobacteria [30]. Interestingly, Scheuer and coworkers succeeded in isolating kulolide-1 from sea hare *Stylocheilus longicaudus*, which suggests that kulolide-1 discovered from *P. speciosa* is possibly accumulated from its prey *Stylocheilus longicaudus*, known to sequester secondary metabolites from its diet of mat-forming cyanobacteria [29]. Thus, similarity among the terminal alkynyl-containing cyclic peptides is suggestive that this intriguing structure family of metabolites in fact originates in cyanobacteria. Interestingly, all the terminal alkynyl fatty acyl moieties identified in the linear peptides were solely discovered as the constituents of metabolites of marine cyanobacteria.

Overall, many of these terminal alkynyl-containing peptides have shown a variety of biological functions as antitumor, antibacterial and antimalarial activities. Intriguingly, some of them with minor structural variations have presented different biological effects. For example, viequeamide A was found to be the most active ($IC_{50} = 60 \pm 10$ nM) against H460 human lung cancer cell line, whereas the other viequeamides with quite similar structures were inactive; hantupeptin A exhibited strong cytotoxicity against the leukemia cell line MOLT-4 with an IC_{50} value of 32 nM, but trungapeptin A was reported to be inactive against KB or LoVo cells at 10 μg/mL. Some cases further indicated that the unsaturated terminal moieties may play an important role in the biological activity, as illustrated by almiramide B and C possessing strong in vitro antiparasitic activity against *L. donovani*, whereas almiramide A was completely inactive.

Another research area to exploit marine peptides as a source of new therapeutics is to harness the genetic versatility of its biosynthetic gene clusters. Acetylenases, a special family of desaturases that catalyze O_2-dependent dehydrogenation of C–C bonds, have been considered to be responsible for formation of terminal alkynes of many natural products [15]. In 2015, Zhu and Zhang et al. reported a thorough characterization of terminal alkyne biosynthetic enzymes responsible for the synthesis of jamaicamide A and B (**65–66**) and carmabins [51,52], which demonstrated the in vitro formation of a short-chain alkynoic starter unit by a three-gene operon, *jamABC*, where *jamA*, *jamB* and *jamC* encode a homolog of fatty acyl-CoA ligase, a membrane-bound fatty acid desaturase and an acyl carrier protein (ACP), respectively [53]. Therefore, the biosynthetic evidences have further shown that the fatty acyl starter unit and the extender units could be engineered using *jamABC* and other modular assembly lines of PKS/NRPS enzymatic machinery to form the terminal alkyne-containing natural product.

A well-known reaction referred to as the "click reaction" (the triazole forming via azide-alkyne cyclo addition), has been quite often used in selective imaging and study of azide- or alkyne-labeled macromolecule interaction. In our opinion, the azide-alkyne click chemistry may serve as a powerful tool to study the drug mechanism of the terminal alkyne-containing peptides as well as to explore their structure activity relationship (SAR). Not surprisingly, it is highly expected to see application of the "click reaction" in combination with the biosynthetically engineered alkynyl-containing peptides playing a role in drug discovery research in the near future.

Acknowledgments: The authors acknowledge the National Natural Science Fund of China (No. 41476121, 81402844, 81302691, 81373321, 41106127, 81172978, 81072573, and 81001394).

Author Contributions: Bing-Nan Han and Qiu-Ye Chai were responsible for writing the review. Zhen Yang assisted in providing references and the final editing the manuscript. Hou-Wen Lin and Bing-Nan Han were in charge of the financial support of this project.

Conflicts of Interest: The authors declare no conflict of interest.

References

1. Costa, M.; Costa-Rodrigues, J.; Fernandes, M.H.; Barros, P.; Vasconcelos, V.; Martins, R. Marine cyanobacteria compounds with anticancer properties: A review on the implication of apoptosis. *Mar. Drugs* **2012**, *10*, 2181–2207. [CrossRef] [PubMed]
2. Cheung, R.C.; Ng, T.B.; Wong, J.H. Marine peptides: Bioactivities and applications. *Mar. Drugs* **2015**, *13*, 4006–4043. [CrossRef] [PubMed]
3. Jo, C.; Khan, F.F.; Khan, M.I.; Iqbal, J. Marine bioactive peptides: Types, structures, and physiological functions. *Food Rev. Int.* **2016**, *33*, 44–61. [CrossRef]
4. Zheng, L.H.; Wang, Y.J.; Sheng, J.; Wang, F.; Zheng, Y.; Lin, X.K.; Sun, M. Antitumor peptides from marine organisms. *Mar. Drugs* **2011**, *9*, 1840–1859. [CrossRef] [PubMed]
5. Sipkema, D.; Franssen, M.C.; Osinga, R.; Tramper, J.; Wijffels, R.H. Marine sponges as pharmacy. *Mar. Biotechnol.* **2005**, *7*, 142–162. [CrossRef] [PubMed]
6. Andavan, G.S.; Lemmens-Gruber, R. Cyclodepsipeptides from marine sponges: Natural agents for drug research. *Mar. Drugs* **2010**, *8*, 810–834. [CrossRef] [PubMed]
7. Rawat, D.S.; Joshi, M.C.; Joshi, P.; Atheaya, H. Marine peptides and related compounds in clinical trial. *Anti-Cancer Agents Med. Chem.* **2006**, *6*, 33–40. [CrossRef]
8. Blunt, J.W.; Copp, B.R.; Keyzers, R.A.; Munro, M.H.; Prinsep, M.R. Marine natural products. *Nat. Prod. Rep.* **2014**, *31*, 160–258. [CrossRef] [PubMed]
9. Park, Y.J.; Jeong, J.-K.; Choi, Y.M.; Lee, M.S.; Choi, J.H.; Cho, E.J.; Song, H.; Park, S.J.; Lee, J.-H.; Hong, S.S. Dolastatin-10 derivative method of producing the same and anticancer drug composition containing the same. *J. Am. Chem. Soc.* **1987**, *109*, 6883–6885.
10. Luesch, H.; Moore, R.E.; Paul, V.J.; Mooberry, S.L.; Corbett, T.H. Isolation of dolastatin 10 from the marine cyanobacterium symploca species vp642 and total stereochemistry and biological evaluation of its analogue symplostatin 1. *J. Nat. Prod.* **2001**, *64*, 907–910. [CrossRef] [PubMed]
11. Rodríguez, J.; Fernández, R.; Quiñoá, E.; Riguera, R.; Debitus, C.; Bouchetj, P. Onchidin: A cytotoxic depsipeptide with C$_2$ symmetry from a marine mollusc. *Tetrahedron Lett.* **1994**, *35*, 9239–9242. [CrossRef]
12. Sitachitta, N.; Williamson, R.T.; Gerwick, W.H. Yanucamides a and b, two new depsipeptides from an assemblage of the marine cyanobacteria *Lyngbya majuscula* and *Schizothrix* species. *J. Nat. Prod.* **2000**, *63*, 197–200. [CrossRef] [PubMed]
13. Mevers, E.; Liu, W.T.; Engene, N.; Mohimani, H.; Byrum, T.; Pevzner, P.A.; Dorrestein, P.C.; Spadafora, C.; Gerwick, W.H. Cytotoxic veraguamides, alkynyl bromide-containing cyclic depsipeptides from the marine cyanobacterium cf. Oscillatoria margaritifera. *J. Nat. Prod.* **2011**, *74*, 928–936. [CrossRef] [PubMed]
14. Sanchez, L.M.; Lopez, D.; Vesely, B.A.; Della Togna, G.; Gerwick, W.H.; Kyle, D.E.; Linington, R.G. Almiramides a–c: Discovery and development of a new class of leishmaniasis lead compounds. *J. Med. Chem.* **2010**, *53*, 4187–4197. [CrossRef] [PubMed]
15. Minto, R.E.; Blacklock, B.J. Biosynthesis and function of polyacetylenes and allied natural products. *Prog. Lipid Res.* **2008**, *47*, 233–306. [CrossRef] [PubMed]
16. Yamaguchi, M.; Park, H.-J.; Ishizuka, S.; Omata, K.; Hirama, M. Chemistry and antimicrobial activity of caryoynencins analogs. *J. Med. Chem.* **1995**, *38*, 5015–5022. [CrossRef] [PubMed]
17. Nagarajan, M.; Maruthanayagam, V.; Sundararaman, M. A review of pharmacological and toxicological potentials of marine cyanobacterial metabolites. *J. Appl. Toxicol.* **2012**, *32*, 153–185. [CrossRef] [PubMed]
18. Raja, R.; Hemaiswarya, S.; Ganesan, V.; Carvalho, I.S. Recent developments in therapeutic applications of cyanobacteria. *Crit. Rev. Microbiol.* **2016**, *42*, 394–405. [CrossRef] [PubMed]
19. Chapman, A.D. *Numbers of Living Species in Australia and the World*; Departmwnt of the Environment: Canberra, Australia, 2010.
20. Aneiros, A.; Garateix, A. Bioactive peptides from marine sources: Pharmacological properties and isolation procedures. *J. Chromatogr. B Anal. Technol. Biomed. Life Sci.* **2004**, *803*, 41–53. [CrossRef] [PubMed]
21. Suarez-Jimenez, G.M.; Burgos-Hernandez, A.; Ezquerra-Brauer, J.M. Bioactive peptides and depsipeptides with anticancer potential: Sources from marine animals. *Mar. Drugs* **2012**, *10*, 963–986. [CrossRef] [PubMed]
22. Luesch, H.; Pangilinan, R.; Yoshida, W.Y.; Moore, R.E.; Paul, V.J. Pitipeptolides a and b, new cyclodepsipeptides from the marine cyanobacterium *Lyngbya majuscula*. *J. Nat. Prod.* **2001**, *64*, 304–307. [CrossRef] [PubMed]

23. Han, B.; Gross, H.; McPhail, K.L.; Goeger, D.; Maier, C.S.; Gerwick, W.H. Wewakamide a and guineamide g, cyclic depsipeptides from the marine cyanobacteria *Lyngbya semiplena* and *Lyngbya majuscula*. *J. Microbiol. Biotechnol.* **2011**, *21*, 930–936. [CrossRef] [PubMed]

24. Wan, F.; Erickson, K.L. Georgamide, a new cyclic depsipeptide with an alkynoic acid residue from an australian cyanobacterium. *J. Nat. Prod.* **2001**, *64*, 143–146. [CrossRef] [PubMed]

25. Liu, W.-T.; Ng, J.; Meluzzi, D.; Bandeira, N.; Gutierrez, M.; Simmons, T.L.; Schultz, A.W.; Linington, R.G.; Moore, B.S.; Gerwick, W.H.; et al. Interpretation of tandem mass spectra obtained from cyclic nonribosomal peptides. *Anal. Chem.* **2009**, *81*, 4200–4209. [CrossRef] [PubMed]

26. Montaser, R.; Paul, V.J.; Luesch, H. Pitipeptolides c-f, antimycobacterial cyclodepsipeptides from the marine cyanobacterium *Lyngbya majuscula* from guam. *Phytochemistry* **2011**, *72*, 2068–2074. [CrossRef] [PubMed]

27. Gunasekera, S.P.; Owle, C.S.; Montaser, R.; Luesch, H.; Paul, V.J. Malyngamide 3 and cocosamides a and b from the marine cyanobacterium *Lyngbya majuscula* from cocos lagoon, guam. *J. Nat. Prod.* **2011**, *74*, 871–876. [CrossRef] [PubMed]

28. Boudreau, P.D.; Byrum, T.; Liu, W.T.; Dorrestein, P.C.; Gerwick, W.H. Viequeamide a, a cytotoxic member of the kulolide superfamily of cyclic depsipeptides from a marine button cyanobacterium. *J. Nat. Prod.* **2012**, *75*, 1560–1570. [CrossRef] [PubMed]

29. Reese, M.T.; Gulavita, N.K.; Nakao, Y.; Hamann, M.T.; Yoshida, W.Y.; Coval, S.J.; Scheuer, P.J. Kulolide: A cytotoxic depsipeptide from a cephalaspidean mollusk, philinopsis speciosa1. *J. Am. Chem. Soc.* **1996**, *118*, 11081–11084. [CrossRef]

30. Nakao, Y.; Yoshida, W.Y.; Szabo, C.M.; Baker, B.J.; Scheuer, P.J. More peptides and other diverse constituents of the marine mollusk philinopsis speciosa. *J. Org. Chem.* **1998**, *63*, 3272–3280. [CrossRef]

31. Han, B.; Goeger, D.; Maier, C.S.; Gerwick, W.H. The wewakpeptins, cyclic depsipeptides from a papua new guinea collection of the marine cyanobacterium *Lyngbya semiplena*. *J. Org. Chem.* **2004**, *70*, 3133–3139. [CrossRef] [PubMed]

32. Horgen, F.D.; Yoshida, W.Y.; Scheuer, P.J. Malevamides a–c, new depsipeptides from the marine cyanobacterium symploca laete-viridis. *J. Nat. Prod.* **2000**, *63*, 461–467. [CrossRef] [PubMed]

33. Tan, L.T.; Sitachitta, N.; Gerwick, W.H. The guineamides, novel cyclic depsipeptides from a papua new guinea collection of the marine cyanobacterium *Lyngbya majuscula*. *J. Nat. Prod.* **2002**, *66*, 764–771. [CrossRef] [PubMed]

34. Williams, P.G.; Yoshida, W.Y.; Quon, M.K.; Moore, R.E.; Paul, V.J. Ulongapeptin, a cytotoxic cyclic depsipeptide from a palauan marine cyanobacterium *Lyngbya* sp. *J. Nat. Prod.* **2003**, *66*, 651–654. [CrossRef] [PubMed]

35. Vining, O.B.; Medina, R.A.; Mitchell, E.A.; Videau, P.; Li, D.; Serrill, J.D.; Kelly, J.X.; Gerwick, W.H.; Proteau, P.J.; Ishmael, J.E.; et al. Depsipeptide companeramides from a panamanian marine cyanobacterium associated with the coibamide producer. *J. Nat. Prod.* **2015**, *78*, 413–420. [CrossRef] [PubMed]

36. Fernández, R.; Rodríguez, J.; Quiñoá, E.; Riguera, R.; Muñoz, L.; Fernández-Suárez, M.; Debitus, C. Onchidin b: A new cyclodepsipeptide from the mollusc *Onchidium* sp. *J. Am. Chem. Soc.* **1996**, *118*, 11635–11643. [CrossRef]

37. Nogle, L.M.; Gerwick, W.H. Isolation of four new cyclic depsipeptides, antanapeptins a–d, and dolastatin 16 from a madagascan collection of *Lyngbya majuscula*. *J. Nat. Prod.* **2001**, *65*, 21–24. [CrossRef]

38. Bunyajetpong, S.; Yoshida, W.Y.; Sitachitta, N.; Kaya, K. Trungapeptins A-C, cyclodepsipeptides from the marine cyanobacterium *Lyngbya majuscula*. *J. Nat. Prod.* **2006**, *69*, 1539–1542. [CrossRef] [PubMed]

39. Tripathi, A.; Puddic, J.; Prinsep, M.R.; Lee, P.P.F.; Tan, L.T. Hantupeptin a, a cytotoxic cyclic depsipeptide from a singapore collection of *Lyngbya majuscula*. *J. Nat. Prod.* **2009**, *72*, 29–32. [CrossRef] [PubMed]

40. Salvador, L.A.; Biggs, J.S.; Paul, V.J.; Luesch, H. Veraguamides a–g, cyclic hexadepsipeptides from a dolastatin 16-producing cyanobacterium symploca cf. Hydnoides from guam. *J. Nat. Prod.* **2011**, *74*, 917–927. [CrossRef] [PubMed]

41. Williams, P.G.; Yoshida, W.Y.; Quon, M.K.; Moore, R.E.; Paul, V.J. The structure of palau'amide, a potent cytotoxin from a species of the marine cyanobacterium *Lyngbya*. *J. Nat. Prod.* **2003**, *66*, 1545–1549. [CrossRef] [PubMed]

42. Pettit, G.R. Isolation and Stuctural Elucidation of the Cytostatic Linear and Cyclo-Depsipeptides Dolastatin 16, Dolastatin 17, and Dolastatin 18. U.S. Patent 6,239,104 B1, 29 May 2001.

43. Luesch, H.; Yoshida, W.Y.; Moore, R.E.; Paul, V.J. Apramides a–g, novel lipopeptides from the marine cyanobacterium *Lyngbya majuscula*. *J. Nat. Prod.* **2000**, *63*, 1106–1112. [CrossRef] [PubMed]

44. Jiménez, J.I.; Scheuer, P.J. New lipopeptides from the caribbean cyanobacterium *Lyngbya majuscula*. *J. Nat. Prod.* **2001**, *64*, 200–203. [CrossRef] [PubMed]

45. McPhail, K.L.; Correa, J.; Linington, R.G.; González, J.; Ortega-Barría, E.; Capson, T.L.; Gerwick, W.H. Antimalarial linear lipopeptides from a panamanian strain of the marine cyanobacterium *Lyngbya majuscula*. *J. Nat. Prod.* **2007**, *70*, 984–988. [CrossRef] [PubMed]

46. Balunas, M.J.; Linington, R.G.; Tidgewell, K.; Fenner, A.M.; Ureña, L.-D.; Togna, G.D.; Kyle, D.E.; Gerwick, W.H. Dragonamide e, a modified linear lipopeptide from *Lyngbya majuscula* with antileishmanial activity. *J. Nat. Prod.* **2010**, *73*, 60–66. [CrossRef] [PubMed]

47. Gunasekera, S.P.; Ross, C.; Paul, V.J.; Matthew, S.; Luesch, H. Dragonamides c and d, linear lipopeptides from the marine cyanobacterium brown *Lyngbya polychroa*. *J. Nat. Prod.* **2008**, *71*, 887–890. [CrossRef] [PubMed]

48. Quintana, J.; Bayona, L.M.; Castellanos, L.; Puyana, M.; Camargo, P.; Aristizabal, F.; Edwards, C.; Tabudravu, J.N.; Jaspars, M.; Ramos, F.A. Almiramide d, cytotoxic peptide from the marine cyanobacterium *oscillatoria nigroviridis*. *Bioorg. Med. Chem.* **2014**, *22*, 6789–6795. [CrossRef] [PubMed]

49. Simmons, T.L.; Engene, N.; Ureña, L.D.; Romero, L.I.; Ortega-Barría, E.; Gerwick, L.; Gerwick, W.H. Viridamides a and b, lipodepsipeptides with antiprotozoal activity from the marine cyanobacterium *oscillatoria nigro-wiridis*. *J. Nat. Prod.* **2008**, *71*, 1544–1550. [CrossRef] [PubMed]

50. Hooper, G.J.; Orjala, J.; Schatzman, R.C.; Gerwick, W.H. Carmabins a and b, new lipopeptides from the caribbean cyanobacterium *Lyngbya majuscula*. *J. Nat. Prod.* **1998**, *61*, 529–533. [CrossRef] [PubMed]

51. Edwards, D.J.; Marquez, B.L.; Nogle, L.M.; McPhail, K.; Goeger, D.E.; Roberts, M.A.; Gerwick, W.H. Structure and biosynthesis of the jamaicamides, new mixed polyketide-peptide neurotoxins from the marine cyanobacterium *Lyngbya majuscula*. *Chem. Biol.* **2004**, *11*, 817–833. [CrossRef] [PubMed]

52. Jones, A.C.; Monroe, E.A.; Podell, S.; Hess, W.R.; Klages, S.; Esquenazi, E.; Niessen, S.; Hoover, H.; Rothmann, M.; Lasken, R.S.; et al. Genomic insights into the physiology and ecology of the marine filamentous cyanobacterium *Lyngbya majuscula*. *Proc. Natl. Acad. Sci. USA* **2011**, *108*, 8815–8820. [CrossRef] [PubMed]

53. Zhu, X.; Liu, J.; Zhang, W. De novo biosynthesis of terminal alkyne-labeled natural products. *Nat. Chem. Boil.* **2015**, *11*, 115–120. [CrossRef] [PubMed]

54. Cruz-Rivera, E.; Paul, V.J. Chemical deterrence of a cyanobacterial metabolite against generalized and specialized grazers. *J. Chem. Ecol.* **2007**, *33*, 213–217. [CrossRef] [PubMed]

55. Xu, Z.; Peng, Y.; Ye, T. The total synthesis and stereochemical revision of yanucamide a. *Org. Lett.* **2003**, *5*, 2821–2824. [CrossRef] [PubMed]

56. Vorde, C.; Hogberg, H.-E.; Hedenström, E. Resolution of 2-methylalkanoic esters: Enantioselective aminolysis by (*R*)-l-phenylethylamine of ethyl 2-methyloctanoate catalysed by lipase B from *Candida antarctica*. *Tetrahedron Asymmetry* **1996**, *7*, 1507–1513. [CrossRef]

57. Engel, K.-H. Lipase-catalyzed enantioselective esterification of 2-methylalkanoic acids. *Tetrahedron Asymmetry* **1991**, *2*, 165–168. [CrossRef]

58. Berglund, P.; Holmquist, M.; Hedenstrom, E.; Hult, K.; Hiigberg, H.-E. 2-Methylalkanoic acids resolved by esterification catalysed by lipase from candida rugosa: Alcohol chain length and enantioselectivity. *Tetrahedron Asymmetry* **1993**, *4*, 1869–1878. [CrossRef]

marine drugs

Article

Anti-Fatigue Effect by Peptide Fraction from Protein Hydrolysate of Croceine Croaker (*Pseudosciaena crocea*) Swim Bladder through Inhibiting the Oxidative Reactions including DNA Damage

Yu-Qin Zhao, Li Zeng, Zui-Su Yang, Fang-Fang Huang, Guo-Fang Ding * and Bin Wang *

Zhejiang Provincial Engineering Technology Research Center of Marine Biomedical Products,
School of Food and Pharmacy, Zhejiang Ocean University, 1st Haidanan Road, Changzhi Island, Lincheng,
Zhoushan 316022, China; zhaoy@hotmail.com (Y.-Q.Z.); 9001000@163.com (L.Z.); yangzs87@163.com (Z.-S.Y.);
gracegang@126.com (F.-F.H.)
* Correspondence: dinggf2007@163.com (G.-F.D.); wangbin4159@hotmail.com (B.W.);
  Tel.: +86-580-229-9809 (G.-F.D.); +86-580-255-5085 (B.W.);
  Fax: +86-580-229-9809 (G.-F.D.); +86-580-255-4781 (B.W.)

Academic Editor: Se-Kwon Kim
Received: 17 September 2016; Accepted: 24 November 2016; Published: 13 December 2016

Abstract: The swim bladder of the croceine croaker (*Pseudosciaena crocea*) was believed to have good curative effects in various diseases, including amnesia, insomnia, dizziness, anepithymia, and weakness after giving birth, in traditional Chinese medicine. However, there is no research focusing on the antioxidant and anti-fatigue peptides from croceine croaker swim bladders at present. Therefore, the purpose of this study was to investigate the bioactivities of peptide fractions from the protein hydrolysate of croceine croaker related to antioxidant and anti-fatigue effects. In the study, swim bladder peptide fraction (SBP-III-3) was isolated from the protein hydrolysate of the croceine croaker, and its antioxidant and anti-fatigue activities were measured using in vitro and in vivo methods. The results indicated that SBP-III-3 exhibited good scavenging activities on hydroxyl radicals (HO•) (EC_{50} (the concentration where a sample caused a 50% decrease of the initial concentration of HO•) = 0.867 mg/mL), 2,2-diphenyl-1-picrylhydrazyl radicals (DPPH•) (EC_{50} = 0.895 mg/mL), superoxide anion radical ($O_2^- •$) (EC_{50} = 0.871 mg/mL), and 2,2′-azino-bis-3-ethylbenzothiazoline-6-sulfonic acid radical (ABTS$^+$•) (EC_{50} = 0.346 mg/mL). SBP-III-3 also showed protective effects on DNA damage in a concentration-effect manner and prolonged the swimming time to exhaustion of Institute of Cancer Research (ICR) mice by 57.9%–107.5% greater than that of the control. SBP-III-3 could increase the levels of muscle glucose (9.4%–115.2% increase) and liver glycogen (35.7%–157.3%), and decrease the levels of blood urea nitrogen (BUN), lactic acid (LA), and malondialdehyde (MDA) by 16.4%–22.4%, 13.9%–20.1%, and 28.0%–53.6%, respectively. SBP-III-3 also enhanced the activity of lactic dehydrogenase to scavenge excessive LA for slowing the development of fatigue. In addition, SBP-III-3 increased the activities superoxide dismutase, catalase, and glutathione peroxidase to reduce the reactive oxygen species (ROS) damage in mice. In conclusion, SBP-III-3 possessed good anti-fatigue capacities on mice by inhibiting the oxidative reactions and provided an important basis for developing the swim bladder peptide functional food.

Keywords: croceine croaker (*Pseudosciaena crocea*); swim bladder; peptide; antioxidant activity; anti-fatigue activity

1. Introduction

Fatigue is one of the most common and disabling non-motor problems, which generally leads to negative effects on physical and cognitive function. Therefore, fatigue is best defined as the difficulty in initiating or sustaining voluntary activities, and classified into mental and physical fatigue [1]. Exercise-induced fatigue usually associates with increased stress levels caused by modern lifestyles, along with a decline in exercise performance [2]. At present, several theories including "exhaustion theory" and "radical theory" have been put forward to explain the mechanisms of exercise-induced fatigue [1]. Among them, the "exhaustion theory" speculates that energy sources, including glucose and liver glycogen, will be exhausted during exercise, which is accompanied by physical fatigue. In this theory, some studies indicated that post-exercise nutrition through the administration of proteins, saccharides, and amino acids can eliminate the accumulated harmful metabolites, repair the damage of organisms, and facilitate recovery from fatigue [3]. Compared with these nutrient substances, protein hydrolysates and peptides have been widely studied due to their potential health benefits associated with high bioactivities, low molecular weight (MW), easy absorption, and less toxicity [4,5]. Ding et al. reported that jellyfish collagen hydrolysate could promote climbing endurance and had anti-fatigue effects in rats [2]. Wang et al. reported that the spleen-derived peptide CMS001 had anti-fatigue effects in mice. Therefore, bioactive protein hydrolysates and peptides are believed to be helpful for counteracting and ameliorating physical fatigue [6].

Except for exhaustion of energy sources, high-intensity exercise often destroys the balance between the oxidation system and anti-oxidation system of human body. The accumulated reactive oxygen species (ROS) will put the body in a state of oxidative stress and bring injury to the body by attacking biomacromolecules and cell organs [6,7]. Some reports found that exogenous dietary antioxidants can decrease the contribution of exercise-induced oxidative stress and improve the animal's physiological condition [5]. You et al. reported that loach peptide could scavenge hydroxyl radical (HO•) and, 2,2-diphenyl-1-picrylhydrazyl radical (DPPH•) in vitro and increase the activities of superoxide dismutase (SOD), catalase (CAT), and glutathione peroxidase (GSH-Px) in vivo [1]. Wei et al. reported that high Fischer ratio oligopeptides derived from food sources such as corn, tuna, and *pinctada martensii*, could scavenge free radicals in vitro and increase the activities of SOD, CAT, and GSH-Px in vivo [8]. In addition, high Fischer ratio oligopeptides could prolong the swimming time, increase liver glucogen contents, and lower blood urea nitrogen (BUN) and lactic acid (LA) levels of exercised mice. However, the anti-fatigue and anti-oxidation mechanisms of protein hydrolysates and peptides have not been fully elucidated. Therefore, more detailed research should be done seeking more high-efficiency antioxidant peptides used in the daily diet to reduce oxidative damage and fight against fatigue.

At present, large quantities of byproducts, accounting for 50%–70% of the original raw material, are generated during the aquatic products processing, and optimal use of these byproducts are an effective approach to protect the environment and produce value-added products to increase the revenue of fish processors [9]. Therefore, preparation of protein hydrolysates and peptides from fish byproducts are extensively researched [5,10]. Proteins in food resources possess a variety of active peptide sequences, and enzymatic hydrolysis is thought as an effective method to release those active fragments without destroying their nutritional value and adding hazardous substances, including residual organic solvents and toxic chemicals in the final products [4,5]. HO• attacks almost every molecule in living cells and is demonstrated to be a highly damaging species in free radical pathology. Thus, the removal of HO• is probably one of the most effective defenses against various diseases for a living body [10]. In our previous research, three antioxidant peptides including Tyr–Leu–Ser–Met–Ser–Arg (YLSMSR), Val–Leu–Tyr–Glu–Glu (VLYEE), and Met–Ile–Leu–Met–Arg (MILMR) were isolated from proteins hydrolysate of croceine croaker (*Pseudosciaena crocea*) muscle and showed strong DPPH•, HO•, superoxide anion radical (O_2^-•), and 2,2'-azino-bis-3-ethylbenzothiazoline-6-sulfonic acid radical (ABTS$^+$•) scavenging activities [11]. Acid and pepsin-soluble collagens from croceine croaker scales were prepared and characterized [12], and antioxidant peptides including Gly–Phe–Arg–Gly–Thr–Ile–Gly–Leu–Val–Gly (GFRGTIGLVG), Gly–Pro–Ala–Gly–Pro–Ala–Gly (GPAGPAG), and Gly–Phe–Pro–Ser–Gly (GFPSG) from the acid-soluble collagen showed strong DPPH•, HO•, ABTS$^+$•, and O$_2^-$• scavenging activities [13]. Traditional Chinese medicine considers that swim bladder of the croceine croaker to have good curative effects in various diseases, including amnesia, insomnia, dizzy, anepithymia, and weakness after giving a birth, and present researchers also suggest that it could serve to remove free radicals and ward against inflammation and cancer [14]. However, there was no research focusing on the antioxidant and anti-fatigue peptides from croceine croaker swim bladders. Therefore, the objectives of the present study were to prepare the active peptide fraction from croceine croaker swim bladders, and its bioactivities related to antioxidant and anti-fatigue effects of prepared fraction were also evaluated.

2. Results and Discussion

2.1. Preparation of Protein Hydrolysates of Swim Bladder and Their HO• Scavenging Activities

In the experiment, four proteases, including alcalase, papain, pepsin, and trypsin were used to hydrolyze the proteins of croceine croaker swim bladders, respectively. The degree of hydrolysis (DH) and the HO• scavenging activities were used to screen the most suitable enzymes for subsequent experiments, and HO• scavenging activities was expressed as EC$_{50}$ (the concentration where a sample caused a 50% decrease of the initial concentration of HO•) (Table 1). The DH (%) of alcalase hydrolysate was 22.32% ± 0.74%, which was significantly higher than those of papain hydrolysate (17.84% ± 0.71%), pepsin hydrolysate (19.52% ± 0.49%), trypsin hydrolysate (16.21% ± 0.37%), and neutrase hydrolysate (21.37% ± 0.67%) ($p < 0.05$). The EC$_{50}$ value of alcalase hydrolysate was 8.85 mg/mL, which was significantly lower than those of papain hydrolysate (11.76 mg/mL), pepsin hydrolysate (13.68 mg/mL), and trypsin hydrolysate (10.02 mg/mL) ($p < 0.05$). The result was in accordance with the previous reports that high DH and low MW of hydrolysates made a great contribute to their antioxidant activities including HO• scavenging activity [15,16]. Proteases digest long protein chains into shorter fragments by splitting the peptide bonds that link amino acid residues. Due to the specificity of enzymes reactions, protein hydrolysates from the same proteins hydrolyzed using different proteases exhibit different DH and bioactivities because the obtained peptides are diverse in terms of chain length and amino acid sequence [5]. Therefore, alcalase hydrolysate (designated as SBP) was selected for further study.

Table 1. HO• scavenging activities of protein hydrolysate of scalloped hammerhead cartilage using different proteases (c = 15 mg protein/mL). All of the values were mean ± standard deviation (SD) (n = 3).

Protease	Enzymolysis Condition	Degree of Hydrolysis (DH%)	HO• Scavenging Rate (%)
Papain	pH 7.0, 60 °C, 4 h, total enzyme dose 2.5%	17.84 ± 0.71 [a]	34.85 ± 1.05 [a]
Alcalase	pH 8.0, 50 °C, 4 h, total enzyme dose 2.5%	22.32 ± 0.74 [b]	54.76 ± 1.94 [b]
Trypsin	pH 8.0, 40 °C, 4 h, total enzyme dose 2.5%	16.21 ± 0.37 [c]	62.38 ± 1.67 [c]
Pepsin	pH 2.0, 37 °C, 4 h, total enzyme dose 2.5%	19.52 ± 0.49 [d]	55.47 ± 2.02 [b]
Neutrase	pH 6.0, 50 °C, 4 h, total enzyme dose 2.5%	21.37 ± 0.67 [b]	50.67 ± 1.85 [d]

[a–d] Values with different letters indicated significant differences at the same concentration ($p < 0.05$).

2.2. Preparation of Antioxidant Peptides from SBP

2.2.1. Fractionation of SBP by Ultrafiltration

Protein hydrolysate is composed of peptides with different molecular sizes and free amino acids, and their bioactivities are influenced by the molecular size of peptides. Therefore, ultrafiltration is a popular method for fractionation and enrichment concentration of peptides with specific molecular sizes from hydrolysates [17]. To obtain purified swim bladder peptide, SBP was fractionated by ultrafiltration with three molecular weight (MW) cut-off membranes of 10, 5, and 3 kDa in turn, respectively, and the resulting four fractions were prepared and named as SBP-I (MW < 3 kDa), SBP-II (3 kDa < MW < 5 kDa), SBP-III (5 kDa < MW < 10 kDa),, and SBP-IV (MW > 10 kDa), respectively. The yields of SBP-I, SBP-II, SBP-III, and SBP-IV were 10.08, 17.30, 23.37, and 28.97 mg protein/g swim bladder, respectively.

For acquiring the fraction with high antioxidant activity, HO• scavenging activities of four prepared fractions were measured, and the results indicated that HO• scavenging activities of SBP, SBP-I, SBP-II, SBP-III, and SBP-IV were 50.32% ± 0.90%, 34.89% ± 2.01%, 40.63% ± 1.59%, 58.36% ± 1.72%, and 52.57% ± 2.20%, respectively, at the concentration of 5 mg/mL. SBP-III showed significantly higher antioxidant activity than SBP and the other three fractions at the tested concentrations ($p < 0.05$). In addition, the EC_{50} of SBP-III was 3.579 mg/mL and significantly less than that of SBP (8.85 mg/mL). The result was in line with the previous report that that samples with lower average molecular weights possibly contained more substrates, which were electron donors and could react with free radicals to convert them to more stable products and terminate the radical chain reactions [9]. From the data, it could be concluded that SBP-III contained more effective antioxidant peptides and could be chosen for subsequent separation.

2.2.2. Gel Filtration Chromatography of SBP-III

Gel filtration chromatography is an effective separation technique on the basis of molecule size and widely applied to separate components in a mixture [10]. As shown in Figure 1, SBP-III was separated into four subfractions (SBP-III-1 to SBP-III-4) using a Sephadex G-25 column. From the linear equation (Log MW = $-0.2036R_t$ + 7.6164, R^2 = 0.9766), the MWs of four subfractions were 9.48 kDa (SBP-III-1), 7.74 kDa (SBP-III-2), 6.78 kDa (SBP-III-3), and 5.79 kDa (SBP-III-4), respectively. HO• scavenging activities of SBP-III-3 was 58.53% ± 2.17% at the concentration of 1 mg/mL, which was significantly higher than those of SBP-III-1 (21.15% ± 1.03%), SBP-III-2 (42.67% ± 1.94%), and SBP-III-4 (33.27% ± 2.11%), respectively ($p < 0.05$). The data indicated that SBP-III-3 could effectively restrain the production of HO• and terminate the radical chain reaction. Pan et al. reported that the hydrolysate subfraction from the skate (*Raja porosa*) cartilage protein using a Sephadex G-15 column had higher radical scavenging activity than other subfractions with smaller molecular size [10]. These results indicated that some factors, including amino acid composition and sequence, may also influence the activities of peptides in addition to the MW. Therefore, SBP-III-3 was chosen for further evaluation on in vitro antioxidant activity and the in vivo anti-fatigue effect.

2.3. In Vitro Antioxidant Activity of SBP-III-3

2.3.1. HO• Scavenging Activity

HO• is a highly reactive radical to the organism because it can destroy virtually all types of macromolecules including carbohydrates, nucleic acids (mutations), lipids (lipid peroxidation), and amino acids (e.g., conversion of Phe to m-Tyrosine and o-Tyrosine). As shown in Figure 2A, the HO• scavenging rate of SBP-III-3 showed a dose-response relationship, and the EC_{50} of SBP-III-3 was 0.867 mg/mL, which was lower than those of Pro–Ser–Tyr–Val (PSYV) (2.64 mg/mL) [18], Pro–Ser–Lys–Tyr–Glu–Pro–Phe–Val (PSKYEPFV) (2.86 mg/mL) [19], Pro–Tyr–Ser–Phe–Lys (PYSFK) (2.283 mg/mL), Gly–Phe–Gly–Pro–Leu (GFGPL) (1.612 mg/mL), Val–Gly–Gly–Arg–Pro (VGGRP)

(2.055 mg/mL) [20], Phe–Ile–Met–Gly–Pro–Tyr (FIMGPY) (3.037 mg/mL), Gly–Pro–Ala–Gly–Asp–Tyr (GPAGDY) (3.92 mg/mL), and Ile–Val–Ala–Gly–Pro–Gln (IVAGPQ) (5.03 mg/mL) [10] from protein hydrolysates of weatherfish loach muscle, grass carp muscle and skin, and skate cartilages. SBP-III-3 showed good HO• scavenging activity, which demonstrated that it could serve as a scavenger for reducing the damage induced by HO• in biological systems.

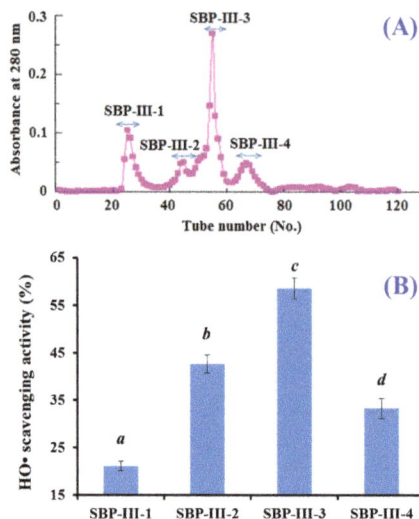

Figure 1. (**A**) Gel filtration chromatography of SBP-III on a Sephadex G-25 column; and (**B**) HO• scavenging activities of subfractions from SBP-III. All of the values were mean ± SD (n = 3). a–d Columnwise values with the same superscripts of this type indicated no significant difference ($p > 0.05$).

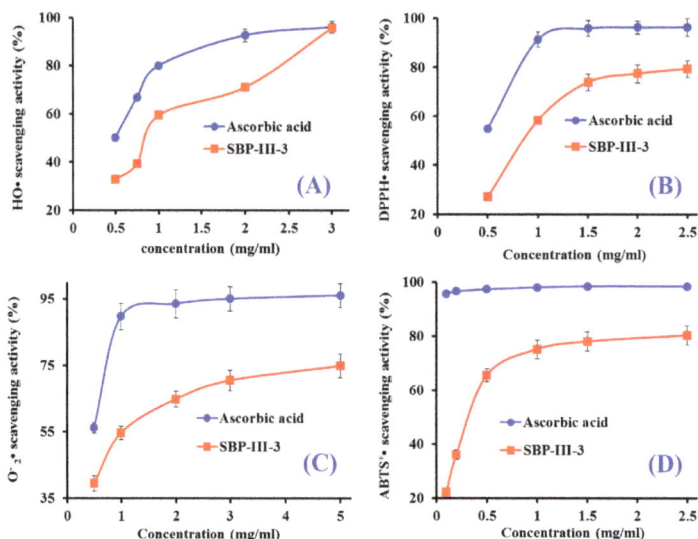

Figure 2. HO• (**A**), DPPH• (**B**), O_2^- • (**C**), and ABTS$^+$• (**D**) scavenging activities of SBP-III-3. All of the values were mean ± SD (n = 3).

2.3.2. DPPH• Scavenging Activity

DPPH• scavenging assay is popular and efficient in predicting the antioxidant activities of protein hydrolysates and peptides. It has a deep violet colour in solution and generates a strong absorption band at about 517 nm. The solution becomes colourless or pale yellow following the reduction of the absorption value at 517 nm when the radicals are neutralized [21]. Therefore, DPPH• scavenging activity of SBP-III-3 was measured and shown in Figure 2B. SBP-III-3 scavenged DPPH• in a concentration-effect manner with EC_{50} of 0.895 mg/mL, but its activity was lower than the positive control of ascorbic acid. In addition, the EC_{50} of SBP-III-3 was lower than those of PSYV (17.0 mg/mL) [18], Phe–Leu–Asn–Glu–Phe–Leu–His–Val (FLNEFLHV) (4.950 mg/mL) [22], Thr–Thr–Ala–Asn–Ile–Glu–Asp–Arg–Arg (TTANIEDRR) (2.503 mg/mL) [23], FIMGPY (2.60 mg/mL), GPAGDY (3.48 mg/mL) and IVAGPQ (3.93 mg/mL) [10], PYSFK (1.575 mg/mL) [20], and Leu–Leu–Pro–Phe (LLPF) (1.084 mg/mL) [24] from hydrolysates of loach, blue mussel, bluefin leatherjacket, salmon pectoral fin, skate cartilage, grass carp skin, and corn gluten meal, but it was higher than those of Gly–Ser–Gln (GSQ) (0.61 mg/mL) [25], Pro–Ile–Ile–Val–Tyr–Trp–Lys (PIIVYWK) (0.713 mg/mL) [20], His–Phe–Gly–Asp–Pro–Phe–His (HFGBPFH) (0.20 mg/mL) [26], Phe–Leu–Pro–Phe (FLPF) (0.789 mg/mL), and Leu–Pro–Phe (LPF) (0.777 mg/mL) [24] from protein hydrolysates of Chinese leek, blue mussel, grass carp skin, mussel sauce and corn gluten meal. Therefore, these results indicated that SBP-III-3 had the strong ability to donate an electron or hydrogen radical for inhibiting the DPPH• reaction.

2.3.3. O_2^- • Scavenging Activity

O_2^- • can promote oxidative reaction to generate H_2O_2 and HO• to damage the biomacromolecule because it can release protein-bound metals and form perhydroxyl radicals which initiate lipid oxidation. The O_2^- • scavenging activity of SBP-III-3 was increased with increasing concentration ranged from 0.5 mg/mL to 5.0 mg/mL (Figure 2C). The IC_{50} value of SBP-III-3 was 0.871 mg/mL, which was lower than those of MILMR (0.993 mg/mL) [11], FIMGPY (1.61 mg/mL), GPAGDY (1.66 mg/mL), and IVAGPQ (1.82 mg/mL) [10] from protein hydrolysates of croceine croaker muscle and skate cartilage. Therefore, SBP-III-3 might have ability to remove O_2^- • damage in biological systems.

2.3.4. ABTS$^+$• Scavenging Activity

The blue/green ABTS$^+$• produced by oxidation of ABTS with $K_2S_2O_8$ has an absorption maximum of 734 nm and can be converted back to its colorless neutral form by antioxidants following the decrease of the absorption. As shown in Figure 2D, SBP-III-3 showed strong ABTS$^+$• scavenging activity in a dose-effect manner with EC_{50} value of 0.346 mg/mL, which was lower than those of FLNEFLHV (1.548 mg/mL) [22], FLPF (1.497 mg/mL), LPF (1.013 mg/mL), LLPF (1.031 mg/mL) [24], FIMGPY (1.04 mg/mL), GPAGDY (0.77 mg/mL), and IVAGPQ (1.29 mg/mL) [10], VGGRP (0.465 mg/mL) [20], Trp–Glu–Gly–Pro–Lys (WEGPK) (5.407 mg/mL), Gly–Pro–Pro (GPP) (2.472 mg/mL), and Gly–Val–Pro–Leu–Thr (GVPLT) (3.124 mg/mL) [17] from protein hydrolysates of salmon, corn gluten meal, skate cartilage, grass carp skin, and bluefin leatherjacket heads. These results indicated that SBP-III-3 could convert ABTS$^+$• to its colorless neutral form and block the free radical reaction.

2.3.5. Protective Activity against Free Radical-Induced DNA Damage

In the organism, the excessive production of ROS may cause a quantity of degenerative processes such as cancer, premature aging, and cardiovascular and neurodegenerative diseases, while DNA damage is a key step in these ROS-induced effects [13,27]. Therefore, the protective activity of SBP-III-3 against oxidative damage of DNA induced by H_2O_2 was also evaluated and showed in Figure 3. The damage of plasmid DNA results in a cleavage of one of the phosphodiester chains and produces a relaxed open circular form. Further cleavage near the first breakage leads to linear double stranded DNA molecules. The formation of circular form of DNA is indicative of single-strand breaks and the formation of linear form of DNA is indicative of double-strand breaks [28]. The plasmid DNA (pBR322DNA) was mainly of the supercoiled form in the absence of $FeSO_4$ and H_2O_2 (Figure 3, lane 1, control). HO• would be generated from iron-mediated decomposition of H_2O_2 when $FeSO_4$ and H_2O_2 were added into the sample, and it subsequently broke the supercoiled DNA and converted the supercoiled form into the open circular form (Figure 3, lane 5). In the experiment, the linear form of DNA was not observed, which indicated that the generated HO• from iron-mediated decomposition of H_2O_2 might be too small and could not break the double-strand of DNA. As shown in Figure 3 (lanes 2, 3, and 4), the contents of supercoiled form of DNA was obvious higher than that of Figure 3 (lane 5). In addition, the contents of supercoiled form of DNA in Figure 3 (lane 2) were higher than that of Figure 3 (lane 4). These data indicated that both SBP-III-3 and the positive control of ascorbic acid could have protective effects on DNA damage in a concentration-effect manner. Therefore, SBP-III-3 could prevent the reaction of Fe^{2+} with H_2O_2 and directly scavenge HO• by donating a hydrogen atom or electron and, therefore, protecting the supercoiled plasmid DNA from HO• dependent strand breaks. This finding was in line with the result that SBP-III-3 could effectively scavenge HO• in HO• scavenging assay in vitro.

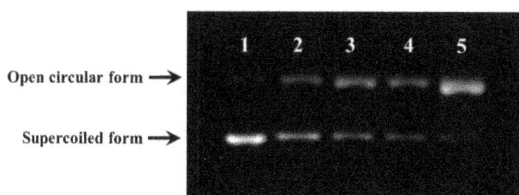

Figure 3. DNA damage protective effect of SBP-III-3. Lane 1, the native pBR322DNA; lanes 2, the DNA treated with $FeSO_4$, H_2O_2 and SBP-III-3 (3.0 mg/mL); lane 3, the DNA treated with $FeSO_4$, H_2O_2, and ascorbic acid (1.0 mg/mL); lane 4, the DNA treated with $FeSO_4$, H_2O_2, and SBP-III-3 (1.0 mg/mL); lane 5, the pBR322DNA treated with $FeSO_4$ and H_2O_2.

2.4. In Vivo Anti-Fatigue Effects of SBP-III-3

2.4.1. SBP-III-3 Prolonged Exhaustive Swimming Time

Exercise tolerance assay is the most direct and objective indicators of reflecting physical fatigue. Swimming to exhaustion is an experimental exercise model to evaluate anti-fatigue; it works well for evaluating the endurance capacity of mice, and gives a high reproducibility [29]. The improvement of exercise endurance was the most powerful representation of anti-fatigue effect. In the experiment, the anti-fatigue effect of SBP-III-3 was investigated through the weight-loaded swimming test, and the length of the swimming time to exhaustion indicated the degree of fatigue. As shown in Figure 4, the mean exhaustion time of the SBP-III-3-HG was 33.41 ± 2.40 min (107.5% greater than that of NCG); the mean exhaustion time of the SBP-III-3-MG was 28.86 ± 1.01 min (79.2% greater than that of NCG); and the mean exhaustion time of the SBP-III-3-LG was 25.43 ± 1.91 min (57.9% greater than that of NCG) ($p < 0.05$ or $p < 0.01$). Therefore, the average loaded swimming time of mice was significantly longer in the SBP-III-3 treatment group (SBP-III-3-LG, SBP-III-3-MG, SBP-III-3-HG) than that of the normal control group (NCG) (16.1 ± 1.46 min) ($p < 0.05$ or $p < 0.01$), and these results indicate that SBP-III-3 has significant effects on movement and endurance in mice, thereby postponing the fatigue.

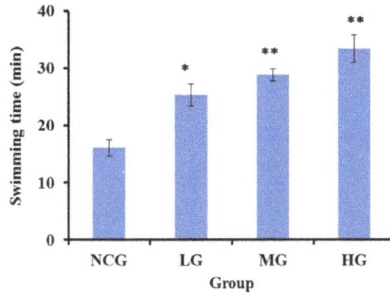

Figure 4. Effects of SBP-III-3 on loaded swimming time of mice. All the values were mean ± SD (n = 12). * p < 0.05, ** p < 0.01 compared with NCG.

2.4.2. Biologic Parameters Determination

SBP-III-3 Decreased Blood Urea Nitrogen (BUN)

Urea is formed in the liver as the end product of protein metabolism. During digestion, protein is broken down into small peptides and amino acids. The amino acid nitrogen is removed as NH_4, while the rest of the molecule is used to produce energy or other substances needed by the cell [30]. Thus, BUN is the metabolic outcome of amino acids and protein, and is one of the sensitive parameters related to fatigue. Therefore, it is usually applied to evaluate the tolerance capability when an animal suffers from a weight load. In other words, the less an animal is adapted to exercise, the more the BUN level increases [1]. As shown in Table 2, the BUN levels of the mice were significantly lower by 16.38%, 16.59%, and 22.06% in the SBP-III-3-LH, SBP-III-3-MG and SBP-III-3-HG compared to the NCG (p < 0.05 or p < 0.01), respectively. It was clear that SBP-III-3 treatment weakened the increase in BUN levels induced by catabolism of amino acids and proteins, which indicated that SBP-III-3 could reduce decomposition of proteins for energy, enhance adaptive capacity to physical load, and eventually improve tolerance capacity. The reduced protein metabolism of SBP-III-3 treatment groups is indicative of enhanced endurance.

Table 2. Effects of SBP-III-3 on BUN, LA, LDH, liver glycogen, muscle glycogen, SOD, GSH-Px, CAT, and MDA in mice (n = 3).

	NCG	SBP-III-3-LG	SBP-III-3-MG	SBP-III-3-HG
BUN (mmol/L)	9.34 ± 0.39	7.81 ± 0.61 **	7.79 ± 0.47 **	7.28 ± 0.43 **,a
LA (mmol/L)	3.08 ± 0.21	2.65 ± 0.47 *	2.56 ± 0.35 *	2.46 ± 0.34 *
LDH (U/gprot)	2784.95 ± 322.92	3397.10 ± 215.90	3605.87 ± 315.21 *	3690.76 ± 337.18 *
Liver glycogen (mg/g)	8.32 ± 0.47	11.29 ± 2.31 *	17.36 ± 1.16 **	21.41 ± 5.23 **,b
Muscle glycogen (mg/g)	2.23 ± 0.56	2.44 ± 0.36	3.39 ± 0.35 *	4.80 ± 1.12 *
SOD (U/mg prot)	68.82 ± 6.17	71.74 ± 2.52	79.63 ± 7.40 *	99.24 ± 4.38 **,b
GSH-Px (IU)	43.22 ± 4.09	71.89 ± 2.34	102.05 ± 5.78 **	147.16 ± 12.80 **,b
CAT (U/g prot)	186.14 ± 2.26	325.27 ± 1.52 *	349.75 ± 4.09 *	483.00 ± 5.87 **,a
MDA in liver (mmol/L)	2.39 ± 0.55	1.72 ± 0.25 **	1.23 ± 0.31 **,a	1.11 ± 0.23 **,b
MDA in plasma (mmol/L)	19.92 ± 2.87	11.75 ± 2.62 **	9.97 ± 1.31 **	9.50 ± 0.55 **

* p < 0.05, ** p < 0.01 compared with the control; [a] p < 0.05, [b] p < 0.01 compared with the low group.

SBP-III-3 Decreased Lactic Acid (LA)

The response to exercise in mammals begins with an increase in aerobic muscular activity, which switches over to anaerobic metabolism if the exercise is intense, which leads to the accumulation of LA [31]. Thus, the accumulation of blood serum LA is an important cause of fatigue. The increased content of LA will lower the pH in muscle tissue and blood, and induce some side effects of various physiological and biochemical processes, which affect both the cardio-circulating system and the skeletal muscle system function, and then do harm to the body performance [6]. The decrease in the contractive strength of the muscle eventually induces fatigue [32]. Therefore, LA was measured as another index to evaluate the level of fatigue. Table 2 showed the LA levels of the mice were significantly lower by 13.96%, 16.88%, and 20.13% in the SBP-III-3-LH, SBP-III-3-MG, and SBP-III-3-HG compared to the NCG ($p < 0.05$), respectively. The LA values of mice from these three groups had a similar trend in the BUN levels. The results indicated that reducing the LA levels might be an anti-fatigue pathway of SBP-III-3.

SBP-III-3 Increased the Activity of Lactic Dehydrogenase (LDH)

Serum LDH is known to be an accurate indicator of muscle damage, the normal function of LDH in cells is to catalyse the interconversion of pyruvate and lactate, thereby reducing the accumulation of LA in muscle [30]. As shown in Table 3, LDH activity was significantly higher in the SBP-III-3-MG (3605.87 ± 315.21 U/gprot) and SBP-III-3-HG (3690.76 ± 337.18U/gprot) compared to the NCG (2784.95 ± 322.92 U/gprot) ($p < 0.05$). LDH activity in the SBP-III-3-LG (3397.10 ± 215.90 U/gprot) was higher than that of the NCG (2784.95 ± 322.92 U/gprot), but the LDH activity showed no significant difference between SBP-III-3-LG and the NCG ($p < 0.05$). The present results suggested that SBP-III-3, especially high-dose groups could scavenge excessive LA by enhancing the activity of LDH, thereby slowing the development of fatigue.

Table 3. The amino acid composition of SBP-III-3 ($n = 3$).

Amino Acid	Concentration (μmol/L)	Composition (%)
Asp	871.33 ± 18.56	4.60 ± 0.10
Glu	1317.30 ± 35.14	6.86 ± 0.18
Ser	540.61 ± 20.47	2.82 ± 0.11
Gly	6744.51 ± 143.47	35.10 ± 0.75
His	86.39 ± 2.34	0.45 ± 0.04
Arg	718.84 ± 12.38	3.74 ± 0.06
Thr	450.29 ± 13.54	2.34 ± 0.07
Ala	2343.97 ± 34.56	13.46 ± 0.19
Pro	2165.95 ± 40.68	11.28 ± 0.21
Hyp	1629.48 ± 26.87	8.49 ± 0.14
Tyr	60.96 ± 1.35	0.32 ± 0.01
Val	499.64 ± 12.58	2.60 ± 0.07
Met	244.28 ± 8.51	1.27 ± 0.05
Ile	207.49 ± 8.34	1.08 ± 0.04
Leu	388.61 ± 12.97	2.02 ± 0.07
Phe	233.10 ± 9.08	1.21 ± 0.05
Lys	453.21 ± 14.58	2.36 ± 0.08
Essential amino acid (EAA)	3017.23 ± 86.37	15.70 ± 0.45
Total	18,955.96 ± 315.20	100%

SBP-III-3 Increased Liver and Muscle Glycogens

The level of energy stored as glycogen is of great importance in evaluating the capacity for high-intense exercise. Energy providing for exercise is derived initially from the decomposition of glycogen, and then from circulation glycogen released by the muscle and liver [33]. When the muscle glycogen level decreases during exercise, the reduction of liver glycogen may be the limiting factor in the capacity of endurance exercise. Thus, increasing the liver and muscle glycogen storage contributes to elevating the tolerance capacity and athletic capacity [33]. Fatigue will happen when the liver and muscle glycogen is mostly consumed [32]. Table 2 showed the liver glycogen levels of the mice in SBP-III-3-LH, SBP-III-3-MG and SBP-III-3-HG groups were increased by 35.70%, 108.65%, and 157.33%, respectively, and the liver glycogen levels in three treated groups were significantly higher than that in the NCG ($p < 0.05$ or $p < 0.01$). Muscle glycogen levels of the mice were significantly higher in the SBP-III-3-MG (3.39 ± 0.35 mg/g) and SBP-III-3-HG (4.80 ± 1.12 mg/g) compared to the NCG (2.23 ± 0.56 mg/g) ($p < 0.05$). Muscle glycogen levels of the mice in the SBP-III-3-LG (2.44 ± 0.36 mg/g) were increased by 9.42% compared to the NCG, but the muscle glycogen level was no significant difference between SBP-III-3-LG and the NCG ($p < 0.05$). These results show that the anti-fatigue activity of SBP-III-3 may be related to the improvement in the metabolic control of exercise and the activation of energy metabolism [34].

2.4.3. SBP-III-3 Enhanced the Antioxidant Enzymes and Decreased the Malondialdehyde (MDA)

Growing evidence indicates that reactive oxygen species (ROS) are responsible for exercise-induced protein oxidation, and contribute strongly to muscle fatigue [35]. ROS, including free radicals such as peroxyl radicals (ROO•), hydroxyl radicals (HO•), nitric oxide radicals (NO•), and superoxide radicals ($O_2^- \bullet$), are physiological metabolites formed during aerobic life as a result of the metabolism of oxygen [5]. Under normal conditions, ROS are effectively eliminated by antioxidant defense systems, such as antioxidant enzymes and non-enzymatic factors. However, the balance between the generation and elimination of ROS is broken under pathological conditions, as a result of these events; bio-macromolecules are damaged by ROS-induced oxidative stress [36]. Muscle cells contain complex endogenous cellular defense mechanisms to clear up ROS to protect the body from exercise-induced oxidative injuries and DNA damages. For example: GSH-Px accelerates the reaction between H_2O_2 and glutathione (GSH) and converts them into H_2O and oxidized GSH, SOD scavenges the $O_2^- \bullet$, and CAT decomposes the HO•. Thus, the improvement in the activities of these defense mechanisms can help to fight against fatigue. Therefore, the present study investigated the activity of GSH-Px, SOD, and CAT to evaluate the anti-fatigue effects of SBP-III-3 on mice. As shown in Table 2, the SOD activities of SBP-III-3-HG and SBP-III-3-MG significantly increased by 44.2% and 15.7%, respectively, compared to the NCG ($p < 0.05$), while the SOD activity of SBP-III-3-LG increased by 4.2% and was without statistical significance compared to that of the NCG ($p > 0.05$). Moreover, the SOD activities between SBP-III-3-HG and SBP-MG showed a statistical difference ($p < 0.01$). The activities of GSH-Px and CAT showed a similar trend in the SOD activities. The GSH-Px activities of mice treated with SBP-III-3-HG, SBP-III-3-MG, and SBP-III-3-LG increased by 240.5%, 136.1%, and 66.3%, respectively, compared to that of the NCG. In addition, the CAT activities of SBP-III-3-HG, SBP-III-3-MG, and SBP-III-3-LG significantly increased by 159.5%, 87.9% and 74.7%, respectively, compared to that of the NCG ($p < 0.05$ or $p < 0.01$). These results suggested that SBP-III-3 could exert its anti-fatigue effects by enhancing the activities of antioxidant enzymes for eliminating the superfluous free radicals in organism.

Fatigue results in the release of ROS which cause lipid peroxidation of membrane structure. MDA, an oxidative degradation product of cell membrane lipids, is generally considered as an indicator of lipid peroxidation. In fatigue conditions, MDA level is increased and accompanied with a decrease in levels of the antioxidant enzymes GSH-Px and SOD [37]. Therefore, MDA is used as a biomarker to measure the level of oxidative stress in an organism. As shown in Table 2, the MDA levels in mice liver and plasma were decreased with the increase of SBP-III-3 dosage. In mice liver, the MDA levels of SBP-III-3-HG, SBP-III-3-MG, and SBP-III-3-LG significantly decreased by 53.6%, 49.9%, and 28.0%, respectively, compared to that of the NCG ($p < 0.01$). In addition, the MDA levels of SBP-III-3-HG, SBP-III-3-MG, and SBP-III-3-LG significantly lessen by 52.3%, 48.5%, and 41.0%, respectively, compared to that of the NCG in mice plasma ($p < 0.01$). The present data were in line with the decrease in levels of the antioxidant enzymes GSH-Px and SOD. These results indicated that SBP-III-3 might induce the MDA level by inhibiting lipid peroxidation of cell membrane lipids.

2.5. Amino Acid Composition of SBP-III-3

The bioactivities of protein hydrolysates and peptides were directly influenced by the amino acid compositions. Chen, Chi, Zhao, and Lv reported that Gly residue may contribute significantly to antioxidant activity since the single hydrogen atom in the side chain of Gly serves as a proton-donator and neutralizes active free radical species [38]. Nimalaratne, Bandara, and Wu also reported that the single hydrogen atom of Gly residue could provide a high flexibility to the peptide backbone and positively influence the antioxidant properties [39]. Pan et al. reported that hydrophobic amino acid residues including Ala, Leu, and Met played an important role in scavenging free radicals because the large hydrophobic group could help them make contact with hydrophobic radical species [10]. The pyrimidine ring of the Pro residue can increase the flexibility of the peptides and also be capable of quenching singlet oxygen due to its low ionization potential [38]. As shown in Table 2, the contents of Gly, Ala, and Pro residues in SBP-III-3 were 35.10%, 13.46%, and 11.28%, respectively, which reached to 59.84% of the total content of amino acids. In addition, the aromatic amino acid , such as Phe, Trp, His, and Tyr, with imidazole groups were also proved to have the ability to quench free radicals by direct electron transfer [11], and acid and basic amino acid residues, including Glu and Asp, were identified to have strong abilities to chelate metal ions, as well as scavenge HO•. Therefore, SBP-III-3 was rich in amino acids with antioxidant activity, which should be one of the important factors of its high antioxidant activity.

Amino acids were also found to play a key role in the regulatory metabolism involved in muscular activity. Vineyard et al. reported that feeding horses with an amino acid-based supplement every day might support muscle development during exercise and promote exercise metabolism and recovery [40]. The essential amino acid (EAA) ingestion during the exercise could attenuate the degradation of myofibrillar protein, thereby enhancing the exercise capability [41]. The EAA content of SBP-III-3 is 15.7%, which might be beneficial to its anti-fatigue capability. Glu residue was proved to have positive influence to the nervous system and would also be beneficial during exercise [42]. Marquezi et al. reported that Asp residue was advantageous in the oxidative deamination and could induce the blood ammonia concentration and postpone the occurrence of fatigue [43]. Therefore, SBP-III-3 contains 6.86% Glu residue and 4.54% Asp residue, suggesting that it might have a potential anti-fatigue effect. Furthermore, Bazzarre et al. reported that the content of amino acid residues, especially Gly, Ala, Val, Thr, Ile, Ser, and Tyr in the plasma will quickly reduce during an endurance test [44]. Table 3 showed that SBP-III-3 contained 57.72% of the above amino acid residues, which was higher than that of loach peptides (32.2%) with antioxidant activity and anti-fatigue effect [1]. The data indicated that these amino acids could enhance the exercise capability of SBP-III-3. In addition, the result of amino acid composition provided a basis for the good antioxidant and anti-fatigue capacities of SBP-III-3.

3. Experimental Section

3.1. Chemicals and Reagents

Frozen swim bladders of the croceine croaker (*P. crocea*) with an average body weight of 300–350 g were obtained from Zhejiang Dahaiyang Sci-Tech Co., Ltd. (Zhoushan, China). Alcalase, Sephadex G-25, and trifluoroacetic acid (TFA) were purchased from Ythx biotechnology Co., Ltd. (Beijing, China). All of the kits, including SOD, CAT, GSH-Px, MDA, BUN, lactic acid (LA), lactic dehydrogenase (LDH), liver glycogen, and muscle glycogen, were purchased from Nanjing Jiancheng Bioengineering Institute (Nanjing, China). Plasmid DNA (pBR322DNA) was purchased from TaKaRa Biotechnology Co., Ltd. (Dalian, China). Other reagents were of analytical grade and purchased from Sinopharm Chemical Reagent Co., Ltd. (Shanghai, China).

3.2. Preparation Protein Hydrolysate of Swim Bladders

Frozen swim bladders were unfrozen, cut into small pieces (0.5 × 0.5 cm) and soaked in 0.1 M NaOH with a solid/solvent ratio of 1:10 (*w/v*) to remove non-collagenous proteins. The mixture was continuously stirred for 24 h at 4 °C, and the NaOH solution was changed every 4 h. Thereafter, the residues were washed with cold distilled water to achieve the neutral pH. Washed samples were then suspended in 10% butyl alcohol for 12 h with a change of solution every 3 h. Defatted samples were thoroughly washed with cold water, homogenized, and suspended in phosphate buffer solution (PBS) with a solid/solvent of 1:3 and hydrolyzed for 8 h separately using alcalase at pH 9.0, 45 °C, papain at pH 7.0, 50 °C, pepsin at pH 3.0, 37 °C, and trypsin at pH 8.0, 60 °C with a total enzyme dose 2%. Enzymatic hydrolysis was stopped by heating for 10 min in boiling water, and hydrolysate was centrifuged at $8000 \times g$ for 10 min. The resulting supernatants using alcalase were lyophilized and named as SBP.

3.3. Isolation and Purification of Antioxidant Peptide from SBP

SBP was fractionated using ultrafiltration (8400, Millipore, Hangzhou, China) with MW cutoffs of 10 kDa, 5 kDa, and 3 kDa membranes, respectively, (Millipore, Hangzhou, China) for the lab scale at 0.30 MPa, 20 °C. Four fractions termed SBP-I (MW < 3 kDa), SBP-II (3 kDa < MW < 5 kDa), SBP-III (5 kDa < MW < 10 kDa), and SBP-IV (MW > 10 kDa) were pooled, concentrated, and lyophilized. SBP-III with the highest HO• scavenging activity among all ultrafiltration fractions was redissolved in distilled water and separated using a Sephadex G-25 gel filtration chromatography column (Φ2.6 cm × 60 cm, Huanyu Glass Co., Ltd., Xuchang, China) eluted with distilled water at a flow rate of 1.5 mL/min, and the eluate was monitored at 280 nm. Four peaks (SBP-III-1 to SBP-III-4) were collected and measured their HO• scavenging activities and SBP-III-3 was chosen for further analysis.

3.4. Degree of Hydrolysis (DH)

DH analysis was performed according to the previously described method [13]. The hydrolysate (50 µL) was mixed with 0.5 mL of 0.2 M phosphate buffer, pH 8.2, and 0.5 mL of 0.05% 2,4,6-trinitrobenzenesulfonic acid (TNBS) reagent. TNBS was freshly prepared before use by diluting with deionized water. The mixture was incubated at 50 °C for 1 h in a water bath. The reaction was stopped by adding 1 mL of 0.1 M HCl, incubating at room temperature for 30 min. The absorbance was monitored at 420 nm. L-leucine was used as a standard. To determine the total amino acid content, mungbean meal was completely hydrolysed with 6 M HCl with a sample to acid ratio of 1:100 at 120 °C for 24 h. DH (%) was calculated using the following equation:

$$DH = ((A_t - A_0)/(A_{max} - A_0)) \times 100$$

where A_t was the amount of a-amino acids released at time t, A_0 was the amount of a-amino acids in the supernatant at 0 h, and A_{max} was the total amount of a-amino acids obtained after acid hydrolysis at 120 °C for 24 h.

3.5. MW Distribution

MW distribution of SBP-III-1 to SBP-III-4 were determined by high-performance size exclusion chromatography (HPSEC) on a TSK-G3000SWXL column (TOSOH Corporation, Tokyo, Japan) using a high-pressure liquid chromatography system (Agilent 1200 HPLC, Agilent Ltd., Santa Rosa, CA, USA) [45]. The mobile phase consisted of 0.1 M sodium phosphate buffer (pH 7.0). A sample (20 µL) was eluted at a flow rate of 0.5 mL/min, and measured by monitoring the absorbance at 230 nm. The approximate MW was determined using standard protein samples (Sigma-Aldrich Co., LLC., St. Louis, MO, USA) as reference: thyroglobulin (670 kDa), γ-globulin (150 kDa), ovalbumin (44 kDa), trypsin inhibitor (20.1 kDa), ribonuclease A (14.7 kDa), Pro–Tyr–Phe–Asn–Lys (667 Da), and Trp–Asp–Arg (475 Da).

The calibration curve showed that the column separated the standard proteins well. The fitted linear equation between MW (logMW) and the retention time (R_t, min) was calculated by the method of least squares, as Log MW = $-0.2036R_t + 7.6164$ ($R^2 = 0.9766$). The MW of SBP-III-1 to SBP-III-4 was calculated by the elution time.

3.6. Amino Acid Composition Analysis

The amino acid composition of SBP-III-3 was analyzed according to the previous method [46]. To determine the amino acid composition, freeze-dried SBP-III-3 was dissolved in distilled water to obtain a concentration of 1 mg/mL, and an aliquot of 50 mL was dried and hydrolyzed in a vacuum-sealed glass tube at 110 °C for 24 h in the presence of 6 M HCl, which contained 0.1% phenol. Norleucine (Sigma Aldrich, Inc., St. Louis, MO, USA) was used as an internal standard. After hydrolysis, the samples were again vacuum-dried, dissolved in application buffer, and injected into an automated amino acid analyzer (HITACHI 835-50 Amino Acid Analyzer, Tokyo, Japan). Determination of tryptophan was also performed by HPLC analysis after alkaline hydrolysis [47]. Briefly, samples (5 mg) were dissolved in 3 mL of 4 N NaOH, sealed in hydrolysis tubes under nitrogen, and incubated in an oven at 100 °C for 4 h. Hydrolysates were cooled on ice, neutralized to pH 7 using 12 N HCl, and diluted to 25 mL with 1 M sodium borate buffer (pH 9). Aliquots of these solutions were filtered through a 0.45 μm Millex filter (Millipore, Hangzhou, China) prior to injection. Standard solutions of tryptophan were prepared by dilution of a stock solution (0.51 mg tryptophan/mL 4 N sodium hydroxide). They were diluted to 3 mL with 4 N sodium hydroxide and incubated as above. Finally, 20 μL samples and tryptophan solutions were determined by an HPLC system (Agilent 1260 HPLC, Agilent Ltd.).

3.7. Antioxidant Activity

3.7.1. Radical Scavenging Activities

The DPPH•, HO•, O_2^-•, and ABTS⁺• scavenging activities were measured by the previous method [13], and the half elimination ratio (EC_{50}) was defined as the concentration where a sample caused a 50% decrease of the initial concentration of DPPH•, HO•, O_2^-•, and ABTS⁺•, respectively.

HO• Scavenging Activity

First, 1.0 mL of a 1.865 mM 1,10-phenanthroline solution and 2.0 mL of the sample were added to a screw-capped tube and mixed. Then, 1.0 mL of a $FeSO_4 \cdot 7H_2O$ solution (1.865 mM) was added to the mixture. The reaction was initiated by adding 1.0 mL of H_2O_2 (0.03%, v/v). After incubating at 37 °C for 60 min in a water bath, the absorbance of the reaction mixture was measured at 536 nm against a reagent blank. The reaction mixture without any antioxidant was used as the negative control, and a mixture without H_2O_2 was used as the blank. The HO• scavenging activity was calculated using the following formula:

$$\text{HO• scavenging activity (\%)} = ((A_s - A_n)/(A_b - A_n)) \times 100$$

where A_s, A_n, and A_b are the absorbance values determined at 536 nm of the sample, the negative control, and the blank after the reaction, respectively.

DPPH• Scavenging Activity

Two millilitres of samples consisting of distilled water and different concentrations of the analytes were placed in cuvettes, and 500 μL of an ethanolic solution of DPPH (0.02%) and 1.0 mL of ethanol were added. A control sample containing the DPPH solution without the sample was also prepared. In the blank, the DPPH solution was substituted with ethanol. The antioxidant activity of the sample was evaluated using the inhibition percentage of the DPPH• with the following equation:

$$\text{DPPH• scavenging activity (\%)} = (A_0 + A' - A)/A_0 \times 100$$

where A is the absorbance rate of the sample, A_0 is the control group absorbance, and A' is the blank absorbance.

$O_2^- \bullet$ Scavenging Activity

In the experiment, superoxide anions were generated in 1 mL of nitrotetrazolium blue chloride (NBT) (2.52 mM), 1 mL of NADH (624 mM), and 1 mL of different sample concentrations. The reaction was initiated by adding 1 mL of phenazinemethosulphate (PMS) solution (120 μM) to the reaction mixture. The absorbance was measured at 560 nm against the corresponding blank after 5-min incubation at 25 °C. The scavenging capacity of the $O_2^- \bullet$ was calculated using the following equation:

$$O_2^- \bullet \text{ scavenging activity } (\%) = ((A_{control} - A_{sample})/A_{control}) \times 100\%,$$

where $A_{control}$ is the absorbance without the sample and A_{sample} is the absorbance with the sample.

ABTS$^+\bullet$ Scavenging Activity

ABTS$^+\bullet$ was generated by mixing an ABTS stock solution (7 mM) with potassium persulphate (2.45 mM). The mixture was left in the dark for 16 h at room temperature. The ABTS$^+\bullet$ solution was diluted in 5 mM phosphate-buffered saline (PBS, pH 7.4) to an absorbance of 0.70 ± 0.02 at 734 nm. One millilitre of diluted ABTS$^+\bullet$ solution was mixed with one millilitre of the different sample concentrations. Ten minutes later, the absorbances were measured at 734 nm against the corresponding blank. The ABTS$^+\bullet$ scavenging activities of the samples were calculated using the same equation as indicated in $O_2^- \bullet$ scavenging activity (%).

3.7.2. DNA Damage Protective Effect

The ability of SBP-III-3 to protect supercoiled pBR322 plasmid DNA was measured by the previous method with a slight modification [28]. The reaction mixtures (15 μL) contained 5 μL of PBS (10 mM, pH 7.4), 1 μL of plasmid DNA (0.5 μg), 5 μL of the SBP-III-3, 2 μL of 1 mM FeSO$_4$, and 2 μL of 1 mM H$_2$O$_2$ were incubated at 37 °C for 30 min. After incubation, 2 μL of a loading buffer (50% glycerol (v/v), 40 mM EDTA and 0.05% bromophenol blue) was added to stop the reaction and the reaction mixtures were electrophoresed on 1% agarose gel containing 0.5 μg/mL ethidium bromide in Tris/acetate/EDTA gel buffer for 50 min (60 V), and the DNA in the gel was visualized and photographed under ultraviolet light. Ascorbic acid was used as a positive control.

3.8. Animals and Experimental Diets

Male Institute of Cancer Research (ICR) mice with an average body weight of 20–25 g were purchased from the Zhejiang Academy of Medical Sciences (China). All of the in vivo tests were carried by the School of Food and Pharmacy of Zhejiang Ocean University (China), which obtained the permission for performing the research protocols and all animal experiments conducted during the present study from the ethics committee of Zhejiang Ocean University. All experimental procedures were conducted under the oversight and approval of the Academy of Experimental Animal Center of Zhejiang Ocean University and in strict accordance with the NIH Guide for the Care and Use of Laboratory Animals (NIH, 2002).

In vivo anti-fatigue activity was determined on the previous method with a slight modification [3,48]. Male ICR mice were housed in a SPF level laboratory under controlled temperature of 21 ± 1 °C with moderate humidity of $55\% \pm 5\%$, and air flow conditions in a 12 h light/dark cycle; noise was <60 dB. Mice had free access to the standard diet and water during the experiments. After one week adaptation, 48 mice were randomly divided into four groups (12 mice per group): the normal control group (NCG) and three swim bladder peptide (SBP-III-3) treatment groups. Mice in NCG were administered with physiological saline (0.1 mL/10 g body weight per day) by gastrogavage; mice in SBP-III-3 treatment groups were respectively fed with SBP-III-3 in three different doses (50, 100, and 200 mg/kg body weight per day) by gastrogavage, and the three groups were accordingly named as low-dose group (SBP-III-3-LG), middle-dose group (SBP-III-3-MG), and high-dose group (SBP-III-3-HG).

3.9. In Vivo Anti-Fatigue Effect of SBP-III-3

3.9.1. Weight-Loaded Swimming Test in ICR Mice

Weight-loaded swimming test was according to the previous method with some modifications [49,50]. Physiological saline/SBP-III-3 were administrated orally (8:00 a.m.) to mice of the normal control group (NCG) and three swim bladder peptide (SBP-III-3) treatment groups (SBP-III-3-LG, SBP-III-3-MG, and SBP-III-3-HG) once daily for 28 days. After the treatment with SBP-III-3 or physiological saline for 30 min at the last time point, the loaded swimming experiment was carried out. All of the mice were weighed and loaded with an iron ring equaling 5% of each mouse's body weight on the tail root of each mouse; they were placed in a swimming pool (50 cm × 50 cm × 40 cm) with 30 cm deep water with $25 \pm 1°C$. The swimming time of mice was calculated from the time they began to swim up to the time the exhibited exhaustion, which was determined as a loss of coordinated movements and failure to return to the surface within 10 s. The length of the swimming time to exhaustion was evaluated as the degree of fatigue.

3.9.2. Biochemical Parameter Determination on Anti-Fatigue

Biochemical parameters were determined according to the method described by You et al. [1], and were performed according to the recommended procedures provided by the kits (Nanjing Jiancheng Bioengineering Institute, Nanjing, China). Briefly, two days later after weight-loaded swimming test, all the mice were forced to swim for 90 min without a load after gavage for 30 min, then anesthetized with 10% chloral hydrate after 60 min rest, and the mice were sacrificed by cervical dislocation.

For the serum assays, 250 L of arterial blood was respectively collected from left femoral artery to determine MDA, BUN, and LA content and LDH, SOD, CAT, and GSH-Px activity using commercial diagnostic kits. The GSH-Px has the ability to decompose hydrogen peroxide (H_2O_2) and other organic hydroperoxides (ROOH). The reaction uses GSH to complete the reaction using H_2O_2, as the substrate. The consumption of nicotinamide adenine dinucleotide phosphate (NADPH) is used to determine the GSH-Px activity. The catalase activity was determined colorimetrically with a CAT assay kit, which is based on the decomposition of the H_2O_2 optical density at 415 nm by CAT.

For the hepatic and muscular assays, the muscles and livers of the mice were also taken to determine the content of MDA and glycogen. The muscles and livers of the mice were dissected immediately after removal, washed with 0.9% saline, and blotted dry with filter paper. Liver samples (~100 mg) were accurately weighed, and homogenized in 8 mL of homogenization buffer. The content of MDA and glycogen was determined according to the recommended procedures.

3.10. Statistical Analysis

The data are reported as the mean $\pm$ standard deviation (SD). An ANOVA test using SPSS 19.0 (Statistical Program for Social Sciences, SPSS Corporation, Chicago, IL, USA) was used to analyze the experimental data. Duncan's multiple range test was used to measure the differences among the parameters means. The differences were considered significant if $p < 0.05$ or $p < 0.01$.

4. Conclusions

Our study firstly prepared and evaluated the high antioxidant and anti-fatigue activities of the peptide fraction (SBP-III-3) from the croceine croaker (*P. crocea*) swim bladder. The results showed that SBP-III-3 could effectively scavenge HO•, DPPH•, $O_2^- •$, and ABTS$^+$•, prolong the swimming time to exhaustion of mice, decreased the BUN, LA, and MDA levels, and increase the LDH, liver glycogen, and muscle glycogen levels in mice. In addition, SBP-III-3 could improve the activities of SOD, GSH-Px, and CAT in vivo. These results confirmed that SBP-III-3 possessed good antioxidant and anti-fatigue capacities in mice and provided an important basis for developing the swim bladder peptide served as a novel functional food.

Protein hydrolysates are composed of free amino acids and short-chain peptides that exhibit numerous advantages as nutraceuticals, functional foods, or medicines. Structure-activity studies of antioxidant peptides reported that peptides and protein hydrolysates display different activities depending on the peptide size, the amino acid sequence, and the presence of amino acids involved in oxidative reactions. Therefore, further research should be done in order to purify and identify antioxidant and anti-fatigue peptides from SBP-III-3, and more detailed studies on physiological functions, pharmacological effects, and structure-activity relationship of SBP-III-3 and the purified peptides will also be needed.

Acknowledgments: This work was funded by the National Natural Science Foundation of China (NSFC) (81673349), Zhejiang Provincial Natural Science Foundation (LY15C200016), the State-level Spark Program (2015GA700044), Zhejiang Ocean University R & D start funding (Q1203) and Key Scientific and Technological Project of Zhejiang Ocean University (X12ZD09).

Author Contributions: Bin Wang and Guo-Fang Ding conceived and designed the experiments. Li Zeng, Zui-Su Yang, Fang-Fang Huang and Yu-Qin Zhao performed the experiments. Yu-Qin Zhao and Bin Wang analyzed the data. Guo-Fang Ding and Bin Wang contributed the reagents/materials/analytical tools and wrote the paper.

Conflicts of Interest: The authors declare no conflicts of interest.

References

1. You, L.; Zhao, M.; Regenstein, J.M.; Ren, J. In vitro antioxidant activity and in vivo anti-fatigue effect of loach (*Misgurnus anguillicaudatus*) peptides prepared by papain digestion. *Food Chem.* **2011**, *124*, 188–194. [CrossRef]
2. Ding, J.F.; Li, Y.Y.; Xu, J.J.; Su, X.Y.; Gao, X.; Yue, F.P. Study on effect of jellyfish collagen hydrolysate on anti-fatigue and anti-oxidation. *Food Hydrocoll.* **2011**, *25*, 1350–1353. [CrossRef]
3. Xu, C.; Lv, J.; Lo, Y.M.; Cui, S.W.; Hu, X.; Fan, M. Effects of oat β-glucan on endurance exercise and its anti-fatigue properties in trained rats. *Carbohyd. Polym.* **2013**, *92*, 1159–1165. [CrossRef] [PubMed]
4. Agyei, D.; Ongkudon, C.M.; Wei, C.Y.; Chan, A.S.; Danquah, M.K. Bioprocess challenges to the isolation and purification of bioactive peptides. *Food Bioprod. Process* **2016**, *98*, 244–256. [CrossRef]
5. Sila, A.; Bougatef, A. Antioxidant peptides from marine by-products: Isolation, identification and application in food systems. A review. *J. Funct. Foods* **2016**, *21*, 10–26. [CrossRef]
6. Wang, L.; Zhang, H.L.; Lu, R.; Zhou, Y.J.; Ma, R.; Lv, J.Q.; Li, X.L.; Chen, L.J.; Yao, Z. The decapeptide CMS001 enhances swimming endurance in mice. *Peptides* **2008**, *29*, 1176–1182. [CrossRef] [PubMed]
7. Chi, C.F.; Hu, F.Y.; Wang, B.; Li, Z.R.; Luo, H.Y. Influence of amino acid compositions and peptide profiles on antioxidant capacities of two protein hydrolysates from skipjack tuna (*Katsuwonus pelamis*) dark muscle. *Mar. Drugs* **2015**, *13*, 2580–2601. [CrossRef] [PubMed]
8. Wei, R.; Huang, C.; Luo, H.; Song, R. Progress in preparation and application of high Fischer ratio oligopeptides derived from food protein source. *Food Sci.* **2014**, *35*, 289–294.
9. Li, Z.R.; Wang, B.; Chi, C.F.; Zhang, Q.H.; Gong, Y.D.; Tang, J.J.; Luo, H.Y.; Ding, G.F. Isolation and characterization of acid soluble collagens and pepsin soluble collagens from the skin and bone of spanish mackerel (*Scomberomorous niphonius*). *Food Hydrocoll.* **2013**, *31*, 103–113. [CrossRef]
10. Pan, X.; Zhao, Y.Q.; Hu, F.Y.; Wang, B. Preparation and identification of antioxidant peptides from protein hydrolysate of skate (*Raja porosa*) cartilage. *J. Funct. Foods* **2016**, *25*, 220–230. [CrossRef]
11. Chi, C.F.; Hu, F.Y.; Wang, B.; Ren, X.J.; Deng, S.G.; Wu, C.W. Purification and characterization of three antioxidant peptides from protein hydrolyzate of croceine croaker (*Pseudosciaena crocea*) muscle. *Food Chem.* **2015**, *168*, 662–667. [CrossRef] [PubMed]
12. Wu, Q.Q.; Li, T.; Wang, B.; Ding, G.F. Preparation and characterization of acid and pepsin-soluble collagens from scales of croceine and redlip croakers. *Food Sci. Biotechnol.* **2015**, *24*, 2003–2010. [CrossRef]

13. Wang, B.; Wang, Y.M.; Chi, C.F.; Luo, H.Y.; Deng, S.G.; Ma, J.Y. Isolation and characterization of collagen and antioxidant collagen peptides from scales of croceine croaker (*Pseudosciaena crocea*). *Mar. Drugs* **2013**, *11*, 4641–4661. [CrossRef] [PubMed]
14. Zhao, X.; Qian, Y.; Li, G.J.; Tan, J. Preventive effects of the polysaccharide of *Larimichthys crocea* swim bladder on carbon tetrachloride (CCl_4)-induced hepatic damage. *Chin. J. Nat. Med.* **2015**, *13*, 521–528. [CrossRef]
15. Abeyrathne, E.D.; Lee, H.Y.; Jo, C.; Suh, J.W.; Ahn, D.U. Enzymatic hydrolysis of ovomucin and the functional and structural characteristics of peptides in the hydrolysates. *Food Chem.* **2016**, *192*, 107–113. [CrossRef] [PubMed]
16. Chi, C.; Hu, F.; Li, Z.; Wang, B.; Luo, H. Influence of different hydrolysis processes by trypsin on the physicochemical, antioxidant, and functional properties of collagen hydrolysates from *Sphyrna lewini*, *Dasyatis akjei*, and *Raja porosa*. *J. Aquat. Food Prod. Technol.* **2016**, *25*, 616–632. [CrossRef]
17. Chi, C.F.; Wang, B.; Wang, Y.M.; Zhang, B.; Deng, S.G. Isolation and characterization of three antioxidant peptides from protein hydrolysate of bluefin leatherjacket (*Navodon septentrionalis*) heads. *J. Funct. Foods* **2015**, *12*, 1–10. [CrossRef]
18. You, L.; Zhao, M.; Regenstein, J.M.; Ren, J. Purification and identification of antioxidative peptides from loach (*Misgurnus anguillicaudatus*) protein hydrolysate by consecutive chromatography and electrospray ionization-mass spectrometry. *Food Res. Int.* **2010**, *43*, 1167–1173. [CrossRef]
19. Ren, J.; Zhao, M.; Shi, J.; Wang, J.; Jiang, Y.; Cui, C.; Kakuda, Y.; Xue, S.J. Purification and identification of antioxidant peptides from grass carp muscle hydrolysates by consecutive chromatography and electrospray ionization-mass spectrometry. *Food Chem.* **2008**, *108*, 727–736. [CrossRef] [PubMed]
20. Cai, L.; Wu, X.; Zhang, Y.; Li, X.; Ma, S.; Li, J. Purification and characterization of three antioxidant peptides from protein hydrolysate of grass carp (*Ctenopharyngodon idella*) skin. *J. Funct. Foods* **2015**, *16*, 234–242. [CrossRef]
21. Wang, B.; Li, L.; Chi, C.F.; Ma, J.H.; Luo, H.Y.; Xu, Y.F. Purification and characterisation of a novel antioxidant peptide derived from blue mussel (*Mytilus edulis*) protein hydrolysate. *Food Chem.* **2013**, *138*, 1713–1719. [CrossRef] [PubMed]
22. Ahn, C.B.; Kim, J.G.; Je, J.Y. Purification and antioxidant properties of octapeptide from salmon byproduct protein hydrolysate by gastrointestinal digestion. *Food Chem.* **2014**, *147*, 78–83. [CrossRef] [PubMed]
23. Park, S.Y.; Kim, Y.S.; Ahn, C.B.; Je, J.Y. Partial purification and identification of three antioxidant peptides with hepatoprotective effects from blue mussel (*Mytilus edulis*) hydrolysate by peptic hydrolysis. *J. Funct. Foods* **2016**, *20*, 88–95. [CrossRef]
24. Zhuang, H.; Tang, N.; Yuan, Y. Purification and identification of antioxidant peptides from corn gluten meal. *J. Funct. Foods* **2013**, *5*, 1810–1821. [CrossRef]
25. Hong, J.; Chen, T.T.; Hu, P.; Yang, J.; Wang, S.Y. Purification and characterization of an antioxidant peptide (GSQ) from Chinese leek (*Allium tuberosum* Rottler) seeds. *J. Funct. Foods* **2014**, *10*, 1–10. [CrossRef]
26. Rajapakse, N.; Mendis, E.; Jung, W.K.; Je, J.Y.; Kim, S.K. Purification of a radical scavenging peptide from fermented mussel sauce and its antioxidant properties. *Food Res. Int.* **2005**, *38*, 175–182. [CrossRef]
27. Apostolou, A.; Stagos, D.; Galitsiou, E.; Spyrou, A.; Haroutounian, S.; Portesis, N.; Trizoglou, I.; Wallace Hayes, A.; Tsatsakis, A.M.; Kouretas, D. Assessment of polyphenolic content, antioxidant activity, protection against ROS-induced DNA damage and anticancer activity of *Vitisvinifera* stem extracts. *Food Chem. Toxicol.* **2013**, *61*, 60–68. [CrossRef] [PubMed]
28. Gao, C.Y.; Lu, Y.H.; Tian, C.R.; Xu, J.G.; Guo, X.P.; Zhou, R.; Hao, G. Main nutrients, phenolics, antioxidant activity, DNA damage protective effect and microstructure of *Sphallerocarpus gracilis* root at different harvest time. *Food Chem.* **2011**, *127*, 615–622. [CrossRef] [PubMed]
29. Jiang, D.Q.; Guo, Y.; Xu, D.H.; Huang, Y.S.; Yuan, K.; Lv, Z.Q. Antioxidant and anti-fatigue effects of anthocyanins of mulberry juice purification (MJP) and mulberry marc purification (MMP) from different varieties mulberry fruit in China. *Food Chem. Toxicol.* **2013**, *59*, 1–7. [CrossRef] [PubMed]
30. Koo, H.N.; Lee, J.K.; Hong, S.H.; Kim, H.M. Herbkines increases physical stamina in mice. *Biol. Pharm. Bull.* **2004**, *27*, 117–119. [CrossRef] [PubMed]
31. Evans, D.A.; Subramoniam, A.; Rajasekharan, S.; Pushpangadan, P. Effect of Trichopus zeylanicus leaf extract on the energy metabolism in mice during exercise and at rest. *Indian J. Pharmacol.* **2002**, *34*, 32–37.
32. Jia, J.M.; Wu, C.F. Antifatigue activity of tissue culture extracts of *Saussurea involucrate*. *Pharm. Biol.* **2008**, *46*, 433–436. [CrossRef]

33. Van-Loon, L.J.; Saris, W.H.; Kruijshoop, M.; Wagenmakers, A.J. Maximizing postexercise muscle glycogen synthesis: carbohydrate supplementation and the application of amino acid or protein hydrolysate mixtures. *Am. J. Clin. Nutr.* **2000**, *72*, 106–111. [PubMed]

34. Wang, J.J.; Shieh, M.J.; Kuo, S.L.; Lee, C.L.; Pan, T.M. Effect of red mold rice on antifatigue and exercise-related changes in lipid peroxidation in endurance exercise. *Appl. Microbiol. Biotechnol.* **2006**, *70*, 247–253. [CrossRef] [PubMed]

35. Powers, S.K.; DeRuisseau, K.C.; Quindry, J.; Hamilton, K.L. Dietary antioxidants and exercise. *J. Sport Sci.* **2004**, *22*, 81–94. [CrossRef] [PubMed]

36. Ko, S.C.; Kim, D.; Jeon, Y.J. Protective effect of a novel antioxidative peptide purified from a marine *Chlorella ellipsoidea* protein against free radical-induced oxidative stress. *Food Chem. Toxicol.* **2012**, *50*, 2294–2302. [CrossRef] [PubMed]

37. Wang, J.; Li, S.; Fan, Y.; Chen, Y.; Liu, D.; Cheng, H.; Gao, X.; Zhou, Y. Anti-fatigue activity of the water-soluble polysaccharides isolated from *Panax ginseng C. A. Meyer. J. Ethnopharmacol.* **2010**, *130*, 421–423. [CrossRef] [PubMed]

38. Chen, C.; Chi, Y.J.; Zhao, M.Y.; Lv, L. Purification and identification of antioxidant peptides from egg white protein hydrolysate. *Amino Acids* **2012**, *43*, 457–466. [CrossRef] [PubMed]

39. Nimalaratne, C.; Bandara, N.; Wu, J. Purification and characterization of antioxidant peptides from enzymatically hydrolyzed chicken egg white. *Food Chem.* **2015**, *188*, 467–472. [CrossRef] [PubMed]

40. Vineyard, K.R.; Gordonm, M.E.; Graham-Thiersm, P.; Jerina, M. Effects of daily administration of an amino acid-based supplement on muscle and exercise metabolism in working horses. *J. Equine Vet. Sci.* **2013**, *33*, 321–399. [CrossRef]

41. Bird, S.P.; Tarpenning, K.M.; Marino, F.E. Liquid carbohydrate/essential amino acid ingestion during a short-termbout of resistance exercise suppresses myofibrillar protein degradation. *Metabolism* **2006**, *55*, 570–577. [CrossRef] [PubMed]

42. Guezennec, C.Y.; Abdelmalki, A.; Serrurier, B.; Merino, D.; Bigard, X.; Berthelot, M.; Pierard, C.; Peres, M. Effect of prolonged exercise on brain ammonia and amino acids. *Int. J. Sports Med.* **1998**, *19*, 323–327. [CrossRef] [PubMed]

43. Marquezi, M.L.; Roschel, H.A.; Costa, A.D.S.; Sawada, L.A.; Lancha, A.H., Jr. Effect of aspartate and asparagine supplementation on fatigue determinants in intense exercise. *Int. J. Sport Nutr. Exerc. Metab.* **2003**, *13*, 65–75. [CrossRef] [PubMed]

44. Bazzarre, T.L.; Murdoch, S.D.; Wu, S.M.; Herr, D.G.; Snider, I.P. Plasma amino acid responses of trained athletes to two successive exhaustion trials with and without interim carbohydrate feeding. *J. Am. Coll. Nutr.* **1992**, *11*, 501–511. [CrossRef] [PubMed]

45. Li, Z.; Wang, B.; Chi, C.; Gong, Y.; Luo, H.; Ding, G. Influence of average molecular weight on antioxidant and functional properties of cartilage collagen hydrolysates from *Sphyrna lewini*, *Dasyatis akjei* and *Raja porosa*. *Food Res. Int.* **2013**, *51*, 283–293. [CrossRef]

46. Chi, C.F.; Wang, B.; Li, Z.R.; Luo, H.Y.; Ding, G.F.; Wu, C.W. Characterization of acid-soluble collagen from the skin of hammerhead shark (*Sphyrna lewini*). *J. Food Biochem.* **2014**, *38*, 236–247. [CrossRef]

47. Yust, M.M.; Pedroche, J.; Girón-Calle, J.; Vioque, J.; Millán, F.; Alaiz, M. Determination of tryptophan by high-performance liquid chromatography of alkaline hydrolysates with spectrophotometric detection. *Food Chem.* **2004**, *85*, 317–320. [CrossRef]

48. Liu, J.; Jia, L.; Kan, J.; Jin, C.H. In vitro and in vivo antioxidant activity of ethanolic extract of white button mushroom (*Agaricus bisporus*). *Food Chem. Toxicol.* **2013**, *51*, 310–316. [CrossRef] [PubMed]

49. Tan, W.; Yu, K.Q.; Liu, Y.Y.; Ouyang, M.Z.; Yan, M.H.; Luo, R.; Zhao, X.S. Anti-fatigue activity of polysaccharides extract from *Radix Rehmanniae Preparata*. *Int. J. Biol. Macromol.* **2012**, *50*, 59–62. [CrossRef] [PubMed]

50. Yu, B.; Lu, Z.X.; Bie, X.M.; Lu, F.X.; Huang, X.Q. Scavenging and anti-fatigue activity of fermented defatted soybean peptides. *Eur. Food Res. Technol.* **2008**, *226*, 415–421. [CrossRef]

marine drugs

MDPI

Article

Antimicrobial and Antitumor Activities of Novel Peptides Derived from the Lipopolysaccharide- and β-1,3-Glucan Binding Protein of the Pacific Abalone *Haliotis discus hannai*

Bo-Hye Nam [1,*], Ji Young Moon [1], Eun Hee Park [1], Hee Jeong Kong [1], Young-Ok Kim [1], Dong-Gyun Kim [1], Woo-Jin Kim [1], Chul Min An [1] and Jung-Kil Seo [2,*]

[1] Biotechnology Research Division, National Institute of Fisheries Science, 216, Gijanghaean-ro, Gijang-eup, Gijang-gun, Busan 46083, Korea; moonjy@gmail.com (J.Y.M.); jeh8478@naver.com (E.H.P.); heejkong@korea.kr (H.J.K.); yobest12@korea.kr (Y.-O.K.); combikola@korea.kr (D.-G.K.); wj2464@korea.kr (W.-J.K.); ancm@korea.kr (C.M.A.)

[2] Department of Food Science and Biotechnology, Kunsan National University, Kunsan 54150, Korea

* Correspondence: nambohye@korea.kr (B.-H.N.); jungkileun@kunsan.ac.kr (J.-K.S.); Tel.: +82-51-720-2452 (B.-H.N.); +82-63-469-1827 (J.-K.S.)

Academic Editor: Se-Kwon Kim
Received: 24 September 2016; Accepted: 23 November 2016; Published: 14 December 2016

Abstract: Antimicrobial peptides are a pivotal component of the invertebrate innate immune system. In this study, we identified a lipopolysaccharide- and β-1,3-glucan-binding protein (LGBP) gene from the pacific abalone *Haliotis discus hannai* (HDH), which is involved in the pattern recognition mechanism and plays avital role in the defense mechanism of invertebrates immune system. The HDH-LGBP cDNA consisted of a 1263-bp open reading frame (ORF) encoding a polypeptide of 420 amino acids, with a 20-amino-acid signal sequence. The molecular mass of the protein portion was 45.5 kDa, and the predicted isoelectric point of the mature protein was 4.93. Characteristic potential polysaccharide binding motif, glucanase motif, and β-glucan recognition motif were identified in the LGBP of HDH. We used its polysaccharide-binding motif sequence to design two novel antimicrobial peptide analogs (HDH-LGBP-A1 and HDH-LGBP-A2). By substituting a positively charged amino acid and amidation at the C-terminus, the pI and net charge of the HDH-LGBP increased, and the proteins formed an α-helical structure. The HDH-LGBP analogs exhibited antibacterial and antifungal activity, with minimal effective concentrations ranging from 0.008 to 2.2 μg/mL. Additionally, both were toxic against human cervix (HeLa), lung (A549), and colon (HCT 116) carcinoma cell lines but not much on human umbilical vein cell (HUVEC). Fluorescence-activated cell sorter (FACS) analysis showed that HDH-LGBP analogs disturb the cancer cell membrane and cause apoptotic cell death. These results suggest the use of HDH-LGBP analogs as multifunctional drugs.

Keywords: antimicrobial peptide; cytotoxic peptide; lipopolysaccharide- and β-1,3-glucan binding protein; *Haliotis discus hannai*

1. Introduction

Invertebrates lack antibodies and an adaptive immune system; instead, they rely on innate immunity to defend themselves against invading pathogens. The innate immune system of marine invertebrates allows them to survive and grow in their microbe-rich benthic environment.

The first stage of the immune response is the recognition of invasive pathogens. Microbial cell-wall components referred to as pathogen-associated molecule patterns (PAMPs), such as LPS, β-1,3-glucan, and peptidoglycans, are recognized by a specific pattern recognition receptors (PRRs) or pattern recognition proteins (PRPs). PRPs bind to PAMPs to form complexes that subsequently activate immune responses such as phagocytosis, nodule formation, encapsulation, activation of proteinase cascades, and synthesis of antimicrobial peptides. To date, various types of invertebrate PRPs, such as peptidoglycan recognition proteins (PGRPs), C-type lectins, lipopolysaccharide (LPS)-binding proteins, and β-glucan binding proteins (βGBPs), have been reported.

Lipopolysaccharide- and β-1,3-glucan-binding proteins (LGBPs) consist of two polysaccharide recognition motifs for polysaccharide binding and a β-glucan recognition motif that recognizes bacterial antigens (saccharide moieties) such as LPS, peptidoglycan, and β-1,3-glucan, a major cellular component of yeast and fungi [1]. Several LGBPs have been cloned and characterized in aquatic animals such as crayfish (*Pacifastacus leniusculus*) [2], kuruma shrimp (*Marsupenaeus japonicas*) [3], Chinese shrimp (*Fenneropenaeus chinensis*) [4], Zhikong scallop (*Chlamys farreri*) [5], disk abalone (*Haliotis discus*) [6], and pearl oyster (*Pinctada fucata*)LGBPs [7]. The LGBP of *Pacifastacus leniusculus* was shown to play an important role in prophenoloxidase activation [2].

The LPS-binding or recognition domain has been used to design new antimicrobial peptides (AMPs). For example, the corresponding synthetic LPS-binding domain peptides of anti-LPS factor (ALF) from several crustacean species were shown to exhibit antimicrobial activities [8–13]. Lactoferrin is a non-hemic iron-binding glycoprotein with antimicrobial activity via its LPS-binding domain (reviewed by [14]). The recombinant *N*-terminal domain of gram-negative binding protein 3 (GNBP3) binds β-1,3-glucan and shows antimicrobial activity [15]. These studies demonstrated that the antimicrobial properties of the polysaccharide recognition motif can be used to develop novel AMPs. Moreover, recent studies of AMPs have shown that they possess other biological properties, including antiviral and cytotoxic activities [16,17]. In particular, cationic antimicrobial peptides, which are toxic to bacteria but not to normal animal cells, possess a broad spectrum of cytotoxic activity against cancer cells (reviewed by [18]).

In the present study, we identified and designed two novel AMPs based on the polysaccharide-binding domain of the β-1,3-glucan-binding protein of *Haliotis discus hannai*. The antimicrobial activities of these peptides against gram-positive and gram-negative bacteria, as well as yeast, and their cytotoxic activities against three tumor cell lines were examined.

2. Results

2.1. Identification of the Antimicrobial Peptide and cDNA Sequences

By using expressed sequencing tags of *Haliotis discus hannai*, a clone with an incomplete open reading frame (ORF) that showed high similarity to the *Haliotis discus discus* LGBP was isolated. A 632-bp sequence was obtained from clone DGT-151, and the *N*-terminal coding sequence was obtained using the Rapid amplification of cDNA-end (RACE) method and gene-specific primers. The sequence of the 380-bp fragment amplified by 5′-RACE overlapped with an EST sequence to generate the full-length cDNA sequence of the *Haliotis discus hannai* LPS- and β-1,3-glucan binding protein (HDH-LGBP) (Figure 1). The complete sequence of the HDH-LGBP cDNA consisted of a 31-bp 5′-untranslated region (5′-UTR), a 162-bp 3′-UTR with a poly-(A) tail, and a 1263-bp ORF encoding a polypeptide of 420 amino acids with an estimated molecular mass of 47.8 kDa and a theoretical pI of 5.27. The HDH-LGBP gene also encodes a 20-amino-acid putative signal sequence. Therefore, the mature HDH-LGBP consists of 400 amino acid residues with a calculated molecular mass of the protein portion of 45.5 kDa and a predicted pI of 4.93 for the mature protein.

Simple Modular Architecture Research Tool (SMART) analysis revealed that the region corresponding to amino acids 164–301 was similar to that of proteins in the glycoside hydrolase family. Five putative glycosylation sites (Asn–Xaa–Ser/Thr, NXS/T) for *N*-linked carbohydrate chains were identified in the mature protein sequence, at Asn-28, -99, -265, -310, and -350. One of the *N*-linked glycosylation sites was located in proximity to the β-glucan recognition motif, suggesting that glycosylation at this site influences the β-glucan-binding capacity. A short putative cell adhesion site and an integrin binding site, Arg/Lys–Gly–Asp (R/KGD), were also detected in the sequence of the mature protein from Lys-189 to Asp-191. The HDH-LGBP also contained a β-1,3-glucanase site, with Trp-209, Glu-214, Ile-215, and Asp-216 as the active residues (Figure 1).

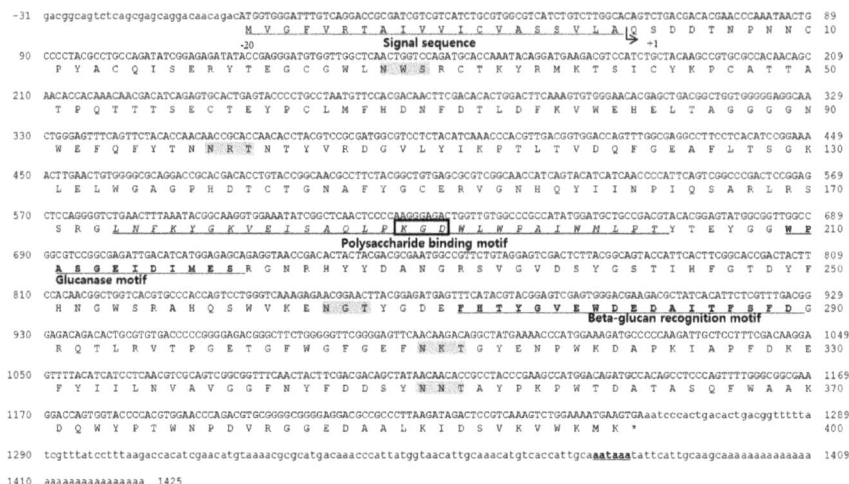

```
 -31  gacggcagtctcagcgagcaggacaacagacATGGTGGGATTTGTCAGGACCGCGATCGTCGTCATCTGCGTGGCGTCATCTGTCTTGGCACAGTCTGACGACACGAACCCAAATAACTG   89
                                   M  V  G  F  V  R  T  A  I  V  V  I  C  V  A  S  S  V  L  A  Q  S  D  D  T  N  P  N  N  C    10
                                       -20                       Signal sequence                    +1
  90  CCCCTACGCCTGCCAGATATCGGAGGATATACCGAGGGATGTGGTTGGCTCAACTGGTCCAGATGCACCAAATACAGGATGAAGACGTCCATCTGCTACAAGCCGTGCGCCACAACAGC   209
       P  Y  A  C  Q  I  S  E  R  Y  T  E  G  C  G  W  L  N  W  S  R  C  T  K  Y  R  M  K  T  S  I  C  Y  K  P  C  A  T  T  A    50
 210  AACACCACAAACAACGACATCAGAGTGCACTGAGTACCCCTGCCTAATGTTCCACGACAACTTCGACACACTGGACTTCAAAGTGTGGGAACACGAGCTGACGGCTGGTGGGGGAGGCAA   329
       T  P  Q  T  T  T  S  E  C  T  E  Y  P  C  L  M  F  H  D  N  F  D  T  L  D  F  K  V  W  E  H  E  L  T  A  G  G  G  G  N    90
 330  CTGGGAGTTTCAGTTCTACACCAACAACCGCACCAACACCTACGTCCGCGATGGCGTCCTCTACATCAAACCCACGTTGACGGTGGACCAGTTTGGCGAGGCCTTCCTCACATCCGGAAA   449
       W  E  F  Q  F  Y  T  N  N  R  T  N  T  Y  V  R  D  G  V  L  Y  I  K  P  T  L  T  V  D  Q  F  G  E  A  F  L  T  S  G  K    130
 450  ACTTGAACTGTGGGGCGCAGGACCGCACGACACCTGTACCGGCAACGCCTTCTACGGCTGTGAGCGCGTCGGCAACCATCAGTACATCATCAACCCCATTCAGTCGGCCCGACTCCGGAG   569
       L  E  L  W  G  A  G  P  H  D  T  C  T  G  N  A  F  Y  G  C  E  R  V  G  N  H  Q  Y  I  I  N  P  I  Q  S  A  R  L  R  S    170
 570  CTCCAGGGGTCTGAACTTTAAATACGGCAAGGTGGAAATATCGGCTCAACTCCCCAAGGGAGACTGGTTGTGGCCCGCATATGGATGCTGCCGACGTACACGGAGTATGGCGGTTGGCC   689
       S  R  G  L  N  F  K  Y  G  K  V  E  I  S  A  Q  L  P  K  G  D  W  L  W  P  A  I  W  M  L  P  T  Y  T  E  Y  G  G  W  P    210
                                              Polysaccharide binding motif
 690  GGCGTCCGGCGAGATTGACATCATGGAGAGCAGAGGTAACCGACACTACTACGACGCGAATGGCCGTTCTGTAGGAGTCGACTCTTACGGCAGTACCATTCACTTCGGCACCGACTACTT   809
       A  S  G  E  I  D  I  M  E  S  R  G  N  R  H  Y  Y  D  A  N  G  R  S  V  G  D  S  Y  G  S  T  I  H  F  G  T  D  Y  F    250
         Glucanase motif
 810  CCAGAACGGCTGGTCACGTGCCCACCAGTCCTGGGTCAAAGAGAACGGAACTTACGGAGATGAGTTTCATACGTACGGAGTCGAGTGGGACGAAGACGCTATCACATTCTCGTTTGACGG   929
       H  N  G  W  S  R  A  H  Q  S  W  V  K  E  N  G  T  Y  G  D  E  F  H  T  Y  G  V  E  W  D  E  D  A  I  T  F  S  F  D  G    290
                                                                           Beta-glucan recognition motif
 930  GAGACAGACACTGCGTGTGACCCCCGGGGACGGGGCTTCTGGGGGTTCGGGGAGTTCAACAAGCAGGCTATGAAAACCCATGGAAAGATGCCCCCAAGATTGCTCCTTTCGACAAGGA   1049
       R  Q  T  L  R  V  T  P  G  E  T  G  F  W  G  F  G  E  F  N  K  T  G  Y  E  N  P  W  K  D  A  P  K  I  A  P  F  D  R  E    330
1050  GTTTTACATCATCCTCAACGTCGCAGTCGGCGGTTTCAACTACTTCGACGACAGCTATAACAACACCGCCTACCCGAAGCCATGGACAGATGCCACAGCCTCCCAGTTTTGGGCGGCGAA   1169
       F  Y  I  I  L  N  V  A  V  G  G  F  N  Y  F  D  D  S  Y  N  N  T  A  Y  P  K  P  W  T  D  A  T  A  S  Q  F  W  A  A  K    370
1170  GGACCAGTGGTACCCCACGTGGAACCCAGACGTGCGGGGCGGGGAGGACGCCGCCCTTAAGATAGACTCCGTCAAAGTCTGGAAAATGAAGTGAaatcccactgacactgacggttttta   1289
       D  Q  W  Y  P  T  W  N  P  D  V  R  G  G  E  D  A  A  L  K  I  D  S  V  K  V  W  K  M  K  *                              400
1290  tcgtttatccttaagaccacatcgaacatgtaaaacgcgcatgacaaacccattatggtaacattgcaaacatgtcaccattgcaaataaatattcattgcaagcaaaaaaaaaaaaaa   1409
1410  aaaaaaaaaaaaaaaaa   1425
```

Figure 1. Nucleotide and deduced amino acid sequence of *Haliotis discus hannai* lipopolysaccharide- and β-1,3-glucan binding protein (HDH-LGBP). The sequences are numbered at the right margin of each line. The signal peptide is underlined, and the poly-(A) signal site is bold and underlined. The integrin-binding motif and *N*-glycosylation sites are boxed and highlighted in gray, respectively. The polysaccharide-binding motif is shown in italics and underlined.

2.2. Peptide Design and Synthesis

To develop a novel AMP, we designed a synthetic peptide analog of HDH-LGBP based on the amino acids sequences located in its polysaccharide-binding motif. One native peptide and two analogs were predicted to show antimicrobial activity. The predicted pI, net positive charge, hydrophobicity, and Boman index are listed in Table 1. Peptide activity is influenced by factors such as hydrophobicity, net charge, and the Boman index, which is an estimate of the potential of peptides to bind to other proteins, including receptors. It is defined as the sum of the free energies of the amino acid residue side chains, divided by the total number of amino acid residues. The native parental peptide WLWPAIWKLPT, rich in W and P residues, has an acidic pI value (5.52) and a zero net charge, but its Boman index is low (−2.21). Schiffer-Edmundson helical wheel projections were used to predict the hydrophobic and hydrophilic regions in the secondary structure of the synthetic peptides HDH-LGBP-A1 (WLWKAIWKLLT) and HDH-LGBP-A2 (WLWKAIWKLLK) (Figure 2).

Table 1. Sequences and physicochemical properties of the peptides used in this study.

Peptide Name	Sequence	Length	M.W.	pI	Hydrophobicity	Hydrophobicmoment	Charge	Boman Index (kcal/mol)	Structure
HDH-LGBP-N	WLWPAIWMLPT-OH	11	1413.7	5.52	−1.62	0.11	0	−2.12	T & R
HDH-LGBP-A1	WLWKAIWKLLT-NH₂	11	1457.8	10.0	−1.12	0.86	+3	−1.34	H
HDH-LGBP-A2	WLWKAIWKLLK-NH₂	11	1484.8	10.3	−0.81	1.07	+4	−1.07	H

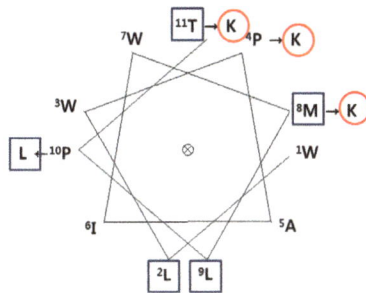

Figure 2. A Schiffer-Edmundson helical wheel representation of HDH-LGBP. The arrows indicate the amino acid residues substituted in the peptide. The hydrophobic and hydrophilic residues are shown in a rectangular box and a circle, respectively.

2.3. Antimicrobial Activity of HDH-LGBP Analogs

The antimicrobial activity of the two HDH-LGBP analogs was determined by measuring their minimum effective concentrations (MECs) against gram-positive bacteria, gram-negative bacteria, and the yeast *C. albicans* using URDA (Table 2). The HDH-LGBP analogs showed antimicrobial activity against the gram-positive bacteria *B. cereus*, *S. aureus*, *S. mutans*, and *S. iniae* (MECs 0.008–1.92 μg/mL) and the gram-negative bacterium *P. aeruginosa* (MECs 1.92–2.12 μg/mL), with maximal killing activity at a peptide concentration of 5 μg/mL. By contrast, the antimicrobial activity of the native peptide (HDH-LGBP-N) was low (data not shown). The two analogs also showed potent activity against *C. albicans* (MECs 2.11–2.16 μg/mL). In the liquid culture bacterial growth inhibition test, the curve clearly showed that growth of microorganisms (Gram negative bacteria: *B. cereus*; *S. auresus*; *S. iniae*; *S. mutans*, Gram positive bacteria: *P. aeruginosa*; *V. anguillarum*; *V. harveyi*) was suppressed at 1 μg/mL HDH-LGBP-A1 or -A2, with greater suppression by the two analogs of up to 5 μg/mL (Figure 3). The results demonstrated that the HDH-LGBP analogs have a broad spectrum of antimicrobial activity.

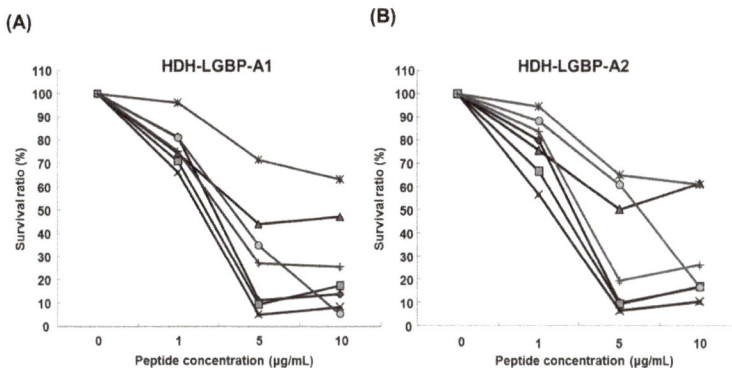

Figure 3. Antimicrobial activity of HDH-LGBP analogs using the broth dilution assay. (**A**) HDH- LGBP-A1; (**B**) HDH-LGBP-A2. Bacterial growth is expressed as a percentage of the maximum optical density (OD) measured in the absence of peptide. Bacterial-killing curve of HDH-LGBP analogs against *B. cereus* (♦), *S. aureus* (■), *S. iniae* (▲), *S. mutans* (×), *P. aeruginosa* (*), *V. anguillarum* (●), and *V. harveyi* (+). The data were obtained from three independent experiments, each performed in triplicate, and are reported as the mean ± SD.

Table 2. Antimicrobial activities of the two HDH-LGBP analogs.

Microbe	Minimal Effective Concentration (μg/mL)		
	Gram	HDH-LGBP-A1	HDH-LGBP-A2
B. cereus	+	1.9	1.8
S. aureus RM4220	+	1.08	1.37
S. iniae FP5229	+	0.57	1.79
S. mutans	+	0.008	1.7
P. aeruginosa KCTC2004	−	2.12	1.92
V. anguillarum	−	>125	>125
V. harveyi	−	>125	>125
C. albicans KCTC7965	Yeast	2.11	2.16

2.4. Thermal Stability of HDH-LGBP Analogs

To investigate thermal stability, 5 μg of the synthetic HDH-LGBP peptides/mL were incubated at 100 °C for 10 min and then cooled before they were used in an URDA against gram-positive and gram-negative bacteria and the yeast *C. albicans*. The antimicrobial activity of the peptides was not greatly altered by heat treatment (Table 3), as evidenced by their strong antimicrobial activities against the tested strains (*S. aureus*, *P. aeruginosa*, and *C. albicans*).

Table 3. Thermal stability of HDH-LGBP analogs against *S. aureus*, *P. aeruginosa*, and *C. albicans*. The upper and lower panels show the radial diffusion assay results of non-heated peptides (N) and of peptides heated for 10 min at 100 °C (H), respectively. Scale bar = 2.3 mm.

Peptide Name	Microbe	*S. aureus*	*P. aeroginosa*	*C. albicans*
HDH-LGBP-A1	N			
	H			
HDH-LGBP-A2	N			
	H			

2.5. Cytotoxicity of HDH-LGBP Analogs

We investigated the cytotoxicity of HDH-LGBP-A1 and HDH-LGBP-A2 on three human cancer cell lines (HeLa, A549, HCT 116 cells) and on a normal cell line, HUVEC, using the MTS assay, which labels live cells based on their mitochondrial dehydrogenase activities, and phase-contrast microscopy. The untreated control cells showed a typical monolayer appearance and had no significant effect on cell viability in the presence of 1–5 µg peptides/mL. However, when the cells were treated for 24 h with 10 µg HDH-LGBP peptides/mL, a decrease in cell number, an increase in the number of rounded cells, and cell shrinkage were observed (Figure 4A,C). In HeLa, A549, and HCT 116 cells treated with 50 µg peptides/mL, cell detachment, swelling, and damage were detected within 5 min (data not shown). This result indicated that higher concentrations (50 µg/mL) of HDH-LGBP-A1 and -A2 directly disrupt the cell membrane. A dose-response experiment showed that treatment of the three cancer cell lines with 1, 5, 10, 25, and 50 µg HDH-LGBP-A1 or -A2/mL for 24 h decreased their viability in a dose-dependent manner (Figure 4B,D). The cytotoxicity of HDH-LGBP-A1 against HeLa cells resulted in 12.4, 98.7, and 99% non-viable cells in cultures exposed to peptide concentrations of 10, 25, and 50 µg/mL, respectively (Figure 4B). The same concentrations also yielded cytotoxic effects in A549 cells (15, 98.5, and 99%) and in HCT 116 cells (22.57, 93.96, and 99%) (Figure 4B). For HDH-LGBP-A2, the corresponding values were 34.4, 99, and 95% in HeLa cells; 24.3, 98.8, and 96.9% in A549 cells; and 29.4, 93.6, and 92% in HCT 116 cells (Figure 4D). At the highest concentration of 50 µg/mL, however, the viability of normal cells was decreased to 32.8% and 47.9% by HDH-LGBP-A1 and HDH-LGBP-A2, respectively (Figure 4D).

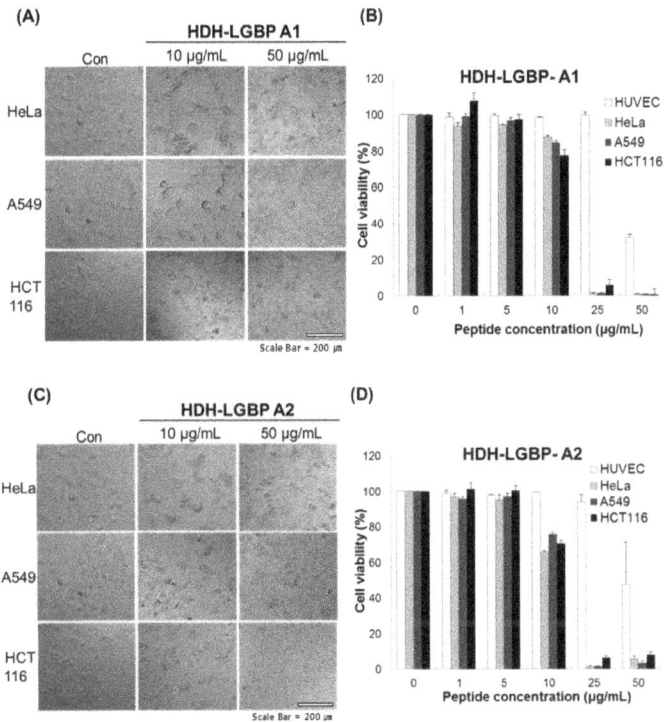

Figure 4. In vitro cytotoxicity of HDH-LGBP analogs. HUVEC, HeLa, A549, and HCT 116 cells were treated with the indicated concentrations of HDH-LGBP-A1 and HDH-LGBP-A2 at 37 °C for 24 h. Cell morphology of HeLa, A549, and HCT116 treated with 10 or 50 µg/mL HDH-LGBP analogs were observed by microscopy (**A** and **C**). Cell viability was measured by an MTS assay after exposure to 0, 1, 5, 10, 25, or 50 µg/mL for 24 h (**B** and **D**). Values represent the mean ± SD (*n* = 3). Scale Bar = 200 µm.

2.6. Effect of HDH-LGBP on Cancer Cell Membranes

Cell death induced by AMPs is thought to involve membrane disruption [19]. In this study, the cell-membrane effects of the HDH-LGBP analogs were investigated using Annexin V-FITC/PI staining. Figure 5 shows the dose-dependent decreases in the proportion of viable HeLa cells (quadrant Q3) and the corresponding increases in damaged and dead HeLa cells (quadrants Q2 and Q4). The percentage of viable HeLa cells decreased from 90.5% (control) to 86.13 (1 μg/mL), 73.33 (5 μg/mL), 68.01 (10 μg/mL), and 40.06% (20 μg/mL) after HDH-LGBPA1 treatment; and to 86.89 (1 μg/mL), 75.21 (5 μg/mL), 51.55 (10 μg/mL), and 29.76% (20 μg/mL) after HDH-LGBPA2 treatment. These results showed that HDH-LGBP analogs disrupt membrane integrity (increased PS exposure) and increase membrane permeability (increase cellular uptake of PI), thereby inducing cell death.

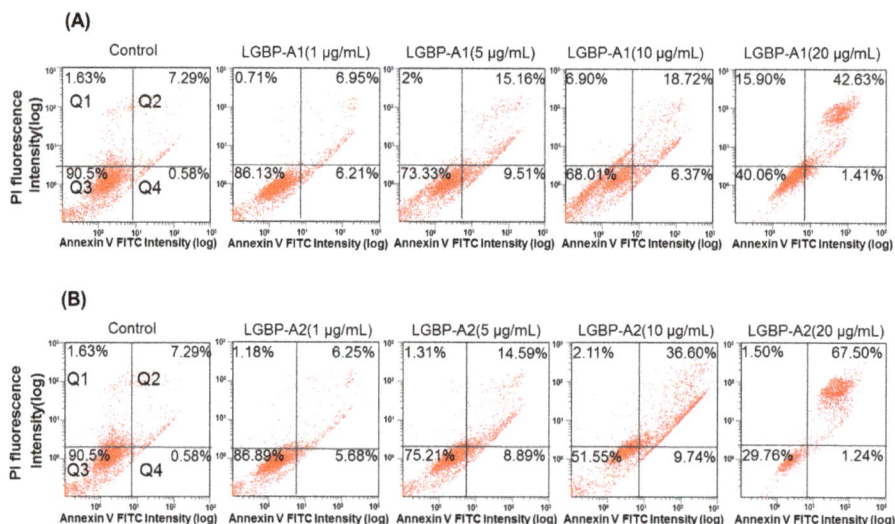

Figure 5. Quantitative analysis of HeLa cells apoptosis and necrosis induced by treatments with abalone HDH-LGBP-A1 (**A**) and HDH-LGBP-A2 (**B**). The cells were incubated with different concentration of 1, 5, 10, and 20 μg HDH-LGBP-A1 and HDH-LGBP-A2/mL for 24 h and then stained with Annexin-V-FITC/PI. Fluorescence intensity was determined using FACS analysis. The upper left part (Q1) represents necrotic cells and the upper right part (Q2) represents secondary necrotic and late apoptotic cells and the lower left part (Q3) represents viable cells and the lower right part (Q4) represents early apoptotic cells.

3. Discussion

AMPs are generally cationic and amphipathic, which enables them to interact with and disrupt lipid membranes. They are also typically very short (5–40 amino acid residues) and contain relatively large (≥30%) proportions of charged (e.g., Lys and Arg) and hydrophobic residues. Some AMPs, such as lactoferricins and indolicidin, are rich in Trp and Arg residues [20]. Unlike currently available conventional antibiotics, which typically interact with a specific target protein, cationic AMPs tend to target the cell membrane of invading microorganisms, leading to cell lysis and death [21]. Thus, AMPs may provide a new class of therapeutic agents whose activities are complementary to those of existing antibiotics. Moreover, bacteria are unlikely to develop resistance to AMPs.

To develop a novel AMP, we designed cationic analogs corresponding to the polysaccharide-binding domain sequence of the abalone β-1,3-glucan-binding protein. The LPS-binding domain is conserved in some PRPs, and is a useful template for designing a mimetic peptide with potential antimicrobial activity. The putative LPS-binding domain of the anti-LPS factor, a small protein with broad-spectrum antimicrobial activities, is pivotal in its antibacterial activity [22]. The synthetic loop of the LPS-binding domain from the ALFs of mud crab [8], shrimp [9,11,12], and Indian mud crab [10] inhibit both gram-negative and gram-positive bacteria, while that from the ALF of black tiger shrimp protects hematopoietic cell cultures from white spot syndrome virus infection [23]. Li et al. [13] had compared antibacterial and antiviral activities of the LPS-binding domain of seven ALF isoforms from the Chinese shrimp and revealed that an identical Lys residue site was specifically conserved in peptide with antimicrobial activity, suggesting that a certain Lys residue is a key residue in antimicrobial activity.

In the present study, to increase the antimicrobial activity of the LGBP derived peptide, we modified the Pro215, Met219, and Pro221 residues of the parent peptide (HDH-LGBP-N) were substituted with Lys, Lys, and Leu to create HDH-LGBP A1; by substituting the Try222 of HDH-LGBP-A1 with Lys, HDH-LGBP-A2 was created (Table 1 and Figure 2). Unlike the parent peptide, the synthetic peptide analogs exhibited inhibitory activities not only against gram-negative and gram-positive bacteria but also against the yeast *C. albicans*. This may have been due to the increased cationicity (net charge) and hydrophobicity of HDH-LGBP-A1 and HDH-LGBP-A2, which facilitated their penetration of the bacterial membrane. The positively charged region of AMPs presumably interacts with the negatively charged bacterial membrane bilayer to form pores via "barrel-stave", "carpet", "toroidal-pore", or "detergent" mechanisms [21,24,25]. Schiffer-Edmandson helical wheel modeling indicated that our LGBP analogs had a hydrophobic area positioned on one side and a positive region on the opposite side (Figure 2). However, the substitution of Lys residue at *C*-terminus of HDH-LGBP-A2 did not increase the antimicrobial activity (Table 2 and Figure 3).

The antimicrobial and cytotoxic activities of AMPs are mediated by targeting the membrane. To determine the effects of HDH-LGBP on mammalian cells, we investigated the toxic effects of HDH-LGBP-A1 and HDH-LGBP-A2 on normal HUVECs and on three cancer cell lines (HeLa, A549, and HCT 116) (Figure 4). The two peptides showed greater cytotoxic than normal-cell toxicity, as determined by comparison of the number of lysed cells. Flow cytometry showed that the two analogs bind to cancer cells and interrupt the cell membrane; thus, the mechanism of the peptides' cytotoxic effects are similar to that underlying their antimicrobial activities. Like the bacterial cell membrane, the membrane of a cancer cell is rich in negatively charged components such as PS, glycoproteins, and glycosaminoglycans [26]. Accordingly, these negatively charged membrane components favor the binding of positively charged AMPs. Further studies are needed to examine the direct interaction of LGBPs with bacterial and cancer cell membranes and to understand the mechanism underlying the cytotoxic effects of these peptides. In our laboratory, we are currently investigating the cytotoxic mechanisms and activities of HDH-LGBP-A1 and HDH-LGBP-A2.

The therapeutic application of AMPs has been hindered by problems such as toxicity, low stability, and high production costs. Furthermore, the salt sensitivity and thermal stability of AMPs pose major obstacles in their development as novel antibiotics, as many of these peptides lose their antimicrobial activities under physiological salt concentrations and high temperatures [27]. HDH-LGBP-A1 and HDH-LGBP-A2, by contrast, maintained their antimicrobial activities after high-temperature treatment. Therefore, these two analogs may be of value in therapeutic applications.

In conclusion, we successfully designed novel AMPs with high thermal stability and anti-cancer activity using peptide mimetics based on the polysaccharide binding motif of the LGBP of *Haliotis discus hannai*. Synthetic, stable HDH-LGBP-A1 and HDH-LGBP-A2 showed potent antimicrobial activity against bacteria and fungi as well as specific cytotoxicity against cancer, but not normal cells, at concentrations <50 μg/mL. Importantly, because HDH-LGBP-A1 and HDH-LGBP-A2 do not contain non-natural or chemically modified amino acids, they can be produced in a cost-effective manner in biological expression systems. Low in vivo stability, toxicity to mammalian cells, and the high cost of production of most AMPs have prevented their clinical use. However, the absence of these features combined with the antimicrobial and cytotoxic effects of HDH-LGBP-A1 and -A2 demonstrated in this study recommend their further exploration for clinical applications.

4. Materials and Methods

4.1. Cloning and Sequencing the Full-Length cDNA of Abalone LGBP

cDNA libraries were constructed from seven tissues obtained from three-year-old disk abalones (*Haliotis discus hannai*), and the expressed sequence tags were analyzed as described in a previous study [28]. The sequence of the 632-bp EST clone DGT-151, isolated from the cDNA library prepared from digestive tract tissues, was homologous to the sequences of the LGBPs of other species. To obtain the full-length cDNA of the LGBP gene, digestive tract cDNAs for the 5′- and 3′-random amplification of cDNA ends (RACE) were synthesized using a SMART RACE cDNA amplification kit (BD Bioscience, San Jose, CA, USA) according to the manufacturer's instructions. Gene-specific primers for 5′- and 3′-RACE were designed based on the partial sequences of the DGT-151 clone (Table 1). The amplified fragments were subcloned into pGEM-T Easy vector (Promega, Madison, WI, USA) and sequenced using an ABI3130 automatic DNA sequencer (Applied Biosystems, Carlsbad, CA, USA). To complete the full-length sequence of LGBP cDNA, the partial sequences of the 5′- and 3′-ends and the partial sequence of DGT-151 were combined and aligned using GENETYX version 8.0 (SDC Software Development, Tokyo, Japan).

4.2. Computational Sequence Analysis

The amino acid sequence was deduced from the obtained cDNA, and the molecular mass and isoelectric point were calculated using GENETYX version 8.0 (SDC Software Development, Tokyo, Japan). Sequence similarities with other known sequences were identified using the BLASTP program from the NCBI [29]. The presence of signal peptides was predicted using SignalP 3.0 [30], and domain searches were conducted in the CD-search in NCBI and Pfam sequence search [31].

4.3. Structure Prediction

The secondary structure of the peptides was predicted using the GOR method (ExPASy). The theoretical isoelectric point (pI) and net charge were estimated using the ExPASy server [32]. Helical wheel diagrams were produced using EMBOSS Pepwheel (European Bioinformatics Institute, Cambridge, UK) [33]. The Boman index [34] was calculated according to the online Antimicrobial Peptide Database [35].

4.4. Peptide Design and Synthesis

A peptide with the amino acid sequence WLWPAIWMLPT, corresponding to the polysaccharide-binding domain of HDH-LGBP and named HDH-LGBP-N, and two modified analogs (HDH-LGBP-A1 and HDH-LGBP-A2) were designed and synthesized commercially by Peptron, Inc. (Daejeon, Korea); the purity grade was >95%. Briefly, the peptide was synthesized using Fmoc solid-phase peptide synthesis (SPPS) with ASP48S (Peptron, Inc., Daejeon, Korea) and purified using reverse-phase high-performance liquid chromatography with a Vydac Everest C18 column (250 mm × 22 mm, 10 μm; Grace, Deerfield, IL, USA). The fractions were eluted with a water-acetonitrile linear gradient (3%–40% (v/v) of acetonitrile) containing 0.1% (v/v) trifluoroacetic acid. The molecular masses of the purified peptides were confirmed using liquid chromatography/mass spectrometry (HP1100 series; Agilent, Santa Clara, CA, USA). All synthetic peptides were dissolved in 0.01% acetic acid to obtain stock solutions of 1000 μg/mL.

4.5. Ultrasensitive Radial Diffusion Assay (URDA) for Antimicrobial Potency

The antimicrobial activity of the purified peptide was assessed as described previously [31]. The antimicrobial activities of the synthetic peptides were tested against the gram-positive bacteria, *Bacillus cereus*, *Staphylococcus aureus* RM4220, *Streptococcu siniae* FP5229, and *S. mutans*; the gram-negative bacteria, *Pseudomonas aeruginosa* KCTC2004, *Vibrio anguillarum*, and *Vibrio harveyi* KCCM40866; and the yeast, *Candida albicans* KCTC7965. The bacterial strains were grown in brain-heart infusion medium (BHI; BD Biosciences, San Jose, CA, USA) at the appropriate temperature (25 °C for *P. aeruginosa* and *S. iniae*, and 37 °C for the other strains). The yeast strain *C. albicans* KCTC7965 was grown in yeast medium (YM) at 25 °C. After 16–18 h of incubation, the bacterial and *C. albicans* suspensions were diluted to a McFarland turbidity standard of 0.5 (Vitek Colorimeter #52-1210; Hach, Loveland, CO, USA) corresponding to ~10^8 CFU/mL for bacteria and ~10^6 CFU/mL for *C. albicans*. A 500-mL aliquot of the diluted bacterial or *C. albicans* suspension was added to 9.5 mL of underlay gel containing 5×10^6 CFU/mL or 5×10^4 CFU/mL in 10 mM phosphate-buffered saline (PBS; pH 6.6) with 0.03% Tryptic Soy Broth (TSB) or 0.03% Sabouraud Dextrose Broth (SDB) and 1% type I low-EEO agarose. The purified peptide was serially diluted twofold in 5 μL of acidified water (0.01% HAc), and each dilution was added to 2.5-mm-diameter wells made in the 1-mm-thick underlay gels. After a 3 h incubation at either 25 °C (*P. aeruginosa*, *S. iniae*, and *C. albicans*) or 37 °C (the other strains), the bacterial or yeast suspension was overlaid with 10 mL of double-strength overlay gel containing 6% BHI or 6% YM prepared in 10 mM PBS (pH 6.6) and using 1% agarose. The plates were incubated for an additional 18–24 h, after which, the clearing zone diameters were measured. After subtracting the diameter of the well, the clearing zone diameter was expressed in units (0.1 mm = 1 U).

4.6. Minimal Effective Concentration of the GBP-Derived Analogs

All tested bacteria and yeast were prepared as described above. The minimal effective concentration (MEC, μg/mL) of the synthetic peptides was calculated as the *x*-intercept of a plot of the above-described units against the log10 of the peptide concentration [36,37]. The antimicrobial assay was performed in triplicate, and the results were averaged.

4.7. Effect of Temperature on Antimicrobial Activity

To explore thermal stability, the LGBP analogs were incubated at 100 °C for 10 min, cooled, and then used in the above-described URDA against the bacteria, *B. cereus*, *S. aureus*, *S. iniae*, and *P. aeruginosa*; and the yeast, *C. albicans*.

4.8. Cell Culture

Primary umbilical vein endothelial cells (HUVEC; normal human cells), HeLa (human cervical adenocarcinoma), A549 (human lung adenocarcinoma), and HCT 116 (human colorectal carcinoma) cell lines were purchased from the American Type Culture Collection (ATCC; Rockville, MD, USA). HUVEC cells were maintained in vascular cell basal medium (ATCC PCS-100-030) containing Plus One endothelial cell growth factor (ATCC PCS-100-040), and 100 U antibiotics-antimycotics/mL (Life Technologies, Carlsbad, CA, USA) at 37 °C in a 5% CO_2 incubator (SANYO, Moriguchi, Osaka, Japan). The three cancer cell lines were maintained in DMEM (Welgene, Gyeongsan, Korea) containing 10% fetal bovine serum (Gibco, Grand Island, NY, USA) and 100 U antibiotics–antimycotics/mL (Life Technologies, Carlsbad, CA, USA) at 37 °C in a 5% CO_2 incubator.

4.9. Cell Viability

The cytotoxicity of the AMPs in HUVEC, HeLa, A549, and HCT 116 cells was determined individually using an MTS assay, according to the manufacturer's instructions of CellTiter 96® Aqueous One Solution Cell Proliferation Assay (Promega, Mannheim, Germany). Briefly, HUVEC, HeLa, A549, and HCT 116 cells (4×10^3 cells/well) were cultured at 37 °C in 96-well plates (Corning, New York, NY, USA) overnight and then incubated for an additional 24 h with 1, 5, 10, 25, or 50 μg/mL of HDH-LGBP-A1 or -A2. Cells in the control group were incubated with 0.01% acetic acid. At the end of the treatment period, 20 μL of a mixture of MTS and the electron-coupling reagent phenazinemethosulfate (Promega, Mannheim, Germany) was added, and the cells were incubated for 4 h at 37 °C. A microtiter plate reader (Perkin Elmer, Waltham, MA, USA) was used to measure the absorbance at 490 nm. The experiment was performed in triplicate and in three independent experiments. The results are expressed as the percentage inhibition of viable cells. Negative control (0.01% acetic acid) values were subtracted from the experimental results.

4.10. FITC-Annexin V and Propidium Iodide (PI) Staining

To evaluate the effects of HDH-LGBP on cell membrane integrity and cell-surface phosphatidylserine (PS) exposure, HeLa cells seeded in a 35-mm dish (3.5×10^5 cells/dish) (Corning, New York, NY, USA) and incubated at 37 °C for 24 h were treated with HDH-LGBP-A1 or -A2 at concentrations of 1–50 μg/mL or with 0.01% acetic acid (negative control). After 24 h, the cells were harvested by tryptic digestion, washed with cold PBS, resuspended in binding buffer (0.01 M Hepes/NaOH (pH 7.4), 0.14 M NaCl, 2.5 mM $CaCl_2$), and stained according to the manufacturer's instructions with FITC-annexin V and PI (FITC-Annexin V apoptosis detection kit, BD Biosciences). The stained cells were gently mixed and evaluated by flow cytometry (FC500, Beckman Coulter). The results were analyzed using Cell Quest software (BD Biosciences, San Jose, CA, USA). During the early stage of apoptosis, PS shifts from the inner to the outer layer of the plasma membrane. Annexin V, a calcium-dependent, phospholipid-binding protein, binds to PS with high affinity, providing a marker of cell apoptosis. Viable cells with an intact membrane exclude PI, whereas the disrupted membranes of damaged or dead cells are permeable to the dye. The Q1, Q2, Q3, and Q4 gates represented dead cells, the late stage of cell apoptosis, normal cells, and the early stage of cell apoptosis, respectively.

Acknowledgments: This work was supported by grants from the National Institute of Fisheries Science (R2016024) and Marine Biotechnology Program (PJT200620) funded by Ministry of Oceans and Fisheries, Korea.

Author Contributions: B.-H.N. and J.-K.S. conceived and designed the experiments, and wrote the paper; J.Y.M. and E.H.P. performed the experiments; H.J.K., Y.-O.K., D.-G.K., W.-J.K. and C.M.A. revised and edited the manuscript.

Conflicts of Interest: The authors declare no conflict of interest.

References

1. Hoffman, O.A.; Olson, E.J.; Limper, A.H. Fungal beta glucans modulate macrophage release of tumor necrosis factor-alpha in response to bacterial lipopolysaccharide. *Immunol. Lett.* **1993**, *37*, 19–25. [CrossRef]
2. Lee, S.Y.; Wang, R.; Söderhäll, K. A lipopolysaccharide- and beta-1,3-glucan-binding protein from hemocytes of the freshwater crayfish *Pacifastacus leniusculus*. Purification, characterization, and cDNA cloning. *J. Biol. Chem.* **2000**, *275*, 1337–1343. [CrossRef] [PubMed]
3. Lin, Y.C.; Vaseeharan, B.; Chen, J.C. Identification and phylogenetic analysis on lipopolysaccharide and beta-1,3-glucan binding protein (LGBP) of kuruma shrimp Marsupenaeus japonicus. *Dev. Comp. Immunol.* **2008**, *32*, 1260–1269. [CrossRef] [PubMed]
4. Liu, F.; Li, F.; Dong, B.; Wang, X.; Xiang, J. Molecular cloning and characterisation of a pattern recognition protein, lipopolysaccharide and beta-1,3-glucan binding protein (LGBP) from Chinese shrimp *Fenneropenaeus chinensis*. *Mol. Biol. Rep.* **2009**, *36*, 471–477. [CrossRef] [PubMed]
5. Su, J.; Ni, D.; Song, L.; Zhao, J.; Qiu, L. Molecular cloning and characterization of a short type peptidoglycan recognition protein (CfPGRP-S1) cDNA from Zhikong scallop *Chlamys farreri*. *Fish Shellfish Immunol.* **2007**, *23*, 646–656. [CrossRef] [PubMed]
6. Nikapitiya, C.; De Zoysa, M.; Lee, J. Molecular characterization and gene expression analysis of a pattern recognition protein from disk abalone, *Haliotis discus discus*. *Fish Shellfish Immunol.* **2008**, *25*, 638–647. [CrossRef] [PubMed]
7. Zhang, D.; Ma, J.; Jiang, J.; Qiu, L.; Zhu, C.; Su, T.; Li, Y.; Wu, K.; Jiang, S. Molecular characterization and expression analysis of lipopolysaccharide and β-1,3-glucan-binding protein (LGBP) from pearl oyster *Pinctada fucata*. *Mol. Biol. Rep.* **2010**, *37*, 3335–3343. [CrossRef] [PubMed]
8. Imjongjirak, C.; Amparyup, P.; Tassanakajon, A.; Sittipraneed, S. Antilipopolysaccharide factor (ALF) of mud crab *Scylla paramamosain*: Molecular cloning, genomic organization and the antimicrobial activity of its synthetic LPS binding domain. *Mol. Immunol.* **2007**, *44*, 3195–3203. [CrossRef] [PubMed]

9.   Pan, C.Y.; Chao, T.T.; Chen, J.C.; Chen, J.Y.; Liu, W.C.; Lin, C.H.; Kuo, C.M. Shrimp (*Penaeus monodon*) anti-lipopolysaccharide factor reduces the lethality of *Pseudomonas aeruginosa* sepsis in mice. *Int. Immunopharmacol.* **2007**, *7*, 687–700. [CrossRef] [PubMed]

10. Sharma, S.; Yedery, R.D.; Patgaonkar, M.S.; Selvaakumar, C.; Reddy, K.V. Antibacterial activity of a synthetic peptide that mimics the LPS binding domain of Indian mud crab, *Scylla serrata* anti-lipopolysaccharide factor (SsALF) also involved in the modulation of vaginal immune functions through NF-kB signaling. *Microb. Pathog.* **2011**, *50*, 179–191. [CrossRef] [PubMed]

11. Guo, S.Y.; Li, S.H.; Li, F.H.; Zhang, X.J.; Xiang, J.H. Modification of a synthetic LPS-binding domain of anti-lipopolysaccharide factor from shrimp reveals strong structure-activity relationship in their antimicrobial characteristics. *Dev. Comp. Immunol.* **2014**, *45*, 227–232. [CrossRef] [PubMed]

12. Li, S.H.; Guo, S.Y.; Li, F.H.; Xiang, J.H. Characterization and function analysis of an anti-lipopolysaccharide factor (ALF) from the Chinese shrimp *Fenneropenaeus chinensis*. *Dev. Comp. Immunol.* **2014**, *46*, 349–355. [CrossRef] [PubMed]

13. Li, S.H.; Guo, S.Y.; Li, F.H.; Xiang, J.H. Functional diversity of anti-lipopolysaccharide factor isoforms in shrimp and their characters related to antiviral activity. *Mar. Drugs* **2015**, *13*, 2602–2616. [CrossRef] [PubMed]

14. Drago-Serrano, M.E.; de la Garza-Amaya, M.; Luna, J.S.; Campos-Rodriguez, R. Lactoferrin-lipopolysaccharide (LPS) binding as key to antibacterial and antiendotoxic effects. *Int. Immunopharmacol.* **2012**, *12*, 1–9. [CrossRef] [PubMed]

15. Lee, H.; Kwon, H.M.; Park, J.W.; Kurokawa, K.; Lee, B.L. *N*-terminal GNBP homology domain of Gram-negative binding protein 3 functions as a beta-1,3-gluganc binding motif in *Tenebrio molitor*. *BMB Rep.* **2009**, *42*, 506–510. [CrossRef] [PubMed]

16. Zasloff, M. Antimicrobial peptides of multicellular organisms. *Nature* **2002**, *415*, 389–395. [CrossRef] [PubMed]

17. Hancock, R.E.; Diamond, G. The role of cationic antimicrobial peptides in innate host defences. *Trends Microbiol.* **2000**, *8*, 402–410. [CrossRef]

18. Gaspar, D.; Veiga, A.S.; Castanho, M.A.R.B. From antimicrobial to cytotoxic peptides. A review. *Front. Microbiol.* **2013**, *4*, 1–16. [CrossRef] [PubMed]

19. Ausbacher, D.; Svineng, G.; Hansen, T.; Strom, M.B. Cytotoxic mechanisms of action of two small amphipathic beta(2,2)-amino acid derivatives derived from antimicrobial peptides. *Biochim. Biophys. Acta* **2012**, *1818*, 2917–2925. [CrossRef] [PubMed]

20. Reddy, K.V.R.; Yedery, R.D.; Aranha, C. Antimicrobial peptides: Premises and promises. *Int. J. Antimicrob. Agents* **2004**, *24*, 536–547. [CrossRef] [PubMed]

21. Oren, Z.; Hong, J.; Shai, Y. A comparative study on the structure and function of a cytolytic alpha-helical peptide and its antimicrobial beta-sheet diastereomer. *Eur. J. Biochem.* **1999**, *259*, 360–369. [CrossRef] [PubMed]

22. Rosa, R.D.; Vergnes, A.; de Lorgeril, J.; Goncalves, P.; Perazzolo, L.M.; Sauné, L.; Romestand, B.; Fievet, J.; Gueguen, Y.; Bachère, E.; et al. Functional divergence in shrimp anti-lipopolysaccharide factors (ALFs): From recognition of cell wall components to antimicrobial activity. *PLoS ONE* **2013**, *8*, e67937. [CrossRef] [PubMed]

23. Tharntada, S.; Ponprateep, S.; Somboonwiwat, K.; Liu, H.; Söderhäll, I.; Söderhäll, K.; Tassanakajon, A. Role of anti-lipopolysaccharide factor from the black tiger shrimp, *Penaeus monodon*, in protection from white spot syndrome virus infection. *J. Gen. Virol.* **2009**, *90*, 1491–1498. [CrossRef] [PubMed]

24. Pouny, Y.; Shai, Y. Interaction of D-amino acid incorporated analogs of pardaxin with membranes. *Biochemistry* **1992**, *31*, 9482–9490. [CrossRef] [PubMed]

25. Schweizer, F. Cationic amphiphilic peptides with cancer-selective toxicity. *Eur. J. Pharmacol.* **2009**, *625*, 190–194. [CrossRef] [PubMed]

26. Hoskin, D.W.; Ramamoorthy, A. Studies on cytotoxic activities of antimicrobial peptides. *Biochim. Biophys. Acta Biomembr.* **2008**, *1778*, 357–375. [CrossRef] [PubMed]

27. Goldman, M.J.; Anderson, G.M.; Stolzenberg, E.D.; Kari, U.P.; Zasloff, M.; Wilson, J.M. Human b-defensin-1 is a salt-sensitive antibiotic in lung that is inactivated in cystic fibrosis. *Cell* **1997**, *88*, 553–560. [CrossRef]

28. Park, E.M.; Nam, B.H.; Kim, Y.O.; Kong, H.J.; Kim, W.J.; Lee, S.J.; Kong, I.S.; Choi, T.J. EST-based survey of gene expression in seven tissue types from the abalone *Haliotis discus hannai*. *J. Fish. Sci. Technol.* **2007**, *10*, 119–126. [CrossRef]

29. National Center for Biotechnology Information. Available online: http://ncbi.nlm.nih.gov/blast/ (accessed on 8 January 2013).

30. SignalP 4.1 Server. Available online: http://www.cbs.dtu.dk/services/SignalP/ (accessed on 11 April 2013).

31. Pfam Protein Family Database. Available online: http://pfam.xfam.org/search/ (accessed on 30 June 2013).

32. SIB Bioinformatics Resources Portal. Available online: http://web.expasy.org/peptide_mass/ (accessed on 30 June 2013).

33. Ramachandran, G.N.; Sasisekharan, V. Conformation of polypeptides and proteins. *Adv. Prot. Chem.* **1968**, *23*, 283–437.

34. Boman, H. Antibacterial peptides, basic facts and emerging concepts. *J. Int. Med.* **2003**, *254*, 197–215. [CrossRef]

35. The Antimicrobial Peptide Database. Available online: http://aps.unmc.edu/Ap/main.php/ (accessed on 30 June 2013).

36. Seo, J.K.; Crawford, J.M.; Stone, K.L.; Noga, E.J. Purification of a novel arthropod defensin from the American oyster, *Crassostrea virginica*. *Biochem. Biophys. Res. Commun.* **2005**, *338*, 1998–2004. [CrossRef] [PubMed]

37. Lehrer, R.I.; Rosenman, M.; Harwig, S.S.L.; Jackson, R.; Eisenhaur, P. Ultrasensitive assay for endogenous antimicrobial polypeptides. *J. Immunol. Methods* **1991**, *137*, 167–173. [CrossRef]

marine drugs

MDPI

Article

Protective Effects of Hydrolyzed Nucleoproteins from Salmon Milt against Ethanol-Induced Liver Injury in Rats

Akiko Kojima-Yuasa [1],*, Mayu Goto [1], Eri Yoshikawa [1], Yuri Morita [1], Hirotaka Sekiguchi [2], Keita Sutoh [2], Koji Usumi [2] and Isao Matsui-Yuasa [1]

[1] Department of Food and Human Health Sciences, Graduate School of Human Life Science, Osaka City University, Osaka 558-8585, Japan; v-oooogue@ezweb.ne.jp (M.G.); eri92011@gmail.com (E.Y.); morimaru.y@gmail.com (Y.M.); yuasa-i@hotmail.co.jp (I.M.-Y.)

[2] Life Science Institute Co., Ltd., Tokyo 103-0012, Japan; sekiguchi@life-science.co.jp (H.S.); sutoh@life-science.co.jp (K.S.); usumi@life-science.co.jp (K.U.)

* Correspondence: kojima@life.osaka-cu.ac.jp; Tel.: +81-6-6605-2865

Academic Editor: Se-Kwon Kim

Received: 17 October 2016; Accepted: 15 December 2016; Published: 19 December 2016

Abstract: Dietary nucleotides play a role in maintaining the immune responses of both animals and humans. Oral administration of nucleic acids from salmon milt have physiological functions in the cellular metabolism, proliferation, differentiation, and apoptosis of human small intestinal epithelial cells. In this study, we examined the effects of DNA-rich nucleic acids prepared from salmon milt (DNSM) on the development of liver fibrosis in an in vivo ethanol-carbon tetrachloride cirrhosis model. Plasma aspartate transaminase and alanine transaminase were significantly less active in the DNSM-treated group than in the ethanol plus carbon tetrachloride (CCl_4)-treated group. Collagen accumulation in the liver and hepatic necrosis were observed histologically in ethanol plus CCl_4-treated rats; however, DNSM-treatment fully protected rats against ethanol plus CCl_4-induced liver fibrosis and necrosis. Furthermore, we examined whether DNSM had a preventive effect against alcohol-induced liver injury by regulating the cytochrome p450 2E1 (CYP2E1)-mediated oxidative stress pathway in an in vivo model. In this model, CYP2E1 activity in ethanol plus CCl_4-treated rats increased significantly, but DNSM-treatment suppressed the enzyme's activity and reduced intracellular thiobarbituric acid reactive substances (TBARS) levels. Furthermore, the hepatocytes treated with 100 mM ethanol induced an increase in cell death and were not restored to the control levels when treated with DNSM, suggesting that digestive products of DNSM are effective for the prevention of alcohol-induced liver injury. Deoxyadenosine suppressed the ethanol-induced increase in cell death and increased the activity of alcohol dehydrogenase. These results suggest that DNSM treatment represents a novel tool for the prevention of alcohol-induced liver injury.

Keywords: DNA-rich nucleic acid prepared from salmon milt (DNSM); in vivo ethanol-carbon tetrachloride cirrhosis model; plasma aminotransferases (AST and ALT); collagen accumulation; CYP2E1 activity; alcohol-induced liver injury; rats

1. Introduction

Alcoholic liver disease is a pathological process characterized by progressive liver damage that leads to steatosis, steatohepatitis, fibrosis, and ultimately cirrhosis, which may further progress to hepatocellular cancer [1–3]. Oxidative stress plays an important role in this process [4,5]. Alcohol-induced oxidative stress associated with ethanol metabolism in the liver plays a major role in ethanol-induced liver injury. Alcohol dehydrogenase is the major enzyme responsible for oxidizing ethanol to aldehyde in alcohol metabolism. Heavy consumption of ethanol induces cytochrome p450 2E1 (CYP2E1) activity in hepatocytes. This enzyme complements the activity of a constitutively expressed alcohol dehydrogenase in oxidizing ethanol to acetaldehyde [6]. However, CYP2E1 generates reactive oxygen species (ROS) quite efficiently, which appears to play a major role in ethanol-induced liver injury [7–10]. Therefore, possible strategies for preventing the production of these ROS may be effective in attempts to minimize the hepatotoxicity of ethanol in humans.

Animal models of liver fibrosis are important for understanding the underlying mechanisms of the treatments used to combat this disease. Currently, two models have been developed for administering alcohol to animals: the Lieber-De Carli liquid diet [11] and the Tsukamoto-French gastric model [12]. In the Lieber-De Carli liquid diet, ethanol replaces the carbohydrates of a normal diet. Tsukamoto and French developed an in vivo animal model in which enteral ethanol is continuously administered to the animal via intragastric infusion. However, neither the Lieber De Caarli nor the Tsukamoto/French feeding protocol results in cirrhosis in rats. Tipoe et al. reported another model for administering dietary alcohol and fish oil (30% of calories) and showed that an increase in profibrogenic mediators was not associated with the presence of histological evidence of fibrosis [13]. Contrastingly, Siegers et al. developed a model for the administration of low-dose carbon tetrachloride (CCl_4) and a 5% ethanol solution that produced histological changes in rats similar to those found in human alcoholic cirrhosis within four weeks [14]. We have also shown that hepatic histological changes occurred within four weeks of the administration of low-dose CCl_4 and a 5% ethanol solution [15,16].

Salmon milts contain mainly nucleic acids, protamine, and polyamine. These components play an important role in the diet. Dietary nucleic acids are particularly important for the development and growth of tissues. Nucleic acids are partly degraded by nucleases in the intestine and absorbed as nucleosides and nucleobases. Dietary nucleotides play a role in maintenance of immune responses in both animals and humans [17–19]. Several researchers have also reported that orally administered nucleic acids from salmon milt play a role in physiological functions such as cellular metabolism, proliferation, differentiation, and apoptosis in human small intestinal epithelial cells [17,20,21]. Additionally, Sakai et al. have reported that dietary ribonucleic acid (RNA) suppresses inflammation in adipose tissue and improves glucose intolerance in mice fed a high-fat diet [22].

In this study, we examined the effect of DNA-rich nucleic acids prepared from salmon milt (DNSM) on the development of liver fibrosis in an in vivo ethanol-CCl_4-induced cirrhosis rat model and in an in vitro alcohol-injury hepatocytes model, and we found that DNSM protected hepatocytes against ethanol induced liver injury.

2. Results

Figure 1 shows changes in body weight during the experimental period. The body weights of ethanol plus CCl_4-treated rats and 0.12% DNSM diet- and ethanol plus CCl_4-treated rats tended to be lower than those rats fed the control diet or CCl_4 alone. However, these differences were not statistically significant.

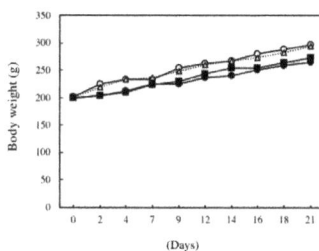

Figure 1. Changes in body weight. ○: Control diet, ●: Control diet with 5% ethanol plus CCl_4, △: Control diet with CCl_4, ■: 0.12% DNSM diet with 5% ethanol plus CCl_4.

We examined the effect of nucleoprotein treatment on plasma aspartate aminotransferase (AST) and alanine aminotransferase (ALT) activities. As shown in Figure 2, in the ethanol plus CCl_4 (0.1 mL/kg of body weight)-treated group, plasma AST and ALT activities increased by 1.8- and 3.5-fold, respectively, compared to the control group. However, these same enzymes were significantly less active in the DNSM-treated group compared to the ethanol plus CCl_4-treated group.

(A)   (B)

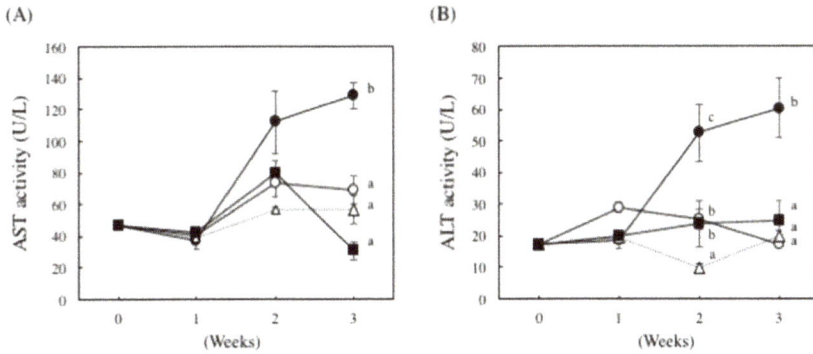

Figure 2. The effect of DNSM on serum AST and ALT activity in ethanol plus CCl_4-treated rats. Effect of DNSM on (**A**) serum AST activity; and (**B**) serum ALT activity. Data are presented as the mean ± S.E. of the activity of five rats. Values without a common letter are significantly different ($p < 0.01$). ○: Control diet, ●: Control diet with 5% ethanol plus CCl_4, △: Control diet with CCl_4, ■: 0.12% DNSM diet with 5% ethanol plus CCl_4.

Histological analysis was performed by hematoxylin and eosin staining as well as elastic van Gieson (EVG) staining and Mason's trichrome staining to assess liver damage (Figure 3). No histological abnormalities were observed in the control rats or CCl_4-treated rats, but collagen accumulation in the liver and hepatic necrosis were observed in ethanol plus CCl_4-treated rats. However, treatment with DNSM fully protected the rats from liver fibrosis and necrosis induced by ethanol plus CCl_4.

(A)

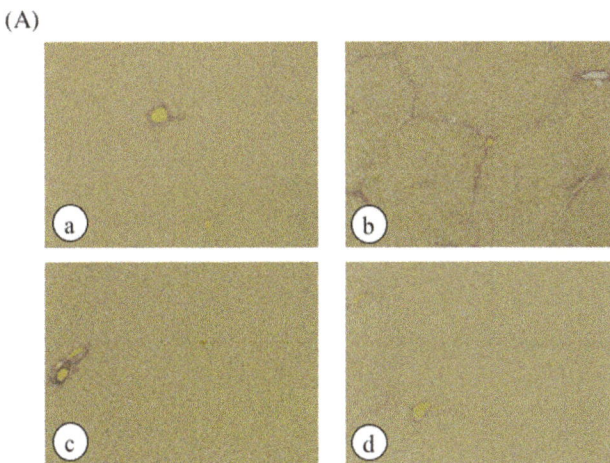

Figure 3. *Cont.*

(B)

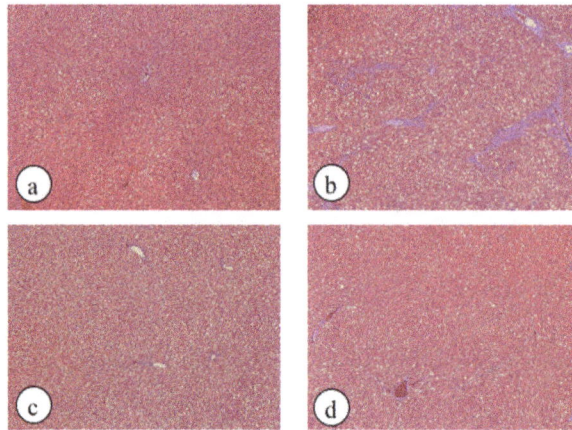

Figure 3. The effect of DNSM on the changes in liver morphology. Liver sections were processed for (**A**) EVG staining; and (**B**) Masson's trichrome staining. (**a**) Control diet; (**b**) Control diet with 5% ethanol plus CCl_4; (**c**) Control diet with CCl_4; (**d**) 0.12% DNSM diet with 5% ethanol plus CCl_4.

Ethanol-induced oxidative stress from CYP2E1 appears to play a major role in ethanol-induced liver injury [7–10]. We have previously demonstrated that Yerba mate extract suppressed the CTP2E1 activity induced by ethanol in both in vitro and in vivo models [16]. Therefore, we examined whether DNSM treatment could also have a preventive effect against alcohol-induced liver injury by regulating the CYP2E1 enzyme in an in vivo model. In ethanol plus CCl_4-treated rats, CYP2E1 activity increased 2.1-fold compared to the control group. However, DNSM-treatments suppressed CYP2E1 activity (Figure 4). We examined the effect of DNSM on the increase in intracellular lipid peroxidation using the thiobarbituric acid reactive substances (TBARS) assay. In ethanol plus CCl4-treated rats, hepatic TBARS levels were significantly increased. However, DNSM-treatment maintained the intracellular TBARS levels at the lower levels of the control rats (Figure 5).

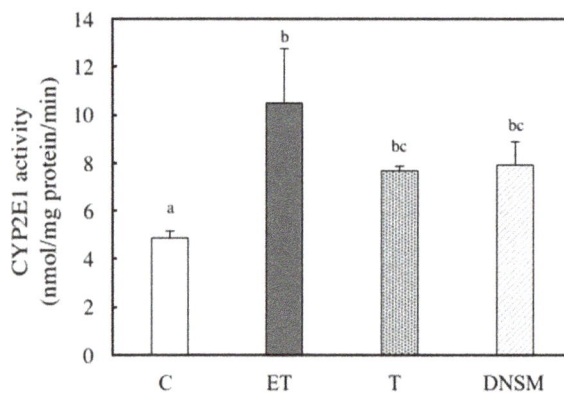

Figure 4. The effect of DNSM on CYP2E1 activity in the livers of ethanol plus CCl_4-treated rats. CYP2E1 activity was determined using the ϱ-nitrophenol (PNP) assay, as described in the Materials and Methods. (C) Control diet; (ET) Control diet with 5% ethanol plus CCl_4; (T) Control diet with CCl_4; (DNSM) 0.12% DNSM diet with 5% ethanol plus CCl_4. Data are presented as the mean ± S.E. of five animals. Values without a common letter are significantly different ($p < 0.01$).

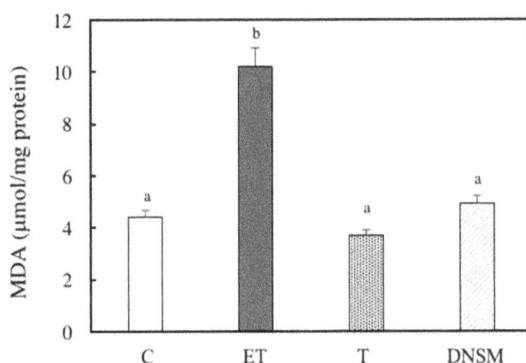

Figure 5. The effect of DNSM on lipid peroxidation in the liver. The measurement of lipid peroxidation using a colorimetric reaction with thiobarbitric acid (TBA) was carried out according to the method described by Ohkawa. The measured lipid peroxidation was expressed as malondialdehyde (MDA). (C) Control diet; (ET) Control diet with 5% ethanol plus CCl_4; (T) Control diet with CCl_4; (DNSM) 0.12% DNSM diet with 5% ethanol plus CCl_4. Each bar is the mean ± S.E. of five animals. Values without a common letter are significantly different ($p < 0.01$).

Furthermore, we measured the weights of various organs after the experimental period. As shown in Table 1, there were no differences among the weights of liver, kidney, and spleen in rats of the four groups. However, the masses of epididymal fat and visceral fat of ethanol plus CCl_4-treated rats and DNSM diet- and ethanol plus CCl_4-treated rats were lower than those rats fed the control diet or CCl_4 alone. These results suggest that loss of body weight in ethanol plus CCl_4-treated rats and DNSM diet- and ethanol plus CCl_4-treated rats may depend on the loss of fat and the loss may be induced by ethanol ingestion.

Table 1. Changes in organ weight of rats.

Groups	Organ Weight (g)				
	Liver	Kidney	Spleen	Visceral Fat	Epididymal Fat
C	10.82 ± 0.38	1.60 ± 0.03	0.76 ± 0.04	6.52 ± 0.55	7.49 ± 0.61
ET	10.54 ± 0.46	1.58 ± 0.05	0.75 ± 0.03	4.54 ± 0.48	4.74 ± 0.37
T	12.35 ± 0.47	1.70 ± 0.08	0.72 ± 0.02	6.22 ± 0.52	6.02 ± 0.20
DNSM	10.76 ± 0.34	1.66 ± 0.02	0.70 ± 0.03	4.79 ± 0.52	5.44 ± 0.52

(C) Control diet; (ET) Control diet with 5% ethanol plus CCl_4; (T) Control diet with CCl_4; (DNSM) 0.12% DNSM diet with 5% ethanol plus CCl_4.

To elucidate whether the protective effect of DNSM is dependent on digestion, we measured the effect of DNSM treatment in an in vitro alcohol-induced injury model in hepatocytes. We previously demonstrated that a treatment of 100 mM ethanol for 24 h significantly decreased cell viability of hepatocytes compared with control cells [15]. Here, we measured the cell viability of hepatocytes treated with 100 mM ethanol with or without various concentrations of DNSM. As shown in Figure 6A, DNSM did not prevent cell death. These results suggest that digestive products of DNSM are effective for the prevention of alcohol-induced liver injury. Therefore, the following experiments were carried out with the presence of deoxyadenosine. Adenosine which is the digestive product of RNA also was measured. Treatments of deoxyadenosine and adenosine prevented cell death in the hepatocytes treated with 100 mM ethanol (Figure 6B). Furthermore, the effect of deoxyadenosine and adenosine on the activity of alcohol dehydrogenase (ADH)—a main pathway of alcohol metabolism—was examined in the cells incubated for 4 h with 100 mM ethanol. The treatment of deoxyadenosine and adenosine increased the activities of ADH compared with the cells treated with 100 mM ethanol (Figure 7).

(A)

(B)

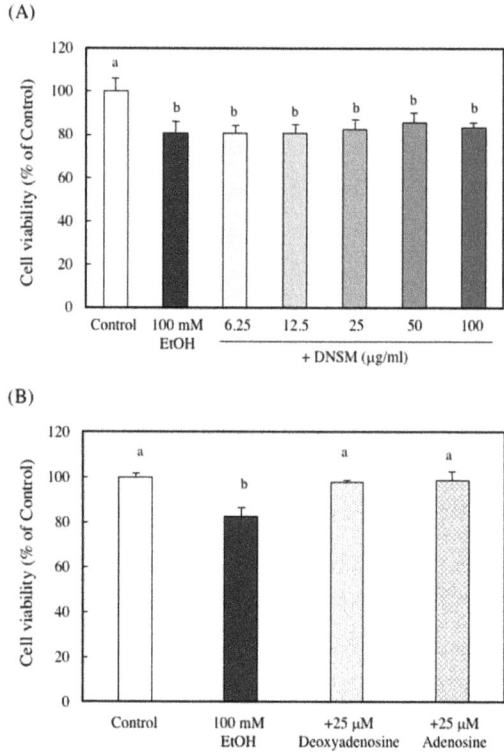

Figure 6. The effect of DNSM on ethanol-treated hepatocyte cell viability. Hepatocytes were incubated with 100 mM ethanol with or without (**A**) various DNSM concentrations; and (**B**) deoxyadenosine or adenosine for 24 h. Cell viability was measured by the Neutral Red assay, as described in the Materials and Methods section. Data are presented as the mean ± S.E. of three experiments. Values without a common letter are significantly different ($p < 0.01$).

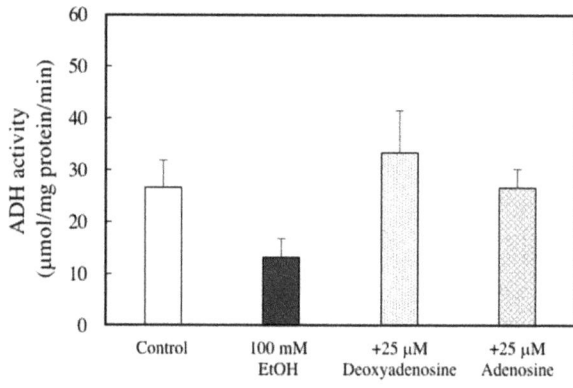

Figure 7. Effects of adenosine or deoxyadenosine on ADH activity in ethanol-treated hepatocytes. Hepatocytes were incubated for 4 h with 100 mM ethanol, with or without 25 µM deoxyadenosine or adenosine. ADH activity analysis was performed as described in the Materials and Methods section. Data are presented as the mean ± S.E.

3. Discussion

This study has shown that DNSM protects against ethanol-induced liver injury. Here, we demonstrated that the liver damage biomarkers, ALT and AST, were increased in a rat model given ethanol plus CCl_4 to induce liver damage, but this increase was reduced by nucleic acid supplementation. Furthermore, the DNSM treatment did not affect cell viability in ethanol-treated hepatocytes, suggesting the digestive products, which are created by the degradation of DNSM by nucleases in the intestine, are effective against alcohol-induced liver injury.

Alcohol-induced liver injury is induced by heavy drinking and is accompanied by the degeneration or necrosis of hepatocytes, which disrupts normal liver function via oxidative stress. The CYP2E1 enzyme is one of the major producers of ethanol-induced ROS. Therefore, decreasing or inhibiting CYP2E1 activity may be a feasible strategy for minimizing the hepatotoxicity of ethanol. Recently, we demonstrated that *Ecklonia cava* polyphenol-treatment maintained CYP2E1 activity in ethanol-treated hepatocytes below that of control cells [23]. We also reported that treatment with an extract of Yerba Mate tea suppressed ethanol-induced increases in CYP2E1 activity to the level of the control cells in an in vitro, alcohol-induced hepatocyte model and an in vivo ethanol plus CCl_4-induced liver-injury model [16].

In the present study, we have shown that DNSM treatment suppressed ethanol-induced increases in CYP2E1 activity to the activity levels observed in the control rats and that the treatment of deoxyadenosine increased the ADH activity compared with the cells treated with ethanol in an in vitro alcohol injury hepatocyte model.

There are several reports that adenosine and deoxyadenosine has a protective effect against various diseases. Modis et al. reported that adenosine and its metabolite, inosine, exerted cytoprotective effects in an in vitro model of liver ischemia-reperfusion injury [24]. Hasemi et al. have shown that adenosine and deoxyadenosine induces apoptosis in human breast cancer cells via the activation of the mitochondria/intrinsic apoptotic pathway [25]. Furthermore, Lee reported that adenosine protected Sprague-Dawley rats from a high-fat diet and repeated acute restraint stress-induced intestinal inflammation and altered expression of nutrient transporters [26].

In the present study, the precise mechanism of the DNSM protection against ethanol-induced liver injuries in rats is not clear. However, there are two possible explanations for this protection. The first is via the cyclic AMP (cAMP)/protein kinase A (PKA) pathway. There are some reports that the cAMP/PKA signaling pathway regulates the activities of alcohol dehydrogenase and CYP2E1 in ethanol metabolism [27,28]. The administration of theophylline to rats, which inhibits cAMP phosphodiesterase and thus increases endogenous cAMP levels, or the addition of dibutyryl cAMP to hepatocyte cultures, both increased ADH activity [27]. Contrastingly, cAMP-dependent phosphorylation of CYP2E1 lead to a reduction in CYP2E1 activity [28]. We have shown that the treatment of *Ecklonia cava* polyphenol with ethanol increased the activity of alcohol dehydrogenase and inhibits CYP2E1 activity [23]. The changes in CYP2E1 and alcohol dehydrogenase activity were suppressed by treatment with H89, an inhibitor of PKA, suggesting that *Eclonia cava* polyphenol has a protective effect against ethanol-induced liver injury in a cAMP-dependent manner.

Charest et al. showed that adenosine and AMP increased cAMP concentration by interacting with the adenosine receptor [29,30]. These results suggest that DNSM was digested by nucleases in the intestine and absorbed as nucleosides and nucleobases. Adenosine and AMP subsequently bind to the adenosine receptor, which activates adenylate cyclase.

Another possibility is the involvement of the adenosine monophosphate-activated protein kinase (AMPK) signaling pathway. AMPK is a sensor that regulates cellular metabolism and oxidative stress [31,32]. Chronic alcohol consumption results in inhibition of the hepatic AMPK signaling pathway by ethanol, which leads to steatosis [33]. Fat accumulation in hepatocytes leads to the development of fatty liver. With continued alcohol consumption, fatty liver may progress to hepatitis and cirrhosis. Therefore, it is important to enhance AMPK pathway signaling to protect the liver against diseases induced by ethanol. Shin et al. have reported that β-lapachone, a naturally occurring quinone, activated the AMPK pathway in ethanol-fed rats [34]. Wang et al. also indicated that oligomeric proanthocyanidines, a class of flavonoid compounds, alleviated liver steatosis and damage through AMPK activation against alcohol-induced liver steatosis and injury [35]. On the other hand, Stenesen et al. reported that dietary adenine controlled adult lifespan via adenosine nucleotide biosynthesis and AMPK activation [36]. Dietary adenine feeding increases the ratio of AMP:ATP and ADP:ATP and activates AMPK. These results suggest that nucleoproteins activate the AMPK signaling pathway in the liver.

It is important to know the precise mechanism of the DNSM protection against ethanol-induced liver injuries in rats. Especially, the involvement of cAMP/PKA pathway or AMPK pathway in an in vivo ethanol-carbon tetrachloride cirrhosis model need to be elucidated further.

4. Materials and Methods

4.1. Materials

DNA-rich nucleic acids from salmon milt (DNSM) were water-solubilized using nuclease and protease. The nucleotide and amino acid composition of the DNSM is shown in Table 2. DNSM were provided by Fordays Co., Ltd. (Tokyo, Japan) and L•S Corporation. Williams' Medium E and β-nicotinamide adenine dinucleotide hydrate were obtained from Sigma-Aldrich Co. (St. Louis, MO, USA). Fetal bovine serum (FBS) was purchased from Nichirei Biosciences, Inc. (Tokyo, Japan). The other chemicals used in this study were special-grade commercial products purchased from WAKO Pure Chemical Co., Ltd. (Osaka, Japan).

Table 2. Composition of nucleotides and amino acids in DNSM.

Nucleotides *	Amount (g/100 g)
5'-dCMP	6.01
5'-dAMP	9.15
5'-dTMP	9.26
5'-dGMP	6.93
Total	31.35
Amino Acids	**Amount (g/100 g)**
Arg	17.80
Lys	2.66
His	0.65
Phe	0.89
Tyr	0.88
Leu	1.95
Ile	1.25
Met	0.60
Val	2.12
Ala	1.95
Gly	4.11
Pro	2.62
Glu	3.48
Ser	2.49
Thr	1.27
Asp	2.24
Trp	0.20
Cys	0.25
Total	47.41

* The amounts of nucleotides were analyzed after treatment of nuclease P_1.

4.2. Animals

Male Wistar rats were purchased from Japan SLC Inc. (Shizuoka, Japan). The rats were housed at a constant temperature and were allowed free access to water and standard rat chow (LaboMR stock, Japan SLC, Inc. Shizuoka, Japan). All animal experiments followed our institution's criteria for the care and use of laboratory animals in research, which meet the guidelines for animal experimentation at Osaka City University.

4.3. Animal Experiments

Male Wistar rats weighing 180–210 g were fed a standard laboratory diet and water ad libitum until three days prior to the experiment. The rats were then fed a control diet for three days and divided into four groups. Group 1 was the control; Group 2 was treated with ethanol and CCl_4; Group 3 was treated with CCl_4 alone; Group 4 was treated with ethanol, CCl_4, and 0.12% DNSM. The composition of each diet is presented in Table 3. CCl_4 (0.1 mL/kg of body weight diluted with olive oil to 25%) was administered by intraperitoneal injection twice a week (on Mondays and Thursdays), and 5% ethanol was administered in the drinking water ad libitum. The rats were euthanized after three weeks.

Table 3. Composition of diets.

Components (g)	Control	0.12% DNSM
Casein	200	200
L-Cystine	3	3
Cornstarch	397.486	396.286
α-Cornstarch	132	132
Sucrose	100	100
Soybean oil	70	70
Cellulose powder	50	50
Mineral mix (AIN-93G-MX) [1]	35	35
Vitamin mix (AIN-93VX) [2]	10	10
Choline hydrogen tartrate	2.5	2.5
t-Butylhydroquinone	0.014	0.014
DNSM	0	1.2
Total	1000	1000

[1] Composition in g/kg diet: Calcium Carbonate, 357; Potassium Phosphate, Monobasic, 196; Potassium Citrate·H_2O, 70.78; Sodium Chloride, 74; Potassium Sulfate, 46.6; Magnesium Oxide, 24; Ferric Citrate, 6.06; Zinc Carbonate, 1.65; Manganese Carbonate, 0.63; Cupric Carbonate, 0.324; Potassium Iodate, 0.01; Sodium Selenate, 0.01025; Chromium K Sulfate·$12H_2O$, 0.275; Ammonium Molybdate·$4H_2O$, 0.00795; Sodium Silicate·$9H_2O$, 1.45; Lithium Chloride, 0.0174; Boric Acid, 0.0815; Sodium Fluoride, 0.0635; Nickel Carbonate·$4H_2O$, 0.0306; Ammonium Vanadate, 0.0066; Sucrose, 221.0032; [2] Composition in g/kg diet: Vitamin A Acetate (500,000 IU/g), 0.8; Vitamin D3 (400,000 IU/g), 0.25; Vitamin E Acetate (500 IU/g), 15; Phylloquinone, 0.075; Biotin, 2; Cyanocobalamin, 2.5; Folic Acid, 0.2; Nicotinic Acid, 3; Calcium Pantothenate, 1.6; Pyridoxine-HCl, 0.7; Riboflavin, 0.6; Thiamin HCl, 0.6; Sucrose, 974.655.

4.4. Histological Analysis

Liver samples were collected from each rat, fixed in 10% buffered formalin fixative, and then dehydrated in a graded alcohol series. Following xylene treatment, the specimens were embedded in paraffin blocks and cut into 5-μm sections. Consecutive sections were stained with EVG and Masson's trichrome staining. The pathologist was blinded to the rats' group assignments.

4.5. Liver Damage Biomarkers

The activity of plasma AST and ALT were estimated using a Transaminase CII-test kit (Wako, Japan).

4.6. CYP2E1 Activity Analysis

Livers were homogenized in nine volumes of tris HCl buffer (containing 0.25 M sucrose, pH 7.4) using a Polytron 1600E (Central Science Trade Co., Inc., Tokyo, Japan). The homogenates were centrifuged at $700 \times g$ for 10 min at 4 °C. The supernatant was collected as an S9 fraction. The activity of CYP2E1 was determined by the rate of hydroxylation of PNP measured at 546 nm [37]. The S9 fraction was added to 100 mM KH_2PO_4 (containing 0.2 mM PNP and 2.0 mM NADPH, pH 6.8) and incubated in a 37 °C water bath for 20 min. The reaction was stopped using 0.6 M perchloric acid (250 µL) and 10 M NaOH (75 µL) was added to the remaining supernatant. The results were expressed as the amount of *p*-nitrophenol pmols/min/mg protein formed and the determined concentration of 4-nitrocatechol ($\varepsilon = 10.28$ $mM^{-1} \cdot cm^{-1}$).

4.7. Measurement of Lipid Peroxidation

Lipid peroxidation was measured according to the method described by Ohkawa using a colorimetric reaction with thiobarbitric acid (TBA) and the measured lipid peroxidation was expressed as malondialdehyde (MDA) [38]. The frozen liver samples were excised and homogenized in nine volumes of ice-cold 1.15% KCl. Samples consisting of less than 0.2 mL of 10% (w/v) tissue homogenate were then added to 0.2 mL of 8.1% sodium dodecyl sulfate, 1.5 mL of 20% acetic acid solution adjusted to pH 3.5 with NaOH, and 1.5 mL of a 0.8% aqueous solution of TBA. The mixture was brought to a total volume of 4.0 mL using distilled water, mixed with 5.0 mL of a mixture of n-butanol and pyridine (15:1, v/v), and shaken vigorously. After centrifugation at 4000 rpm for 10 min, the organic layer was collected and its absorbance was measured at 532 nm. 1,1,3,3-Terramethoxypropane was used as an external standard. The level of lipid peroxidation was expressed in nmol of MDA.

4.8. Hepatocyte Preparation and Culture

Hepatocytes were isolated by collagenase perfusion following their removal from 10-week-old male Wistar rats anesthetized with sodium pentobarbital [39]. The viability of the isolated hepatocytes was greater than 90%, as determined by 0.2% trypan blue exclusion. The cells were plated on 35-mm plastic dishes at a density of 2.5×10^5 cells/mL in 2 mL of Williams' Medium E supplemented with 10% FBS. The cells were cultured in a humidified atmosphere (5% CO_2/95% air) at 37 °C overnight. After pre-incubation, the cells were cultured in 10% FBS containing fresh Williams' Medium E with different concentrations of ethanol, with or without DNSM, adenosine, or deoxyadenosine for 0–24 h.

4.9. Cell Viability Assay

The cell viability of the hepatocytes was measured by the Neutral Red assay, as previously described [40]. Neutral Red stock solution (0.4% Neutral Red in water) was diluted 1:80 in phosphate-buffered saline (PBS). Hepatocytes were incubated with the Neutral Red solution for 2 h at 37 °C to allow for the uptake of the lysosomal dye into viable cells. The Neutral Red solution was then removed, and the cultures were washed rapidly (in under 2.5 min) with a mixture of 1% formaldehyde-1% calcium chloride. A mixture of 1% acetic acid-50% ethanol was added to the cells at room temperature for 30 min to extract the Neutral Red from the hepatocytes. The optical density of each sample was then measured at 540 nm with a spectrophotometer. Cell viability was estimated as a percentage of the value obtained for untreated controls.

4.10. Assay of ADH Activities

After incubation, the cells were washed twice and then dissolved with cold PBS. The debris was obtained by centrifugation at $2600 \times g$ for 1 min at 4 °C, and then buffer (50 mM HEPES pH 7.5, 0.25 M sucrose, 1 mM EDTA, 1 mM dithiothreitol (DTT), 3 mM $MgCl_2$, 1 mM phenylmethylsulphonyl fluoride) was added. After two freeze-thaw cycles using liquid nitrogen, the cells were sonicated and centrifuged at $12,000 \times g$ for 20 min at 4 °C. Finally, the supernatant was collected. ADH activity was determined at 25 °C in a 1.5 mL volume (50 mM HEPES pH 8.0, 10 mM MgCl$_2$, 1 mM DTT, 300 μM NAD$^+$) in the presence or absence of ethanol (50 μL). The reaction was started by adding ethanol, and the absorbance at 340 nm was followed with a spectrophotometer. The linear initial increase in absorbance was used to determine specific enzyme activities with an absorption coefficient of 6.2 mM·cm^{-1}.

4.11. Statistical Analysis

Statistical comparisons were performed between groups using one-way analysis of variance and post hoc multiple comparisons using Tukey's test. A p-value less than 0.05 was considered significant.

5. Conclusions

In conclusion, we found that DNSM had protective effects against ethanol-induced liver injury. Although its precise mechanisms need to be elucidated further, DNSM may represent a novel tool for preventing alcohol-induced liver injury.

Acknowledgments: This study was partially supported by a Grant-in-Aid for Scientific Research from the Japan Society for the Promotion of Science (24500987).

Author Contributions: A.K-Y. and I.M-Y. contributed to the concept of the study and the manuscript preparation. A.K-Y., I.M-Y., H.S., K.S. and K.U. wrote the paper. A.K-Y., M.G., E.Y., Y.M., H.S. and K.S. performed experiments and analyzed data.

Conflicts of Interest: The authors declare no conflict of interest.

References

1. Lieber, C.S. Ethanol metabolism, cirrhosis and alcoholism. *Clin. Chim. Acta* **1997**, *257*, 59–84. [CrossRef]
2. Tsukamoto, H.; Lu, S.C. Current concepts in the pathogenesis of alcoholic liver injury. *FASEB J.* **2001**, *15*, 1335–1349. [CrossRef] [PubMed]
3. Lucey, M.R.; Mathurin, P.; Morgan, T.R. Alcoholic hepatitis. *N. Engl. J. Med.* **2009**, *360*, 2758–2769. [CrossRef] [PubMed]
4. Zima, T.; Fialova, L.; Mestek, O.; Janebova, M.; Crikovska, J.; Malbonan, I.; Stipek, S.; Milkulikova, L.; Popov, P. Oxidative stress, metabolism of ethanol and alcohol-related disease. *J. Biomed. Sci.* **2001**, *8*, 59–70. [CrossRef] [PubMed]
5. Albano, E. Alcohol, oxidative stress and free radical damage. *Proc. Nutr. Soc.* **2006**, *65*, 278–290. [CrossRef] [PubMed]
6. Lieber, C.S.; DeCarli, L.M. Ethanol oxidation by hepatic microsomes-adaptive increase after ethanol feeding. *Science* **1968**, *162*, 912–918. [CrossRef]
7. Lieber, C.S. CYP2E1: From ASH to NASH. *Hepatol. Res.* **2004**, *28*, 1–11. [CrossRef] [PubMed]
8. Guengrich, F.P. Oxidative cleavage of carboxylic esters by cytochrome-P-450. *J. Biol. Chem.* **1987**, *262*, 8459–8462.
9. Porter, T.D.; Coon, M.J. Cytochrome P-450: Multiplicity of isoforms, substrates, and catalytic and regulatory mechanisms. *J. Biol. Chem.* **1991**, *266*, 13469–13472. [PubMed]
10. Robertson, G.; Leclercq, I.; Farrell, G.C. Nonalcoholic p-450 enzymes and oxidative stress. *Am. J. Physiol. Gastrointest. Liver Physiol.* **2001**, *281*, G1135–G1139. [PubMed]
11. Lieber, C.A.; De Carli, L.M.; Sorrel, M.F. Experimental methods of ethanol administration. *Hepatology* **1989**, *10*, 501–510. [CrossRef] [PubMed]
12. Tsukamoto, H.; Horne, W.; Kamimura, S.; Niemela, O.; Parkkla, S.; Ylaherttula, S.; Brinttenham, G.M. Experimental liver cirrhosis induced by alcohol and iron. *J. Clin. Investig.* **1995**, *96*, 620–630. [CrossRef] [PubMed]
13. Tipoe, G.L.; Liong, E.C.; Casey, C.A.; Donohue, T.M., Jr.; Eagon, P.K.; So, H.; Leung, Y.M.; Fogt, F.; Nanji, A.A. A voluntary oral ethanol-feeding rat model associated with necroinflammatory liver injury. *Alcohol Clin. Exp. Res.* **2008**, *32*, 669–682. [CrossRef] [PubMed]
14. Siegers, C.P.; Pauli, V.; Korb, G.; Younes, M. Hepatoprotection by malotilate against carbon tetrachloride-alcohol-induced liver fibrosis. *Agents Actions* **1986**, *18*, 600–603. [CrossRef] [PubMed]

15. Takahashi, M.; Satake, N.; Yamashita, H.; Tamura, A.; Sasaki, M.; Matsui-Yuasa, I.; Tabuchi, M.; Akahoshi, Y.; Terada, M.; Kojima-Yuasa, A. Ecklonia-cava polyphenol protects the liver against ethanol-induced injury in rats. *Biochim. Biophysica. Acta* **2012**, *1820*, 978–988. [CrossRef] [PubMed]
16. Tamura, A.; Sasaki, M.; Yamashita, H.; Matsui-Yuasa, I.; Saku, T.; Hikima, T.; Tabuchi, M.; Munakata, H.; Kojima-Yuasa, A. Yerba-mate (*Ilex paraguarienesis*) extract prevents ethanol-induced liver injury in rats. *J. Funct. Food.* **2013**, *5*, 1714–1723. [CrossRef]
17. Carver, J.D. Dietary nucleotides: Effects on the immune and gastrointestinal systems. *Acta Paediatr. Suppl.* **1999**, *88*, 83–88. [CrossRef] [PubMed]
18. Jyonouchi, H.; Sun, S.; Abiru, T.; Winship, T.; Kuchan, M.J. Dietary nucleotides modulate antigen-specific type 1 and type 2 T-cell responses in young c57bl/6 mice. *Nutrition* **2000**, *16*, 442–446. [CrossRef]
19. Sudo, N.; Aiba, Y.; Oyama, N.; Yu, X.N.; Matsunaga, M.; Koga, Y.; Kudo, C. Dietary nucleic acid and intestinal microbiota synergistically promote a shift in the Th1/Th2 balance toward Th1-dominant immunity. *Clin. Exp. Allergy* **2004**, *135*, 132–135.
20. He, Y.; Sanderson, I.R.; Walker, W.A. Uptake, transport and metabolism of exogenous nucleosides in intestinal epithelial cell culture. *J. Nutr.* **1994**, *124*, 1942–1949. [PubMed]
21. Tanaka, M.; Lee, K.; Martinez-Augustin, O.; He, Y.; Sanderson, L.R.; Walker, W.A. Exogenous nucleotides after the proliferation, differentiation and apoptosis of human small intestinal epitherium. *J. Nutr.* **1996**, *126*, 424–433. [PubMed]
22. Sakai, T.; Taki, T.; Nakamoto, A.; Tazaki, S.; Arakawa, M.; Nakamoto, M.; Tsutsumi, R.; Shuto, E. Dietary ribonucleic acid suppresses inflammation of adipose Tissue and improves glucose intolerance that is medicated by immune cells in C57BL/6 Mice fed a high-fat diet. *J. Nutr. Sci. Vitaminol.* **2015**, *61*, 73–78. [CrossRef] [PubMed]
23. Yamashita, H.; Goto, M.; Matsui-Yuasa, I.; Kojima-Yuasa, A. *Ecklonia cava* polyphenol has a protective effect against ethanol-induced liver injury in a cyclic AMP-dependent manner. *Mar. Drugs* **2015**, *13*, 3877–3891. [CrossRef] [PubMed]
24. Módis, K.; Geró, D.; Stangl, R.; Rosero, O.; Szijártó, A.; Lotz, G.; Mohácsik, P.; Szoleczky, P.; Coletta, C.; Szabó, C. Adenisine and inosine exert cytoprotective effects in an in vitro model of liver ischemia-reperfusion injury. *Int. J. Mol. Med.* **2013**, *31*, 437–446. [PubMed]
25. Hasemi, M.; Karami-Tehrani, F.; Ghavami, S.; Maddika, S. Adenosine and deoxyadenosine induced apoptosis in oestrogen receptor-positive and -negative human breast cancer cells via the intrinsic pathway. *Cell Prolif.* **2005**, *38*, 269–285. [CrossRef] [PubMed]
26. Lee, C.Y. Adenosine protects Sprague Dawley rats from high-fat diet and repeated acute restraint stress-induced intestinal inflammation and altered expression of nutrient transporters. *J. Anim. Physiol. Anim. Nutr.* **2015**, *99*, 317–325. [CrossRef] [PubMed]
27. Potter, J.J.; MacDougalg, O.A.; Mezey, E. Regulation of rat alcohol dehydrogenase by cyclic AMP in primary hepatocyte culture. *Arch. Biochem. Biophys.* **1995**, *321*, 329–335. [CrossRef] [PubMed]
28. Oesch-Bartlomowicz, B.; Padma, P.R.; Becker, R.; Richter, B.; Hengstler, J.G.; Freeman, J.E.; Wolf, C.R.; Oesch, F. Differential modulation of CYP2E1 activity by cAMP-dependent protein kinase upon Ser129 replacement. *Exp. Cell Res.* **1998**, *242*, 294–302. [CrossRef] [PubMed]
29. Charest, R.; Blckmore, P.F.; Exton, J.H. Characterization of responses of isolated rat hepatocytes to ATP and ADP. *J. Biol. Chem.* **1985**, *260*, 15789–15794. [PubMed]
30. Okajima, F.; Tokumitsu, Y.; Kondo, Y.; Ui, M. P2-Purinergic receptors are coupled to two signal transduction systems leading to inhibition of cAMP generation and to production of inositol trisphosphate in rat hepatocytes. *J. Biol. Chem.* **1987**, *262*, 13483–13490. [PubMed]
31. Mandrekar, P.; Szabo, G. Signalling pathways in alcohol-induced liver inflammation. *J. Hepatol.* **2009**, *50*, 1258–1266. [CrossRef] [PubMed]
32. Long, Y.C. AMP-activated protein kinase signaling in metabolic regulation. *J. Clin. Investig.* **2006**, *116*, 1776–1783. [CrossRef] [PubMed]
33. Sid, B.; Verrax, J.; Calderon, P.B. Role of AMPK activation in oxidative cell damage: Implications for alcohol-induced liver disease. *Biochem. Pharmacol.* **2013**, *86*, 200–209. [CrossRef] [PubMed]
34. Shin, S.; Park, J.; Li, Y.; Min, K.N.; Kong, G.; Hur, G.M.; Kim, J.M.; Shong, M.; Jung, M.-S.; Park, J.K.; et al. β-Lapachone alleviates alcoholic fatty liver disease in rats. *Cell. Signal.* **2014**, *26*, 295–305. [CrossRef] [PubMed]

35. Wang, Z.; Su, B.; Fan, S.; Fei, H.; Zhao, W. Protective effect of oligomeric proanthocyanidins against alcohol-induced liver steatosis and injury in mice. *Biochem. Biophys. Res. Commun.* **2015**, *458*, 757–762. [CrossRef] [PubMed]

36. Stenesen, D.; Suh, J.M.; Seo, J.; Yu, K.; Lee, K.-S.; Kim, J.-S.; Min, K.-J.; Graff, J.M. Dietary adenine controls adult lifespan via adenosine nucleotide biosynthesis and AMPK, and regulates the longevity benefit of caloric restriction. *Cell Metab.* **2013**, *8*, 101–112. [CrossRef] [PubMed]

37. Sapone, A.; Affatato, A.; Canistro, D.; Broccoli, M.; Trespidi, S.; Pozzetti, L.; Biagi, G.L.; Cantelli-Forti, G.; Paoline, M. Induction and suppression of cytochrome P450 isoenzymes and generation of oxygen radicals by procymidone in liver, kidney and lung of CD1 mice. *Mutat. Res.* **2003**, *527*, 67–80. [CrossRef]

38. Ohkawa, H.; Ohnishi, N.; Yagi, K. Assay for lipidperoxide in animal tissues by thiobarbituric acid reaction. *Anal. Biochem.* **1979**, *95*, 351–358. [CrossRef]

39. Moldéus, P.; Högber, J.; Orrenius, S. Isolation and use of liver cells. *Methods Enzymol.* **1978**, *52*, 60–71. [PubMed]

40. Zhang, S.Z.; Lipsky, M.M.; Trump, B.F.; Hsu, I.C. Neutral red (NR) assay for cell viability and xenobiotic-induced cytotoxicity in primary culture of human and rat hepatocytes. *Cell Biol. Toxicol.* **1990**, *6*, 219–234. [CrossRef] [PubMed]

marine drugs

MDPI

Article

Novel Peptide with Specific Calcium-Binding Capacity from *Schizochytrium* sp. Protein Hydrolysates and Calcium Bioavailability in Caco-2 Cells

Xixi Cai [1,2], Jiaping Lin [2] and Shaoyun Wang [2,*]

[1]   The Key Lab of Analysis and Detection Technology for Food Safety of the MOE, College of Chemistry, Fuzhou University, Fuzhou 350108, China; caixx_0123@163.com
[2]   College of Biological Science and Technology, Fuzhou University, Fuzhou 350108, China; kathleen369@163.com
*   Correspondence: shywang@fzu.edu.cn; Tel.: +86-591-2286-6375

Academic Editor: Se-Kwon Kim
Received: 1 October 2016; Accepted: 20 December 2016; Published: 27 December 2016

Abstract: Peptide-calcium can probably be a suitable supplement to improve calcium absorption in the human body. In this study, a specific peptide Phe-Tyr (FY) with calcium-binding capacity was purified from *Schizochytrium* sp. protein hydrolysates through gel filtration chromatography and reversed phase HPLC. The calcium-binding capacity of FY reached 128.77 ± 2.57 µg/mg. Results of ultraviolet spectroscopy, fluorescence spectroscopy, and infrared spectroscopy showed that carboxyl groups, amino groups, and amido groups were the major chelating sites. FY-Ca exhibited excellent thermal stability and solubility, which were beneficial to be absorbed and transported in the basic intestinal tract of the human body. Moreover, the calcium bioavailability in Caco-2 cells showed that FY-Ca could enhance calcium uptake efficiency by more than three times when compared with $CaCl_2$, and protect calcium ions against dietary inhibitors, such as tannic acid, oxalate, phytate, and Zn^{2+}. Our findings further the progress of algae-based peptide-calcium, suggesting that FY-Ca has the potential to be developed as functionally nutraceutical additives.

Keywords: *Schizochytrium* sp.; protein hydrolysate; calcium-binding peptide; structure; bioavailability

1. Introduction

Marine algae, which have traditionally formed part of the diet for centuries, especially in Asian countries such as China, Korea, and Japan, have become a popular research topic because of their biological implication [1]. *Schizochytrium* sp., belonging to marine fungi, possesses a large number of bioactive substances beneficial to the human body, such as unsaturated fatty acids, pigments, and proteins [2]. *Schizochytrium* sp. has been widely used in the industrial production of docosahexaenoic acid. The remaining by-products, containing about 41% protein, are usually used for biological baits or just discarded as industrial waste. Preparation of bioactive peptides from proteins through enzymatic hydrolysis has been a hot topic [3,4]. Therefore, the utilization of protein from the defatted *Schizochytrium* sp. by-products presents an opportunity.

Calcium is the most abundant mineral in the human body, mostly stored in the bones and supporting their structure and function. Calcium deficiency may result in many diseases, such as osteoporosis, kidney stones, and arterial hypertension [5,6]. Therefore, numerous calcium-fortified medicines and foods have come to market. However, calcium deficiency is still widespread due to insufficient absorption of the intake calcium. Ionized calcium is the primary calcium supplement for humans, but intestinal absorption of ionized calcium could be easily affected by the presence of dietary factors, such as tannin, phytate, oxalate, and other divalent metal ions [7]. Thus, a new class of calcium-enriched nutrients that can overcome these shortcomings has the potential to improve calcium nutrition. Organic calcium supplements show their superiority. Calcium-binding peptides, one of the organic calcium supplements, such as casein ophosphopeptides (CPPs) [8], soybean protein hydrolysates [9], whey protein hydrolysates [10], and serum protein hydrolysates [11], have been reported to be capable of promoting calcium uptake. Among these, CPPs were known as excellent mineral carriers with a significant role in promoting calcium ion absorption through the formation of CPP-Ca aggregates and maintaining the solubility [8,12]. CPPs induced calcium uptake in Caco-2 cells involved the transient receptor potential cation of the vanilloid subfamily V member 6, TRPV6 channel, also designated as calcium transporter-1, or CaT1 [13]. In the previous study, the nanocomposites of *Schizochytrium* sp. protein hydrolysate (SPH) chelated with calcium ions were prepared and the characterization of nano-composites was investigated by our group [14]. However, none has been reported about purified peptide with specific calcium-binding capacity from *Schizochytrium* sp. protein hydrolysate and calcium bioavailability. The research on the purified peptide is necessary to further understand the relationship between structure and function, and action mechanism.

The objectives of this study were, therefore, to isolate and characterize specific calcium-binding peptides from *Schizochytrium* sp. protein hydrolysate (SPH) and explore the possible chelating mechanism. Additionally, the Caco-2 cell monolayer model was used to determine the calcium bioavailability. This study could provide a new train of thought of the calcium-binding peptide from *Schizochytrium* sp. protein hydrolysate for the potential to be developed as a new kind of functionally nutraceutical supplements to improve bone health in the human body.

2. Results and Discussion

2.1. Purification of Calcium-Binding Peptide

Schizochytrium sp. protein hydrolysates consisted of various peptides were confirmed to possess calcium-binding capacity [14]. Systematic investigation on the calcium-binding properties of various peptides in SPH is of great importance. For this purpose, a specific peptide with calcium-binding capacity was first purified.

As shown in Figure 1a, SPH was divided into three size-dependent fraction through Sephadex G-25 chromatography. The calcium-binding capacities of F2 and F3 were similar and remarkably higher than F1 and SPH. Many studies have shown that peptides with lower molecular mass exhibited higher chelating capacity [15–17]. Therefore, the active fraction F3 with lower molecular mass was pooled and loaded onto semi-preparative C18 RP-HPLC. Twenty-two distinct fractions were collected and all of them exhibited different degrees of calcium-binding capacities (Figure 1b). Among them, activities of fraction 13 and fraction 17 were significantly higher than other fractions and fraction F3 from Sephadex G-25 chromatography ($p < 0.05$). Fraction 17, which showed the highest chelating capacity, was first selected for further purification by analytic RP-HPLC. Finally, fraction A, with the highest calcium-binding activity (128.77 ± 2.57 µg/mg), was collected and lyophilized for further studies (Figure 1c).

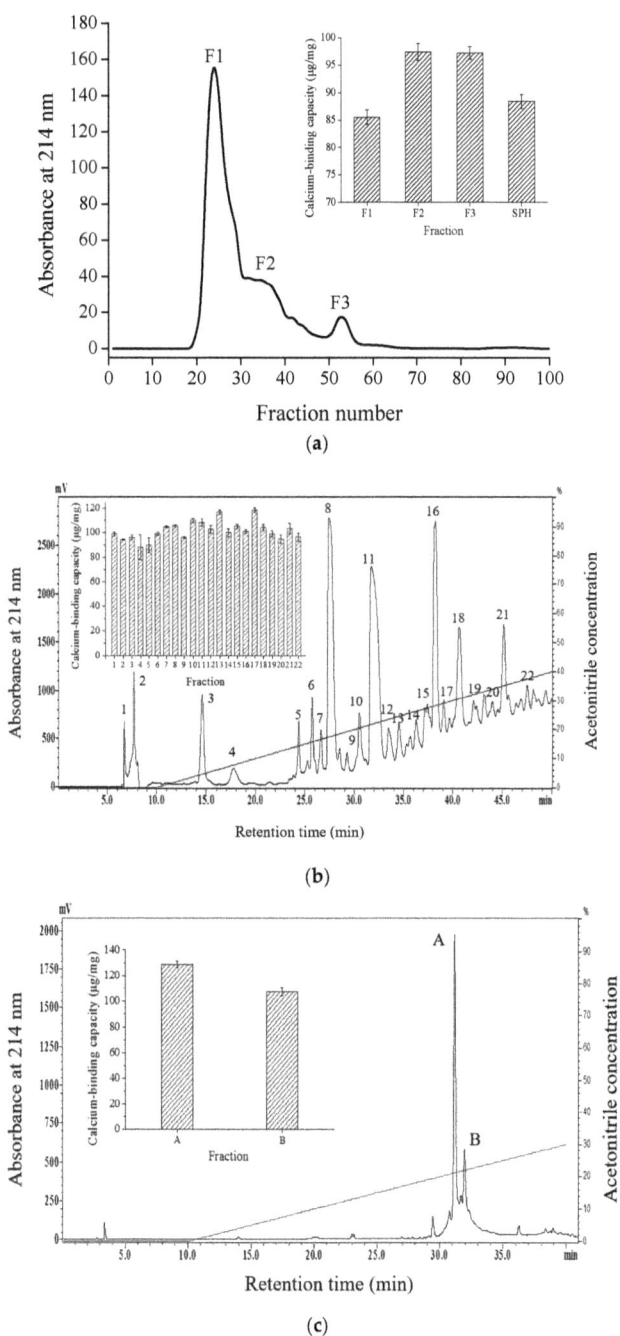

Figure 1. Elution profiles and calcium-binding capacities of calcium-binding peptides. (**a**) Sephadex G-25 gel filtration chromatography of SPH; (**b**) semi-preparative C18 RP-HPLC of fraction F3; and (**c**) analytic RP-HPLC of fraction 17 from semi-preparative HPLC.

2.2. Identification of the Calcium-Binding Peptide

The amino acid sequence of fraction A was determined to be Phe-Tyr (FY) with a molecular weight (MW) of 328.17 Da using liquid chromatography-electrospray ionization-tandem mass spectrometry (LC-ESI-MS/MS) (Figure 2). Subsequently, the identified peptide was chemically synthesized and the calcium-binding capacity was determined to be 125.91 ± 1.63 μg/mg, which was equivalent to the purified sample. Calcium-binding peptides from various sources with different MW and sequences have been isolated. Jeon reported that a peptide purified from *Chlorella* protein hydrolysates had a calcium binding activity of 0.166 mM and was determined to be 700.48 Da [18]. In our previous works, four dipeptides or tripeptides from whey protein hydrolysates were confirmed to possess 70–80 μg/mg calcium-binding capacity [19–22]. Not only the differences in the length and net charge of peptides, but also the different amino acid composition and sequence, could affect the extent of chelate formation with divalent metal cations [15,23]. Previous reports showed that the phosphorylation of tyrosine residues could provide appropriate chelating sites for positively charged metals, like calcium, zinc, and iron [24]. Kim indicated that an iron-binding peptide separated from heated whey hydrolysates contained 16.58% of phenylalanine residues, which was higher than other amino acids [25]. Moreover, dipeptide or tripeptide was deemed to promote metal ion absorption more effectively than higher MW peptides in intestinal epithelial cells [26]. Consequently, both of the Phe and Tyr residues in the purified peptide might contributed to chelation with metal cations.

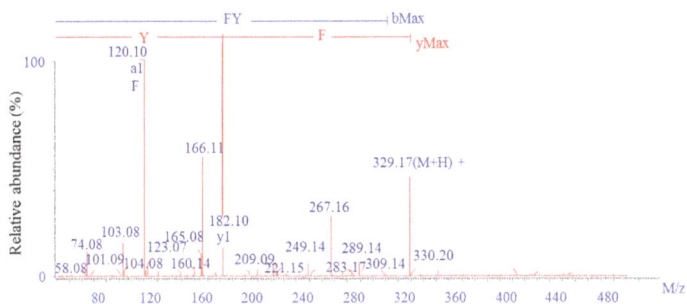

Figure 2. Identification of the amino acid sequence of the calcium-binding peptide using LC-ESI-MS/MS.

2.3. Structural Characterization of Peptide-Calcium Chelate

2.3.1. Ultraviolet Spectroscopy Analysis

Aromatic amino acids including tryptophan, phenylalanine, and tryptophan residues, could produce different UV spectra because of different chromophores. Phenylalanine has a specific absorption peak at 260 nm, and tyrosine at 280 nm approximately [27]. Therefore, the UV spectra was utilized to discuss the chelating mechanism of FY. As shown in Figure 3, the UV absorption spectra of FY-Ca chelates presented distinct differences from that of FY, which implied that a new substance was formed when FY interacted with calcium ions. Dipeptide FY had a maximum UV absorption peak at about 196 nm. With the increase of calcium ion concentration, the absorbance of the maximum absorption peak gradually increased from 1.937 to 2.149, showing a hyperchromic effect and redshift phenomenon. The results indicated that the chromophore groups (-C=O, -COOH) and auxochrome groups (-OH, -NH$_2$) generated polarizing changes when the ligands bound with calcium ions in the chelating process [28,29]. In addition, both FY and FY-Ca chelate had specific absorption peaks near 280 nm with the same intensity, suggesting that the tyrosine structure remained unchanged and the phenolic hydroxyl group of Tyr in FY was not involved in the chelation reaction because of the steric hindrance of the benzene ring. Hence, it could be presumed that the nitrogen atom of -NH- and -NH$_2$ and the oxygen atom of -C=O and -COOH participated in the chelation.

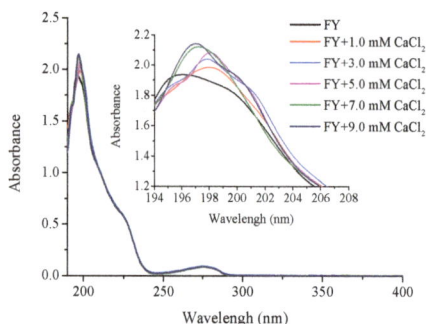

Figure 3. UV spectra of FY with different CaCl$_2$ concentrations over the wavelength range from 190 to 400 nm.

2.3.2. Fluorescence Spectroscopy Analysis

The specific calcium-binding peptide FY included Phe and Tyr, which could generate endogenous fluorescence at an excitation wavelength of 280 nm, and the corresponding emission peaks of Phe and Tyr were 303 nm and 313 nm, respectively. The fluorescence spectra of FY and FY-Ca chelate were shown in Figure 4. With the increase of calcium ion concentration, the intensity of endogenous fluorescence at 310 nm was reduced, which implied that calcium ions could be chelated by aromatic amino acids and lead to fluorescence quenching. Particularly, obvious endogenous fluorescence quenching appeared as soon as 1.0 mM of CaCl$_2$ was introduced. However, when the concentration of calcium ion reached 5.0 mM, no further changes were observed. This potentially manifested that changes in the fluorescence occurred when calcium chelated with the peptides and excess free calcium made no difference. Similar results has been reported by Zhou that fluorescence quenching was observed when calcium ions combined with the calcium-chelating peptide [30]. Moreover, Wu proved that reduced fluorescence intensity was a classic indicator of peptide folding when ferrous ions chelated with sturgeon protein peptide, and ferrous ions closed to tryptophan residues in the folding process [31]. Therefore, the results demonstrated that the calcium ions chelated with FY might cause folding of the peptide and form a compact structure, which contributed to the decrease in fluorescence intensity.

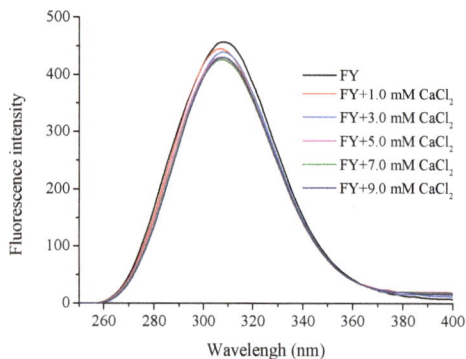

Figure 4. Fluorescence spectra of FY with different CaCl$_2$ concentration over the wavelength range from 295 to 500 nm.

2.3.3. Fourier Transform Infrared Spectroscopy (FTIR) Measurement

The specific FTIR absorption peak changes of the amides and carboxylates in FY could reflect the interaction of calcium ions and organic ligand groups of the peptides. As shown in Figure 5, displacement and intensity changes of main absorption peaks could be observed when calcium ions bound with the amino acid residues. The two most important vibrational modes of amides are the amide-I vibration and amide-II vibration, the amide-I vibration is primarily caused by the stretching of C=O bonds, amide-II vibration is assigned to deformation of N-H bonds and stretching of C–N bonds [21,32]. The absorption band of FY at 1668.17 cm^{-1} for the amide I band shifted to a higher frequency (1680.58 cm^{-1}) after chelating with calcium, manifesting that the -COO- group participated in the covalent combining reaction with the metal cations [33]. Additionally, the amide II band at 1516.43 cm^{-1} in FY also shifted to 1587.11 cm^{-1} in the FY-Ca chelate. The characteristic peak of amide-A stretching vibration of FY shifted from 3394.61 cm^{-1} to 3422.96 cm^{-1} might be due to the replacement of N-OH bonds (hydrogen bonds) with Ca-N bonds after calcium chelation [22]. After chelation, the spectrum shifted towards high-frequency wavenumbers (3500–2800 cm^{-1}), indicating that dipole field effect or induced effect led to the electron cloud density and frequency increased [14]. In the fingerprint region, the absorption intensity at 1187.51 cm^{-1} decreased and moved towards 1214.78 cm^{-1} simultaneously when FY chelated with calcium. A reasonable explanation was that FY bound with calcium ion and form C–O–Ca [14]. Furthermore, the absorption intensity observably reduced at lower frequency 837.24 cm^{-1} in FY-Ca chelate, it might attributed to the changes of -C–H group and -N–H group of FY in the chelating procedure. Previous research showed that the carboxyl group loss of protons and negative electricity (-COO-) was also potential binding site. Additionally, the amino group (-NH$_2$) and imino group of the peptide bond (-NH) were also likely to be involved in the formation of chelate [34]. The results of FTIR proved that oxygen atoms of the carboxyl group and nitrogen atoms of the amino group might be involved in the chelating reaction and generated a new substance.

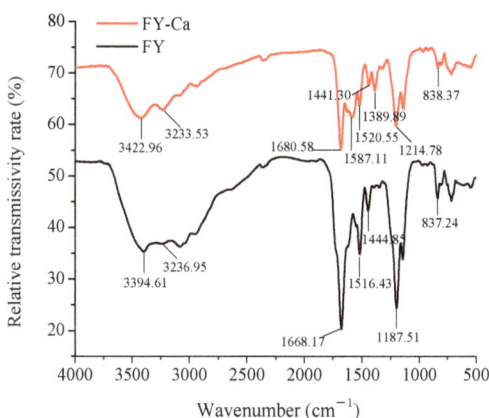

Figure 5. Fourier transform infrared (FTIR) spectra of FY and FY-Ca chelate in the regions from 4000 to 400 cm^{-1}.

2.4. Thermal and pH Stability Analysis of Peptide-Calcium Chelate

2.4.1. Thermogravimetry-Differential Scanning Calorimetry (TG-DCS) Analysis

The difference of thermostability between FY and FY-Ca chelate was explored through TG-DSC analysis. As shown in Figure 6a, the TG curve of dipeptide FY revealed that the thermal decomposition reaction of FY involved three stages in the whole process of 76.35% weight loss, and the thermal transition temperature was 155.16 °C, 161.26 °C, 298.66 °C, and 386.34 °C, respectively according to DSC analysis. The endothermic peaks were mainly caused by the destruction of C–N bonds in different positions of FY [20]. However, the TG curve of FY-Ca performed only two stages and lost 43.42% of its weight entirely (Figure 6b). The temperature of endothermic peaks significantly shifted to 265.12 °C, 335.75 °C, and 417.82 °C after the calcium ion chelated with FY, suggesting that FY-Ca chelate was less sensitive to thermal denaturation and performed better thermostability than FY, which was advantageous for application in medicine and functional food.

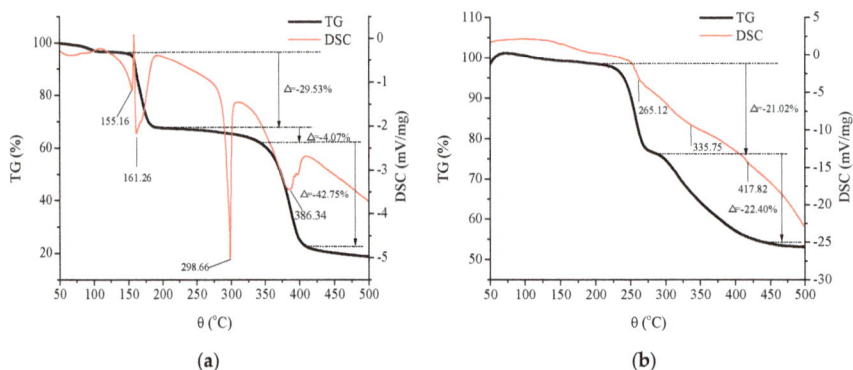

Figure 6. Typical TG-DSC thermograms of (**a**) FY and (**b**) FY-Ca chelate.

2.4.2. Calcium-Releasing Percentage Assay

The calcium-releasing percentages of the FY-Ca chelate and $CaCl_2$ at different pH were shown in Figure 7. The solubility of FY-Ca and $CaCl_2$ was obviously different. The calcium-releasing percentage of $CaCl_2$ exhibited a distinctly downward trend with the increase of pH value, and was reduced to 75.7% at pH 8.0, which could deduce that the free calcium ions and OH^- formed precipitates and led to a decline in the percentage. In contrast, the calcium-releasing percentage of FY-Ca chelate was always apparently higher than that of $CaCl_2$ at pH 2.0–8.0, and it maintained a relatively stable value of about 95% as well. The pH value in human intestinal tract is approximately pH 7.2, and FY-Ca chelate had higher solubility and better bioavailability in the gastrointestinal tract, which implied that FY-Ca chelate could be effectively absorbed and transported by intestinal epithelial cells than $CaCl_2$ [35].

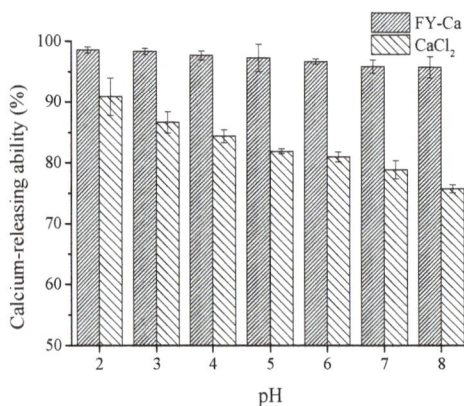

Figure 7. Calcium-releasing percentage of FY-Ca chelate and $CaCl_2$ at different pH.

2.5. Calcium Bioavailability in Human Intestinal Caco-2 Cell Lines

2.5.1. Cell Uptake of FY-Ca

For the uptake studies, Caco-2 cells were pre-incubated with FY-Ca chelate at different concentrations with $CaCl_2$ used as control. The effect of FY-Ca on the intracellular calcium concentration increased dose-dependently and then approximately trended to stable when the calcium concentration reached 9 mM, according to results in Figure 8. Additionally, the absorption-enhanced effects of FY-Ca were more than three times that of $CaCl_2$ at the same calcium concentration. Similar findings were also reported for desalted duck egg white peptides [36], soybean protein hydrolysates-calcium complex [9], and CPPs [37], which might act as calcium carriers and interact with the plasma membrane to transport calcium to the cytosol and ultimately significantly promote calcium uptake.

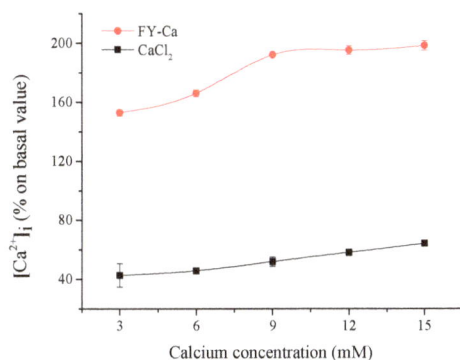

Figure 8. Cell uptake of FY-Ca chelate and $CaCl_2$ in Caco-2 cell by Fluo-3-AM loading for fluorescence analysis.

2.5.2. Calcium Bioavailability under the Action of Dietary Inhibition Factors

Well-established dietary factors, such as tannic acid, oxalate, phytate, and zinc ions, were chosen to evaluate whether the typical inhibitors from food would affect the uptake of calcium chelated by FY, with $CaCl_2$ as control. As expected, the addition of zinc ions, oxalate, phytate, and tannic acid severely decreased the calcium uptake efficiency of $CaCl_2$ by 39.7%, 84.4%, 74.9%, and 86.6%. FY-Ca, by contrast, could protect calcium ions from precipitation caused by oxalate, phytate, and tannic acid, and retain 83.0%, 65.2%, and 36.5% of calcium uptake efficiency, which were 5.3, 2.6, and 2.7 times higher than $CaCl_2$, respectively (Figure 9). Furthermore, the addition of Zn ions had little impact on the calcium uptake efficiency of FY-Ca.

Divalent metal ions, such as zinc and ferrous ions, have negative interactions with calcium nutrients and inhibit their uptake since the common receptors for these metal ions, DMT1, are located in the intestine [38]. In this study, the addition of FY could significantly attenuate the inhibition effect of zinc ions on calcium uptake, indicating that FY-Ca might pass through the cell membrane through specific pathways other than the DMT1 receptor. Organic phosphates, such as oxalate and phytate, greatly inhibit calcium uptake due to the formation of insoluble and indigestible complexes [7,39]. In the present study, the calcium uptake efficiency of FY-Ca was superior to $CaCl_2$ in the same condition, obviously, which might be due to the stronger chelating power of FY than organic phosphate and prevention of calcium precipitation. Tannin is another dietary factor belonging to polyphenols that exhibits extremely strong protein degeneration and metal ions complexing actions [40]. The addition of tannic acid also decreased the absorptivity of FY-Ca in Caco-2, which might be attributed to the peptide denaturation under high-dose tannic acid. Despite all of these, the calcium uptake efficiency of FY-Ca was remarkably higher than $CaCl_2$. These results demonstrated that FY could prevent a great amount of calcium from being precipitated by certain substances, thus improving calcium uptake. The present study provides powerful evidence for the idea that some proteins/peptides could be considered as mineral carriers because of their ability to bind and solubilize calcium with the possible role in increasing calcium transport across intestinal epithelial cells [41].

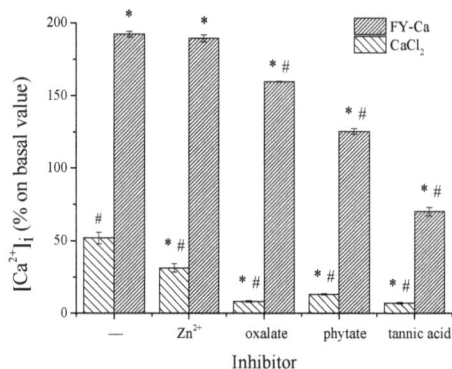

Figure 9. Effect of FY-Ca chelate on calcium bioavailability under the action of dietary inhibition factors. The concentration of calcium was 10 mM and tannic acid/Ca, oxalate/Ca, phytate/Ca, or Zn/Ca = 20:1. * Statistical significance $p < 0.05$, compared with the $CaCl_2$ control group. # Statistical significance $p < 0.05$, compared with the FY-Ca control group.

3. Materials and Methods

3.1. Materials

The defatted *Schizochytrium* sp. was kindly provided by Fisheries Research Institute of Fujian, China. The commercial protease, Alcalase (EC. 3.4.21.62, 2.2×10^5 U/g) and Flavourzyme (EC. 3.4.11.1, 7.8×10^4 U/g) were products of Novozymes (Copenhagen, Denmark). Sephadex G-25 was purchased from Amersham Pharmacia Co. (Uppsala, Sweden). Methanol and acetonitrile used in liquid chromatography were of HPLC grade. All of the other chemicals and solvents were of analytical grade and commercially available.

3.2. Preparation of Schizochytrium sp. Protein Hydrolysates

Schizochytrium sp. protein was prepared from *Schizochytrium* sp. by alkali extraction and acid precipitation, and *Schizochytrium* sp. protein hydrolysate possessing high calcium-binding capacity was prepared through stepwise enzymatic hydrolysis with Alcalase and Flavourzyme, as described in our previous work [14].

3.3. Purification of Specific Calcium-Binding Peptides

The lyophilized SPH dissolved in deionized water was loaded onto a Sephadex G-25 column (100 × 2.0 cm) and then eluted with deionized water at a flow rate of 0.3 mL/min. The absorbance of the elution was monitored at 214 nm and the calcium-binding capacity of the fractions was determined. The fraction with the highest calcium-binding activity from Sephadex G-25 chromatography was pooled and further purified by semi-preparation reversed phase HPLC on a C18 reversed-silica gel chromatograph (Gemini 5 μ C18, 250 × 10 mm; Phenomenex Inc.; Torrance, CA, USA). Elution was performed with solution A (0.05% trifluoroacetic acid (TFA) in water) and solution B (0.05% TFA in acetonitrile) with a gradient of 0%–40% B at a flow rate of 2 mL/min for 50 min. The elution was monitored at 214 nm, and the fractions were collected for calcium-binding capacity analysis. The most active fraction was further purified by analytic HPLC. Buffers A and B were the same as those used in semi-preparative RP-HPLC. Runs were conducted with a liner gradient of 0%–30% solvent B at a flow rate of 1 mL/min.

3.4. Identification of Purified Calcium-Binding Peptide

The molecular mass and amino acid sequence of the purified calcium-binding peptide were determined using LC-ESI-MS/MS (Delta Prep 4000, Waters Co., Milford, MA, USA) over the m/z range of 300–3000.

3.5. Synthesis of the Purified Peptide

The purified peptide (Phe-Tyr, FY) was synthesized by GL Biochem Corporation. Ltd. (Shanghai, China) through a solid-phase procedure. The purity of the synthesized peptide was 99.22% by HPLC analysis and the structure of peptide was confirmed by mass spectrometry analysis.

3.6. Analysis of Calcium-Binding Capacity

The calcium-binding capacity was measured with ortho-cresolphthalein complexone reagent according to the method described by Wang [35] with some modifications. One milliliter of 9 mM $CaCl_2$ was mixed with 2 mL of 0.2 M sodium phosphate buffer (pH 8.0), and then 1 mL of 1 mg/mL of peptides was added to create a competitive environment. The mixture was stirred at 37 °C for 2 h. Afterward, the insoluble calcium phosphate salts was removed by centrifugation at 10,000 rpm for 10 min and the calcium contents in the supernatant were determined by the absorbance at 570 nm after introducing the working solution to the samples.

3.7. Structural Characterization of Peptide-Calcium Chelate

3.7.1. Fabrication of Peptide-Calcium Chelate

One-hundred milligrams of lyophilized peptide was dissolved in 10 mL of distilled water, and $CaCl_2$ solution was introduced subsequently to a 3:1 ratio of peptide to calcium (w/w) at pH 6.0. The reaction solution was placed in a shaking water bath at 140 rpm and 37 °C for 20 min. Peptide-calcium chelate was precipitated after introducing absolute ethanol and collected by centrifugation at 10,000 rpm for 20 min.

3.7.2. Ultraviolet Spectroscopy

The ultraviolet spectra of calcium-binding peptide and its calcium chelate were monitored over the wavelength range from 190 nm to 400 nm using an ultraviolet spectrophotometer (UV-2600, UNICO Instrument Co. Ltd., Shanghai, China) as the method described in our previous work with some modifications [14]. For determinations, 20 μg/mL of peptide solution was prepared and the pH was adjusted to 6.5. Then 0, 0.5, 1.0, 1.0, 1.0, and 1.0 μL of 2 M $CaCl_2$ was constantly introduced every 10 min and the UV spectra were recorded.

3.7.3. Fluorescence Spectroscopy

Fluorescence spectroscopy was utilized to investigate the conformational changes of the peptide chelating with calcium ions by a Hitachi F-4600 fluorescence spectrophotometer (Hitachi Co., Tokyo, Japan). The excitation wavelength was 285 nm and the emission wavelengths between 250 and 400 nm were recorded. The slit width of excitation and emission was 20 and 30 nm respectively, and the sensitivity was 1. The preparation of sample was the same as that of ultraviolet spectroscopy analysis.

3.7.4. FTIR

One milligram of lyophilized sample and 100 mg of dried KBr were fully mixed and ground in an agate mortar. After tableting, FTIR spectra were recorded at room temperature by an infrared spectrophotometer (360 Intelligent, Thermo Nicolet Co., Madison, WI, USA) from 4000 to 400 cm^{-1}. For each spectrum, 64 scans were acquired at 4 cm^{-1} resolution. The peak signals in the spectra were analyzed using OMNIC 8.2 software (Thermo Nicolet Co., Madison, WI, USA).

3.8. Thermal and pH Stability Analysis of Peptide-Calcium Chelate

3.8.1. TG-DCS Analysis

A TG-DSC simultaneous thermal analyzer (STA449C, NETZSCH, Bavaria, Germany) was used to analyze the thermostability of the samples. The lyophilized powder samples (5 mg) were set in hermetic pans and heated from 30 °C to 500 °C with programmed heating rate of 10 °C/min and argon flow rate of 30 mL/min.

3.8.2. Calcium Releasing Assay

The calcium ions releasing percentages of peptide-calcium chelate and $CaCl_2$ (50 µg/mL in deionized water) were determined at pH ranges of 2.0–8.0. After incubation in a water bath shaking at 140 rpm and 37 °C for 2 h, the reaction solutions were centrifuged at 10,000 rpm for 10 min. The calcium content of the supernatant and the total calcium in the solution were measured using a colorimetric method with ortho-cresolphthalein complexone reagent. The calcium-releasing percentage was calculated as follows:

$$\text{Calcium releasing } (\%) = \frac{\text{Calcium in supernatant}}{\text{Total calcium in solution}} \times 100 \tag{1}$$

3.9. The Effect of Peptide-Calcium Chelate on the Cellular Uptake of Calcium

3.9.1. Cell Culture

The human colon adenocarcinoma cells, Caco-2, were grown in dulbecco's modified eagle medium (DMEM) supplemented with 15% (v/v) fetal bovine serum (FBS), 1% non-essential amino acid, 100 units/mL penicillin, and 100 µg/mL streptomycin and maintained at 37 °C in a humidified atmosphere with 5% CO_2. At 80%–90% confluence, cells were seeded on 12-well plastic cell culture clusters at a density of 1×10^4 cells/cm^2 for seven days.

3.9.2. Fluorescence Analysis for Calcium Bioavailability

Caco-2 cells were pre-incubated with peptide-calcium chelate, $CaCl_2$, at different concentrations, and tannic acid/phytate/oxalate/Zn^{2+} plus chelate, and tannic acid/phytate/oxalate/Zn^{2+} plus $CaCl_2$, respectively, for 1 h after cells were grown in 12-well plastic cell culture clusters for seven days. The cells were then washed with Hank's balanced salt solution (HBSS, without calcium and magnesium) three times followed by treatment with 10 µM Fluo-3-AM. After incubation for 1 h, cells were washed with HBSS and harvested for analysis by a F-4600 FL spectrophotometer. Intracellular calcium concentrations $[Ca^{2+}]_i$ are expressed as an increase in fluorescence intensity compared to the baseline, which is the original fluorescence intensity without the addition of exogenous calcium.

3.10. Statistical Analyses

All data were presented as means ± standard deviations (SDs) in three replicates. Statistical analysis was performed adopting SPSS 17.0 (SPSS, Chicago, IL, USA). Analysis of variance (ANOVA) was done to determine the significance of the main effects. A confidence level of $p < 0.05$ was considered statistically significant.

4. Conclusions

In summary, a specific dipeptide Phe-Tyr (FY) with strong calcium-chelating capacity from *Schizochytrium* sp. protein hydrolysates was purified and the chelating mechanism was investigated.

It showed that calcium ions could form dative bonds with carboxyl oxygen atoms and amino nitrogen atoms, as well as nitrogen and oxygen atoms of amido bonds, inducing conformational changes of the dipeptide, and ultimately a new and stable peptide-calcium chelate was formed. The calcium bioavailability of FY-Ca was superior to $CaCl_2$, suggesting the potential of FY-Ca to be used as functionally nutraceutical additives.

Acknowledgments: This work was supported by Natural Science Foundation of China (No. 31571779), High & New project of Fujian Marine Fisheries Department (No. [2015]20), Fujian Production & Study project of Provincial Science & Technology Hall (No. 2016N5006), China.

Author Contributions: Xixi Cai, Jiaping Lin and Shaoyun Wang conceived and designed the experiments; Jiaping Lin and Xixi Cai carried out the experiments and analyzed the data; Xixi Cai drafted the paper; Shaoyun Wang reviewed the manuscript and provided useful suggestion to improve the manuscript. All authors read and approved the final manuscript.

Conflicts of Interest: The authors declare no conflict of interest.

References

1. Ngo, D.H.; Wijesekara, I.; Vo, T.S.; Van Ta, Q.; Kim, S.K. Marine food-derived functional ingredients as potential antioxidants in the food industry: An overview. *Food Res. Int.* **2011**, *44*, 523–529. [CrossRef]
2. Yaguchi, T.; Tanaka, S.; Yokochi, T.; Nakahara, T.; Higashihara, T. Production of high yields of docosahexaenoic acid by *Schizochytrium* sp. strain SR21. *J. Am. Oil Chem. Soc.* **1997**, *74*, 1431–1434. [CrossRef]
3. Opheim, M.; Šližytė, R.; Sterten, H.; Provan, F.; Larssen, E.; Kjos, N.P. Hydrolysis of Atlantic salmon (*Salmo salar*) rest raw materials—Effect of raw material and processing on composition, nutritional value, and potential bioactive peptides in the hydrolysates. *Process Biochem.* **2015**, *50*, 1247–1257. [CrossRef]
4. Coscueta, E.R.; Amorim, M.M.; Voss, G.B.; Nerli, B.B.; Picó, G.A.; Pintado, M.E. Bioactive properties of peptides obtained from Argentinian defatted soy flour protein by Corolase PP hydrolysis. *Food Chem.* **2016**, *198*, 36–44. [CrossRef] [PubMed]
5. Adluri, R.S.; Zhan, L.J.; Bagchi, M.; Maulik, N.; Maulik, G. Comparative effects of a novel plant-based calcium supplement with two common calcium salts on proliferation and mineralization in human osteoblast cells. *Mol. Cell. Biochem.* **2010**, *340*, 73–80. [CrossRef] [PubMed]
6. Singh, G.; Muthukumarappan, K. Influence of calcium fortification on sensory, physical and rheological characteristics of fruit yogurt. *LWT—Food Sci. Technol.* **2008**, *41*, 1145–1152. [CrossRef]
7. Amalraj, A.; Pius, A. Bioavailability of calcium and its absorption inhibitors in raw and cooked green leafy vegetables commonly consumed in India—An in vitro study. *Food Chem.* **2015**, *170*, 430–436. [CrossRef] [PubMed]
8. Bennett, T.; Desmond, A.; Harrington, M.; McDonagh, D.; FitzGerald, R.; Flynn, A.; Cashman, K.D. The effect of high intakes of casein and casein phosphopeptide on calcium absorption in the rat. *Br. J. Nutr.* **2000**, *83*, 673–680. [CrossRef] [PubMed]
9. Lv, Y.; Bao, X.L.; Yang, B.C.; Ren, C.G.; Guo, S.T. Effect of soluble soybean protein hydrolysate-calcium complexes on calcium uptake by Caco-2 cells. *J. Food Sci.* **2008**, *73*, H168–H173. [CrossRef] [PubMed]
10. Pan, D.D.; Lu, H.Q.; Zeng, X.Q. A newly isolated Ca binding peptide from whey protein. *Int. J. Food Prop.* **2013**, *16*, 1127–1134. [CrossRef]
11. Choi, D.W.; Lee, J.H.; Chun, H.H.; Song, K.B. Isolation of a calcium-binding peptide from bovine serum protein hydrolysates. *Food Sci. Biotechnol.* **2012**, *21*, 1663–1667. [CrossRef]
12. Cosentino, S.; Donida, B.M.; Marasco, E.; Del Favero, E.; Cantù, L.; Lombardi, G.; Colombini, A.; Iametti, S.; Valaperta, S.; Fiorilli, A.; et al. Calcium ions enclosed in casein phosphopeptide aggregates are directly involved in the mineral uptake by differentiated HT-29 cells. *Int. Dairy J.* **2010**, *20*, 770–776. [CrossRef]
13. Perego, S.; Zabeo, A.; Marasco, E.; Giussani, P.; Fiorilli, A.; Tettamanti, G.; Ferraretto, A. Casein phosphopeptides modulate calcium uptake and apoptosis in Caco-2 cells through their interaction with the TRPV6 calcium channel. *J. Funct. Foods* **2013**, *5*, 847–857. [CrossRef]
14. Lin, J.P.; Cai, X.X.; Tang, M.R.; Wang, S.Y. Preparation and evaluation of the chelating nanocomposite fabricated with marine algae *Schizochytrium* sp. protein hydrolysate and calcium. *J. Agric. Food Chem.* **2015**, *63*, 9704–9714. [CrossRef] [PubMed]

15. Chaud, M.V.; Izumi, C.; Nahaal, Z.; Shuhama, T.; de Lourdes Pires Bianchi, M.; de Freitas, O. Iron derivatives from casein hydrolysates as a potential source in the treatment of iron deficiency. *J. Agric. Food Chem.* **2002**, *50*, 871–877. [CrossRef] [PubMed]

16. Huang, G.G.; Ren, L.; Jiang, J.X. Purification of a histidine-containing peptide with calcium binding activity from shrimp processing byproducts hydrolysate. *Eur. Food Res. Technol.* **2010**, *232*, 281–287. [CrossRef]

17. Huang, G.R.; Ren, Z.Y.; Jiang, J.X. Separation of iron-binding peptides from shrimp processing by-products hydrolysates. *Food Bioprocess Technol.* **2010**, *4*, 1527–1532. [CrossRef]

18. Jeon, S.J.; Lee, J.H.; Song, K.B. Isolation of a calcium-binding peptide from Chlorella protein hydrolysates. *J. Food Sci. Nutr.* **2010**, *15*, 282–286. [CrossRef]

19. Zhao, L.N.; Cai, X.X.; Huang, S.L.; Wang, S.Y.; Huang, Y.F.; Hong, J.; Rao, P.F. Isolation and identification of a whey protein-sourced calcium-binding tripeptide Tyr-Asp-Thr. *Int. Dairy J.* **2015**, *40*, 16–23. [CrossRef]

20. Zhao, L.N.; Huang, Q.M.; Huang, S.L.; Lin, J.P.; Wang, S.Y.; Huang, Y.F.; Hong, J.; Rao, P.F. Novel peptide with a specific calcium-binding capacity from whey protein hydrolysate and the possible chelating mode. *J. Agric. Food Chem.* **2014**, *62*, 10274–10282. [CrossRef] [PubMed]

21. Zhao, L.N.; Huang, S.L.; Cai, X.X.; Hong, J.; Wang, S.Y. A specific peptide with calcium chelating capacity isolated from whey protein hydrolysate. *J. Funct. Foods* **2014**, *10*, 46–53. [CrossRef]

22. Huang, S.L.; Zhao, L.N.; Cai, X.X.; Wang, S.Y.; Huang, Y.F.; Hong, J.; Rao, P.F. Purification and characterisation of a glutamic acid-containing peptide with calcium-binding capacity from whey protein hydrolysate. *J. Dairy Res.* **2015**, *82*, 29–35. [CrossRef] [PubMed]

23. Lv, Y.; Liu, Q.; Bao, X.L.; Tang, W.X.; Yang, B.C.; Guo, S.T. Identification and characteristics of iron-chelating peptides from soybean protein hydrolysates using IMAC-Fe^{3+}. *J. Agric. Food Chem.* **2009**, *57*, 4593–4597. [CrossRef] [PubMed]

24. Miquel, E.; Farré, R. Effects and future trends of casein phosphopeptides on zinc bioavailability. *Trends Food Sci. Technol.* **2007**, *18*, 139–143. [CrossRef]

25. Kim, S.B.; Seo, I.S.; Khan, M.A.; Ki, K.S.; Lee, W.S.; Lee, H.J.; Shin, H.S.; Kim, H.S. Enzymatic hydrolysis of heated whey: Iron-binding ability of peptides and antigenic protein fractions. *J. Dairy Sci.* **2007**, *90*, 4033–4042. [CrossRef] [PubMed]

26. Wang, C.; Li, B.; Ao, J. Separation and identification of zinc-chelating peptides from sesame protein hydrolysate using IMAC-Zn^{2+} and LC-MS/MS. *Food Chem.* **2012**, *134*, 1231–1238. [CrossRef] [PubMed]

27. Aitken, A.; Learmonth, M. Protein Determination by UV Absorption. In *The Protein Protocols Handbook*; Walker, J.M., Ed.; Humana Press: Totowa, NJ, USA, 1996; pp. 3–6.

28. Houser, R.P.; Fitzsimons, M.P.; Barton, J.K. Metal-dependent intramolecular chiral Induction: The Zn^{2+} complex of an ethidium-peptide conjugate. *Inorg. Chem.* **1999**, *38*, 1368–1370. [CrossRef] [PubMed]

29. Armas, A.; Sonois, V.; Mothes, E.; Mazarguil, H.; Faller, P. Zinc(II) binds to the neuroprotective peptide humanin. *J. Inorg. Biochem.* **2006**, *100*, 1672–1678. [CrossRef] [PubMed]

30. Zhou, J.; Wang, X.; Ai, T.; Cheng, X.; Guo, H.Y.; Teng, G.X.; Mao, X.Y. Preparation and characterization of β-lactoglobulin hydrolysate-iron complexes. *J. Dairy Sci.* **2012**, *95*, 4230–4236. [CrossRef] [PubMed]

31. Wu, H.H.; Liu, Z.Y.; Zhao, Y.H.; Zeng, M.Y. Enzymatic preparation and characterization of iron-chelating peptides from anchovy (*Engraulis japonicus*) muscle protein. *Food Res. Int.* **2012**, *48*, 435–441. [CrossRef]

32. Nara, M.; Morii, H.; Tanokura, M. Coordination to divalent cations by calcium-binding proteins studied by FTIR spectroscopy. *Biochim. Biophys. Acta* **2013**, *1828*, 2319–2327. [CrossRef] [PubMed]

33. Liu, F.R.; Wang, L.; Wang, R.; Chen, Z.X. Calcium-binding capacity of wheat germ protein hydrolysate and characterization of peptide-calcium complex. *J. Agric. Food Chem.* **2013**, *61*, 7537–7544. [CrossRef] [PubMed]

34. Wang, X.L.; Li, K.; Yang, X.D.; Wang, L.L.; Shen, R.F. Complexation of Al(III) with reduced glutathione in acidic aqueous solutions. *J. Inorg. Biochem.* **2009**, *103*, 657–665. [CrossRef] [PubMed]

35. Wang, X.; Zhou, J.; Tong, P.S.; Mao, X.Y. Zinc-binding capacity of yak casein hydrolysate and the zinc-releasing characteristics of casein hydrolysate-zinc complexes. *J. Dairy Sci.* **2011**, *94*, 2731–2740. [CrossRef] [PubMed]

36. Hou, T.; Wang, C.; Ma, Z.L.; Shi, W.; Lui, W.W.; He, H. Desalted duck egg white peptides: Promotion of calcium uptake and structure characterization. *J. Agric. Food Chem.* **2015**, *63*, 8170–8176. [CrossRef] [PubMed]

37. Cosentino, S.; Gravaghi, C.; Donetti, E.; Donida, B.M.; Lombardi, G.; Bedoni, M.; Fiorilli, A.; Tettamanti, G.; Ferraretto, A. Caseinphosphopeptide-induced calcium uptake in human intestinal cell lines HT-29 and Caco-2 is correlated to cellular differentiation. *J. Nutr. Biochem.* **2010**, *21*, 247–254. [CrossRef] [PubMed]

38. Hallberg, L.; Brune, M.; Erlandsson, M.; Sandberg, A.S.; Rossander-Hultén, L. Calcium: Effect of different amounts on nonheme- and heme-iron absorption in humans. *Am. J. Clin. Nutr.* **1991**, *53*, 112–119. [PubMed]

39. Li, M.L.; Zhang, T.; Yang, H.X.; Zhao, G.H.; Xu, C.S. A novel calcium supplement prepared by phytoferritin nanocages protects against absorption inhibitors through a unique pathway. *Bone* **2014**, *64*, 115–123. [CrossRef] [PubMed]

40. Ma, Z.H.; Lu, Z.B.; Shi, B. Chemical properties and application of tannic acid. *Nat. Prod. Res. Dev.* **2003**, *15*, 87–91.

41. Daengprok, W.; Garnjanagoonchorn, W.; Naivikul, O.; Pornsinlpatip, P.; Issigonis, K.; Mine, Y. Chicken eggshell matrix proteins enhance calcium transport in the human intestinal epithelial cells, Caco-2. *J. Agric. Food Chem.* **2003**, *51*, 6056–6061. [CrossRef] [PubMed]

marine drugs

MDPI

Article

In vitro Anti-Thrombotic Activity of Extracts from Blacklip Abalone (*Haliotis rubra*) Processing Waste

Hafiz Ansar Rasul Suleria [1,2], **Barney M. Hines** [2], **Rama Addepalli** [2], **Wei Chen** [2], **Paul Masci** [1], **Glenda Gobe** [1,*] **and Simone A. Osborne** [2,*]

[1] School of Medicine, The University of Queensland, Translational Research Institute, Kent Street, Woolloongabba 4102, Australia; hafiz.suleria@uqconnect.edu.au (H.A.R.S.); p.masci@uq.edu.au (P.M.)
[2] CSIRO Agriculture, 306 Carmody Road, St Lucia 4067, Australia; barney.hines@csiro.au (B.M.H.); rama.addepalli@csiro.au (R.A.); wei.chen@csiro.au (W.C.)
*   Correspondence: g.gobe@uq.edu.au (G.G.); simone.osborne@csiro.au (S.A.O.)

Academic Editor: Se-Kwon Kim
Received: 17 November 2016; Accepted: 28 December 2016; Published: 31 December 2016

Abstract: Waste generated from the processing of marine organisms for food represents an underutilized resource that has the potential to provide bioactive molecules with pharmaceutical applications. Some of these molecules have known anti-thrombotic and anti-coagulant activities and are being investigated as alternatives to common anti-thrombotic drugs, like heparin and warfarin that have serious side effects. In the current study, extracts prepared from blacklip abalone (*Haliotis rubra*) processing waste, using food grade enzymes papain and bromelain, were found to contain sulphated polysaccharide with anti-thrombotic activity. Extracts were found to be enriched with sulphated polysaccharides and assessed for anti-thrombotic activity *in vitro* through heparin cofactor-II (HCII)-mediated inhibition of thrombin. More than 60% thrombin inhibition was observed in response to 100 µg/mL sulphated polysaccharides. Anti-thrombotic potential was further assessed as anti-coagulant activity in plasma and blood, using prothrombin time (PT), activated partial thromboplastin time (aPTT), and thromboelastography (TEG). All abalone extracts had significant activity compared with saline control. Anion exchange chromatography was used to separate extracts into fractions with enhanced anti-thrombotic activity, improving HCII-mediated thrombin inhibition, PT and aPTT almost 2-fold. Overall this study identifies an alternative source of anti-thrombotic molecules that can be easily processed offering alternatives to current anti-thrombotic agents like heparin.

Keywords: blacklip abalone; processing waste; bioactive molecules; anti-thrombotic activity

1. Introduction

Marine organisms are increasingly being investigated as sources of bioactive molecules with therapeutic applications as nutraceuticals and pharmaceuticals [1]. Accordingly, processing waste from these organisms is an important source of bioactive molecules [2]. Abalone, a marine gastropod, contains a variety of bioactive molecules with reported anti-oxidant, anti-thrombotic, anti-inflammatory, anti-microbial and anti-cancer activities [3]. For thousands of years, different cultures have used abalone as a traditional functional food, believing that its consumption provides health benefits [4]. Recent research has revealed that abalone is composed of many bioactive molecules like sulphated polysaccharides, proteins and fatty acids that provide health benefits beyond basic nutrition [3]. In recent years abalone has been investigated as source of sulphated polysaccharides with anti-thrombotic activity with the potential to reduce thrombosis [5].

Thrombosis involves local clotting of blood in the vessel system that often leads to severe health related disorders like heart attack and stroke. The risk factors for thrombosis are abnormally high blood lipids, high blood glucose, elevated plasma fibrinogen, hypertension and cancer insurgence [6]. In the last several decades, prevention and treatment of thrombosis has been achieved with drugs, including heparin and warfarin. Heparin, a highly sulphated glycosaminoglycan (GAG) present in many mammalian tissues, is used commercially as an anti-coagulant or anti-thrombotic drug [7]. Heparin is administered intravenously, with frequent laboratory monitoring needed to prevent unwanted and sometimes life-threatening bleeding [8]. Heparin-induced thrombocytopenia (a low platelet count) is another serious complication following heparin therapy, particularly in some cardiac patients [9]. Heparin also has other disadvantages as it is extracted and purified from bovine and porcine internal organs making its production difficult and prone to contamination by other GAGs present in these sources [10].

These disadvantages have necessitated a field of research aimed at discovering novel anti-thrombotic and anti-coagulant agents with fewer side effects than heparin. Heparin-like molecules are present in lower invertebrates [11], lobster [12], ascidians and tunicates [13]. Many other marine species including molluscs, that are rich in sulphated polysaccharides, contain GAG-like molecules that have comparable biological activity to heparin [14]. These uniquely sulphated polysaccharides have complex structures composed of galactose, fucose, glucuronic acid and galactosamine. Sulphated polysaccharides isolated from molluscs were found to contain anti-thrombin and anti-coagulant bioactive molecules with unique 3-O-sulphated glycosamine residues [11]. In general, these sulphated polysaccharides significantly vary between species with respect to their composition. Furthermore, bioactivity has also been found to differ depending upon the degree of sulphation, molecular weight, types of saccharides present and glycosidic branching [15].

Many studies have demonstrated abalone viscera, gonads and pleopods are sources of potent anti-thrombotic and anti-coagulant polysaccharides, however limited research has been conducted *in vitro* using plasma and blood to investigate the anti-thrombotic and anti-coagulant mechanisms [8,11,14,15]. Several well-established analyses are used to indicate anti-thrombotic activity including prothrombin time (PT), activated partial thrombin time (aPTT) and thromboelastography (TEG). These assays help to indicate if molecules act in both the intrinsic and extrinsic pathways of the blood coagulation cascade [16] and what the impact on platelets might be. The aim of the current research was to extract, purify and characterise GAG-like sulphated polysaccharides from offal samples from processed wild caught *H. rubra* and assess anti-thrombotic and anti-coagulant activity in the extracts using *in vitro* plasma and blood assays.

2. Results and Discussion

2.1. Protein and Sulphated Polysaccharide Content of Extracts from Blacklip Abalone Processing Waste

Table 1 shows the protein and sulphated polysaccharides that were estimated in all crude extracts and expressed as mg sulphated polysaccharides or protein per gram of starting abalone processing waste material (wet weight). Sulphated polysaccharides were similar across all samples, however protein content was found to be significantly higher in canned abalone extracts compared to the liquid abalone extracts, particularly following enzymatic treatment with papain, and with papain and bromelain combined.

There were no sulphated polysaccharides detected in any enzyme (empty digest) control. However as expected, protein was detected in all enzyme controls (data not shown). Simultaneous digestion with papain and bromelain produced higher contents of sulphated polysaccharides in both canned and liquid samples compared to single enzyme digestions, however these differences were not statistically significant. Initial screening of anti-thrombin activity mediated by heparin cofactor-II (HCII) depicted that most of the thrombin inhibition arose directly from the abalone samples with enzyme control contributing less than 10% of the total thrombin inhibition.

The presence of sulphated polysaccharides in abalone processing waste has been observed previously in other abalone species [17]. The extraction and purification of these sulphated polysaccharides can be achieved by different processes, including enzyme digestion. Different enzymes have different hydrolysing activities that vary in efficiency depending upon sample type, time of incubation, pH and buffer [18]. Kechaou et al. [19] digested cuttlefish and sardine viscera with several commercial proteases, including papain. In this study the degree of hydrolysis for cuttlefish was higher than that obtained for sardine. The authors speculated that the differences may be due to a difference in protein composition of the tissues and nature of the samples influencing the digestion profile. In short, enzyme digestion varies depending upon the nature of the sample and hydrolysing conditions, however it appears as though papain is commonly used in the digestion of various marine samples.

Table 1. Protein and sulphated polysaccharide content of extracts from abalone processing waste expressed in mg per gram (on a wet weight basis).

Abalone Waste	Treatments	Protein (mg/g)	Sulphated Polysaccharides (mg/g)	Anti-Thrombin HCII (% Inhibition)
Canned	Papain	29.95 ± 0.51 [b]	1.36 ± 0.09 [a]	92.1 ± 1.31 [b]
Canned	Bromelain	25.06 ± 1.79 [c]	1.39 ± 0.91 [a]	89.9 ± 2.09 [c]
Canned	Papain + Bromelain	36.10 ± 0.72 [a]	1.46 ± 0.38 [a]	96.8 ± 1.12 [a]
Liquid	Papain	18.82 ± 0.10 [e]	1.27 ± 0.82 [a]	97.1 ± 0.08 [a]
Liquid	Bromelain	23.38 ± 2.09 [d]	1.03 ± 0.13 [a]	95.4 ± 2.13 [a]
Liquid	Papain + Bromelain	18.90 ± 0.80 [e]	1.41 ± 0.68 [a]	91.1 ± 0.79 [b]

Alphabetic letters shows the difference between different samples and treatments. HCII, heparin cofactor-II.

In the study presented here, combined enzymatic treatment with 0.5% *w/v* papain and 0.5% *w/v* bromelain liberated the highest levels of sulphated polysaccharides and protein from canned abalone processing waste compared to separate treatments with 1% *w/v* papain or 1% *w/v* bromelain. These results suggest that the different protease activities of bromelain and papain are required to produce optimal release of sulphated polysaccharides and protein from abalone processing waste. Regardless of enzyme treatment, all extracts were investigated for anti-thrombotic and anti-coagulant activity.

2.2. Separation of Abalone Extracts Using Anion Exchange Chromatography-Fast Performance Liquid Chromatography (AEC-FPLC)

Based on previous studies it was hypothesised that sulphated polysaccharides present in the abalone processing waste may confer both anti-thrombotic and anti-coagulant activity. To produce fractions from the abalone extracts with enhanced sulphated polysaccharide content and anti-thrombotic and anti-coagulant activity, all extracts were subjected to an anion exchange chromatography-fast performance liquid chromatography (AEC-FPLC) system. As shown in Figure 1, separation into sulphated polysaccharide-containing fractions was monitored through their interactions with 1,9-dimethylene blue (DMMB) dye. Similar elution profiles were observed for almost all extracts where sulphated polysaccharide concentration was mostly lower in first 10 eluted fractions (0.4 M NaCl) and generally increased until fraction 30 (1.6 M NaCl). For analysis, seven AEC pools were prepared. Five AEC pools were prepared from the NaCl gradient (AEC pools 1–5), whilst the initial unbound material and a final column wash were also collected and analyzed for bioactivity along linear gradient pools. All three enzyme controls showed similar elution profiles. Most of the peptides eluted during sample application while few peptides were detected in AEC Pool 1. However, there was no protein or sulphated polysaccharides detected after AEC Pool 1 in all three enzyme controls indicating no binding with column. Finally, to decrease NaCl concentration in all pools, 3 kDa spin columns and deionised water washes were performed until the NaCl concentration was below 60 mM.

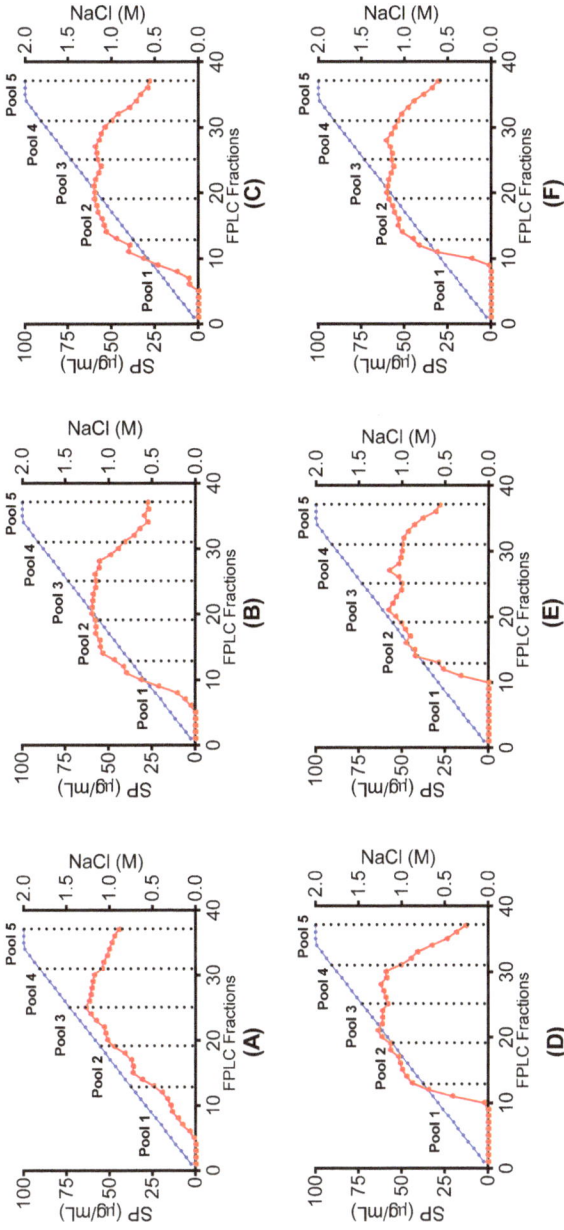

Figure 1. AEC-FPLC chromatograms showing separation of abalone extracts through interaction with DMMB. Concentration of sulphated polysaccharide (SP) ▲ is shown in relation to linear NaCl gradient ● in canned abalone processing waste digested with (**A**) papain; (**B**) bromelain; (**C**) papain + bromelain, and liquid abalone processing waste digested with (**D**) papain; (**E**) bromelain and (**F**) papain + bromelain.

Protein and sulphated polysaccharides concentrations (via Blyscan™ Sulphated GAG assay) were estimated in all the AEC pools and are shown in Table 2. The highest sulphated polysaccharide concentrations were mostly measured in the gradient AEC pools 3 and 4 (0.8–1.0 M NaCl) whereas the highest protein concentrations were measured in the unbound material because of almost complete elution of enzymes. Some of the enzyme peptides were also detected in early AEC pool 1 during ion exchange chromatography of empty control/enzyme controls.

Table 2. Protein and sulphated polysaccharide concentration of pooled fractions from AEC-FPLC abalone extracts.

Sample Descriptions	Protein (mg/mL)	Sulphated Polysaccharides (mg/mL)
Can_Ab_Pap_Unbound material	3.40 ± 1.1	1.12 ± 0.9
Can_Ab_Pap_AEC Pool 1	0.76 ± 0.7	0.19 ± 0.4
Can_Ab_Pap_AEC Pool 2	0.59 ± 0.4	1.04 ± 0.2
Can_Ab_Pap_AEC Pool 3	1.87 ± 0.2	1.82 ± 1.2
Can_Ab_Pap_AEC Pool 4	0.37 ± 0.1	1.49 ± 0.2
Can_Ab_Pap_AEC Pool 5	0.23 ± 0.9	1.15 ± 1.3
Can_Ab_Pap_Final column wash	0.36 ± 2.3	0.03 ± 2.1
Can_Ab_Bro_Unbound material	5.77 ± 1.4	0.56 ± 0.9
Can_Ab_Bro_AEC Pool 1	0.74 ± 0.4	0.30 ± 0.3
Can_Ab_Bro_AEC Pool 2	0.64 ± 0.1	1.02 ± 0.8
Can_Ab_Bro_AEC Pool 3	0.16 ± 0.2	1.11 ± 0.4
Can_Ab_Bro_AEC Pool 4	0.11 ± 0.1	1.21 ± 0.1
Can_Ab_Bro_AEC Pool 5	0.05 ± 1.2	0.22 ± 1.1
Can_Ab_Bro_Final column wash	0.06 ± 2.1	0.02 ± 2.1
Can_Ab_Pap+Bro_Unbound material	8.92 ± 1.2	0.65 ± 0.4
Can_Ab_Pap+Bro_AEC Pool 1	1.31 ± 0.3	0.42 ± 0.2
Can_Ab_Pap+Bro_AEC Pool 2	1.90 ± 0.1	2.01 ± 1.1
Can_Ab_Pap+Bro_AEC Pool 3	0.37 ± 0.7	2.45 ± 0.9
Can_Ab_Pap+Bro_AEC Pool 4	0.33 ± 0.1	2.00 ± 0.1
Can_Ab_Pap+Bro_AEC Pool 5	0.05 ± 0.3	0.38 ± 0.7
Can_Ab_Pap+Bro_Final column wash	0.67 ± 0.9	0.54 ± 1.2
Liquid_Ab_Pap_Unbound material	2.52 ± 1.1	0.03 ± 0.4
Liquid_Ab_Pap_AEC Pool 1	0.17 ± 0.2	0.04 ± 0.1
Liquid_Ab_Pap_AEC Pool 2	0.44 ± 0.8	0.70 ± 0.8
Liquid_Ab_Pap_AEC Pool 3	0.33 ± 0.3	1.25 ± 0.7
Liquid_Ab_Pap_AEC Pool 4	0.13 ± 0.1	1.64 ± 0.2
Liquid_Ab_Pap_AEC Pool 5	0.02 ± 1.2	0.16 ± 0.7
Liquid_Ab_Pap_Final column wash	0.39 ± 2.1	0.01 ± 1.1
Liquid_Ab_Bro_Unbound material	6.12 ± 1.1	0.40 ± 0.9
Liquid_Ab_Bro_AEC Pool 1	0.00 ± 2.1	0.01 ± 0.1
Liquid_Ab_Bro_AEC Pool 2	0.00 ± 0.5	0.02 ± 0.9
Liquid_Ab_Bro_AEC Pool 3	0.91 ± 0.1	1.04 ± 0.2
Liquid_Ab_Bro_AEC Pool 4	1.41 ± 0.5	3.10 ± 0.7
Liquid_Ab_Bro_AEC Pool 5	0.21 ± 1.2	0.61 ± 1.1
Liquid_Ab_Bro_Final column wash	0.11 ± 0.9	0.47 ± 1.9
Liquid_Ab_Pap+Bro_Unbound material	10.34 ± 0.9	0.95 ± 0.5
Liquid_Ab_Pap+Bro_AEC Pool 1	0.45 ± 0.7	0.06 ± 0.6
Liquid_Ab_Pap+Bro_AEC Pool 2	0.39 ± 0.9	2.07 ± 0.7
Liquid_Ab_Pap+Bro_AEC Pool 3	0.56 ± 0.3	2.06 ± 0.2
Liquid_Ab_Pap+Bro_AEC Pool 4	0.27 ± 0.2	1.43 ± 0.1
Liquid_Ab_Pap+Bro_AEC Pool 5	0.11 ± 0.1	0.53 ± 0.9
Liquid_Ab_Pap+Bro_Final column wash	0.10 ± 1.1	0.05 ± 0.1

"Ab" stands for abalone, "Pap" stands for papain enzyme, "Bro" stands for bromelain enzyme while "Pap+Bro" reflects the combination of both papain and bromelain enzymes.

In similar separation studies, different types of sulphated polysaccharides have been obtained from visceral portions and gonads of abalone [20]. Wang et al. [21] isolated and characterized three sulphated polysaccharides, AAP, AVAP I and AVAP II, from the pleopods and viscera of the abalone *H. discus hannai* Ino. The crude polysaccharide extract was initially separated by AEC on a diethylaminoethyl-cellulose (DEAE-cellulose) column with the main polysaccharide fraction from the pleopods eluted with 0.42–0.60 M NaCl, whilst two fractions from the viscera eluted with 0.28–0.40 M NaCl and 0.44–0.56 M NaCl.

Zhu et al. [22] demonstrated that sulphated polysaccharides isolated from pleopods of abalone consist of 1-1,4-, 1,6-, or 1,4,6-linked glucose, and in accretion 1-, 1,3-, 1,6-, and 1,4,6-linked galactose. Prior to this, She [23] also proposed that sulphated polysaccharides from abalone pleopods are comprised of galactose, glucose, fructose and xylose. The acidic polysaccharide content has not been fully determined, however several GAG-like structures have been defined in abalone. The chemical structure of sulphated polysaccharides, isolated by Li et al. [17], contains a galactosamine and glucuronic acid backbone linked to sulphated-fucose and galactose, considered to be similar to the fucosylated chondroitin sulphate present in the sea cucumber [24]. In abalone, even though these sulphated polysaccharides are linked with galactose to the fucose branch, it is still considered a chondroitin-like polysaccharide [17].

Based on the assumption that the sulphated polysaccharides present in abalone processing waste are also GAG-like molecules, the anti-thrombotic and anti-coagulant activities were investigated using several *in vitro* assays.

2.3. Anti-Thrombotic Activity Measured through HCII-Mediated Thrombin Inhibition

To determine the anti-thrombotic activity of the different abalone processing waste extracts and AEC pooled fractions, HCII-mediated thrombin inhibition was measured using the chromogenic substrate Chromozym TH. All samples were initially screened without dilution to determine which extracts and AEC pooled fractions contained anti-thrombotic molecules (data not shown). No activity was observed in the enzyme controls, AEC Pool 1 and in the final column washes (data not shown). The extracts and AEC pooled fractions that produced anti-thrombotic activity were then examined further in order to compare the *in vitro* HCII-mediated thrombin inhibition between the different samples. *In vitro* HCII-mediated anti-thrombin activity, expressed as percentage inhibition of thrombin, is presented in Table 3 and shows that AEC pools 4 or 5 displayed the highest activity. AEC Pool 4 of all abalone extracts showed significantly higher inhibition of thrombin ($p < 0.05$) relative to other pools at a sulphated polysaccharide concentration of 100 µg/mL.

It was also observed that most of the time AEC pool 4 displayed highest activity at 10 and 1 µg/mL sulphated polysaccharide concentration. Moreover, specific enzyme digestion also showed a marked effect on percentage inhibition; papain treatment alone or in combination appeared to be the most efficient enzyme with respect to release of sulphated polysaccharides, whilst bromelain digestion inconsistently inhibited thrombin and generally produced the lowest percentage inhibition. The activity of the AEC pool 4 of the liquid abalone sample digested with papain and bromelain showed thrombin inhibition only 2–3 times less compared to the heparin standard (on a similar sulphated polysaccharides basis).

Other studies involving sulphated polysaccharides from marine sources support the anti-thrombotic effects observed in this study [25]. These studies also proposed different mechanisms of action involving factors Xa and thrombin (IIa) mediated by HCII and anti-thrombin III (ATIII) [26]. The specific pattern of sulphation and the position of glycosidic linkages varied among different species and may contribute to difference in activity [27]. For example, sea cucumber polysaccharides displayed weaker anti-factor Xa and anti-thrombin activities mediated by ATIII compare to both heparin and low molecular weight heparin, suggesting that their anti-coagulant mechanisms are different from those of heparin-like drugs. Thus, the structural interaction of these polysaccharides with coagulation cofactors (HCII and ATIII) and their target proteases may be influenced by the conformation and length of repetitive sulphated units [28].

Based on previous reports and on the findings presented here, extracts as well as the AEC pooled fractions 3–5 were selected for further assessment using PT, aPTT and TEG assays to confirm the anti-thrombotic and anti-coagulant effect of these samples in blood and plasma and to help elucidate the role of these molecules in the coagulation cascade.

Table 3. Heparin cofactor II-mediated thrombin inhibition by abalone samples expressed as percentage inhibition of thrombin activity.

	Percentage Inhibition of Thrombin Mediated by HCII at 10 min			
Sample Description	Sulphated Polysaccharide Concentration (μg/mL)			
	100	**50**	**10**	**1**
Can_Ab_Pap	93.1 ± 0.8 [d]	26 ± 7.5 [j]	11.4 ± 10.2 [l]	0
Can_Ab_Pap_Unbound material	56 ± 0.4 [k]	26.9 ± 0.8 [j]	0	0
Can_Ab_Pap_AEC Pool 2	8.5 ± 1.9 [s]	8.7 ± 3.4 [q]	0	0
Can_Ab_Pap_AEC Pool 3	87.6 ± 3.2 [e]	81.8 ± 0.8 [f]	13.9 ± 6.6 [k]	0
Can_Ab_Pap_AEC Pool 4	96.1 ± 0.4b [c]	92.7 ± 1.7 [c]	43 ± 3.9 [f]	9.4 ± 6.6 [d]
Can_Ab_Pap_AEC Pool 5	97.5 ± 0.2 [ab]	94.6 ± 1.8 [b]	78.2 ± 2.7 [b]	20.7 ± 4.2 [b]
Can_Ab_Bro	72.3 ± 1.2 [g]	13.2 ± 2.9 [o]	2.4 ± 1.1 [p]	0
Can_Ab_Bro_Unbound material	14.2 ± 5.1 [r]	0	0	0
Can_Ab_Bro_AEC Pool 2	56.6 ± 1.9 [jk]	38.5 ± 6.1 [h]	0	0
Can_Ab_Br_AEC Pool 3	26.4 ± 3.7 [q]	21.5 ± 3.6 [m]	3.6 ± 4.5 [o]	0
Can_Ab_Bro_AEC Pool 4	93.1 ± 0.5 [d]	89.4 ± 0.2 [e]	27.7 ± 3.8 [g]	0
Can_Ab_Bro_AEC Pool 5	45.3 ± 2.1 [n]	17.9 ± 3.8 [n]	0	0
Can_Ab_Pap + Bro	82.4 ± 0.6 [f]	24 ± 2.2 [l]	10.3 ± 6.2 [m]	0
Can_Ab_Pap+Bro_Unbound material	61.27 ± 4.8 [i]	13.21 ± 3.9 [o]	0	0
Can_Ab_Pap+Bro_AEC Pool 2	34.3 ± 1.9 [o]	25.8 ± 2.4 [kl]	0	0
Can_Ab_Pap+Bro_AEC Pool 3	93.6 ± 0.3 [d]	90.9 ± 0.5 [d]	19.4 ± 2 [j]	0
Can_Ab_Pap+Bro_AEC Pool 4	96.4 ± 0.3b [c]	94.9 ± 0.3 [b]	56.9 ± 2.1 [d]	14.3 ± 4.2 [c]
Can_Ab_Pap+Bro_AEC Pool 5	58.45 ± 2.8 [j]	21.29 ± 1.2 [m]	9.8 ± 2.7 [mn]	0
Liquid_Ab_Pap	93.4 ± 1.1 [d]	65.2 ± 2.1 [g]	21.4 ± 0.9 [i]	5.4 ± 1.1 [e]
Liquid_Ab_Pap_Unbound material	12.43 ± 1.8 [r]	0	0	0
Liquid_Ab_Pap_AEC Pool 2	28.3 ± 0.9 [q]	11.2 ± 0.7 [p]	0	0
Liquid_Ab_Pap_AEC Pool 3	94.9 ± 0.1 [cd]	88.5 ± 1.8 [e]	24.8 ± 2.3 [h]	0
Liquid_Ab_Pap_AEC Pool 4	98.5 ± 0.1 [a]	96.3 ± 0.2 [a]	69.1 ± 2.2 [c]	13 ± 3.5 [c]
Liquid_Ab_Pap_AEC Pool 5	57.12 ± 0.7 [jk]	21.23 ± 0.1 [m]	10.2 ± 1.1 [m]	0
Liquid_Ab_ Bro	64.32 ± 1.9 [h]	21.4 ± 0.4 [m]	10.2 ± 1.9 [mn]	0
Liquid_Ab_Bro_Unbound material	14.26 ± 4.9 [r]	2.7 ± 1.9 [r]	0	0
Liquid_Ab_Bro_AEC Pool 3	34.68 ± 0.8 [o]	11.2 ± 2.9 [p]	0	0
Liquid_Ab_Bro_AEC Pool 4	52.9 ± 0.7 [l]	34.7 ± 1.9 [i]	10.21 ± 0.8 [mn]	0
Liquid_Ab_Bro_AEC Pool 5	64.98 ± 1.8 [h]	38.9 ± 0.7 [h]	9.7 ± 1.9 [n]	1.1 ± 1.8 [f]
Liquid_Ab_Pap+Bro	93.1 ± 0.8 [d]	26 ± 3.5 [jk]	11.4 ± 10.2 [l]	0
Liquid_Ab_Pap+Bro_AEC Pool 2	32.1 ± 0.6 [p]	20.6 ± 3.5 [m]	3.1 ± 4.1 [o]	0
Liquid_Ab_Pap+Bro_AEC Pool 3	95.3 ± 0.3 [cd]	93.6 ± 1 [c]	47.3 ± 0.8 [e]	2.8 ± 2.2 [ef]
Liquid_Ab_Pap+Bro_AEC Pool 4	98.4 ± 0.1 [a]	96.1 ± 2.2 [a]	92.4 ± 1.2 [a]	25.7 ± 0.4 [a]
Liquid_Ab_Pap+Bro_AEC Pool 5	47.4 ± 2.9 [m]	21.49 ± 4.1 [m]	1.4 ± 2.9 [q]	0
Heparin Standard	16	4	2	0.5
	91.5 ± 0.6 [a]	75.0 ± 1.3 [b]	48.0 ± 2.1 [c]	27.6 ± 1.2 [d]

"Ab" stands for abalone, "Pap" stands for papain enzyme, "Bro" stands for bromelain enzyme while "Pap+Bro" reflects the combination of both papain and bromelain enzymes. Alphabetic letters shows the difference between different samples and treatments, ANOVA $p < 0.05$.

2.4. Anti-Thrombotic and Anti-Coagulant Activity in Blood and Plasma

2.4.1. Prothrombin and Activated Partial Thromboplastin Time

The *in vitro* clotting assays PT and aPTT showed positive responses to all abalone extracts and pools 3 and 4. Figure 2 demonstrates that PT increased with sulphated polysaccharide concentration, and that pools 3 and 4 generally prolonged PT time more effectively than the original extracts. In particular, pool 4 from both the canned and liquid abalone samples significantly increased PT compared with other pools. Comparing to heparin standard, both extract and pools showed an increase in PT activity. PT and aPTT were not observed in the enzyme controls. The aPTT also showed an increase with sulphated polysaccharide concentration. In Figure 3, both canned and liquid extracts, digested with different enzymes, prolonged aPTT significantly compared to the saline control.

Furthermore, all three enzyme controls were also subjected to aPTT analysis and they were unable to prolong aPTT. Moreover, heparin standard was also subjected to aPTT analysis. Even at a very low concentration of 0.02 mg/mL, no clot formed, demonstrating that heparin standard has a very strong aPTT activity and it can prolong aPTT many fold higher compared with our extract. However, from the results it appeared that AEC pools 3 and 4 prolonged aPTT more compared with the original extract. Furthermore, AEC pool 4 from both the canned and liquid abalone processing waste appeared to prolong aPPT more when compared to the remaining AEC pools.

The results presented here are supported by several other studies that report the pharmaceutical importance of abalone. Previously, Li et al. [17] isolated a GAG-like polysaccharide from abalone, and conducted *in vitro* investigations on its anti-coagulant activity. Usually *in vitro* anti-coagulant assays thrombin time (TT), PT and aPTT and help to indicate anti-coagulant activity with respect to the intrinsic and extrinsic pathways in the blood coagulation cascade. PT reflects the extrinsic pathway of the coagulation cascade whilst aPTT reflects changes in the intrinsic pathway of the blood coagulation cascade [16]. Li et al. [17] found that the GAG-like polysaccharide could prolong aPTT as well as TT.

In another study by Li et al. [29], different types of extracts were prepared from abalone viscera and it was found that water extracts were associated with higher PT, aPTT and TT when compared to extracts prepared using different solvents. The heparin control displayed higher specific activity in processing waste from abalone offal. Other research also demonstrated that GAG-like molecules isolated from different sea cucumbers did not show a prominent effect on PT but efficiently improved the aPTT and TT. It also indicated that the type of sulphated polysaccharides present in sea cucumber may affect the intrinsic but not the extrinsic and common coagulation process [24] compared with abalone. This suggests abalone sulphated polysaccharides are involved in both intrinsic and extrinsic pathways. In order to confirm this mode of action and further demonstrate a role for abalone sulphated polysaccharides in both intrinsic and extrinsic pathway, an *in vitro* blood assay was performed using TEG.

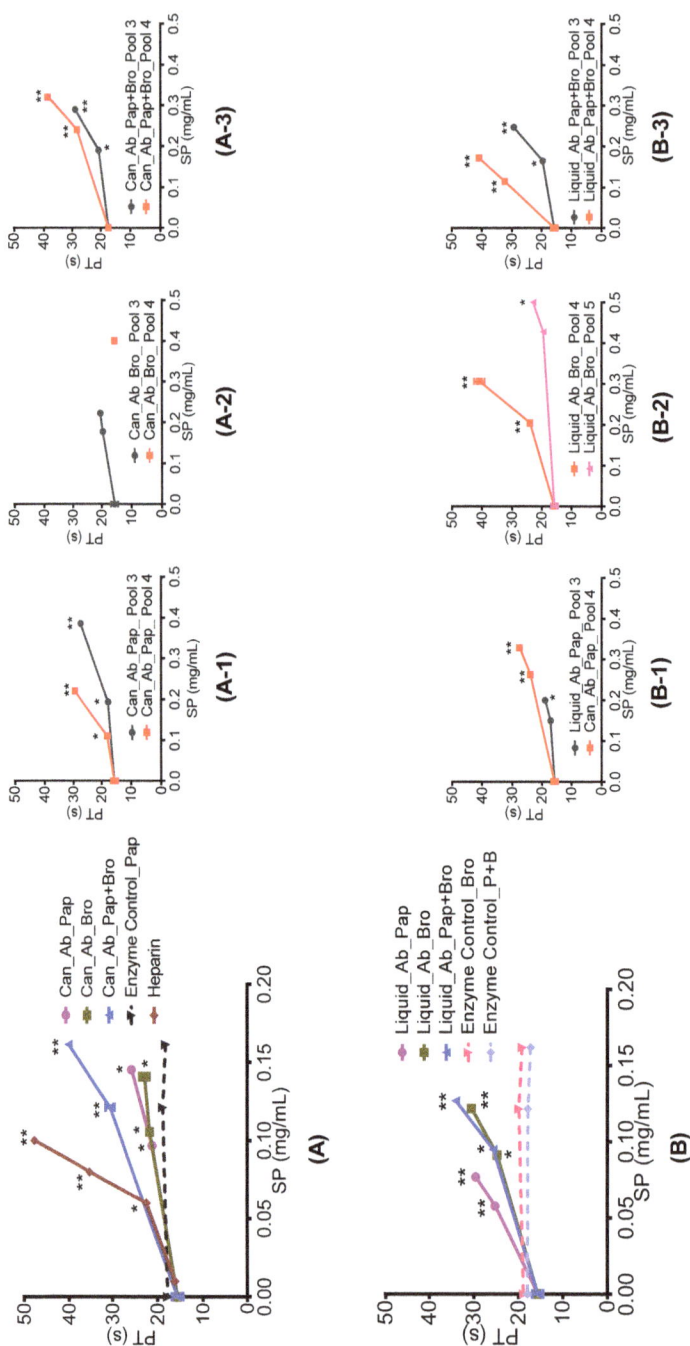

Figure 2. Prothrombin time (PT) of abalone processing waste extracts and AEC pooled fractions. Graphs (**A**, **A-3**): canned abalone processing waste digested with papain, bromelain and a combination of papain and bromelain and their respective AEC pools 3 and 4; (**B**, **B-3**): liquid abalone processing waste digested with papain, bromelain and a combination of papain and bromelain and their respective AEC pools 3 and 4. * = Statistical significance determined using a one way ANOVA with Dunnett's Multiple Comparison Test compared to saline control with * *p* < 0.05 and ** *p* < 0.01.

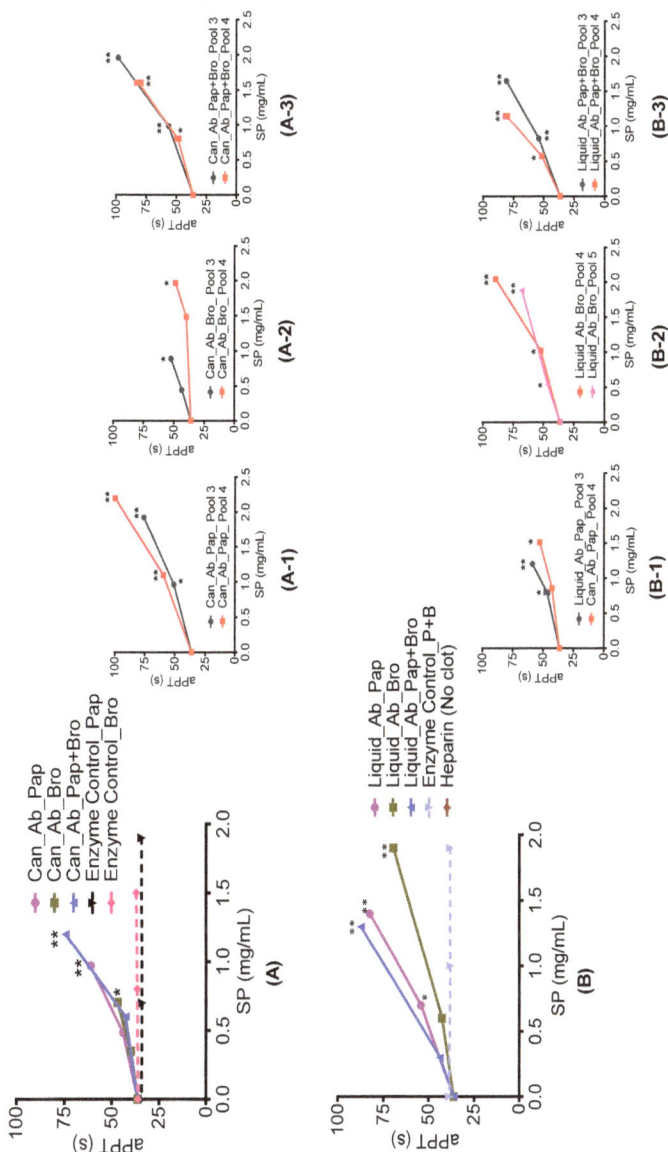

Figure 3. Activated partial thrombin time (aPTT) of abalone processing waste extracts and AEC pooled fractions. (**A**, **A-3**) canned abalone digested with papain, bromelain and a combination of papain and bromelain and their respective AEC pool 3 and 4 while (**B**, **B-3**), liquid abalone samples digested with papain, bromelain and a combination of papain and bromelain and their respective AEC pool 3 and 4. * Statistical significance determined using a one way ANOVA with Dunnett's Multiple Comparison Test compared to saline control with * $p < 0.05$ and ** $p < 0.01$. Heparin standard did not form a clot even at very low concentration. It showed that the heparin standard is many times stronger in aPTT activity compare to abalone extract and anion exchanged pools.

2.4.2. Thromboelastography (TEG)

TEG is a global assessment of haemostatic function investigating the interaction of platelets with the coagulation cascade from the time of initial fibrin formation through to platelet aggregation, clot strengthening, fibrin cross linkage, and to eventual clot lysis. To assess anti-coagulant activity in this study, TEG parameters including R-time, α-angle and MA value were measured.

All extracts and AEC pooled fractions were freeze dried, resuspended in deionised water and added to whole blood in the TEG assay. Table 4 demonstrates that both liquid and canned abalone samples have anti-coagulant activity. In particular, R time is prolonged significantly when sulphated polysaccharide concentrations are increases compared to the saline control. Higher anti-coagulant activity was associated with stronger effects on α-angle and MA values that decreased significantly compared to saline control. These results suggest that molecules present in abalone extracts have an effect on clot strength and platelet function indicating that the kinetics of fibrin polymerization and networking are also affected.

Table 4. Abalone extracts and AEC pools and thromboelastography.

Sample Description	SP Conc. (μg/mL)	R (s)	MA (mm)	α (Degree)
Control Saline	0	445 ± 14.5	55.2 ± 1.2	45.2 ± 0.5
Can_Ab_Pap	20	760 ± 20.5 **	37.6 ± 2.4 **	12 ± 1.5 **
	80	1115 ± 21.8 **	33.3 ± 1.7 **	22.2 ± 5.4 **
Can_Ab_Pap_AEC Pool 3	20	770 ± 8.5 **	35.3 ± 2.3 **	12 ± 1.7 **
	30	1245 ± 12.2 **	34.8 ± 0.5 **	15.4 ± 0.9 **
Can_Ab_Pap_AEC Pool 4	22	930 ± 10.6 **	34.4 ± 1.9 **	12.8 ± 0.8 **
	34	1475 ± 25.5 **	24.9 ± 1.5 **	10 ± 0.9 **
Can_Ab_Bro	10	635 ± 10.2 **	49.1 ± 2.5 **	23.1 ± 3.7 **
	60	1010 ± 24.7 **	29.8 ± 1.9 **	5.6 ± 4.1 **
Can_Ab_Bro_AEC Pool 3	7	515 ± 15.6 **	40.2 ± 3.6 **	31.7 ± 2.1 **
	35	915 ± 20.3 **	33.4 ± 2.8 **	12.5 ± 1.7 **
Can_Ab_Bro_AEC Pool 4	8	495 ± 4.9 **	27 ± 0.5 **	18.1 ± 0.6 **
	38	645 ± 15.6 **	33.6 ± 1.2 **	26 ± 0.9 **
Can_Ab_Pap+Bro	20	845 ± 12.8 **	41.6 ± 1.4 **	19 ± 2.1 **
	70	1340 ± 24.7 **	37.3 ± 2.5 **	14.3 ± 1.9 **
Can_Ab_Pap+Bro_AEC Pool 3	15	810 ± 20.3 **	37.3 ± 4.9 **	17.4 ± 1.1 **
	31	1180 ± 25.9 **	35.6 ± 1.6 **	16.6 ± 0.4 **
Can_Ab_Pap+Bro_AEC Pool 4	13	690 ± 6.5 **	28.5 ± 1.4 **	17.2 ± 0.2 **
	25	855 ± 12.5 **	21.7 ± 1.8 **	20.8 ± 0.4 **
Liquid_Ab_Pap	30	940 ± 16.5 **	34.5 ± 4.4 **	12.5 ± 4.7 **
	50	1410 ± 35.2 **	31.2 ± 1.2 **	14.9 ± 1.2 **
Liquid_Ab_Pap_AEC Pool 3	8	670 ± 14.7 **	46.3 ± 6.5 **	26.9 ± 4.9 **
	39	1540 ± 23.6 **	33.6 ± 3.9 **	13.6 ± 3.4 **
Liquid_Ab_Pap_AEC Pool 4	10	500 ± 6.9 **	33.8 ± 2.1 **	28.8 ± 0.2 **
	51	705 ± 11.8 **	25.1 ± 0.5 **	10.2 ± 0.9 **
Liquid_Ab_Bro	40	710 ± 7.5 **	38.9 ± 4.7 **	25.9 ± 2.8 **
	90	1075 ± 12.5 **	31.5 ± 1.2 **	12.2 ± 3.1 **
Liquid_Ab_Bro_AEC Pool 4	32	990 ± 4.2 **	36.1 ± 1.9 **	35.7 ± 1.7 **
	64	1295 ± 10.9 **	26.9 ± 2.9 **	17.5 ± 2.1 **
Liquid_Ab_Bro_AEC Pool 5	4	620 ± 27.1 **	36.7 ± 5.9 **	32.3 ± 3.5 **
	19	1115 ± 12.8 **	33.8 ± 7.6 **	13.7 ± 4.7 **
Liquid_Ab_Pap+Bro	20	890 ± 8.5 **	39.6 ± 3.2 **	23.5 ± 1.9 **
	40	1495 ± 24.5 **	36.1 ± 1.9 **	12.4 ± 2.2 **
Liquid_Ab_Pap+Bro_AEC Pool 3	13	685 ± 14.3 **	43 ± 2.9 **	29.6 ± 4.1 **
	26	1865 ± 20.5 **	31.8 ± 3.9 **	8.5 ± 3.2 **
Liquid_Ab_Pap+Bro_AEC Pool 4	9	650 ± 10.3 **	27.3 ± 1.2 **	27.7 ± 0.8 **
	45	1420 ± 15.8 **	35.6 ± 0.9 **	15.2 ± 0.1 **

"SP" stands for Sulphated Polysaccharides "Ab" stands for abalone, "Pap" stands for papain enzyme, "Bro" stands for bromelain enzyme while "Pap+Bro" reflects the combination of both papain and bromelain enzymes. ** Using the Dunnett's Multiple Comparison Test, all the treatment are significantly different to each other by comparing with saline control.

The AEC pool fractions prolonged R time more than the original extracts. Generally, AEC pool 3 and 4 appeared to increase R time more than other AEC pools (data not shown). In these samples, R time increased with an increasing sulphated polysaccharide concentration.

However it was also observed that both fractions from AEC pool 3 and 4 affected MA values and α-angles more than the extracts, suggesting a greater effect on clot strength and platelet function because α-angles indicate fibrin build up and cross-linking while MA values reflect clot formation, firmness and platelet function.

The relationship between structure and anti-coagulant activity has been previously investigated in detail for heparin and fucosylated chondroitin sulphate [24]. Both molecules inhibit the intrinsic and/or common pathways of coagulation and thrombin activity or conversion of fibrinogen to fibrin, as observed in the study presented here. This is in agreement with a TEG study reported by Fischer et al. [30] who demonstrated that a variety of glucosamine-based biopolymers including the marine-derived poly-N-acetyl glucosamine could decrease the R time and increase maximal clot strength in plasma. The haemostatic properties were highly dependent on the chemical nature and tertiary/quaternary structure of these biomaterials.

3. Materials and Methods

3.1. Chemicals

All chemicals were reagent grade and unless otherwise indicated were from Sigma-Aldrich (St Louis, MO, USA). Food grade enzymes papain 30,000 (30,000 PU/gram, Papain Units) and bromelain concentrate (2400 GDU/gram, Gelatin Digestion Units) were from Enzyme Solutions Pty Ltd. (Croydon South, Australia). The Blyscan™ Glycosaminoglycan Assay was from Biocolor (Carrickfergus, County Antrim, UK). Other suppliers are listed here: Pierce BCA Protein Assay kit (Quantum Scientific, Murarrie, Australia); Q Sepharose™ Big Beads (GE Healthcare Life Science, Chicago, IL, USA); Thromborel S® (Dade Behring Inc. Newark, NJ, USA); Triniclot (Haemostasis, Wicklow, Ireland); Chromozym TH (Roche Diagnostics, Basel, Switzerland); and human alpha-thrombin and human Heparin Cofactor II (HCII) (US Biologicals, Salem, MA, USA).

3.2. Preparation of Extracts

Processed wild caught *H. rubra* byproducts (comprised of viscera and gonads) were provided by Lonimar Australia Pty Ltd (formerly of Melbourne, Australia). Samples were received either as a paste (canned) or liquid (hot-filled bag). All samples were stored at −20 °C.

Extracts were prepared using the food grade proteases papain and bromelain either separately or combined as a 1:1 ratio. Each digest contained 20 g abalone processing waste and 1% *w/v* papain or bromelain, or a mixture of 0.5% *w/v* papain and 0.5% *w/v* bromelain, in water (final volume 100 mL). For the purpose of determining the contribution of the two enzymes to the subsequent assay measurements, enzyme-only control digests were prepared using the same concentrations and conditions but with no added abalone extract. Digests were incubated overnight (14–16 h) at 50 °C, inactivated by heating at 95 °C for 10 min, cooled on ice and centrifuged (Beckman Coulter, Avanti® J-26XP1, Brea, CA, USA) at 5940× *g* for 10 min to remove undigested material (pellet). Approximately 0.6 g of undigested material was usually discarded. Supernatants were clarified using 2, 1 μm (Whatman™, GE Healthcare, Life Science, Chicago, IL, USA) and 0.45 μm filtration (mixed cellulose ester, Merck Millipore, Bayswater VIC, Australia) and extracts were stored at −20 °C. Due to likely interference with bioassays, salt ions were removed from all abalone extracts and pooled fractions using 3 kDa molecular weight cut off (MWCO) spin columns (Centrifuge Filter Unit, Merck Millipore, Billerica, MA, USA). 10 mL samples were added into the spin column and centrifuged at 3270× *g* for 30 min. This process was repeated using deionised water until salt was at or below 60 mM. Conductivity was measured using a Metler Toledo-AG conductivity meter (VWR International Pty, Ltd., Dietikon, Switzerland). Salt concentration was extrapolated using a NaCl standard curve.

3.3. Estimation of Sulphated Glycosaminoglycan Content

3.3.1. Dimethyl-Methylene Blue (DMMB) Assay

Sulphated polysaccharide concentration was initially estimated in samples and extracts through interaction with DMMB dye. 200 µL DMMB dye solution was added to 25 µL samples, blanks or standards (chondroitin sulphate from bovine trachea, Sigma-Aldrich, Castle Hill, NSW, Australia) in triplicate, in a 96 well plate (Nunclon Delta Surface, Thermo Fisher Scientific, Waltham, MA, USA). The plate was mixed for 1 min using a plate mixer (IKA® MS digital 96 Well Plate Mixer, Staufen im Breisgau, Germany) and absorbance was measured at 525 nm using a Spectra-Max M3 System spectrophotometer (Molecular Devices, Sunnyvale, CA, USA). Sulphated polysaccharides were calculated from the standard curve using SoftMax-Pro 6.1 software (Molecular Devices, Sunnyvale, CA, USA).

3.3.2. Blyscan Sulphated Glycosaminoglycan (GAG) Assay

Sulphated polysaccharide concentration was measured in all samples and pooled fractions using the Blyscan™ Sulphated Glycosaminoglycan assay according to manufacturer's instructions (Biocolor Ltd., Carrickfergus, County Antrim, UK), with modifications. Blyscan Dye Reagent (250 µL) was added to 25 µL sample, blank or standard (chondroitin sulphate from bovine trachea, Sigma-Aldrich, Castle Hill, NSW, Australia) in triplicate, in a 96 well V-bottom plate (Stor Plate-96 V-bottom, Perkin Elmer, Waltham, MA, USA). The plate was placed on an orbital shaker for 30 min (500 rpm) followed by centrifugation at $3270 \times g$ (Beckman Coulter, Allegra™ X-12R, Lane Cove, NSW, Australia) for 10 min. Supernatant was removed without disturbing the pellet using a vacuum before 250 µL Blyscan Dye Dissociation Reagent was added to each well. The plate was again placed on the orbital shaker for 30 min (500 rpm) or until complete dissociation of the pellet. 200 µL of the resuspended solution was transferred from each well into a 96 Well Plate to enable absorption to be measured at 656 nm in a Spectra-Max M3 System spectrophotometer. Sulphated polysaccharides were calculated from the standard curve using SoftMax-Pro 6.1 software.

3.4. Estimation of Protein Content

Protein content was estimated in all samples and extracts using the Pierce BCA Protein Assay Kit (Thermo Fisher Scientific, Waltham, MA, USA) with bovine serum albumin (BSA) as a protein standard, according to manufacturer's instructions. Absorbance was measured at 562 nm using a Spectra-Max M3 System spectrophotometer. Protein concentration was calculated from the standard curve using SoftMax-Pro 6.1 software.

3.5. Separation of Extracts Using Anion Exchange Chromatography-Fast Performance Liquid Chromatography (AEC-FPLC)

An empty 200 mm × 16 mm column (GE-XK 16/20) packed with 13.5 mL Q Sepharose™ Big Beads and connected to fast protein liquid chromatography (FPLC) system (ÄKTA Lab-Scale Systems, GE Healthcare Life Science, Chicago, IL, USA) was used to fractionate the abalone extracts on the basis of their anionic interactions. The column was equilibrated with deionised water (Buffer A) before approximately 28 mg sample (based on sulphated polysaccharides as measured by DMMB assay) was loaded onto the column. Flow rate was set at 5 mL/min with a column pressure of 1 MPa. Fractions (2 mL each) were collected using a 0–2 M NaCl linear gradient over 20 min. AEC fractions were collected and pooled (7 pools for each sample) on the basis of their interactions with DMMB dye. For further analysis, all AEC pooled samples were desalted using 3 kDa MWCO spin columns and washed using deionised water.

3.6. Assessment of Anti-Thrombotic and Anti-Coagulant Activity

3.6.1. Heparin Cofactor II (HCII) Mediated Thrombin Inhibition Assay

In vitro HCII-mediated thrombin inhibition was measured in the extracts and AEC pooled fractions using a kinetic assay as previously described by Dupouy et al. [31] with modifications by Hines et al. [32]. Briefly, 12.9 μM HCII in 0.02 M Tris-HCl pH 7.4/0.15 M NaCl/1 mg/mL polyethylene glycol (PEG) and 1 μL serially diluted extract, sample or standard (heparin from porcine intestinal mucosa, Sigma-Aldrich) was added to a 384 well plate (SpectraPlate-384 TC, clear, tissue culture with lid, Perkin Elmer, Waltham, MA, USA) by an epMotion® 5075L/epMotion® 5075TMX automated pipetting system (Eppendorf, Hamburg, Germany), mixed and incubated at room temperature for 22 min. Thrombin was then added (0.45 μM) prior to the final addition of 83 μM Chromozym TH. The assay was incubated at 37 °C for 40 min with absorbance measured at 405 nm every 2 min in a Spectra-Max M3 System spectrophotometer. HCII-mediated thrombin inhibition was measured at 10 min in triplicate and expressed as the mean percentage inhibition of thrombin activity $\pm$ standard error.

3.6.2. Prothrombin Time (PT) Assay

To measure PT, 100 μL citrated plasma was added to a glass clotting tube and incubated at 37 °C on the heating block of a Hyland-Clotek clotting machine for 5 min. 50 μL saline (negative control), heparin/Clexane (positive control) or diluted abalone extracts or AEC pooled fractions were added to the tube. The volume was adjusted to 150 μL with plasma before the final addition of 100 μL Thromborel S® (Dade Behring Inc. Newark, NJ, USA) to initiate clotting. Time in seconds until clot formation was measured in triplicate and expressed as the mean $\pm$ standard error.

3.6.3. Activated Partial Thromboplastin Time (aPTT) Assay

To measure aPTT, 100 μL citrated plasma, 100 μL Triniclot (Haemostasis, Wicklow, Ireland) and diluted abalone extracts and AEC pooled fractions were added to a clotting tube. The final volume was adjusted to 250 μL with saline. The clotting tube was incubated at 37 °C in a heating block (Hyland-Clotek clotting machine) for 5 min before 50 μL 50 mM calcium solution in saline were added to initiate clotting. Time in seconds until clot formation was measured in triplicate and expressed as the mean $\pm$ standard error.

3.6.4. Thromboelastography (TEG)

To measure clot dynamics in the presence of abalone extracts and AEC pooled fractions, TEG (Haemonetics, Braintree, MA, USA) analyses were undertaken. These analyses provide measurements of whole blood hemostasis that help to assess bleeding and thrombotic risks, as well as monitor anti-thrombotic therapies by investigating the shear elasticity of a clot as it forms or lyses through the following parameters: reaction time (R) which is the time from the start of a sample run until the first detectable clot formation (this is the point at which most traditional plasma clotting assays reach their end point); α-Angle (α) which is the measurement of the rapidity of fibrin build-up and cross-linking (or clot strengthening); and maximal amplitude (MA) which is the maximal stiffness or strength (maximal shear modulus) of the developed clot. A typical TEG tracing, generated by the TEG companion software is shown in Figure 4 and depicts R, α and MA. The parameters are as labelled and described by the TEG® (Haemostasis Analyser, Braintree, MA, USA) 5000 Series Manual.

In this method, 280 μL citrated whole blood and 20 μL 0.2 M CaCl$_2$, along with 20 μL of the negative (saline) or positive (heparin) controls or the abalone test compounds, were added into a disposable TEG cup (Haemonetics, Braintree, MA, USA).

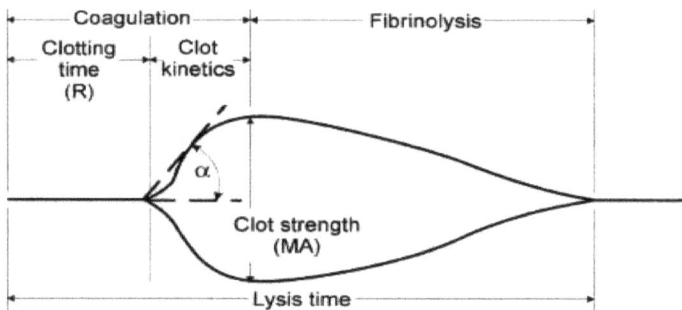

Figure 4. Representation of a typical TEG trace obtained during the clotting of citrated whole blood [33].

3.7. Statistical Analyses

All statistical analyses were conducted using a one-way ANOVA with post hoc comparisons using Tukey's multiple comparison test. These calculations were carried out using GraphPad Prism 5 Software for Windows (GraphPad 5 Software, San Diego, CA, USA, www.graphpad.com).

4. Conclusions

Extracts were prepared from blacklip abalone processing waste using food grade proteases and separated by AEC to produce fractions enriched with sulphated polysaccharides. These sulphated polysaccharides appeared to display properties similar to GAG-like molecules previously characterised in other abalone species by inhibiting thrombin activity through HCII and displaying significant anti-coagulant activity in plasma and blood. Further studies are needed to improve our understanding of the anti-coagulant mechanism *in vitro* and the critical structures required for the mechanism/s behind the anti-coagulant activity.

Acknowledgments: Hafiz Ansar Rasul Suleria was awarded an International Postgraduate Research Scholarship (IPRS), Australia Postgraduate Award (APA) and Postgraduate Studentship (Commonwealth Scientific and Industrial Research Organisation, CSIRO) from the Australian Government at the University of Queensland, Australia. This research received funding from the Fisheries Research and Development Corporation for initial work, which is sponsored by the Australian Government.

Author Contributions: Hafiz Ansar Rasul Suleria, Simone A. Osborne and Paul Masci conceived and designed the experiments; Hafiz Ansar Rasul Suleria performed all the experiments; Barney M. Hines and Rama Addepalli helped in analyzing the data; Glenda Gobe and Wei Chen contributed reagents/materials/analysis tools; Hafiz Ansar Rasul Suleria wrote the original draft while edited by Simone Osborne and Glenda Gobe.

Conflicts of Interest: The authors declare no conflict of interest.

References

1. Ngo, D.H.; Vo, T.S.; Ngo, D.N.; Wijesekara, I.; Kim, S.K. Biological activities and potential health benefits of bioactive peptides derived from marine organisms. *Int. J. Biol. Macromol.* **2012**, *51*, 378–383. [CrossRef] [PubMed]

2. Hickey, R.M. Extraction and characterization of bioactive carbohydrates with health benefits from marine resources: macro- and microalgae, cyanobacteria, and invertebrates. In *Marine Bioactive Compounds: Sources, Characterization and Applications*; Hayes, M., Ed.; Springer: Boston, MA, USA, 2012; pp. 159–172.

3. Suleria, H.A.R.; Masci, P.P.; Gobe, G.C.; Osborne, S.A. Therapeutic potential of abalone and status of bioactive molecules: A comprehensive review. *Crit. Rev. Food Sci. Nutr.* **2015**. [CrossRef] [PubMed]

4. Gates, K.W. Marine nutraceuticals: prospects and perspectives Edited by Se-Kwon Kim. *J. Aquat. Food Prod. Technol.* **2014**, *23*, 522–527. [CrossRef]

5. Harnedy, P.A.; FitzGerald, R.J. Bioactive peptides from marine processing waste and shellfish: A review. *J. Funct. Foods* **2012**, *4*, 6–24. [CrossRef]

6. Leopold, J.A.; Loscalzo, J. Oxidative risk for atherothrombotic cardiovascular disease. *Free Radic. Biol. Med.* **2009**, *47*, 1673–1706. [CrossRef] [PubMed]
7. David, J.C. *Pharmacology and Pharmaco-Therapeutics*, 7th ed.; Vikas Publishing House: New Delhi, India, 1979.
8. Franchini, M.; Liumbruno, G.M.; Bonfanti, C.; Lippi, G. The evolution of anticoagulant therapy. *Blood Transfus.* **2016**, *14*, 175–184. [PubMed]
9. Ahmed, I.; Majeed, A.; Powell, R. Heparin induced thrombocytopenia: Diagnosis and management update. *Postgrad. Med. J.* **2007**, *83*, 575–582. [CrossRef] [PubMed]
10. Shanmugam, M.; Mody, K.H.; Oza, R.M.; Ramavat, B.K. Blood anticoagulant activity of a green marine alga *Codium dwarkense* (Codiaceae, Chlorophyta) in relation to its growth stages. *Indian J. Geo-Mar. Sci.* **2001**, *30*, 49–52.
11. Kim, Y.S.; Jo, Y.Y.; Chang, I.M.; Toida, T.; Park, Y.; Linhardt, R.J. A new glycosaminoglycan from the giant African snail *Achatina fulica*. *J. Biol. Chem.* **1996**, *271*, 11750–11755. [CrossRef] [PubMed]
12. Mauro, M.C.; Toutain, S.; Walter, B.; Pinck, L.; Otten, L.; Coutos-Thevenot, P.; Barbier, P. High efficiency regeneration of grapevine plants transformed with the GFLV coat protein gene. *Plant Sci.* **1995**, *112*, 97–106. [CrossRef]
13. Santos, J.C.; Mesquita, J.M.F.; Belmiro, C.L.R.; da Silveira, C.B.M.; Viskov, C.; Mourier, P.A.; Pavão, M.S.G. Isolation and characterization of a heparin with low antithrombin activity from the body of *Styela plicata* (Chordata-Tunicata). Distinct effects on venous and arterial models of thrombosis. *Thromb. Res.* **2007**, *121*, 213–223. [CrossRef] [PubMed]
14. Lopes-Lima, M.; Ribeiro, I.; Pinto, R.A.; Machado, J. Isolation, purification and characterization of glycosaminoglycans in the fluids of the mollusc *Anodonta cygnea*. *Comp. Biochem. Physiol. A* **2005**, *141*, 319–326. [CrossRef] [PubMed]
15. Qi, H.; Zhang, Q.; Zhao, T.; Chen, R.; Zhang, H.; Niu, X.; Li, Z. Antioxidant activity of different sulfate content derivatives of polysaccharide extracted from *Ulva pertusa* (Chlorophyta) *in vitro*. *Int. J. Biol. Macromol.* **2005**, *37*, 195–199. [CrossRef] [PubMed]
16. Sikka, P.; Bindra, V.K. Newer antithrombotic drugs. Indian journal of critical care medicine: Peer-reviewed. *Off. Publ. Indian Soc. Crit. Care Med.* **2010**, *14*, 188–195. [CrossRef] [PubMed]
17. Li, G.Y.; Chen, S.G.; Wang, Y.M.; Xue, Y.; Chang, Y.G.; Li, Z.J.; Xue, C.H. A novel glycosaminoglycan-like polysaccharide from abalone *Haliotis discus hannai* Ino: Purification, structure identification and anticoagulant activity. *Int. J. Biol. Macromol.* **2001**, *49*, 1160–1166. [CrossRef] [PubMed]
18. Zhou, D.Y.; Tang, Y.; Zhu, B.W.; Qin, L.; Li, D.M.; Yang, J.F.; Murata, Y. Antioxidant activity of hydrolysates obtained from scallop (*Patinopecten yessoensis*) and abalone (*Haliotis discus hannai* Ino) muscle. *Food Chem.* **2012**, *132*, 815–822. [CrossRef]
19. Kechaou, E.S.; Dumay, J.; Donnay-Moreno, C.; Jaouen, P.; Gouygou, J.P.; Bergé, J.P.; Amar, R.B. Enzymatic hydrolysis of cuttlefish (*Sepia officinalis*) and sardine (*Sardina pilchardus*) viscera using commercial proteases: Effects on lipid distribution and amino acid composition. *J. Biosci. Bioeng.* **2009**, *107*, 158–164. [CrossRef] [PubMed]
20. Zhu, B.W.; Zhou, D.Y.; Li, T.; Yan, S.; Yang, J.F.; Li, D.M.; Murata, Y. Chemical composition and free radical scavenging activities of a sulphated polysaccharide extracted from abalone gonad (*Haliotis discus hannai* Ino). *Food Chem.* **2010**, *121*, 712–718. [CrossRef]
21. Wang, Y.M.; Wu, F.J.; Du, L.; Li, G.Y.; Takahashi, K.; Xue, Y.; Xue, C.H. Effects of polysaccharides from abalone (*Haliotis discus hannai* Ino) on HepG2 cell proliferation. *Int. J. Biol. Macromol.* **2014**, *66*, 354–361. [CrossRef] [PubMed]
22. Zhu, B.W.; Zhou, D.Y.; Yang, J.F.; Li, D.M.; Yin, H.L.; Tada, M. A neutral polysaccharide from the abalone pleopod, *Haliotis discus hannai* Ino. *Eur. Food Res. Technol.* **2008**, *228*, 591–595.
23. She, Z.G.; Hu, G.P.; Wu, Y.W. Study on the methanolysis of the sulphated polysaccharide Hal-A from *Haliotis diverisicolor* Reeve. *Chin. J. Org. Chem.* **2002**, *22*, 367–370.
24. Chen, S.; Xue, C.; Yin, L.A.; Tang, Q.; Yu, G.; Chai, W. Comparison of structures and anticoagulant activities of fucosylated chondroitin sulfates from different sea cucumbers. *Carbohydr. Polym.* **2011**, *83*, 688–696. [CrossRef]
25. Bordbar, S.; Anwar, F.; Saari, N. High-value components and bioactives from sea cucumbers for functional foods—A review. *Mar. Drugs* **2011**, *9*, 1761–1805. [CrossRef] [PubMed]

26. Pomin, V.H.; Pereira, M.S.; Valente, A.-P.; Tollefsen, D.M.; Pavão, M.S.G.; Mourão, P.A.S. Selective cleavage and anticoagulant activity of a sulfated fucan: Stereospecific removal of a 2-sulfate ester from the polysaccharide by mild acid hydrolysis, preparation of oligosaccharides, and heparin cofactor II–dependent anticoagulant activity. *Glycobiology* **2005**, *15*, 369–381. [CrossRef] [PubMed]
27. Pereira, M.S.; Melo, F.R.; Mourão, P.A.S. Is there a correlation between structure and anticoagulant action of sulfated galactans and sulfated fucans? *Glycobiology* **2002**, *12*, 573–580. [CrossRef] [PubMed]
28. Wu, M.; Wen, D.; Gao, N.; Xiao, C.; Yang, L.; Xu, L.; Lian, W.; Peng, W.; Jiang, J.; Zhao, J. Anticoagulant and antithrombotic evaluation of native fucosylated chondroitin sulfates and their derivatives as selective inhibitors of intrinsic factor Xase. *Eur. J. Med. Chem.* **2015**, *92*, 257–269. [CrossRef] [PubMed]
29. Li, J.; Tong, T.; Ko, D.O.; Kang, S.G. Antithrombotic potential of extracts from abalone, *Haliotis discus hannai* Ino: and animal studies. *Food Sci. Biotechnol.* **2013**, *22*, 471–476. [CrossRef]
30. Fischer, T.H.; Bode, A.P.; Demcheva, M.; Vournakis, J.N. Hemostatic properties of glucosamine-based materials. *J. Biomed. Mater. Res. A* **2007**, *80*, 167–174. [CrossRef] [PubMed]
31. Dupouy, D.; Sié, P.; Dol, F.; Boneu, B. A simple method to measure dermatan sulfate at sub-microgram concentrations in plasma. *Thromb. Haemost.* **1988**, *60*, 236–239. [PubMed]
32. Hines, B.M.; Suleria, H.A.R.; Osborne, S.A. Automated screening potential thrombin inhibitors using the epMotion® 5075. *Eppendorf* **2016**, *377*, 1–6.
33. Rivard, G.E.; Brummel-Ziedins, K.E.; Mann, K.G.; Fan, L.; Hofer, A.; Cohen, E. Evaluation of the profile of thrombin generation during the process of whole blood clotting as assessed by thromboelastography. *J. Thromb. Haemost.* **2005**, *3*, 2039–2043. [CrossRef] [PubMed]

marine drugs

MDPI

Article

Biochemical and Structural Insights into a Novel Thermostable β-1,3-Galactosidase from *Marinomonas* sp. BSi20414

Haitao Ding, Qian Zeng, Lili Zhou, Yong Yu and Bo Chen *

SOA Key Laboratory for Polar Science, Polar Research Institute of China, Shanghai 200136, China; htding@outlook.com (H.D.); zengqianmu@126.com (Q.Z.); lilizhou1199@163.com (L.Z.); yuyong@pric.org.cn (Y.Y.)
* Correspondence: chenbo@pric.org.cn; Tel.: +86-21-5871-1026

Academic Editor: Se-Kwon Kim
Received: 1 November 2016; Accepted: 24 December 2016; Published: 8 January 2017

Abstract: A novel β-1,3-galactosidase, designated as MaBGA (β-galactosidase from *Marinomonas* sp. BSi20414), was successfully purified to homogeneity from *Marinomonas* sp. BSi20414 isolated from Arctic sea ice by ammonium sulfate precipitation and anion exchange chromatography, resulting in an 8.12-fold increase in specific activity and 9.9% recovery in total activity. MaBGA displayed its maximum activity at pH 6.0 and 60 °C, and maintained at least 90% of its initial activity over the pH range of 5.0–8.0 after incubating for 1 h. It also exhibited considerable thermal stability, which retained 76% of its initial activity after incubating at 50 °C for 6 h. In contrast to other β-galactosidases, MaBGA displayed strict substrate specificity, not only for the glycosyl group, but also for the linkage type. To better understand the structure–function relationship, the encoding gene of MaBGA was obtained and subject to bioinformatics analysis. Multiple alignments and phylogenetic analysis revealed that MaBGA belonged to the glycoside hydrolase family 42 and had closer genetic relationships with thermophilic β-galactosidases of extremophiles. With the aid of homology modeling and molecular docking, we proposed a reasonable explanation for the linkage selectivity of MaBGA from a structural perspective. On account of the robust stability and 1,3-linkage selectivity, MaBGA would be a promising candidate in the biosynthesis of galacto-oligosaccharide with β1–3 linkage.

Keywords: β-galactosidase; *Marinomonas*; thermostable; purification; gene cloning; linkage selectivity

1. Introduction

The enzyme β-galactosidases (EC 3.2.1.23, BGA), which are widely distributed in various organisms, including animals, plants, bacteria, archaea, yeasts and fungi, are capable of catalyzing the hydrolysis of molecules containing the β-glycosidic bond, to release their terminal non-reducing galactose molecules. In some cases, β-galactosidases can catalyze the reverse reaction of the hydrolysis, transglycosylation, when receptors of galactosyl are monosaccharides, disaccharides or oligosaccharides, instead of water molecules [1]. Due to the catalytic characteristic, β-galactosidases are important for the dairy industry to produce milk with low/no lactose for people who suffer from lactose intolerance [2]. Moreover, β-galactosidases are also widely utilized for enzymatic synthesis of galacto-oligosaccharides, which can be employed to stimulate the growth of beneficial bacteria selectively in the gut, as prebiotics [3].

Based on the similarity of the amino acid sequences, β-galactosidases are mainly divided into four glycoside hydrolase (GH) families [4]—GH1, GH2, GH35 and GH42—according to the carbohydrate-active enzymes database (CAZy) [5]. All of these four families belong to the GH-A superfamily, of which members have two glutamic acid residues as catalytic active sites located in an $(\alpha/\beta)_8$ TIM (the triosephosphate isomerase) barrel domain [6]. Generally, GH1 and GH2 β-galactosidases are mainly found in mesophiles and display high lactase activity [7]. GH35 β-galactosidases are usually found in pathogens such as Streptococcus pneumoniae [8–10], with specific activity toward β-1,3-linkages. β-galactosidases belonging to the GH42 family are mostly stemmed from extremophiles, including thermophiles [11–15], halophiles [16,17] and alkaliphiles [18].

Owing to the attractive properties such as heat resistance and salt tolerance, GH42 β-galactosidases have received extensive attention in recent years. It is expected to obtain new enzymes with excellent properties from microorganisms living in extreme environments [19]. The Arctic is one of the most extreme regions to be inhabited by plenty of microorganisms, which have been proven to be the natural treasure house for screening novel enzymes [20,21]. In our previous study, a strain designated as BSi20414 with high β-galactosidase activity was isolated from Arctic sea ice and identified as *Marinomonas* [22]. The optimal catalytic temperature of the crude enzyme was determined as 60 °C, indicating that it might be a thermophilic enzyme. Generally, robust thermal-stability is indispensable for the practical application of enzymes. Thus, to obtain a promising thermal-stable β-galactosidase and provide a comprehensive evaluation of its potential in practical application, the enzyme that possessed β-galactosidase activity from *Marinomonas* sp. BSi20414 was purified to homogeneity and characterized extensively in the present work. In addition to biochemical characterization, the encoding gene of MaBGA was cloned by degenerate PCR and chromosome walking, and was further subject to bioinformatics analysis to investigate its structure–function relationships.

2. Results

2.1. Purification of Wild-Type MaBGA

The crude enzyme was concentrated by 60% of ammonium sulfate and then separated into five components, peak I–V (Figure 1a), by anion exchange chromatography. Among these five peaks, only peak IV exhibited β-galactosidase activity toward *o*-nitrophenyl-β-galactoside (ONPG). The purity of peak IV was examined by SDS-PAGE (sodium dodecyl sulfate polyacrylamide gel electrophoresis) analysis, which showed a single band corresponding to about 70 kDa (Figure 1b), indicating that MaBGA had been successfully purified. As shown in Table 1, the two-step purification procedure yielded an 8.12-fold increase in specific activity and a recovery of 9.9% in total activity.

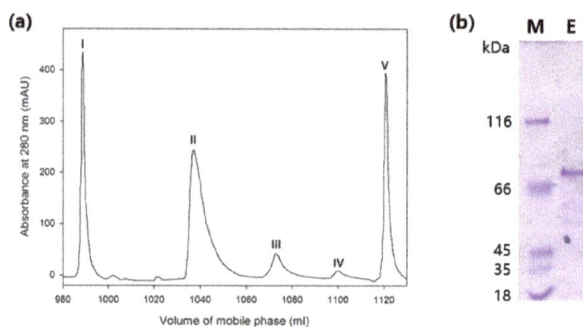

Figure 1. Purification of wild-type MaBGA (β-galactosidase from *Marinomonas* sp. BSi20414). (**a**) Ion exchange chromatography. Peak I was unbound proteins; Peak II and III were proteins eluted by 0.1–0.2 M of NaCl; Peak IV was protein eluted by 0.20–0.24 M of NaCl; Peak V was protein eluted by 0.24–0.6 M of NaCl; (**b**) SDS-PAGE (sodium dodecyl sulfate polyacrylamide gel electrophoresis) analysis of purified MaBGA. Lane M: protein molecular weight marker; Lane E: purified MaBGA.

Table 1. Purification of MaBGA.

Purification Steps	Total Protein (mg)	Total Activity (U)	Specific Activity (U/mg)	Recovery (%)	Purification Fold
Cell lysate	162.54	1818.18	11.19	100	1
Ammonium sulfate precipitation	29.9	927.27	31.01	51	2.77
HiTrap DEAE FF	1.01	91.80	90.89	9.9	8.12

2.2. Enzymatic Characterization of MaBGA

2.2.1. Effect of pH on the Activity and Stability of MaBGA

The optimum pH of MaBGA was determined as 6.0, and it exhibited more than 80% of its maximum activity over the pH range of 5.0–7.0, outside of which the activity decreased sharply (Figure 2a). The stability of MaBGA showed a similar pattern with that of the activity response to pH, which was stable around the neural condition, and could maintain at least 90% of its initial activity over the pH ranging from 5.0 to 8.0, after incubating in Britton–Robinson buffer with different pH values for 1 h (Figure 2b).

2.2.2. Effect of Temperature on the Activity and Stability of MaBGA

MaBGA exhibited the highest activity at 60 °C, and less than 50% of the maximum activity was measured at temperatures below 45 °C (Figure 2c). Generally, an enzyme with a relatively high optimal reaction temperature often possessed superior thermal stability. With no exception, MaBGA was stable at 50 °C, which could maintain 76% of its initial activity after incubating for 6 h (Figure 2d). In addition, the half-life of MaBGA at 50 °C was determined as 16 h.

2.2.3. Effect of NaCl on the Activity and Stability of MaBGA

MaBGA showed the highest activity with 0.5 M NaCl contained in the reaction buffer. Although the activity decreased along with the increase in the concentration of NaCl, MaBGA still displayed 55% of its maximum activity with 5 M NaCl added (Figure 2e). MaBGA was unstable while incubated in buffers containing NaCl above 0.5 M, and it could only maintain 30% of its initial activity after incubating in buffer with 5 M NaCl added for 1 h (Figure 2f).

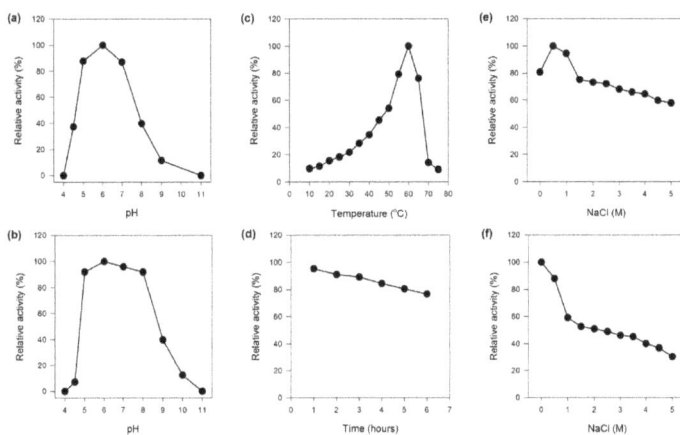

Figure 2. Effects of pH, temperature and NaCl on the activity and stability of MaBGA. (**a**) Effect of pH on the activity of MaBGA; (**b**) Effect of pH on the stability of MaBGA; (**c**) Effect of temperature on the activity of MaBGA; (**d**) Effect of time on the stability of MaBGA; (**e**) Effect of NaCl on the activity of MaBGA; (**f**) Effect of NaCl on the stability of MaBGA.

2.2.4. Effects of Metal Ions and Chemicals on the Activity of MaBGA

As shown in Table 2, K^+, Na^+ and Mn^{2+} displayed no significant effects on the activity of MaBGA, as well as EDTA. Interestingly, Fe^{2+} is capable of improving the activity of MaBGA by 111%, whereas other bivalent cations—Mg^{2+}, Co^{2+}, Ni^{2+} and Zn^{2+}—slightly inhibited the activity of the enzyme. Moreover, reducing agents, such as L-cysteine, L-glutathion and dithiotreitol showed no notable effect on the activity of MaBGA, indicating that no disulfide bond was indispensable to the enzyme.

Table 2. Effects of metal ions and chemicals on the activity of MaBGAL.

Metal Ions	Relative Activity (%)	Chemicals	Relative Activity (%)
K^+	96	EDTA	98
Na^+	95	L-Cysteine	110
Fe^{2+}	211	L-Glutathion	103
Mn^{2+}	98	Dithiotreitol	106
Mg^{2+}	89		
Co^{2+}	88		
Ni^{2+}	76		
Zn^{2+}	74		

2.2.5. Substrate Specificity and Steady-State Kinetic Analysis

MaBGA possessed a narrow substrate spectrum, which showed no activity toward *p*-nitrophenyl-β-D-glucopyranoside, *p*-nitrophenyl-β-D-xylopyranoside, *p*-nitrophenyl-β-D-lactopyranoside, *p*-nitrophenyl-β-D-glucuronide, *p*-nitrophenyl-a-D-galactopyranoside, *p*-nitrophenyl-β-L-arabinopyranoside and *p*-nitrophenyl-β-D-cellobioside. Moreover, MaBGA showed not only group selectivity, but also showed linkage selectivity in the substrate recognition process, of which the activity toward *p*-nitrophenyl-β-D-galactopyranoside was 4.22-fold greater than that of ONPG (Table 3).

Table 3. Substrate specificity of MaBGA.

Substrate	Relative Activity (%)
o-nitrophenyl-β-D-galactopyranoside	100
p-nitrophenyl-β-D-galactopyranoside	422
p-nitrophenyl-β-D-glucopyranoside	<1
p-nitrophenyl-β-D-xylopyranoside	<1
p-nitrophenyl-β-D-lactopyranoside	<1
p-nitrophenyl-β-D-glucuronide	<1
p-nitrophenyl-a-D-galactopyranoside	<1
p-nitrophenyl-β-L-arabinopyranoside	<1
p-nitrophenyl-β-D-cellobioside	<1

The steady-state kinetic constants of MaBGA were determined by using a nonlinear fitting plot. The apparent Michaelis–Menten constant K_m and the maximum reaction velocity V_{max} were calculated as 14.19 mM and 1.049 μM·min^{-1}, respectively.

2.2.6. Linkage Selectivity Analysis

As shown in Figure 3, the chromatograms of Galβ1–4GlcNAc showed no change before and after the reaction catalyzed by MaBGA, as well as Galβ1–6GlcNAc, suggesting that MaBGA was unable to hydrolyze both Galβ1–4GlcNAc and Galβ1–6GlcNAc. With regard to Galβ1–3GlcNAc, the product chromatogram generated a new peak corresponding to the standard of β-galactose, with an identical retention time of 10.1 min, indicating that MaBGA was capable of degrading Galβ1–3GlcNAc selectively.

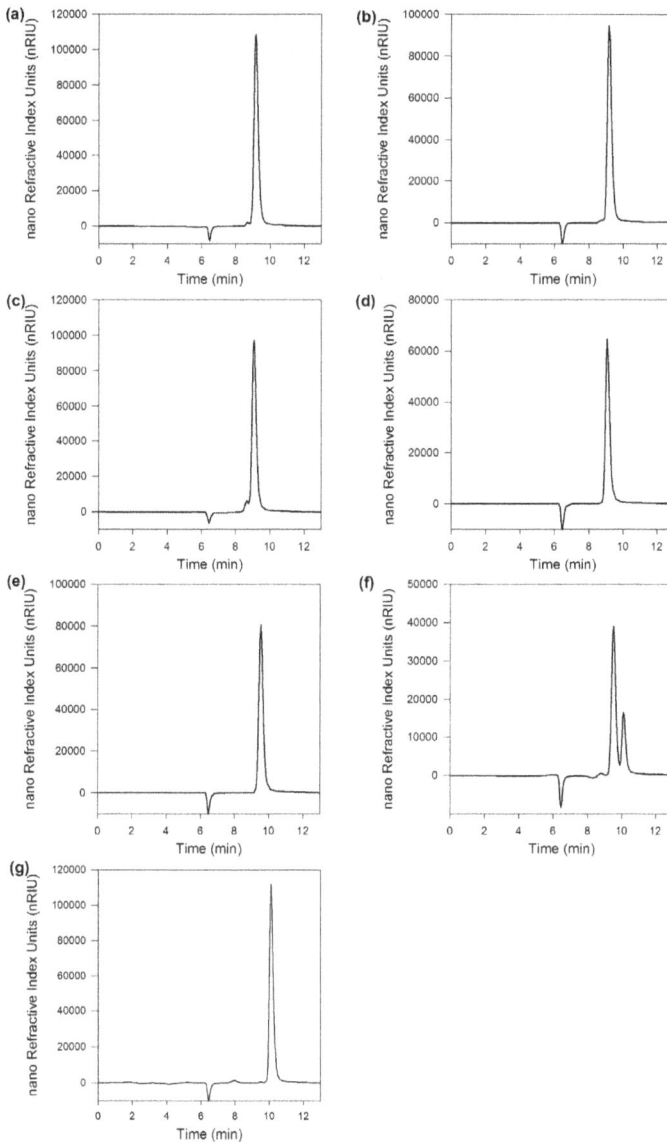

Figure 3. Linkage selectivity analysis. (**a**), (**c**), (**e**) and (**g**) were chromatograms of Galβ1–4GlcNAc, Galβ1–6GlcNAc, Galβ1–3GlcNAc and galactose, respectively; (**b**), (**d**) and (**f**) were chromatograms of Galβ1–4GlcNAc, Galβ1–6GlcNAc and Galβ1–3GlcNAc hydrolyzed by MaBGA, respectively.

2.3. Gene Cloning and Sequence Analysis

2.3.1. Gene Cloning

A 500-bp fragment was amplified from the genomic DNA of *Marinomonas* sp. BSi20414, by using the degenerate primers F1 and R2 (Figure 4a). The nucleotide sequence of the fragment showed an identity of 84% with a putative β-galactosidase gene of *Marinomonas* sp. MWYL1, suggesting that the partial β-galactosidase gene sequence of *Marinomonas* sp. BSi20414 had been successfully obtained. After chromosome walking, 1800-bp and a 1600-bp DNA fragments (Figure 4b), corresponding to the upstream and downstream sequence of the target gene, were amplified and sequenced. The intermediate and flanking sequences were utilized for assembly and subsequent amplification of the intact *mabga* gene, with a size of 2000 bp (Figure 4c).

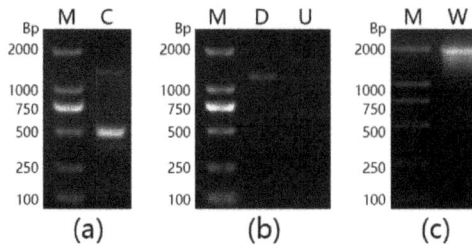

Figure 4. Agarose gel electrophoresis analysis of DNA fragments amplified by PCR. (**a**) Degenerated PCR; (**b**) Chromosome walking PCR; (**c**) Full length amplification. Lane M: DNA marker; Lane C: Fragment amplified by F1/R2; Lane D: Downstream fragment; Lane U: Upstream fragment; Lane W: Full-length fragment of gene *mabga*.

2.3.2. Sequence Analysis

Gene *mabga* encodes a peptide consisting of 656 amino acids with a calculated molecular weight of 74.28 kDa in accordance with the results of SDS-PAGE. Significant Pfam-A matches revealed that MaBGA belonged to the glycoside hydrolase family 42. The protein sequence of MaBGA showed 54.6%, 54.0%, 52.5%, 52.0%, 40.6%, and 38.2% identities with the well-characterized galactosidases from *Thermus* sp. IB-21 (Q8GEA9) [13], *Thermus thermophilus* A4 (O69315) [23], *Thermus* sp. T2 (O54315) [14], *Thermus brockianus* ITI360 (Q9X6C6) [24], *Halorubrum lacusprofundi* ATCC 49239 (B9LW38) [16], and *Haloferax lucentense* DSM 14919 (P94804) [17], respectively. In addition, MaBGA exhibited the highest identity of 90% with a putative BGA from *Marinomonas* sp. MWYL1.

Multiple alignments of protein sequences of structure-solved GH42 β-galactosidase showed that MaBGA shared the conserved catalytic residues, Glu142 and Glu314, as well as other GH42 BGAs (Figure 5). Phylogenetic analysis of characterized BGAs showed that these BGAs diverged into two clusters, and MaBGA affiliated to the branch consisting of thermophilic BGAs (Figure 6) that exhibited considerable stability against heat in previous studies, suggesting that these BGAs, including MaBGA, might be originated from the same ancestral sequence.

2.4. Structural Analysis of MaBGA

2.4.1. Homology Modelling

The tertiary structures of MaBGA were constructed by various software or online servers, including SWISS-MODEL [25], Robetta [26], MODELLER [27] and I-TASSER [28], then evaluated by ProSA [29] and PROCHECK [30]. Both of the Z-score and Ramachandran plot statistics indicated that the three-dimensional structures of MaBGA had been modeled reasonably (Table 4), especially for the model constructed by MODELLER, which scored highest and was selected for the docking study. The superposition of the MaBGA monomer structure onto the structure of β-galactosidase from *Thermus thermophilus* A4 [23] demonstrated the relatively high similarity between them, with a root mean square deviation value of 0.17 (Figure 7a). As with other GH42 BGAs [11,18,23], the quaternary structure of MaBGA was predicted as a homo-trimer, which resembles a flowerpot, with a cone-shaped tunnel in the center of the flowerpot surrounded by three subunits (Figure 7b).

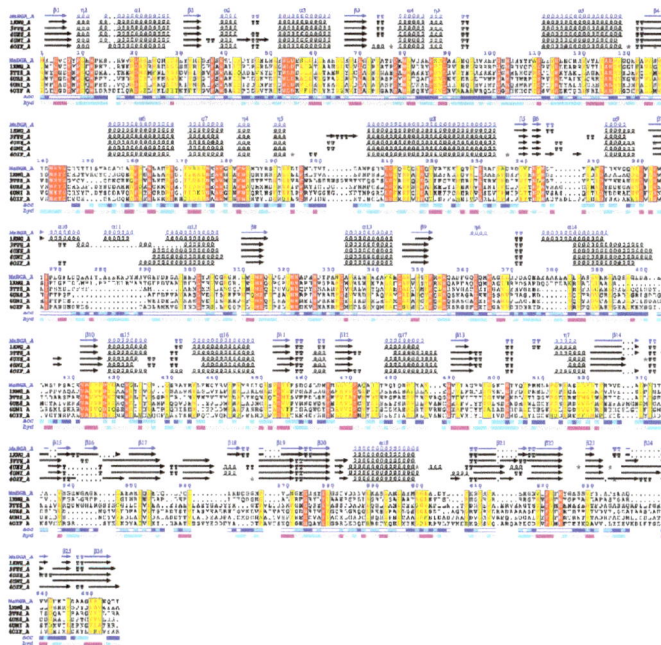

Figure 5. Multiple alignment of structure-solved β-galactosidases of the GH42 family (glycoside hydrolase 42 family). Identical residues and conserved substitutions are shaded red and yellow, respectively. Secondary structures of β-galactosidases are presented on the top: helices with squiggles, β-strands with arrows, turns with TT letters.

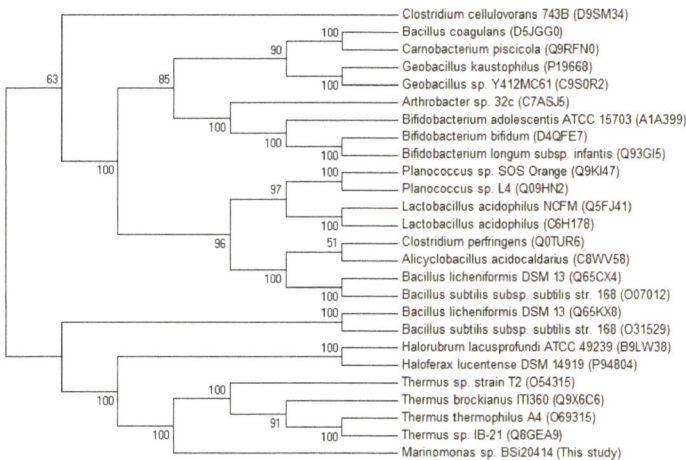

Figure 6. Unrooted phylogenetic tree of β-galactosidase belonging to the GH42 family. The phylogenetic tree was built using the neighbor joining method [31] in MEGA 6 [32], with a bootstrap test of 1000 replicates. The GenBank accession numbers were provided in the bracket followed by the species names.

Table 4. Evaluation of models generated by different modeling approaches.

Model	Z-Score [1]	Ramachandran Plot [2]			
		Most Favored (%)	Additional Allowed (%)	Generously Allowed (%)	Disallowed (%)
Template (4oif)	−12.19	88.3	10.7	0.6	0.3
Robetta	−10.19	87.9	10.2	1.6	0.4
Template (1kwk)	−12.24	90.6	8.2	0.7	0.4
SWISS-MODEL	−9.93	89.7	8.1	1.5	0.7
MODELLER	−10.08	91.9	6.8	0.4	0.9
Template (1kwg)	−12.27	91.2	7.9	0.6	0.4
I-TASSER	−10.14	79.1	15.8	3.2	1.9

[1] Calculated by ProSA-web; [2] Calculated by PROCHECK.

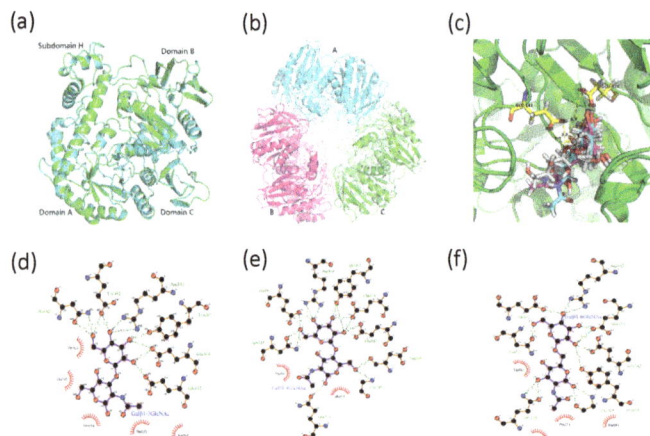

Figure 7. Three-dimensional structures of MaBGA. (**a**) Superposition of the MaBGA monomer structure (green) on the structures of β-galactosidase from Thermus thermophilus A4 (cyan; PDB entry 1kwk); (**b**) Ribbon representation of the trimer structure of MaBGA; (**c**) Ball and stick representation of the docking models of MaBGA with Galβ1–3GlcNAc (white), Galβ1–4GlcNAc (cyan) and Galβ1–6GlcNAc (magenta), respectively; (**d**) Schematic diagram of MaBGA/Galβ1–3GlcNAc interactions; (**e**) Schematic diagram of MaBGA/Galβ1–4GlcNAc interactions; (**f**) Schematic diagram of MaBGA/Galβ1–6GlcNAc interactions.

2.4.2. Molecular Docking Analysis

The model of MaBGA generated by MODELLER [27] was subject to GROMACS [33] software packages for energy minimization, to remove steric clashes. The refined model was employed for molecular docking with Galβ1–3GlcNAc, Galβ1–4GlcNAc and Galβ1–6GlcNAc by Autodock 4.2 [34], respectively. Cluster analysis was performed on different conformations with a root mean square deviation (RMSD) tolerance of 2.0 Å. Conformation with the lowest estimated binding free energy was utilized for analysis. As shown in Figure 7c, the galactosyl group of these three substrates adopts similar conformations, including the oxygen atom which links the acetylglucosamine group. However, the acetylglucosamine group of the substrates adopts a varied conformation corresponding to their lowest free energy. The two-dimensional projection of the interaction of the enzyme/substrate complex showed that no hydrogen bond was generated between the enzyme and the glucosyl group of Galβ1–3GlcNAc (Figure 7d), in contrast to those of Galβ1–4GlcNAc (Figure 7e) and Galβ1–6GlcNAc (Figure 7f), which formed three and four pairs with the enzyme, respectively.

3. Discussion

In the present study, a thermostable β-1,3-galactosidase MaBGA was successfully purified to homogeneity from *Marinomonas* sp. BSi20414 isolated from Arctic sea ice by ammonium sulfate precipitation and anion exchange chromatography, resulting in an 8.12-fold increase in specific activity and 9.9% recovery in total activity. The purification results showed that the two-step purification method is efficient for separating MaBGA from the wild-type strain of *Marinomonas* sp. BSi20414, which also provides a reference for extracting other proteins from strains belonging to the genus of *Marinomonas*.

Interestingly, as an enzyme stemmed from a strain living in permanently low-temperature marine environments, MaBGA displayed extraordinary stability against heat, with the half-life determined as 16 h at 50 °C. Phylogenetic analysis of characterized GH42 BGAs also revealed that MaBGA had closer genetic relationships with thermophilic BGAs derived from extremophiles, including thermophiles [11–15] and halophiles [16,17]. On account of the enzymatic and phylogenetic analyses, MaBGA was considered as a thermophilic enzyme, although the thermal stability of MaBGA is weaker than those of its thermophilic counterparts. Additionally, MaBGA only shared high identity (>70%) with BGAs of the genus *Marinomonas*, and no sequence with identity more than 55% was found in their related marine species. On the basis of the above evidence, a putative explanation was proposed to illustrate the mismatch of enzyme stability and circumstance. It is supposed that the ancestor of the genus *Marinomonas* acquired the gene encoding thermophilic β-galactosidase from other thermophiles by occasional horizontal transfer, then experienced adaptive evolution under low-temperature marine environments for a long period, which led to a decrease in thermal stability without selection pressures.

Another point worth mentioning is that MaBGA has a strict substrate specificity, unlike other GH42 BGAs. Furthermore, it displayed not only group selectivity, but also linkage selectivity in the substrate recognition process. As indicated above, MaBGA was able to hydrolyze Galβ1–3GlcNAc, but was unable to hydrolyze Galβ1–4GlcNAc and Galβ1–6GlcNAc. To better understand the linkage selectivity of MaBGA, it is essential to put MaBGA against its structural contexts. Thus, the three-dimensional structure of MaBGA was constructed and subject to docking analysis after energy minimization by molecular dynamics. As shown in Figure 7c, for all these three substrates, although the galactosyl group adopts similar lowest energy conformations, the distance is a bit long for the reaction between the oxygen atom linking the acetylglucosamine group and the carboxyl group of catalytic residues (Glu142/Glu314). Therefore, the substrate molecule needs to fine-tune its geometry to shorten the distance mentioned above by overcoming the energy barrier. However, the planar representation of the interaction of the enzyme/substrate complex indicated that the strong interaction between the glucosyl group of Galβ1–4GlcNAc (Figure 7e)/Galβ1–6GlcNAc (Figure 7f) and the enzyme might lead to the failure of the substrates to adjust their conformation for an optimal fit. Therefore, we proposed that the favored binding conformation with lowest free binding energy of the substrate is not close enough to the catalytic residues to let the reaction occur, thus the substrate might be fine-tuning its conformation to achieve an optimal geometry for the reaction. However, due to the different binding energy between the glucosyl group and enzyme, Galβ1–4GlcNAc and Galβ1–6GlcNAc cannot readily overcome the energy barrier, other than Galβ1–3GlcNAc. In general, further experiments, such as enzyme/substrate complex co-crystallization and site-directed mutagenesis, are still needed to test the hypothesis.

A previous study had proven that galacto-oligosaccharides with β1–3 linkage have a stronger bifidogenic effect than those with β1–4 and β1–6 linkages [35], indicating that the former would be more popular as prebiotics than the latter two. Generally, the production of galacto-oligosaccharides is implemented by the transglycosylation activity of β-galactosidase [36], therefore, the linkage of galacto-oligosaccharides will depend on the linkage recognition ability of β-galactosidase. Since β-galactosidases that existed in the nature which are capable of recognizing β1–3 galactoside linkage are very few, the β-1,3-galactosidase MaBGA studied in the present work not only could provide a promising candidate for the biosynthesis of galacto-oligosaccharides with β1–3 linkage, but also would offer a good model for research on the substrate recognition mechanism of β-galactosidase.

4. Materials and Methods

4.1. Strains, Plasmids, and Culture Conditions

Strain BSi20414, used as the source of β-galactosidase, was isolated from a core sample of sea ice collected from Canada Basin, Arctic Ocean, and was characterized as *Marinomonas* in our previous study [22]. The strain was cultivated in medium (pH 7.0) containing $MgCl_2$ (0.5%, w/v), $MgSO_4 \cdot 7H_2O$ (0.4%, w/v), KCl (0.1%, w/v), $CaCl_2$ (0.06%, w/v), lactose (1.5%, w/v) and Tryptone (0.5%, w/v), on a shaking incubator at 180 rev·min^{-1} at 30 °C for 96 h. *Escherichia coli* DH5α used for gene cloning was cultivated at 37 °C in Luria–Bertani medium. Plasmid pMD18-T (Takara) was used to conduct TA cloning for sequencing. All chemicals used in this study were of analytical grade.

4.2. Purification of Wild-Type MaBGA

All purification steps were conducted at 4 °C. Cells were harvested by centrifugation at 10,000× g for 10 min. The pellet was washed three times with normal saline and was suspended by pre-cooling PBS buffer (pH 7.0, 50 mM). The suspension was lysed by sonication (burst of 2 s followed by intervals of 5 s for 30 min). The cell debris was removed by centrifugation at 10,000× g for 15 min and the supernatant was precipitated with ammonium sulfate (60%, w/v). The precipitate was collected by centrifugation at 10,000× g for 10 min, then dissolved and dialyzed using PBS buffer (pH 7.0, 50 mM) overnight. Subsequently, the protein solution was filtered by cellulose acetate film with pore size of 0.22 μm, and the filtrate was loaded onto an anion exchange column HiTrap DEAE FF, which was pre-equilibrated with PBS buffer (pH 8.0, 50 mM). The column was first washed with PBS buffer (pH 8.0, 50 mM) for tenfold resin volumes, then was eluted by PBS buffer (pH 8.0, 50 mM) containing NaCl with a linear gradient from 0.1 M to 0.6 M. Every eluting peak was collected and measured by standard activity assay. The protein concentration was assayed by the method of Bradford using BSA (bovine serum albumin) as a standard [37].

4.3. SDS-PAGE Analysis

The purified MaBGA was analyzed by denaturing discontinuous SDS-PAGE on a 5% stacking gel and a 10% separating gel as described by Laemmli [38]. Gels were stained with Coomassie Brilliant Blue R-250. The molecular weight of MaBGA was determined by comparing its electrophoretic mobility with Protein Molecular Weight Marker (MBI).

4.4. β-galactosidase Activity Assay

The β-galactosidase activity was assayed by measuring the absorbance of ONP (*o*-nitrophenyl) at 420 nm in 50 mM PBS buffer (pH 7.0) with 10 mM ONPG as substrate. The ONP concentration was calculated from the standard curve obtained under the same experimental condition. One unit of enzyme activity was defined as the amount of the enzyme that catalyzed the formation of 1 μmol of ONP per minute.

4.5. Effect of pH on the Activity and Stability of MaBGA

The optimum pH for MaBGA was determined by measuring the activity in Britton–Robinson buffer with different pH ranging from 3.0 to 12.0. The pH stability was assayed by measuring the residual activity after incubating MaBGA in different pH buffers at 37 °C for 1 h.

4.6. Effect of Temperature on the Activity and Stability of MaBGA

To study the effect of temperature on the activity of MaBGA, the enzyme activity was assayed at different temperatures from 10 to 70 °C with 5 °C intervals at pH 7.0. The thermal stability was determined by assaying the residual activity after incubating the enzyme at 50 °C for 6 h with 1 h intervals.

4.7. Effect of NaCl on the Activity and Stability of MaBGA

In order to determine the effects of NaCl on the activity of MaBGA, the enzyme activity was assayed with 0.5, 1, 1.5, 2, 2.5, 3, 3.5, 4, 4.5, 5 M NaCl added individually. The NaCl tolerance of MaBGA was determined by measuring the residual activity after incubating the enzyme in buffers containing diverse concentration of NaCl from 0.5 M to 5M at 37 °C for 1 h.

4.8. Effect of Metal Ions and Chemicals on the Activity of MaBGA

To investigate the effects of metal ions and chemicals on the MaBGA activity, 1 mM of KCl, NaCl, $FeCl_2$, $MnCl_2$, $MgCl_2$, $CoCl_2$, $NiCl_2$, $ZnCl_2$, EDTA and 10 mM of L-cysteine, L-glutathion and dithiotreitol were added to the reaction system individually, and the activity of MaBGA was then measured under the standard assay as described above. No chemical was added in the control.

4.9. Substrate Specificity

The substrates' specificity of MaBGA was measured by the standard assay, except that ONPG was replaced by *p*-nitrophenyl-β-D-galactopyranoside, *p*-nitrophenyl-β-D-glucopyranoside, *p*-nitrophenyl-β-D-xylopyranoside, *p*-nitrophenyl-β-D-lactopyranoside, *p*-nitrophenyl-β-D-glucuronide, *p*-nitrophenyl-a-D-galactopyranoside, *p*-nitrophenyl-β-L-arabinopyranoside, *p*-nitrophenyl-β-D-cellobioside, respectively.

4.10. Steady-State Kinetic Analysis

For steady-state kinetic analysis, the activity of MaBGA was measured by using various concentrations of ONPG from 0.1 mM to 19 mM. The kinetic constants of the enzyme were determined by using a nonlinear fitting of the Michaelis–Menten equation: $v = V_{max} \cdot [S]/(K_m + [S])$, where $[S]$ and K_m are the concentration and Michaelis constants of ONPG, respectively.

4.11. Linkage Selectivity Analysis

To determine the activity of MaBGA toward different linkage types, Galβ1–3GlcNAc, Galβ1–4GlcNAc and Galβ1–6GlcNAc were used as substrates, respectively. The reaction products were filtered by nitrocellulose membrane with pore size of 0.22 μm, in advance of being subject to detection by HPLC equipped with an Aminex HPX-87P column and differential detector. The column temperature and flow rate were set as 85 °C and 0.5 mL·min^{-1}.

4.12. Gene Cloning and Sequence Analysis

The partial sequence of gene *mabga* was amplified by using degenerate primer pairs F1/R1, F1/R2, F1/R3, F2/R1, F2/R2, F2/R3, F3/R1, F3/R2, F3/R3, A208/B1, A208/C1, A208/C2, A76/B1, A76/C1, A76/C2, A195/B1, A195/C1 and A195/C2 (Table 1), respectively, which were designed according to the conservative regions of the protein sequence of β-galactosidases. PCR was performed as follows: 95 °C for 4 min; followed by 30 cycles of 95 °C for 1 min, 50 °C for 1 min, and 72 °C for 2 min; with a final extension at 72 °C for 10 min. The amplified fragment was inserted into the pMD18-T vector and then transformed into *E. coli* DH5α for sequencing.

The 5′ and 3′ flanking regions of the known partial sequence were obtained by using DNA Walking *SpeedUp*™ Kit of Seegene, which adopted the thermal asymmetric interlaced PCR (TAIL-PCR) strategy [39]. The nested specific primers for upstream and downstream regions were designed based on the obtained partial sequence (Table 5). TAIL-PCR was performed as described by the kit. The amplified fragments were purified and ligated into the pMD18-T vector for sequencing. The upstream, downstream and obtained partial sequences were assembled to obtain a full-length *mabga* gene.

Table 5. Primers used for gene cloning.

Primers	Sequence (5′ to 3′)	Purpose
F1	GCNTGGGGNAAYGTNTTYT	Degenerated PCR
F2	TNTGGACNTGGGARGCNTT	Degenerated PCR
F3	GGARCARCARCCNGGNCCNGT	Degenerated PCR
R1	CCARCANGCRTCRTARTCRAA	Degenerated PCR
R2	RAANGCYTCCCANGTCCA	Degenerated PCR
R3	GGRTTRTGNGGNGCCARTT	Degenerated PCR
A208	TGGATHATGGAGGAGCCC	Degenerated PCR
A76	CGGGACCTGGTGCAYAAYTAY	Degenerated PCR
A195	CAYAAYTAYATGGGCTTCTTC	Degenerated PCR
B1	CAGACCCAGAACGAGTAYKGN	Degenerated PCR
C1	GCACCACAAGTACCACCARGA	Degenerated PCR
C2	GTYCTRDWNCTGCACCGGCCG	Degenerated PCR
U1	CCGTAAAGAATCCCATGAGT	DNA Walking (1st-round upstream)
D1	GGACATTTTGCGTGCG	DNA Walking (1st-round downstream)
U2	AACGCTGAAAGTCCAACCCGAT	DNA Walking (2nd-round upstream)
D2	GGACACTTATCCGCTGGGTTT	DNA Walking (2nd-round downstream)
U3	GATTGGCTTCGGTCACGGT	DNA Walking (3rd-round upstream)
D3	CCCGATTTTGGTGCTTTTCA	DNA Walking (3rd-round downstream)
DW-ACP 1	ACP-AGGTC	DNA Walking (1st-round)
DW-ACP 2	ACP-TGGTC	DNA Walking (1st-round)
DW-ACP 3	ACP-GGGTC	DNA Walking (1st-round)
DW-ACP 4	ACP-CGGTC	DNA Walking (1st-round)
DW-ACP N	ACPN-GGTC	DNA Walking (2nd-round)
Uni-primer	TCACAGAAGTATGCCAAGCGA	DNA Walking (3rd-round)
MaBGA-F	CGGAATTCAAGTTAGGTGTATGTTACTACCCAG	Full-length amplification
MaBGA-R	GTTCGCGCTCGAGGATTTCTTGCCAAATGGC	Full-length amplification

Homologous search in GenBank was performed using the BLAST server (http://www.ncbi.nlm.nih.gov/BLAST). Alignment of multiple protein sequences was conducted using the Clustal X 2.0 program [40] and rendered by ESPript [41]. A phylogenetic tree of multiple β-galactosidase was constructed using the neighbor-joining method [31] in MEGA6 [32], with a bootstrap test of 1000 replicates.

4.13. Homology Modelling and Molecular Docking Analysis

The three-dimensional model of MaBGA was constructed by using MODELLER [27], Robetta [26], I-TASSER [28] and SWISS-MODEL [25], respectively. Precise evaluation of the model quality was performed using ProSA-web [29] and PROCHECK [30]. To remove steric clashes, the constructed model was subject to an energy minimization process in vacuum by using the steepest descent method for about 5000 iterations in GROMACS 4.5 [33].

The refined model was used for docking with Galβ1–3GlcNAc, Galβ1–4GlcNAc and Galβ1–6GlcNAc, respectively, using Autodock 4.2 [34] with default parameters. The representation of the protein structure was achieved using the program PyMOL (The PyMOL Molecular Graphics System, Version 1.7 Schrödinger, LLC., New York, NY, USA).

5. Conclusions

In this work, a thermostable β-1,3-galactosidase MaBGA derived from *Marinomonas* sp. BSi20414, was first purified to homogeneity and characterized extensively. MaBGA displayed robust stability against heat and strict substrate specificity toward both the glycosyl group and the linkage type. Although further experiments are required to decipher its substrate recognition mechanism, our study provided an attractive alternative for biosynthesis of galacto-oligosaccharide with β1–3 linkage and laid the groundwork for the protein engineering to modify the linkage preference of β-galactosidase.

Acknowledgments: This study was supported by Chinese Polar Environment Comprehensive Investigation and Assessment Program (CHINARE-01-06, CHINARE-04-02, CHINARE-02-01), Open Fund of Key Laboratory of Biotechnology and Bioresources Utilization (Dalian Minzu University, KF2015009), State Ethnic Affairs Commission & Ministry of Education of China, and National Natural Science Foundation of China (31200599).

Author Contributions: Y.Y. and B.C. conceived and designed the experiments; Q.Z. and L.Z. performed the experiments; H.D., Q.Z. and L.Z. analyzed the data; Y.Y. and B.C. contributed reagents/materials/analysis tools; H.D. and B.C. wrote the paper.

Conflicts of Interest: The authors declare no conflict of interest.

References

1. Juers, D.H.; Matthews, B.W.; Huber, R.E. LacZ β-galactosidase: Structure and function of an enzyme of historical and molecular biological importance. *Protein Sci.* **2012**, *21*, 1792–1807. [CrossRef] [PubMed]
2. Oliveira, C.; Guimaraes, P.M.; Domingues, L. Recombinant microbial systems for improved β-galactosidase production and biotechnological applications. *Biotechnol. Adv.* **2011**, *29*, 600–609. [CrossRef] [PubMed]
3. Park, A.R.; Oh, D.K. Galacto-oligosaccharide production using microbial β-galactosidase: Current state and perspectives. *Appl. Microbiol. Biotechnol.* **2010**, *85*, 1279–1286. [CrossRef] [PubMed]
4. Henrissat, B.; Davies, G. Structural and sequence-based classification of glycoside hydrolases. *Curr. Opin. Struct. Biol.* **1997**, *7*, 637–644. [CrossRef]
5. Lombard, V.; Ramulu, H.G.; Drula, E.; Coutinho, P.M.; Henrissat, B. The carbohydrate-active enzymes database (cazy) in 2013. *Nucleic Acids Res.* **2014**, *42*, D490–D495. [CrossRef] [PubMed]
6. Davies, G.; Henrissat, B. Structures and mechanisms of glycosyl hydrolases. *Structure* **1995**, *3*, 853–859. [CrossRef]
7. Adam, A.C.; Rubio-Texeira, M.; Polaina, J. Lactose: The milk sugar from a biotechnological perspective. *BFSN* **2005**, *44*, 553–557. [CrossRef]
8. Singh, A.K.; Pluvinage, B.; Higgins, M.A.; Dalia, A.B.; Woodiga, S.A.; Flynn, M.; Lloyd, A.R.; Weiser, J.N.; Stubbs, K.A.; Boraston, A.B.; et al. Unravelling the multiple functions of the architecturally intricate streptococcus pneumoniae β-galactosidase, BgaA. *PLoS Pathog.* **2014**, *10*, e1004364. [CrossRef] [PubMed]
9. Hu, D.; Zhang, F.; Zhang, H.; Hao, L.; Gong, X.; Geng, M.; Cao, M.; Zheng, F.; Zhu, J.; Pan, X.; et al. The β-galactosidase (BgaC) of the zoonotic pathogen streptococcus suis is a surface protein without the involvement of bacterial virulence. *Sci. Rep.* **2014**, *4*, 4140. [CrossRef] [PubMed]
10. Cheng, W.; Wang, L.; Jiang, Y.L.; Bai, X.H.; Chu, J.; Li, Q.; Yu, G.; Liang, Q.L.; Zhou, C.Z.; Chen, Y. Structural insights into the substrate specificity of streptococcus pneumoniae β(1,3)-galactosidase BgaC. *J. Biol. Chem.* **2012**, *287*, 22910–22918. [CrossRef] [PubMed]
11. Solomon, H.V.; Tabachnikov, O.; Lansky, S.; Salama, R.; Feinberg, H.; Shoham, Y.; Shoham, G. Structure-function relationships in Gan42B, an intracellular GH42 β-galactosidase from geobacillus stearothermophilus. *Acta Crystallogr. Sect. D Biol. Crystallogr.* **2015**, *71*, 2433–2448. [CrossRef] [PubMed]
12. Solomon, H.V.; Tabachnikov, O.; Feinberg, H.; Govada, L.; Chayen, N.E.; Shoham, Y.; Shoham, G. Crystallization and preliminary crystallographic analysis of ganb, a GH42 intracellular β-galactosidase from geobacillus stearothermophilus. *Acta Crystallogr. Sect. F Struct. Biol. Cryst. Commun.* **2013**, *69*, 1114–1119. [CrossRef] [PubMed]
13. Kang, S.K.; Cho, K.K.; Ahn, J.K.; Bok, J.D.; Kang, S.H.; Woo, J.H.; Lee, H.G.; You, S.K.; Choi, Y.J. Three forms of thermostable lactose-hydrolase from *Thermus* sp. IB-21: Cloning, expression, and enzyme characterization. *J. Biotechnol.* **2005**, *116*, 337–346. [CrossRef] [PubMed]
14. Vian, A.; Carrascosa, A.V.; Garcia, J.L.; Cortes, E. Structure of the β-galactosidase gene from thermus sp. Strain T2: Expression in escherichia coli and purification in a single step of an active fusion protein. *Appl. Environ. Microbiol.* **1998**, *64*, 2187–2191. [PubMed]
15. Ohtsu, N.; Motoshima, H.; Goto, K.; Tsukasaki, F.; Matsuzawa, H. Thermostable β-galactosidase from an extreme thermophile, *Thermus* sp. A4: Enzyme purification and characterization, and gene cloning and sequencing. *Biosci. Biotechnol. Biochem.* **1998**, *62*, 1539–1545. [CrossRef] [PubMed]
16. Karan, R.; Capes, M.D.; DasSarma, P.; DasSarma, S. Cloning, overexpression, purification, and characterization of a polyextremophilic β-galactosidase from the antarctic haloarchaeon halorubrum lacusprofundi. *BMC Biotechnol.* **2013**, *13*, 3. [CrossRef] [PubMed]

17. Holmes, M.L.; Scopes, R.K.; Moritz, R.L.; Simpson, R.J.; Englert, C.; Pfeifer, F.; Dyall-Smith, M.L. Purification and analysis of an extremely halophilic β-galactosidase from haloferax alicantei. *Biochim. Biophys. Acta* **1997**, *1337*, 276–286. [CrossRef]

18. Maksimainen, M.; Paavilainen, S.; Hakulinen, N.; Rouvinen, J. Structural analysis, enzymatic characterization, and catalytic mechanisms of β-galactosidase from bacillus circulans sp. Alkalophilus. *FEBS J.* **2012**, *279*, 1788–1798. [CrossRef] [PubMed]

19. Ferrer, M.; Golyshina, O.; Beloqui, A.; Golyshin, P.N. Mining enzymes from extreme environments. *Curr. Opin. Microbiol.* **2007**, *10*, 207–214. [CrossRef] [PubMed]

20. Lee, H.S.; Kwon, K.K.; Kang, S.G.; Cha, S.-S.; Kim, S.-J.; Lee, J.-H. Approaches for novel enzyme discovery from marine environments. *Curr. Opin. Biotechnol.* **2010**, *21*, 353–357. [CrossRef] [PubMed]

21. Ferrer, M.; Beloqui, A.; Timmis, K.N.; Golyshin, P.N. Metagenomics for mining new genetic resources of microbial communities. *J. Mol. Microbiol. Biotechnol.* **2008**, *16*, 109–123. [CrossRef] [PubMed]

22. Zeng, Q.; Wang, Y.; Sun, K.; Yu, Y.; Chen, B. Preliminary studies on the screening, identification and optimum fermentative conditions of a strain *Marinomonas* sp. Bsi20414 isolated from arctic sea ice producing β-galactosidase. *Chin. J. Polar Res.* **2011**, 108–114.

23. Hidaka, M.; Fushinobu, S.; Ohtsu, N.; Motoshima, H.; Matsuzawa, H.; Shoun, H.; Wakagi, T. Trimeric crystal structure of the glycoside hydrolase family 42 β-galactosidase from thermus thermophilus a4 and the structure of its complex with galactose. *J. Mol. Biol.* **2002**, *322*, 79–91. [CrossRef]

24. Fridjonsson, O.; Watzlawick, H.; Gehweiler, A.; Rohrhirsch, T.; Mattes, R. Cloning of the gene encoding a novel thermostable α-galactosidase from thermus brockianus ITI360. *Appl. Environ. Microbiol.* **1999**, *65*, 3955–3963. [PubMed]

25. Biasini, M.; Bienert, S.; Waterhouse, A.; Arnold, K.; Studer, G.; Schmidt, T.; Kiefer, F.; Gallo Cassarino, T.; Bertoni, M.; Bordoli, L.; et al. Swiss-model: Modelling protein tertiary and quaternary structure using evolutionary information. *Nucleic Acids Res.* **2014**, *42*, W252–W258. [CrossRef] [PubMed]

26. Kim, D.E.; Chivian, D.; Baker, D. Protein structure prediction and analysis using the robetta server. *Nucleic Acids Res.* **2004**, *32*, W526–W531. [CrossRef] [PubMed]

27. Webb, B.; Sali, A. Comparative protein structure modeling using Modeller. *Curr. Protoc. Bioinform.* **2014**. [CrossRef]

28. Yang, J.; Yan, R.; Roy, A.; Xu, D.; Poisson, J.; Zhang, Y. The I-tasser suite: Protein structure and function prediction. *Nat. Methods* **2015**, *12*, 7–8. [CrossRef] [PubMed]

29. Wiederstein, M.; Sippl, M.J. Prosa-web: Interactive web service for the recognition of errors in three-dimensional structures of proteins. *Nucleic Acids Res.* **2007**, *35*, W407–W410. [CrossRef] [PubMed]

30. Laskowski, R.A.; MacArthur, M.W.; Moss, D.S.; Thornton, J.M. Procheck: A program to check the stereochemical quality of protein structures. *J. Appl. Crystallogr.* **1993**, *26*, 283–291. [CrossRef]

31. Saitou, N.; Nei, M. The neighbor-joining method: A new method for reconstructing phylogenetic trees. *Mol. Biol. Evol.* **1987**, *4*, 406–425. [PubMed]

32. Tamura, K.; Stecher, G.; Peterson, D.; Filipski, A.; Kumar, S. MEGA6: Molecular Evolutionary Genetics Analysis version 6.0. *Mol. Biol. Evol.* **2013**, *30*, 2725–2729. [CrossRef] [PubMed]

33. Pronk, S.; Páll, S.; Schulz, R.; Larsson, P.; Bjelkmar, P.; Apostolov, R.; Shirts, M.R.; Smith, J.C.; Kasson, P.M.; van der Spoel, D. Gromacs 4.5: A high-throughput and highly parallel open source molecular simulation toolkit. *Bioinformatics* **2013**, *29*, 845–854. [CrossRef] [PubMed]

34. Morris, G.M.; Huey, R.; Lindstrom, W.; Sanner, M.F.; Belew, R.K.; Goodsell, D.S.; Olson, A.J. Autodock4 and autodocktools4: Automated docking with selective receptor flexibility. *J. Comput. Chem.* **2009**, *30*, 2785–2791. [CrossRef] [PubMed]

35. Depeint, F.; Tzortzis, G.; Vulevic, J.; I'Anson, K.; Gibson, G.R. Prebiotic evaluation of a novel galactooligosaccharide mixture produced by the enzymatic activity of bifidobacterium bifidum ncimb 41171, in healthy humans: A randomized, double-blind, crossover, placebo-controlled intervention study. *Am. J. Clin. Nutr.* **2008**, *87*, 785–791. [PubMed]

36. Legler, G. Glycoside hydrolases: Mechanistic information from studies with reversible and irreversible inhibitors. *Adv. Carbohydr. Chem. Biochem.* **1990**, *48*, 319–384. [PubMed]

37. Bradford, M.M. A rapid and sensitive method for the quantitation of microgram quantities of protein utilizing the principle of protein-dye binding. *Anal. Biochem.* **1976**, *72*, 248–254. [CrossRef]

38. Laemmli, U.K. Cleavage of structural proteins during the assembly of the head of bacteriophage t4. *Nature* **1970**, *227*, 680–685. [CrossRef] [PubMed]

39. Liu, Y.G.; Whittier, R.F. Thermal asymmetric interlaced pcr: Automatable amplification and sequencing of insert end fragments from p1 and yac clones for chromosome walking. *Genomics* **1995**, *25*, 674–681. [CrossRef]

40. Larkin, M.A.; Blackshields, G.; Brown, N.P.; Chenna, R.; McGettigan, P.A.; McWilliam, H.; Valentin, F.; Wallace, I.M.; Wilm, A.; Lopez, R.; et al. Clustal W and clustal X version 2.0. *Bioinformatics* **2007**, *23*, 2947–2948. [CrossRef] [PubMed]

41. Gouet, P.; Robert, X.; Courcelle, E. Espript/endscript: Extracting and rendering sequence and 3d information from atomic structures of proteins. *Nucleic Acids Res.* **2003**, *31*, 3320–3323. [CrossRef] [PubMed]

marine drugs

MDPI

Article

Preparation of Antioxidant Peptides from Salmon Byproducts with Bacterial Extracellular Proteases

Ribang Wu [1,†], Leilei Chen [2,†], Dan Liu [1], Jiafeng Huang [1], Jiang Zhang [1], Xiao Xiao [1], Ming Lei [1], Yuelin Chen [1] and Hailun He [1,*]

[1] School of Life Science, State Key Laboratory of Medical Genetics, Central South University, Changsha 410013, China; ribang.wu@gmail.com (R.W.); liudan.forever@163.com (D.L.); 1608110217@csu.edu.cn (J.H.); zhangjiang915@163.com (J.Z.); 162511013@csu.edu.cn (X.X.); leiming@csu.edu.cn (M.L.); lumc96@hotmail.com (Y.C.)
[2] Institute of Agro-Food Science and Technology & Shandong Provincial Key Laboratory of Agro-Products Processing Technology, Shandong Academy of Agricultural Sciences, Jinan 250100, China; chenleilei8210@163.com
* Correspondence: helenhe@csu.edu.cn; Tel.: +86-731-8265-0230
† These authors contributed equally to this work.

Academic Editor: Se-Kwon Kim
Received: 26 September 2016; Accepted: 16 December 2016; Published: 11 January 2017

Abstract: Bacterial extracellular proteases from six strains of marine bacteria and seven strains of terrestrial bacteria were prepared through fermentation. Proteases were analyzed through substrate immersing zymography and used to hydrolyze the collagen and muscle proteins from a salmon skin byproduct, respectively. Collagen could be degraded much more easily than muscle protein, but it commonly showed weaker antioxidant capability. The hydrolysate of muscle proteins was prepared with crude enzymes from *Pseudoalteromonas* sp. SQN1 displayed the strongest activity of antioxidant in DPPH and hydroxyl radical scavenging assays (74.06% ± 1.14% and 69.71% ± 1.97%), but did not perform well in Fe^{2+} chelating assay. The antioxidant fractions were purified through ultrafiltration, cation exchange chromatography, and size exclusion chromatography gradually, and the final purified fraction U2-S2-I displayed strong activity of antioxidant in DPPH, hydroxyl radical scavenging assays (IC_{50} = 0.263 ± 0.018 mg/mL and 0.512 ± 0.055 mg/mL), and oxygen radical absorption capability assay (1.960 ± 0.381 mmol·TE/g). The final purified fraction U2-S2-I possessed the capability to protect plasmid DNA against the damage of hydroxyl radical and its effect was similar to that of the original hydrolysis product. It indicated that U2-S2-I might be the major active fraction of the hydrolysate. This study proved that bacterial extracellular proteases could be utilized in hydrolysis of a salmon byproduct. Compared with collagen, muscle proteins was an ideal material used as an enzymatic substrate to prepare antioxidant peptides.

Keywords: bacterial extracellular proteases; antioxidant peptide; enzymatic hydrolysis; peptide purification; evaluation of antioxidant activity

1. Introduction

Oxidative stress is an imbalance between oxidation and antioxidation, caused by external or internal stimulation. Reactive oxygen species (ROS), such as superoxide anion radicals ($O_2^{\bullet-}$), hydrogen peroxide (H_2O_2), hydroxyl radicals ($^{\bullet}OH$), and peroxyl radicals ($ROO^{\bullet}$), can damage DNA, proteins, and membrane systems, which is the significant nosogenesis of chronic diseases [1], including cancer [2], heart disease [3], and Alzheimer's [4]. In addition, free-radical-mediated lipid peroxidation can lead to food spoilage and the generation of potentially toxic products. Although an endogenous antioxidant system containing antioxidant enzymes (superoxide dismutase, glutathione S-transferase, and catalase) and antioxidant substrates (glutathine and victamin C) can scavenge ROS to protect cells in vivo, the excessive ROS that cannot be removed promptly still damage cells and tissues. In the food industry, synthetic antioxidants, such as butylated-hydroxytoluene (BHT), butylated-hydroxyanisole (BHA), tertbutyl-hydroquinone (TBHQ), and propyl gallate (PG), have been used during food processing, but existing potentially toxic effect on health. Antioxidant peptides are a series of oligopeptides with specific amino acid sequences, which can scavenge free radicals or inhibit the generation of ROS.

A lot of studies have reported that the enzymatic hydrolysis of protein is an effective way to prepare novel bioactive peptides, including antioxidant peptides. Numerous protein resources were found to be ideal materials of antioxidant peptides preparation, such as blacktip shark skin [5], blue mussel protein [6], cod protein [7], skate skin [8], oysters [9], and so on. Meanwhile, during fish processing, a great amount of fish byproducts, such as fish skin, bone frame, and fins, were processed into animal feed, which were not utilized comprehensively and therefore caused a severe waste of protein resources. Recent studies have reported that byproducts with abundant protein can be recycled and used to prepare bioactive peptides. Series of peptides can be released from the parental protein by enzymatic hydrolysis, and they may possess different kinds of biological activities, such as antioxidant, inhibition of angiotensin I-converting enzyme (ACE), antibiotic, anti-freezing, and so on [5,10]. Salmon is a valuable and popular edible fish worldwide [11]. Previous studies have reported that salmon byproducts could be hydrolyzed by various proteases to prepare bioactive peptides. Sathiel et al. extracted functional and nutritional peptides from salmon head hydrolysates with different enzymes [12]. Ahn et al. used six kinds of proteases to hydrolyze protein from the salmon pectoral fin, and found that peptic hydrolysates exhibited antioxidant and anti-inflammatory activities [13]. Most of the enzymes used in these researches are commercial proteases, such as pepsin, trypsin, chymotrypsin, papain, flavourzyme, and so on. However, non-commercial proteases were seldom reported in peptide preparation.

Recently, marine bacteria have become important sources for the selection of novel enzymes. The proteases secreted by marine bacteria play an important role in the decomposition of organic nitrogen in oceans, and therefore they have incomparable advantages in hydrolyzing marine-sourced protein [14]. Compared with land proteases, marine proteases usually have higher catalytic efficiency in hydrolyzing marine protein, such as fish skin, muscle, and bone frame [15,16]. The cleavage site of proteases from different bacteria also varies widely. With these enzymes, peptides obtained from protein hydrolysate may possess different amino acid sequences. It would be beneficial to discover more new bioactive peptides. This study aimed at investigating the effect of several extracellular proteases from bacteria on hydrolyzing the protein of salmon byproduct. The antioxidant activity of each hydrolysate and purified fraction was evaluated in order to verify the possibility of application.

2. Results and Discussion

2.1. Substrate-Immersing Zymography of Bacterial Extracellular Proteases

The bacterial fermentation products were obtained every 24 h for five days. Then the enzymatic activity and concentration of total protein were quantified with the Folin's phenol and BCA method to calculate specific activity and the time course of protease activity, respectively (Figure 1). It was obvious that *Pseualtermonas* sp. SQN1 produced a protease with much higher specific activity (347.27 ± 5.58 U/mg) compared with other tested bacteria from sea water. The enzymatic activity could become stable after 72 h fermentation, which was faster than *Vibrio* sp. SQS2-3 and *Vibrio* sp. SWN2. Bacteria from fresh water commonly display lower specific activity, such as *Bacillus* sp. MH12, *Aeromonas* sp. ZM3, and *Aeromonas* sp. ZM7. In addition, *Exiguobacterium* sp. MH2 and *Pseudomonas* sp. ZM9 could produce proteases in a shorter time, but the specific activity decreased rapidly after two and three days, respectively.

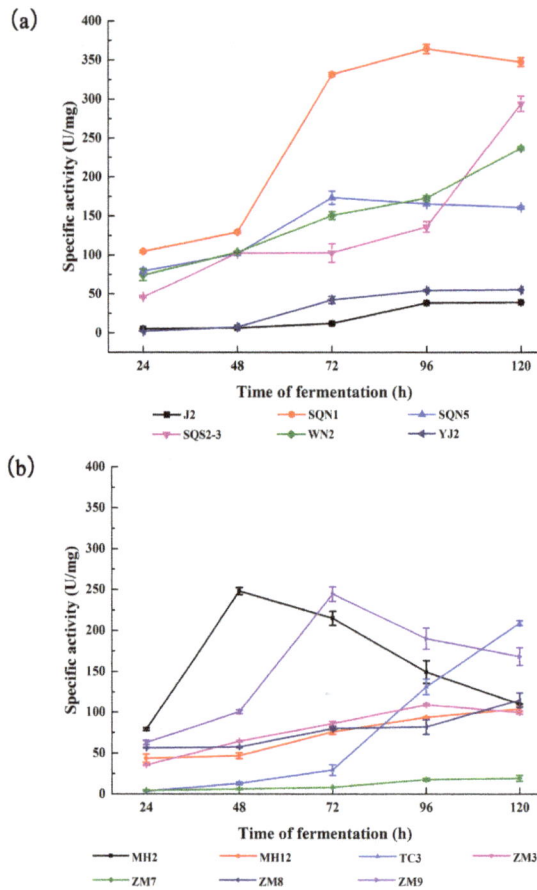

Figure 1. Time course specific activity of bacterial crude enzymes came from marine bacteria (**a**) and fresh water bacteria (**b**) during five-day fermentation. Values are displayed as means ± SD (*n* = 3).

The enzymatic activity and protease composition of each bacterial fermentation products after five days were detected through substrate immersing zymography. The result of zymography was summarized in Table 1, including the amount of proteases, the total protein concentration of crude enzyme, specific activity, and the molecular weight of each active band. As shown in Figure 2, crude enzymes from *Pseudoalteromonas* sp. J2, *Pseudoalteromonas* sp. SQN1, *Vibrio* sp. SQS2-3, and *Bacillus* sp. MH12 contained several proteases. Only one kind of protease was displayed in the lanes of *Photobacterium* sp. YJ2, *Bacillus* sp. TC3, and *Pseudomonas* sp. ZM9. The enzymes from *Aeromonas* sp. ZM3 and *Aeromona* sp. ZM7 showed the strongest enzymatic activity on casein. The crude enzyme from *Bacillus* sp. SQN5, *Vibrio* sp. SWN2, *Exiguobacterium* sp. MH2, and *Paenibacillus* sp. ZM8 did not display the ability of degrading casein in zymography. Theoretically, multi-proteases possess more enzymatic cleavage sites, which could hydrolyze protein more effectively. The antioxidant peptides were commonly separated from the hydrolysates with a relatively higher hydrolysis degree, so multi-proteases might be more suitable for antioxidant peptide preparation. Wang et al. reported that hydrolyzing collagen from croceine croaker with pepsin and trypsin could obtain peptides with higher antioxidant activity compared with preparation with pepsin or trypsin alone [17]. Meanwhile, other factors should be considered when choosing enzymes, such as enzymatic activity and substrate binding capability. Higher enzymatic activity and substrate binding capability could shorten the reaction time and release antioxidant peptides more easily. On the other hand, some crude proteases contain a lot of extracellular polysaccharide, which makes the samples sticky and impedes the migration of proteases in gel, just like the enzymes from *Aeromonas* sp. ZM3 and *Aeromonas* sp. ZM7. The byproducts of fermentation would increase cost and difficulty in purification, so these two enzymes were not selected. Zymography analysis of bacterial proteases was necessary and helpful to choose enzymes by screening before the preparation of antioxidant peptides. With zymography analysis, the enzymatic activity, the amount of proteases, purification difficulty, or even the type of each protease could be forecasted to a certain degree.

Table 1. Summary of crude enzymes from bacterial fermentation.

Bacteria Strain	Concentration of Total Protein (mg/mL)	Specific Activity (U/mg)	Amount of Proteases	Molecular Weight (kD)
Bacteria from sea water				
Pseudoalteromonas sp. J2	2.63	39.06	3	90, 40, 30~40
Pseudoalteromonas sp. SQN1	1.26	347.27	2	90~120, 40
Bacillus sp. SQN5,	1.98	161.06		
Vibrio sp. SQS2-3	1.18	293.95	3	60, 40, 30
Vibrio sp. SWN2	1.65	236.92		
Photobacterium sp. YJ2	2.00	54.93	1	14
Bacteria from fresh water				
Bacillus sp. TC3	1.03	110.04	1	<14
Exiguobacterium sp. MH2	2.20	103.67		
Bacillus sp. MH12	3.84	208.78	2	<14
Aeromonas sp. ZM3	5.35	99.47	Proteases cannot be separated	
Aeromonas sp. ZM7	5.26	19.25	Proteases cannot be separated	
Paenibacillus sp. ZM8	1.02	114.68		
Pseudomonas sp. ZM9	1.06	168.12	1	30~40

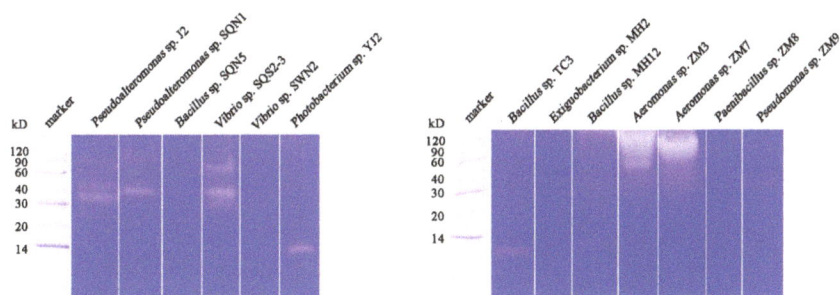

Figure 2. Substrate immersing zymography of different bacterial extracellular proteases.

2.2. Hydrolysis of Salmon Protein

The collagen and muscle proteins from salmon byproducts were both hydrolyzed with different kinds of bacterial proteases, respectively. The hydrolysis results of 5 min were analyzed by sodium dodecyl sulfate polyacrylamide gel electrophoresis (SDS-PAGE) (Figure 3) and the rate of hydrolysis of each bacterial crude enzyme was also measured (Figure 4). The antioxidant activities of each hydrolysis product after 30 min were measured. As depicted in Figure 3, collagen obtained from salmon skin through hot water treatment consisted of proteins of similar molecular weight. The triple α-helical structure of marine-sourced collagen was unstable at high temperature; it could convert from tight form to relaxed form, and part of the collagen fiber would be broken during hot water treatment. That might be the reason why the salmon collagen extracted in hot water did not display as several specific bands. It could be observed that crude enzymes from *Pseudoalteromonas* sp. J2, *Pseudoalteromonas* sp. SQN1, *Vibrio* sp. SQS2-3, *Photobacterium* sp. YJ2, *Bacillus* sp. TC3, *Bacillus* sp. MH12, *Aeromonas* sp. ZM3, and *Aeromonas* sp. ZM7 could hydrolyze collagen into smaller pieces more effectively. Compared with the results of substrate-immersing zymography, those crude enzymes that degraded collagen quickly commonly contained several proteases or formed brilliant bands in gel, which indicated high enzymatic activity. The antioxidant activity of each group was measured through DPPH, hydroxyl radical scavenging assays, and ferrous ion chelating assay (Figure 3). The products with a higher hydrolysis degree tend to display strong activity in DPPH and hydroxyl radical scavenging assays, such as J2-C (47.77% $\pm$ 1.78% in DPPH and 32.75% $\pm$ 3.49 in $^{\bullet}$OH), SQS2-3-C (44.08% $\pm$ 1.77% in DPPH and 25.67% $\pm$ 3.38% in $^{\bullet}$OH), and ZM3-C (42.36% $\pm$ 1.61% in DPPH and 44.88% $\pm$ 1.70% in $^{\bullet}$OH). These results indicated that those antioxidant peptides scavenging free radicals directly were closely related to the degree of hydrolysis or molecular weight. Peptides with small molecular weight could react with free radicals more easily and displayed stronger antioxidant activity [17]. However, collagen hydrolysates with higher hydrolysis degree did not show significant activity in Fe^{2+} chelating assay. Contrarily, the MH2-C and ZM8-C groups with lower hydrolysis degree displayed relatively stronger Fe^{2+} chelating activity (21.75% $\pm$ 2.87% and 20.96% $\pm$ 2.44%).

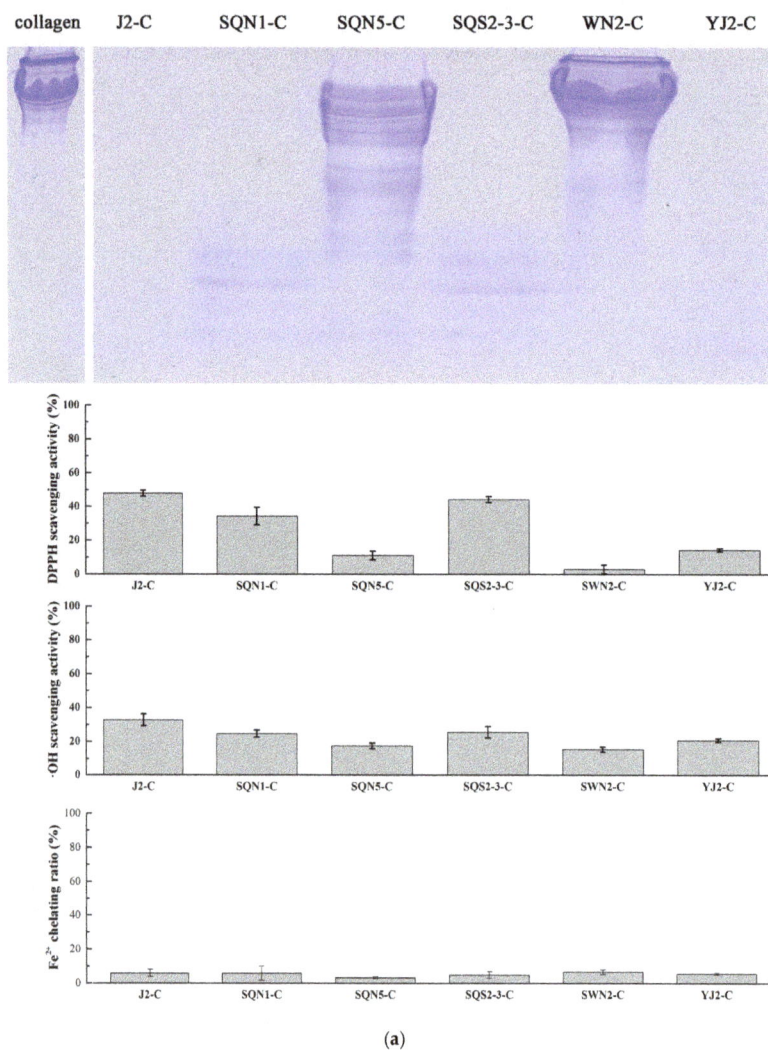

(a)

Figure 3. *Cont.*

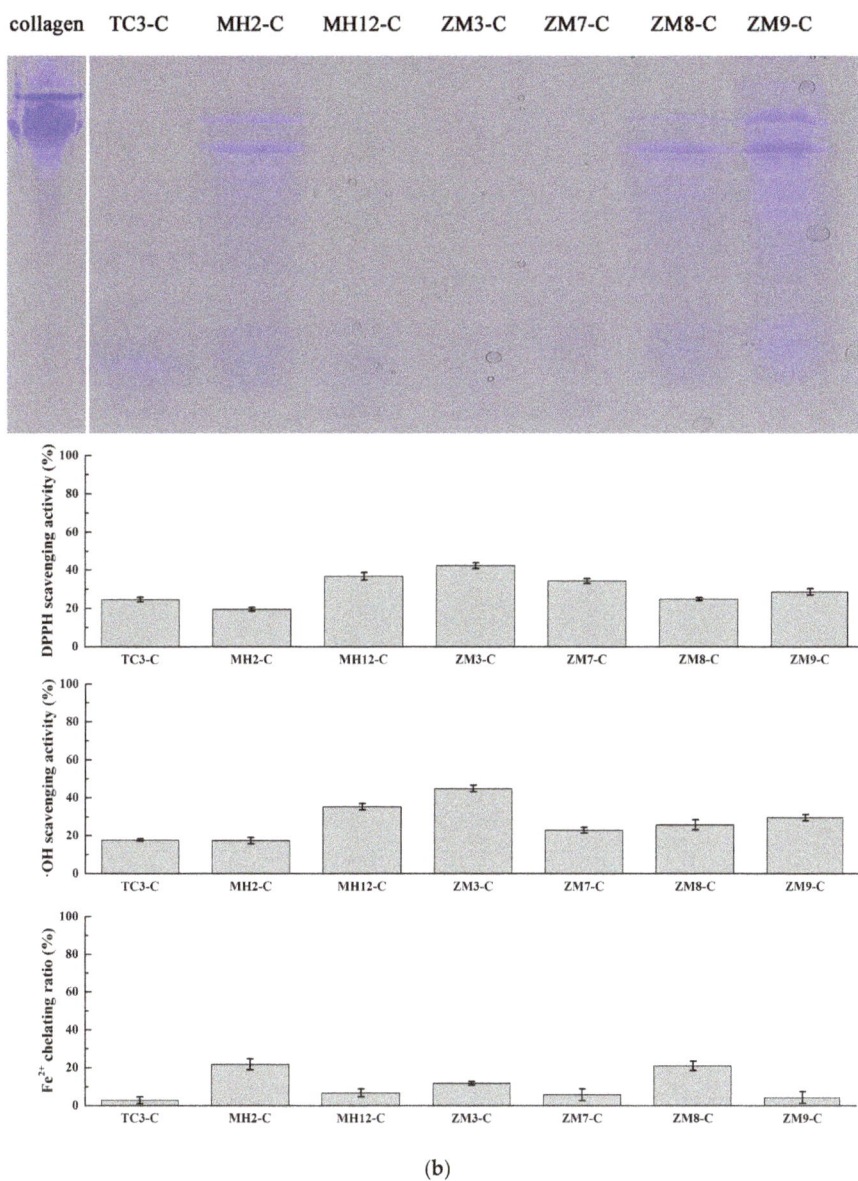

Figure 3. *Cont.*

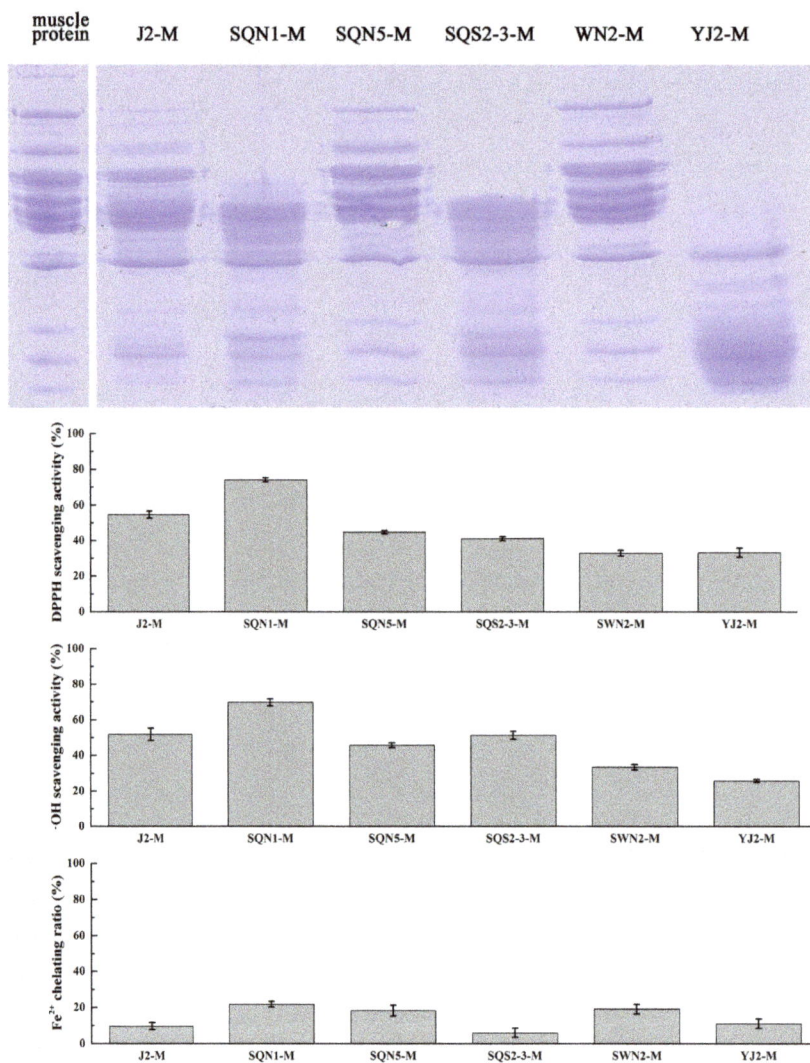

Figure 3. *Cont.*

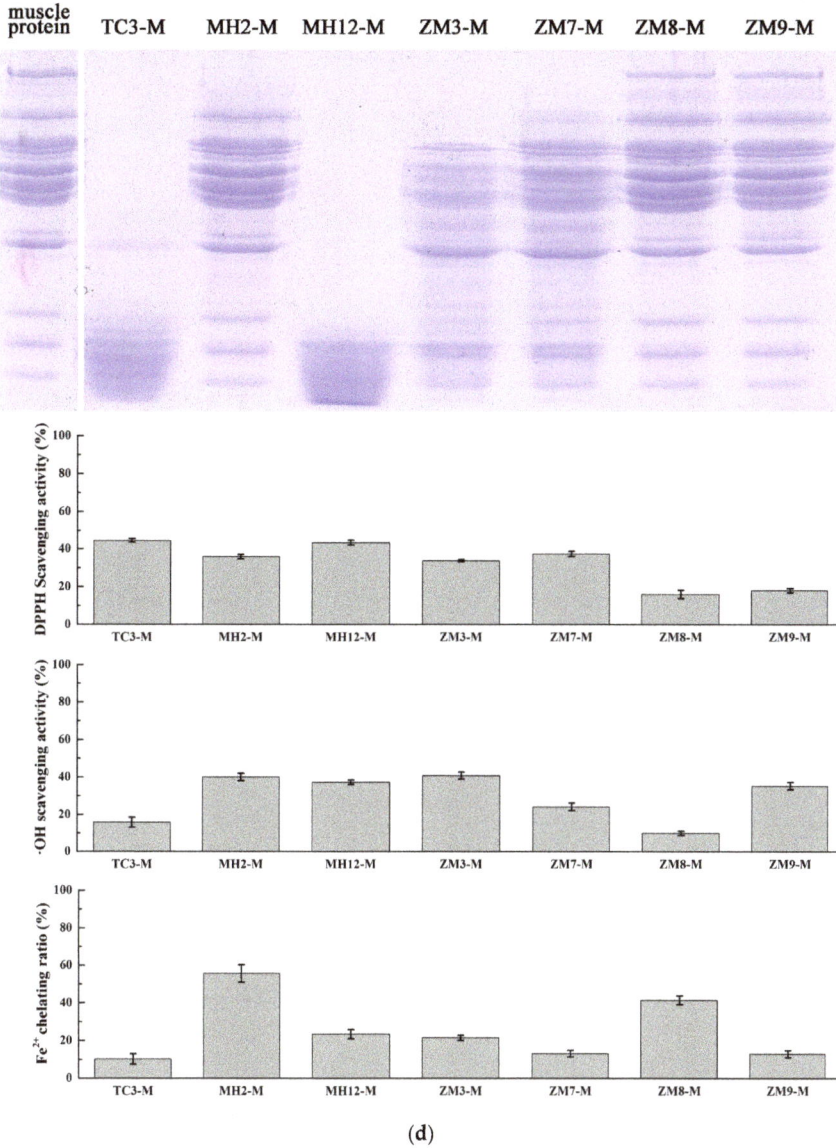

Figure 3. Hydrolysis results of (**a**) salmon collagen with sea water bacterial proteases; (**b**) salmon collagen with fresh water bacterial proteases; (**c**) salmon muscle protein with sea water bacterial proteases; and (**d**) salmon muscle proteins with fresh water bacterial proteases in SDS-PAGE, and the antioxidant activity measured with DPPH, hydroxyl radical scavenging assays, and ferrous ion chelating assay (*n* = 3).

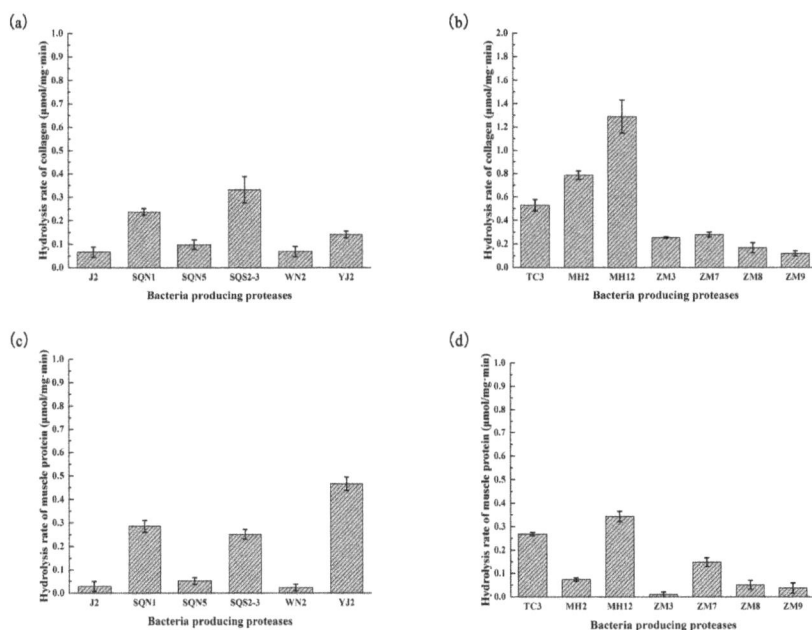

Figure 4. Hydrolysis rate of (**a**) marine bacterial proteases towards collagen; (**b**) fresh water bacterial proteases towards collagen; (**c**) marine bacterial proteases towards muscle protein; and (**d**) fresh water bacterial proteases towards muscle protein. Values were displayed as means ± SD (*n* = 3).

Previous studies commonly hydrolyzed collagen with commercial proteases. For example, Wang et al. used trypsin and pepsin to degrade collagen from a croceine croaker for 4 h, and obtained products with hydroxyl radical scavenging activity (53.11% ± 0.97% and 44.96% ± 1.97%, at a concentration of 10 mg/mL, respectively) [18]. Yang et al. reported that bromelain or papain could finish the digestion of retorted gelatin from cobia skin within 0.5 h, producing antioxidant fractions, while pancreatin or trypsin needed at least 2 h [19]. Mendis et al. obtained antioxidant peptides from the skin gelatin of jumbo squid with trypsin for 4 h [20]. Compared with the crude enzymes used in this study, most of the commercial proteases took longer to produce antioxidant peptides. It was common that commercial proteases were single-enzyme, while crude enzymes obtained from bacterial fermentation were multi-enzymes. Multi-proteases possessed more cleavage sites. More potential bioactive peptides would be released and the reaction time could be shortened.

The protein component from salmon muscle was more complicated than collagen, which contained a series of proteins with different molecular weights (Figure 3). In addition, it was much more difficult for enzymes to hydrolyze muscle proteins compared with collagen. In the first five minutes, *Bacillus* sp. TC3 and *Bacillus* sp. MH12 could degrade most of the protein from salmon muscle. *Pseudoalteromonas* sp. SQN1, *Vibrio* sp. SQS2-3, *Photobacterium* sp. YJ2, *Bacillus* sp. MH12, *Aeromonas* sp. ZM3, and *Aeromonas* sp. ZM7 could hydrolyze part of muscle protein. In DPPH and hydroxyl radical scavenging assays, the antioxidant activity of muscle hydrolysate was much stronger than that of collagen hydrolysate. The product of muscle hydrolyzed with enzyme from *Pseudoalteromonas* sp. SQN1 showed the strongest activity after 30 min hydrolysis (74.06% ± 1.14% in DPPH and 69.71% ± 1.97% in $\bullet$OH). Compared with the Fe^{2+} chelating assay results of collagen hydrolysates, muscle protein hydrolysates generally displayed stronger activity, especially the MH2-M and ZM-8 groups, which exhibited significant activity (55.52% ± 4.51% and 41.42% ± 2.29%, respectively). Similar to collagen hydrolysates, those muscle protein hydrolysates with a lower hydrolysis degree displayed higher Fe^{2+} chelating activity. It was possible that a suitable hydrolysis degree was an important factor in preparing peptides with better ion chelating activity. It was also reported that the amino acid residues of source protein could affect the antioxidant activity of peptides [17]. Nazeer et al. used gastrointestinal digestive enzymes to hydrolyze croaker (*Otolithes ruber*) muscle proteins and prepared a peptide Lys-Thr-Phe-Cys-Gly-Arg-His with strong DPPH and hydroxyl radical scavenging activity (84.5% $\pm$ 1.2% and 62.4% $\pm$ 2.9%) [21]. Chi et al. used trypsin to hydrolyze monkfish muscle proteins and prepared three peptides Glu-Trp-Pro-Ala-Gln, Phe-Leu-His-Arg-Pro, and Leu-Met-Gly-Gln-Trp. All of these peptides displayed strong activities in DPPH (EC_{50} 2.408, 3.751, and 1.399 mg/mL), hydroxyl radical (EC_{50} 0.269, 0.114, and 0.040 mg/mL), and superoxide anion radical (EC_{50} 0.624, 0.101, and 0.042 mg/mL) scavenging assays [20]. Most of these reported antioxidant peptides contained specific amino acid, such as cysteine, tyrosine, histidine, and so on, and they made a great contribution to remove free radical or chelate oxidation-related ions [17]. The Gly-X-Y repeating sequence made the amino acid composition of collagen simple and rich in glycine and proline, which do not possess strong active sites against free radicals. The amino acid composition of muscle proteins was more complicated than that of collagen, which may exist more potential antioxidant peptide sequences. In addition, the muscle proteins were much more stable than collagen. Extracted with homogenization, the structure of the muscle proteins could be kept intact, which ensures that the activity of the products was very similar even though they were prepared in different batches. Therefore the muscle was more suitable than collagen to be used in antioxidant peptide preparation.

2.3. Optimization of Hydrolysis Condition

A single factor analysis towards the hydrolysis of muscle protein with enzyme from *Pseudoalteromonas* sp. SQN1 was carried out, including time of hydrolysis, temperature of hydrolysis, and the ratio of [E]/[S] (Figure 5). Then the time, temperature, and ratio of [E]/[S] were selected to be 25 min, 45 °C, and 1:50 (g/g). The central composite design was designed with Design Expert 8.0. The experimental conditions and DPPH scavenging activity are listed in Table 2. Variance analysis of linear model with ANOVA is displayed in Table 3, which indicates that the ratio of [E]/[S] was a significant factor influencing the DPPH scavenging activity of muscle protein hydrolysate.

Table 2. Optimization of muscle protein hydrolysis condition through central composite design (CCD).

Group	Time (min)	Temperature (°C)	Ratio of [E]/[S] (g/g)	DPPH Scavenging Activity (%)
1	29.2	45	1:50	68.13
2	25	45	1:50	68.70
3	25	45	1:50	69.27
4	22.5	47.5	1.5:50	71.19
5	20.8	45	1:50	72.91
6	25	45	1:50	70.80
7	25	40.8	1:50	69.46
8	27.5	42.5	0.5:50	70.99
9	25	45	1.8:50	74.44
10	27.5	42.5	1.5:50	76.54
11	25	49.2	1:50	69.66
12	27.5	47.5	0.5:50	67.93
13	25	45	1:50	73.67
14	25	45	1:50	71.95
15	25	45	1:50	73.48
16	22.5	42.5	1.5:50	73.29
17	22.5	47.5	0.5:50	69.08
18	22.5	42.5	0.5:50	70.42
19	27.5	47.5	1.5:50	70.80
20	25	45	0.2:50	70.43

Table 3. Variance analysis of linear model with ANOVA.

Source	Sum of Square	df	Mean Square	F Value	p-Value (Prob > F)
Model	0.390×10^{-3}	3	1.230×10^{-3}	6.53	0.0043
A-Time	9.880×10^{-5}	1	9.880×10^{-5}	0.52	0.4793
B-Temperature	3.781×10^{-4}	1	3.781×10^{-4}	2.01	0.1756
C-Ratio of [E]/[S]	3.213×10^{-3}	1	3.213×10^{-3}	17.07	0.0008
Residual	3.013×10^{-3}	16	1.883×10^{-4}		
Lack of Fit	2.197×10^{-3}	11	1.998×10^{-4}	1.23	0.4375
Pure Error	8.152×10^{-4}	5	1.630×10^{-4}		
Cor Total	6.703×10^{-3}	19			

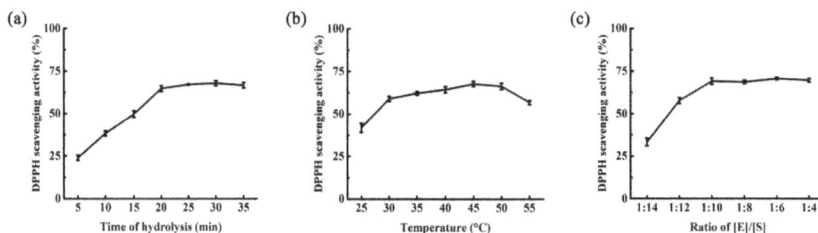

Figure 5. Single-factor analysis of muscle protein hydrolysis with SQN1: (**a**) incubated at 50 °C with a ratio of [E]/[S] in 1:10 for 5, 10, 15, 20, 25, 30, and 35 min; (**b**) incubated at 25, 30, 35, 40, 45, 50, and 55 °C with a ratio of [E]/[S] in 1:10 for 30 min; (**c**) incubated at 50 °C for 30 min with a ratio of [E]/[S] in 1:4, 1:6, 1:8, 1:10, 1:12, and 1:14. Values were displayed as means ± SD (*n* = 3).

2.4. Purification of Antioxidant Peptides from Hydrolysate of Muscle Proteins

Since the crude enzyme from *Pseudoalteromonas* sp. SQN1 could hydrolyze salmon muscle proteins to release peptides with strong antioxidant activity, the active fractions were further purified with ultrafiltration, cation exchange chromatography, and size exclusion chromatography gradually. Ultrafiltration tubes with 3 kDa molecular weight cutoff (MWCO) were selected to separate small peptides after 30 min hydrolysis. As shown in Table 4, the DPPH and hydroxyl radical scavenging activity of muscle hydrolysis product (IC_{50} 0.721 ± 0.024 mg/mL and 1.371 ± 0.178 mg/mL) was higher than the activity of smaller fraction U2 (IC_{50} 0.377 ± 0.013 mg/mL and 0.882 ± 0.127 mg/mL) but lower than U1, which was the fraction with a larger size (IC_{50} 0.972 ± 0.031 mg/mL and 1.495 ± 0.214 mg/mL). This indicated that peptides with smaller size were the major active fraction in this product, and therefore U2 was chosen to be further purified with cation exchange chromatography. As shown in Figure 6a, U2 was separated into three major fraction peaks, which were collected, lyophilized, and detected. The first eluted peak (U2-S2) showed the strongest antioxidant activity (IC_{50} 0.289 ± 0.022 mg/mL and 0.681 ± 0.078 in DPPH and $^{\bullet}$OH). This fraction was further purified with a Sephadex G-15 size exclusion column (Uppsala, Sweden), and two fractions (U2-S2-I and U2-S2-II) were obtained (Figure 6b). U2-S2-I accounted for 99.03% of U2-S2 according to the integral area calculation of Bio-Rad ChromLab software (Hercules, CA, USA). Furthermore, this fraction also displayed similar DPPH and hydroxyl radical scavenging activity (IC_{50} 0.263 ± 0.018 mg/mL and 0.512 ± 0.055 mg/mL) compared with U2-S2. This result showed that U2-S2-I was the major active fraction. As shown in Table 5, the fraction U2-S2-I displayed higher DPPH scavenging activity compared with other reported antioxidant peptides purified from different muscle protein hydrolysate. This indicated that salmon muscle hydrolyzed by protease from *Pseualtermonas* sp. SQN1 would be a feasible method to prepare antioxidant peptides.

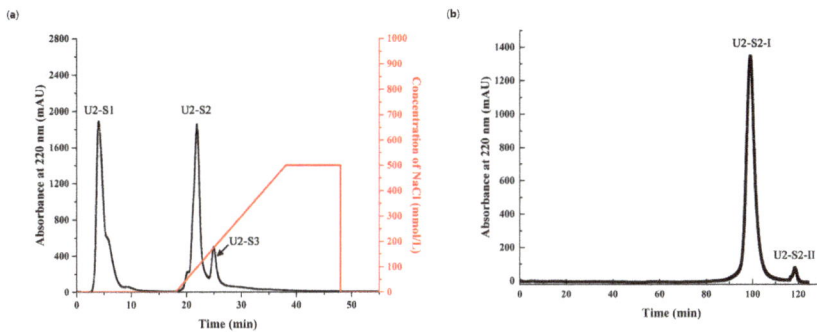

Figure 6. Purification of antioxidant fractions with fast protein liquid chromatography on (**a**) Macro-Prep High S column and (**b**) Sephadex G-15 column.

Table 4. DPPH scavenging activity of purified fractions in each step.

Preparation Step	Fractions	IC$_{50}$ Value (mg/mL)		Yield (%)
		DPPH	•OH	
Enzymatic hydrolysis	Hydrolysate	0.721 ± 0.024	1.371 ± 0.178	100
Ultrafiltration	U1	0.972 ± 0.151	1.495 ± 0.214	35.18
	U2	0.377 ± 0.013	0.882 ± 0.127	
Cation exchange chromatography	U2-S1	1.781 ± 0.048	1.689 ± 0.118	12.81
	U2-S2	0.289 ± 0.022	0.681 ± 0.078	
	U2-S3	0.972 ± 0.053	0.920 ± 0.093	
Size exclusion chromatography	U2-S2-I	0.263 ± 0.018	0.512 ± 0.055	12.68
	U2-S2-II	4.832 ± 0.552	3.191 ± 0.323	

Table 5. Comparing with reported antioxidant peptides sourced from muscle protein.

Source	Enzyme	Antioxidant Activity (DPPH)	Reference
Salmon muscle	Protease from *Pseudoalteromonas* sp. SQN1	IC$_{50}$ = 0.51 mg/mL	
Scorpaena notata muscle	neutral serine protease	IC$_{50}$ = 0.60 mg/mL	[22]
Croceine croaker muscle	pepsin and alcalase	IC$_{50}$ = 1.35 mg/mL	[23]
Monkfish muscle	trypsin	IC$_{50}$ = 1.40 mg/mL	[24]
Smooth hound muscle	gastrointestinal proteases	IC$_{50}$ = 0.60 mg/mL	[25]
Sphyrna ewini muscle	papain	IC$_{50}$ = 3.06 mg/mL	[26]

2.5. Oxygen Radical Absorption Capability (ORAC) Assay

Oxygen radical absorption capability assay was used to detect the antioxidant activity against peroxyl radicals [27]. Peroxyl radical was considered to be the major free radical generated during the auto-oxidation process of lipid and fatty acid. Peptides with strong absorption capability against peroxyl radical could be a potential antioxidant additive in the food processing industry. The activities of two fractions separated in size exclusion chromatography were detected with this method. The decay speed of the fluorescence curve reflected the reaction speed of the peptides and the area under the curve reflected the quantity of the peroxyl radical removed by antioxidant. As shown in Figure 7, U2-S2-I displayed its effect in decreasing the decay of fluorescence and its antioxidant activity (1.960 ± 0.381 mmol·TE/g) was much stronger than U2-S2-II (0.344 ± 0.079 mmol·TE/g). Antioxidant-donating hydrogen could block the radical chain reaction caused by peroxyl radical in ORAC assay. The result indicated that U2-S2-I might contain active peptides working as hydrogen donors. Specific amino acid residues, such as cysteine, tyrosine, and histidine, could provide hydrogen for free radicals from the sulfydryl group (-SH), the phenolic hydroxyl group, and the iminazole circle, respectively. Meanwhile, specific amino acid residues could also form a stable structure to stop the radical chain reaction.

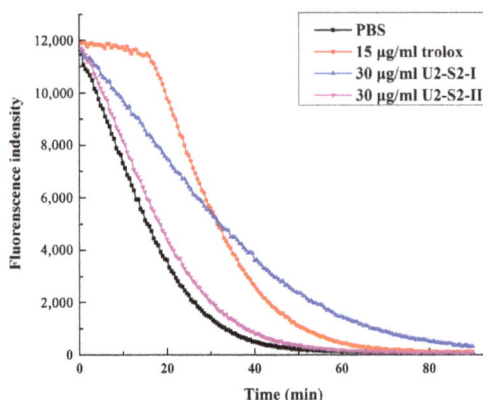

Figure 7. Oxygen radical absorption capability of U2-S2-I and U2-S2-II compared with PBS control group and trolox-positive group.

2.6. DNA Protection Effect against Oxidation-Induced Damage

Hydroxyl radicals are known for causing oxidative breaks in DNA strands. The DNA protection effect of U2-S2-I was examined using plasmid DNA in vitro compared with the initial hydrolysate. The results in Figure 8 showed that the concentration of supercoil DNA significantly decreased, and the open circle DNA appeared in damage group. When the plasmid DNA was exposed to U2-S2-I or initial hydrolysate, the concentration of supercoil DNA was still high. This indicated that both the hydrolysis product and the final purified fractions have an antioxidant effect. In addition, the antioxidant effect of these two groups was similar, which also indicates that the final fraction U2-S2-I might be the major active fraction of this hydrolysis product. DNA damage is a typical phenomenon of cytopathy caused by oxidative stress in vivo. Peptides with a DNA protection effect against oxidation could be further developed as a functional supplement to prevent diseases related to oxidation. Sheih et al. prepared an antioxidant peptide with DNA protection effect from algae protein hydrolysate, and the peptide could increase the viability of AGS cells [28]. Karawita et al. found that the enzymatic extracts of microalgae could effectively inhibit DNA damage and repair H_2O_2-induced DNA damage in mouse lymphoma L5178 cells [29].

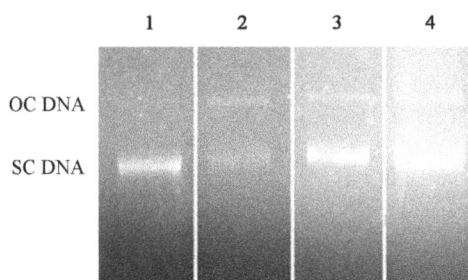

Figure 8. DNA protection effect of U2-S2-I against oxidation-induced damage. Lane 1: plasmid DNA pET-22b without oxidation damage; Lane 2: plasmid DNA pET-22b without antioxidant was attacked by hydroxyl radical; Lane 3: plasmid DNA pET-22b was protected by U2-S2-I against the attack of hydroxyl radical; Lane 4: plasmid DNA pET-22b was protected by SQN1-M hydrolysate against the attack of hydroxyl radical.

3. Experimental Section

3.1. Materials

Fresh salmon skin with muscle was purchased from a seafood market in Shanghai, China, and was stored at -20 °C prior to use. The soybean meal, corn powder, and wheat bran were purchased from a supermarket in Changsha, Hunan province, China. Tryptone and yeast extraction were purchased from Thermo Fisher Oxoid (Basingstoke, Hamshire, UK). Sephadex G-15 size exclusion gel was purchased from GE Healthcare Life Sciences (Uppsala, Sweden). The other regents used are commercially available.

Pseudoalteromonas sp. J2, *Pseudoalteromonas* sp. SQN1, *Bacillus* sp. SQN5, *Vibrio* sp. SQS2-3, *Vibrio* sp. SWN2, and *Photobacterium* sp. YJ2 were from the inshore environment of the South China Sea. *Bacillus* sp. TC3, *Exiguobacterium* sp. MH2, *Bacillus* sp. MH12, *Aeromonas* sp. ZM3, *Aeromonas* sp. ZM7, *Paenibacillus* sp. ZM8, and *Pseudomonas* sp. ZM9 were from the lakes on the Yungui plateau.

3.2. Preparation of Bacterial Extracellular Proteases

3.2.1. Preparation of Bacterial Proteases from Fermentation

The method of proteases preparation was modified according to Liu's study [30]. The protease-producing bacteria were activated in a 2216E medium with shaking at 200 rpm and 18 °C. When the OD_{600} value reached 0.6, the bacteria were incubated in a fermentation broth (0.5% corn powder, 0.5% bean powder, 0.25% wheat bran, 0.1% $CaCl_2$, 0.4% Na_2HPO_4, and 0.03% KH_2PO_4, prepared with sea water) [31] and fermented at 200 rpm and 18 °C for 5fivedays. The supernatant was centrifuged at $12,000 \times g$ and 4 °C for 30 min to collect the crude proteases.

3.2.2. Detection of Proteases with Substrate-Immersing Zymography

Substrate immersing zymography was modified to detect the proteolytic activities according to the method developed by Liu [30]. Crude proteases (16 µL) were loaded in a 12.5% SDS-PAGE gel with constant voltage at 100 V for 10 min and 160 V for 50 min in proper order. The gel was washed three times with 2.5% Triton X-100 for 15 min to remove SDS after electrophoresis. Then the gel was washed with 50 mM Tris-HCl (pH 8.0) and immersed in 0.1% casein at 37 °C for 60 min. Subsequently, the gel was stained with 0.1% Coomassie Brilliant Blue R-250 for 3 h, and destained by ethanol/acetic acid/H_2O (2:1:7) mixture with shaking until the bands of proteolytic activity became visible.

3.3. Protein Hydrolysis of Salmon Byproducts Using Bacterial Extracellular Proteases

The salmon byproducts consisted of fish skin and subcutaneous muscle, which were separated and pretreated in different ways. The total protein concentration of crude enzymes from different bacteria was diluted to 1 mg/mL.

3.3.1. Hydrolysis of Salmon Collagen Using Bacterial Extracellular Proteases

Fish skin was cut into pieces and washed with cold distilled water three times. Then 5 g fish skin was cooked in 50 mL distilled water at 75 °C for 30 min. The supernatant was collected by centrifugation at 12,000× *g* for 15 min. Salmon skin collagen was lyophilized for 24 h until the protein changed into a solid. Then the collagen was dissolved into ddH$_2$O to the concentration of 10 mg/mL, then mixed with different kinds of bacterial extracellular proteases at an enzyme/substrate ratio of 1:90 (*g*/*g*), and the mixtures were incubated at 50 °C. After 5 min, 20 μL of hydrolysates were sampled and inactivated at 90 °C for 10 min. The hydrolysis of collagen was detected by 12.5% SDS-PAGE. The hydrolysis rates of collagen with different bacterial crude enzyme were quantified as follows:

$$\text{Rate of hydrolysis (μmol/min·g)} = (c_t - c_0) \times 0.18 \text{ mL}/(5 \text{ min} \times 1 \text{ mg/mL} \times 0.02 \text{ mL}),$$

where c_0 and c_t were defined as the concentration of peptides product before and after hydrolysis, respectively. After 25 min, the mixtures of hydrolysis were inactivated at 90 °C for 10 min.

3.3.2. Hydrolysis of Salmon Muscle Using Bacterial Extracellular Proteases

The salmon muscle was homogenized at a speed of 10,000 rpm for 2 min in distilled water. Then the supernatant was collected by centrifugation at 13,000× *g* for 30 min. The concentration of muscle proteins was quantified and diluted to 5 mg/mL with BCA method. Then the muscle protein was mixed with different proteases at an enzyme/substrate ratio of 1:45 (*g*/*g*), and the mixtures were incubated at 50 °C. After 5 min, 20 μL of hydrolysates were sampled and inactivated at 90 °C for 10 min. The hydrolysis of collagen was detected by 12.5% SDS-PAGE. The hydrolysis rates of muscle protein with different bacterial crude enzyme were quantified using the method described in Section 3.3.1. After 25 min, the mixtures of hydrolysis were inactivated at 90 °C for 10 min.

3.4. Isolation of Antioxidant Peptides from Muscle Hydrolysate

3.4.1. Isolation of Muscle Hydrolysate by Ultrafiltration

After being hydrolyzed by proteases from *Pseudoalteromonas* sp. SQN1, the hydrolysate of salmon muscle proteins was centrifuged in ultrafiltration tube with 3 kDa molecular weight cutoff (Millipore, Temecula, CA, USA) at 5000× *g* for 30 min to isolate fractions above 3 kDa (MUF-1) and below 3 kDa (MUF-2). The antioxidant activities of these two fractions were detected by hydroxyl radical scavenging assay.

3.4.2. Cation Exchange Chromatography

The active fraction MUF-2 was further purified in a Bio-Rad Macro-Prep High S column (5 × 1.26 cm) with NGC chromatography system (Bio-Rad, Hercules, CA, USA). The column was equilibrated with distilled water. Then 5 mL MUF-2 were loaded into the pre-equilibrated column at a flow rate of 1.5 mL/min, and eluted with distilled water for 18 min at a flow rate of 1.5 mL/min. Then the column was eluted with a linear gradient of 1 M NaCl (0%–50%) at a flow rate of 1.5 mL/min for 20 min. Peptide fractions were monitored at 220 nm and collected at a volume of 5 mL. All the fractions were lyophilized.

3.4.3. Size Exclusion Chromatography

After cation exchange chromatography, the fraction with the strongest activity was further purified by Sephadex G-15 size exclusion column (1.6 × 80 cm) with NGC chromatography system (Bio-Rad). Fraction was dissolved in 1 mL distilled water and loaded into a column pre-equilibrated with distilled water. The column was eluted with distilled water at a flow rate of 0.75 mL/min. Peptide fractions were monitored at 220 nm and collected at a volume of 5 mL. All the fractions were lyophilized and the antioxidant activities were evaluated by DPPH radical scavenging assay.

3.5. Evaluation of Antioxidant Activity

3.5.1. DPPH Radical Scavenging Activity

The DPPH radicals scavenging activity assay was measured according to the method of Shimada et al. [32]. One hundred microliters of DPPH solution (0.1 mM in 95% ethanol) were mixed with 20 μL of purified fraction solution in an Eppendorf tube to initiate the reaction, which was incubated at room temperature for 60 min. Then the reaction mixture was transferred into a 96-well microplate. The absorbance of the resulting solution was measured at 517 nm using an Enspire 2300 microplate reader (Perkin Elmer, Waltham, MA, USA). For the blank, the purified fraction was replaced with distilled water. The DPPH radicals scavenging activity was calculated using the following formula:

$$\text{DPPH radical scavenging activity (\%)} = [1 - ABS_{\text{sample}}/ABS_{\text{blank}}] \times 100\%.$$

3.5.2. Hydroxyl Radical Scavenging Activity

Hydroxyl radical scavenging activity was measured according to the method developed by Wang et al. [33]. Hydroxyl radicals were generated by the Fenton reaction. Ferrous ion (Fe^{2+}) could combine with 1,10-phenanthroline to form red compounds with a maximum absorbance at 536 nm. The absorbance value would decrease when ferrous ion was oxidized into ferric ions by a hydroxyl radical, which reflected the concentration of hydroxyl radicals. In this system, 1,10-phenanthroline (40 μL, 2 mM) and the sample (80 μL) were added into an Eppendorf tube and mixed. The $FeSO_4$ solution (40 μL, 2 mM) was then pipetted into the mixture. The reaction was initiated by adding 40 μL H_2O_2 (0.03% *v/v*). After being incubated at 37 °C for 60 min, the reaction mixture was transferred into a 96-well microplate. The absorbance of the resulting solution was measured at 536 nm using an Enspire 2300 Multimode Plate Reader (Perkin Elmer). The group without any antioxidant was used as a negative control, while the mixture without H_2O_2 was used as the blank. The hydroxyl radical scavenging activity (HRSA) was calculated as follows:

$$\text{HRSA (\%)} = [(A_s - A_n)/(A_b - A_n)] \times 100\%,$$

where A_s, A_n, and A_b were the absorbance values of the sample, the negative control, and the blank determined at 536 nm after the reaction, respectively.

3.5.3. Ferrous Ion Chelating Assay

The ferrous ion chelating activity was measured according to the method of Thiansilakul [34]. Fresh ferrous sulfate (2 mM, 40 μL) and a 400-μL sample were mixed together, and the ferrous ion was detected by 80 μL 5 mM ferrozine. Distilled water was added into the mixture until the total volume reached 2 mL. Then the mixture was incubated at 37 °C for 20 min. The absorbance of the reaction product was measured at 562 nm, and the chelating ratio was calculated as follows:

$$Fe^{2+} \text{ chelating ratio (\%)} = 1 - (A_s/A_c) \times 100\%,$$

where A_s and A_c were the absorbance values of the sample group and the control group determined at 562 nm, respectively.

3.5.4. Oxygen Radical Absorbance Capability (ORAC) Assay

The ORAC assay was measured according to the method developed by Alberto et al. [35]. Sample solution (20 µL) and fluorescein (100 µL, 96 nM) were added into a 96-well microplate and pre-incubated at 37 °C in Enspire 2300 Multimode Plate Reader (Perkin Elmer). The reaction was initiated by adding 30 µL pre-warmed AAPH (320 mM). The reaction was performed at 37 °C. The fluorescence intensity was measured every 30 s for 180 cycles with excitation and emission wavelengths of 485 nm and 538 nm, respectively. Trolox was used as a positive control. The ORAC was defined as the trolox equivalent (mmol·TE/g) according to the area under the curve (AUC), and was calculated as follows:

$$\text{ORAC} = (AUC_{\text{sample}} - AUC_{\text{control}})/(AUC_{\text{Trolox}} - AUC_{\text{control}}) \times (M_{\text{Trolox}}/M_{\text{sample}}),$$

where AUC_{sample}, AUC_{control}, and AUC_{Trolox} were the integral areas under the fluorescence decay curve of the peptide, 75 mM PBS (pH 7.4), and Trolox, respectively. M_{Trolox} and M_{sample} are the concentrations of trolox and peptide.

3.5.5. Protection Effect on Oxidation-Induced DNA Damage

The protection effect on oxidation-induced DNA damage assay was modified according to the method described by Qian et al. [36]. Plasmid DNA displays different structures and a different running rate in agarose gel electrophoresis according to the damage degree. The reaction system included 6 µL of pET-22b DNA, 3 µL of 2 mM FeSO$_4$, 6 µL of antioxidant, and 5 µL of 0.3% H$_2$O$_2$. The mixture was incubated at 37 °C for 10 min, and then analyzed by 1% agarose gel electrophoresis at a constant voltage (130 V) for 30 min. The gel was then monitored in a gel imaging system (Bio-Rad) after being immersed in ethidium bromide for 15 min.

3.6. Statistical Analysis

All experiments were conducted in triplicate (n = 3). The values were expressed as mean ± standard deviation, which were calculated with the Origin 9.1 software. An ANOVA test was used to analyze data in the SPSS 19.0 software.

4. Conclusions

Marine protein resources are considered to be a huge treasury of bioactive peptides. More and more researchers are attempting to prepare novel bioactive peptides from marine protein through enzymatic hydrolysis. This study proved that muscle proteins from salmon byproducts were more suitable to be used as a preparation material of antioxidant peptides compared with collagen from salmon byproducts. Moreover, the antioxidant peptide fraction exhibited a DNA protection effect, which could be developed as a potential dietary supplement to prevent oxidation-related diseases. In addition, commercial proteases were the first choice to be used in protein hydrolysis in previous studies, because these proteases are relatively thoroughly studied and are easily obtained from supermarkets. However, non-commercial proteases from bacteria also possess great potential in bioactive peptide preparation, due to their high efficiency and low cost.

Acknowledgments: The work was supported by the National Natural Science Foundation of China (31070061, 31370104, 21205142), the National Sparking Plan Project (2013GA770009), the Opening Foundation of the Chinese National Engineering Research Center for Control and Treatment of Heavy Metal Pollution (No. 2015CNERC-CTHMP-07), the Open-End Fund for the Valuable and Precision Instruments of Central South University (CSUZC201640), the Young Talents Training Program of Shandong Academy of Agricultural Sciences, and the Fundamental Research Funds for the Central Universities of Central South University (2015zzts273).

Author Contributions: R.W., L.C. and H.H. conceived and designed this study. D.L. and J.Z. provided the bacteria separated from sea water and fresh water, respectively. R.W., J.H. and M.L. performed the experiments. R.W., L.C. and H.H. wrote this article. X.X. and Y.C. reviewed and edited the manuscript. All authors read and approved the manuscript.

Conflicts of Interest: The authors declare no conflict of interest.

References

1. Rizzo, A.M.; Berselli, P.; Zava, S.; Montorfano, G.; Negroni, M.; Corsetto, P.; Berra, B. Endogenous antioxidants and radical scavengers. *Adv. Exp. Med. Biol.* **2010**, *698*, 52–67. [PubMed]

2. Spolarics, Z. Endotoxemia, Pentose cycle, and the oxidant/antioxidant balance in the hepatic sinusoid. *J. Leukoc. Biol.* **1998**, *63*, 534–541. [PubMed]

3. Diaz, M.N.; Frei, B.; Vita, J.A.; Keaney, J.F. Antioxidant and atherosclerotic heart disease. *N. Engl. J. Med.* **1997**, *337*, 408–416. [PubMed]

4. Stohs, S.J. The role of free radicals in toxicity and disease. *J. Basic Clin. Physiol. Pharmacol.* **1995**, *6*, 205–228. [CrossRef] [PubMed]

5. Phanat, K.; Soottawat, B.; Wonnop, V.; Fereidoon, S. Gelatin hydrolysate from blacktip shark skin prepared using papaya latex enzyme: Antioxidant activity and its potential in model systems. *Food Chem.* **2002**, *135*, 1118–1126.

6. Wang, B.; Li, L.; Chi, C.F.; Luo, H.Y.; Xu, Y.F. Purification and characterisation of a novel antioxidant peptide derived from blue mussel (*Mytilus edulis*) protein hydrolysate. *Food Chem.* **2013**, *138*, 1713–1719. [CrossRef] [PubMed]

7. Abraham, T.G.; He, R.; Fida, M.H.; Chibuike, C.U.; Tom, A.G.; Rotimi, E.A. Evaluation of the in vitro antioxidant properties of a cod (*Gadus morhua*) protein hydrolysate and peptide fractions. *Food Chem.* **2015**, *173*, 652–659.

8. Ngo, D.H.; Ryu, B.; Kim, S.K. Active peptides from skate (*Okamejei kenojei*) skin gelatin diminish angiotensin-I converting enzyme activity and intracellular free radical-mediated oxidation. *Food Chem.* **2014**, *143*, 246–255. [CrossRef] [PubMed]

9. Wang, Q.K.; Li, W.; He, Y.H.; Ren, D.D.; Felicia, K.; Song, L.L.; Yu, X.J. Novel antioxidative peptides from the protein hydrolysate of oysters (*Crassostrea talienwhanensis*). *Food Chem.* **2014**, *145*, 991–996. [CrossRef] [PubMed]

10. Mirari, Y.A.; Ailén, A.; Marta, M.C.; López-Caballero, M.E.; Pilar, M.; Gómez-Guillén, M.C. Antimicrobial and antioxidant chitosan solutions enriched with active shrimp (*Litopenaeus vannamei*) waste materials. *Food Hydrocoll.* **2014**, *35*, 710–717.

11. Yang, R.; Wang, J.; Liu, Z.; Pe, X.; Han, X.; Li, Y. Antioxidant effect of a marine oligopeptide preparation from chum salmon (*Oncorhynchus keta*) by enzymatic hydrolysis in radiation injured mice. *Mar. Drugs* **2011**, *9*, 2304–2315. [CrossRef] [PubMed]

12. Subramaniam, S.; Scott, S.; Witoon, P.; Peter, J.B. Functional and nutritional properties of red salmon (*Oncorhynchus nerka*) enzymatic hydrolysates. *J. Food Sci.* **2005**, *70*, 401–406.

13. Ahn, C.B.; Kim, J.G.; Je, J.Y. Purification and antioxidant properties of octapeptide from salmon byproduct protein hydrolysate by gastrointestinal digestion. *Food Chem.* **2014**, *147*, 78–83. [CrossRef] [PubMed]

14. Hunter, E.M.; Mills, H.J.; Kostka, J.E. Microbial Community diversity associated with carbon and nitrogen cycling in permeable marine sediments. *Appl. Environ. Microbiol.* **2006**, *72*, 5689–5701. [CrossRef] [PubMed]

15. Gómez-Estaca, J.; Bravo, L.; Gómez-Guillén, M.C.; Alemán, A.; Montero, P. Antioxidant properties of tuna-skin and bovine-hide gelatin films induced by the addition of oregano and rosemary extracts. *Food Chem.* **2009**, *112*, 18–25. [CrossRef]

16. Kim, S.Y.; Je, J.Y.; Kim, S.K. Purification and characterization of antioxidant peptide from hoki (*Johnius belengerii*) frame protein by gastrointestinal digestion. *J. Nutr. Biochem.* **2007**, *18*, 31–38. [CrossRef] [PubMed]

17. Wu, R.B.; Wu, C.L.; Liu, D.; Yang, X.H.; Huang, J.F.; Zhang, J.; Liao, B.Q.; He, H.L.; Li, H. Overview of antioxidant peptides derived from marine resources: The sources, characteristic, purification, and evaluation methods. *Appl. Biochem. Biotechnol.* **2015**, *176*, 1815–1833. [CrossRef] [PubMed]

18. Wang, B.; Wang, Y.M.; Chi, C.F.; Luo, H.Y.; Deng, S.G.; Ma, J.Y. Isolation and characterization of collagen and antioxidant collagen peptides from scales of croceine croaker (*Pseudosciaena crocea*). *Mar. Drugs* **2013**, *11*, 4641–4661. [CrossRef] [PubMed]
19. Yang, J.I.; Ho, H.Y.; Chu, Y.J.; Chow, C.J. Characteristic and antioxidant activity of retorted gelatin hydrolysates from cobia (*Rachycentron canadum*) skin. *Food Chem.* **2008**, *110*, 128–136. [CrossRef] [PubMed]
20. Mendis, E.; Rajapakse, N.; Byun, H.G.; Kim, S.K. Investigation of jumbo squid (*Dosidicus gigas*) skin gelatin peptides for their in vitro antioxidant effects. *Life Sci.* **2005**, *77*, 2166–2178. [CrossRef] [PubMed]
21. Nazeer, R.A.; Sampath, K.N.S.; Jai, G.R. In vitro and in vivo studies on the antioxidant activity of fish peptide isolated from the croaker (*Otolithes ruber*) muscle proteins hydrolysate. *Peptides* **2012**, *35*, 261–268. [CrossRef] [PubMed]
22. Ferid, A.; Neyssene, A.; Chobert, J.M.; Thomas, H.; Mohamed, N.M. Neutral serine protease from *Penicillium italicum*. purification, biochemical characterization, and use for antioxidative peptide preparation from *Scorpaena notata* muscle. *Appl. Biochem. Biotechnol.* **2014**, *174*, 186–205.
23. Chi, C.F.; Hu, F.Y.; Wang, B.; Ren, X.J.; Deng, S.G.; Wu, C.W. Purification and characterization of three antioxidant peptides from protein hydrolyzate of croceine croaker (*Pseudosciaena crocea*) muscle. *Food Chem.* **2015**, *168*, 662–667. [CrossRef] [PubMed]
24. Chi, C.F.; Wang, B.; Deng, Y.Y.; Wang, Y.M.; Deng, S.G.; Ma, J.Y. Isolation and characterization of three antioxidant pentapeptides from protein hydrolysate of monkfish (*Lophius litulon*) muscle. *Food Res. Int.* **2014**, *55*, 222–228. [CrossRef]
25. Ali, B.; Mohamed, H.; Rafik, B.; Imen, L.; Yosra, T.E.; Moncef, N. Antioxidant and free radical-scavenging activities of smooth hound (*Mustelus mustelus*) muscle protein hydrolysates obtained by gastrointestinal proteases. *Food Chem.* **2009**, *114*, 1198–1205.
26. Luo, H.Y.; Wang, B.; Li, Z.R.; Chi, C.F.; Zhang, Q.H.; He, G.Y. Preparation and evaluation of antioxidant peptide from papain hydrolysate of *Sphyrna lewini* muscle protein. *LWT Food Sci. Technol.* **2013**, *51*, 281–288. [CrossRef]
27. Elias, R.J.; Kellerby, S.S.; Decker, E.A. Antioxidant activity of proteins and peptides. *Crit. Rev. Food Sci. Nutr.* **2008**, *48*, 430–441. [CrossRef] [PubMed]
28. Sheih, I.C.; Wu, T.K.; Fang, T.J. Antioxidant properties of a new antioxidative peptide from algae protein waste hydrolysate in different oxidation systems. *Bioresour. Technol.* **2009**, *100*, 3419–3425. [CrossRef] [PubMed]
29. Karawita, R.; Senevirathne, M.; Athukorala, Y.; Affan, A.; Lee, Y.J.; Kim, S.K.; Lee, J.B.; Jeon, Y.J. Protective effect of enzymatic extracts from microalgae against DNA damage induced by H_2O_2. *Mar. Biotechnol.* **2007**, *9*, 479–490. [CrossRef] [PubMed]
30. Liu, D.; Yang, X.H.; Huang, J.F.; Wu, R.B.; Wu, C.L.; He, H.L.; Li, H. In situ demonstration and characteristic analysis of the protease components from marine bacteria using substrate immersing zymography. *Appl. Biochem. Biotechnol.* **2014**, *175*, 489–501. [CrossRef] [PubMed]
31. He, H.L.; Guo, J.; Chen, X.L.; Xie, B.B.; Zhang, X.Y.; Yu, Y.; Chen, B.; Zhang, Y.Z.; Zhou, B.C. Structural and functional characterization of mature forms of metalloprotease E495 from arctic sea-ice bacterium *Pseudoalteromonas* sp. SM495. *PLoS ONE* **2012**, *7*, e35442. [CrossRef] [PubMed]
32. Shimada, K.; Fujikawa, K.; Yahara, K.; Nakamura, T. Antioxidative properties of xanthan on the antioxidation of soybean oil in cyclodextrin emulsion. *J. Agric. Food Chem.* **1992**, *40*, 945–948. [CrossRef]
33. Wang, B.; Li, Z.R.; Chi, C.F.; Zhang, Q.H.; Luo, H.Y. Preparation and evaluation of antioxidant peptides from ethanol-soluble proteins hydrolysate of *Sphyrna lewini* Muscle. *Peptides* **2012**, *36*, 240–250. [CrossRef] [PubMed]
34. Thiansilakul, Y.; Benjakul, S.; Shahidi, F. Compositions functional properties and antioxidative activity of protein hydrolysates prepared from round scad (*Decapterus maruadsi*). *Food Chem.* **2007**, *103*, 1385–1394. [CrossRef]

35. Alberto, D.; Carmen, G.C.; Begona, B. Extending applicability of the oxygen radical absorbance capacity (ORAC-fluorescein) assay. *J. Agric. Food Chem.* **2004**, *52*, 48–54.

36. Qian, Z.J.; Jung, W.K.; Byun, H.G.; Kim, S.K. protective effect of an antioxidative peptide purified from gastrointestinal digests of oyster (*Crassostrea gigas*) against free radical induced DNA damage. *Bioresour. Technol.* **2008**, *99*, 3365–3371. [CrossRef] [PubMed]

marine drugs

MDPI

Review

Marine Antifreeze Proteins: Structure, Function, and Application to Cryopreservation as a Potential Cryoprotectant

Hak Jun Kim [1,*], Jun Hyuck Lee [2,*], Young Baek Hur [3], Chang Woo Lee [2], Sun-Ha Park [2] and Bon-Won Koo [1]

[1] Department of Chemistry, Pukyong National University, Busan 48513, Korea; 89guti14@gmail.com
[2] Unit of Polar Genomics, Korea Polar Research Institute, Incheon 21990, Korea; justay@kopri.re.kr (C.W.L.); psh@kopri.re.kr (S.-H.P.)
[3] Tidal Flat Research Institute, National Fisheries Research and Development Institute, Gunsan, Jeonbuk 54014, Korea; hur0100@korea.kr
* Correspondence: kimhj@pknu.ac.kr (H.J.K.); junhyucklee@kopri.re.kr (J.H.L.); Tel.: +82-51-629-5587 (H.J.K.); +82-32-760-5555 (J.H.L.)

Academic Editor: Keith B. Glaser
Received: 1 December 2016; Accepted: 20 January 2017; Published: 27 January 2017

Abstract: Antifreeze proteins (AFPs) are biological antifreezes with unique properties, including thermal hysteresis (TH), ice recrystallization inhibition (IRI), and interaction with membranes and/or membrane proteins. These properties have been utilized in the preservation of biological samples at low temperatures. Here, we review the structure and function of marine-derived AFPs, including moderately active fish AFPs and hyperactive polar AFPs. We also survey previous and current reports of cryopreservation using AFPs. Cryopreserved biological samples are relatively diverse ranging from diatoms and reproductive cells to embryos and organs. Cryopreserved biological samples mainly originate from mammals. Most cryopreservation trials using marine-derived AFPs have demonstrated that addition of AFPs can improve post-thaw viability regardless of freezing method (slow-freezing or vitrification), storage temperature, and types of biological sample type.

Keywords: antifreeze proteins; ice-binding proteins; ice recrystallization inhibition; cryoprotectant; slow-freezing; vitrification

1. Introduction

Antifreeze proteins (AFPs) are biological antifreeze materials originally found in polar fish; AFPs can bind to ice and subsequently inhibit the growth of the ice crystals. Fish can inhabit ice-laden or cold seawater below the freezing point (-0.7 °C) of their blood serum by virtue of AFPs [1–4]. However, in a literal sense, the term AFP is a misnomer since AFP does not stop freezing of the blood serum or solution containing AFP. Hence, the term ice-binding protein (IBP) has been proposed to include any protein that binds to ice including AFPs [5]. The term IBP has a bit more nuance than the term AFP. The term ice structuring protein (ISP), which is not used frequently, is synonymous with AFP. However, AFPs are a subset of the larger class of IBPs that includes ice nucleating proteins. In short, all AFPs are IBPs, but not all IBPs are AFPs. In this review, the terms AFP and IBP are used interchangeably.

Marine organisms known to possess or express AFPs, as shown in Figure 1, include bacteria [6–9], fungi [10–12], crustacean [13], microalgae [14–19], and fish [20]. Propelled by next-generation sequencing (NGS) technologies, identification of antifreeze genes from marine organisms has advanced rapidly within the last few years. However, until now, other than fish AFPs, only a few AFPs have been thoroughly characterized from *Colwellia* sp. [21], *Flavobacterium frigoris* [7], *Glaciozyma antarctica* [12,22], *Navicula glaciei* [16], *Fragilariopsis cylindrus* [23,24], and *Chaetocero neoglacile* [15]. The unique function of AFPs, i.e., enabling fish to survive in subfreezing environments, has inspired the researchers in academia and industries to examine the potential applications of AFPs as a potential cryoprotective agents or cryoprotectants (CPAs) in the cryopreservation of biological samples [25–31]. In this review, we discuss the biophysical and biochemical aspects of marine-derived AFPs as well as investigate past and current research of the practical applications of AFPs in cryopreservation. We also describe the possible role of AFPs in cryopreservation.

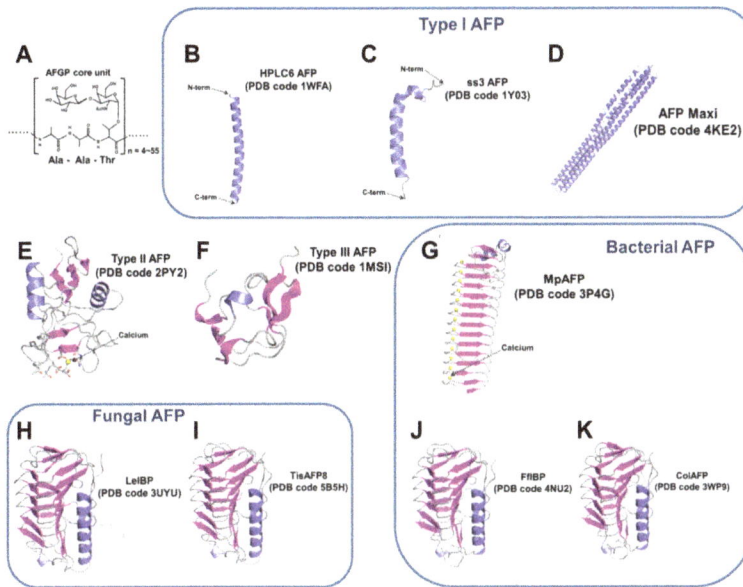

Figure 1. Structural diversity of AFPs: (**A**) core unit structure of antifreeze glycoproteins (AFGPs); (**B**) Type I HPLC6 AFP structure; (**C**) Type I ss3 AFP structure; (**D**) the structure of AFP Maxi from winter flounder, *Pseudopleuronectes americanus*; (**E**) calcium-dependent type II AFP structure; (**F**) Type III HPLC12 AFP structure; (**G**) the structure of MpAFP from *Marinomonas primoryensis*; (**H**) the structure of LeIBP from *Glaciozyma* sp. AY30; (**I**) the structure of TisAFP8 from *Typhula ishikariensis*; (**J**) the structure of FfIBP from *Flavobacterium frigoris* PS I; and (**K**) the structure of ColAFP from *Colwellia* sp. strain SLW05.

2. AFP Properties: Thermal Hysteresis (TH), Ice Recrystallization Inhibition (IRI), and Interaction with Biological Membranes

Generally, AFPs can be characterized based on two properties: TH and IRI. However, interaction of AFPs with membranes should not be ruled out. In this section, the unique features of AFPs and their contribution to cryopreservation are discussed briefly (for more an in-depth biophysical discussion on these properties, refer to recent reviews [32,33]).

TH refers to the difference between melting and freezing points of a solution. In AFP-containing solution, the temperature gap can be created by irreversible binding of AFPs to ice crystals and subsequent inhibition of their growth until the temperature decreases to the non-equilibrium freezing point [32,34–39]. Below the non-equilibrium freezing point, the burst of the ice crystal can be observed (Figure 2A). During the TH gap, AFPs bind to the specific planes of ice crystals, shaping a unique ice morphology. For example, type I AFPs bind to the prism plane of ice and creates a hexagonal bipyramidal shape [35,40], whereas hyperactive FfIBP binds to prism and basal planes and generates a lemon shape [41]. Moderately active AFPs bind to prism and/or pyramidal planes [40,42,43], whereas hyperactive AFPs are able to bind to the basal plane of ice crystals [42,44]. The ice morphology shaped by AFPs is a hallmark of binding of AFPs to ice (inset of Figure 2A) [20]. TH has been used to describe the activity of AFPs quantitatively. For most fish AFPs, the observed TH activity is approximately 1 °C [20,45]. This temperature gap provides enough cushion against seawater during the winter season (−1.9 °C) for polar fish to survive in cold environments. In addition to fish AFPs, many marine AFPs are associated with sea ice [7,13,14,16,21,23,46]. Unlike the blood plasma of polar fish, seawater in brine channels in sea ice undergoes freezing to ice. Hence, AFPs from sea-ice associated bacteria, microalgae, and eukaryotic protists are secreted into the surrounding environment to protect themselves from freezing, and some of them are hyperactive (Figure 2B) [7,21,41].

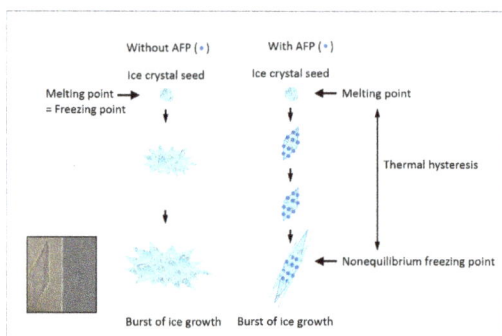

(A)

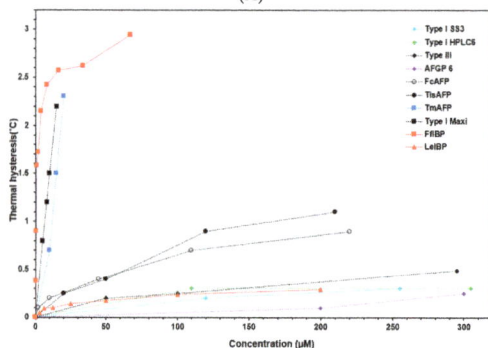

(B)

Figure 2. (**A**) Cartoon illustration of TH phenomenon. In the left panel, the ice starts to grow rapidly as temperature drops. However, as shown in the right panel, AFPs adsorb irreversibly on to the specific planes of ice surface, inhibiting the further growth of ice until the temperature reaches nonequilibrium freezing point. This adsorption-inhibition mechanism by AFPs separates melting and freezing points of solution. The inset shows the bipyramidal and lemon ice morphologies created by moderately active type I AFP (left) and hyperactive FfIBP (right), respectively; (**B**) Comparison of TH activities of AFPs from various organisms. TH activity of marine-derived FfIBP (from *Flavobacterium frigoris*), and type I-Hyp (from *Pseudopleuronectes americanus*) are comparable to hyperactive insect and fungal AFPs, TmAFP and TisAFP, respectively, of non-marine origin. Other marine AFPs are moderately. Abbreviations are as follows: TmAFP, *Tenebrio molitor* AFP; TisAFP, *Thyphula ishikariensis* AFP; FcAFP, *Fragilariopsis cylindrus* AFP; and LeIBP, *Glaciozyma* (formerly known as *Leucosporidium*) sp. IBP.

The second function of AFPs, which may be more useful for cryopreservation, is IRI. Ice recrystallization (IR), as depicted in Figure 3, explains a thermodynamically favorable process in which the formation of larger ice grains takes place at the expense of smaller ones with a high internal energy [47,48]. Eventually, the larger ice crystals formed because of this phenomenon can be fatal to the cryopreserved cells as well as the organisms inhabiting polar or cold regions [49,50]. Fortunately for these organisms, AFPs can inhibit IR at very low concentrations. The AFP-dependent IRI mechanism remains to be elucidated; however, similar to TH activity, IRI is attributed to the binding of AFPs to ice [5,45,51]. AFPs at the interface between the grain boundaries bind to the surface of ice grains and inhibit the growth process [50]. The IRI is more likely to be a key property for cold-tolerant organisms to survive in extremely cold environments [47,52–55]. To this end, IRI is eventually thought to defend membranes against freezing injury [27–31,56,57]. IRI activity was first analyzed using a splat cooling assay developed by Knight [58]. In splat assays, a small droplet of a solution is expelled from a height of 2 m onto a very cold (−70 °C) metal plate and freezes instantaneously as a polycrystalline wafer. The ice is then annealed at −6 °C over a certain period of time, allowing ice recrystallization to occur. Modified methods have been proposed wherein the ice grains are generated from a few-microliter sample placed between coverslips by flash freezing [54,55] or where the sample inside 10 µL glass capillary undergoes freezing and annealing [59]. However, the IRI result was only semi-quantitatively reported by presenting the IRI endpoint, expressed as mg/mL or µM, where IRI is no longer observed [54,55,58,59]. To assess IRI activity quantitatively, Jackman et al. employed domain recognition software to measure and report the mean grain size (MGS) of the 10 largest ice crystals after the annealing period [60]. This method displayed percent MGS as a function of AFP relative to the control. Very recently, Voets' group adopted an automated image analysis using the circle Hough transform (CHT) algorithm with a modified splat assay [55] to quantitatively evaluate IRI [61,62]. The CHT is a basic technique for detecting circular objects in a digital image. They attempted to include all ice crystal images instead of only the 10 largest ice crystals in the calculation, which obviously makes the quantitative evaluation of IR kinetics more statistically significant. In this method, the inflection point of the kinetic curve was presented as an IRI endpoint.

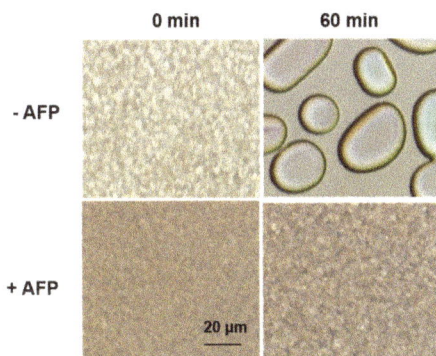

Figure 3. Results of ice recrystallization inhibition (IRI) assay using modified splat assay. In this assay, AFP containing solution was mixed with 30% sucrose in a 1:1 ratio. The mixed solution was spotted between two coverslips and flash frozen. Then, the sample was placed at −6 °C stage and the changes were observed over a specific period of time-in this case 30 min. As in upper panel, larger ice grains grow as expense of smaller ice crystals, while the growth was halted in lower panel in the presence of AFPs. All subfigures are drawn in the same scale.

Both TH and IRI properties are based on the affinity of AFPs for ice. Intriguingly, however, TH activity is not always proportional to IRI activity (Figure 4), which remains to be elucidated [63]. In the comparison of TH and IRI activities of hyperactive insect, bacterial, and fish AFPs with moderately active fish AFPs,

Yu et al. reported that the TH hyperactivity of AFPs was not reflected in their IRI activity [63]. This was corroborated by other marine-derived AFPs, i.e., LeIBP and FfIBP [7,10,41,44,64]. Olijve et al. also demonstrated that type III AFP and its mutant showed different TH values but almost the same IRI activity [61]. The hyperactive FfIBP showed less activity in IRI, compared to the moderately active LeIBP [7,44]. These results implied that TH activity was not necessarily translated into improvements in the cryopreservation efficiency of biological samples [65–69]. Therefore, the utilization of AFPs in cryopreservation cannot be considered from their TH activity only.

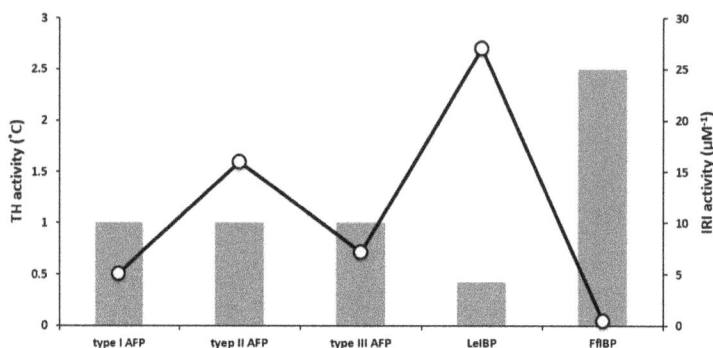

Figure 4. A graph of TH and IRI activities of marine-derived AFPs. TH values, represented as a bar, are from Figure 2B. The IRI activities (O) are expressed as the reciprocal of endpoint of each AFP. The endpoint indicates the lowest concentration at which the AFP shows IRI activity. Higher IRI value means more effective in IRI. The LeIBP is weaker in TH but higher in IRI activity, but vice versa in FfIBP. This plot demonstrates that the TH values are not proportional to IRI activities.

Along with the IRI feature, the interaction of AFPs with membranes (or proteins in membranes) also may also ameliorate the cryoinjury of cells. In the early study of Rubinsky and his colleagues, fish AFPs were found to protect cell membranes during hypothermic storage [70]. As membranes are cooled to low temperatures, one mechanism of injury is often thermotropic phase transition partly due to weakened hydrophobic interactions [71–75]. During the transition from liquid crystalline to gel phase, membranes become leaky, resulting in the loss of intracellular contents [76]. It is not entirely clear what causes leakage during the phase transition; however, this process may be related to defects in packing of the hydrocarbon chains during the coexistence of gel and liquid crystalline domains [71]. Since the phase transition temperature of each lipid depends on the degree of unsaturation of lipid tails and the number of carbons in the lipid alkane chains, model membranes with diverse compositions, such as dielaidoylphosphatidylcholine (DEPC), dielaidoylphos-phatidylethanolamine (DEPE), and dielaidoylphosphatidylglycerol (DEPG), have been used in order to better understand the nature of the interactions between AFPs and cell membranes [72,73,75,77–82]. The results showed that these interactions were lipid specific, i.e., the lipid composition of the bilayer dictates whether or not a certain AFP or antifreeze glycoprotein (AFGP) will protect/interact with the membranes [61,62,64–69]. Other reports have indicated that the cryoprotective effects of AFPs arose from their interaction with membrane proteins, such as potassium and calcium ion channels [83–86]. However, in some cases, the addition of AFPs in cryopreservation medium induces leakage from cryopreserved cells [87–95]. These results implied that protection against freezing damage by AFPs depends on the type of membrane and the type of AFP [80,96].

3. Marine-Derived AFPs

3.1. Fish AFPs

Two scientists, Scholander and DeVries, first observed that some fish inhabiting the polar oceans could survive in cold water that occasionally reached sub-zero temperatures [1,3,97]. Following this observation, they attempted to elucidate how these fish could survive in icy water, reaching temperatures below the freezing point of fish blood. When cooling was increased even further, they observed that the growth of ice crystals was sluggish and delayed due to the presence of glycoproteins that depress the noncolligative freezing point of solutions [98]. These proteins were designated AFGPs [3]. Thereafter, nonglycosylated AFPs (type I AFPs) were found in the winter flounder, *Pseudopleuronectes americanus* [99]. In addition, several types of AFPs have been discovered and classified within distinct groups (classified into types I, II, III, and IV) in the Arctic and Antarctic regions. Even though the AFP types are fundamentally different in terms of their primary sequences and three-dimensional structures, they all have equivalent properties allowing them to bind to ice and depress the freezing point of the solutions. Moreover, these different types of AFPs do not seem to share any common ancestor genes.

AFGPs contain a three amino-acid (Ala-Ala-Thr) repeating sequence motif with a disaccharide connected to the hydroxyl group of the threonine residue [100]. However, there are sequence variations at the first residue position; sometimes, the first Ala residue is replaced by a Pro, Thr, or Arg. There are eight AFGPs (AFGP1-8), named according to the number of repeating units. AFGP1 has about 50 repeating units and therefore the highest molecular weight (33.7 kDa), whereas AFGP8 has the lowest molecular weight (2.6 kDa), with only four repeating units. Typically, the antifreeze activities of AFGPs are proportional to the number of repeating units. It is thought that high-molecular-weight AFGPs cover a wider ice surface and inhibit ice growth more efficiently than smaller AFGPs [47,101–104]. Recent studies have also shown that carbohydrate moieties are important for AFGP activity. Structural studies using nuclear magnetic resonance (NMR) have revealed that carbohydrate moieties and Ala residues are located on opposite sides. This feature confers AFGPs with a helical shape and amphipathic characteristics. Consequently, AFGPs show strong recrystallization properties. However, there are several limitations regarding their commercial utilization toward cryopreservation. Natural polar fish sources are not sufficient to prepare large quantities of AFGPs, and chemical synthesis is difficult to establish in large-scale mass production systems. In contrast, AFPs can be prepared in large quantities by recombinant protein expression techniques. For that reason, AFPs have been more broadly used for application studies than AFGPs. This review focuses on marine AFPs used for cryopreservation applications.

3.1.1. Type I AFPs

Type I AFPs are found in many flounders and sculpins. Type I AFP HPLC6 from winter flounder has been the most extensively studied. This protein possesses 37 amino acids and its sequences are composed of 11 amino acid repeating units [20,105]. Moreover, this protein also has a high Ala residue content, making up 23 of 37 residues. The molecular structure of HPLC6 (PDB code 1WFA) was determined using the X-ray crystallography, which showed that HPLC6 AFP is an α-helical protein with amphipathic characteristics. Another type I AFP has been isolated from the shorthorn sculpin (*Myoxocephalus Scorpius*; ss3 AFP), also displaying a high Ala content (21 Ala residues among a total of 33 residues). The structure of the ss3 AFP (PDB code 1Y03) was determined by NMR spectroscopy. The overall structure of ss3 AFP is similar to that of HPLC6 AFP; however, ss3 AFP contains a Pro residue at position 4, inducing a helix kink. Recently, the four-repeat containing isoform AFP9 and a much larger type I AFP (a 195-residue protein, AFP Maxi) were discovered in winter flounder (*Pseudopleuronectes americanus*). These two proteins exhibit significantly higher TH activities than HPLC6 AFP. Moreover, the increased size of the AFP may induce higher antifreeze activity by facilitating binding to multiple ice crystals and increasing coverage of the ice surface.

Furthermore, analysis of the AFP Maxi structure (PDB code 4KE2) revealed that this protein folds into a dimeric four-helix bundle and that its ordered water may be involved in ice binding, thereby enhancing its antifreeze activity.

The ice-binding mechanism of type I AFP was previously investigated through an ice-etching experiment, which is used to identify AFP binding sites. A simple crystal growth and etching technique allows the identification of the crystallographic planes where the binding occurs [40]. Furthermore, ice etching has also been used to identify the ice-binding planes of AFPs and enhanced green fluorescent protein (EGFP) fusion constructs allow their clear visualization. In 1991, Knight et al. reported that type I AFPs from winter flounder (*Pseudopleuronectes americanus*) and Alaskan plaice (*Pleuronectes quadritaberulatus*) adsorb onto the {2 0 −2 1} pyramidal planes of ice, whereas the sculpin (*Myoxocephalus scorpius*) AFP adsorbs onto {2 −1 −1 0}, the secondary prism planes [40]. This finding suggests that each type I AFP has a unique ice-binding mechanism depending on its sequence length and composition. Moreover, ice-binding sites of type I AFPs were analyzed by site-directed mutagenesis, truncated variants, and molecular docking studies [106–111]. Currently, it is generally accepted that ice-binding sites of type I AFPs are located on their Ala-rich hydrophobic faces.

3.1.2. Type II AFPs

Type II AFPs are found in sea raven, smelt, herring, and long snout poacher. Type II AFPs are globular cysteine-rich fish AFPs with molecular weights ranging from 11 to 24 kDa. The overall structure of type II AFPs shows numerous similarities with C-type lectin-like domains (CTLDs). Type II AFPs have two α-helices and nine β-strands with specific cysteines forming disulfide bonds. Those disulfide bonds are known for their capacity to increase the structural stability of type II AFPs [112–116]. Structural comparison studies between various groups of type II AFPs showed that even if their amino acid sequence similarity is low, overall, their structures are similar, and they display the same functions. These results indicate that type II AFPs evolved from the backbones of CTLDs [117].

Type II AFPs are distinguished by their dependence on calcium ions to enable their antifreeze activities. Herring and two types of smelts produce Ca^{2+}-dependent type II AFPs. Herring type II AFP (hAFP) has close structural similarities with lithostathine (PDB code: 1qdd; root mean square deviation [RMSD] = 1.7 Å for 122 Cα atoms) and mannose-binding protein (PDB code: 1sl6; RMSD = 2.2 Å for 124 Cα atoms). However, these two proteins have no ice-binding activities. Likewise, hAFP has no carbohydrate-binding activity. Thus, this high similarity in carbon backbone structure along with different activities indicates a divergent evolutionary pattern. Another difference between hAFP and C-type lectin protein is the number of cysteine bonds. hAFP has five disulfide bonds, whereas C-type lectin only possesses three or four. Thr96, Leu97, Thr98, and Thr115 residues are important for ice-binding. Interestingly, all of these residues are located near the Ca^{2+} binding site. Therefore, the results obtained from these investigations suggest that Ca^{2+} binding in hAFP is critical for forming an ice-binding state structure and increasing ice-binding activity [115]. Sea raven and long snout poacher produce Ca^{2+}-independent type II AFPs. Structural comparisons between Ca^{2+}-dependent and -independent type II AFPs showed that several residues near the Ca^{2+} binding site are different. Gln92, Asp94, Glu99, and Asn113 residues of hAFP are substituted with Lys95, Asn97, Asp102, and Asp116 residues, respectively, in long snout poacher AFP (lpAFP). Through these studies, the critical amino acids for Ca^{2+} binding were identified. These amino acids could be important indicators allowing the distinction between Ca^{2+}-dependent and-independent type II AFPs [116]. Additionally, a type II AFP was found in Japanese smelt (*Hypomesus nipponensis*; HniAFP), which does not inhabit polar regions, but instead is found in fresh waters in regions near the middle latitudes. Interestingly, HniAFP can bind to Ca^{2+}, but its ice binding activity does not depend on this feature; indeed, despite adding ethylenediaminetetraacetic acid (EDTA) to remove Ca^{2+}, its antifreeze activity was not affected [118].

3.1.3. Type III AFPs

Type III AFPs are small globular proteins with an average molecular weight of 6.5 kDa, found in Antarctic eelepout (*Macrozoarces americanus*) and wolf fish [119,120]. Type III AFPs can be divided into two groups, quaternary-amino-ethyl (QAE) and sulfopropyl (SP) sephadex-binding isoforms, based on both their sequence similarities and affinities for SP and QAEs [121]. QAEs can be further categorized into QAE1 and QAE2 subgroups [122]. According to some studies, QAE1 isoforms have higher TH activities than the QAE2 and SP isoforms. SP and QAE2 isoforms are incapable of stopping ice growth [123,124]. The structures of type III AFPs have been extensively studied, and about 40 models have been solved and deposited in the protein data bank (http://www.rcsb.org/pdb/) to date. Among these, the three-dimensional structure of HPLC12 AFP, belonging to the QAE1 subgroup, was the first to be determined, showing a globular β-sandwich consisting of two antiparallel triple-stranded β-sheets [125–127]. Although type III AFPs are mainly composed of several loops, they form stable structures through hydrophobic interactions and a number of hydrogen bonds at the center of the structure. Type III AFPs were found to be active over a broad pH range (2–11), indicating that the protein fold is stable even at extreme pH, which would normally cause protein denaturation [125]. Recent studies have shown that temperature treatment at 80 °C and pressure treatment at 400 MPa (duration of 1 min for both treatments) did not influence the IRI activity of type III AFPs [128]. Interestingly, sialic acid synthase (SAS) has a C-terminal antifreeze-like domain similar to that of type III AFPs. However, these two homologous proteins have very different temperature-dependent stabilities, activities, and backbone dynamics. While type III AFPs are mostly rigid, with a few residues showing slow motions, SAS is remarkably flexible at low temperature [129,130]. These two proteins, displaying different functions, may have evolved from a common structural ancestor.

The most widely accepted hypothesis to describe the mechanism through which type III AFPs interact with ice crystals involves the Thr18 residue located on the flat surface; this residue is thought to be responsible for the recognition and interaction with the primary prism planes of ice. AFPs cover water-accessible ice surfaces, thereby inhibiting ice growth. Several reports have shown that putative ice-binding residues (Gln9, Asn14, Thr15, Ala16, Thr18, and Gln44) are capable of significantly altering TH activity and ice crystal morphology [125,126,131,132]. Notably, the replacement of Thr18 by Asn causes a significant loss of TH activity (90% loss). Computer simulation studies have emphasized that hydrophobic interactions within ice-binding sites are also important for the antifreeze activity of the protein [132,133]. When hydrophobic residues, such as Leu19, Val20, and Val41, were replaced with Ala, a 20% loss in activity was observed. Double mutants (L19A/V41A and L10A/I13A) showed more than 50% loss of activity compared with the activity of the wild-type protein [124]. Ice-etching studies revealed a more complex ice-binding mechanism within type III AFPs, showing that they could interact with both the primary prism and a pyramidal plane of ice [1]. While the QAE1 isoform is able to bind both the primary prism and a pyramidal plane of ice, the SP and QAE2 isoforms can only bind pyramidal ice planes [134]. Interestingly, a triple mutant of the inactive QAE2 isoform (V9Q/V19L/G20V) is able to bind to the primary prism ice plane and shows full TH activity, similar to the QAE1 isoform [135]. More recently, NMR experiments with inactive QAE2-like mutants containing the V20G mutation were reported. These experiments showed that the mutants exhibited increased conformational flexibility and were incapable of binding to the primary prism plane of ice crystals. These results suggested that inactive type III AFPs may be unable to anchor water molecules via H-bond interactions in the first 3_{10} helix (residues 18–22) and therefore have no antifreeze activity [136].

Interestingly, two almost identical type III AFP domains tied by linker residues, designated RD3, were found in nature in the Antarctic eelpout, *Rhigophila dearborni* [137–139]. RD3 possesses 5.9-fold higher activity than a single domain in the range of 0 to 0.5 mM. This high activity at low concentrations may be related to the need for much smaller concentrations of AFP for cryopreservation, as mentioned below.

3.2. Fungal AFPs

To date, various mushrooms and Basidiomycetous psychrophilic yeast species have been screened and reported to have antifreeze activities. Only two mushrooms (enoki and shiitake), one snow mold fungus (*Typhula ishikariensis*), and two yeast organisms (*Glaciozyma antarctica* and *Glaciozyma* sp. AY30) have been characterized both genomically and for their antifreeze properties [9,10,140,141].

Lee et al. were the first to report the antifreeze activity of a protein isolated from the psychrophilic yeast *Glaciozyma* sp. AY30, itself isolated from an ice core sample of a freshwater pond near the Dasan station, Ny-Ålesund, Svalbard archipelago, Norway, and named LeIBP [10]. LeIBP contains a right-handed β-helical structure, which provides the advantage of a broad-range interaction surface for ice binding [44,64]. The ice-binding site of LeIBP was determined to be a B-face using site-directed mutagenesis experiments [64]. Moreover, the codon-optimized LeIBP (pLeIBP) was constructed and subjected to high-level expression in the *Pichia pastoris* system [142]. In pilot-scale fermentation (700 L), pLeIBP was secreted into culture medium, and the yield was 300 mg/L. The TH activity of pLeIBP was about 0.42 °C, which was similar to that of LeIBP expressed in *E. coli*. The availability of large quantities of pLeIBP allowed us to use this protein in further application studies [65–67,143–145].

Snow mold fungus (*Typhula ishikariensis*) secretes seven antifreeze protein isoforms composing the TisAFPs [141]. Among them, the structures of TisAFP6 (PDB code 3VN3) and TisAFP8 (PDB code 5B5H) were determined and their antifreeze mechanisms were characterized [146]. The results suggested that TisAFP8 has a more adapted shape and higher hydrophobicity to allow ice binding than TisAFP6, which may possess a higher TH activity. Notably, the overall structures of LeIBP (PDB code 3UYU), TisAFP6, and TisAFP8 are very similar, with RMSD values within 0.73 Å when superimposed.

Glaciozyma antarctica AFP (Afp1), described by Hashim et al., possesses both TH and RI activities and shows 30% sequence similarity with TisAFPs [12]. Amino-acid sequence analysis showed that Afp1 contains four α-helices. Shah et al. confirmed the antifreeze activity of each helical peptide [22]. In addition, the NMR structures of the peptides were determined and the ice-binding model was generated using a molecular dynamics method. The results indicated that the Afp1 peptides work like type I AFPs. In 2014, another *Glaciozyma antarctica* AFP (Afp4) was identified and characterized [147]. The Afp4 sequence shows the highest amino acid similarity (93%) to LeIBP. A recombinant Afp4 protein changed ice crystals into hexagonal shaped crystals and showed a TH value of 0.8 °C at a protein concentration of 5 mg/mL.

3.3. Diatom AFPs

Studies aiming to identify new AFP genes from polar sea diatoms (*Chaetoceros neogracile*, *Berkeleya* sp., *Navicula* sp., *Fragilariopsis* sp., and *Nitzschia frustulum*) have been performed, and further gene expression studies have shown that the expression of AFP genes is regulated in response to stress conditions, such as cold temperature and high salinity [16,17,23,24,148]. Thus, AFP genes may play an important role in the environmental adaptation of diatoms. In 2009, Gwak et al. first produced recombinant antifreeze protein (Cn-AFP) from a marine diatom, *C. neogracile*, and characterized its antifreeze activity [15]. The TH value of the mature form of Cn-AFP is 0.8 °C, whereas pre-mature Cn-AFP has a 16-fold lower TH activity, indicating that the signal peptide induces improper folding of Cn-AFP or masks the ice-binding site.

3.4. Bacterial AFPs

In 2004, Gilbert et al. published an interesting finding showing bacterial AFP screening results obtained from Antarctic lake bacteria [149]. The authors managed to culture 866 bacterial isolates from an Antarctic lake and found RI activity in 19 of these isolates. The first bacterial IBP gene (~25 kDa) was identified, and the protein purified through ice affinity purification, in the sea ice gram-negative bacterium *Colwellia* strain SLW05 [8]. In 2008, other bacterial IBPs (54 kDa) were isolated from a deep Antarctic ice core of the subglacial Lake Vostok, at a depth of 3519 m (GenBank EU694412) [140].

The sequence of the protein is similar to those of IBPs previously found in sea ice habitats, even though the protein is longer. In addition, uncharacterized proteins similar to IBPs were found in sea ice bacteria *Polaribacter irgensii* (ZP_01118128; sequence identity: 61%, sequence similarity: 75%), *Psychromonas ingrahamii* (ZP_01349469; sequence identity: 59%, sequence similarity: 71%), and marine bacterium *Shewanella frididimarina* (YP_749708; sequence identity: 52%, sequence similarity: 69%).

The first bacterial AFP structure was solved using a protein isolated from an Antarctic lake bacterium (*Marinomonas primoryensis;* MpAFP) [150]. MpAFP is a 1.5-MDa protein with calcium-dependent antifreeze activity [6]. The solved MpAFP structure (PDB code 3P4G) shows a calcium-bound β-helical fold and bound water molecules, which fit well onto the ice crystal lattice. Therefore, this structure may explain the anchored clathrate mechanism of AFPs when binding to ice.

Recently, another IBP (FfIBP) from the Antarctic bacterium *F. frigoris* PS1 was identified from sea ice on the shore of McMurdo Sound (GenBank accession no. AHKF00000000.1) and characterized [41,151]. FfIBP shares 56% sequence similarity with LeIBP, but displays an antifreeze activity that is up to 10-fold higher than that of LeIBP. Structural and functional characterization of FfIBP revealed that this protein displays regular motifs (T-A/G-X-T/N motif) and more regularly aligned ice-binding residues on its IBS than LeIBP [7]. These structural differences may confer FfIBP with higher TH activity.

In 2014, structural and biochemical data on an AFP from *Colwellia* sp. strain SLW05 (ColAFP) were published [21]. Interestingly, the ColAFP structure is similar to those of LeIBP, TisAFP, and FfIBP, displaying a β-helical structure. In addition, the alignment of sequences and phylogenetic trees of the bacterial AFPs with those of other AFPs and IBPs suggests that eukaryotic IBPs could have been acquired from bacteria by horizontal gene transfer (HGT) [151]. One theory in favor of HGT is "restricted occurrence", which suggests that the same small set of organisms can be found in different locations [152]. IBPs seem to satisfy this criterion because hundreds of organisms have IBPs or IBP-like genes. Another potential explanation involves virus-mediated transformation of IBP genes. For example, Arctic cryoconite holes are built on snow, glaciers, or ice caps where viruses are abundant; these viruses are able to infect a broad range of bacterial species and other organisms, suggesting that viruses in the environment may play a role in the exchange of genetic material [153].

Furthermore, a new bacterial AFP [154] with high IRI activity [155] was reported very recently. Metagenomic sequencing of the Antarctic psychrophilic marine ciliate *Euplotes focardii* revealed two sequences encoding IBPs, designated as EFsymbAFP and EFsymbIBP, obtained from its putative bacterial symbiont [154,155]. These IBPs seem to be structurally similar to TisIBP, LeIBP, and FfIBP [154]. Of these, N-terminal 23 residue-deleted EFsymbAFP was recombinantly expressed in *E. coli* and characterized [155]. Its TH activity was 0.53 °C at 50 µM, but its IRI activity was in the nanomolar range, as determined by Voets method. This value is the lowest observed to date. The recombinant protein also effectively protected bacterial cells from freezing damage. Further investigations of this IBP will provide more insight into the relationships among IRI and TH and the evolution of IBP.

4. Cryopreservation Using AFPs as a Potential Cryoprotectants (CPAs)

4.1. Cryopreservation and Ice Recrystallization

Cryopreservation is an important technique used to store various types of cells, tissues, and organs at very low temperature, usually in liquid nitrogen (−196 °C) [156], and has become crucial in cell biology and regenerative medicine [157,158]. However, cells are not always viable after thawing [145,159]. The freezing and thawing process during cryopreservation causes cryo-injury to cells (Figure 5). Currently, two methods, i.e., slow-freezing [156] and vitrification [160], are commonly adopted in cryopreservation. Prior to addressing the role of AFPs in cryopreservation, we will discuss the association of cryo-injury with freezing with regard to methods other than decreased temperature.

During the slow-freezing process, since the solute concentration inside a cell is higher than that in the medium, the cell is supercooled and ice forms extracellularly [156]. The growth of extracellular ice leaves the unfrozen fraction highly concentrated with salt, leading to dehydration of the cell and destabilization of cellular membranes simultaneously due to osmotic pressure. Incomplete dehydration inside the cell allows intracellular ice formation, which is believed to be detrimental to cells. Eventually the further growth of extracellular ice may cause rupture of the cell membrane. In addition, recrystallization of intracellular and extracellular ice during the thawing process may further damage the cryopreserved cells. Since cell-penetrating CPAs, such as dimethylsulfoxide (DMSO) and glycerol, reduce ice formation by replacing water outside and within the cell as well as stabilize the membranes, the addition of CPAs can increase the post-thaw viability of cryopreserved cells.

Vitrification is a process in which a liquid turns into an amorphous glass solid in the absence of crystallization [160]. Vitrification of cells requires very high concentrations of CPAs and ultrafast cooling rates to completely avoid fatal intracellular and extracellular ice formation [160,161]. In addition to the osmotic stress and chemical toxicity caused by high CPA concentrations, however, vitrification is also associated with ice recrystallization during thawing. In both cases, ice recrystallization during thawing seems to be one of major cold damages. In this context, AFPs are believed to play a crucial role in inhibiting ice recrystallization, improving the cryopreservation efficiency.

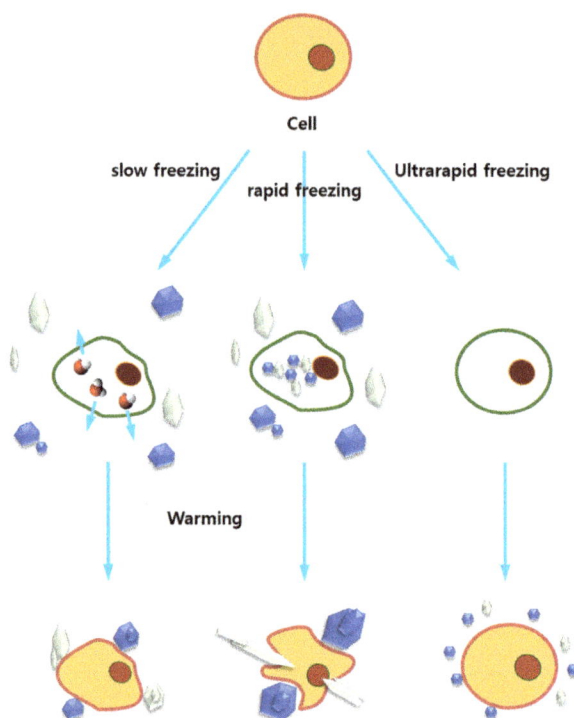

Figure 5. Schematic illustration of freezing rate and ice recrystallization during warming. In slow freezing process, the extracellular ice starts to form below the equilibrium freezing point. Subsequently, water is expelled from inside the cell by osmotic pressure, eventually eliminating the intracellular ice formation. Fast freezing process, however, causes the intracellular ice formation since water cannot leave the cell quickly. In ultrarapid cooling, such as vitrification process, theoretically no bulk ice will form in the presence of higher concentration of CPAs. The ice formed during freezing will become problematic, when the cryopreserved cells are thawed (or warmed). They start to grow bigger: a process known as ice recrystallization. This process is fatal to the cells. Even in vitrification, ice can form during the warming. Therefore, freezing rate should be optimized depending of cell type, CPAs used, etc. The addition of AFPs in freezing media seems to alleviate the ice formation and recrystallization.

4.2. AFPs in Cryopreservation

The first application of marine AFPs to the protection of membranes at hypothermic temperatures was made in 1990 using AFGP from Antarctic and Arctic fishes [83]. Since then, marine-derived AFPs have been tested for cryopreservation on numerous occasions. Almost all reports of cryopreservation using AFPs are summarized in Table 1. Of eight AFPs, including nonmarine insect DcAFP, as shown in Figure 6A, type III AFP has been tested most in cryopreservation, followed by type I AFP, AFGP, and LeIBP. This is because type III AFP is easy to produce recombinantly compared with other fish AFPs and because it has been studied longer than other marine-derived AFPs, such as LeIBP and FfIBP. The results listed in Table 1 also showed that hyperactive AFPs do not always ensure better cryopreservation efficiency [65,66,145]. For example, moderately active LeIBP protects mouse ovarian tissue more effectively than 10-fold hyperactive FfIBP [65]. The same result was obtained in human cell line cryopreservation (Hak Jun Kim, unpublished result), consistent with the observation that hyperactive AFPs do not ensure increased IRI activity (Figure 4) [63]. The AFP concentration used in cryopreservations was also determined empirically (Table 1). The IRI endpoint, sometimes expressed as mg/mL, does not indicate the effective amount of AFP in cryopreservation, and the solubility of AFPs and the molar concentration in the freezing medium should also be considered [145]. Quite frequently, higher concentrations of AFPs lead to a decrease in the post-thaw survival of cryopreserved cells, which may be due to the formation of destructive needle-like ice at high AFP concentrations [65,66,68,69,92,162,163].

Cryopreserved biological samples are relatively diverse ranging from diatoms and reproductive cells to embryos and organs (Figure 6B). Most cryopreserved biological samples originated from mammals. Most cryopreservation trials using AFPs have demonstrated that the addition of AFPs could improve post-thaw viability, regardless of the freezing method (slow-freezing or vitrification), storage temperature, and biological sample, but several reports showed no beneficial effects [68,87–94,164,165].

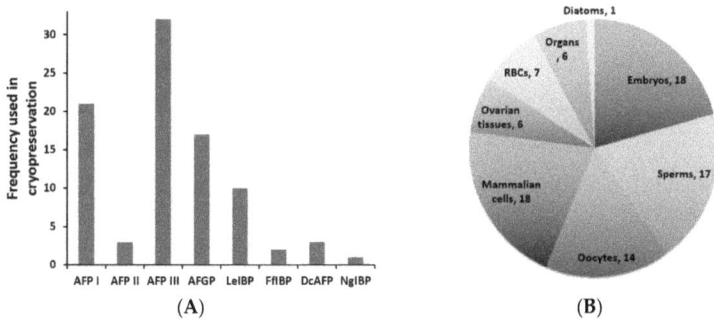

Figure 6. Cryopreservation research using AFPs: (**A**) frequency of AFPs used in cryopreservation; and (**B**) types and frequency of biological samples in cryopreservation using AFPs.

Table 1. Lists of AFPs used in cryopreservation of biological samples.

AFPs	Origin Species of AFPs	Cryopreserved Biological Samples		AFP Quantities Used	Freezing Methods	References
		Organisms	Sample Types			
III	Fish	Turbot (*Scophthalmus maximus*)	Embryos	20 nL (final conc. 0.77 mg/mL) of 10 mg/mL type III AFP injected in yolk sac 0.23914 mm³)	Vitrification	[166]
I/III	Fish	Gilthead seabream (*Sparus aurata*)	Sperm	1 µg/mL	Vitrification	[167]
I/III/AFGP	Fish	Bull	Sperm	0.1, 1, 10, and 100 µg/mL	Cryopreservation	[168]
LeIBP	Yeast	Boar	Sperm	0.01, 0.1, and 1 mg/mL	Cryopreservation	[169]
III FfIBP LeIBP	Fish Bacteria Yeast	Mouse	Ovarian tissue	0.1, 1, and 10 mg/mL	Vitrification	[65]
AFGP	Fish	Buffalo	Sperm	0.1, 1, and 10 µg/mL	Vitrification	[170]
III	Fish	Buffalo	Sperm	0.01, 0.1, 1, and 10 mg/mL	Cryopreservation	[171]
III	Fish	Bovine	Embryos	10 mg/mL	Hypothermic	[172]
AFGP	Fish	Pig	Oocyte	40 mg/mL	Hypothermic	[83]
I II III	Winter flounder Sea raven Eel pout	Bovine	Oocyte	20 mg/mL	Hypothermic	[70]
III	Notched-fin eelpout	Human	HepG2	2–10 mg/mL	Hypothermic	[173]
I/III/AFGP	Fish	Rat	RIN-5F cells (insulinoma)	10 mg/mL	Hypothermic	[174]
I III	Winter flounder Ocean pout	Sheep	Embryo	1 or 10 mg/mL	Hypothermic	[175]
III	Fish	Rabbit	Sperm Embryo	0.1, 1, 10, and 100 µg/mL 100, 500, and 1000 µg/mL	Vitrification	[176]
III	Fish	Mouse	Oocyte	500 ng/mL	Vitrification	[177]
LeIBP	Yeast	Diatom	Diatom	0.1 mg/mL	Cryopreservation	[143]
AFGP	Fish	Carp	Spermatozoa	2–10 mg/mL	Hypothermic	[178]
I III	Fish	Mouse	Pronuclear embryos, 4-cell embryos	0.1 and 1.0 mg/mL 0.1 mg/mL	Vitrification	[92]
I/III	Fish	Sea bream	Spermatozoa	0.1, 1, and 10 µg/mL	Cryopreservation	[165]
AFGP	Fish	Equine	Embryos	20 mg/mL	Hypothermic/ Cryopreservation	[91]
DcAFP	Insect	Mouse	A10 smooth muscle cell	1 µg/mL	Cryopreservation	[179]

Table 1. *Cont.*

AFPs	Origin Species of AFPs	Cryopreserved Biological Samples		AFP Quantities Used	Freezing Methods	References
		Organisms	Sample Types			
AFGP	*Gadus morhua*	Mouse	Embryos	20 mg/mL	Vitrification	[180]
III	Fish	Mouse	Ovarian tissue	0, 5, and 20 mg/mL	Vitrification	[67]
III	Fish	Mouse	Mature oocyte	2.5 mg/mL	Vitrification	[181]
I	Fish	Rat	Hippocampal slice cultures	10 mg/mL	Hypothermic	[182]
AFGP	Fish	Pig	Oocyte	40 mg/mL	Vitrification	[183]
AFGP	Fish	Mouse	Embryos	20 mg/mL	Vitrification	[183]
LeIBP	Yeast	Human	Red blood cells	0.4–0.8 mg/mL	Cryopreservation	[144]
III	Fish	Rat	Heart	3, 5, and 15 mg/mL	Hypothermic	[184]
AFGP	Fish	Rat	Cardiomyoctes	0.5–10 mg/mL	Hypothermic (−4 °C)	[162]
III	Fish	Mouse	Oocytes	500 ng/mL	Cryopreservation	[185]
AFGP	Fish	Rat	Cardiac	10 μg/mL, 10 and 15 mg/mL	Hypothermic	[164]
I/III	Fish	Zebra fish	Embryo	40 μg/mL	Hypothermic	[186]
NgIBP	Diatom	Human	Red blood cells	25, 50, and 77 μg/mL	Cryopreservation	[18]
I/III	Fish	Zebra fish	Embryo	40 μg/mL	Vitrification/Cryopreservation	[187]
DcAFP	Insect	Centipede	Gut cells	0.02 mg/mL	Cryopreservation	[188]
III	Fish	Rat	Hearts	15 mg/mL	Hypothermic	[189]
AFGP	Fish	Mouse	Oocytes	1 mg/mL	Cryopreservation	[190]
I/III	Fish	Mouse	Blastocysts	0.1, 1.0 mg/mL	Cryopreservation	[191]
I/III/AFGP	Fish	Mouse	Spermatozoa	1–100 μg/mL	Cryopreservation	[192]
AFGP	Synthetic	Rat	Islet cell	500 μg/mL	Cryopreservation	[193]
I/II/III/AFGP	Fish	Mouse	Oocytes	20 mg/mL	Vitrification	[194]
I	Fish	Human	Myelogenous leukemia cells	0–1000 μg/mL	Cryopreservation	[163]
III	Ocean pout	Chimpanzee (*Pan troglodytes*)	Spermatozoa	1, 10, and 100 μg/mL	Cryopreservation	[195]

Table 1. *Cont.*

AFPs	Origin Species of AFPs	Cryopreserved Biological Samples		AFP Quantities Used	Freezing Methods	References
		Organisms	Sample Types			
I	Fish	Seabream	Embryos	20 nL of 10 mg/mL type I AFP injected	Vitrification	[196]
III	Fish	Turbot	Embryos	10 mg/mL	Hypothermic	[166]
I	Fish	Rat	Liver	1 mg/mL	Hypothermic	[94]
I/III	Fish	Rat	Hearts	10, 15, and 20 mg/mL	Hypothermic	[69]
I/II/III	Fish	Human	Red blood cells	0–1.54 mg/mL	Cryopreservation	[197]
I	Fish	Human	Red blood cells	5–160 µg/mL	Cryopreservation	[68]
III FfIBP LeIBP	Eel pout Bacteria Yeast	Mouse	Oocyte	0.1 mg/mL 0.05 mg/mL 0.1 mg/mL	Vitrification	[66]
III	Eel pout	Bovine	Oocyte	0.5–1 µg/mL	Vitrification	[198]
AFGP8	Fish	Bovine	Oocyte	1 mM (2.6 mg/mL)	Vitrification	[199]
DcAFP	Beetle *Dendroides canadensis*	Buffalo	Semen	0.1, 1.0, and 10 µg/mL	Cryopreservation	[200]
LeIBP	Yeast	Human	Cell lines	0.1 mg/mL	Cryopreservation	[145]

In the cryopreservation of cell lines, AFPs have been used as additives to conventional freezing medium to reduce the high amount of cytotoxic CPAs and reduce freezing damage [31]. Some of the cell types tested for cryopreservation with the addition of AFPs include sperms [167–171,176,178,200], oocytes [66,70,83,177,181,183,185,190,198,199], human liver cells [173], RIN-5F insulin tumor cells [174], diatoms [143], red blood cells [18,144,197], muscle cells [162,179], gut cell [188], islet cells [193], *E. coli* [83], and human cell lines [145] including HeLa cells, NIH/3T3 cells, preosteoblasts (MC3T3-E1 cells), and human ketatinocytes (HaCaT cells). Thus, the addition of AFPs seems to mainly enhance the cryopreservation efficiency regardless of cell type and freezing method, with a handful of exceptions [68,89,90,162,164]. Notably, these exceptions appear to be related to the concentration of AFPs used; indeed, at higher concentrations, AFPs form needle-like ice, which penetrates and destroys cells during freezing [68,143–145,162,193,201]. The amount used in cryopreservation also differs between AFPs. LeIBP, which shows lower TH activity, but higher IRI activity than fish AFPs has been used in the range of 0.1–0.8 mg/mL in red blood cells [144], diatoms [143], oocytes [66], and mammalian cell lines [145], whereas fish AFPs have been used at concentrations lower than 0.1 mg/mL, depending on the cell type (Table 1). Interestingly, in the vitrification of mouse oocytes, 0.05 mg/mL FfIBP is more effective at maintaining in murine oocyte quality and embryo development than 0.1 mg/mL LeIBP and 0.1 mg/mL type III AFP [66]. Since the results obtained vary between studies, the utilization of AFPs in cryopreservation needs fine-tuning depending on the type of AFPs, cells, freezing media, and storage temperature.

Embryos from fish [166,186,187,196,202], cows [172,203], sheep [175], rabbits [176], mice [67,92,183,191], and horses [91] were preserved in the presence of AFPs. Early attempts with equine and mouse embryos demonstrated that fish AFPs had negligible effects [91,92]; however, fish embryos subjected to microinjection or incubation in type I AFP solution showed significantly increased survival after chilling at 4 °C or −10 °C. Vitrified 5-somite embryos in type I AFP solution showed similar survival to that of cells recovered from unfrozen embryos [187]. Similarly, AFPs can help improve the survival of embryos preserved at hypothermic temperatures [150,153,154]. These promising results may fuel research in not only hypothermic storage but also vitrification of other embryos such as mammalian embryos.

Lee and colleagues evaluated the beneficial effects of AFPs in vitrification of mouse ovarian tissues [65,67]. Ovarian tissues treated with type III AFPs showed significantly higher intact follicle ratios and lower apoptotic follicle rates than control tissues. The transplanted vitrified-warmed ovaries showed higher intact follicle ratios [65]. In another attempt, all AFP-treated groups had significantly improved follicle preservation with decreasing efficiency in the order of LeIBP > FfIBP > type III AFP [65].

Few studies have evaluated the potential use of AFPs in the hypothermic storage of organs [94,164,184,204]. The TH activity of AFPs has been exploited for subzero preservation of organs. As anticipated, the presence of AFPs decreases cold-induced injury during the hypothermic storage of rat livers [204] and mammalian hearts [69,205] by decreasing the ice formation [189,204,206]. In contrast, Wang et al. reported that higher concentrations of AFGPs have adverse effects on heart preservation [164].

5. Conclusions and Perspective

Thanks to their unique properties as biological antifreezes, AFPs have attracted interest from researchers in academia and biomedical fields. In this review, we surveyed the past and current trends in cryopreservation applications of AFPs. The first property of freezing point depression, termed TH, has typically been utilized primarily in the hypothermic storage of tissues and organs. Due to the complexity and size of tissues and organs, more advancement is needed to achieve effective hypothermic storage of these biological materials. The ability to inhibit ice recrystallization is known to neutralize the catastrophic large icy environment for the cryopreserved cells during freezing and/or warming. The third and less characterized function of AFPs is the interaction with cellular membranes and/or integral membrane proteins. It is not likely that these interactions themselves can confer the cryopreserved cells with post-thaw viability. However, AFPs are thought to augment the viability or cryopreservation efficiency of the cells together with the other two features, particularly IRI.

Our physicochemical understanding of unique binding of AFPs to ice crystal has been the main focus of scientists within the last five decades [51,207]. Relatively few studies have evaluated the application of AFPs in cryopreservation. This is mainly because AFPs are expensive to obtain. Therefore, prior and current cryopreservation research has been limited only to moderately active fish AFPs. Additionally, the applications of AFPs has still only partly characterized based on empirical features, similar to other CPAs [31]. In other words, researchers still need to determine the optimal working concentrations of AFPs in cryopreservation; neither TH nor IRI can provide this information.

For the application of AFPs to be practical, a few questions should be addressed. First, mass production of AFPs should be established. Currently, only type III AFP has been produced on the industrial scale owing to its use as an ingredient in ice cream. However, advancements in molecular biology and genomics have improved our ability to produce genes and proteins easily, expanding AFP-related research. Indeed, a mass production system for LeIBP, *Glaciozyma* IBP, has been reported [142]. Additionally, the LeIBP has been shown to yield better post-thaw viability in several studies compared with that of other marine-derived AFPs [65,66,143–145]. Second, the behaviors of AFPs in freezing medium should be characterized thoroughly. Typically, freezing medium contains high concentrations of chemicals, such as DMSO, ethylene glycol, polyvinylpyrrolidone, and polyethylene glycol, which may destabilize AFPs, leading to loss of function [208]. Third, functionalized AFPs should be engineered and developed to overcome the limitations of natural counterparts. Mother nature has suggested the use of RD3 and an IBP from Vostok glacial bacterium [137,209]. In both cases, connecting two almost homologous domains increases the TH value cooperatively compared with their monomeric AFPs [139,209]. Studies from the laboratories of Tsuda, Davies, and Holland have demonstrated that the multimerization of native type III AFP can increase TH activity [210–212]. Recently, Steven et al. claimed the dendrimer-like AFPs showed higher TH values [213], and Phippen et al. demonstrated 12 AFP-fused protein cage nanoparticles that increased the TH value to more than 50-fold that of monomeric AFP [214]. A few groups have attempted to synthesize AFP or AFGP derivatives to elucidate the underlying mechanism of action and to develop practical applications [215–221]. Another interesting approach is the development of cell-internalizable or -penetrating AFPs. AFPs are usually nonpenetrating, such that the internal ice formation should be inhibited by high amounts of cytotoxic CPAs. Cell-internalizable AFPs may also reduce the amount of CPAs in freezing medium, eventually increasing the efficiency of cryopreservation.

Finally, it is encouraging that many research groups studying AFP worldwide have started expanding their research into cryopreservation using AFPs. We hope these concerted efforts will accelerate the development of biomedical application of AFPs.

Acknowledgments: We thank the anonymous reviewers for their careful reading of our manuscript and their insightful comments and suggestions. This work is supported by the Polar Genomics 101 Project: Genome analysis of polar organisms and establishment of application platform (PE17080) funded by the Korea Polar Research Institute (KOPRI), and National Fisheries Research and Development Institute (R2017013 to Y.B.H).

Conflicts of Interest: The authors declare no conflict of interest.

References

1. DeVries, A.L.; Wohlschlag, D.E. Freezing resistance in some Antarctic fishes. *Science* **1969**, *163*, 1073–1075. [CrossRef] [PubMed]
2. DeVries, A.L.; Komatsu, S.K.; Feeney, R.E. Chemical and physical properties of freezing point-depressing glycoproteins from Antarctic fishes. *J. Biol. Chem.* **1970**, *245*, 2901–2908. [PubMed]
3. DeVries, A.L. Glycoproteins as biological antifreeze agents in antarctic fishes. *Science* **1971**, *172*, 1152–1155. [CrossRef] [PubMed]

4. DeVries, A.L. Freezing resistance in fishes of the Antarctic penninsula. *Antarct. J. US* **1969**, *4*, 104–105.
5. Davies, P.L. Ice-binding proteins: A remarkable diversity of structures for stopping and starting ice growth. *Trends Biochem. Sci.* **2014**, *39*, 548–555. [CrossRef] [PubMed]
6. Gilbert, J.A.; Davies, P.L.; Laybourn-Parry, J. A hyperactive, Ca^{2+}-dependent antifreeze protein in an Antarctic bacterium. *FEMS Microbiol. Lett.* **2005**, *245*, 67–72. [CrossRef] [PubMed]
7. Do, H.; Kim, S.J.; Kim, H.J.; Lee, J.H. Structure-based characterization and antifreeze properties of a hyperactive ice-binding protein from the Antarctic bacterium *Flavobacterium frigoris* PS1. *Acta Crystallogr. D Biol. Crystallogr.* **2014**, *70*, 1061–1073. [CrossRef] [PubMed]
8. Raymond, J.A.; Fritsen, C.; Shen, K. An ice-binding protein from an Antarctic sea ice bacterium. *FEMS Microbiol. Ecol.* **2007**, *61*, 214–221. [CrossRef] [PubMed]
9. Singh, P.; Hanada, Y.; Singh, S.M.; Tsuda, S. Antifreeze protein activity in Arctic cryoconite bacteria. *FEMS Microbiol. Lett.* **2014**, *351*, 14–22. [CrossRef] [PubMed]
10. Lee, J.K.; Park, K.S.; Park, S.; Park, H.; Song, Y.H.; Kang, S.H.; Kim, H.J. An extracellular ice-binding glycoprotein from an Arctic psychrophilic yeast. *Cryobiology* **2010**, *60*, 222–228. [CrossRef] [PubMed]
11. Boo, S.Y.; Wong, C.M.V.L.; Rodrigues, K.F.; Najimudin, N.; Murad, A.M.A.; Mahadi, N.M. Thermal stress responses in Antarctic yeast, *Glaciozyma antarctica* PI12, characterized by real-time quantitative PCR. *Polar Biol.* **2013**, *36*, 381–389. [CrossRef]
12. Hashim, N.H.; Bharudin, I.; Nguong, D.L.; Higa, S.; Bakar, F.D.; Nathan, S.; Rabu, A.; Kawahara, H.; Illias, R.M.; Najimudin, N.; et al. Characterization of Afp1, an antifreeze protein from the psychrophilic yeast *Glaciozyma antarctica* PI12. *Extremophiles* **2013**, *17*, 63–73. [CrossRef] [PubMed]
13. Kiko, R. Acquisition of freeze protection in a sea-ice crustacean through horizontal gene transfer? *Polar Biol.* **2010**, *33*, 543–556. [CrossRef]
14. Jung, W.; Gwak, Y.; Davies, P.L.; Kim, H.J.; Jin, E. Isolation and characterization of antifreeze proteins from the antarctic marine microalga *Pyramimonas gelidicola*. *Mar. Biotechnol.* **2014**, *16*, 502–512. [CrossRef] [PubMed]
15. Gwak, I.G.; Jung, W.; Kim, H.J.; Kang, S.H.; Jin, E. Antifreeze protein in Antarctic marine diatom, *Chaetoceros neogracile*. *Mar. Biotechnol.* **2009**, *12*, 630–639. [CrossRef] [PubMed]
16. Janech, M.; Krell, A.; Mock, T.; Kang, J.-S.; Raymond, J. Ice-binding proteins from sea ice diatoms (bacillariophyceae). *J. Phycol.* **2006**, *42*, 410–416. [CrossRef]
17. Krell, A.; Beszteri, B.; Dieckmann, G.; Glöckner, G.; Valentin, K.; Mock, T. A new class of ice-binding proteins discovered in a salt-stress-induced cDNA library of the psychrophilic diatom *Fragilariopsis cylindrus* (Bacillariophyceae). *Eur. J. Phycol.* **2008**, *43*, 423–433. [CrossRef]
18. Kang, J.S.; Raymond, J.A. Reduction of freeze-thaw-induced hemolysis of red blood cells by an algal ice-binding protein. *Cryo Lett.* **2004**, *25*, 307–310.
19. Raymond, J.A.; Janech, M.; Fritsen, C. Novel ice-binding proteins from a psychrophilic antarctic alga (Chlamydomonadaceae, chlorophyceae). *J. Phycol.* **2009**, *45*, 130–136. [CrossRef] [PubMed]
20. Davies, P.L.; Hew, C.L. Biochemistry of fish antifreeze proteins. *FASEB J.* **1990**, *4*, 2460–2468. [PubMed]
21. Hanada, Y.; Nishimiya, Y.; Miura, A.; Tsuda, S.; Kondo, H. Hyperactive antifreeze protein from an Antarctic sea ice bacterium *Colwellia* sp. has a compound ice-binding site without repetitive sequences. *FEBS J.* **2014**, *281*, 3576–3590. [CrossRef] [PubMed]
22. Shah, S.H.H.; Kar, R.K.; Asmawi, A.A.; Rahman, M.B.A.; Murad, A.M.A.; Mahadi, N.M.; Basri, M.; Rahman, R.N.Z.A.; Salleh, A.B.; Chatterjee, S.; et al. Solution structures, dynamics, and ice growth inhibitory activity of peptide fragments derived from an antarctic yeast protein. *PLoS ONE* **2012**, *7*, e49788. [CrossRef] [PubMed]
23. Bayer-Giraldi, M.; Weikusat, I.; Besir, H.; Dieckmann, G. Characterization of an antifreeze protein from the polar diatom *Fragilariopsis cylindrus* and its relevance in sea ice. *Cryobiology* **2011**, *63*, 210–219. [CrossRef] [PubMed]
24. Uhlig, C.; Kabisch, J.; Palm, G.J.; Valentin, K.; Schweder, T.; Krell, A. Heterologous expression, refolding and functional characterization of two antifreeze proteins from *Fragilariopsis cylindrus* (Bacillariophyceae). *Cryobiology* **2011**, *63*, 220–228. [CrossRef] [PubMed]
25. Hew, C.L.; Davies, P.L.; Fletcher, G. Antifreeze protein gene transfer in Atlantic salmon. *Mol. Mar. Biol. Biotechnol.* **1992**, *1*, 309–317. [PubMed]

26. Wohrmann, A.P. Antifreeze glycopeptides of the high-Antarctic silverfish *Pleuragramma antarcticum* (Notothenioidei). *Comp. Biochem. Physiol. C Pharmacol. Toxicol. Endocrinol.* **1995**, *111*, 121–129. [CrossRef]

27. Barrett, J. Thermal hysteresis proteins. *Int. J. Biochem. Cell Biol.* **2001**, *33*, 105–117. [CrossRef]

28. Ben, R.N. Antifreeze glycoproteins-preventing the growth of ice. *Chembiochem* **2001**, *2*, 161–166. [CrossRef]

29. Bouvet, V.; Ben, R.N. Antifreeze glycoproteins. *Cell Biochem. Biophys.* **2003**, *39*, 133–144. [CrossRef]

30. Harding, M.M.; Anderberg, P.I.; Haymet, A.D. "Antifreeze" glycoproteins from polar fish. *Eur. J. Biochem.* **2003**, *270*, 1381–1392. [CrossRef] [PubMed]

31. Fuller, B.J. Cryoprotectants: The essential antifreezes to protect life in the frozen state. *Cryo Lett.* **2004**, *25*, 375–388.

32. Kristiansen, E.; Zachariassen, K.E. The mechanism by which fish antifreeze proteins cause thermal hysteresis. *Cryobiology* **2005**, *51*, 262–280. [CrossRef] [PubMed]

33. Bar Dolev, M.; Braslavsky, I.; Davies, P.L. Ice-binding proteins and their function. *Annu. Rev. Biochem.* **2016**, *85*, 515–542. [CrossRef] [PubMed]

34. Raymond, J.A.; DeVries, A.L. Freezing behavior of fish blood glycoproteins with antifreeze properties. *Cryobiology* **1972**, *9*, 541–547. [CrossRef]

35. Raymond, J.A.; DeVries, A.L. Adsorption inhibition as a mechanism of freezing resistance in polar fishes. *Proc. Natl. Acad. Sci. USA* **1977**, *74*, 2589–2593. [CrossRef] [PubMed]

36. Wilson, P.W.; Beaglehole, D.; Devries, A.L. Antifreeze glycopeptide adsorption on single crystal ice surfaces using ellipsometry. *Biophys. J.* **1993**, *64*, 1878–1884. [CrossRef]

37. Wilson, P.W. Explaining thermal hysteresis by the Kelvin effect. *Cryo Lett.* **1993**, *14*, 31–36.

38. Wilson, P.W.; Leader, J.P. Stabilization of supercooled fluids by thermal hysteresis proteins. *Biophys. J.* **1995**, *68*, 2098–2107. [CrossRef]

39. Celik, Y.; Drori, R.; Pertaya-Braun, N.; Altan, A.; Barton, T.; Bar-Dolev, M.; Groisman, A.; Davies, P.L.; Braslavsky, I. Microfluidic experiments reveal that antifreeze proteins bound to ice crystals suffice to prevent their growth. *Proc. Natl. Acad. Sci. USA* **2013**, *110*, 1309–1314. [CrossRef] [PubMed]

40. Knight, C.A.; Cheng, C.C.; DeVries, A.L. Adsorption of alpha-helical antifreeze peptides on specific ice crystal surface planes. *Biophys. J.* **1991**, *59*, 409–418. [CrossRef]

41. Do, H.; Lee, J.H.; Lee, S.G.; Kim, H.J. Crystallization and preliminary X-ray crystallographic analysis of an ice-binding protein (FfIBP) from *Flavobacterium frigoris* PS1. *Acta Crystallogr. Sect. F Struct. Biol. Cryst. Commun.* **2012**, *68*, 806–809. [CrossRef] [PubMed]

42. Drori, R.; Celik, Y.; Davies, P.L.; Braslavsky, I. Ice-binding proteins that accumulate on different ice crystal planes produce distinct thermal hysteresis dynamics. *J. R. Soc. Interface* **2014**, *11*, 2014526. [CrossRef] [PubMed]

43. Pertaya, N.; Marshall, C.B.; Celik, Y.; Davies, P.L.; Braslavsky, I. Direct visualization of spruce budworm antifreeze protein interacting with ice crystals: Basal plane affinity confers hyperactivity. *Biophys. J.* **2008**, *95*, 333–341. [CrossRef] [PubMed]

44. Park, K.S.; Do, H.; Lee, J.H.; Park, S.I.; Kim, E.J.; Kim, S.J.; Kang, S.H.; Kim, H.J. Characterization of the ice-binding protein from Arctic yeast *Leucosporidium* sp. AY30. *Cryobiology* **2012**, *64*, 286–296. [CrossRef] [PubMed]

45. Fletcher, G.L.; Hew, C.L.; Davies, P.L. Antifreeze proteins of teleost fishes. *Annu. Rev. Physiol.* **2001**, *63*, 359–390. [CrossRef] [PubMed]

46. Jung, W.; Campbell, R.L.; Gwak, Y.; Kim, J.I.; Davies, P.L.; Jin, E. New cysteine-rich ice-binding protein secreted from Antarctic microalga, *Chloromonas* sp. *PLoS ONE* **2016**, *11*, e0154056. [CrossRef] [PubMed]

47. Knight, C.A.; DeVries, A.L.; Oolman, L.D. Fish antifreeze protein and the freezing and recrystallization of ice. *Nature* **1984**, *308*, 295–296. [CrossRef] [PubMed]

48. Knight, C.A.; Wen, D.; Laursen, R.A. Nonequilibrium antifreeze peptides and the recrystallization of ice. *Cryobiology* **1995**, *32*, 23–34. [CrossRef] [PubMed]

49. Raymond, J.A.; Fritsen, C.H. Semipurification and ice recrystallization inhibition activity of ice-active substances associated with Antarctic photosynthetic organisms. *Cryobiology* **2001**, *43*, 63–70. [CrossRef] [PubMed]

50. Raymond, J.A.; Knight, C.A. Ice binding, recrystallization inhibition, and cryoprotective properties of ice-active substances associated with Antarctic sea ice diatoms. *Cryobiology* **2003**, *46*, 174–181. [CrossRef]

51. Jia, Z.; Davies, P.L. Antifreeze proteins: An unusual receptor-ligand interaction. *Trends Biochem. Sci.* **2002**, *27*, 101–106. [CrossRef]

52. Knight, C.A.; Duman, J.G. Inhibition of recrystallization of ice by insect thermal hysteresis proteins: A possible cryoprotective role. *Cryobiology* **1986**, *23*, 256–262. [CrossRef]

53. Worrall, D.; Elias, L.; Ashford, D.; Smallwood, M.; Sidebottom, C.; Lillford, P.; Telford, J.; Holt, C.; Bowles, D. A carrot leucine-rich-repeat protein that inhibits ice recrystallization. *Science* **1998**, *282*, 115–117. [CrossRef] [PubMed]

54. Smallwood, M.; Worrall, D.; Byass, L.; Elias, L.; Ashford, D.; Doucet, C.J.; Holt, C.; Telford, J.; Lillford, P.; Bowles, D.J. Isolation and characterization of a novel antifreeze protein from carrot (*Daucus carota*). *Biochem. J.* **1999**, *340*, 385–391. [CrossRef] [PubMed]

55. Sidebottom, C.; Buckley, S.; Pudney, P.; Twigg, S.; Jarman, C.; Holt, C.; Telford, J.; McArthur, A.; Worrall, D.; Hubbard, R.; et al. Heat-stable antifreeze protein from grass. *Nature* **2000**, *406*, 256. [CrossRef] [PubMed]

56. Hew, C.L.; Yang, D.S. Protein interaction with ice. *Eur. J. Biochem.* **1992**, *203*, 33–42. [CrossRef] [PubMed]

57. Wohrmann, A. Antifreeze glycoproteins in fishes: Structure, mode of action and possible applications. *Tierarztl. Prax.* **1996**, *24*, 1–9. [PubMed]

58. Knight, C.A.; Hallett, J.; DeVries, A.L. Solute effects on ice recrystallization: An assessment technique. *Cryobiology* **1988**, *25*, 55–60. [CrossRef]

59. Tomczak, M.M.; Marshall, C.B.; Gilbert, J.A.; Davies, P.L. A facile method for determining ice recrystallization inhibition by antifreeze proteins. *Biochem. Biophys. Res. Commun.* **2003**, *311*, 1041–1046. [CrossRef] [PubMed]

60. Jackman, J.; Noestheden, M.; Moffat, D.; Pezacki, J.P.; Findlay, S.; Ben, R.N. Assessing antifreeze activity of AFGP 8 using domain recognition software. *Biochem. Biophys. Res. Commun.* **2007**, *354*, 340–344. [CrossRef] [PubMed]

61. Olijve, L.L.C.; Oude Vrielink, A.S.; Voets, I.K. A simple and quantitative method to evaluate ice recrystallization kinetics using the circle Hough Transform algorithm. *Cryst. Growth Des.* **2016**, *16*, 4190–4195. [CrossRef]

62. Olijve, L.L.C.; Meister, K.; DeVries, A.L.; Duman, J.G.; Guo, S.; Bakker, H.J.; Voets, I.K. Blocking rapid ice crystal growth through nonbasal plane adsorption of antifreeze proteins. *Proc. Natl. Acad. Sci. USA* **2016**, *113*, 3740–3745. [CrossRef] [PubMed]

63. Yu, S.O.; Brown, A.; Middleton, A.J.; Tomczak, M.M.; Walker, V.K.; Davies, P.L. Ice restructuring inhibition activities in antifreeze proteins with distinct differences in thermal hysteresis. *Cryobiology* **2010**, *61*, 327–334. [CrossRef] [PubMed]

64. Lee, J.H.; Park, A.K.; Do, H.; Park, K.S.; Moh, S.H.; Chi, Y.M.; Kim, H.J. Structural basis for the antifreeze activity of an ice-binding protein from an Arctic yeast. *J. Biol. Chem.* **2012**, *287*, 11460–11468. [CrossRef] [PubMed]

65. Lee, J.; Kim, S.K.; Youm, H.W.; Kim, H.J.; Lee, J.R.; Suh, C.S.; Kim, S.H. Effects of three different types of antifreeze proteins on mouse ovarian tissue cryopreservation and transplantation. *PLoS ONE* **2015**, *10*, e0126252. [CrossRef] [PubMed]

66. Lee, H.H.; Lee, H.J.; Kim, H.J.; Lee, J.H.; Ko, Y.; Kim, S.M.; Lee, J.R.; Suh, C.S.; Kim, S.H. Effects of antifreeze proteins on the vitrification of mouse oocytes: Comparison of three different antifreeze proteins. *Hum. Reprod.* **2015**, *30*, 2110–2119. [CrossRef] [PubMed]

67. Lee, J.R.; Youm, H.W.; Lee, H.J.; Jee, B.C.; Suh, C.S.; Kim, S.H. Effect of antifreeze protein on mouse ovarian tissue cryopreservation and transplantation. *Yonsei Med. J.* **2015**, *56*, 778–784. [CrossRef] [PubMed]

68. Carpenter, J.F.; Hansen, T.N. Antifreeze protein modulates cell survival during cryopreservation: Mediation through influence on ice crystal growth. *Proc. Natl. Acad. Sci. USA* **1992**, *89*, 8953–8957. [CrossRef] [PubMed]

69. Amir, G.; Rubinsky, B.; Kassif, Y.; Horowitz, L.; Smolinsky, A.K.; Lavee, J. Preservation of myocyte structure and mitochondrial integrity in subzero cryopreservation of mammalian hearts for transplantation using antifreeze proteins-an electron microscopy study. *Eur. J. Cardiothorac. Surg.* **2003**, *24*, 292–297. [CrossRef]

70. Rubinsky, B.; Arav, A.; Fletcher, G.L. Hypothermic protection-a fundamental property of "antifreeze" proteins. *Biochem. Biophys. Res. Commun.* **1991**, *180*, 566–571. [CrossRef]

71. Quinn, P.J. A lipid-phase separation model of low-temperature damage to biological membranes. *Cryobiology* **1985**, *22*, 128–146. [CrossRef]

72. Hays, L.M.; Feeney, R.E.; Crowe, L.M.; Crowe, J.H.; Oliver, A.E. Antifreeze glycoproteins inhibit leakage from liposomes during thermotropic phase transitions. *Proc. Natl. Acad. Sci. USA* **1996**, *93*, 6835–6840. [CrossRef] [PubMed]

73. Tomczak, M.M.; Hincha, D.K.; Estrada, S.D.; Wolkers, W.F.; Crowe, L.M.; Feeney, R.E.; Tablin, F.; Crowe, J.H. A mechanism for stabilization of membranes at low temperatures by an antifreeze protein. *Biophys. J.* **2002**, *82*, 874–881. [CrossRef]

74. Tomczak, M.M.; Vigh, L.; Meyer, J.D.; Manning, M.C.; Hincha, D.K.; Crowe, J.H. Lipid unsaturation determines the interaction of AFP type I with model membranes during thermotropic phase transitions. *Cryobiology* **2002**, *45*, 135–142. [CrossRef]

75. Tablin, F.; Oliver, A.E.; Walker, N.J.; Crowe, L.M.; Crowe, J.H. Membrane phase transition of intact human platelets: Correlation with cold-induced activation. *J. Cell. Physiol.* **1996**, *168*, 305–313. [CrossRef]

76. Crowe, J.H.; Crowe, L.M. Water and carbohydrate interactions with membranes: Studies with infrared spectroscopy and differential scanning calorimetry methods. *Methods Enzymol.* **1986**, *127*, 696–703. [PubMed]

77. Hays, L.M.; Crowe, J.H.; Wolkers, W.; Rudenko, S. Factors affecting leakage of trapped solutes from phospholipid vesicles during thermotropic phase transitions. *Cryobiology* **2001**, *42*, 88–102. [CrossRef] [PubMed]

78. Tomczak, M.M.; Hincha, D.K.; Estrada, S.D.; Feeney, R.E.; Crowe, J.H. Antifreeze proteins differentially affect model membranes during freezing. *Biochim. Biophys. Acta* **2001**, *1511*, 255–263. [CrossRef]

79. Kun, H.; Byk, G.; Mastai, Y. Effects of antifreeze protein fragments on the properties of model membranes. *Adv. Exp. Med. Biol.* **2009**, *611*, 85–86. [PubMed]

80. Wu, Y.; Fletcher, G.L. Efficacy of antifreeze protein types in protecting liposome membrane integrity depends on phospholipid class. *Biochim. Biophys. Acta* **2001**, *1524*, 11–16. [CrossRef]

81. Kun, H.; Mastai, Y. Isothermal calorimetry study of the interactions of type I antifreeze proteins with a lipid model membrane. *Protein Pept. Lett.* **2009**, *17*, 739–743. [CrossRef]

82. Tomczak, M.M.; Hincha, D.K.; Crowe, J.H.; Harding, M.M.; Haymet, A.D. The effect of hydrophobic analogues of the type I winter flounder antifreeze protein on lipid bilayers. *FEBS Lett.* **2003**, *551*, 13–19. [CrossRef]

83. Rubinsky, B.; Arav, A.; Mattioli, M.; Devries, A.L. The effect of antifreeze glycopeptides on membrane potential changes at hypothermic temperatures. *Biochem. Biophys. Res. Commun.* **1990**, *173*, 1369–1374. [CrossRef]

84. Negulescu, P.A.; Rubinsky, B.; Fletcher, G.L.; Machen, T.E. Fish antifreeze proteins block Ca entry into rabbit parietal cells. *Am. J. Physiol.* **1992**, *263*, C1310–C1313. [PubMed]

85. Arav, A.; Rubinsky, B.; Seren, E.; Roche, J.F.; Boland, M.P. The role of thermal hysteresis proteins during cryopreservation of oocytes and embryos. *Theriogenology* **1994**, *41*, 107–112. [CrossRef]

86. Rubinsky, B.; Mattioli, M.; Arav, A.; Barboni, B.; Fletcher, G.L. Inhibition of Ca^{2+} and K^+ currents by "antifreeze" proteins. *Am. J. Physiol.* **1992**, *262*, R542–R545. [PubMed]

87. Wang, J.H.; Bian, H.W.; Huang, C.N.; Ge, J.G. Studies on the application of antifreeze proteins in cryopreservation of rice suspension cells. *Shi Yan Sheng Wu Xue Bao* **1999**, *32*, 271–276. [PubMed]

88. Wang, L.H.; Wusteman, M.C.; Smallwood, M.; Pegg, D.E. The stability during low-temperature storage of an antifreeze protein isolated from the roots of cold-acclimated carrots. *Cryobiology* **2002**, *44*, 307–310. [CrossRef]

89. Ishiguro, H.; Rubinsky, B. Influence of fish antifreeze proteins on the freezing of cell suspensions with cryoprotectant penetrating cells. *Int. J. Heat Mass Transf.* **1998**, *41*, 1907–1915. [CrossRef]

90. Payne, S.R.; Oliver, J.E.; Upreti, G.C. Effect of antifreeze proteins on the motility of ram spermatozoa. *Cryobiology* **1994**, *31*, 180–184. [CrossRef] [PubMed]

91. Lagneaux, D.; Huhtinen, M.; Koskinen, E.; Palmer, E. Effect of anti-freeze protein (AFP) on the cooling and freezing of equine embryos as measured by DAPI-staining. *Equine Vet. J. Suppl.* **1997**, *25*, 85–87. [CrossRef]

92. Shaw, J.M.; Ward, C.; Trounson, A.O. Evaluation of propanediol, ethylene glycol, sucrose and antifreeze proteins on the survival of slow-cooled mouse pronuclear and 4-cell embryos. *Hum. Reprod.* **1995**, *10*, 396–402. [PubMed]

93. Mezhevikina, L.M.; Karanova, M.V. The use of antifreeze glycoproteins in the freezing in liquid nitrogen of early mouse embryos. *Izv. Akad. Nauk. Seriia Biol. Akad. Nauk.* **1994**, *2*, 172–177.

94. Soltys, K.A.; Batta, A.K.; Koneru, B. Successful nonfreezing, subzero preservation of rat liver with 2,3-butanediol and type I antifreeze protein. *J. Surg. Res.* **2001**, *96*, 30–34. [CrossRef] [PubMed]

95. Zhang, E.; Zhang, L.; Wang, B.; Yan, B.; Wang, Q. Cryopreservation of marine diatom algae by encapsulation-vitrification. *Cryo Lett.* **2009**, *30*, 224–231.

96. Larese, A.; Acker, J.; Muldrew, K.; Yang, H.Y.; McGann, L. Antifreeze proteins induce intracellular nucleation. *Cryoletters* **1996**, *17*, 175–182.

97. Scholander, P.F.; Van Dam, L.; Kanwisher, J.W.; Hammel, H.T.; Gordon, M.S. Supercooling and osmoregulation in Arctic fish. *J. Cell. Physiol.* **1957**, *49*, 5–24. [CrossRef]

98. Komatsu, S.K.; DeVries, A.L.; Feeney, R.E. Studies of the structure of freezing point-depressing glycoproteins from an Antarctic fish. *J. Biol. Chem.* **1970**, *245*, 2909–2913. [PubMed]

99. Duman, J.G.; Devries, A.L. Freezing resistance in winter flounder *Pseudopleuronectes americanus*. *Nature* **1974**, *247*, 237–238. [CrossRef]

100. Slaughter, D.; Fletcher, G.L.; Ananthanarayanan, V.S.; Hew, C.L. Antifreeze proteins from the sea raven, *Hemitripterus americanus*. Further evidence for diversity among fish polypeptide antifreezes. *J. Biol. Chem.* **1981**, *256*, 2022–2026. [PubMed]

101. Feeney, R.E.; Yeh, Y. Antifreeze proteins from fish bloods. *Adv. Protein Chem.* **1978**, *32*, 191–282. [PubMed]

102. Burcham, T.S.; Osuga, D.T.; Chino, H.; Feeney, R.E. Analysis of antifreeze glycoproteins in fish serum. *Anal. Biochem.* **1984**, *139*, 197–204. [CrossRef]

103. Kao, M.H.; Fletcher, G.L.; Wang, N.C.; Hew, C.L. The relationship between molecular weight and antifreeze polypeptide activity in marine fish. *Can. J. Zool.* **1986**, *64*, 578–582. [CrossRef]

104. Ahlgren, J.A.; Cheng, C.C.; Schrag, J.D.; DeVries, A.L. Freezing avoidance and the distribution of antifreeze glycopeptides in body fluids and tissues of Antarctic fish. *J. Exp. Biol.* **1988**, *137*, 549–563. [PubMed]

105. Hew, C.L.; Wang, N.C.; Yan, S.; Cai, H.; Sclater, A.; Fletcher, G.L. Biosynthesis of antifreeze polypeptides in the winter flounder. *Eur. J. Biochem.* **1986**, *160*, 267–272. [CrossRef] [PubMed]

106. Chao, H.; Houston, M.E., Jr.; Hodges, R.S.; Kay, C.M.; Sykes, B.D.; Loewen, M.C.; Davies, P.L.; Sönnichsen, F.D. A diminished role for hydrogen bonds in antifreeze protein binding to ice. *Biochemistry* **1997**, *36*, 14652–14660. [CrossRef] [PubMed]

107. Haymet, A.D.; Ward, L.G.; Harding, M.M.; Knight, C.A. Valine substituted winter flounder "antifreeze": Preservation of ice growth hysteresis. *FEBS Lett.* **1998**, *430*, 301–306. [CrossRef]

108. Loewen, M.C.; Chao, H.; Houston, M.E., Jr.; Baardsnes, J.; Hodges, R.S.; Kay, C.M.; Sykes, B.D.; Sönnichsen, F.D.; Davies, P.L. Alternative roles for putative ice-binding residues in type I antifreeze protein. *Biochemistry* **1999**, *38*, 4743–4749. [CrossRef] [PubMed]

109. Zhang, W.; Laursen, R.A. Structure-function relationships in a type I antifreeze polypeptide. The role of threonine methyl and hydroxyl groups in antifreeze activity. *J. Biol. Chem.* **1998**, *273*, 34806–34812. [CrossRef] [PubMed]

110. Vasina, E.N.; Paszek, E.; Nicolau, D., Jr.; Nicolau, D.V. The BAD project: Data mining, database and prediction of protein adsorption on surfaces. *Lab Chip* **2009**, *9*, 891–900. [CrossRef] [PubMed]

111. Baardsnes, J.; Kondejewski, L.H.; Hodges, R.S.; Chao, H.; Kay, C.; Davies, P.L. New ice-binding face for type I antifreeze protein. *FEBS Lett.* **1999**, *463*, 87–91. [CrossRef]

112. Ewart, K.V.; Yang, D.S.; Ananthanarayanan, V.S.; Fletcher, G.L.; Hew, C.L. Ca^{2+}-dependent antifreeze proteins. Modulation of conformation and activity by divalent metal ions. *J. Biol. Chem.* **1996**, *271*, 16627–16632. [PubMed]

113. Gronwald, W.; Loewen, M.C.; Lix, B.; Daugulis, A.J.; Sönnichsen, F.D.; Davies, P.L.; Sykes, B.D. The solution structure of type II antifreeze protein reveals a new member of the lectin family. *Biochemistry* **1998**, *37*, 4712–4721. [CrossRef] [PubMed]

114. Drickamer, K. C-type lectin-like domains. *Curr. Opin. Struct. Biol.* **1999**, *9*, 585–590. [CrossRef]

115. Liu, Y.; Li, Z.; Lin, Q.; Kosinski, J.; Seetharaman, J.; Bujnicki, J.M.; Sivaraman, J.; Hew, C.L. Structure and evolutionary origin of Ca^{2+}-dependent herring type II antifreeze protein. *PLoS ONE* **2007**, *2*, e548. [CrossRef] [PubMed]

116. Nishimiya, Y.; Kondo, H.; Takamichi, M.; Sugimoto, H.; Suzuki, M.; Miura, A.; Tsuda, S. Crystal structure and mutational analysis of Ca^{2+}-independent type II antifreeze protein from longsnout poacher, *Brachyopsis rostratus*. *J. Mol. Biol.* **2008**, *382*, 734–746. [CrossRef] [PubMed]

117. Ewart, K.V.; Lin, Q.; Hew, C.L. Structure, function and evolution of antifreeze proteins. *Cell. Mol. Life Sci.* **1999**, *55*, 271–283. [CrossRef] [PubMed]

118. Yamashita, Y.; Miura, R.; Takemoto, Y.; Tsuda, S.; Kawahara, H.; Obata, H. Type II antifreeze protein from a mid-latitude freshwater fish, Japanese smelt (*Hypomesus nipponensis*). *Biosci. Biotechnol. Biochem.* **2003**, *67*, 461–466. [CrossRef] [PubMed]

119. Yeh, Y.; Feeney, R.E. Antifreeze proteins: Structures and mechanisms of function. *Chem. Rev.* **1996**, *96*, 601–618. [CrossRef] [PubMed]

120. Antson, A.A.; Smith, D.J.; Roper, D.I.; Lewis, S.; Caves, L.S.; Verma, C.S.; Buckley, S.L.; Lillford, P.J.; Hubbard, R.E. Understanding the mechanism of ice binding by type III antifreeze proteins. *J. Mol. Biol.* **2001**, *305*, 875–889. [CrossRef] [PubMed]

121. Hew, C.L.; Wang, N.C.; Joshi, S.; Fletcher, G.L.; Scott, G.K.; Hayes, P.H.; Buettner, B.; Davies, P.L. Multiple genes provide the basis for antifreeze protein diversity and dosage in the ocean pout, *Macrozoarces americanus*. *J. Biol. Chem.* **1988**, *263*, 12049–12055. [PubMed]

122. Nishimiya, Y.; Sato, R.; Takamichi, M.; Miura, A.; Tsuda, S. Co-operative effect of the isoforms of type III antifreeze protein expressed in Notched-fin eelpout, *Zoarces elongatus Kner*. *FEBS J.* **2005**, *272*, 482–492. [CrossRef] [PubMed]

123. DeLuca, C.I.; Chao, H.; Sönnichsen, F.D.; Sykes, B.D.; Davies, P.L. Effect of type III antifreeze protein dilution and mutation on the growth inhibition of ice. *Biophys. J.* **1996**, *71*, 2346–2355. [CrossRef]

124. Baardsnes, J.; Davies, P.L. Contribution of hydrophobic residues to ice binding by fish type III antifreeze protein. *Biochim. Biophys. Acta* **2002**, *1601*, 49–54. [CrossRef]

125. Chao, H.; Sönnichsen, F.D.; DeLuca, C.I.; Sykes, B.D.; Davies, P.L. Structure-function relationship in the globular type III antifreeze protein: Identification of a cluster of surface residues required for binding to ice. *Protein Sci.* **1994**, *3*, 1760–1769. [CrossRef] [PubMed]

126. Jia, Z.; DeLuca, C.I.; Chao, H.; Davies, P.L. Structural basis for the binding of a globular antifreeze protein to ice. *Nature* **1996**, *384*, 285–288. [CrossRef] [PubMed]

127. Sönnichsen, F.D.; DeLuca, C.I.; Davies, P.L.; Sykes, B.D.; Sönnichsen, F.D.; DeLuca, C.I.; Davies, P.L.; Sykes, B.D. Refined solution structure of type III antifreeze protein: Hydrophobic groups may be involved in the energetics of the protein–ice interaction. *Structure* **1996**, *4*, 1325–1337. [CrossRef]

128. Leiter, A.; Rau, S.; Winger, S.; Muhle-Goll, C.; Luy, B.; Gaukel, V. Influence of heating temperature, pressure and pH on recrystallization inhibition activity of antifreeze protein type III. *J. Food Eng.* **2016**, *187*, 53–61. [CrossRef]

129. Hamada, T.; Ito, Y.; Abe, T.; Hayashi, F.; Guntert, P.; Inoue, M.; Kigawa, T.; Terada, T.; Shirouzu, M.; Yoshida, M.; et al. Solution structure of the antifreeze-like domain of human sialic acid synthase. *Protein Sci.* **2006**, *15*, 1010–1016. [CrossRef] [PubMed]

130. Choi, Y.-G.; Park, C.-J.; Kim, H.-E.; Seo, Y.-J.; Lee, A.-R.; Choi, S.-R.; Lee, S.S.; Lee, J.-H. Comparison of backbone dynamics of the type III antifreeze protein and antifreeze-like domain of human sialic acid synthase. *J. Biomol. NMR* **2015**, *61*, 137–150. [CrossRef] [PubMed]

131. DeLuca, C.I.; Davies, P.L.; Ye, Q.; Jia, Z. The effects of steric mutations on the structure of type III antifreeze protein and its interaction with ice. *J. Mol. Biol.* **1998**, *275*, 515–525. [CrossRef] [PubMed]

132. Graether, S.P.; DeLuca, C.I.; Baardsnes, J.; Hill, G.A.; Davies, P.L.; Jia, Z. Quantitative and qualitative analysis of type III antifreeze protein structure and function. *J. Biol. Chem.* **1999**, *274*, 11842–11847. [CrossRef] [PubMed]

133. Chen, G.; Jia, Z. Ice-binding surface of fish type III antifreeze. *Biophys. J.* **1999**, *77*, 1602–1608. [CrossRef]

134. Garnham, C.P.; Natarajan, A.; Middleton, A.J.; Kuiper, M.J.; Braslavsky, I.; Davies, P.L. Compound ice-binding site of an antifreeze protein revealed by mutagenesis and fluorescent tagging. *Biochemistry* **2010**, *49*, 9063–9071. [CrossRef] [PubMed]

135. Garnham, C.P.; Nishimiya, Y.; Tsuda, S.; Davies, P.L. Engineering a naturally inactive isoform of type III antifreeze protein into one that can stop the growth of ice. *FEBS Lett.* **2012**, *586*, 3876–3881. [CrossRef] [PubMed]

136. Choi, S.; Seo, Y.; Kim, M.; Eo, Y.; Ahn, H.; Lee, A.; Park, C.; Ryu, K.; Cheong, H.; Lee, S.S. NMR study of the antifreeze activities of active and inactive isoforms of a type III antifreeze protein. *FEBS Lett.* **2016**, *590*, 4202–4212. [CrossRef] [PubMed]

137. Wang, X.; DeVries, A.L.; Cheng, C.H. Antifreeze peptide heterogeneity in an antarctic eel pout includes an unusually large major variant comprised of two 7 kDa type III AFPs linked in tandem. *Biochim. Biophys. Acta* **1995**, *1247*, 163–172. [CrossRef]

138. Miura, K.; Ohgiya, S.; Hoshino, T.; Nemoto, N.; Suetake, T.; Miura, A.; Spyracopoulos, L.; Kondo, H.; Tsuda, S. NMR analysis of type III antifreeze protein intramolecular dimer. Structural basis for enhanced activity. *J. Biol. Chem.* **2001**, *276*, 1304–1310. [CrossRef] [PubMed]

139. Miura, K.; Ohgiya, S.; Hoshino, T.; Nemoto, N.; Odaira, M.; Nitta, K.; Tsuda, S. Determination of the solution structure of the N-domain plus linker of Antarctic eel pout antifreeze protein RD3. *J. Biochem.* **1999**, *126*, 387–394. [CrossRef] [PubMed]

140. Raymond, J.A.; Christner, B.C.; Schuster, S.C. A bacterial ice-binding protein from the Vostok ice core. *Extremophiles* **2008**, *12*, 713–717. [CrossRef] [PubMed]

141. Xiao, N.; Suzuki, K.; Nishimiya, Y.; Kondo, H.; Miura, A.; Tsuda, S.; Hoshino, T. Comparison of functional properties of two fungal antifreeze proteins from Antarctomyces psychrotrophicus and *Typhula ishikariensis*. *FEBS J.* **2010**, *277*, 394–403. [CrossRef] [PubMed]

142. Lee, J.H.; Lee, S.G.; Do, H.; Park, J.C.; Kim, E.; Choe, Y.H.; Han, S.J.; Kim, H.J. Optimization of the pilot-scale production of an ice-binding protein by fed-batch culture of *Pichia pastoris*. *Appl. Microbiol. Biotechnol.* **2013**, *97*, 3383–3393. [CrossRef] [PubMed]

143. Koh, H.Y.; Lee, J.H.; Han, S.J.; Park, H.; Lee, S.G. Effect of the antifreeze protein from the Arctic yeast *Leucosporidium* sp. AY30 on cryopreservation of the marine diatom *Phaeodactylum tricornutum*. *Appl. Biochem. Biotechnol.* **2015**, *175*, 677–686. [CrossRef] [PubMed]

144. Lee, S.G.; Koh, H.Y.; Lee, J.H.; Kang, S.H.; Kim, H.J. Cryopreservative effects of the recombinant ice-binding protein from the arctic yeast *Leucosporidium* sp. on red blood cells. *Appl. Biochem. Biotechnol.* **2012**, *167*, 824–834. [CrossRef] [PubMed]

145. Kim, H.J.; Shim, H.E.; Lee, J.H.; Kang, Y.-C.; Hur, Y.B. Ice-binding protein derived from *Glaciozyma* can improve the viability of cryopreserved mammalian cells. *J. Microbiol. Biotechnol.* **2015**, *25*, 1989–1996. [CrossRef] [PubMed]

146. Kondo, H.; Hanada, Y.; Sugimoto, H.; Hoshino, T.; Garnham, C.P.; Davies, P.L.; Tsuda, S. Ice-binding site of snow mold fungus antifreeze protein deviates from structural regularity and high conservation. *Proc. Natl. Acad. Sci. USA* **2012**, *109*, 9360–9365. [CrossRef] [PubMed]

147. Hashim, N.H.F.; Sulaiman, S.; Bakar, F.D.A.; Illias, R.M.; Kawahara, H.; Najimudin, N.; Mahadi, N.M.; Murad, A.M.A. Molecular cloning, expression and characterisation of Afp4, an antifreeze protein from *Glaciozyma antarctica*. *Polar Biol.* **2014**, *37*, 1495–1505. [CrossRef]

148. Bayer-Giraldi, M.; Uhlig, C.; John, U.; Mock, T.; Valentin, K.; Bayer-Giraldi, M.; Uhlig, C.; John, U.; Mock, T.; Valentin, K. Antifreeze proteins in polar sea ice diatoms: Diversity and gene expression in the genus *Fragilariopsis*. *Environ. Microbiol.* **2010**, *12*, 1041–1052. [CrossRef] [PubMed]

149. Gilbert, J.A.; Hill, P.J.; Dodd, C.E.; Laybourn-Parry, J. Demonstration of antifreeze protein activity in Antarctic lake bacteria. *Microbiology* **2004**, *150*, 171–180. [CrossRef] [PubMed]

150. Garnham, C.P.; Gilbert, J.A.; Hartman, C.P.; Campbell, R.L.; Laybourn-Parry, J.; Davies, P.L. A Ca^{2+}-dependent bacterial antifreeze protein domain has a novel beta-helical ice-binding fold. *Biochem. J.* **2008**, *411*, 171–180. [CrossRef] [PubMed]

151. Raymond, J.A.; Kim, H.J. Possible role of horizontal gene transfer in the colonization of sea ice by algae. *PLoS ONE* **2012**, *7*, e35968. [CrossRef] [PubMed]

152. Gogarten, J.P.; Doolittle, W.F.; Lawrence, J.G. Prokaryotic evolution in light of gene transfer. *Mol. Biol. Evol.* **2002**, *19*, 2226–2238. [CrossRef] [PubMed]

153. Anesio, A.M.; Mindl, B.; Laybourn-Parry, J.; Hodson, A.J.; Sattler, B. Viral dynamics in cryoconite holes on a high Arctic glacier (Svalbard). *J. Geophys. Res. Biogeosci.* **2007**, *112*, G04S31. [CrossRef]

154. Pucciarelli, S.; Chiappori, F.; Devaraj, R.R.; Yang, G.; Yu, T.; Ballarini, P.; Miceli, C. Identification and analysis of two sequences encoding ice-binding proteins obtained from a putative bacterial symbiont of the psychrophilic Antarctic ciliate *Euplotes focardii*. *Antarct. Sci.* **2014**, *26*, 491–501. [CrossRef]

155. Mangiagalli, M.; Bar-Dolev, M.; Tedesco, P.; Natalello, A.; Kaleda, A.; Brocca, S.; Pascale, D.; Pucciarelli, S.; Miceli, C.; Bravslavsky, I. Cryo-protective effect of an ice-binding protein derived from Antarctic bacteria. *FEBS J.* **2016**. [CrossRef] [PubMed]

156. Mazur, P. Freezing of living cells: Mechanisms and implications. *Am. J. Physiol.* **1984**, *247*, C125–C142. [PubMed]

157. Naaldijk, Y.; Staude, M.; Fedorova, V.; Stolzing, A. Effect of different freezing rates during cryopreservation of rat mesenchymal stem cells using combinations of hydroxyethyl starch and dimethylsulfoxide. *BMC Biotechnol.* **2012**, *12*, 49. [CrossRef] [PubMed]

158. Fowler, A.; Toner, M. Cryo-injury and biopreservation. *Ann. N. Y. Acad. Sci.* **2006**, *1066*, 119–135. [CrossRef] [PubMed]

159. Chaytor, J.L.; Tokarew, J.M.; Wu, L.K.; Leclre, M.; Tam, R.Y.; Capicciotti, C.J.; Guolla, L.; Von Moos, E.; Findlay, C.S.; Allan, D.S.; et al. Inhibiting ice recrystallization and optimization of cell viability after cryopreservation. *Glycobiology* **2012**, *22*, 123–133. [CrossRef] [PubMed]

160. Fahy, G.M.; MacFarlane, D.R.; Angell, C.A.; Meryman, H.T. Vitrification as an approach to cryopreservation. *Cryobiology* **1984**, *21*, 407–426. [CrossRef]

161. Wowk, B.; Leitl, E.; Rasch, C.M.; Mesbah-Karimi, N.; Harris, S.B.; Fahy, G.M. Vitrification enhancement by synthetic ice blocking agents. *Cryobiology* **2000**, *40*, 228–236. [CrossRef] [PubMed]

162. Mugnano, J.A.; Wang, T.; Layne, J.R., Jr.; DeVries, A.L.; Lee, R.E., Jr. Antifreeze glycoproteins promote intracellular freezing of rat cardiomyocytes at high subzero temperatures. *Am. J. Physiol.* **1995**, *269*, R474–R479. [PubMed]

163. Hansen, T.N.; Smith, K.M.; Brockbank, K.G. Type I antifreeze protein attenuates cell recoveries following cryopreservation. *Transpl. Proc.* **1993**, *25*, 3182–3184.

164. Wang, T.; Zhu, Q.; Yang, X.; Layne, J.R., Jr.; Devries, A.L. Antifreeze glycoproteins from antarctic notothenioid fishes fail to protect the rat cardiac explant during hypothermic and freezing preservation. *Cryobiology* **1994**, *31*, 185–192. [CrossRef] [PubMed]

165. Zilli, L.; Beirão, J.; Schiavone, R.; Herraez, M.P.; Gnoni, A.; Vilella, S. Comparative proteome analysis of cryopreserved flagella and head plasma membrane proteins from sea bream spermatozoa: Effect of antifreeze proteins. *PLoS ONE* **2014**, *9*, e99992. [CrossRef] [PubMed]

166. Robles, V.; Cabrita, E.; Anel, L.; Herraez, M.P. Microinjection of the antifreeze protein type III (AFPIII) in turbot (*Scophthalmus maximus*) embryos: Toxicity and protein distribution. *Aquaculture* **2006**, *261*, 1299–1306. [CrossRef]

167. Beirão, J.; Zilli, L.; Vilella, S.; Cabrita, E.; Schiavone, R.; Herraez, M.P. Improving sperm cryopreservation with antifreeze proteins: Effect on gilthead seabream (*Sparus aurata*) plasma membrane lipids. *Biol. Reprod.* **2012**, *86*, 59. [CrossRef] [PubMed]

168. Prathalingam, N.S.; Holt, W.V.; Revell, S.G.; Mirczuk, S.; Fleck, R.A.; Watson, P.F. Impact of antifreeze proteins and antifreeze glycoproteins on bovine sperm during freeze-thaw. *Theriogenology* **2006**, *66*, 1894–1900. [CrossRef] [PubMed]

169. Kim, J.S.; Yoon, J.H.; Park, G.H.; Bae, S.H.; Kim, H.J.; Kim, M.S.; Hwang, Y.J.; Kim, D.Y. Influence of antifreeze proteins on boar sperm DNA damaging during cryopreservation. *Dev. Biol.* **2011**, *356*, 195. [CrossRef]

170. Qadeer, S.; Khan, M.A.; Ansari, M.S.; Rakha, B.A.; Ejaz, R.; Iqbal, R.; Younis, M.; Ullah, N.; DeVries, A.L.; Akhter, S. Efficiency of antifreeze glycoproteins for cryopreservation of Nili-Ravi (*Bubalus bubalis*) buffalo bull sperm. *Anim. Reprod. Sci.* **2015**, *157*, 56–62. [CrossRef] [PubMed]

171. Qadeer, S.; Khan, M.A.; Ansari, M.S.; Rakha, B.A.; Ejaz, R.; Husna, A.U.; Ashiq, M.; Iqbal, R.; Ullah, N.; Akhter, S. Evaluation of antifreeze protein III for cryopreservation of Nili-Ravi (*Bubalus bubalis*) buffalo bull sperm. *Anim. Reprod. Sci.* **2014**, *148*, 26–31. [CrossRef] [PubMed]

172. Ideta, A.; Aoyagi, Y.; Tsuchiya, K.; Nakamura, Y.; Hayama, K.; Shirasawa, A.; Sakaguchi, K.; Tominaga, N.; Nishimiya, Y.; Tsuda, S. Prolonging hypothermic storage (4 °C) of bovine embryos with fish antifreeze protein. *J. Reprod. Dev.* **2015**, *61*, 1–6. [CrossRef] [PubMed]

173. Hirano, Y.; Nishimiya, Y.; Matsumoto, S.; Matsushita, M.; Todo, S.; Miura, A.; Komatsu, Y.; Tsuda, S. Hypothermic preservation effect on mammalian cells of type III antifreeze proteins from notched-fin eelpout. *Cryobiology* **2008**, *57*, 46–51. [CrossRef] [PubMed]

174. Kamijima, T.; Sakashita, M.; Miura, A.; Nishimiya, Y.; Tsuda, S. Antifreeze protein prolongs the life-time of insulinoma cells during hypothermic preservation. *PLoS ONE* **2013**, *8*, e73643. [CrossRef] [PubMed]

175. Baguisi, A.; Arav, A.; Crosby, T.F.; Roche, J.F.; Boland, M.P. Hypothermic storage of sheep embryos with antifreeze proteins: Development in vitro and in vivo. *Theriogenology* **1997**, *48*, 1017–1024. [CrossRef]

176. Nishijima, K.; Tanaka, M.; Sakai, Y.; Koshimoto, C.; Morimoto, M.; Watanabe, T.; Fan, J.; Kitajima, S. Effects of type III antifreeze protein on sperm and embryo cryopreservation in rabbit. *Cryobiology* **2014**, *69*, 22–25. [CrossRef] [PubMed]

177. Jo, J.W.; Jee, B.C.; Lee, J.R.; Suh, C.S. Effect of antifreeze protein supplementation in vitrification medium on mouse oocyte developmental competence. *Fertil. Steril.* **2011**, *96*, 1239–1245. [CrossRef] [PubMed]
178. Karanova, M.V.; Pronina, N.D.; Tsvetkova, L.I. The effect of antifreeze glycoproteins on survival and quality of fish spermatozoa under the conditions of long-term storage at +4 degree C. *Izv. Akad. Nauk. Ser. Biol.* **2002**, *1*, 88–92.
179. Halwani, D.O.; Brockbank, K.G.; Duman, J.G.; Campbell, L.H. Recombinant *Dendroides canadensis* antifreeze proteins as potential ingredients in cryopreservation solutions. *Cryobiology* **2014**, *68*, 411–418. [CrossRef] [PubMed]
180. Karanova, M.V.; Mezhevikina, L.M.; Petropavlov, N.N. Study of cryoprotective properties of antifreeze glycoproteins from the white sea cod *Gadus morhua* on low temperature freezing of mouse embryos. *Biofizika* **1994**, *40*, 1341–1347.
181. Wen, Y.; Zhao, S.; Chao, L.; Yu, H.; Song, C.; Shen, Y.; Chen, H.; Deng, X. The protective role of antifreeze protein 3 on the structure and function of mature mouse oocytes in vitrification. *Cryobiology* **2014**, *69*, 394–401. [CrossRef] [PubMed]
182. Rubinsky, L.; Raichman, N.; Lavee, J.; Frenk, H.; Ben-Jacob, E.; Bickler, P.E. Antifreeze protein suppresses spontaneous neural activity and protects neurons from hypothermia/re-warming injury. *Neurosci. Res.* **2010**, *67*, 256–259. [CrossRef] [PubMed]
183. Rubinsky, B.; Arav, A.; Devries, A.L. The cryoprotective effect of antifreeze glycopeptides from antarctic fishes. *Cryobiology* **1992**, *29*, 69–79. [CrossRef]
184. Amir, G.; Rubinsky, B.; Smolinsky, A.K.; Lavee, J. Successful use of ocean pout thermal hysteresis protein (antifreeze protein III) in cryopreservation of transplanted mammalian heart at subzero temperature. *J. Hear Lung Transplant.* **2002**, *21*, 137. [CrossRef]
185. Jo, J.W.; Jee, B.C.; Suh, C.S.; Kim, S.H. The Beneficial Effects of antifreeze proteins in the vitrification of immature mouse oocytes. *PLoS ONE* **2012**, *7*, e37043. [CrossRef] [PubMed]
186. Martínez-Páramo, S.; Pérez-Cerezales, S.; Robles, V.; Anel, L.; Herraez, M.P. Incorporation of antifreeze proteins into zebrafish embryos by a non-invasive method. *Cryobiology* **2008**, *56*, 216–222. [CrossRef] [PubMed]
187. Martínez-Páramo, S.; Barbosa, V.; Pérez-Cerezales, S.; Robles, V.; Herraez, M.P. Cryoprotective effects of antifreeze proteins delivered into zebrafish embryos. *Cryobiology* **2009**, *58*, 128–133. [CrossRef] [PubMed]
188. Tursman, D.; Duman, J.G. Cryoprotective effects of thermal hysteresis protein on survivorship of frozen gut cells from the freeze-tolerant centipede *Lithobius forficatus*. *J. Exp. Zool.* **1995**, *272*, 249–257. [CrossRef]
189. Amir, G.; Rubinsky, B.; Horowitz, L.; Miller, L.; Leor, J.; Kassif, Y.; Mishaly, D.; Smolinsky, A.K.; Lavee, J. Prolonged 24-hour subzero preservation of heterotopically transplanted rat hearts using antifreeze proteins derived from arctic fish. *Ann. Thorac. Surg.* **2004**, *77*, 1648–1655. [CrossRef] [PubMed]
190. O'Neil, L.; Paynter, S.J.; Fuller, B.J.; Shaw, R.W.; DeVries, A.L. Vitrification of mature mouse oocytes in a 6 M Me$_2$SO solution supplemented with antifreeze glycoproteins: The effect of temperature. *Cryobiology* **1998**, *37*, 59–66. [CrossRef] [PubMed]
191. Shaw, J.M.; Ward, C.; Trounson, A.O. Survival of mouse blastocysts slow cooled in propanediol or ethylene glycol is influenced by the thawing procedure, sucrose and antifreeze proteins. *Theriogenology* **1995**, *43*, 1289–1300. [CrossRef]
192. Koshimoto, C.; Mazur, P. Effects of warming rate, temperature, and antifreeze proteins on the survival of mouse spermatozoa frozen at an optimal rate. *Cryobiology* **2002**, *45*, 49–59. [CrossRef]
193. Matsumoto, S.; Matsusita, M.; Morita, T.; Kamachi, H.; Tsukiyama, S.; Furukawa, Y.; Koshida, S.; Tachibana, Y.; Nishimura, S.; Todo, S. Effects of synthetic antifreeze glycoprotein analogue on islet cell survival and function during cryopreservation. *Cryobiology* **2006**, *52*, 90–98. [CrossRef] [PubMed]
194. Arav, A.; Rubinsky, B.; Fletcher, G.; Seren, E. Cryogenic protection of oocytes with antifreeze proteins. *Mol. Reprod. Dev.* **1993**, *36*, 488–493. [CrossRef] [PubMed]
195. Younis, A.I.; Rooks, B.; Khan, S.; Gould, K.G. The effects of antifreeze peptide III (AFP) and insulin transferrin selenium (ITS) on cryopreservation of chimpanzee (*Pan troglodytes*) spermatozoa. *J. Androl.* **1998**, *19*, 207–214. [PubMed]
196. Robles, V.; Barbosa, V.; Herraez, M.P.; Martinez-Paramo, S.; Cancela, M.L. The antifreeze protein type I (AFP I) increases seabream (*Sparus aurata*) embryos tolerance to low temperatures. *Theriogenology* **2007**, *68*, 284–289. [CrossRef] [PubMed]

197. Chao, H.; Davies, P.L.; Carpenter, J.F. Effects of antifreeze proteins on red blood cell survival during cryopreservation. *J. Exp. Biol.* **1996**, *199*, 2071–2076. [PubMed]

198. Chaves, D.F.; Campelo, I.S.; Silva, M.M.A.S.; Bhat, M.H.; Teixeira, D.I.A.; Melo, L.M.; Souza-Fabjan, J.M.G.; Mermillod, P.; Freitas, V.J.F. The use of antifreeze protein type III for vitrification of in vitro matured bovine oocytes. *Cryobiology* **2016**, *73*, 324–328. [CrossRef] [PubMed]

199. Bouvet, V.R.; Lorello, G.R.; Ben, R.N. Aggregation of antifreeze glycoprotein fraction 8 and its effect on antifreeze activity. *Biomacromolecules* **2006**, *7*, 565–571. [CrossRef] [PubMed]

200. Qadeer, S.; Khan, M.A.; Shahzad, Q.; Azam, A.; Ansari, M.S.; Rakha, B.A.; Ejaz, R.; Husna, A.U.; Duman, J.G.; Akhter, S. Efficiency of beetle (*Dendroides canadensis*) recombinant antifreeze protein for buffalo semen freezability and fertility. *Theriogenology* **2016**, *86*, 1662–1669. [CrossRef] [PubMed]

201. Liu, S.; Wang, W.; Moos, E.; Jackman, J.; Mealing, G.; Monette, R.; Ben, R.N. In vitro studies of antifreeze glycoprotein (AFGP) and a C-linked AFGP analogue. *Biomacromolecules* **2007**, *8*, 1456–1462. [CrossRef] [PubMed]

202. Martinez-Paramo, S.; Perez-Cerezales, S.; Barbosa, V.; Robles, V.; Herraez, M.P. Advances on fish embryo cryopreservation using antifreeze proteins. *Biol. Reprod.* **2008**, *78*, 152.

203. Ideta, A.; Aoyagi, Y.; Tsuchiya, K.; Kamijima, T.; Nishimiya, Y.; Tsuda, S. A simple medium enables bovine embryos to be held for seven days at 4 °C. *Sci. Rep.* **2013**, *3*, 1173. [CrossRef] [PubMed]

204. Lee, C.Y.; Rubinsky, B.; Fletcher, G.L. Hypothermic preservation of whole mammalian organs with antifreeze proteins. *Cryo-Letters* **1992**, *13*, 59–66.

205. Amir, G.; Horowitz, L.; Rubinsky, B.; Yousif, B.S.; Lavee, J.; Smolinsky, A.K. Subzero nonfreezing cryopresevation of rat hearts using antifreeze protein I and antifreeze protein III. *Cryobiology* **2004**, *48*, 273–282. [CrossRef] [PubMed]

206. Rubinsky, B.; Arav, A.; Hong, J.S.; Lee, C.Y. Freezing of mammalian livers with glycerol and antifreeze proteins. *Biochem. Biophys. Res. Commun.* **1994**, *200*, 732–741. [CrossRef] [PubMed]

207. Davies, P.L.; Sykes, B.D. Antifreeze proteins. *Curr. Opin. Struct. Biol.* **1997**, *7*, 828–834. [CrossRef]

208. Arakawa, T.; Kita, Y.; Timasheff, S.N. Protein precipitation and denaturation by dimethyl sulfoxide. *Biophys. Chem.* **2007**, *131*, 62–70. [CrossRef] [PubMed]

209. Wang, C.; Oliver, E.E.; Christner, B.C.; Luo, B.-H. Functional Analysis of a bacterial antifreeze protein indicates a cooperative effect between its two ice-binding domains. *Biochemistry* **2016**, *55*, 3975–3983. [CrossRef] [PubMed]

210. Nishimiya, Y.; Ohgiya, S.; Tsuda, S. Artificial multimers of the type III antifreeze protein. Effects on thermal hysteresis and ice crystal morphology. *J. Biol. Chem.* **2003**, *278*, 32307–32312. [CrossRef] [PubMed]

211. Baardsnes, J.; Kuiper, M.J.; Davies, P.L. Antifreeze protein dimer: When two ice-binding faces are better than one. *J. Biol. Chem.* **2003**, *278*, 38942–38947. [CrossRef] [PubMed]

212. Can, O.; Holland, N.B. Utilizing avidity to improve antifreeze protein activity: A type III antifreeze protein trimer exhibits increased thermal hysteresis activity. *Biochemistry* **2013**, *52*, 8745–8752. [CrossRef] [PubMed]

213. Stevens, C.A.; Drori, R.; Zalis, S.; Braslavsky, I.; Davies, P.L. Dendrimer-linked antifreeze proteins have superior activity and thermal recovery. *Bioconjug. Chem.* **2015**, *26*, 1908–1915. [CrossRef] [PubMed]

214. Phippen, S.W.; Stevens, C.A.; Vance, T.D.R.; King, N.P.; Baker, D.; Davies, P.L. Multivalent display of antifreeze proteins by fusion to self-assembling protein cages enhances ice-binding activities. *Biochemistry* **2016**, *55*, 6811–6820. [CrossRef] [PubMed]

215. Bang, J.K.; Lee, J.H.; Murugan, R.N.; Lee, S.G.; Do, H.; Koh, H.Y.; Shim, H.E.; Kim, H.C.; Kim, H.J. Antifreeze peptides and glycopeptides, and their derivatives: Potential uses in biotechnology. *Mar. Drugs* **2013**, *11*, 2013–2041. [CrossRef] [PubMed]

216. Balcerzak, A.K.; Capicciotti, C.J.; Briard, J.G.; Ben, R.N. Designing ice recrystallization inhibitors: From antifreeze (glyco) proteins to small molecules. *RSC Adv.* **2014**, *4*, 42682–42696. [CrossRef]

217. Liu, S.; Ben, R.N. C-linked galactosyl serine AFGP analogues as potent recrystallization inhibitors. *Org. Lett.* **2005**, *7*, 2385–2388. [CrossRef] [PubMed]

218. Garner, J.; Harding, M.M. Design and synthesis of antifreeze glycoproteins and mimics. *Chembiochem* **2010**, *11*, 2489–2498. [CrossRef] [PubMed]

219. Hachisu, M.; Hinou, H.; Takamichi, M.; Tsuda, S.; Koshida, S.; Nishimura, S. One-pot synthesis of cyclic antifreeze glycopeptides. *Chem. Commun.* **2009**. [CrossRef] [PubMed]

220. Garner, J.; Harding, M.M. Design and synthesis of alpha-helical peptides and mimetics. *Org. Biomol. Chem.* **2007**, *5*, 3577–3585. [CrossRef] [PubMed]
221. Can, O.; Holland, N.B. Conjugation of type I antifreeze protein to polyallylamine increases thermal hysteresis activity. *Bioconjug. Chem.* **2011**, *22*, 2166–2171. [CrossRef] [PubMed]

marine drugs

MDPI

Review

Enzyme-Assisted Discovery of Antioxidant Peptides from Edible Marine Invertebrates: A Review

Tsun-Thai Chai [1,2,*], Yew-Chye Law [1], Fai-Chu Wong [1,2] and Se-Kwon Kim [3,4]

[1] Department of Chemical Science, Faculty of Science, Universiti Tunku Abdul Rahman, 31900 Kampar, Malaysia; yew_chye89@hotmail.com (Y.-C.L.); wongfc@utar.edu.my (F.-C.W.)
[2] Centre for Bio-diversity Research, Universiti Tunku Abdul Rahman, 31900 Kampar, Malaysia
[3] Department of Marine Bio-Convergence Science, Pukyong National University, 48513 Busan, Korea; sknkim@pknu.ac.kr
[4] Institute for Life Science of Seogo (ILSS), Kolmar Korea Co, 137-876 Seoul, Korea
*   Correspondence: chaitt@utar.edu.my; Tel.: +60-5-468-8888

Academic Editor: Orazio Taglialatela-Scafati
Received: 29 November 2016; Accepted: 8 February 2017; Published: 16 February 2017

Abstract: Marine invertebrates, such as oysters, mussels, clams, scallop, jellyfishes, squids, prawns, sea cucumbers and sea squirts, are consumed as foods. These edible marine invertebrates are sources of potent bioactive peptides. The last two decades have seen a surge of interest in the discovery of antioxidant peptides from edible marine invertebrates. Enzymatic hydrolysis is an efficient strategy commonly used for releasing antioxidant peptides from food proteins. A growing number of antioxidant peptide sequences have been identified from the enzymatic hydrolysates of edible marine invertebrates. Antioxidant peptides have potential applications in food, pharmaceuticals and cosmetics. In this review, we first give a brief overview of the current state of progress of antioxidant peptide research, with special attention to marine antioxidant peptides. We then focus on 22 investigations which identified 32 antioxidant peptides from enzymatic hydrolysates of edible marine invertebrates. Strategies adopted by various research groups in the purification and identification of the antioxidant peptides will be summarized. Structural characteristic of the peptide sequences in relation to their antioxidant activities will be reviewed. Potential applications of the peptide sequences and future research prospects will also be discussed.

Keywords: antioxidant peptide; enzymatic hydrolysis; marine invertebrate; peptide identification; peptide purification

1. Introduction

Reactive oxygen species (ROS) and reactive nitrogen species (RNS) are free radicals that play vital roles in the body, such as participating in intracellular signaling cascades and host defense against invading pathogens. Imbalance between free radical production and endogenous antioxidant defense may result in cellular oxidative stress, causing oxidative damage to various cellular components, such as DNA, proteins and membrane lipids. Oxidative damage has been implicated in and is believed to be a key factor causing various pathological conditions, such as heart disease, stroke, arteriosclerosis, diabetes, and cancer [1–4]. Furthermore, accumulation of oxidized proteins underlies the aging process in humans and the development of some age-related diseases [5]. Dietary intake of antioxidants has been associated with reduced risks of some of the aforementioned diseases [6,7]. The effectiveness of antioxidant therapies in preventing and/or managing human pathologies was also highlighted [8–10].

Oxidation in the form of lipid peroxidation is also a deleterious process occurring in foodstuffs. Lipid peroxidation is a major cause of rancidity and reduced shelf-life in foods [11]. Oxidation compromises the nutritive value of food, in addition to causing the loss of flavors and the formation of toxic by-products.

An effective approach to keep oxidation of food constituents in check is by incorporating synthetic food-grade antioxidants (e.g., butylated hydroxytoluene (BHT), butylated hydroxyanisole (BHA), *tert*-butylhydroquinone (TBHQ), and propyl gallate) during food processing [11,12].

Free radicals can be quenched through a number of mechanisms. Antioxidants can directly scavenge free radicals (e.g., via hydrogen atom transfer or electron transfer) or prevent free radical formation by chelating metal ions. Antioxidants can also interrupt the radical chain reactions of lipid peroxidation, thus retarding its progression. There is currently great interest to search for natural antioxidants as alternatives to the synthetic ones for applications in food processing, functional food development, cosmetic formulations, and therapy. One of the factors driving such a trend is the concern about potential side effects of synthetic antioxidants and consumer preference for natural antioxidants, which are perceived as relatively safe, especially those derived from edible sources [11–14].

The last two decades have seen a marked increase worldwide in studies searching for bioactive peptides from edible animals and plants as well as from food products and processing wastes derived from them. Bioactive peptides have a broad range of activities, such as antioxidant, antimicrobial, antiviral, antitumor, antihypertensive, immunomodulatory, analgesic, anti-diabetic, and neuroprotective activities [15]. Such bioactive peptides are potential candidates for development into future peptide drugs. The global market for peptide therapeutics was valued at USD 17.5 billion in 2015, expected to hit USD 47 billion by 2025 [16]. There are more than 60 FDA-approved peptide drugs already on the market [17], with about 400 more peptide therapeutics in different phases of preclinical and clinical development as of February 2016 [16]. Overall, peptide drugs are recognized as one of the fastest growing segment, with enormous future growth potential, in the pharmaceutical industry [16,17].

Bioactive peptides are encrypted in an inactive state within the structure of the parent proteins. Such peptides become active after release from the parent proteins, which can be achieved by means of in vitro enzymatic hydrolysis, gastrointestinal digestion, and food processing (e.g., fermentation) [18–20]. The activation of antioxidant peptides upon their liberation from the parent protein may be due to their less restricted interaction with free radicals, unhindered by their positions within the bulky protein structure or by poor lipid solubility of the parent protein. The aforementioned proposal, however, remains to be experimentally validated. Enzymatic hydrolysis under optimal conditions is the most efficient and reliable strategy for releasing antioxidant and other bioactive peptides from food proteins, including proteins of the marine origin [21–23]. It is also the preferred method for bioactive peptide production in the food and pharmaceutical industries [24]. Antioxidant peptides liberated upon enzymatic hydrolysis of parent proteins are small, ranging from 2 to 20 residues [13,18]. Such peptides usually contain varying percentages of hydrophobic amino acids (e.g., Ala, Leu, and Pro) or aromatic amino acids (e.g., Tyr, His, and Phe) in their sequences. The functionality of antioxidant peptides has been attributed to the ability of such amino acid constituents to donate protons to free radicals, chelate metal ions and/or trap lipid peroxyl radicals [25,26].

Antioxidant peptides are an important area of scientific interest. The input query "antioxidant peptide" OR "antioxidative peptide" in the Scopus database [27], as of November 2016, revealed 542 publications between the years 1992 and 2016. An increasing trend in the number of publications can be seen over the last 24 years, leading to about 80 publications annually between 2013 and 2016 (see Figure S1). A more precise input query "antioxidant peptide" OR "antioxidative peptide" AND "marine" in the Scopus database revealed 22 publications between 2009 and 2016. This comprised of 15 journal article, five book chapters and two reviews. As of September 2016, 531 antioxidant peptide sequences ranging between 2 and 20 residues in length have been deposited into the BIOPEP database [28]. All these point to a growing interest in the research area of antioxidant peptides over the last two decades.

Antioxidant peptides have been isolated and identified from numerous edible marine animals, including various fish species [29,30], and edible marine invertebrates, such as mussels [31,32], clams [33,34], and oysters [35,36]. Identification of antioxidant peptides from food products manufactured from edible marine invertebrates, e.g., fermented mussel sauce [37,38] and shrimp paste [39], has also been reported. The effectiveness of the marine invertebrate peptides in scavenging ROS, chelating metals, suppressing lipid peroxidation, and protecting cells against ROS-induced toxicity has been demonstrated by using chemical and cell-based assays [40–43]. Notably, the in vivo effects of antioxidant peptides identified from the mussel [44] and sea squirt [45] were reported, suggesting that antioxidant peptides identified from edible marine invertebrates can have biological or physiological significance. Collectively, the aforementioned findings suggest that edible marine invertebrates deserve more attention than they have received to date as a promising source of potent antioxidant peptides.

In reviews on marine bioactive peptides, enzymatic hydrolysates and antioxidant peptides of fish and their processing by-products often overshadowed those of edible marine invertebrates [18,29,46,47]. In this review, we focus on antioxidant peptides purified and identified from enzymatic hydrolysates of edible marine invertebrates, including those prepared by using in vitro gastrointestinal digestion. This review presents an overview of enzyme-assisted production, assay-guided purification, and identification of antioxidant peptides, in addition to their structure–activity relationships. Potential applications of the pure antioxidant peptides in food, therapy and cosmetics are discussed. Future research opportunities, in relation to gaps in current knowledge, are highlighted. When preparing for this review, we analyzed the published antioxidant peptide sequences that were identified from edible marine invertebrates for additional functions (e.g., anticancer activity) or properties (e.g., allergenicity) by using a number of in silico tools. The significance of the new information is discussed in this review where relevant. Our emphasis is on studies which have successfully identified potential antioxidant peptides from edible marine invertebrates. Nonetheless, evidence from protein hydrolysate studies may still be referred to where appropriate. We also highlight studies which have taken the additional step of validating the antioxidant activity of the identified peptide sequences by using synthetic peptides. Literature pertaining to the identification of antioxidant peptides naturally occurring in the cells of edible marine invertebrates, or those present in fermented products (e.g., [38,39]) is not within the scope of this review.

2. Enzyme-Assisted Production, Purification, and Identification of Antioxidant Peptides

Antioxidant peptides encrypted in the proteins of edible marine invertebrates have been effectively released by enzymatic proteolysis and identified. Table 1 shows the primary structures of 13 edible marine invertebrate-derived antioxidant peptides whose activities were validated using chemically synthesized peptides. A general workflow employed in the purification and identification of antioxidant peptides from edible marine invertebrates is shown in Figure 1.

Table 1. Primary structures of selected antioxidant peptides identified from edible marine invertebrates.

Antioxidant Peptides	References
VKP, VKCFR	[42]
IKK, FKK, FIKK	[48]
HMSY, PEASY	[49]
LWHTH	[50]
LPHPSF	[45]
PIIVYWK, FSVVPSPK, TTANIEDRR	[32]
GPLGLLGFLGPLGLS	[51]

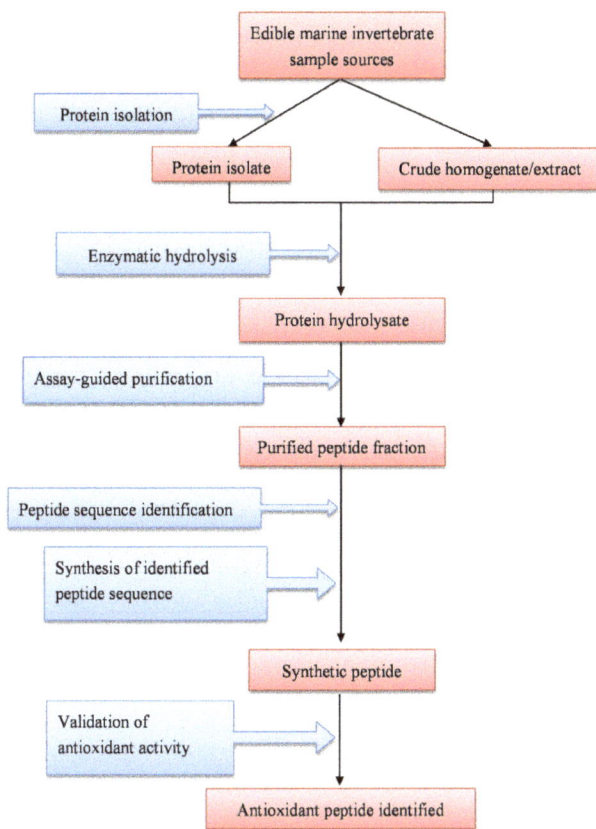

Figure 1. A workflow used for the purification and identification of antioxidant peptides from enzymatic hydrolysates of edible marine invertebrates.

2.1. Production of Antioxidant Peptides

Different forms of protein samples from edible marine invertebrates have been used as the starting material for the isolation of antioxidant peptides. Crude homogenate or mince of samples in cold water as well as pulverized, lyophilized samples, without further protein enrichment, were used for proteolysis and subsequent isolation of antioxidant peptides from the oyster [35], prawn [48], short-neck clam [34], mussels [32,44,52,53], shrimp processing waste [54], scallop female gonad [49], and sea squirt [50]. Isopropanol was used to defat the homogenates of the blood clam [33] and blue mussel [31] prior to their use in the preparation of hydrolysates. To prepare a protein isolate to be used for hydrolysis, Zhou et al. [55] used the trichloroacetic acid/acetone precipitation method to extract proteins from the body wall of the sea cucumber. Nonetheless, most of the aforementioned studies used either crude homogenates or pulverized samples as their starting materials for antioxidant peptide isolation. This suggests that the preparation of a protein isolate or protein-enriched sample is not a requisite for successful isolation of antioxidant peptides.

Enzymatic hydrolysis of protein samples of edible marine invertebrates has been performed by using individual proteases or a combination of digestive proteases with the aim of simulating gastrointestinal digestion. Proteases of animal, plant, and microbial origins that are commercially available in pure form, such as pepsin, trypsin, α-chymotrypsin, papain, alcalase, and neutrase, have been used to prepare protein hydrolysates from edible marine invertebrates. Other commercial proteases that have been used, but less frequently, include kojizyme, flavourzyme, protamex, neutral protease, acid protease, newlase F, pronase and pancreatin [44,45,48,50,53,55].

Owing to the unique cleavage specificities of the proteases, when the same protein sample is treated with different proteases, hydrolysates each consisting of a complex pool of polypeptides hydrolyzed to different extents can be produced. Proteolysis and the generation of various peptides often alter the antioxidant activity of the original protein sample. Hence, one strategy used by some studies was comparing the degree of hydrolysis (DH), yield, and/or antioxidant activity of multiple hydrolysates produced by using different proteases under optimum pH and temperature conditions. A hydrolysate was then chosen and used for the subsequent purification and identification of antioxidant peptides [31,33,41,43–45,48,50,53,55–57]. For example, Rajapakse, Mendis, Byun and Kim [43] hydrolyzed squid muscle separately with pepsin, trypsin and α-chymotrypsin for six hours. Tryptic hydrolysate, which showed the highest DH and inhibitory activity against linoleic acid oxidation, was selected for further purification. This led to the discovery of two potent antioxidant oligopeptides NADFGLNGLEGLA and NGLEGLK [43]. A number of studies successfully identified antioxidant peptides from optimum hydrolysates chosen based on only antioxidant efficacy, without considering DH or yield (e.g., [31,44,45,53]). Ko, Kim, Jung, Kim, Lee, Son, Kim and Jeon [45] hydrolyzed sea squirt with nine proteases. Based on only relative antioxidant efficacies of the hydrolysates, the authors chose tryptic hydrolysate for further purification; this culminated in the discovery of an antioxidant peptide LPHPSF [45]. Thus, unless it is the objective of the study to evaluate the effectiveness of proteolysis or that availability of protein sources is limited, measurements of DH and yield appear omittable when screening for an optimum hydrolysate for purification of antioxidant peptides.

In contrast to the aforementioned studies, some investigations omitted the tedious process of screening for an optimum hydrolysate, focusing directly on a protein hydrolysate generated by using a single protease. The protease treatments in these studies were chosen based on previously reported efficacy on other species or sample matrices [32,34,35,42,49,51,54]. In these studies, the protease treatments may not be considered optimum for the samples anymore. Even though antioxidant peptides were eventually identified in these studies, the questions remain whether a more potent hydrolysate could have been generated from the samples under investigation, and whether antioxidant peptides of greater potency could have been discovered.

Reports of the use of in vitro gastrointestinal digestion to release antioxidant peptides from proteins of edible marine invertebrates, which culminated in the identification of antioxidant peptide sequences, are scarce. Through the use of pepsin, trypsin and α-chymotrypsin under simulated gastrointestinal conditions and followed by subsequent purification work, antioxidant peptides LVGDEQAVPAVCVP and LKQELEDLLEKQE were identified from the mussel [52] and oyster [40], respectively. The use of a single commercial protease in enzymatic hydrolysis is relatively straightforward when compared with a combination of several proteases, as is the case in in vitro gastrointestinal digestion. Single-protease hydrolysis likely allows a better control of the physicochemical conditions of the process, in addition to that of the compositions and molecular weights of the resulting peptides [58]. For single-protease hydrolysis, proteolysis is carried out only under a selected optimum temperature and pH. Optimum hydrolysis duration is often selected by screening a range of different durations to identify the one which produces the highest DH and/or antioxidant activity (e.g., [31,35,50,56,57]). Often, the N- or C-terminal amino acid residues of the resulting peptides can be predicted based on the cleavage specificity of the protease used. On the other hand, in vitro gastrointestinal digestion by using proteases requires more complicated pH control and is more time-consuming. For example, simulated gastrointestinal digestion of mussel muscle was performed by first hydrolyzing the sample with pepsin at 37 °C and pH 2.5 for 120 min. Next, the resulting digest was hydrolyzed by trypsin and α-chymotrypsin at 37 °C and pH 7.0 for 150 min [52]. Due to the use of multiple proteases and the unique cleavage specificity of each protease, prediction of the molecular compositions of the resulting peptides is no longer straightforward. Nevertheless, it is believed that purification of antioxidant peptides from simulated gastrointestinal digests may increase the chance of producing peptides that resist breakdown by gastrointestinal peptidases in vivo [40,59]. Notwithstanding this potential advantage, the latter approach is less commonly adopted than the single-protease hydrolysis approach, possibly owing to the aforementioned difficulties.

2.2. Purification of Antioxidant Peptides

A protein hydrolysate is essentially a mixture comprising of both antioxidant and prooxidant peptides [60]. Further fractionation will aid in eliminating the prooxidant components [22]. Active hydrolysates produced from edible marine invertebrates were usually subjected to assay-guided fractionation to enrich and eventually isolate the most potent antioxidant peptides. Peptide variability in terms of the molecular mass, net charge and polarity/hydrophobicity underlies the basis of separation techniques used, which include membrane ultrafiltration (UF), low-pressure column chromatography, fast protein liquid chromatography (FPLC), and high-performance liquid chromatography (HPLC). Table 2 summarizes the purification techniques, which enabled successful purification and identification of antioxidant peptides from edible marine invertebrates in various studies.

Table 2. Selected antioxidant peptides identified from edible marine invertebrates as reported in the literature between years 2000 and 2016.

Species	Protease Used for Hydrolysis *	Antioxidant Parameters Used to Guide Purification and Characterize Purified Peptides	Purification Techniques	Peptide Sequence Identified	Validated with Synthetic Peptides	Reference
Oyster (*Crassostrea madrasensis*)	Papain	DPPH scavenging / OH• scavenging # / FRAP # / Iron chelating # / LPI #	UF SPE RP-HPLC	ISIGCQPAGRIVM	×	[36]
Oyster (*Crassostrea gigas*)	In vitro gastrointestinal digestion (Pepsin, Trypsin and α-Chymotrypsin)	OH• scavenging / O2•− scavenging / Cellular radical scavenging Protection against OH•-induced DNA damage / LPI	AEC RP-HPLC(×2)	LKQELED/LEKQE	×	[40]
Oyster (*Crassostrea talienwhanensis*)	Subtilisin (Alcalase)	DPPH scavenging / OH• scavenging	UF SEC RP-HPLC(×2)	PVMGA QHGV	×	[35]
Mussel (*Mytilus coruscus*)	Papain	DPPH scavenging / OH• scavenging / O2•− scavenging / Alkyl radical scavenging / In vivo antioxidant defense	UF AEC RP-HPLC(×2) GPC	SLPIGLMIAM	×	[44]
Mussel (*Mytilus coruscus*)	In vitro gastrointestinal digestion (Pepsin, Trypsin and α-Chymotrypsin)	LPI / OH• scavenging / O2•− scavenging / Carbon-centered radical scavenging	AEC SEC RP-HPLC	LVGDEQAVPAVCVP	×	[52]
Blue mussel (*Mytilus edulis*)	Pepsin	DPPH scavenging / ORAC / Protection against H2O2-induced cytotoxicity	UF CEC RP-HPLC(×2)	PIIVYWK TTANIEDRR FSVVPSPK	√	[32]
Blue mussel (*Mytilus edulis*)	Neutrase	DPPH scavenging / OH• scavenging / O2•− scavenging / LPI	UF SEC RP-HPLC	YPPAK	×	[31]
Blood Clam (*Tegillarca granosa*)	Neutrase	DPPH scavenging / ABTS scavenging / OH• scavenging / O2•− scavenging / LPI	UF AEC SEC RP-HPLC	WPP QP	×	[33]
Short-necked Clam (*Ruditapes philippinarum*)	α-Chymotrypsin	DPPH scavenging / OH• scavenging / Alkyl radicalscavenging / O2•− scavenging	UF AEC RP-HPLC(×3)	SVEIQALCDM	×	[53]
Short-necked Clam (*Ruditapes philippinarum*)	Trypsin	DPPH scavenging / Reducing power / Protection against OH•-induced DNA damage	UF SEC RP-HPLC	GDQQK	×	[34]
Scallop (*Patinopecten yessoensis*)	Neutrase	DPPH scavenging # / OH• scavenging / Iron chelating # / Reducing power # / Protection against OH•-induced DNA damage	SEC	HMSY PEASY	√	[49]
Jellyfish (*Rhopilema esculentum*)	Alcalase	OH• scavenging / Protection against H2O2-induced cytotoxicity / Cellular antioxidant enzyme activity	UF AEC RP-HPLC	VKP VKCFR	√	[42]

Table 2. *Cont.*

Species	Protease Used for Hydrolysis *	Antioxidant Parameters Used to Guide Purification and Characterize Purified Peptides	Purification Techniques	Peptide Sequence Identified	Validated with Synthetic Peptides	Reference
Jumbo squid (*Dosidicus gigas*)	Trypsin	LPI OH• scavenging Carbon-centered radical scavenging Iron chelating # Protection against *t*-butyl hydroperoxide-induced cytotoxicity	UF CEC SEC RP-HPLC	FDSGPAGVL NGPLQAGQPGER	×	[41]
Giant squid (*Dosidicus gigas*)	Trypsin	LPI OH• scavenging O₂•⁻ scavenging Carbon-centered radical scavenging Protection against *t*-butyl hydroperoxide-induced cytotoxicity	UF CEC SEC RP-HPLC(×2)	NADFGLNGLEGLA NGLEGLK	×	[43]
Giant squid (*Dosidicus gigas*) (skin)	Alcalase	ABTS scavenging FRAP	UF SEC	GPLGLLGFLGPLGLS	√	[51]
Shortclub cuttlefish (*Sepia brevimana*)	Trypsin	DPPH scavenging ABTS scavenging # O₂•⁻ scavenging # Total antioxidant capacity # Reducing power Iron chelating # LPI Protection against OH•-induced DNA damage	AEC SEC	I/L N I/L CCN	×	[56]
Indian squid (*Loligo duvauceli*)	α-chymotrypsin	DPPH scavenging OH• scavenging O₂•⁻ scavenging # Reducing power Iron chelating LPI Protection against OH•-induced DNA damage Cellular radical scavenging	AEC SEC	WCTSVS	×	[57]
Prawn (*Penaeus japonicus*)	Pepsin	LPI	SEC CEC RP-HPLC	IKK FKK FIKK	√	[48]
Shrimp processing by-products	Alcalase	DPPH scavenging	Methanol extraction CEC SEC RP-HPLC(×2)	SVAMLFH	×	[54]
Sea cucumber (*Stichopus japonicus*)	Trypsin	OH• scavenging O₂•⁻ scavenging	SEC AEC SEC RP-HPLC	GPEPTGPT GAPQWLR	×	[55]
Sea squirt (*Styela clava*)	Pepsin	Peroxyl radical scavenging	SEC RP-HPLC	LWHTH	√	[50]
Sea squirt (*Styela plicata*)	Trypsin	Peroxyl radical scavenging DPPH scavenging OH• scavenging Cellular radical scavenging Protection against APPH-induced cytotoxicity Protection against APPH-induced ROS generation and cell death in zebrafish embryos	AEC SEC RP-HPLC	LPHPSF	√	[45]

* Only the protease which was associated with the peptide sequence identified is listed. # Not tested with purified or synthetic peptides. AEC, anion exchange chromatography; CEC, cation exchange chromatography; FRAP, Ferric Reducing Antioxidant Power; LPI, lipid peroxidation inhibition; RP-HPLC(×2), two-step RP-HPLC; RP-HPLC(×3), three-step RP-HPLC; SEC, size exclusion chromatography; SPE, solid phase extraction; UF, ultrafiltration membrane; ×, not validated; √, validated.

Membrane UF of protein hydrolysates is often the first step in the assay-guided purification of antioxidant peptides from edible marine invertebrates. This technique uses special porous membranes having certain molecular weight cut-off (MWCO) specifications and made of materials such as regenerated cellulose and polyethersulfone to fractionate peptides in a protein hydrolysate. UF membranes with different MWCO values were frequently used in combination for the fractionation of antioxidant peptides. For example, UF membranes of MWCO 3, 5 and 10 kDa were used to separate the hydrolysates of squid muscle, squid skin gelatin, and the short-necked clam into multiple fractions of different molecular mass ranges [34,41,43]. UF membranes of MWCO 1 and 3 kDa were used to separate a jellyfish protein hydrolysate into three fractions [42]. Less frequently, some studies used only one UF membrane of a selected MWCO for initial fractionation of hydrolysates [33,35,51]. When the antioxidant activities of UF fractions were compared, the fraction with the smallest molecular size, e.g., <1 kDa [33,42] or <3 kDa [31,34,41,43], often showed the strongest activity and was therefore chosen for further purification.

Following membrane UF, the most active fraction obtained from marine invertebrate protein hydrolysates was usually further purified by means of size-exclusion chromatography (SEC) and/or ion exchange chromatography (IEC), either driven by FPLC (e.g., [45,52,56,57]), or carried out as low pressure column chromatography (e.g., [33,34,50]). Notably, some studies omitted the UF step and directly purified protein hydrolysates on an SEC and/or IEC column [45,46,56,57]. Sephadex G-15, G-25 and G-50 are SEC stationary phases that have been used in the purification of antioxidant peptides from marine invertebrate hydrolysates. Sephadex G-15 was used in the purification of antioxidant peptides from the oyster [35] and blue mussel [31]. Sephadex G-25 was used more commonly by other researchers [33,34,41,43,45,48–50,54,56,57]. Zhou, Wang and Jiang [55] used both Sephadex G-25 and G-50 stationary phases to purify antioxidant peptides from the sea cucumber. Elution of SEC columns was usually performed by using deionized or distilled water and monitored at 214 nm [36], 220 nm [33,34,49,55], or 280 nm [31,45,54,56,57].

IEC can be divided into two broad classes based on the type of exchangers used: anion exchange chromatography (AEC) and cation exchange chromatography (CEC). For the purification of antioxidant peptides by AEC from the Indian squid [57], shortclub cuttlefish [56], oyster [40], mussel [44,52], sea cucumber [55], short-necked clam [53], blood clam [33], sea squirt [45], and jellyfish [42], weak anion exchangers with diethylaminoethyl (DEAE) exchange groups were used. To purify antioxidant peptides by CEC from the prawn [48], squid [43], shrimp processing by-products [54], and jumbo squid skin gelatin [41], resins with sulphopropyl (SP) strong cation exchange groups were used. Prior to IEC, Zhao, Huang and Jiang [54] extracted lyophilized hydrolysate of shrimp processing by-products with 90% methanol overnight to remove interfering non-protein compounds having high antioxidant activity. Most of the aforementioned studies eluted peptide fractions during AEC and CEC by using a linear NaCl gradient, although step gradient elution with NaCl was also reported for the purification of antioxidant peptides from the blood clam and sea cucumber [33,55]. Elution profiles of IEC were monitored in these studies at 215 nm [43], 220 nm [33,41,55], and 280 nm [40,44,45,53,54,56,57].

Some researchers purified their peptide samples on only SEC [34,50], whereas others on only IEC [32,40], before proceeding to peptide purification by Reversed-Phase HPLC (RP-HPLC). The use of both SEC and IEC to purify antioxidant peptides from the protein hydrolysates of the prawn [48], Indian squid [57], shortclub cuttlefish [56], sea squirt [45], and shrimp processing by-products [54] has also been reported. When purifying antioxidant peptides from the sea cucumber, Zhou, Wang and Jiang [55] adopted the approach of two-step SEC linked by IEC. Their peptide sample was purified on a Sephadex G-50 column, followed by a DEAE cellulose DE-52 column, and lastly on a Sephadex G-25 column, guided by $OH^{\bullet}$ and $O_2^{\bullet-}$ scavenging assays. The Sephadex G-25 step allowed them to desalt their sample prior to subjecting it to RP-HPLC separation [55].

RP-HPLC was almost always the last purification technique performed before the peptides purified from edible marine invertebrates were taken to peptide sequence determination. Peptide samples were often purified with low-pressure column chromatography before they were subjected to RP-HPLC purification. Notwithstanding, Asha, Remya Kumari, Ashok Kumar, Chatterjee, Anandan and Mathew [36] purified an active UF fraction of the oyster protein hydrolysate with four C18 solid phase extraction (SPE) cartridges connected in series prior to subjecting it to RP-HPLC purification. With a few exceptions [49,56,57], most studies which successfully identified antioxidant peptides from edible marine invertebrates purified their peptides by RP-HPLC at least once. RP-HPLC was carried out using a C18 stationary phase with a linear gradient of acetonitrile either containing low percentage of formic acid [51], or trifluoroacetic acid (TFA)(e.g., [31–33,36,45,54]), or without any mobile-phase modifiers [34,41,43,44,50,52,53]. An alternative mobile phase comprising of a linear gradient of methanol with 0.1% TFA was reported by others who successfully purified and identified antioxidant peptides from the jellyfish [42] and sea cucumber [55]. Eluted RP-HPLC fractions were monitored at 214 nm [36,54,55], 215 nm [32,40,41,43,44,52,53], 220 nm [33,34,42,48,50], 230 nm [35], or 280 nm [31,35,45,61].

Multi-step RP-HPLC involving the use of a semi-preparative C18 column followed by one or two analytical C18 columns was used to purify antioxidant peptides from the short-necked clam [53], giant squid [43], oyster [40], hard-shelled mussel [44], and shrimp processing by-products [54]. To isolate antioxidant peptides from the blue mussel, Park, Kim, Ahn and Je [32] purified their sample on the same semi-preparative C18 column twice, each time using a different linear acetonitrile gradient.

2.3. Identification of Antioxidant Peptides

Following RP-HPLC purification, the purified antioxidant peptides were usually taken to amino acid sequence identification by using either liquid chromatography-tandem mass spectrometry (LC-MS/MS) or the Edman degradation method. Typical LC-MS/MS experiments which enabled successful identification of antioxidant peptides from edible marine invertebrates involved the use of a quadrupole time-of-flight tandem mass spectrometer, equipped with an electrospray ionization (ESI) source and run in the positive ion mode (e.g., [42,49]). Alternatively, the use of a hybrid triple quadrupole/linear ion trap mass spectrometer to sequence antioxidant peptides from the oyster [36], shortclub cuttlefish [56], and Indian squid [57] was reported. To identify a radical scavenging peptide from shrimp processing by-products, Zhao, Huang and Jiang [54] used an ESI-triple quadrupole mass spectrometer run in the negative ion mode. Mass spectra data obtained were typically analyzed with de novo sequencing algorithms to identify the amino acid sequences of the peptides isolated. At the same time, information on molecular mass of the isolated peptide can be computed from the mass spectra data [32,42,55].

On the other hand, using a protein sequencer, the sequencing of antioxidant peptides purified from the short-necked clam [34,53], hard-shelled mussel [44], blue mussel [31], blood clam [33], and prawn [48] was carried out based on the Edman degradation reaction. Following peptide sequencing, mass spectrometry was used by some research groups to determine the molecular masses of the antioxidant peptides identified [31,33,48]. The use of SEC to further purify the antioxidant peptide contained in an active RP-HPLC fraction prior to determining the amino acid sequence of the peptide by using the Edman degradation method was also reported [44,53].

After an antioxidant peptide is identified, synthesizing the peptide and validating its bioactivity will provide valuable confirmation of the function of the antioxidant peptide. A search of the literature revealed that of the 22 reports published in the last 16 years of antioxidant peptides identified from edible marine invertebrates, only seven studies have validated the antioxidant peptides identified with chemically synthesized peptides [32,42,45,48–51]. For synthetic peptide production, Fmoc-based solid-phase peptide synthesis was typically carried out. Purity of the synthetic peptide and its molecular mass were analyzed by means of RP-HPLC and mass spectrometry [32,45,49,51].

3. Evaluation of Antioxidant Activities

Different types of chemical and cell-based assays have been used in the assay-guided purification and the characterization of antioxidant peptides from edible marine invertebrates (Table 2). Among the chemical assays used, radical scavenging and lipid peroxidation inhibition assays were common. Concerning the principles as well as the advantages and limitations of the antioxidant assays listed in Table 2, we refer the reader to reviews by Zhong and Shahidi [11], Sila and Bougatef [18], Samaranayaka and Li-Chan [21], and Wu, et al. [46].

The 2,2-diphenyl-1-picrylhydrazyl (DPPH) radical scavenging assay, likely due to its simplicity, was used in quite a number of studies to guide the purification of antioxidant peptides from the oyster [35,36], mussel [31,32], blood clam [33], short-necked clam [34], scallop female gonad [49], shortclub cuttlefish [56], Indian squid [57], and shrimp processing by-products [54]. Besides DPPH scavenging activity, the ability of antioxidant peptides identified from edible marine invertebrates to quench ABTS, $OH^{\bullet}$, $O_2^{\bullet-}$, alkyl, peroxyl, and carbon-centered radicals have been reported (Table 2). Using a linoleic acid model system, the ability of antioxidant peptides identified from the oyster [40], mussel [31,52], blood clam [33], squid [41,43], shortclub cuttlefish [56], Indian squid [57], and prawn [48] to inhibit lipid peroxidation has also been demonstrated. The ability to protect against $OH^{\bullet}$-induced DNA damage has been demonstrated for oyster-derived peptide LKQELEDLLEKQE [40], short-necked clam-derived GDQQK [34], scallop-derived HMSY and PEASY [49], Indian squid-derived WCTSVS [57], and shortclub cuttlefish-derived I/L N I/L CCN [56]. Using the fluorescence probe DCFH-DA, cellular radical scavenging activity of LKQELEDLLEKQE and LPHPSF in mouse macrophage cells (RAW 264.7) [40,45] and that of WCTSVS in human breast adenocarcinoma cells (MCF7) [57] has been demonstrated.

Antioxidant peptides identified from the giant squid [43], blue mussel [32], jellyfish [42], sea squirt [45], and jumbo squid skin gelatin [41] were shown to mitigate radical-induced cytotoxicity (Table 2). Cell types used in these studies were human lung fibroblast cells [41,43], Chang liver cells [32], RAW 264.7 [45], and rat cerebral microvascular endothelial cells (RCMEC) [42]. Notably, the cytoprotective activity of PIIVYWK, FSVVPSPK, VKP, and VKCFR was confirmed using chemically synthesized peptide sequences, rather than purified fractions [32,42]. Mendis, Rajapakse, Byun and Kim [41] found that the efficacy of FDSGPAGVL and NGPLQAGQPGER in protecting human lung cells against *t*-butyl hydroperoxide-induced oxidative cell death was comparable to or surpassed that of α-tocopherol. On the other hand, the concentration range of Indian squid-derived WCTSVS that was non-toxic to MCF7 cells [57], of shortclub cuttlefish-derived I/L N I/L CCN non-toxic to human colorectal adenocarcinoma cells (HT29) [56], and of sea squirt-derived LPHPSF non-toxic to RAW 264.7 cells [45] have been reported. LKQELEDLLEKQE purified from the oyster was also reportedly non-toxic to human embryonic lung fibroblast and RAW 264.7 cells, although the range of peptide concentration tested was not reported [40].

Little work has been done to elucidate the molecular or biochemical basis underlying the cytoprotective effects of antioxidant peptides identified from edible marine invertebrates. The two peptides PIIVYWK and FSVVPSPK derived from the blue mussel protected human liver cells against H_2O_2-induced toxicity by upregulating the protein expression of hemeoxygenase-1 [32]. Protection of RCMEC cells by jellyfish-derived peptides VKP and VKCFR against H_2O_2-induced toxicity, on the other hand, was associated with enhanced enzymatic activities of superoxide dismutase, catalase, and glutathione peroxidase [42]. Current evidence is preliminary, but the observed cytoprotection appears attributable to the ability of the antioxidant peptides to activate the expression of cytoprotective enzymes. To the authors' knowledge, only two animal studies have been conducted to investigate the effects of antioxidant peptides identified from any edible marine invertebrates. Oral administration of peptide SLPIGLMIAM to mice was reported to inhibit the level of malondialdehyde in the liver, thus providing evidence for in vivo antioxidant effect [44]. The treatment enhanced superoxide dismutase activity in vivo but had no effects on catalase or glutathione peroxidase activities [44]. Recently, sea squirt-derived antioxidant peptide LPHPSF was shown to be capable of attenuating radical-induced cell death and ROS production in zebrafish embryos [45]. Figure 2 summarizes modes of action reported for antioxidant peptides identified from edible marine invertebrates.

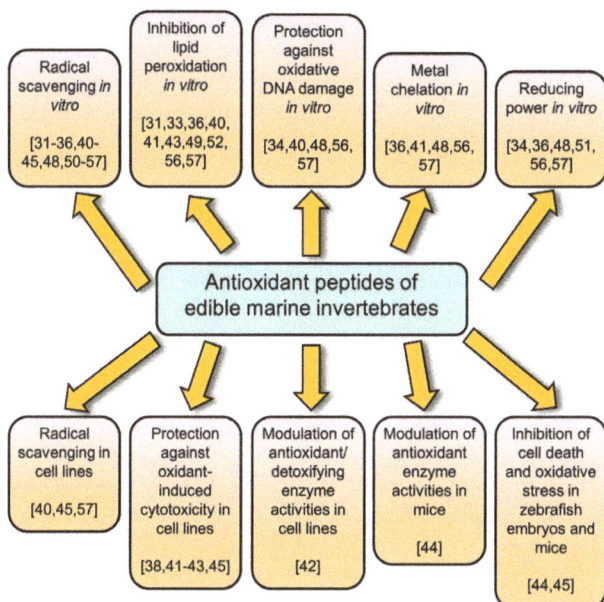

Figure 2. Antioxidant mechanisms reported for antioxidant peptides identified from edible marine invertebrates.

Owing to the diverse types of assays and assay conditions used to characterize the bioactivities of food-derived antioxidant peptides, comparison between studies is often difficult [21]. For example, Zhou, Wang and Jiang [55] and Sudhakar and Nazeer [57] reported markedly different levels of BHT potency in their hydroxyl radical and superoxide anion radical scavenging assays. Such a discrepancy is likely due to the different assay protocols used in the two studies. Chi et al. [33] compared the EC_{50} values of blood clam-derived WPP for DPPH and ABTS scavenging activities with EC_{50} values reported for other antioxidant peptides by other research groups. Such a comparison will be more meaningful if identical assay conditions and ideally an identical positive control were established between studies to check for laboratory-to-laboratory variations. Although Chi et al. [33] and Zhuang et al. [62] both ran the DPPH and ABTS scavenging assays using ascorbic acid as the positive control, comparison of relative potency of peptides between studies remains difficult because EC_{50} for ascorbic acid was reported by only Zhuang et al. [62]. Comparison of antioxidant potential between peptides based on data obtained from different antioxidant assay procedures in different studies (e.g., [31,32]) should therefore be considered with caution.

4. Molecular Characteristics and Structure–Activity Relationship

The structure–activity relationships (SAR) of food-derived antioxidant peptides were recently reviewed [18,25,26]. Comprehension of SAR may contribute towards an effective prediction of potential antioxidant activity in a new peptide and the development of strategies for enzyme-assisted release of antioxidant peptides from food proteins [18]. Generally, structural characteristics such as molecular mass, hydrophobicity, amino acid composition, and peptide sequence are considered determinants of the antioxidant activity of a peptide [25,26]. Despite such generalizations, current knowledge of the SAR of antioxidant peptides is still incomplete.

In the context of molecular mass, a majority of the food-derived antioxidant peptides have molecular masses ranging between 500 and 1800 Da [21]. Concurring with this, we found that 26 of the 32 antioxidant peptides identified from edible marine invertebrates fall within this range (Table 3).

Peptides having smaller molecular masses are generally believed to have greater antioxidant activity than those having larger masses [25,26]. In accordance with this, during the fractionation of marine invertebrate protein hydrolysates with UF membranes, peptidic fractions having the lowest molecular mass range showed the highest DPPH radical scavenging activity [31,33,36] and the highest hydroxyl radical scavenging activity [42,53]. Similarly, UF fractions with the lowest molecular mass range (<3 kDa), which were prepared from the giant squid muscle [43] and jumbo squid skin gelatin [41], had the highest inhibitory activity against linoleic acid oxidation relative to other UF fractions in the same study. By contrast, in SEC or gel filtration chromatography, the peptidic fraction with the lowest molecular mass was not always the most potent antioxidant fraction. SEC fractions with the lowest molecular mass were reported to be the most active fraction in scavenging DPPH [54,57], hydroxyl [35,57], and peroxyl [50] radicals, as well as having the highest reducing power [57]. Furthermore, others reported that SEC fractions with intermediate molecular masses were the most active in scavenging DPPH [31,33,34,49,56] and ABTS [51], as well as having the strongest reducing power [34,51,56]. Zhou, Wang and Jiang [55] had purified antioxidant peptides from a sea cucumber protein hydrolysate by using a two-step SEC. In the first step, the fraction with the lowest molecular mass showed the highest superoxide and hydroxyl scavenging activity. In the second step, the fraction with the highest molecular mass was the most potent [55]. Alemán et al. [51] suggested that in SEC, a peptidic fraction with the lowest molecular mass may contain large number of free amino acids and small peptides lacking antioxidant activity. Taken together, the aforementioned discrepancies imply that the notion of smaller peptides exhibiting greater antioxidant activity may be oversimplified, or is only applicable to the analysis of protein hydrolysates purified by certain techniques. The proposal thus contributes only limited, if any, knowledge about the SAR of antioxidant peptides.

Table 3. Molecular masses of 32 antioxidant peptides identified from edible marine invertebrates.

Antioxidant Peptides	Molecular Mass (Da)	References
QP	243.23	[33]
VKP	342	[42]
IKK	388	[48]
WPP	398.44	[33]
FKK	422	[48]
QHGV	440	[35]
PVMGA	518	[35]
FIKK	535	[48]
HMSY	536.16	[49]
PEASY	565.21	[49]
YPPAK	574	[31]
GDQQK	574.27 *	[34]
VKCFR	651	[42]
I/L N I/L CCN	679.5	[56]
WCTSVS	682.5	[57]
LWHTH	692.2	[50]
LPHPSF	696.3	[45]
NGLEGLK	747	[43]
SVAMLFH	804.4	[54]
FSVVPSPK	860.09	[32]
FDSGPAGVL	880.18	[41]
PIIVYWK	1004.57	[32]
SLPIGLMIAM	1044.57 *	[44]
TTANIEDRR	1074.54	[32]
SVEIQALCDM	1107.49 *	[53]
NGPLQAGQPGER	1241.59	[41]
ISIGGQPAGRIVM	1297.72	[36]
NADFGLNGLEGLA	1307	[43]
GPLGLLGFLGPLGLS	1409.63 **	[51]
GPEPTGPTGAPQWLR	1563	[55]
LVGDEQAVPAVCVP	1590	[52]
LKQELEDLLEKQE	1600	[40]

* Calculated online using PepDraw [63]; ** Calculated from the *m/z* value reported.

Hydrophobicity is an important determinant of antioxidant activity of peptides. The presence of hydrophobic residues allows an antioxidant peptide to interact with lipid-soluble free radicals and retard lipid peroxidation [64]. In agreement with this, 15 antioxidant peptides identified from edible marine invertebrates, which inhibited lipid peroxidation in vitro, contain 28%–100% hydrophobic residues in their sequences (Table 4). In fact, the 12 antioxidant peptides identified from edible marine invertebrates, whose antioxidant activities (radical scavenging, reducing power, and inhibition of lipid peroxidation) were validated by using synthetic peptides, contain 22%–71% hydrophobic residues (Table 5). Alemán et al. [51] compared the reducing power and ABTS radical scavenging activity of three peptides having the same molecular mass, namely GPLGLLGFLGPLGLS, GPOGOOGFOGPOGOS (where O represents hydroxyproline), and GPOGOOGFLGPOGOS. The peptide GPLGLLGFLGPLGLS, the most hydrophobic sequence of the three, was found to have the strongest antioxidant activity [51]. Apparently, hydrophobicity may confer antioxidant activity to peptides not only in a lipid oxidation model, but also via other antioxidant mechanisms.

Table 4. Percentages of hydrophobic residues in 15 edible marine invertebrate-derived antioxidant peptides, which exhibited lipid peroxidation inhibitory activity.

Antioxidant Peptides	Hydrophobic Amino Acid Residue (%) *	References
NGLEGLK	28.57	[43]
LKQELEDLLEKQE	30.77	[40]
NGPLQAGQPGER	33.33	[41]
IKK	33.33	[48]
FKK	33.33	[48]
WCTSVS	33.33	[57]
I/L N I/L CCN	33.33	[56]
FDSGPAGVL	44.44	[41]
NADFGLNGLEGLA	46.15	[43]
QP	50	[33]
FIKK	50	[48]
ISIGGQPAGRIVM	53.85	[36]
YPPAK	60	[31]
LVGDEQAVPAVCVP	64.29	[52]
WPP	100	[33]

* Percentages of hydrophobic residues were computed manually, based on the classification of A, I, L, M, F, P, W, and V as hydrophobic amino acids (The IARCTP53 Database [65]).

Table 5. Percentages of hydrophobic residues in 13 edible marine invertebrate-derived antioxidant peptides, whose activities were confirmed using pure synthetic peptides.

Antioxidant Peptides	Hydrophobic Amino Acid Residue (%) *	References
TTANIEDRR	22.22	[32]
HMSY	25	[49]
IKK	33.33	[48]
FKK	33.33	[48]
PEASY	40	[49]
LWHTH	40	[50]
VKCFR	40	[42]
FIKK	50	[48]
GPLGLLGFLGPLGLS	60	[51]
FSVVPSPK	62.5	[32]
VKP	66.67	[42]
LPHPSF	66.67	[45]
PIIVYWK	71.43	[32]

* Percentages of hydrophobic residues were computed manually, based on the classification of A, I, L, M, F, P, W, and V as hydrophobic amino acids (The IARCTP53 Database [65]).

Amino acid compositions and peptide sequences also influence the antioxidant activity of a peptide [25,26]. Food-derived antioxidant peptides often contain hydrophobic residues at the N-terminus, in addition to having Pro, His, Tyr, Trp, Met, and Cys in the peptide sequences [21]. Examination of the list of 32 antioxidant peptides identified from edible marine invertebrates (Table 3) revealed 18 sequences containing a hydrophobic residue at the N-terminus (FDSGPAGVL, FIKK, FKK, FSVVPSPK, I/L N I/L CCN, IKK, ISIGGQPAGRIVM, LKQELEDLLEKQE, LPHPSF, LVGDEQAVPAVCVP, LWHTH, PEASY, PIIVYWK, PVMGA, VKCFR, VKP, WCTSVS, and WPP). The 14 other peptides in Table 3, although lacking a hydrophobic N-terminal residue, contain at least one hydrophobic residue in their sequences. The ability of some of the aforementioned peptides to inhibit lipid peroxidation (Table 4) may therefore be attributable, at least in part, to the presence of hydrophobic residues. GDQQK identified from the short-necked clam is the only antioxidant peptide which contains no hydrophobic residue in its sequence. At present there is no report of GDQQK having any lipid peroxidation inhibitory activity [34].

His-containing peptides can exert their antioxidant effects through the hydrogen donating, lipid peroxyl radical trapping, and metal ion chelating actions of the His imidazole group [66]. On this score, the presence of His residue may account for, to a certain extent, the radical scavenging activities exhibited by QHGV [35], LWHTH [50], SVAMLFH [54], LPHPSF [45] and HMSY [49]. On the other hand, aromatic amino acid residues (e.g., Tyr, Trp and Phe) in peptides may exert antioxidant effects by donating protons to electron-deficient free radicals [26,67]. Thus, the presence of aromatic amino acid residues may account for the antioxidant activities demonstrated by some peptides identified from edible marine invertebrates (e.g., WPP, FKK, and FIKK).

IKK (388 Da) and FKK (422 Da), two antioxidant tripeptides identified from the prawn, showed different levels of inhibition against linoleic acid oxidation despite both containing one hydrophobic residue. Furthermore, a mixture of the constituent amino acids at the same concentrations as the peptides showed no antioxidant activity [48]. This suggests that the specific amino acid sequences, and more specifically, the identity of the N-terminal residue of the tripeptides, are key to their potency as antioxidants. Wu et al. [49] compared the antioxidant activity of two peptides of similar molecular masses that were identified from scallop female gonads, namely HMSY (536 Da) and PEASY (565 Da). HMSY, despite its lower content of hydrophobic residues (25%), exhibited 4.6-fold stronger hydroxyl radical scavenging activity than PEASY (40% hydrophobic residues) [49]. It was suggested that antioxidant activity of the two peptides may be attributed to the Tyr residue at the C-terminus of the peptides [49]. Nevertheless, considering their distinct difference in antioxidant potency, the specific amino acid sequences in HMSY and PEASY may have imparted different levels of antioxidant effects too.

Several antioxidant peptides identified from edible marine invertebrates which exhibited lipid peroxidation inhibition activity (Table 4) were amphiphilic in nature. For example, NGLEGLK, NGPLQAGQPGER, and FIKK are antioxidant peptides containing one or more hydrophilic, basic amino acids (e.g., K and R), in addition to varying percentages of hydrophobic residues in their sequences. The lipophilic and hydrophilic amino acid residues may have collectively contributed to the overall antioxidant activity of the peptides. It was noted that the amphiphilic nature of peptides may influence their antioxidant activity by facilitating their interaction with a hydrophobic target and also proton exchanges with free radicals [26]. Amphiphilic peptides may reside in the oil-water interface and effectively quench free radicals in both the aqueous and oil phases of a linoleic acid emulsion system [68].

Altogether, despite others' assertions [25,26], molecular mass is not a reliable determinant of antioxidant activity of peptide fractions derived from edible marine invertebrates. On the other hand, our assessment of the hydrophobicity and amino acid compositions of the antioxidant peptides identified from edible marine invertebrates supports the proposal that these two parameters are important determinants of peptide antioxidant efficacy [25,26]. Thus, hydrophobicity and amino acid compositions are two criteria that are useful to future efforts to develop strategies to discover antioxidant peptides from edible marine invertebrates based on SAR knowledge.

5. Potential Applications in Food, Therapy and Cosmetics

The bioactive peptide ingredient market, currently dominated by the soy and dairy industries, is very competitive [69]. Bioactive peptides, in pure form and as an unpurified mixture, have been incorporated into a number of functional foods and dietary supplements already on the market [64,69–71]. At present, a number of peptides with well-established antioxidant effects are commercially available as dietary supplements, e.g., reduced glutathione, carnosine, anserine, and melatonin. Thus there is market demand for antioxidant peptide-based supplements or functional foods. Nevertheless, marine peptide-based functional foods and dietary supplements with approved antioxidant health claims are scarce [12,72]. To the authors' knowledge, health foods or supplements containing pure antioxidant peptides identified from edible marine invertebrates are either unavailable or have not been documented in the literature. Apparently, opportunities abound for the development of novel foods or dietary supplements containing antioxidant peptides identified from edible marine invertebrates. On this score, peptides which exhibited antioxidant potential in cell culture models and in mice, for example, VKP and VKCFR identified from the jellyfish [42] and SLPIGLMIAM identified from the mussel [44] are promising candidates for the development of high-value peptide ingredients for health food or supplements. Furthermore, VKP and VKCFR which demonstrated antioxidant and angiotensin converting enzyme inhibitory activities [42] as well as WPP which showed antioxidant and antiproliferative potential [33] are valuable candidates for the development of multifunctional health foods or supplements.

Antioxidant peptides which can inhibit lipid oxidation are potentially useful in the preservation of lipid-rich foods [18]. Besides food quality preservation during storage, antioxidant peptides incorporated into foodstuffs can provide nutrients in the form of amino acids when consumed, which is an advantage over synthetic antioxidants. Among the antioxidant peptides identified from edible marine invertebrates, 15 of them inhibited in vitro lipid oxidation (Table 4). Peptides isolated from the giant squid (NADFGLNGLEGLA, and NGLEGLK) [43], mussel (LVGDEQAVPAVCVP) [52], and oyster (LKQELEDLLEKQE) [40], for instance, were comparable or superior to lipid-soluble antioxidant alpha-tocopherol in inhibiting lipid peroxidation. Tripeptide WPP isolated from blood clam protein hydrolysate was as effective as glutathione in attenuating lipid peroxidation [33]. Among the 15 peptides, prawn muscle-derived IKK, FKK, and FIKK are the most notable. Their ability to dampen lipid peroxidation, validated using chemically synthesized peptides designed based on the identified sequences, surpassed that of alpha-tocopherol [48]. Currently, the antioxidant effects of edible marine invertebrate-derived peptides in food systems are a gap in knowledge. Nevertheless, some work has been done using protein hydrolysates prepared from edible marine invertebrates. Squid protein hydrolysate prepared using papain was shown to retard lipid oxidation in a sardine mince model system as effectively as ascorbic acid [73]. Cuttlefish skin gelatin hydrolysate prepared with alcalase delayed lipid oxidation in cooked turkey meat sausage during storage at 4 °C [74]. Protein hydrolysates prepared by alcalase hydrolysis of shrimp waste, when applied as a dipping treatment on whole Croaker fish, reduced lipid oxidation and maintained fish skin color during storage at 4 °C for 10 days [75].

Recently, the peptide PAGT isolated from Amur sturgeon skin gelatin was reported to inhibit lipid oxidation when added to the Japanese sea bass mince [76] and to retard both lipid and protein oxidation when applied in combination with caffeic acid to the same mince model [77]. These studies [76,77] suggest that besides protein hydrolysates, pure antioxidant peptides may be developed into novel food additives to be used for delaying oxidative rancidity and maintain food quality. Edible marine invertebrate-derived peptides with the ability to inhibit lipid oxidation in vitro (Table 2) have potential applications as food preservatives, but future work is required to first investigate their efficacy in food systems. Furthermore, issues such as potential unfavorable effects on food organoleptic properties and antioxidant peptide stability following food processing operations (e.g., heat treatment) have to be addressed [21]. To make their application as food antioxidant additives economically feasible, it is crucial that such peptides have considerably higher efficacy than crude protein hydrolysates, thus allowing the peptides to be used at low quantities (0.001%–0.02%) [18]. Otherwise, the use of crude protein hydrolysates, e.g., those prepared from the squid, cuttlefish skin gelatin and shrimp waste [73–75], are likely more cost-effective options as food antioxidant additives. Moreover, certain aromatic and hydrophobic amino acids that confer antioxidant activity on peptides may impart bitterness. Antioxidant peptides having high potency can be added to foodstuffs in small quantities, thus minimizing the issue of food bitterness [78].

Antioxidant peptides have promising future applications as therapeutics or adjunct therapeutics against oxidative damage-related diseases or conditions. In rats, the ability of a mitochondrion-targeted antioxidant peptide SS31 to reduce myocardial lipid peroxidation and infarct size in ischemia-reperfusion injury was reported [79]. The peptide also mitigated kidney injury in diabetic mice [80] and retinal damage in diabetic rats [81]. In a mouse model of burn injury, the antioxidant peptide SS31 alleviated symptoms of burn injury and promoted recovery of skeletal muscle mitochondrial functions [82,83]. An antioxidant peptide identified from ostrich egg white protein hydrolysate was also shown to promote wound healing in rats [84]. At present, the effects of pure antioxidant peptides identified from edible marine invertebrates have not been tested in animal models of human diseases. Future research in this direction should provide a better understanding of the potential of edible marine invertebrate-derived antioxidant peptides as therapeutic agents for targeting oxidative stress-related diseases or conditions. One current focus of the pharmaceutical industry is the development of peptide drugs having multiple pharmacological activities [17]. On this score, some antioxidant peptides identified from edible marine invertebrates (Table 2) may have potential applications in the development of future multifunctional peptide therapeutics and/or adjunct drugs. Promising candidates include blood clam-derived WPP, which in addition to its antioxidant activity, also inhibited the proliferation of different cancer cell lines, showing only marginal cytotoxicity against normal cells [33]. Other multifunctional antioxidant peptides discovered from edible marine invertebrates are VKP and VKCFR (jellyfish) [42] and GPLGLLGFLGPLGLS (squid skin gelatin) [51] which exhibited angiotensin converting enzyme inhibitory activity.

Skin care or cosmetic products are one area in which antioxidant peptides discovered from edible marine invertebrates may have applications. The beneficial effects of bioactive peptides whether as cosmeceuticals or as dermatologic tools which modulate collagen, elastin and melanin synthesis have been highlighted [85–88]. The observations that more than 25 different peptides are used as active ingredients in skin care products manufactured by companies based in Canada, USA, Spain, Switzerland, and France, and that many more skin care-relevant peptides are in development worldwide point to the commercial value of peptides as cosmeceutical ingredients [88]. Antioxidant peptides, exemplified by glutathione and carnosine, are one of the many bioactive peptides used in skin care products. One key consideration in using peptides in the formulation of skin care products is skin penetration. Longer peptides (e.g., containing six or more amino acids) generally do not penetrate well into deeper layers of the skin [85]. Longer peptides are also likely more expensive to produce compared to the shorter ones. Hence, among the antioxidant peptides derived from edible marine invertebrates (Table 2), those with short sequences (e.g., QHGV [35], WPP, QP [33], HMSY [49], VKP [42], IKK, FKK, and FIKK [48]) may have greater potential for application in skin care products compared with the rest. Such short peptides can be used as lead structures for the development of effective peptide ingredients taking into considerations factors such as potency, transcutaneous delivery, stability and compatibility within cosmetic formulations, toxicity, and cost of production [85,88]. In the light of many bioactive peptides used currently in skin care formulations, antioxidant peptides identified from edible marine invertebrates may have to outperform current peptide ingredients in critical areas, such as cost of production, potency and multifunctionality, in order to stand out as candidates for future peptide ingredients. The copper-binding tripeptide GHK-Cu, which is popularly used in skin care products, is an antioxidant peptide with wound healing, anti-inflammatory, and anti-aging properties [89]. Glutathione, in addition to its well-established antioxidant properties, also exhibits skin-lightening effects [90]. Thus, the multifunctional nature of the antioxidant peptides identified from edible marine invertebrates (Table 2) warrants more research.

6. Current Gaps in Knowledge and Future Perspectives

6.1. Biological Significance

One issue to take note of in future research is the biological significance of edible marine invertebrate-derived antioxidant peptides. Among the 32 antioxidant peptides identified from edible marine invertebrates (Table 2), three peptides showed cellular radical scavenging activity [40,45,57]; nine peptides, protective effects against AAPH-, H_2O_2- or *t*-butyl hydroperoxide-induced cytotoxicity [32,41–43,45]; and two peptides, in vivo antioxidant effects [44,45]. It is unclear whether antioxidant effects of the other 19 peptides, discovered by using cell-free or chemical-based methods, have any biological significance. Shen et al. [91] identified 16 antioxidant peptides from ovotransferrin guided by the oxygen radical absorbance capacity (ORAC) assay; antioxidant activity of pure peptides synthesized based on the 16 sequences was validated with the ORAC assay. Recently, Jahandideh et al. [92] reported that these 16 peptides had no antioxidant activities in human umbilical vein endothelial cells. A plausible explanation would be that a chemical- or cell-free antioxidant assay, such as the ORAC assay, fails to reflect the complexity of a biological system when used to assess antioxidant peptides [92]. Whether the same observation will hold had a different cellular antioxidant assay or cell type been used to test the 16 peptides in the study of Jahandideh, Chakrabarti, Davidge and Wu [92] is an open question. In any case, if the targeted applications of an edible marine invertebrate-derived antioxidant peptide involve a biological system, it is highly desirable to have its biological significance ascertained at least in a cell culture model before more time and effort are invested to studying it.

6.2. Stability

Depending on the intended applications for an antioxidant peptide, its stability upon thermal processing and gastrointestinal digestion, its bioactivity in biological tissues or cells, as well as its bioavailability may or may not be crucial. For example, antioxidant peptides to be used for food preservation may not need to exhibit significant cellular radical scavenging activity or gastrointestinal stability, but should remain active when incorporated into food matrices and ideally be thermal-stable. Meanwhile, antioxidant peptides to be used as injectable therapeutics may not need to be thermal-stable or resistant to gastrointestinal digestion, but should not be readily degraded by human plasma peptidases.

To gauge the potential of an antioxidant peptide as orally administered therapeutic agent or as functional food ingredient, the intestinal stability and absorption of the peptide is a key consideration. Monolayers of the human colorectal adenocarcinoma cell (Caco-2) are considered a model of the intestinal epithelium. The Caco-2 permeability assay has been used to demonstrate that CKYVCTCKMS, an antioxidant peptide derived from buffalo milk product, had good stability against brush-border peptidases and was absorbed intact through a Caco-2 monolayer [93]. On the other hand, the in vitro resistance of a lactoferricin B-derived antihypertensive peptide RRWQWR to human plasma peptidases has also been investigated through monitoring their degradation in human serum over time with RP-HPLC [94]. At present, very little is known about the in vivo stability of the antioxidant peptides identified from edible marine invertebrates. Future studies to investigate their bioavailability and stability against intestinal and plasma peptidases are warranted to better appraise their potential applications in biological systems. Certain strategies may be adopted to enhance the intestinal and systemic stability of the edible marine invertebrate-derived antioxidant peptides if necessary. Peptide structural modifications such as cyclisation and replacement of L-amino acid with D-amino acid as well as other related strategies have been reviewed [95,96]. Proline- and hydroxyproline-containing peptides are generally able to resist breakdown by digestive enzymes [97].

On this score, selection of proline-containing peptides from the pool of antioxidant peptides identified from edible marine invertebrates for further study or incorporation of a proline residue into a selected antioxidant peptide may be a promising approach.

Development of efficacious and stable peptide drugs that can be administered orally, thus improving patient convenience, is one of the current foci of the pharmaceutical industry [17]. Notwithstanding, a pertinent point to consider is that in vivo absorption of an antioxidant peptide may not always be required for its application. Ingested materials and pathogens can induce gastrointestinal oxidative injury and inflammation, which may increase the risks of diseases such as peptic ulcers, cancers, and inflammatory bowel disease [98]. Hence, antioxidant peptides that are resistant to gastrointestinal proteolysis following oral administration, even without getting absorbed, may still play a role in maintaining the health of the epithelial lining of the gastrointestinal tract [22].

6.3. Application of In Silico Tools

Integration of in silico and in vitro experiments is a promising research strategy for the discovery of antioxidant peptides from food sources, including edible marine invertebrates. Recently, some research groups have used in silico tools to predict bioactive peptides that can be enzymatically released from selected protein sources and compare relative effectiveness of different proteases. To the authors' knowledge, the applications of in silico tools in the discovery of potential antioxidant peptides from edible marine invertebrates are scarce. BIOPEP is a database which currently houses 3285 bioactive peptides, including 531 antioxidant peptides (accessed in September 2016) [28]. Darewicz et al. [99] used the BIOPEP database and its computational tools to identify carp proteins that are potential sources of bioactive peptides, in addition to predicting whether those peptides can be released during human gastrointestinal digestion. Their in silico analyses predicted that peptides with 11 types of bioactivity, including antioxidant activity, could be released by carp protein proteolysis. Meanwhile, 13 antioxidant peptides could potentially be released from myosin heavy chain after gastro-duodenal digestion [99], suggesting that the protein is an excellent source of antioxidant peptides. Huang et al. [100] used the BIOPEP database tools to predict the number of antioxidant peptides and other biopeptides that could be released from seven tilapia proteins after in silico proteolysis by using 27 proteases. The study identified myosin heavy chain as the best source of antioxidant peptides. Moreover, chymotrypsin C, ficain, and thermolysin were predicted to be the three most effective among 27 proteases for releasing antioxidant peptides from myosin heavy chain, potentially releasing 13, 13, and 17 antioxidant peptides, respectively, from the protein [100]. Meanwhile, Garcia-Mora et al. [101] used the BIOPEP database to identify antioxidant amino acid sequences harbored within the primary structures of 17 peptides they identified from a pinto bean hydrolysate exhibiting antioxidant activity. On the other hand, in silico analyses involving PeptideRanker, BioPep, and PepBank were used to select five candidates from bioactive peptides identified from donkey milk for chemical synthesis. Further validations of the synthetic peptides lead to the discovery of two novel antioxidant peptides from donkey milk [102]. Considering the aforementioned examples, in silico tools may be a useful set of resources for the discovery of antioxidant peptides from edible marine invertebrates in future. These tools may expedite the screening or pre-selection of protein precursors in edible marine invertebrates for potential sources of bioactive peptides. In silico tools can also be used for searching potential strategies to release bioactive peptides from marine invertebrate proteins [99].

6.4. Multifunctionality

Multifunctionality of antioxidant peptides is a potentially productive area of research. As pointed out above, some edible marine invertebrate-derived peptides which showed multiple bioactivities, i.e., WPP, VKP, VKCFR, and GPLGLLGFLGPLGLS [33,42,51], are promising starting points for the design and development of future peptide therapeutics and/or adjunct drugs. When we used antioxidant peptide sequences in Table 2 as queries in the BIOPEP database (accessed in July 2016), we found that QP identified from the blood clam [33] was also reported by others as an inhibitor of dipeptidyl peptidase IV, a potential therapeutic target for the treatment of type 2 diabetes mellitus [103]. SATPdb is a database of therapeutic peptides curated from 22 public domain peptide databases. Analysis of the 19,192 experimentally validated peptide sequences in the database revealed that 39% of the sequences (7512 peptides) have two to three functions [104]. A list of 26 antioxidant peptides with additional experimentally validated bioactivities (e.g., antibacterial, antifungal, anticancer and antihypertensive activities) can be accessed at the SATPdb database (accessed in July 2016) [104]. Thus, it should not be surprising that the antioxidant peptides identified from edible marine invertebrates (Table 2), even many other marine antioxidant peptides yet to be discovered, have multiple functionalities. For example, when we used antioxidant peptide sequences in Table 2 as queries in C2Pred webserver (accessed in July 2016) [105], ten peptide sequences, namely, LPHPSF, YPPAK, WPP, QP, GDQQK, PEASY, VKP, VKCFR, IKK, and FKK, were predicted to be cell-penetrating peptides. Except for GDQQK, the other nine of these peptides were predicted to have anticancer potential by the AntiCP webserver [106]. Despite in silico predictions, experimental validations of the anticancer activity of these peptides and their ability to cross the cellular membrane is necessary in future. Antioxidant peptides that have additional functions, e.g., anticancer and cell-penetrating activity, likely possess greater versatility and commercial value than other antioxidant peptides when it comes to applications in therapy and cosmetics.

6.5. Safety

Safety or toxicity assessment of food-derived antioxidant peptides, including those identified from edible marine invertebrates, requires more attention in future. Even when such peptides are identified from food with a long history of human consumption without adverse effects, absence of toxicity and allergenicity in vivo should still be asserted [107,108] before they are used in food, therapy and cosmetics. For the initial screening of potential allergens in food proteins, the European Food Safety Authority (EFSA) recommended the use of in silico tools [109]. A total 2872 peptides identified from hydrolysates of bovine blood globulins were recently assessed in silico for toxicity and allergenicity [110]. Potential toxicity was predicted using the ToxinPred web server [111]. Potential allergenicity of the peptides was assessed by using AlgPred [112] and AllerTOP [113]. In silico analyses predicted that all the peptides were non-toxic, although 564 peptides were predicted to be potential allergens. Such a large-scale screening of peptides for toxicity and allergenicity in the laboratory is predictably expensive. In silico tools represent a less costly and faster strategy to conduct screenings. Such an approach may also be used to narrow down the choices of bioactive peptides to be used for chemical synthesis and further validation of bioactivity, depending on the research objectives. Such a strategy in initial screening for potential toxicity and allergenicity can be adopted in the future search for safe antioxidant peptides from edible marine invertebrates.

When we used the 32 antioxidant peptide sequences listed in Table 2 as input queries in the ToxinPred server, all were predicted to be non-toxic, except for I/L N I/L CCN (accessed in July 2016). The four sequence variations (i.e., INICCN, LNICCN, INLCCN, and LNLCCN) of shortclub cuttlefish-derived antioxidant peptide I/L N I/L CCN [56] were predicted to be toxic. Interestingly, Sudhakar and Nazeer [56] reported that I/L N I/L CCN was not cytotoxic to HT29 cells, which showed greater than 50% viability when exposed to up to 140 µg/mL peptide. Whether the peptide varies in toxicity in different cell types or biological models is an interesting question to address in future. Meanwhile, prediction by ToxinPred or other related in silico tools should also be considered with caution. Such in silico prediction tools, especially when developed primarily based on datasets of bacterial origin, may not always generate predictions that are relevant to the human body [114]. On the other hand, four antioxidant peptides, i.e., LKQELEDLLEKQE [40], NGPLQAGQPGER [41], NADFGLNGLEGLA [43], and GPLGLLGFLGPLGLS [51], were predicted to be allergenic by AlgPred and AllerTOP servers (accessed in July 2016). Such predictions, although still requiring experimental validations, apparently contradict the general assumption that food-derived peptides are safe and non-toxic. Methodologies for the assessment of food peptide toxicity are beyond the scope of this review. For a review on the in vitro, in vivo and in silico tools used for evaluating toxicology of food, the reader is referred to Gosslau [115]. For a comprehensive discussion on empirical and in silico approaches to designing peptides with low toxicity and to predicting peptide toxicity, we refer the reader to Gupta et al. [114].

6.6. Need for More Intensive Research on Processing Wastes

Antioxidative protein hydrolysates prepared from by-products/processing wastes of edible marine invertebrates (e.g., squid skin [41,51,116], shrimp waste [54,75,117], shrimp processing wastewater [118], sea cucumber viscera and green sea urchin processing waste [119], scallop gonads [49], and cuttlefish processing wastewater [120]) have been reported. Nevertheless, investigations on such bioresources which culminated in the identification of antioxidant peptide sequences are limited (e.g., see [41,49,51,54]). The antioxidant properties of protein hydrolysates prepared from edible marine invertebrate and their processing by-products were previously reviewed [121,122]. In general, more progresses have been made in the discovery of antioxidant peptides from fish processing by-products than from by-products of edible marine invertebrates. Worldwide catching and processing of shellfish (i.e., cephalopods, bivalves, echinoderms and crustaceans) generate enormous amount of by-products and processing wastes annually [123–126]. These are protein-rich raw materials which should be tapped into more intensively for the discovery of antioxidant peptides. Meanwhile, more antioxidant peptides have been identified from mussels and oysters, whereas other edible marine invertebrates, such as the lobster, crab, octopus, jellyfish, scallop, abalone, sea cucumber, and sea squirt are underexplored. Future research to valorize shellfish processing by-products should also pay attention to these underexplored species as promising sources of marine antioxidant peptides.

7. Conclusions

The purpose of this review was to summarize current progress in the discovery of antioxidant peptides from edible marine invertebrates and to discuss potential applications of the peptides. It is clear from the research reviewed that enzyme-assisted release of peptides from edible marine invertebrates is an effective strategy for the purification and identification of marine antioxidant peptides. A number of edible marine invertebrate-derived peptides exhibiting antioxidant effects in vitro were identified; some of which also demonstrated antioxidant effects in animal models and/or additional bioactivities. Notwithstanding, knowledge gaps exist with regards to the multifunctionality, in vivo stability and safety of edible marine invertebrate-derived peptides. Future research in this direction, supported by the application of bioinformatics tools, should contribute towards realizing potential future applications of these antioxidant peptides in the food, pharmaceutical and cosmetics industries.

Supplementary Materials: The following are available online at www.mdpi.com/1660-3397/15/2/42/s1. Figure S1: The trends in the numbers of publications in the field of antioxidant peptides over the past 24 years, based on the Scopus database (accessed in November 2016). Input query used was "antioxidant peptide" OR "antioxidative peptide".

Acknowledgments: Current work on marine bioactive peptides in the laboratory of T.-T.C., Y.-C.L. and F.-C.W. is funded by Fundamental Research Grant Scheme of Ministry of Education, Malaysia.

Author Contributions: T.-T.C. and Y.-C.L. performed search of the literature and wrote the review. S.-K.K. and F.-C.W. commented, revised, and proofread the paper. All authors agreed with the final submitted version.

Conflicts of Interest: The authors declare no conflict of interest.

References

1. Weidinger, A.; Kozlov, A. Biological activities of reactive oxygen and nitrogen species: Oxidative stress versus signal transduction. *Biomolecules* **2015**, *5*, 472–484. [CrossRef] [PubMed]

2. Ye, Z.-W.; Zhang, J.; Townsend, D.M.; Tew, K.D. Oxidative stress, redox regulation and diseases of cellular differentiation. *BBA-Gen. Subj.* **2015**, *1850*, 1607–1621. [CrossRef] [PubMed]

3. Luca, M.; Luca, A.; Calandra, C. The role of oxidative damage in the pathogenesis and progression of Alzheimer's disease and vascular dementia. *Oxid. Med. Cell. Longev.* **2015**, *2015*, 504678. [CrossRef] [PubMed]

4. Chakrabarti, S.; Jahandideh, F.; Wu, J. Food-Derived bioactive peptides on inflammation and oxidative stress. *Biomed. Res. Int.* **2014**, *2014*, 608979. [CrossRef] [PubMed]

5. Reeg, S.; Grune, T. Protein oxidation in aging: Does it play a role in aging progression? *Antioxid. Redox Signal.* **2015**, *23*, 239–255. [CrossRef] [PubMed]

6. Miyake, Y.; Fukushima, W.; Tanaka, K.; Sasaki, S.; Kiyohara, C.; Tsuboi, Y.; Yamada, T.; Oeda, T.; Miki, T.; Kawamura, N.; et al. Dietary intake of antioxidant vitamins and risk of Parkinson's disease: A case–control study in Japan. *Eur. J. Neurol.* **2011**, *18*, 106–113. [CrossRef] [PubMed]

7. Bo, Y.; Lu, Y.; Zhao, Y.; Zhao, E.; Yuan, L.; Lu, W.; Cui, L.; Lu, Q. Association between dietary vitamin C intake and risk of esophageal cancer: A dose-response meta-analysis. *Int. J. Cancer* **2016**, *138*, 1843–1850. [CrossRef] [PubMed]

8. Aboonabi, A.; Singh, I. The effectiveness of antioxidant therapy in aspirin resistance, diabetes population for prevention of thrombosis. *Biomed. Pharmacother.* **2016**, *83*, 277–282. [CrossRef] [PubMed]

9. Tamay-Cach, F.; Quintana-Pérez, J.C.; Trujillo-Ferrara, J.G.; Cuevas-Hernández, R.I.; Del Valle-Mondragón, L.; García-Trejo, E.M.; Arellano-Mendoza, M.G. A review of the impact of oxidative stress and some antioxidant therapies on renal damage. *Ren. Fail.* **2016**, *38*, 171–175. [CrossRef] [PubMed]

10. Bielli, A.; Scioli, M.G.; Mazzaglia, D.; Doldo, E.; Orlandi, A. Antioxidants and vascular health. *Life Sci.* **2015**, *143*, 209–216. [CrossRef] [PubMed]

11. Zhong, Y.; Shahidi, F. Methods for the assessment of antioxidant activity in foods. In *Handbook of Antioxidants for Food Preservation*; Shahidi, F., Ed.; Woodhead Publishing: Oxford, UK, 2015; pp. 287–333.

12. Agyei, D.; Danquah, M.K.; Sarethy, I.P.; Pan, S. Antioxidative peptides derived from food proteins. In *Free Radicals in Human Health and Disease*; Rani, V., Yadav, S.U.C., Eds.; Springer: New Delhi, India, 2015; pp. 417–430.

13. Sarmadi, B.H.; Ismail, A. Antioxidative peptides from food proteins: A review. *Peptides* **2010**, *31*, 1949–1956. [CrossRef] [PubMed]

14. Sampath Kumar, N.S.; Nazeer, R.A.; Jaiganesh, R. Purification and identification of antioxidant peptides from the skin protein hydrolysate of two marine fishes, horse mackerel (*Magalaspis cordyla*) and croaker (*Otolithes ruber*). *Amino Acids* **2011**, *42*, 1641–1649. [CrossRef] [PubMed]

15. Cheung, R.C.F.; Ng, T.B.; Wong, J.H. Marine peptides: Bioactivities and applications. *Mar. Drugs* **2015**, *13*, 4006–4043. [CrossRef] [PubMed]

16. Ghosh, S. Peptide therapeutics market: Forecast and analysis 2015–2025. *Chim. Oggi Chem. Today* **2016**, *34*, 5–7.

17. Fosgerau, K.; Hoffmann, T. Peptide therapeutics: Current status and future directions. *Drug Discov. Today* **2015**, *20*, 122–128. [CrossRef] [PubMed]

18. Sila, A.; Bougatef, A. Antioxidant peptides from marine by-products: Isolation, identification and application in food systems. A review. *J. Funct. Foods* **2016**, *21*, 10–26. [CrossRef]

19. Lemes, A.C.; Sala, L.; Ores, J.d.C.; Braga, A.R.C.; Egea, M.B.; Fernandes, K.F. A review of the latest advances in encrypted bioactive peptides from protein-rich waste. *Int. J. Mol. Sci.* **2016**, *17*, 950. [CrossRef] [PubMed]

20. Ngo, D.-H.; Vo, T.-S.; Ngo, D.-N.; Wijesekara, I.; Kim, S.-K. Biological activities and potential health benefits of bioactive peptides derived from marine organisms. *Int. J. Biol. Macromol.* **2012**, *51*, 378–383. [CrossRef] [PubMed]

21. Samaranayaka, A.G.P.; Li-Chan, E.C.Y. Food-derived peptidic antioxidants: A review of their production, assessment, and potential applications. *J. Funct. Foods* **2011**, *3*, 229–254. [CrossRef]

22. Xiong, Y.L. Antioxidant peptides. In *Bioactive Proteins and Peptides as Functional Foods and Nutraceuticals*; Mine, Y., Li-Chan, E., Jiang, B., Eds.; Wiley-Blackwell: Oxford, UK, 2010; pp. 29–42.

23. Kim, S.-K.; Wijesekara, I. Marine-Derived Peptides: Development and Health Prospects. In *Marine Proteins and Peptides: Biological Activities and Applications*; Kim, S.-K., Ed.; John Wiley & Sons, Ltd.: Chichester, UK, 2013; pp. 1–3.

24. Kim, S.-K.; Wijesekara, I. Development and biological activities of marine-derived bioactive peptides: A review. *J. Funct. Foods* **2010**, *2*, 1–9. [CrossRef]

25. Li, Y.; Yu, J. Research progress in structure-activity relationship of bioactive peptides. *J. Med. Food* **2015**, *18*, 147–156. [CrossRef] [PubMed]

26. Zou, T.-B.; He, T.-P.; Li, H.-B.; Tang, H.-W.; Xia, E.-Q. The structure-activity relationship of the antioxidant peptides from natural proteins. *Molecules* **2016**, *21*, 72. [CrossRef] [PubMed]

27. Scopus. Available online: http://www.scopus.com (accessed on 29 November 2016).

28. Minkiewicz, P.; Dziuba, J.; Iwaniak, A.; Dziuba, M.; Darewicz, M. BIOPEP database and other programs for processing bioactive peptide sequences. *J. AOAC Int.* **2008**, *91*, 965–980. [PubMed]

29. Halim, N.R.A.; Yusof, H.M.; Sarbon, N.M. Functional and bioactive properties of fish protein hydolysates and peptides: A comprehensive review. *Trends Food Sci. Technol.* **2016**, *51*, 24–33. [CrossRef]

30. Senevirathne, M.; Kim, S.-K. Development of bioactive peptides from fish proteins and their health promoting ability. *Adv. Food Nutr. Res.* **2012**, *65*, 235–248. [PubMed]

31. Wang, B.; Li, L.; Chi, C.F.; Ma, J.H.; Luo, H.Y.; Xu, Y.F. Purification and characterisation of a novel antioxidant peptide derived from blue mussel (*Mytilus edulis*) protein hydrolysate. *Food Chem.* **2013**, *138*, 1713–1719. [CrossRef] [PubMed]

32. Park, S.Y.; Kim, Y.S.; Ahn, C.B.; Je, J.Y. Partial purification and identification of three antioxidant peptides with hepatoprotective effects from blue mussel (*Mytilus edulis*) hydrolysate by peptic hydrolysis. *J. Funct. Foods* **2016**, *20*, 88–95. [CrossRef]

33. Chi, C.F.; Hu, F.Y.; Wang, B.; Li, T.; Ding, G.F. Antioxidant and anticancer peptides from the protein hydrolysate of blood clam (*Tegillarca granosa*) muscle. *J. Funct. Foods* **2015**, *15*, 301–313. [CrossRef]

34. Li, R.; Yang, Z.S.; Sun, Y.; Li, L.; Wang, J.B.; Ding, G. Purification and antioxidant property of antioxidative oligopeptide from short-necked clam (*Ruditapes philippinarum*) hydrolysate in vitro. *J. Aquat. Food Prod. Technol.* **2015**, *24*, 556–565. [CrossRef]

35. Wang, Q.; Li, W.; He, Y.; Ren, D.; Kow, F.; Song, L.; Yu, X. Novel antioxidative peptides from the protein hydrolysate of oysters (*Crassostrea talienwhanensis*). *Food Chem.* **2014**, *145*, 991–996. [CrossRef] [PubMed]

36. Asha, K.K.; Remya Kumari, K.R.; Ashok Kumar, K.; Chatterjee, N.S.; Anandan, R.; Mathew, S. Sequence determination of an antioxidant peptide obtained by enzymatic hydrolysis of oyster *Crassostrea madrasensis* (Preston). *Int. J. Pept. Res. Ther.* **2016**, 1–13. [CrossRef]

37. Jung, W.K.; Rajapakse, N.; Kim, S.K. Antioxidative activity of a low molecular weight peptide derived from the sauce of fermented blue mussel, Mytilus edulis. *Eur. Food Res. Technol.* **2005**, *220*, 535–539. [CrossRef]

38. Rajapakse, N.; Mendis, E.; Jung, W.K.; Je, J.Y.; Kim, S.K. Purification of a radical scavenging peptide from fermented mussel sauce and its antioxidant properties. *Food Res. Int.* **2005**, *38*, 175–182. [CrossRef]

39. Kleekayai, T.; Harnedy, P.A.; O'Keeffe, M.B.; Poyarkov, A.A.; CunhaNeves, A.; Suntornsuk, W.; FitzGerald, R.J. Extraction of antioxidant and ACE inhibitory peptides from Thai traditional fermented shrimp pastes. *Food Chem.* **2015**, *176*, 441–447. [CrossRef] [PubMed]

40. Qian, Z.-J.; Jung, W.-K.; Byun, H.-G.; Kim, S.-K. Protective effect of an antioxidative peptide purified from gastrointestinal digests of oyster, *Crassostrea gigas* against free radical induced DNA damage. *Bioresour. Technol.* **2008**, *99*, 3365–3371. [CrossRef] [PubMed]

41. Mendis, E.; Rajapakse, N.; Byun, H.-G.; Kim, S.-K. Investigation of jumbo squid (*Dosidicus gigas*) skin gelatin peptides for their in vitro antioxidant effects. *Life Sci.* **2005**, *77*, 2166–2178. [CrossRef] [PubMed]

42. Li, J.; Li, Q.; Li, J.; Zhou, B. Peptides derived from *Rhopilema esculentum* hydrolysate exhibit angiotensin converting enzyme (ACE) inhibitory and antioxidant abilities. *Molecules* **2014**, *19*, 13587–13602. [CrossRef] [PubMed]

43. Rajapakse, N.; Mendis, E.; Byun, H.-G.; Kim, S.-K. Purification and in vitro antioxidative effects of giant squid muscle peptides on free radical-mediated oxidative systems. *J. Nutr. Biochem.* **2005**, *16*, 562–569. [CrossRef] [PubMed]

44. Kim, E.-K.; Oh, H.-J.; Kim, Y.-S.; Hwang, J.-W.; Ahn, C.-B.; Lee, J.S.; Jeon, Y.-J.; Moon, S.-H.; Sung, S.H.; Jeon, B.-T.; et al. Purification of a novel peptide derived from *Mytilus coruscus* and in vitro/in vivo evaluation of its bioactive properties. *Fish Shellfish Immunol.* **2013**, *34*, 1078–1084. [CrossRef] [PubMed]

45. Ko, S.C.; Kim, E.A.; Jung, W.K.; Kim, W.S.; Lee, S.C.; Son, K.T.; Kim, J.I.; Jeon, Y.J. A hexameric peptide purified from *Styela plicata* protects against free radical-induced oxidative stress in cells and zebrafish model. *RSC Adv.* **2016**, *6*, 54169–54178. [CrossRef]

46. Wu, R.B.; Wu, C.L.; Liu, D.; Yang, X.H.; Huang, J.F.; Zhang, J.; Liao, B.; He, H.L.; Li, H. Overview of antioxidant peptides derived from marine resources: The sources, characteristic, purification, and evaluation methods. *Appl. Biochem. Biotechnol.* **2015**, *176*, 1815–1833. [CrossRef] [PubMed]

47. Ngo, D.-H.; Kim, S.-K. Marine bioactive peptides as potential antioxidants. *Curr. Protein Pept. Sci.* **2013**, *14*, 189–198. [CrossRef] [PubMed]

48. Suetsuna, K. Antioxidant peptides from the protease digest of prawn (*Penaeus japonicus*) muscle. *Mar. Biotechnol.* **2000**, *2*, 5–10. [PubMed]

49. Wu, H.T.; Jin, W.G.; Sun, S.G.; Li, X.S.; Duan, X.H.; Li, Y.; Yang, Y.T.; Han, J.R.; Zhu, B.W. Identification of antioxidant peptides from protein hydrolysates of scallop (*Patinopecten yessoensis*) female gonads. *Eur. Food Res. Technol.* **2016**, *242*, 713–722. [CrossRef]

50. Kang, N.; Ko, S.C.; Samarakoon, K.; Kim, E.A.; Kang, M.C.; Lee, S.C.; Kim, J.; Kim, Y.T.; Kim, J.S.; Kim, H.; et al. Purification of antioxidative peptide from peptic hydrolysates of Mideodeok (*Styela clava*) flesh tissue. *Food Sci. Biotechnol.* **2013**, *22*, 541–547. [CrossRef]

51. Alemán, A.; Giménez, B.; Pérez-Santin, E.; Gómez-Guillén, M.C.; Montero, P. Contribution of Leu and Hyp residues to antioxidant and ACE-inhibitory activities of peptide sequences isolated from squid gelatin hydrolysate. *Food Chem.* **2011**, *125*, 334–341. [CrossRef]

52. Jung, W.-K.; Qian, Z.-J.; Lee, S.-H.; Choi, S.Y.; Sung, N.J.; Byun, H.-G.; Kim, S.-K. Free radical scavenging activity of a novel antioxidative peptide isolated from in vitro gastrointestinal digests of *Mytilus coruscus*. *J. Med. Food* **2007**, *10*, 197–202. [CrossRef] [PubMed]

53. Kim, E.K.; Hwang, J.W.; Kim, Y.S.; Ahn, C.B.; Jeon, Y.J.; Kweon, H.J.; Bahk, Y.Y.; Moon, S.H.; Jeon, B.T.; Park, P.J. A novel bioactive peptide derived from enzymatic hydrolysis of *Ruditapes philippinarum*: Purification and investigation of its free-radical quenching potential. *Process Biochem.* **2013**, *48*, 325–330. [CrossRef]

54. Zhao, J.; Huang, G.; Jiang, J. Purification and characterization of a new DPPH radical scavenging peptide from shrimp processing by-products hydrolysate. *J. Aquat. Food Prod. Technol.* **2013**, *22*, 281–289. [CrossRef]

55. Zhou, X.; Wang, C.; Jiang, A. Antioxidant peptides isolated from sea cucumber *Stichopus Japonicus*. *Eur. Food Res. Technol.* **2012**, *234*, 441–447. [CrossRef]

56. Sudhakar, S.; Nazeer, R.A. Preparation of potent antioxidant peptide from edible part of shortclub cuttlefish against radical mediated lipid and DNA damage. *LWT-Food Sci. Technol.* **2015**, *64*, 593–601. [CrossRef]

57. Sudhakar, S.; Nazeer, R.A. Structural characterization of an Indian squid antioxidant peptide and its protective effect against cellular reactive oxygen species. *J. Funct. Foods* **2015**, *14*, 502–512. [CrossRef]

58. Grienke, U.; Silke, J.; Tasdemir, D. Bioactive compounds from marine mussels and their effects on human health. *Food Chem.* **2014**, *142*, 48–60. [CrossRef] [PubMed]

59. Pérez-Vega, J.A.; Olivera-Castillo, L.; Gómez-Ruiz, J.T.; Hernández-Ledesma, B. Release of multifunctional peptides by gastrointestinal digestion of sea cucumber (*Isostichopus badionotus*). *J. Funct. Foods* **2013**, *5*, 869–877. [CrossRef]

60. Amarowicz, R.; Shahidi, F. Antioxidant activity of peptide fractions of capelin protein hydrolysates. *Food Chem.* **1997**, *58*, 355–359. [CrossRef]

61. Ngo, D.-H.; Wijesekara, I.; Vo, T.-S.; Van Ta, Q.; Kim, S.-K. Marine food-derived functional ingredients as potential antioxidants in the food industry: An overview. *Food Res. Int.* **2011**, *44*, 523–529. [CrossRef]

62. Zhuang, H.; Tang, N.; Yuan, Y. Purification and identification of antioxidant peptides from corn gluten meal. *J. Funct. Foods* **2013**, *5*, 1810–1821. [CrossRef]

63. PepDraw. Available online: http://www.tulane.edu/~biochem/WW/PepDraw/ (accessed on 30 July 2016).

64. Harnedy, P.A.; FitzGerald, R.J. Bioactive peptides from marine processing waste and shellfish: A review. *J. Funct. Foods* **2012**, *4*, 6–24. [CrossRef]

65. IARC TP53 Database. Available online: http://p53.iarc.fr/AAProperties.aspx (accessed on 30 July 2016).

66. Chan, K.M.; Decker, E.A.; Feustman, C. Endogenous skeletal muscle antioxidants. *Crit. Rev. Food Sci. Nutr.* **1994**, *34*, 403–426. [CrossRef] [PubMed]

67. Dhaval, A.; Yadav, N.; Purwar, S. Potential applications of food derived bioactive peptides in management of health. *Int. J. Pept. Res. Ther.* **2016**, 1–22. [CrossRef]

68. Wang, B.; Wang, Y.-M.; Chi, C.-F.; Luo, H.-Y.; Deng, S.-G.; Ma, J.-Y. Isolation and characterization of collagen and antioxidant collagen peptides from scales of croceine croaker (*Pseudosciaena crocea*). *Mar. Drugs* **2013**, *11*, 4641–4661. [CrossRef] [PubMed]

69. Thorkelsson, G.; Kristinsson, H.G. *Bioactive Peptides from Marine Sources. State of Art. Report to the NORA Fund*; Matis Food Research, Innovation and Technology: Reykjavik, Iceland, 2009; pp. 1–19.

70. Hartmann, R.; Meisel, H. Food-Derived peptides with biological activity: From research to food applications. *Curr. Opin. Biotechnol.* **2007**, *18*, 163–169. [CrossRef] [PubMed]

71. Dziuba, B.; Dziuba, M. Milk proteins-derived bioactive peptides in dairy products: Molecular, biological and methodological aspects. *Acta Sci. Pol. Technol. Aliment.* **2014**, *13*, 5–26. [CrossRef] [PubMed]

72. Freitas, A.C.; Andrade, J.C.; Silva, F.M.; Rocha-Santos, T.A.P.; Duarte, A.C.; Gomes, A.M. Antioxidative peptides: Trends and perspectives for future research. *Curr. Med. Chem.* **2013**, *20*, 4575–4594. [CrossRef] [PubMed]

73. Sivaraman, B.; Shakila, R.J.; Jeyasekaran, G.; Sukumar, D.; Manimaran, U.; Sumathi, G. Antioxidant activities of squid protein hydrolysates prepared with papain using response surface methodology. *Food Sci. Biotechnol.* **2016**, *25*, 665–672. [CrossRef]

74. Jridi, M.; Lassoued, I.; Nasri, R.; Ayadi, M.A.; Nasri, M.; Souissi, N. Characterization and potential use of cuttlefish skin gelatin hydrolysates prepared by different microbial proteases. *Biomed. Res. Int.* **2014**, *2014*, 461728:1–461728:14. [CrossRef] [PubMed]

75. Dey, S.S.; Dora, K.C. Antioxidative activity of protein hydrolysate produced by alcalase hydrolysis from shrimp waste (*Penaeus monodon* and *Penaeus indicus*). *J. Food Sci. Technol. (Mysore)* **2014**, *51*, 449–457. [CrossRef] [PubMed]

76. Nikoo, M.; Benjakul, S.; Ehsani, A.; Li, J.; Wu, F.; Yang, N.; Xu, B.; Jin, Z.; Xu, X. Antioxidant and cryoprotective effects of a tetrapeptide isolated from Amur sturgeon skin gelatin. *J. Funct. Foods* **2014**, *7*, 609–620. [CrossRef]

77. Nikoo, M.; Regenstein, J.M.; Ghomi, M.R.; Benjakul, S.; Yang, N.; Xu, X. Study of the combined effects of a gelatin-derived cryoprotective peptide and a non-peptide antioxidant in a fish mince model system. *LWT-Food Sci. Technol.* **2015**, *60*, 358–364. [CrossRef]

78. Aluko, R.E. Amino acids, peptides, and proteins as antioxidants for food preservation. In *Handbook of Antioxidants for Food Preservation*; Shahidi, F., Ed.; Woodhead Publishing: Oxford, UK, 2015; pp. 105–140.

79. Cho, J.; Won, K.; Wu, D.; Soong, Y.; Liu, S.; Szeto, H.H.; Hong, M.K. Potent mitochondria-targeted peptides reduce myocardial infarction in rats. *Coron. Artery Dis.* **2007**, *18*, 215–220. [CrossRef] [PubMed]

80. Hou, Y.; Li, S.; Wu, M.; Wei, J.; Ren, Y.; Du, C.; Wu, H.; Han, C.; Duan, H.; Shi, Y. Mitochondria-Targeted peptide SS-31 attenuates renal injury via an antioxidant effect in diabetic nephropathy. *Am. J. Physiol. Ren. Physiol.* **2016**, *310*, F547–F559. [CrossRef] [PubMed]

81. Huang, J.; Li, X.; Li, M.; Li, J.; Xiao, W.; Ma, W.; Chen, X.; Liang, X.; Tang, S.; Luo, Y. Mitochondria-targeted antioxidant peptide SS31 protects the retinas of diabetic rats. *Curr. Mol. Med.* **2013**, *13*, 935–945. [CrossRef] [PubMed]

82. Righi, V.; Constantinou, C.; Mintzopoulos, D.; Khan, N.; Mupparaju, S.P.; Rahme, L.G.; Swartz, H.M.; Szeto, H.H.; Tompkins, R.G.; Tzika, A.A. Mitochondria-targeted antioxidant promotes recovery of skeletal muscle mitochondrial function after burn trauma assessed by in vivo 31P nuclear magnetic resonance and electron paramagnetic resonance spectroscopy. *FASEB J.* **2013**, *27*, 2521–2530. [CrossRef] [PubMed]

83. Lee, H.Y.; Kaneki, M.; Andreas, J.; Tompkins, R.G.; Martyn, J.A.J. Novel mitochondria-targeted antioxidant peptide ameliorates burn-induced apoptosis and endoplasmic reticulum stress in the skeletal muscle of mice. *Shock* **2011**, *36*, 580–585. [CrossRef] [PubMed]

84. Homayouni-Tabrizi, M.; Asoodeh, A.; Abbaszadegan, M.R.; Shahrokhabadi, K.; Nakhaie Moghaddam, M. An identified antioxidant peptide obtained from ostrich (*Struthio camelus*) egg white protein hydrolysate shows wound healing properties. *Pharm. Biol.* **2015**, *53*, 1155–1162. [CrossRef] [PubMed]

85. Lintner, K. Peptides and proteins. In *Cosmetic Dermatology: Products and Procedures*; Draelos, Z.D., Ed.; John Wiley & Sons, Ltd.: Oxford, UK, 2015; pp. 308–317.

86. Reddy, B.; Jow, T.; Hantash, B.M. Bioactive oligopeptides in dermatology: Part I. *Exp. Dermatol.* **2012**, *21*, 563–568. [CrossRef] [PubMed]

87. Reddy, B.Y.; Jow, T.; Hantash, B.M. Bioactive oligopeptides in dermatology: Part II. *Exp. Dermatol.* **2012**, *21*, 569–575. [CrossRef] [PubMed]

88. Zhang, L.; Falla, T.J. Cosmeceuticals and peptides. *Clin. Dermatol.* **2009**, *27*, 485–494. [CrossRef] [PubMed]

89. Pickart, L.; Vasquez-Soltero, J.; Margolina, A. GHK-Cu may prevent oxidative stress in skin by regulating copper and modifying expression of numerous antioxidant genes. *Cosmetics* **2015**, *2*, 236–247. [CrossRef]

90. Sonthalia, S.; Daulatabad, D.; Sarkar, R. Glutathione as a skin whitening agent: Facts, myths, evidence and controversies. *Indian J. Dermatol. Venereol. Leprol.* **2016**, *82*, 262–272. [CrossRef] [PubMed]

91. Shen, S.; Chahal, B.; Majumder, K.; You, S.-J.; Wu, J. Identification of novel antioxidative peptides derived from a thermolytic hydrolysate of ovotransferrin by LC-MS/MS. *J. Agric. Food. Chem.* **2010**, *58*, 7664–7672. [CrossRef] [PubMed]

92. Jahandideh, F.; Chakrabarti, S.; Davidge, S.T.; Wu, J. Antioxidant peptides identified from ovotransferrin by the ORAC method did not show anti-inflammatory and antioxidant activities in endothelial cells. *J. Agric. Food. Chem.* **2016**, *64*, 113–119. [CrossRef] [PubMed]

93. Tenore, G.C.; Ritieni, A.; Campiglia, P.; Stiuso, P.; Di Maro, S.; Sommella, E.; Pepe, G.; D'Urso, E.; Novellino, E. Antioxidant peptides from "Mozzarella di Bufala Campana DOP" after simulated gastrointestinal digestion: In vitro intestinal protection, bioavailability, and anti-haemolytic capacity. *J. Funct. Foods* **2015**, *15*, 365–375. [CrossRef]

94. Fernández-Musoles, R.; Salom, J.B.; Castelló-Ruiz, M.; Contreras, M.d.M.; Recio, I.; Manzanares, P. Bioavailability of antihypertensive lactoferricin B-derived peptides: Transepithelial transport and resistance to intestinal and plasma peptidases. *Int. Dairy J.* **2013**, *32*, 169–174. [CrossRef]

95. Renukuntla, J.; Vadlapudi, A.D.; Patel, A.; Boddu, S.H.S.; Mitra, A.K. Approaches for enhancing oral bioavailability of peptides and proteins. *Int. J. Pharm.* **2013**, *447*, 75–93. [CrossRef] [PubMed]

96. Bruno, B.J.; Miller, G.D.; Lim, C.S. Basics and recent advances in peptide and protein drug delivery. *Ther. Deliv.* **2013**, *4*, 1443–1467. [CrossRef] [PubMed]

97. Vermeirssen, V.; Camp, J.V.; Verstraete, W. Bioavailability of angiotensin I converting enzyme inhibitory peptides. *Br. J. Nutr.* **2004**, *92*, 357–366. [CrossRef] [PubMed]

98. Bhattacharyya, A.; Chattopadhyay, R.; Mitra, S.; Crowe, S.E. Oxidative stress: An essential factor in the pathogenesis of gastrointestinal mucosal diseases. *Physiol. Rev.* **2014**, *94*, 329–354. [CrossRef] [PubMed]

99. Darewicz, M.; Borawska, J.; Pliszka, M. Carp proteins as a source of bioactive peptides—An in silico approach. *Czech J. Food Sci.* **2016**, *34*, 111–117. [CrossRef]

100. Huang, B.B.; Lin, H.C.; Chang, Y.W. Analysis of proteins and potential bioactive peptides from tilapia (*Oreochromis* spp.) processing co-products using proteomic techniques coupled with BIOPEP database. *J. Funct. Foods* **2015**, *19*, 629–640. [CrossRef]

101. Garcia-Mora, P.; Peñas, E.; Frias, J.; Zieliński, H.; Wiczkowski, W.; Zielińska, D.; Martínez-Villaluenga, C. High-pressure-assisted enzymatic release of peptides and phenolics increases angiotensin converting enzyme I inhibitory and antioxidant activities of pinto bean hydrolysates. *J. Agric. Food. Chem.* **2016**, *64*, 1730–1740. [CrossRef] [PubMed]

102. Zenezini Chiozzi, R.; Capriotti, A.L.; Cavaliere, C.; La Barbera, G.; Piovesana, S.; Samperi, R.; Laganà, A. Purification and identification of endogenous antioxidant and ACE-inhibitory peptides from donkey milk by multidimensional liquid chromatography and nanoHPLC-high resolution mass spectrometry. *Anal. Bioanal. Chem.* **2016**, *408*, 5657–5666. [CrossRef] [PubMed]

103. Hikida, A.; Ito, K.; Motoyama, T.; Kato, R.; Kawarasaki, Y. Systematic analysis of a dipeptide library for inhibitor development using human dipeptidyl peptidase IV produced by a *Saccharomyces cerevisiae* expression system. *Biochem. Biophys. Res. Commun.* **2013**, *430*, 1217–1222. [CrossRef] [PubMed]

104. Singh, S.; Chaudhary, K.; Dhanda, S.K.; Bhalla, S.; Usmani, S.S.; Gautam, A.; Tuknait, A.; Agrawal, P.; Mathur, D.; Raghava, G.P.S. SATPdb: A database of structurally annotated therapeutic peptides. *Nucleic Acids Res.* **2015**, *44*, D1119–D1126. [CrossRef] [PubMed]

105. Tang, H.; Su, Z.-D.; Wei, H.-H.; Chen, W.; Lin, H. Prediction of cell-penetrating peptides with feature selection techniques. *Biochem. Biophys. Res. Commun.* **2016**, *477*, 150–154. [CrossRef] [PubMed]

106. Tyagi, A.; Kapoor, P.; Kumar, R.; Chaudhary, K.; Gautam, A.; Raghava, G.P.S. In silico models for designing and discovering novel anticancer peptides. *Sci. Rep.* **2013**, *3*, 2984. [CrossRef] [PubMed]

107. Hartmann, R.; Wal, J.M.; Bernard, H.; Pentzien, A.K. Cytotoxic and allergenic potential of bioactive proteins and peptides. *Curr. Pharm. Des.* **2007**, *13*, 897–920. [CrossRef] [PubMed]

108. Schaafsma, G. Safety of protein hydrolysates, fractions thereof and bioactive peptides in human nutrition. *Eur. J. Clin. Nutr.* **2009**, *63*, 1161–1168. [CrossRef] [PubMed]

109. Christer, H.; Andersson, S.; Arpaia, D.; Casacuberta, J.; Davies, H.; Jardin, P.; Flachowsky, G. Scientific opinion on the assessment of allergenicity of GM plants and microorganisms and derived food and feed. *EFSA J.* **2010**, *8*, 1700.

110. Lafarga, T.; Wilm, M.; Wynne, K.; Hayes, M. Bioactive hydrolysates from bovine blood globulins: Generation, characterisation, and in silico prediction of toxicity and allergenicity. *J. Funct. Foods* **2016**, *24*, 142–155. [CrossRef]

111. Gupta, S.; Kapoor, P.; Chaudhary, K.; Gautam, A.; Kumar, R.; Raghava, G.P.S. In silico approach for predicting toxicity of peptides and proteins. *PLoS ONE* **2013**, *8*, e0073957. [CrossRef] [PubMed]

112. Saha, S.; Raghava, G.P.S. AlgPred: Prediction of allergenic proteins and mapping of IgE epitopes. *Nucleic Acids Res.* **2006**, *34*, W202–W209. [CrossRef] [PubMed]

113. Dimitrov, I.; Bangov, I.; Flower, D.R.; Doytchinova, I. AllerTOP v.2—A server for in silico prediction of allergens. *J. Mol. Model.* **2014**, *20*, 1–6. [CrossRef] [PubMed]

114. Gupta, S.; Kapoor, P.; Chaudhary, K.; Gautam, A.; Kumar, R.; Raghava, G.P.S. Peptide toxicity prediction. In *Computational Peptidology*; Zhou, P., Huang, J., Eds.; Springer: New York, NY, USA, 2015; pp. 143–157.

115. Gosslau, A. Assessment of food toxicology. *Food Sci. Hum. Wellness* **2016**. [CrossRef]

116. Nakchum, L.; Kim, S.M. Preparation of squid skin collagen hydrolysate as an antihyaluronidase, antityrosinase, and antioxidant agent. *Prep. Biochem. Biotechnol.* **2016**, *46*, 123–130. [CrossRef] [PubMed]

117. Vieira, M.A.; Oliveira, D.D.; Kurozawa, L.E. Production of peptides with radical scavenging activity and recovery of total carotenoids using enzymatic protein hydrolysis of shrimp waste. *J. Food Biochem.* **2016**. [CrossRef]

118. Tonon, R.V.; Dos Santos, B.A.; Couto, C.C.; Mellinger-Silva, C.; Brígida, A.I.S.; Cabral, L.M.C. Coupling of ultrafiltration and enzymatic hydrolysis aiming at valorizing shrimp wastewater. *Food Chem.* **2016**, *198*, 20–27. [CrossRef] [PubMed]

119. Mamelona, J.; Saint-Louis, R.; Pelletier, É. Nutritional composition and antioxidant properties of protein hydrolysates prepared from echinoderm byproducts. *Int. J. Food Sci. Technol.* **2010**, *45*, 147–154. [CrossRef]

120. Amado, I.R.; Vázquez, J.A.; González, M.P.; Murado, M.A. Production of antihypertensive and antioxidant activities by enzymatic hydrolysis of protein concentrates recovered by ultrafiltration from cuttlefish processing wastewaters. *Biochem. Eng. J.* **2013**, *76*, 43–54. [CrossRef]

121. Giménez, B.; López-Caballero, E.M.; Montero, P.M.; Gómez-Guillén, C.M. Antioxidant peptides from marine origin: Sources, properties and potential applications. In *Antioxidant Polymers: Synthesis, Properties, and Applications*; John Wiley & Sons, Inc.: Hoboken, NJ, USA, 2012; pp. 203–257.

122. Lee, J.K.; Jeon, J.K.; Kim, S.K.; Byun, H.G. Characterization of bioactive peptides obtained from marine invertebrates. *Adv. Food Nutr. Res.* **2012**, *65*, 47–72. [PubMed]

123. Yan, N.; Chen, X. Sustainability: Don't waste seafood waste. *Nature* **2015**, *524*, 155–157. [CrossRef] [PubMed]

124. Food and Agriculture Organization of the United Nations (FAO). *The State of World Fisheries and Aquaculture 2014. Opportunities and Challenges*; Food and Agriculture Organization of the United Nations: Rome, Italy, 2014.

125. Food and Agriculture Organization of the United Nations (FAO). *The State of World Fisheries and Aquaculture 2016. Contributing to Food Security and Nutrition for All*; Food and Agriculture Organization of the United Nations: Rome, Italy, 2016.

126. Olsen, R.L.; Toppe, J.; Karunasagar, I. Challenges and realistic opportunities in the use of by-products from processing of fish and shellfish. *Trends Food Sci. Technol.* **2014**, *36*, 144–151. [CrossRef]

marine drugs

MDPI

Article

Purification and Identification of Antioxidant Peptides from Protein Hydrolysate of Scalloped Hammerhead (*Sphyrna lewini*) Cartilage

Xue-Rong Li [1], Chang-Feng Chi [1,*], Li Li [1] and Bin Wang [2,*]

[1] School of Marine Science and Technology, Zhejiang Ocean University, 1st Haidanan Road, Changzhi Island, Lincheng, Zhoushan 316022, China; xuerongl0312@163.com (X.-R.L.); wenwenlili@163.com (L.L.)
[2] School of Food and Pharmacy, Zhejiang Ocean University, 1st Haidanan Road, Changzhi Island, Lincheng, Zhoushan 316022, China
* Correspondence: chichangfeng@hotmail.com (C.-F.C.); wangbin4159@hotmail.com (B.W.);
   Tel.: +86-580-255-4818 (C.-F.C.); +86-580-255-5085 (B.W.);
   Fax: +86-580-255-4818 (C.-F.C.); +86-580-255-4781 (B.W.)

Academic Editor: Paul Long
Received: 26 August 2016; Accepted: 18 February 2017; Published: 1 March 2017

Abstract: The aim of this study was to purify and identify peptides with antioxidant properties from protein hydrolysate of scalloped hammerhead (*Sphyrna lewini*) cartilage. Cartilaginous proteins of the scalloped hammerhead were extracted by guanidine hydrochloride, and three antioxidant peptides, named enzymolysis peptide of scalloped hammerhead cartilage A (SCPE-A), SCPE-B and SCPE-C, were subsequently isolated from the hydrolysate of the cartilaginous proteins using ultrafiltration and chromatography. The amino acid sequences of SCPE-A, SCPE-B and SCPE-C were identified as Gly-Pro-Glu (GPE), Gly-Ala-Arg-Gly-Pro-Gln (GARGPQ), and Gly-Phe-Thr-Gly-Pro-Pro-Gly-Phe-Asn-Gly (GFTGPPGFNG), with molecular weights of 301.30 Da, 584.64 Da and 950.03 Da, respectively. As per in vitro activity testing, SCPE-A, SCPE-B and SCPE-C exhibited strong scavenging activities on 2,2-diphenyl-1-picrylhydrazyl radicals (DPPH•) (half elimination ratio (EC$_{50}$) 2.43, 2.66 and 1.99 mg/mL), hydroxyl radicals (HO•) (EC$_{50}$ 0.28, 0.21 and 0.15 mg/mL), 2,2'-azino-bis-3-ethylbenzothiazoline-6-sulfonic acid radicals (ABTS$^+$•) (EC$_{50}$ 0.24, 0.18 and 0.29 mg/mL), and superoxide anion radicals (O$_2^-$•) (EC$_{50}$ 0.10, 0.14 and 0.11 mg/mL). In addition, SCPE-A showed inhibition activity similar to butylated hydroxytoluene (BHT) in lipid peroxidation in a linoleic acid model system. The amino acid residues of Gly, Pro and Phe could positively influence the antioxidant activities of GPE, GARGPQ and GFTGPPGFNG. These results suggested that GPE, GARGPQ and GFTGPPGFNG might serve as potential antioxidants and be used as food additives and functional foods.

Keywords: scalloped hammerhead (*Sphyrna lewini*); cartilage; peptide; antioxidant activity

1. Introduction

Oxidation is an important factor in the food industry because it causes a loss of nutrition, color and functionality, as well as undesirable off-flavors and toxic compounds, which further induce the deterioration of food. Furthermore, the accumulation of toxic products is dangerous to the health of consumers [1,2]. Therefore, the inhibition of free radical formation and oxidation reactions play an important role in preventing or retarding the autoxidation of food components [3]. Many synthetic antioxidants, including butylated hydroxytoluene (BHT), butylated hydroxyanisole (BHA), and tertiary butylhydroquinone (TBHQ), are widely used in the food industry for preservation and to retard lipid oxidation [4,5]. However, the dosages of the synthetic antioxidants are under strict regulation due to their potential health hazards and toxic effects [6,7]. Therefore, there has been a large amount of interest in researching safe antioxidants from natural sources as an alternative to synthetic antioxidants [8].

In recent years, peptides with different activities, including anticancer, antioxidant, antimicrobial, antihypertensive, and mineral-binding properties, have been isolated from various bioresources, such as byproducts from the fish processing industry [9]. Antioxidant peptides have drawn great attention and have been extensively reported as free radical scavengers, peroxide decomposers, metal inactivators and oxygen inhibitors to protect food and organisms from reactive oxygen species (ROS) [8,10]. Arg-Gln-Ser-His-Phe-Ala-Asn-Ala-Gln-Pro (RQSHFANAQP), with molecular weight (MW) 1155 Da, from the protein hydrolysate of chickpeas showed significant dose-dependent scavenging activities on hydroxyl radicals (HO•) (EC$_{50}$ 2.03 µmol/mL), 2,2-diphenyl-1-picrylhydrazyl radicals (DPPH•) (EC$_{50}$ 3.15 µmol/mL) and 2, 2′-azino-bis-3-ethylbenzothiazoline-6-sulfonic acid radicals (ABTS$^+$•) (EC$_{50}$ 2.31 µmol/mL) [10]. Asp-Leu-Glu-Glu (DLEE), with MW 504.2 Da, was confirmed to be one of the main antioxidant peptides generated in dry-cured Xuanwei ham, and its DPPH• scavenging rate was 74.45% at 0.5 mg/mL [9]. Phe-Ile-Met-Gly-Pro-Tyr (FIMGPY), Gly-Pro-Ala-Gly-Asp-Tyr (GPAGDY) and Ile-Val-Ala-Gly-Pro-Gln (IVAGPQ), with MWs of 726.90 Da, 578.58 Da and 583.69 Da, respectively, showed strong scavenging activities on DPPH• (EC$_{50}$ 3.5768, 6.0147 and 6.733 M), HO• (EC$_{50}$ 4.1821, 6.7752 and 8.6176 M), superoxide anion radicals (O$_2^-$•) (EC$_{50}$ 2.2149, 2.8691 and 3.1181 M) and ABTS$^+$• (EC$_{50}$ 1.4307, 1.3308 and 2.2101 M) [11]. In addition, FIMGPY induced HeLa cell apoptosis by up-regulating the Bax (B-cell lymphoma 2 (Bcl-2) assaciated X protein)/Bcl-2 ratio and caspase-3 activation [12]. Structure–activity studies on the antioxidant peptides suggested that the peptide length and the composition and position of amino acids in a peptide sequence are important determinants of the bioactivity of a specific peptide [8,13,14].

Cartilage is a form of connective tissue that is chemically abundant in bioactive components, and many cartilaginous proteins, low MW proteins, glycoproteins, and peptides have been prepared from various soft bone fish sources, such as bamboo and blacktip sharks [15], *Prionace glauca* [16], Amur sturgeon [17], spotless smoothhound [7], and silvertip shark [18]. Current research has shown that those active substances could be used as angiogenesis inhibitors, tumor cells inhibitory factors, antioxidants, immune regulatory factors, and anti-invasion factors for the treatment of some diseases, especially in tumor therapy and prevention. Scalloped hammerhead (*Sphyrna lewini*), belonging to the family Triakidae, is a commercially valuable fishery resource. At present, large quantities of muscle protein and cartilages (except the cartilages from fins) from scalloped hammerhead are not used efficiently. In our previous research, three antioxidant peptides, Trp-Asp-Arg (WDR), Pro-Tyr-Phe-Asn-Lys (PYFNK) and Leu-Asp-Lys (LDK), were isolated from the hydrolysate of scalloped hammerhead muscle, and WDR, PYFNK and LDK exhibited good scavenging activities on DPPH• (EC$_{50}$ 3.63, 4.11 and 3.06 mg/mL), HO• (EC$_{50}$ 0.15, 0.24 and 0.17 mg/mL), ABTS$^+$• (EC$_{50}$ 0.34, 0.12 and 0.19 mg/mL), and O$_2^-$• (EC$_{50}$ 0.09, 0.11 and 0.12 mg/mL) [19,20]. Acid-soluble collagen and its hydrolysate were prepared from scalloped hammerhead cartilage [21]. However, to the best of our knowledge, there is no research focusing on the antioxidative peptides in scalloped hammerhead cartilage. Thus, in this study, three novel antioxidant peptides were isolated from the protein hydrolysate of scalloped hammerhead cartilage and their antioxidant activities were evaluated by DPPH•, HO•, O$_2^-$ •, and ABTS$^+$• scavenging and lipid peroxidation inhibition assays.

2. Results and Discussion

2.1. Preparation of the Protein Hydrolysate of Scalloped Hammerhead Cartilage (SHCH)

Proteins in raw and processed foods possess antioxidant peptide sequences and structural domains, and enzymatic hydrolysis is considered as an attractive way for releasing those active fragments without impairing their nutritional value and without leaving residual organic solvents and toxic chemicals in the final product [22]. In addition, protein hydrolysates using different proteases exhibit different antioxidant activities against various antioxidant systems because the peptides are different in terms of chain length and amino acid sequence [1,23]. Therefore, five proteases including papain, alcalase, trypsin, pepsin, and neutrase were used to hydrolyze the cartilaginous proteins of scalloped hammerhead. HO• scavenging assay is quick, convenient, and efficient in predicting the antioxidant activities of protein hydrolysates and purified peptides. As a consequence, HO• was used to evaluate the antioxidant activity of compounds to act as free radical scavengers or hydrogen donors, and the results are shown in Table 1.

Table 1. Hydroxyl radical (HO•) scavenging activity of the protein hydrolysate of scalloped hammerhead cartilage using different proteases (c = 15 mg protein/mL).

Protease	Enzymolysis Condition	Yields (g/100 g Cartilage)	Degree of Hydrolysis (DH%)	HO• Scavenging Rate (%)
Papain	pH 7.0, 60 °C, 4 h, total enzyme dose 2.5%	1.93 ± 0.08 [a]	18.33 ± 0.25 [a]	34.85 ± 1.05 [a]
Alcalase	pH 8.0, 50 °C, 4 h, total enzyme dose 2.5%	1.96 ± 0.10 [a,b]	21.37 ± 0.35 [b,c]	54.76 ± 1.94 [b]
Trypsin	pH 8.0, 40 °C, 4 h, total enzyme dose 2.5%	2.11 ± 0.11 [b]	23.72 ± 0.31 [c]	62.38 ± 1.67 [c]
Pepsin	pH 2.0, 37 °C, 4 h, total enzyme dose 2.5%	1.99 ± 0.07 [a,b]	21.58 ± 0.26 [c]	55.47 ± 2.02 [b]
Neutrase	pH 6.0, 50 °C, 4 h, total enzyme dose 2.5%	1.85 ± 0.06 [a]	20.87 ± 0.36 [b]	50.67 ± 1.85 [d]

[a–d] Values with different letters indicate significant differences at the same concentration ($p < 0.05$).

The yield and degree of hydrolysis (DH%) of trypsin hydrolysate was 2.11 ± 0.11 g/100 g cartilage and 23.72% ± 0.31%, respectively, which was higher than for papain hydrolysate, alcalase hydrolysate, pepsin hydrolysate, and neutrase hydrolysate. The result indicated that trypsin could more effectively hydrolyze the proteins from scalloped hammerhead cartilages than the other four proteases. Furthermore, trypsin hydrolysate (SHCH) showed a significantly higher HO• scavenging activity ($p < 0.05$) with 62.38% ± 1.67% at 15 mg/mL, whereas papain hydrolysate showed a significantly lower HO• scavenging activity ($p < 0.05$) at 34.85% ± 1.05%. Based on these data, the protein hydrolysate of scalloped hammerhead cartilage produced by trypsin was named SHCH and was selected for follow-up studies.

2.2. Purification of the Antioxidant Peptides from SHCH

2.2.1. Ultrafiltration

Protein hydrolysate is a complex mixture of active and inactive peptides (of various sizes) and amino acid compositions, and ultrafiltration membrane technology is an important method for the fractionation of protein hydrolysate and the enrichment of peptides with specific MW ranges [1,5]. SHCH was fractionated by ultrafiltration using two molecular weight cut-off (MWCO) membranes (10 and 3 kDa), and three fractions, SHCH-I (MW < 3 kDa), SHCH-II (3 kDa < MW < 10 kDa), and SHCH-III (MW > 10 kDa), were prepared. As shown in Figure 1, the HO• scavenging activity of SHCH-I was 79.10% ± 2.38% at 15 mg protein/mL, which was significantly stronger than those of SHCH, SHCH-II, and SHCH-III ($p < 0.05$). The MW of peptides plays a critical role in bioactivity, and protein hydrolysates with smaller MW usually exhibited higher antioxidant activity than larger MW hydrolysates [4,5]. SHCH-I, which is abundant in smaller MW peptides, showed high HO• scavenging activity, and the result was in agreement with other reports that the ultrafiltration fractions of protein hydrolysates with lower MW could more effectively interact with the free radicals interfering in oxidative processes [6,9].

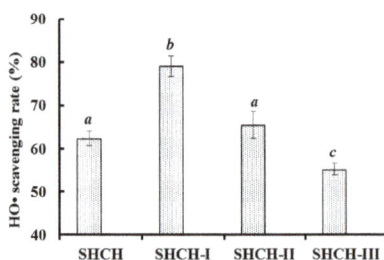

Figure 1. HO• scavenging activities of trypsin hydrolysate (SHCH) and its three fractions at 15 mg protein/mL. All data are presented as the mean ± standard deviation (SD) of triplicate results. [a–c] Values with same letters indicate no significant difference for each group of samples at the same concentration ($p > 0.05$).

2.2.2. Anion-Exchange Chromatography

Ion-exchange chromatography is used to separate the charged molecules based on their affinity to the ion exchanger (anion and/or cation exchange resins), and their interaction was determined by the number and location of the charges on the molecules [5]. SHCH-I was loaded onto a Diethylaminoethyl cellulose 52 (DEAE-52) cellulose anion-exchange column and separated by stepwise elution using deionized water and 0.1, 0.5, and 1.0 M NaCl (Figure 2A). Five separated fractions (Fr.1 to Fr.5) were collected. Their HO• scavenging activities were measured and are shown in Figure 2B. The HO• scavenging rate of Fr.4 reached 72.03% ± 2.64% at 10 mg protein/mL, and it exhibited significantly more efficient antioxidant activity than the other fractions ($p < 0.05$). Peptides with basic and/or hydrophobic amino acid residues, such as His, Lys and Pro, are thought to have strong antioxidant activities [24]. Therefore, anion and cation exchange resins have been widely used to purify antioxidant peptides from protein hydrolysates [25–27]. The present data showed that Fr.4 had the strongest HO• scavenging activity and was selected for further purification.

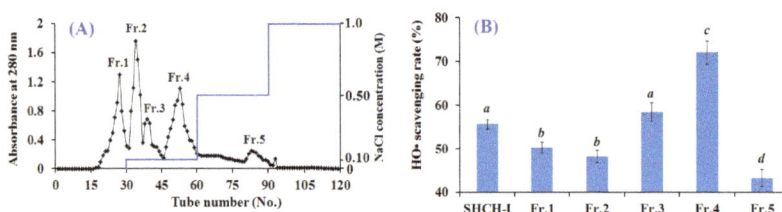

Figure 2. Elution profile of SHCH-I in DEAE-52 cellulose chromatography (**A**); and the HO• scavenging rate (%) of different fractions of SHCH-I at 10 mg protein/mL (**B**). All data are presented as the mean ± standard deviation (SD) of triplicate results. [a–d] Values with same letters indicate no significant difference for each group of samples at the same concentration ($p > 0.05$). Fr: separated fractions.

2.2.3. Gel Filtration Chromatography

Molecular size is an important determinant of the bioactivity of a specific peptide [8]. Therefore, gel filtration chromatography is an important method to purify peptides. Fr.4 was loaded onto a Sephadex G-15 column and separated into two fractions of Fr.4-1 and Fr.4-2 (Figure 3A). Each fraction was collected, lyophilized, and evaluated for HO• scavenging activity. As shown in Figure 3B, the HO• scavenging rate of Fr.4-1 reached 87.80% ± 2.24% at 5 mg protein/mL and was higher than those of Fr.4 (72.03% ± 2.64%) and Fr.4-2 (52.38% ± 1.62%). Therefore, Fr.4-1 was selected for further purification by RP-HPLC.

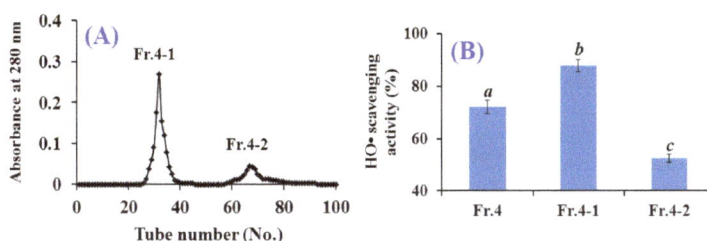

Figure 3. Elution profile of Fr.4 in Sephadex G-15 chromatography (**A**) and HO• scavenging activity of Fr.4 and its fractions at 5 mg protein/mL (**B**). All data are presented as the mean ± SD of triplicate results. [a–c] Values with same letters indicate no significant difference for each group of samples at the same concentration ($p > 0.05$).

2.2.4. Isolation of Peptides from Fr.4-1 by Reversed-Phase High Performance Liquid Chromatography (RP-HPLC)

The hydrophobic and hydrophilic properties of peptides play a key role in their retention time (RT) on an RP-HPLC column, and the RT can be adjusted by changing the ratio of polar (water) and nonpolar (methanol, acetonitrile) solvents [11]. Using an ultrafiltration membrane system, anion-exchange chromatography and gel filtration chromatography, Fr.4-1, which had the highest HO• scavenging activity among all fractions, was separated using RP-HPLC on a Zorbax C-18 column, and the eluted fractions were collected separately according to the chromatographic peaks (Figure 4). Three fractions, referred to as enzymolysis peptide of scalloped hammerhead cartilage A (SCPE-A), SCPE-B and SCPE-C, with RTs of 10.642, 13.605, and 17.979 min, respectively, showed high antioxidant activities, and their HO• scavenging rates reached 80.7% ± 1.22%, 75.4% ± 2.33%, and 92.2% ± 3.44%, respectively, at 3.0 mg/mL. Therefore, SCPE-A, SCPE-B and SCPE-C were collected for further research.

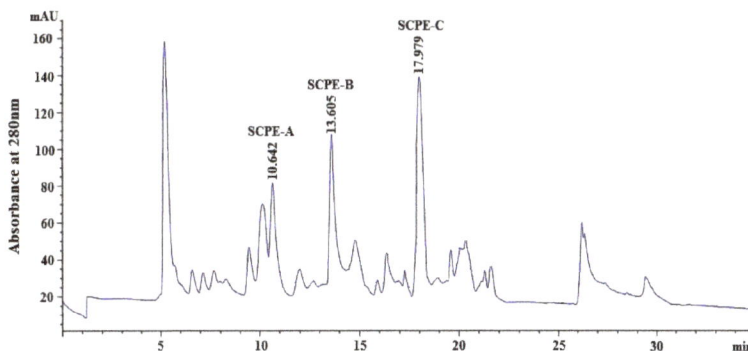

Figure 4. RP-HPLC profile of Fr.4-1 on a Zorbax C18 column with a linear gradient of acetonitrile (0%–50% for 32 min) containing 0.1% trifluoroacetic acid (TFA) at a flow rate of 0.8 mL/min.

2.3. Molecular Mass and Amino Acid Sequences of the Purified Peptides

The properties of peptides are related to their composition, structure, MW, amino acid sequence and hydrophobicity. Considering the radical-scavenging ability, the amino acid sequences and molecular mass of the three isolated peptides were analyzed using a protein sequencer and quadrupole-time of flight mass spectrometry (Q-TOF MS), respectively. The mass spectra of the three isolated peptides were shown in Figure 5. The amino acid sequences of SCPE-A, SCPE-B and SCPE-C were identified as Gly-Pro-Glu (GPE), Gly-Ala-Arg-Gly-Pro-Gln (GARGPQ), and Gly-Phe-Thr-Gly-Pro-Pro-Gly-Phe-Asn-Gly (GFTGPPGFNG), with molecular masses of 301.30 Da, 584.64 Da and 950.03 Da, respectively, which were in agreement with the theoretical masses of 301.27 Da, 584.64 Da, and 949.12 Da, respectively.

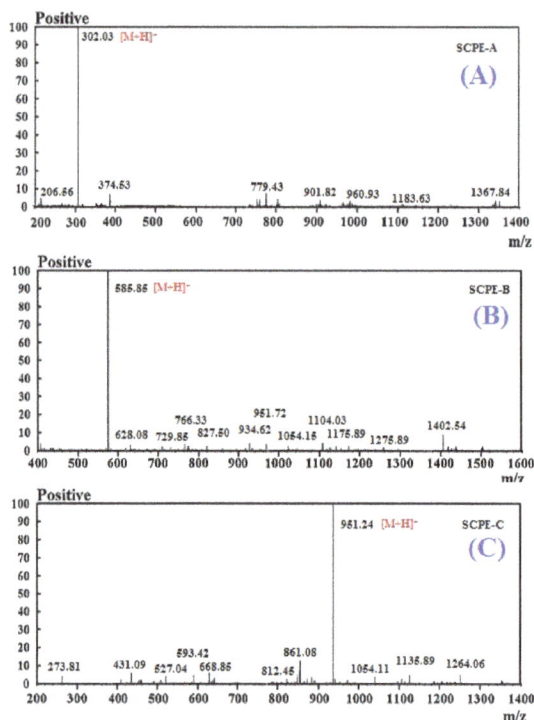

Figure 5. Mass spectrograms of SCPE-A, SCPE-B and SCPE-C.

2.4. Antioxidant Activity of SCPE-A, SCPE-B and SCPE-C

2.4.1. DPPH• Scavenging Activity

DPPH is a relatively stable organic radical that can be scavenged by accepting a proton-donating substance (H^+), which reduces the absorbance at 517 nm because the solution color changes from deep purple to yellow [19]. As shown in Figure 6A, SCPE-A, SCPE-B and SCPE-C showed dose-dependent anti-DPPH• activity, with EC_{50} values of 2.43, 2.43, and 1.99 mg/mL, respectively, and SCPE-C exhibited the highest radical-scavenging activity among all samples, except the positive control of ascorbic acid. The EC_{50} of SCPE-C was lower than that of Pro-Ser-Tyr-Val (PSYV) (17.0 mg/mL) [28], Thr-Thr-Ala-Asn-Ile-Glu-Asp-Arg-Arg (TTANIEDRR) (2.503 mg/mL) [26], Phe-Leu-Asn-Glu-Phe-Leu-His-Val (FLNEFLHV) (4.950 mg/mL) [29], Trp-Glu-Gly-Pro-Lys (WEGPK) (4.438 mg/mL), Gly-Val-Pro-Leu-Thr (GVPLT) (4.541 mg/mL) [5], Gly-Phe-Gly-Pro-Leu (GFGPL) (2.249 mg/mL), Val-Gly-Gly-Arg-Pro (VGGRP) (2.937 mg/mL) [30], FIMGPY (2.60 mg/mL), GPAGDY (3.48 mg/mL), IVAGPQ (3.93 mg/mL) [11], WDR(3.63 mg/mL), PYFNK (4.11 mg/mL) and LDK (3.06 mg/mL) [19,20] from the protein hydrolysates of loach, blue mussel, salmon, bluefin leatherjacket, grass carp skin, skate (*Raja porosa*) cartilage and scalloped hammerhead muscle, but it was higher than that of Gly-Ser-Gln (GSQ) (0.61 mg/mL) [31], Pro-Ile-Ile-Val-Tyr-Trp-Lys (PIIVYWK) (0.713 mg/mL), Phe-Ser-Val-Val-Pro-Ser-Pro-Lys (FSVVPSPK) (0.937 mg/mL) [29], Pro-Tyr-Ser-Phe-Lys (PYSFK) (1.575 mg/mL) [30], His-Phe-Gly-Asp-Pro-Phe-His (HFGDPFH) (0.20 mg/mL) [32],

Phe-Leu-Pro-Phe (FLPF) (0.789 mg/mL), Leu-Pro-Phe (LPF) (0.777 mg/mL) and Leu-Leu-Pro-Phe (LLPF) (1.084 mg/mL) [33] from the protein hydrolysates of Chinese leek, blue mussel, grass carp skin, mussel sauce and corn gluten meal. Therefore, the present results suggested that SCPE-A, SCPE-B and SCPE-C were DPPH• inhibitors and primary antioxidants that reacted with free radicals.

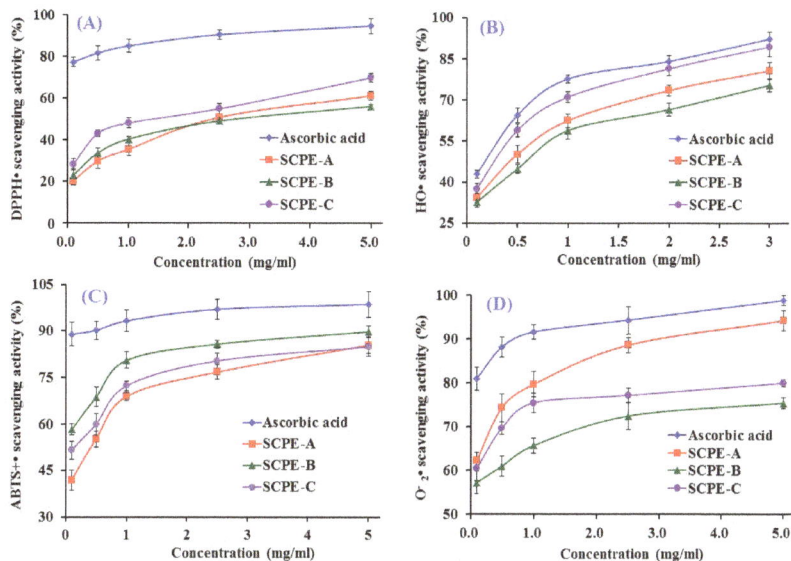

Figure 6. DPPH• (**A**); HO• (**B**); ABTS+• (**C**); and O_2^-• (**D**) scavenging activities of SCPE-A, SCPE-B and SCPE-C. All data are presented as the mean ± SD of triplicate results. DPPH•: 2,2-diphenyl-1-picrylhydrazyl radicals; ABTS+•: 2, 2′-azino-bis-3-ethylbenzothiazoline-6-sulfonic acid radicals; O_2^- •: superoxide anion radicals.

2.4.2. HO• Scavenging Activity

HO• is highly reactive and consequently short-lived and can damage virtually all types of macromolecules, including carbohydrates, nucleic acids, lipids, and proteins [5]. The HO• scavenging activity of SCPE-A, SCPE-B and SCPE-C was dose-dependent at the test concentrations, as shown in Figure 6B. The EC_{50} values of SCPE-A, SCPE-B and SCPE-C were 0.28, 0.21, and 0.15 mg/mL, respectively, and SCPE-C exhibited the highest HO• scavenging activity. The EC_{50} of SCPE-C was lower than that of PYFNK (0.24 mg/mL), LDK (0.17 mg/mL) [19,20], Leu-Gly-Leu-Asn-Gly-Asp-Asp-Val-Asn (LGLNGDDVN) (0.687 mg/mL) [34], PSYV (2.64 mg/mL) [28], HFGDPFH (0.50 mg/mL) [32], Phe-Pro-Glu-Leu-Leu-Ile (FPELLI) (0.57 mg/mL) and Val-Phe-Ala-Ala-Leu (VFAAL) (0.31 mg/mL) [4], as well as that of Tyr-Pro-Pro-Ala-Lys (YPPAK) (0.228 mg/mL) [23], Pro-Ser-Lys-Tyr-Glu-Pro-Phe-Val (PSKYEPFV) (2.86 mg/mL) [35], PYSFK (2.283 mg/mL), GFGPL (1.612 mg/mL), VGGRP (2.055 mg/mL) [30], Tyr-Leu-Gly-Ala-Lys (YLGAK) (scavenging rate: 45.14% at 0.5 mg/mL), Gly-Gly-Leu-Glu-Pro-Ile-Asn-Phe-Gln (GGLEPINFQ) (scavenging rate: 41.07% at 0.5 mg/mL) [36], Asn-Gly-Leu-Glu-Gly-Leu-Lys (NGLEGLK) (0.313 mg/mL), Asn-Ala-Asp-Phe-Gly-Leu-Asn-Gly-Leu-Glu-Gly-Leu-Ala (NADFGLNGLEGLA) (0.612 mg/mL) [32], FIMGPY (3.04), GPAGDY (3.92 mg/mL) and IVAGPQ (5.03 mg/mL) [11] from the protein hydrolysates of scalloped hammerhead muscle, conger eel, weatherfish loach, mussel sauce, Chinese cherry seeds, blue mussel, grass carp, egg white, giant squid and skate (*R. porosa*) cartilage. The three isolated peptides, especially SCPE-C, revealed good HO• scavenging activity, which demonstrated that it could serve as a scavenger to reduce or eliminate the damage induced by HO• in foods and biological systems.

2.4.3. ABTS$^+\bullet$ Scavenging Activity

The ABTS$^+\bullet$ scavenging assay is a sensitive method to determine the antioxidant capacity of bioactive compounds, in which sodium persulfate converts ABTS to its radical cation with a blue color and an absorption maximum of 734 nm, and the blue ABTS$^+\bullet$ is converted back to its colorless neutral form when ABTS$^+\bullet$ is reactive towards an antioxidant [10,37,38]. The abilities of SCPE-A, SCPE-B and SCPE-C to scavenge ABTS$^+\bullet$ in comparison with ascorbic acid were investigated, and dose-related effects were observed at different peptide concentrations ranging from 0 to 5.0 mg/mL (Figure 6C). SCPE-B, with an EC$_{50}$ of 0.18 mg/mL, showed the strongest scavenging activity on ABTS$^+\bullet$ among the protein hydrolysate, fractions, and prepared peptides at all tested concentrations. The EC$_{50}$ of SCPE-B was lower than those of WDR (0.34 mg/mL) [19], LDK (0.19 mg/mL) [20], FLNEFLHV (1.548 mg/mL) [29], FPELLI (0.40 mg/mL) and VFAAL (0.38 mg/mL) [4], FLPF (1.497 mg/mL), LPF (1.013 mg/mL), LLPF (1.031 mg/mL) [33], GFGPL (0.328 mg/mL), VGGRP (0.465 mg/mL) [30], WEGPK (5.407 mg/mL), Gly-Pro-Pro (GPP) (2.472 mg/mL), GVPLT (3.124 mg/mL) [6], FIMGPY (1.04 mg/mL), GPAGDY (0.77 mg/mL) and IVAGPQ (1.29 mg/mL) [11] from the protein hydrolysates of scalloped hammerhead muscle, salmon, Chinese cherry seeds, corn gluten meal, grass carp skin, bluefin leatherjacket heads and skate cartilage. The present results indicated that SCPE-A, SCPE-B and SCPE-C could strongly donate electrons or hydrogen atoms to inactivate ABTS$^+\bullet$.

2.4.4. O$_2^-\bullet$ Scavenging Activity

O$_2^-\bullet$ is the most common free radical and can produce hydrogen peroxide and hydroxyl radicals through dismutation and other reactions in vivo, which can cause damage to DNA, proteins and cell membranes. The O$_2^-\bullet$ scavenging activities of SCPE-A, SCPE-B and SCPE-C were studied, and the dose–effect relations were observed as the concentration gradually increased from 0.1 to 5.0 mg/mL (Figure 6D). The EC$_{50}$ values of SCPE-A, SCPE-B and SCPE-C were 0.08, 0.14, and 0.11 mg/mL, respectively. SCPE-A showed stronger O$_2^-\bullet$ scavenging activity than SCPE-B and SCPE-C and reached 91.7% ± 2.58% scavenging activity at 5.0 mg/mL. The EC$_{50}$ of SCPE-A was lower than that of WDR (0.09 mg/mL), PYFNK (0.11 mg/mL), LDK(0.12 mg/mL) [19,20], HFGDPFH (0.20 mg/mL) [32], GSQ (0.70 mg/mL) [31], Ser-Leu-Pro-Ile-Gly-Leu-Met-Ile-Ala-Met (SLPIGLMIAM) (0.3168 mg/mL) [39], YLGAK (scavenging rate: 36.27% at 1.0 mg/mL), GGLEPINFQ (scavenging rate: 32.05% at 1.0 mg/mL) [36], His-Asp-His-Pro-Val-Cys (HDHPVC) (0.265 mg/mL) and His-Glu-Lys-Val-Cys (HEKVC) (0.235 mg/mL) [40], Tyr-Leu-Met-Arg (YLMR) (0.450 mg/mL), Val-Leu-Tyr-Glu-Glu (VLYEE) (0.693 mg/mL), Met-Ile-Leu-Met-Arg (MILMR) (0.993 mg/mL) [41], FIMGPY (1.61 mg/mL), GPAGDY (1.66 mg/mL) and IVAGPQ (1.82 mg/mL) [11] from the protein hydrolysates of scalloped hammerhead muscle, mussel sauce, Chinese leek seeds, *Mytilus coruscus*, egg white, round scad, croceine croaker muscle and skate cartilage. O$_2^-\bullet$ could be catalyzed into hydrogen peroxide and oxygen by superoxide dismutase (SOD) with a reaction rate 10,000-fold higher than that of spontaneous dismutation in an organism [19]. Therefore, SCPE-A, SCPE-B and SCPE-C might have high antioxidant activity similar to SOD and could be applied as O$_2^-\bullet$ scavengers in biological systems.

2.4.5. Lipid Peroxidation Inhibition Assay

Scavenging activities on DPPH$\bullet$, ABTS$^+\bullet$, HO$\bullet$ and O$_2^-\bullet$ have been widely used to assess the antioxidant capacities of protein hydrolysates and peptides. However, each of these assays only measures an antioxidant property representing a different mechanism, which does not reflect the multiple mechanisms by which samples may act as antioxidants to retard or inhibit lipid oxidation in a food system [42]. Therefore, in this section, the ability of the soluble samples to suppress lipid peroxidation in a linoleic acid model system was investigated. Lipid peroxidation is a complex process that involves the formation and propagation of lipid radicals and lipid hydroperoxides, which are formed as the primary oxidation products in the presence of oxygen [43]. The inhibitory activities of SCPE-A, SCPE-B and SCPE-C were measured by lipid peroxidation in an in vitro model and were compared with the commercially available antioxidant BHT for 7 days. Figure 7 shows that SCPE-A had a similar inhibitory effect on lipid peroxidation as BHT and significantly retarded the lipid peroxidation compared with the control (without sample), SCPE-B and SCPE-C. In previous research, SCPE-A showed excellent scavenging activity on DPPH•, HO•, ABTS$^+$• and O$_2^-$ •, with EC$_{50}$ values of 2.43, 0.28, 0.24, and 0.08 mg/mL, respectively. Therefore, the inhibition effect of lipid peroxidation caused by SCPE-A could be attributed to its radical-scavenging activity. In addition, SCPE-A may have potential applications in the food industry for retarding the production of unwanted off-flavors and toxic products.

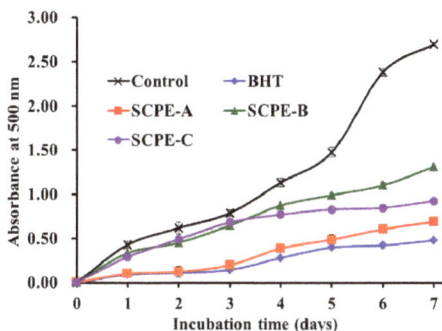

Figure 7. Lipid peroxidation inhibition assays of SCPE-A, SCPE-B and SCPE-C. All data are presented as the mean ± SD of triplicate results.

2.5. Structure-Antioxidant Activity Relationship of Peptides

The structural characteristics of peptides provide guides for the evaluation of food-derived proteins as potential precursors of antioxidant peptides and predict the possible release of bioactive peptides from various proteins using an appropriate protease [1].

Many researchers found that the antioxidant activity of peptides was highly dependent on their amino acid sequence and composition. Chen et al. (2012) reported that the Gly residue may contribute significantly to antioxidant activity since the single hydrogen atom in the side chain of Gly serves as a proton-donating source and neutralizes active free radical species [44]. In addition, Nimalaratne et al. (2015) reported that the single hydrogen atom of Gly (G) can provide high flexibility to the peptide backbone and positively influence the antioxidant properties [8]. Therefore, Gly residues might be important contributors to the antioxidant activity of SCPE-A, SCPE-B, and SCPE-C because there are one, two and four Gly residues in their amino acid sequences, respectively.

The pyrrolidine ring of proline (P) can interact with the secondary structure of the peptide, thereby increasing the flexibility, and it is also capable of quenching singlet oxygen due to its low ionization potential [6]. Samaranayaka and Li-Chan (2011) reported that the Pro residue plays an important role in the antioxidant activity of the peptide purified from *Saccharomyces cerevisiae* protein hydrolysate [45]. Therefore, the one, one and two Pro residues in the amino acid sequences of SCPE-A, SCPE-B, and SCPE-C should enhance the radical-scavenging activities of the three peptides.

Aromatic amino acids, such as Phe, Tyr, His, and Trp, and hydrophobic amino acids, including Ala, Val, and Leu, have been reported to be critical to the antioxidant activities of peptides [1]. Huang et al. (2005) reported that amino acids with aromatic residues, such as Phe, Tyr and Trp, can quench free radicals by direct electron transfer [46]. The results from Guo et al. (2015) indicated that hydrophobic amino acids (e.g., Val, Ala, Leu) and aromatic amino acids (Phe, His, Tyr and Trp) can enhance the radical-scavenging abilities of peptides from Chinese cherry seeds [4]. Therefore, the presence of the one Ala residue and two Phe residues in the sequences of SCPE-B and SCPE-C, respectively, should have a positive impact on their radical-scavenging and lipid peroxidation inhibitory activities.

The presence of acidic and basic amino acids plays a critical role in the metal ion chelating activity, which is related to the carboxyl and amino groups in their side chains [47]. Similar results were reported by Memarpoor-Yazdi et al. (2012), who found that the basic (Arg) and acidic (Asp and Glu) amino acid residues in the sequences of Asn-Thr-Asp-Gly-Ser-Thr-Asp-Tyr-Gly-Ile-Leu-Gln-Ile-Asn-Ser-Arg (NTDGSTDYGILQINSR) and Leu-Asp-Glu-Pro-Asp-Pro-Leu-Ile (LDEPDPLI) were critical to their antioxidant activities [48]. Díaz, et al. (2003) found that Glu is an effective cation chelator that forms complexes with calcium, iron and zinc and may contribute to the antioxidant activity [49]. Therefore, Glu in SCPE-A and Arg in SCPE-B might be favorable to their antioxidant activities.

In addition, the antioxidant activities of peptides are dependent on their molecular size, and shorter peptides, especially peptides with 2–10 amino acid residues, have stronger radical-scavenging and lipid peroxidation inhibition activities than their parent native proteins or long-chain peptides [1,46]. SCPE-A, SCPE-B, and SCPE-C exhibited good antioxidant activities in the radical scavenging and lipid peroxidation inhibition assays, which suggested that the short SCPE-A, SCPE-B, and SCPE-C could interact more effectively and easily with free radicals and inhibit the propagation cycles of lipid peroxidation in the radical scavenging and lipid peroxidation model system [50]. However, SCPE-A had the strongest $O_2^- \bullet$ scavenging and lipid peroxidation inhibition activities, SCPE-B had the strongest scavenging activity on $ABTS^+ \bullet$, and SCPE-C exhibited the highest DPPH$\bullet$ and HO$\bullet$ scavenging activities among all samples and fractions. The results indicated no consistent trends in the antioxidant capacities of SCPE-A, SCPE-B, and SCPE-C in different antioxidant assays. Therefore, more detailed study should be performed to clarify the relationship between the activity and structure of the three isolated peptides.

3. Experimental Section

3.1. Chemicals and Reagents

Scalloped hammerhead (*S. lewini*) was purchased from Fengmao market in Zhoushan City, Zhejiang Province, China. DEAE-52 cellulose and Sephadex G-15 were purchased from Shanghai Source Poly Biological Technology Co., Ltd. (Shanghai, China). Acetonitrile was of liquid chromatography (LC) grade and was purchased from Thermo Fisher Scientific Co., Ltd. (Shanghai, China). All other reagents used in the experiment were of analytical grade and were purchased from Sinopharm Chemical Reagent Co., Ltd. (Shanghai, China).

3.2. Preparation of the Protein Hydrolysate from Scalloped Hammerhead Cartilage

Frozen scalloped hammerhead cartilage was thawed, minced to homogenate and soaked in 1.0 M guanidine hydrochloride with a solid-to-solvent ratio of 1:5 (*w*/*v*) for 48 h with continuous stirring, and the liquid supernatant was collected by centrifugation at 12,000× *g* at 4 °C for 10 min. The resulting supernatant was dialyzed (MW 5 kDa) against 25 volumes of distilled water for 12 h, with the solution changed every 4 h, and the resulting dialysate was freeze-dried.

The freeze-dried sample was dissolved (5% *w*/*v*) in 0.2 M phosphate buffer solution (PBS, pH 7.2) and hydrolyzed for 4 h using neutrase at pH 7.0, 60 °C; alcalase at pH 8.0, 50 °C; trypsin at pH 8.0, 40 °C; pepsin at pH 2.0, 37 °C; or papain at pH 6.0, 50 °C, with a total enzyme dose of 2.5%. Enzymatic hydrolysis was stopped by heating for 10 min in boiling water, and the hydrolysate was centrifuged at 9000× *g* for 15 min. The supernatant was freeze-dried and stored at −20 °C for further analysis. The protein hydrolysate of scalloped hammerhead cartilage using trypsin was named SHCH.

3.3. Isolation of Peptides from SHCH

3.3.1. Fractionation of SHCH by Ultrafiltration

SHCH was fractionated using ultrafiltration (8400, Millipore, Hangzhou, China) with 10 kDa and 3 kDa MW cutoff (MWCO) membranes (Millipore, Hangzhou, China). Three peptide fractions, named SHCH-I (MW < 3 kDa), SHCH-II (3 kDa < MW < 10 kDa) and SHCH-III (MW > 10 kDa), were collected and lyophilized.

3.3.2. Anion-Exchange Chromatography

SHCH-I (5 mL, 40.0 mg/mL) was injected onto a DEAE-52 cellulose column (1.6 × 70 cm) that was pre-equilibrated with deionized water and stepwise eluted with 150 mL distilled water and 0.1, 0.5, and 1.0 M NaCl solution at a flow rate of 1.0 mL/min. Each eluted fraction (5 mL) was collected and measured at 280 nm, and five fractions (Fr.1–5) were pooled and lyophilized.

3.3.3. Gel filtration Chromatography

Fr.4 (5 mL, 10.0 mg/mL) was fractionated on a Sephadex G-15 column (2.6 × 160 cm) eluted with deionized water at a flow rate of 0.6 mL/min. Each eluate (3 mL) was collected and monitored at 280 nm, and two fractions (Fr.4-1 and Fr.4-2) were lyophilized.

3.3.4. RP-HPLC

Fr.4-2 was separated by RP-HPLC (Agilent 1260 HPLC, Agilent Ltd., Santa Rosa, CA, USA) on a Thermo C-18 column (4.6 × 250 mm, 5 μm) (Thermo Co., Ltd., Yokohama, Japan) using a linear gradient of acetonitrile (0%–50% in 0–32 min) in 0.1% trifluoroacetic acid at a flow rate of 0.8 mL/min. The eluate was analyzed at 280 nm, and three peptides (SCPE-A, SCPE-B and SCPE-C) were isolated and lyophilized.

3.4. Determination of the Amino Acid Sequence and Molecular Mass

The amino acid sequences of SCPE-A, SCPE-B and SCPE-C were determined on an Applied Biosystems 494 protein sequencer (Perkin Elmer/Applied Biosystems Inc., Foster City, CA, USA). The molecular masses were determined using a Q-TOF mass spectrometer coupled to an electrospray ionization source (ESI) (Micromass, Waters, Los Angeles, CA, USA).

3.5. Degree of Hydrolysis (DH)

DH analysis was performed according to the previously described method [49]. The hydrolysate (50 μL) was mixed with 0.5 mL of 0.2 M phosphate buffer, pH 8.2 and 0.5 mL of 0.05% trinitrobenzenesulfonic acid (TNBS) reagent. TNBS was freshly prepared before use by diluting with DI water. The mixture was incubated at 50 °C for 1 h in a water bath. The reaction was stopped by adding 1 mL of 0.1 M HCl and incubating at room temperature for 30 min. The absorbance was monitored at 420 nm. L-leucine was used as a standard. To determine the total amino acid content, mungbean meal was completely hydrolyzed with 6 M HCl with a sample to acid ratio of 1:100 at 120 °C for 24 h. DH (%) was calculated using the following equation:

$$DH = [(A_t - A_0)/(A_{max} - A_0)] \times 100$$

where A_t was the amount of a-amino acids released at time t, A_0 was the amount of a-amino acids in the supernatant at 0 h, and A_{max} was the total amount of a-amino acids obtained after acid hydrolysis at 120 °C for 24 h.

3.6. Antioxidant Activity

The radical (DPPH•, HO•, O_2^-•, and ABTS$^+$•) scavenging activity and lipid peroxidation inhibition assays were performed according to previously reported methods [19,51], and the half elimination ratio (EC_{50}) was defined as the concentration of a sample that caused a 50% decrease in the initial concentration of DPPH•, O_2^-•, HO•, and ABTS$^+$•. The EC_{50} was calculated based on the linear relationship of the radical-scavenging rate and concentration of the samples.

3.6.1. HO• Scavenging Activity

In this system, hydroxyl radicals are generated by the Fenton reaction. Hydroxyl radicals can oxidize Fe^{2+} into Fe^{3+}, and only Fe^{2+} can combine with 1,10-phenanthroline to form a red compound (1,10-phenanthroline-Fe^{2+}) with the maximum absorbance at 536 nm. The concentration of hydroxyl radical is reflected by the degree of decolorization of the reaction solution. Briefly, 1,10-phenanthroline solution (1.0 mL, 1.865 mM) and the sample (2.0 mL) were added into a screw-capped tube and mixed. The $FeSO_4 \cdot 7H_2O$ solution (1.0 mL, 1.865 mM) was then pipetted into the mixture. The reaction was initiated by adding 1.0 mL H_2O_2 (0.03% v/v). After being incubated at 37 °C for 60 min in a water bath, the absorbance of the reaction mixture was measured at 536 nm against a reagent blank. The reaction mixture without any antioxidant was used as the negative control, and mixture without H_2O_2 was used as the blank. The HO• scavenging activity was calculated by the following formula:

$$\text{HO• scavenging activity (\%)} = [(A_s - A_n)/(A_b - A_n)] \times 100\%$$

where A_s, A_n, and A_b were the absorbance values determined at 536 nm of the sample, the negative control, and the blank after reaction, respectively.

3.6.2. DPPH• Scavenging Activity

Two milliliters of deionized water containing different concentrations of samples were placed in cuvettes, and then 500 μL of ethanol solution of DPPH (0.02%) and 1.0 mL of ethanol were added into. A control sample containing DPPH solution without sample was also prepared. For the blank absorbance, DPPH solution was substituted with ethanol. The antioxidant activity of the sample was evaluated by the inhibition percentage of DPPH• with the following equation:

$$\text{DPPH• scavenging activity (\%)} = (A_0 + A' - A)/A_0 \times 100\%$$

where A was sample absorbance rate; A_0 was the absorbance of control group; A' was the blank absorbance.

3.6.3. O_2^-• Scavenging Activity

In the experiment, superoxide anions were generated in 1 mL of nitrotetrazolium blue chloride (NBT) (2.52 mM), 1 mL of nicotinamide adenine dinucleotide (NADH) (624 mM) and 1 mL of different concentrations of samples. The reaction was initiated by adding 1 mL of phenazine methosulfate (PMS) solution (120 μg) to the reaction mixture. The absorbance was measured at 560 nm against the corresponding blank after 5 min of incubation at 25 °C. The capacity of scavenging the O_2^-• was calculated using the following equation:

$$O_2^-\text{• scavenging activity (\%)} = \left[\left(A_{control} - A_{sample}\right)/A_{control}\right] \times 100\%$$

where $A_{control}$ was the absorbance without sample and A_{sample} was the absorbance with sample.

3.6.4. ABTS$^{+}\bullet$ Scavenging Activity

The ABTS radical cation was generated by mixing ABTS stock solution (7 mM) with potassium persulphate (2.45 mM). The mixture was left in the dark at room temperature for 16 h. The ABTS radical solution was diluted in 5 mM phosphate buffered saline (PBS) pH 7.4, to an absorbance of 0.70 ± 0.02 at 734 nm. One milliliter of diluted ABTS radical solution was mixed with one milliliter of different concentrations of samples. Ten minutes later, the absorbance was measured at 734 nm against the corresponding blank. The ABTS$^{+}\bullet$ scavenging activity of samples was calculated using the following equation:

$$\text{ABTS}^{+}\bullet \text{ scavenging activity (\%)} = [(A_{control} - A_{sample})/A_{control}] \times 100\%$$

where $A_{control}$ was the absorbance without sample and A_{sample} was the absorbance with sample.

3.6.5. Lipid Peroxidation Inhibition Assay

A sample (5.0 mg) was dissolved in 10 mL of 50 mM phosphate buffer (pH 7.0), and added to a solution of 0.13 mL of linoleic acid and 10 mL of 99.5% ethanol. Then, the total volume was adjusted to 25 mL with deionized water. The mixture was incubated in a conical flask with a screw cap at $40 \pm 1\,^{\circ}\text{C}$ in a dark room and the degree of oxidation was evaluated by measuring the ferric thiocyanate values. The reaction solution (100 µL) incubated in the linoleic acid model system was mixed with 4.7 mL of 75% ethanol, 0.1 mL of 30% ammonium thiocyanate, and 0.1 mL of 20 mM ferrous chloride solution in 3.5% HCl. After 3 min, the thiocyanate value was measured by reading the absorbance at 500 nm following color development with $FeCl_2$ and thiocyanate at different intervals during the incubation period at $40 \pm 1\,^{\circ}\text{C}$.

3.7. Statistical Analysis

All experiments were performed in triplicate ($n = 3$), and the data are expressed as the mean $\pm$ standard deviation (SD). ANOVA was applied to analyze the data using SPSS 19.0 (SPSS Corporation, Chicago, IL, USA). Duncan's multiple range test was used to measure the differences between the parameter means. The differences were considered significant if $p < 0.05$.

4. Conclusions

In this study, three new antioxidant peptides (SCPE-A, SCPE-B and SCPE-C) were isolated from the protein hydrolysate of scalloped hammerhead (*S. lewini*) cartilage by ultrafiltration and chromatography, and their amino acid sequences were identified as Gly-Pro-Glu (GPE), Gly-Ala-Arg-Gly-Pro-Gln (GARGPQ), and Gly-Phe-Thr-Gly-Pro-Pro-Gly-Phe-Asn-Gly (GFTGPPGFNG). SCPE-A, SCPE-B and SCPE-C exhibited strong radical scavenging and lipid peroxidation inhibition activities. These results suggested that the purified peptides from the protein hydrolysate of scalloped hammerhead cartilage may be applied as ingredients in functional foods in bioactive food products. Our subsequent studies will focus on the molecular mechanisms and the relationship between the antioxidant activity and structure of the three isolated peptides.

Acknowledgments: This work was funded by the National Natural Science Foundation of China (NSFC) (No. 81673349) and the International Science and Technology Cooperation Program of China (No. 2012DFA30600).

Author Contributions: Bin Wang and Chang-Feng Chi conceived and designed the experiments. Xue-Rong Li and Li Li performed the experiments. Xue-Rong Li and Bin Wang analyzed the data. Chang-Feng Chi and Bin Wang contributed the reagents, materials, and analytical tools and wrote the paper.

Conflicts of Interest: The authors declare no conflicts of interest.

References

1. Sila, A.; Bougatef, A. Antioxidant peptides from marine by-products: Isolation, identification and application in food systems. A review. *J. Funct. Foods* **2016**, *21*, 10–26. [CrossRef]

2. Jang, H.L.; Liceaga, A.M.; Yoon, K.Y. Purification, characterisation and stability of an antioxidant peptide derived from sandfish (*Arctoscopus japonicus*) protein hydrolysates. *J. Funct. Foods* **2016**, *20*, 433–442. [CrossRef]

3. Wattanasiritham, L.; Theerakulkait, C.; Wickramasekara, S.; Maier, C.S.; Stevens, J.F. Isolation and identification of antioxidant peptides from enzymatically hydrolyzed rice bran protein. *Food Chem.* **2016**, *192*, 156–162. [CrossRef] [PubMed]

4. Guo, P.; Qi, Y.; Zhu, C.; Wang, Q. Purification and identification of antioxidant peptides from Chinese cherry (*Prunus pseudocerasus* Lindl.) seeds. *J. Funct. Foods* **2015**, *19*, 394–403. [CrossRef]

5. Chi, C.F.; Hu, F.Y.; Wang, B.; Li, Z.R.; Luo, H.Y. Influence of amino acid compositions and peptide profiles on antioxidant capacities of two protein hydrolysates from skipjack tuna (*Katsuwonus pelamis*) dark muscle. *Mar. Drugs* **2015**, *13*, 2580–2601. [CrossRef] [PubMed]

6. Mirzaei, M.; Mirdamadi, S.; Ehsani, M.R.; Aminlari, M.; Hosseini, E. Purification and identification of antioxidant and ACE-inhibitory peptide from *Saccharomyces cerevisiae* protein hydrolysate. *J. Funct. Foods* **2015**, *19*, 259–268. [CrossRef]

7. Wang, B.; Gong, Y.D.; Li, Z.R.; Yu, D.; Chi, C.F.; Ma, J.Y. Isolation and characterisation of five novel antioxidant peptides from ethanol-soluble proteins hydrolysate of spotless smoothhound (*Mustelus griseus*) muscle. *J. Funct. Foods* **2014**, *6*, 176–185. [CrossRef]

8. Nimalaratne, C.; Bandara, N.; Wu, J. Purification and characterization of antioxidant peptides from enzymatically hydrolyzed chicken egg white. *Food Chem.* **2015**, *188*, 467–472. [CrossRef] [PubMed]

9. Xing, L.; Hu, Y.; Hu, H.; Ge, Q.; Zhou, G.; Zhang, W. Purification and identification of antioxidative peptides from dry-cured Xuanwei ham. *Food Chem.* **2016**, *194*, 951–958. [CrossRef] [PubMed]

10. Xue, Z.; Wen, H.; Zhai, L.; Yu, Y.; Li, Y.; Yu, W.; Cheng, A.; Wang, C.; Kou, X. Antioxidant activity and anti-proliferative effect of a bioactive peptide from chickpea (*Cicer arietinum* L.). *Food Res. Int.* **2015**, *77*, 75–81. [CrossRef]

11. Pan, X.; Zhao, Y.Q.; Hu, F.Y.; Wang, B. Preparation and identification of antioxidant peptides from protein hydrolysate of skate (*Raja porosa*) cartilage. *J. Funct. Foods* **2016**, *25*, 220–230. [CrossRef]

12. Pan, X.; Zhao, Y.Q.; Hu, F.Y.; Chi, C.F.; Wang, B. Anticancer activity of a hexapeptide from skate (*Raja porosa*) cartilage protein hydrolysate in HeLa Cells. *Mar. Drugs* **2016**, *14*, 153. [CrossRef] [PubMed]

13. Guo, H.; Kouzuma, Y.; Yonekura, M. Structures and properties of antioxidative peptides derived from royal jelly protein. *Food Chem.* **2009**, *113*, 238–245. [CrossRef]

14. You, L.; Zhao, M.; Regenstein, J.M.; Ren, J. Purification and identification of antioxidative peptides from loach (*Misgurnus anguillicaudatus*) protein hydrolysate by consecutive chromatography and electrospray ionizationmass spectrometry. *Food Res. Int.* **2010**, *43*, 1167–1173. [CrossRef]

15. Kittiphattanabawon, P.; Benjakul, S.; Visessanguan, W.; Shahidi, F. Isolation and characterization of collagen from the cartilages of brownbanded bamboo shark (*Chiloscyllium punctatum*) and blacktip shark (*Carcharhinus limbatus*). *LWT Food Sci. Technol.* **2010**, *43*, 792–800. [CrossRef]

16. Zheng, L.; Ling, P.; Wang, Z.; Niu, R.; Hu, C.; Zhang, T.; Lin, X. A novel polypeptide from shark cartilage with potent anti-angiogenic activity. *Cancer Biol. Ther.* **2007**, *6*, 775–780. [CrossRef] [PubMed]

17. Liang, Q.; Wang, L.; Sun, W.; Wang, Z.; Xu, J.; Ma, H. Isolation and characterization of collagen from the cartilage of Amur sturgeon (*Acipenser schrenckii*). *Process Biochem.* **2014**, *49*, 318–323. [CrossRef]

18. Jeevithan, E.; Bao, B.; Bu, Y.S.; Zhou, Y.; Zhao, Q.B.; Wu, W.H. Type II collagen and gelatin from silvertip shark (*Carcharhinus albimarginatus*) cartilage: Isolation, purification, physicochemical and antioxidant properties. *Mar. Drugs* **2014**, *12*, 3852–3873. [CrossRef] [PubMed]

19. Wang, B.; Li, Z.R.; Chi, C.F.; Zhang, Q.H.; Luo, H.Y. Preparation and evaluation of antioxidant peptides from ethanol-soluble proteins hydrolysate of *Sphyrna lewini* muscle. *Peptides* **2012**, *36*, 240–250. [CrossRef] [PubMed]

20. Luo, H.Y.; Wang, B.; Li, Z.R.; Chi, C.F.; Zhang, Q.H.; He, G.Y. Preparation and evaluation of antioxidant peptide from papain hydrolysate of *Sphyrna lewini* muscle protein. *LWT Food Sci. Technol.* **2013**, *51*, 281–288. [CrossRef]

21. Li, Z.; Wang, B.; Chi, C.; Gong, Y.; Luo, H.; Ding, G. Influence of average molecular weight on antioxidant and functional properties of cartilage collagen hydrolysates from *Sphyrna lewini*, *Dasyatis akjei* and *Raja porosa*. *Food Res. Int.* **2013**, *51*, 283–293. [CrossRef]

22. Najafian, L.; Babji, A.S. Production of bioactive peptides using enzymatic hydrolysis and identification antioxidative peptides from patin (*Pangasius sutchi*) sarcoplasmic protein hydolysate. *J. Funct. Foods* **2014**, *9*, 280–289. [CrossRef]

23. Wang, B.; Li, L.; Chi, C.F.; Ma, J.H.; Luo, H.Y.; Xu, Y.F. Purification and characterisation of a novel antioxidant peptide derived from blue mussel (*Mytilus edulis*) protein hydrolysate. *Food Chem.* **2013**, *138*, 1713–1719. [CrossRef] [PubMed]

24. Delgado, M.C.O.; Nardo, A.; Pavlovic, M.; Rogniaux, H.; Añón, M.C.; Tironi, V.A. Identification and characterization of antioxidant peptides obtained by gastrointestinal digestion of amaranth proteins. *Food Chem.* **2016**, *197*, 1160–1167. [CrossRef] [PubMed]

25. Liu, K.; Zhao, Y.; Chen, F.; Fang, Y. Purification and identification of Se-containing antioxidative peptides from enzymatic hydrolysates of Se-enriched brown rice protein. *Food Chem.* **2015**, *187*, 424–430. [CrossRef] [PubMed]

26. Park, S.Y.; Kim, Y.S.; Ahn, C.B.; Je, J.Y. Partial purification and identification of three antioxidant peptides with hepatoprotective effects from blue mussel (*Mytilus edulis*) hydrolysate by peptic hydrolysis. *J. Funct. Foods* **2016**, *20*, 88–95. [CrossRef]

27. Wu, Q.; Du, J.; Jia, J.; Kuang, C. Production of ACE inhibitory peptides from sweet sorghum grain protein using alcalase: Hydrolysis kinetic, purification and molecular docking study. *Food Chem.* **2016**, *199*, 140–149. [CrossRef] [PubMed]

28. You, S.J.; Wu, J.P. Angiotensin-I converting enzyme inhibitory and antioxidant activities of egg protein hydrolysates produced with gastrointestinal and nongastrointestinal enzymes. *J. Food Sci.* **2011**, *76*, 801–807. [CrossRef] [PubMed]

29. Ahn, C.B.; Kim, J.G.; Je, J.Y. Purification and antioxidant properties of octapeptide from salmon byproduct protein hydrolysate by gastrointestinal digestion. *Food Chem.* **2014**, *147*, 78–83. [CrossRef] [PubMed]

30. Cai, L.; Wu, X.; Zhang, Y.; Li, X.; Ma, S.; Li, J. Purification and characterization of three antioxidant peptides from protein hydrolysate of grass carp (*Ctenopharyngodon idella*) skin. *J. Funct. Foods* **2015**, *16*, 234–242. [CrossRef]

31. Hong, J.; Chen, T.T.; Hu, P.; Yang, J.; Wang, S.Y. Purification and characterization of an antioxidant peptide (GSQ) from Chinese leek (*Allium tuberosum* Rottler) seeds. *J. Funct. Foods* **2014**, *10*, 1–10. [CrossRef]

32. Rajapakse, N.; Mendis, E.; Jung, W.K.; Je, J.Y.; Kim, S.K. Purification of a radical scavenging peptide from fermented mussel sauce and its antioxidant properties. *Food Res. Int.* **2005**, *38*, 175–182. [CrossRef]

33. Zhuang, H.; Tang, N.; Yuan, Y. Purification and identification of antioxidant peptides from corn gluten meal. *J. Funct. Foods* **2013**, *5*, 1810–1821. [CrossRef]

34. Ranathunga, S.; Rajapakse, N.; Kim, S.K. Purification and characterization of antioxidative peptide derived from muscle of conger eel (*Conger myriaster*). *Eur. Food Res. Technol.* **2006**, *222*, 310–315. [CrossRef]

35. Ren, J.; Zhao, M.; Shi, J.; Wang, J.; Jiang, Y.; Cui, C.; Kakuda, Y.; Xue, S.J. Purification and identification of antioxidant peptides from grass carp muscle hydrolysates by consecutive chromatography and electrospray ionization-mass spectrometry. *Food Chem.* **2008**, *108*, 727–736. [CrossRef] [PubMed]

36. Chen, C.; Chi, Y.J.; Zhao, M.Y.; Lv, L. Purification and identification of antioxidant peptides from egg white protein hydrolysate. *Amino Acids* **2012**, *43*, 457–466. [CrossRef] [PubMed]

37. Re, R.; Pellegrini, N.; Proteggente, A.; Pannala, A.; Yang, M.; Rice-Evans, C. Antioxidant activity applying an improved ABTS radical cation decolorization assay. *Free Radic. Biol. Med.* **1999**, *26*, 1231–1237. [CrossRef]

38. Zheng, L.; Zhao, M.; Xiao, C.; Zhao, Q.; Su, G. Practical problems when using ABTS assay to assess the radical-scavenging activity of peptides: Importance of controlling reaction pH and time. *Food Chem.* **2016**, *192*, 288–294. [CrossRef] [PubMed]

39. Kim, E.K.; Oh, H.J.; Kim, Y.S.; Hwang, J.W.; Ahn, C.B.; Lee, J.S.; Jeon, Y.J.; Moon, S.H.; Sung, S.H.; Jeon, B.T.; et al. Purification of a novel peptide derived from *Mytilus coruscus* and in vitro/in vivo evaluation of its bioactive properties. *Fish Shellfish Immunol.* **2013**, *34*, 1078–1084. [CrossRef] [PubMed]

40. Jiang, H.; Tong, T.; Sun, J.; Xu, Y.; Zhao, Z.; Liao, D. Purification and characterization of antioxidative peptides from round scad (*Decapterus maruadsi*) muscle protein hydrolysate. *Food Chem.* **2014**, *154*, 158–163. [CrossRef] [PubMed]

41. Chi, C.F.; Hu, F.Y.; Wang, B.; Ren, X.J.; Deng, S.G.; Wu, C.W. Purification and characterization of three antioxidant peptides from protein hydrolyzate of croceine croaker (*Pseudosciaena crocea*) muscle. *Food Chem.* **2015**, *168*, 662–667. [CrossRef] [PubMed]
42. Conway, V.; Gauthier, S.F.; Pouliot, Y. Antioxidant activities of buttermilk proteins, whey proteins, and their enzymatic hydrolysates. *J. Agric. Food Chem.* **2013**, *61*, 364–372. [CrossRef] [PubMed]
43. Hu, F.Y.; Chi, C.F.; Wang, B.; Deng, S.G. Two novel antioxidant nonapeptides from protein hydrolysate of skate (*Raja porosa*) muscle. *Mar. Drugs* **2015**, *13*, 1993–2009. [CrossRef] [PubMed]
44. Chen, C.; Chi, Y.J.; Zhao, M.Y.; Xu, W. Influence of degree of hydrolysis on functional properties, antioxidant and ACE inhibitory activities of egg white protein hydrolysate. *Food Sci. Biotechnol.* **2012**, *21*, 27–34. [CrossRef]
45. Samaranayaka, A.G.P.; Li-Chan, E.C.Y. Food-derived peptidic antioxidants: A review of their production, assessment, and potential applications. *J. Funct. Foods* **2011**, *3*, 229–254. [CrossRef]
46. Huang, D.; Ou, B.; Prior, R.L. The chemistry behind antioxidant capacity assays. *J. Agric. Food Chem.* **2005**, *53*, 1841–1856. [CrossRef] [PubMed]
47. Gimenez, B.; Aleman, A.; Montero, P.; Gomez-Guillen, M.C. Antioxidant and functional properties of gelatin hydrolysates obtained from skin of sole and squid. *Food Chem.* **2009**, *114*, 976–983. [CrossRef]
48. Memarpoor-Yazdi, M.; Asoodeh, A.; Chamani, J. A novel antioxidant and antimicrobial peptide from hen egg white lysozyme hydrolysates. *J. Funct. Foods* **2012**, *4*, 278–286. [CrossRef]
49. Díaz, M.; Dunn, C.M.; McClements, D.J.; Decker, E.A. Use of caseinophosphopeptides as natural antioxidants in oil-in-water emulsions. *J. Agric. Food Chem.* **2003**, *51*, 2365–2370. [CrossRef] [PubMed]
50. Wiriyaphan, C.; Xiao, H.; Decker, E.A.; Yongsawatdigul, J. Chemical and cellular antioxidative properties of threadfin bream (*Nemipterus* spp.) surimi byproduct hydrolysates fractionated by ultrafiltration. *Food Chem.* **2015**, *167*, 7–15. [CrossRef] [PubMed]
51. Wang, B.; Wang, Y.M.; Chi, C.F.; Hu, F.Y.; Deng, S.G.; Ma, J.Y. Isolation and characterization of collagen and antioxidant collagen peptides from scales of croceine croaker (*Pseudosciaena crocea*). *Mar. Drugs* **2013**, *11*, 4641–4661. [CrossRef] [PubMed]

marine drugs

Review

Bioactive Peptide of Marine Origin for the Prevention and Treatment of Non-Communicable Diseases

Ratih Pangestuti [1] and Se-Kwon Kim [2,3,*]

1   Research Center for Oceanography, Indonesian Institute of Sciences (LIPI), Jakarta 14430, Indonesia;
     ratih.pangestuti@lipi.go.id or pangestuti.ratih@gmail.com
2   Department of Marine-bio Convergence Science, Pukyong National University, Busan 608-737, Korea
3   Institute for Life Science of Seogo (ILSS), Kolmar Korea Co., Seoul 137-876, Korea
*   Correspondence: sknkim@pknu.ac.kr; Tel.: +82-51-629-7550

Academic Editor: Keith B. Glaser
Received: 1 December 2016; Accepted: 6 March 2017; Published: 9 March 2017

Abstract: Non-communicable diseases (NCD) are the leading cause of death and disability worldwide. The four main leading causes of NCD are cardiovascular diseases, cancers, respiratory diseases and diabetes. Recognizing the devastating impact of NCD, novel prevention and treatment strategies are extensively sought. Marine organisms are considered as an important source of bioactive peptides that can exert biological functions to prevent and treatment of NCD. Recent pharmacological investigations reported cardio protective, anticancer, antioxidative, anti-diabetic, and anti-obesity effects of marine-derived bioactive peptides. Moreover, there is available evidence supporting the utilization of marine organisms and its bioactive peptides to alleviate NCD. Marine-derived bioactive peptides are alternative sources for synthetic ingredients that can contribute to a consumer's well-being, as a part of nutraceuticals and functional foods. This contribution focus on the bioactive peptides derived from marine organisms and elaborates its possible prevention and therapeutic roles in NCD.

Keywords: bioactive peptide; marine; prevention; treatment; non-communicable diseases

1. Introduction

Non-communicable diseases (NCD), sometimes referred to as chronic diseases, are the leading cause of death and disability globally [1,2]. NCD are not passed from person to person, and these diseases are of long duration and slow progression. Many of the NCD are strongly associated with lifestyle-related choices (unhealthy diet, physical inactivity, and tobacco and alcohol use), and environmental and genetic factors [3]. The four main leading causes of NCD deaths are cardiovascular diseases (CVD), cancers, respiratory diseases and diabetes [4]. In 2012, CVD was responsible for around 17.5 million deaths (46.2% of NCD deaths), while cancers around 8.2 million deaths (21.7% of NCD deaths) (Figure 1).

NCD are increase rapidly poses one of the major health challenges of the 21st century. Of the 56 million global deaths in 2012, 68% or 38 million were attributed to NCD and projected to rise further worldwide. It has been predicted by the World Health Organization of the United Nations (WHO) that NCD will be responsible for a significant increase total number of deaths in the next decade. The greatest NCD increase is expected to be seen in low and middle income countries where 80% of NCD deaths occur. Notably, NCD are projected to surpass communicable, maternal, perinatal and nutritional diseases as the most common cause of death by 2030 in Africa [5]. The rapidly growing burden of NCD in low and middle income countries is not only accelerated by population aging, but also by the negative impact of globalization [2].

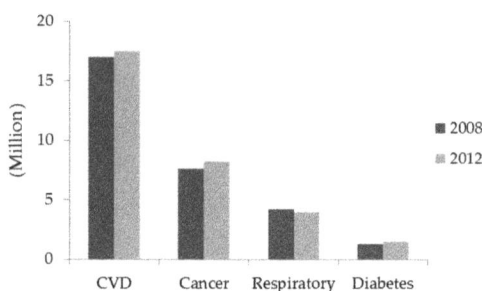

Figure 1. Top four cause of death attributed to non-communicable diseases in the world (References: [1,2]).

Recognizing the devastating impact of NCD, novel preventive and therapeutic strategies are extensively sought. Many research groups have combed both terrestrial and marine natural resources for NCD remedies [4,6–8]. Marine organisms are consistently exposed to biotic and abiotic pressures, which exert an influence on the organisms physiology, leading to the production of metabolites to survive and thrive [4,9]. Therefore, marine organisms are reservoirs of structurally diverse bioactive materials with numerous biological effects for human's body. These bioactive materials include polysaccharides (agar, alginates, carrageenan, fucoidan, ulvan, laminarin, porphyran, and fulcellaran), pigments (chlorophyll, carotenoids, and phycobillins), protein and peptides, polyunsaturated fatty acids (PUFA), polyphenols, and other bioactive compounds. Among marine-derived bioactive materials, much attention has been paid to unraveling the structural and biological properties of bioactive peptides. Depending on the structural and sequence of amino acids, these peptides can exhibit diverse activities for NCD remedies, including cardio protective, antihypertensive, anticancer, anti-diabetic, and antioxidative. Not restricted to one activity, many of the bioactive peptides are multifunctional and can exert more than one of the effects mentioned. For above reasons, marine-derived bioactive peptides are considered prominent candidates for NCD prevention and treatment.

This article focuses on bioactive peptides reported from fish, mollusks, crustaceans, and seaweeds. It highlights and compiles the most relevant studies on the structural diversity of peptides found in these marine organisms and outlines their potential as candidate raw materials for the generation of bioactive peptides. Notably, their possible biological role with potential utilization as NCD prevention and remedy will be briefly discussed. Furthermore, some purification and isolation technique of marine-derived bioactive peptides will be outlined.

2. Marine-Derived Cardio Protective Peptides

The CVD is the leading cause of death and diseases burden in many countries [10,11]. The major independent risk factor for CVD is hypertension. In 2000, the estimated total number of adults with hypertension was nearly one billion or equal to 25% of the total adult population worldwide. The total number of adult with hypertension was predicted to increase to a total of 1.56 billion (60% of the total adult population) in 2025 [12].

The important regulator of blood pressure homeostasis in mammals is renin-angiotensin system (RAS). Renin (EC 3.4.23.15) converts angiotensinogen to angiotensin I, and it will be converted to biologically active angiotensin II by angiotensin-I converting enzyme (ACE, peptidyldipeptide hydrolase, EC 3.4.15.1), which ultimately leads to hypertension. In addition, ACE regulates the inactivation of bradykinin [13]. Therefore, ACE and renin inhibitor makes a positive contribution to hypertension treatment and specific inhibitors are currently used in pharmaceuticals. Synthetic hypertension drugs such as captopril, enalapril, and lisinopril are remarkably effective; however, they are known to cause adverse side effects. Hence, search for natural antihypertensive as alternative to synthetic inhibitors are of interest. Marine-derived anti-hypertensive peptides have shown potent renin and ACE inhibitory activities (Figure 2) and, therefore, potential to be used and developed as cardio protective peptides.

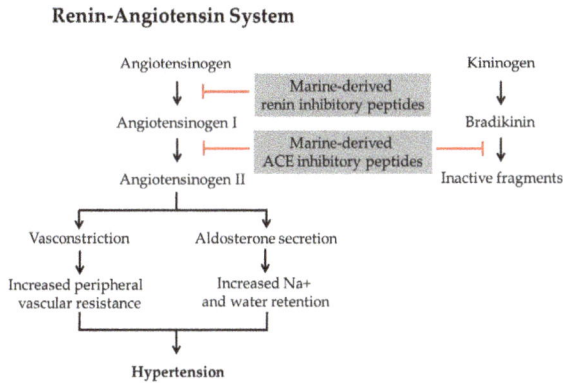

Figure 2. Potent renin and angiotensin-I converting enzyme inhibitory activity of marine-derived anti-hypertensive peptides.

2.1. Marine-Derived Renin Inhibitory Peptides

Renin has long been recognized as the key regulator of RAS, which has an established role in controlling blood volume, arterial pressure, and cardiac and vascular function [14]. The first new class of orally active, non-peptide, low molecular weight renin inhibitors was discovered in Switzerland. The renin inhibitor was named Aliskiren (formerly CGP 60536) [15]. Afterwards, many studies have identified renin inhibitory substances derived from plant sources.

In 2012, Fitzgerald and his colleagues had successfully isolated and characterized renin inhibitory peptides derived from marine red algae *Palmaria palmata* papain hydrolysates [16]. The tridecapeptide sequence was identified as Ile-Arg-Leu-Ile-Ile-Val-Leu-Met-Pro-Ile-Leu-Met-Ala. In vivo result showed that *P. palmata* hydrolysate and tridecapeptide reduced spontaneously hypertensive rat (SHR) blood pressure when administered orally after a 24 h period. After 24 h, SHR group fed the *P. palmata* hydrolysate recorded a drop of 34 mm Hg in systolic blood pressure (SBP), while the group fed the tridecapeptide presented a drop of 33 mm Hg in blood pressure compared to the SBP recorded at time zero [17]. It was concluded that the potential active form of the peptide is dipeptides originated along the passage through gastrointestinal tract [18]. Further, *P. palmata* protein hydrolysate was formulated in wheat bread. Four percent *P. palmata* protein hydrolysate content in wheat bread did not affect the texture or sensory properties of the bread to a large degree. Interestingly, wheat bread containing the hydrolysate retained renin inhibitory bioactivity after the baking process; therefore, baked products may be one of the suitable delivery vehicles for bioactive peptides as renin inhibitor [19].

2.2. Marine-Derived ACE Inhibitory Peptides

It was revealed that ACE inhibitors significantly reduced the mortality of heart failure patients. Marine-derived ACE inhibitory peptides have been studied intensively and the first one was isolated from sardine by a Japanese scientist [20]. Afterwards, many other marine-derived ACE inhibitory peptides have been discovered. Up to now, more than 125 ACE-inhibitory peptides sequences have been isolated and identified from marine organisms. The potency of marine-derived ACE inhibitory peptides are normally expressed as half maximal inhibitory concentration (IC_{50}) value, which is the ACE inhibitor concentration leading to 50% inhibition of ACE activity [8]. The ACE inhibition patterns of marine-derived ACE inhibitory peptides were analyzed by Lineweaver–Burk plot and the competitive inhibitions are the more frequent reported pattern compared to non-competitive inhibition [21]. Competitive inhibition means that marine-derived ACE inhibitory peptides can bind to the active site to block it or to the inhibitor binding site that is remote from the active site to alter the enzyme conformation such as that the substrate no longer binds to the active site [22].

As summarized in Table 1, peptides derived from algae, tuna, shark and salmon showed stronger ACE inhibitory activity compared to other marine organisms such as oyster, sipuncula, and jellyfish. The ACE inhibitory activity of marine-derived bioactive peptides were higher compared to ACE inhibitory peptide-derived from terrestrial food source (i.e., milk, chicken muscle and bovine) [23,24]. Marine-derived ACE inhibitory peptides are generally short chain peptides [18,25–27]. It was reported that amino acid residues with bulky side chain as well as hydrophobic side chains were more active for dipeptides [28]. Meanwhile, for tripeptides, the most favorable residue for the C-terminus was aromatic amino acids, positively charged amino acid in the middle and hydrophobic amino acid in the N-terminus [29]. Molecular weight is also an important factor on ACE inhibitory activity of peptides. Generally, ACE inhibitory peptides are short sequences of hydrophobic amino acids, and have low molecular weights.

Table 1. ACE inhibitory activity of marine-derived bioactive peptides.

Source	Extraction	Sequence	Inhibition (IC$_{50}$)	References
Seaweed (*Undaria pinnatifida*)	Hot water extraction; Chromatography	Ile-Tyr	2.7 μM	[18]
	Enzymatic hydrolysis (Protease S); Chromatography	Ile-Trp	1.5 μM	[30]
Seaweed (*P. yezoensis*)	Chromatography	Ala-Lys-Tyr-Ser-Tyr	1.52 μM	[31]
Microalgae (*Spirulina platensis*)	Enzymatic hydrolysis (Pepsin); Chromatography	Ile-Ala-Pro-Gly	11.4 μM	[32]
Yellowfin tuna (*Neothunnus macropterus*)	Chromatography	Pro-Thr-His-Ile-Lys-Trp-Gly-Asp	2 μM	[33]
Skipjack tuna (*Katsuwonus pelamis*) bowels	Chromatography	Leu-Arg-Pro	1 μM	[34]
Alaska Pollack skin (*Theragra chalcogramma*)	Enzymatic hydrolysis (serial protease); Chromatography	Gly-Pro-Leu	2.6 μM	[35]
Chum salmon (*Oncorhynchus keta*) muscle	Enzymatic hydrolysis (Thermolysin); Chromatography	Val-Trp	2.5 μM	[36]
Pink salmon (*Oncorhynchus gorbuscha*)	Enzymatic hydrolysis (papain); Chromatography	Ile-Trp	1.2 μM	[37]
Skate skin (*Okamejei kenojei*)	Enzymatic hydrolysis (alkalase/protease); Chromatography	Met-Val-Gly-Ser-Ala-Pro-Gly-Val-Leu	3.09 μM	[38]
Small-spotted catshark (*Scyliorhinus canicula*)	Enzymatic hydrolysis (Trypsin, subtilisin); Chromatography	Val-Ala-Met-Pro-Phe	0.44 μM	[39]
Pelagic thresher (*Alopias pelagicus*) muscle	Enzymatic hydrolysis (thermolysin); Chromatography	Ile-Lys-Trp	0.54 μM	[26]
Marine shrimp (*Acetes chinensis*)	Enzymatic hydrolysis (Protease); Chromatography	Ile-Phe-Val-Pro-Ala-Phe	3.4 μM	[40]
	Fermentation; Chromatography	Asp-Pro	2.15 μM	[41]
	Enzymatic hydrolysis (Pepsin); Chromatography	Leu-His-Pro	3.4 μM	[42]
Izumi shrimp (*Plesionika izumiae* Omori, 1971)	Enzymatic hydrolysis (Protease); Chromatography	Ser-Thr	4.03 μM	[43]
Jellyfish (*Rhopilema esculentum*)	Enzymatic hydrolysis (pepsin, papain); ultrafiltration; Chromatography	Gln-Pro-Gly-Pro-Thr	80.67 μM	[44]
Sipuncula (*Phascolosoma esculenta*)	Enzymatic hydrolysis (Pepsin); Chromatography	Ala-Trp-Leu-His-Pro-Gly-Ala-Pro-Lys-Val-Phe	135 M	[45]
Pearl oyster (*Pinctada fucata martensii*)	Enzymatic hydrolysis (Pepsin); Chromatography	Ala-Leu-Ala-Pro-Glu	167.5 μM	[46]

Many in vivo studies in SHR and hypertensive human volunteers demonstrated that marine-derived ACE inhibitory peptides significantly reduce blood pressure. For example, bonito oligopeptide (at a dose of 3 mg/day) decreased blood pressure in human subjects with borderline or mild hypertension.

More recently, the purified oligopeptide from bonito was optimized by ultrafiltration methods. The optimized bonito peptide (at a dose of 1.5 mg/day) showed anti-hypertensive effects in a double-blind, randomized, cross-over study in 61 human subjects with borderline or mild hypertension without any side-effects [47,48]. Subsequent report indicated that bonito oligopeptide played a direct action on relaxation of vascular smooth muscle in addition to the ACE-inhibitory activity [49].

Anti-hypertensive effect of peptides-derived from fish gelatin has already been reported in SHR. Peptides-derived from *O. kenojei* inhibited vasoconstriction via PPAR-c expression, activation and phosphorylation of eNOS in lungs. The peptides also involved in the expression levels of endothelin-1, RhoA, a-smooth muscle actin, cleaved caspase 3 and MAPK were decreased by SAP in lungs. SP1 (Leu-Gly-Pro-Leu-Gly-Val-Leu, molecular weight (MW): 720 Da) and SP2 (Met-Val-Gly-Ser-Ala-Pro-Gly-Val-Leu, MW: 829 Da) showed potent ACE inhibition with IC_{50} values of 4.22 and 3.09 µM, respectively [38]. Peptide from tuna and chum salmon (*O. keta*) also showed potent anti-hypertensive activity as tested in SHR [50,51]. Oral administration of tuna peptides (Gly-Asp-Leu-Gly-Lys-Thr-Thr-Thr-Val-Ser-Asn-Trp-Ser-Pro-Pro-Lys-Try-Lys-Asp-Thr-Pro, MW: 2480 Da) in SHR decreased SBP of 21 mmHg. Lee et al. (2014) demonstrated that oral administration (20 mg/kg) of chum salmon peptides showed a strong suppressive effect on SBP of SHR. They claimed that antihypertensive activity of chum salmon peptide was similar with captopril [50].

The ACE inhibitory activities of brown and red seaweed-derived bioactive peptides have been confirmed in SHR. More than one decade ago, Suetsuna et al. (2000) successfully characterized di- and tetrapeptides derived from the brown algae, *U. pinatifida* and showed that administration of those peptides in SHR significantly decreased blood pressure in SHR [25]. Marine microalgae (*C. ellipsoidea*) tetrapeptides (Val-Glu-Gly-Tyr) also showed a potent anti-hypertensive activity. Oral administration of *C. ellipsoidea* tetrapeptides at a dose of 10 mg/kg significantly decrease SBP in SHR [52].

Due to their effectiveness in regulating blood pressure, marine-derived bioactive peptides have prospective use as high quality diets for the prevention and treatment of CVD as well as other NCD. In Japan, some of the marine-derived peptides and hydrolysates have been approves as "foods for specified health uses" (FOSHU) by Japanese Ministry of Health, Labor, and Welfare. Presently, bonito oligopeptide are incorporated in blood pressure lowering capsules and sold as nutraceuticals worldwide. However, generally, marine-derived anti-hypertensive peptides are short sequences of hydrophobic amino acids, which normally give bitter taste. Therefore, to increase consumer's acceptance, flavor manipulation needs to be used when developing marine-derived peptides as functional foods products.

3. Marine-Derived Anti-Cancer Peptides

Cancer is a condition of uncontrolled growth of cells which interferes with the normal functioning of the body and has undesirable systematic effects [53]. It is a dreadful NCD which increases with changing lifestyle, unhealthy diet and global warming [54]. Therefore, fruitful approaches are needed for the prevention and treatment of these diseases. Current cancer available treatments such as chemotherapy many times causing disastrous side effect; and most anticancer drugs currently used in chemotherapy are giving toxic effects to the normal cells which cause immunotoxicity and, hence, aggravate patient's recovery [55]. In this context, a variety of ingredients of traditional medicines are being widely investigated to analyze their potential as cancer therapeutic agents. Presently, more than 60% of the used anticancer agents are derived from natural sources [56]. Although marine resources are still underrepresented in current pharmacopeia, it is anticipated that marine environment will become the invaluable source for cancer therapeutic agents in the future [57]. Many studies reported that marine-derived bioactive peptides could induce cancer cell death by different mechanisms such as apoptosis, affecting the tubulin-microtubule equilibrium, or inhibiting angiogenesis [57,58].

3.1. Anti-Cancer Peptides Derived from Sponges

Marine sponges (Porifera) are the oldest metazoan group, having an outstanding importance as a living fossil. There are approximately 8000 described species of sponges and perhaps twice as many un-described species. Sponges inhabit every type of marine environment, from polar seas to temperate and tropical waters and also thrive and prosper at all depths. Marine sponges have been renowned and ranked at the top with respect to the discovery of bioactive compounds with the diversity in chemical structures being related to an equally diverse pattern of activities. The chemical diversity of sponge bioactive metabolites is remarkable, including unusual nucleosides, bioactive terpenes, sterols, peptides, alkaloids, fatty acids, peroxides, and amino acid derivatives (which are frequently halogenated). In recent years, anticancer peptides have been isolated from marine sponges.

Discodermins is the first head-to-side chain novel cyclodepsipeptides isolated from marine sponge *Discodermia kiiensis*. Discodermins A–H contain 13–14 known and rare amino acids as a chain, with a macrocyclic ring constituted by lactonization of a threonine unit with the carboxy terminal. All the discodermins types are cytotoxic against murine leukemia (P388) cells, human lung (A549) cell with IC_{50} values from 0.02 to 20 μg/mL. It was demonstrated that macrolactone ring is also essential for the cytotoxic activity. Furthermore, Fusetani and co-workers (1995) reported the isolation and structure of Halicylindramides A–C, which are cyclic depsipeptides isolated from the Japanese marine sponge *Halichondria cylindrata*. Further, the structures of halicylindramide D and halicylindramide E have also been reported. Halicylindramide E is a truncated and linear version of Halicylindramide B amidated at the C-terminus. Compared to other type of Halicylindramide, Halicylindramide E loses cytotoxicity and shows low antifungal activity; suggesting that "head to side chain" arrangement are crucial for the bioactivity of these peptides.

Jaspamide (also known as Jasplakinolide) is a cyclic depsipeptide with 15-carbon macrocyclic ring containing three amino acid residues (L-alanine, *N*-methyl-2-bromotryptophan, and β-tyrosine). Jasplakinolide was originally isolated from the marine sponge *Jaspis johnstoni* [59]. These cyclic depsipeptides have been extensively investigated as a potential cancer therapeutic agent. Jaspamide has been demonstrated to have growth inhibitory effect on PC-3, prostate carcinoma (DU-145), and Lewis lung carcinoma (LNCaP) cells [60]. It is unique anti-cancer agents that stabilizes actin filaments in vitro, and disrupts actin filaments and induce polymerization of monomeric actin into amorphous masses in vivo. In recent years, several analogs of jaspamides have been isolated from *J. splendens* and many of them possess anticancer activity [61]. Another sponge-derived cyclic depsipeptide, Geodiomolides A, B, H and I, also showed anti-proliferative activity against breast cancer (T47D and MCF-7) cells via actin depolymerization. Geodiamolides were previously isolated and characterized from the Carribean sponge *Geodia* sp. (order Astrophorida; family Geodidae). Further experiments demonstrated that geodiamolide H induces striking phenotypic modifications in human breast cancer (Hs578T) cells [62]. Geodiamolide H decreases Hs578T cell migration and invasion which probably mediated through modifications in the actin cytoskeleton. Interestingly, Geodiamolides H was not cytotoxic for human mammary epithelial (MCF 10A) cell lines [63].

Hemiasterlins comprise a small family of naturally occurring *N*-methylated tripeptide with highly alkylated unnatural amino acids, was originally isolated from the sponge *Hemiasterella minor* (class, Demospongiae; order, Hadromedidia; family, Hemiasterllidae). Hemiasterlins act as potent tumor growth inhibitors. It was reported that Hemiasterlins exhibit antimitotic activity and thus are useful for the treatment of certain cancers. Synthetic analog of hemiasterlins, taltobulin (HTI-286) was a potent inhibitor of proliferation in 18 human tumor cell lines and had substantially less interaction with multidrug resistance protein 1 than currently used antimicrotubule agents, including vinblastine, paclitaxel, docetaxel, or vinorelbine [64]. HTI-286 and another hemiasterlin analog (E7974) are recently being evaluated in clinical trials [65].

Arenastatin A, also known as cryptophycin-24, is potent cytotoxic cyclodepsipeptide isolated from the Okinawan marine sponge *Dysidea arenaria* [66]. Arenastatin A showed extremely potent cytotoxicity against an epidermal carcinoma [67] tumor cell line. Further experiments of cryptophycin-24 showed only marginal in vivo antitumor activity, making it ineligible for further clinical trials [68]. Phakellistatins, a group of proline rich cyclopeptides, have been isolated from *Phakellia* sp. (class Demospongiae, order Axinellida). Up to now, 19 phakellistatins have been isolated [69–72]. Of all the phakellistatins, four comprise the distinctive Pro-Pro track, which represents a considerable synthetic challenge. Phakellistatin 3 represents a new type of cyclopeptide containing an amino acid unit apparently derived from a photooxidation product of tryptophan. Interestingly, all phakellistatins exhibited cancer cell growth inhibitory activities [73]. Reniochalistatins is another group of cyclopeptides rich in proline residues from an extract of a tropical marine sponge, *Reniochalina stalagmitis* Lendenfeld (class Demospongiae, order Halichondria, family Axnellidae) [74]. Recently, Zhan et al. successfully isolated reniochalistatin [75] and reported that only octapeptide (reniochalistatin) was effective inhibited growth different tumor cell lines (RPMI-8226, MGC-803, HL-60, HepG2, and HeLa). Notably, owing to conflicting reports of naturally occurring, proline-rich cyclopeptides that were initially described as having anti-proliferative activity, but subsequent synthetic samples were not active; it is premature to draw any general conclusions regarding a structure–activity relationship among the proline-rich cyclic peptides.

Mostly, anti-cancer activities of peptides-derived from sponge were investigated in vitro, therefore further detailed animal studies and clinical human trials are highly needed to evaluate the physiological anti-cancer activities of these peptides. It is important to note that sponges are susceptible to over exploitation due to their richness in bioactive compounds, hence management and conservations issue of sponge also need to be addressed. Once isolated and characterized, bioactive peptides derived from sponges can be synthesized by peptide synthesis. Synthesis of anticancer peptides derived from sponges can be used for further steps of clinical trials and may provide an alternative to the overexploitation of sponges as for medicinal purposes

3.2. Anti-Cancer Peptides Derived from Fish

The medicinal use of shark cartilage originated from the basic science and observational studies. Early theories regarding the use of shark cartilage for cancer stemmed from the belief that sharks are not afflicted by cancer. In 1992, William Lane published a book entitled "Sharks Don't Get Cancer" [76]. Additionally, cartilage is often recommended by natural medicine experts for cancer, psoriasis, and inflammatory joint diseases [77]. Those traditional remedies and studies have gained attention to develop commercialized anti-cancer agents derived from shark cartilage.

Neovastat (AE-941) is a standardized liquid extract comprising the <500 kDa fraction from the cartilage of shark, *Squalus acanthias* [78]. In vitro and in vivo studies of AE-941 have demonstrated anti-tumor, anti-angiogenic and anti-inflammatory properties. AE-941 could inhibit matrix metalloproteinases (MMP)-2, MMP-9, and MMP-12, and stimulate tissue plasminogen activator enzymatic activities. AE-941 also selectively competes for the binding of vascular endothelial growth factor (VEGF) to its receptor (VEGFR), causing disruption of the signaling pathway which finally induces apoptotic activities in endothelial cells [79]. Further, AE-941 has been tested in a randomized phase III trial in patients with advanced solid tumors (prostate, lung, breast and kidney). However, the result showed that AE-941 was inactive in patients with advanced-stage cancers. AE-941 failed to meet endpoint in the phase III trial, and hence the development was stopped [80].

In 2007, Zheng et al. purified a linear polypeptide with (PG155) from the cartilage of blue shark (*Prionace glauca*). The isolated peptide could inhibit VEGF induced migration and tubulogenesis of human umbilical vein endothelial cells (HUVECs) [81]. As summarized in Table 2, anti-cancer peptides from other marine fish such as pipefish, Red Sea Moses sole, tuna, anchovy and grouper have also been isolated and purified [82–86]. The peptides isolated from marine fish showed anti-cancer activity in human breast cancer (MCF-7), human lung carcinoma (A549), human leukemic lymphoblasts (CCRF-CEM), hepatocellular carcinoma (HA59T/VGH), cervical cancer [87], human liver cancer (HepG2), human fibrosarcoma (HT1080), human myeloid leukemia (U937), human prostate cancer (PC-3), and oral squamous cell carcinoma (OSCC) cells. Pardaxin, a cell-penetrating peptide with cytotoxicity against cancer cells has been isolated from the marine fish Red Sea Moses sole (*Pardachirus marmoratus*) [88]. Pardaxin anti-cancer activity was mediated by apoptosis, as demonstrated by an increase in the externalization of plasma membrane phosphatidylserine and the presence of chromatin condensation. Cancer cells treated with pardaxin also showed elevation of caspase-3/7 activities, disruption of the mitochondrial membrane potential, and accumulation of reactive oxygen species (ROS) production [89]. However, compared to the snake-derived venom peptide; IC_{50} value of anti-cancer effects of marine-derived bioactive peptides is relatively higher (Table 2).

Anticancer peptide has also been isolated from half-fin anchovy (*Setipinna taty*), the peptide sequence was identified as Tyr-Ala-Leu-Pro-Ala-His. The peptide was found to be active inhibiting prostate cancer cells proliferation. Further, three modified peptide were synthesized in order to disclose the contribution of specific amino acid residue to the anti-proliferative activity. The authors concluded that hydrogen-bond formation of the guanidine moiety in arginine (R) with phosphates, sulfates, and carboxylates on cellular components was proposed to be appreciated for cell-permeation efficacy and crucial for the anti-cancer activity. However, the underlying mechanisms of anti-cancer activities are yet clarified.

Table 2. Anti-cancer effects of bioactive peptides derived from marine fish and other organisms.

Name	Source	Anti-Cancer Activity	References
Neovastat (AE-941)	Spiny dogfish shark (*Squalus acanthias*)	Inhibition of metastatic activity on HUVEC, BAEC cells; inhibition of matrix metalloproteinase; Anti-angiogenic effects; Pro-apoptotic on BAEC cells	[78,90]
Pardaxin	Red Sea Moses sole (*Pardachirus marmoratus*)	Pro-apoptotic on HT1080 (IC_{50}: 14.52–15.74 μg/mL), HeLa, OSCC cells	[89,91–94]
PG155	Blue shark (*Prionace glauca*)	Anti-angiogenic effects on HUVECs	[81]
Syngnathusin	Pipefish (*Syngnathus acus*)	Pro-apoptotic on A549 (IC_{50}: 84.9 μg/mL), and CCRF-CEM (IC_{50}: 215.3 μg/mL), cells	[86]
Epinecidin-1	Grouper (*Epinephelus coioides*)	Anti-angiogenic effects on A549, HA59T/VGH, HeLa, HepG2, and HT1080 cells Pro-apoptotic on U937 cells	[82,83]
PAB 1; PAB2	Long tail tuna (*Thunnus tonggol*)	Pro-apoptotic on MCF-7 cells (IC_{50}: 8.1; 8.8 μM)	[84]
YALRAH	Half-fin anchovy (*Setipinna taty*)	Pro-apoptotic on PC-3 cells (IC_{50}: 11.1 μM)	[85]
Rusvinoxidase	Venom of *Daboia russelii russelii*	Pro-apoptotic on MCF-7 cells (IC_{50}: 83 nM)	[95]

The Food and Agriculture Organization of the United Nations (FAO) estimates that world global fishery capture in 2014 was 93.4 million tons, 81.5 million tons from marine waters and 11.9 million tons from inland waters [96]. These numbers are estimated to rise every year due to the increasing consumer knowledge about health benefits of fish. It was estimated that in high-risk populations, consumption of 40–60 g fish per day leads to 50% reduction in death from NCD (i.e., CVD, and cancer) [97]. Supporting those epidemiological studies, anti-cancer effects of fish-derived bioactive peptides in several cell lines also has been reported (Table 2). Unfortunately, fish consumption is very low even in some countries known for their large fish stock, such as in the north African region; hence, nutraceuticals derived from fish peptide can be develop in order to alleviate NCD. For many years, a great deal of interest has been developed by many research groups towards identification of anti-cancer peptides from fish. To develop fish-derived anti-cancer peptides as bioactive materials in food and pharmaceutical industries, large further research is needed. In addition, the potential value of fish by-product is still being ignored. It was estimated that almost half of the fish is commonly discarded to prepared seafood industrially. The amount of fisheries by-products varies depending on species, size, season, and the fishing grounds [98]. Assuming 25% of the animal weight is wasted, the total amount of waste generated from marine capture can be as high as 20.4 million tons per year. These huge amounts of fish by-product harbor useful source of anti-cancer and other bioactive peptides. Scientists should find sustainable ways to refine fish and fish by-products, and governments and industry should invest in using this marine resource in sustainable ways.

3.3. Anti-Cancer Peptides Derived from Urochordata

The urochordata, also known as tunicates and ascidians, have emerged as a rich source of metabolites with potent anticancer activities [99]. Chemical studies of Caribbean tunicates, *Trididemnum solidum*, led to the discovery of the didemnin depsipeptides. Of the didemnins that have been isolated, didemnin B is the most well-known member. Early studies reported that didemnin B possesses in vitro and in vivo antitumor activity against melanoma (B-16), sarcoma (M5076), prostatic, and leukemia (P388) cell lines [100,101]. Based on the significant activity and low toxicity of didemnin B in pre-clinical models, this peptide has been submitted to clinical trials, making it the first marine natural product evaluated in clinical trials [102]. Didemnin B has been tested in clinical phase I and phase II trials against several human tumors. In a clinical phase II trial, patients with non-Hodgkin's lymphoma were given a short intravenous infusion of didemnin B every 28 days, and antitumor effects were observed [103]. Didemnin B has shown modest activity in patients with advanced pretreated non-Hodgkin's lymphoma, and advances epithelial ovarian cancer [100,103]. Nausea, vomiting and anemia are the most frequent reported toxicities due to didemnin B. However, didemnin B clinical trials were stopped, owing to the onset of severe fatigue in patients. An analog of didemnin B that appears to be more active in preclinical models is aplidine (plitidepsin, degydrodidemnin B, DDB or aplidin). Aplidine, a cyclic depsipeptides isolated from the tunicates *Aplidium albicans*, has a pyruvyl group instead of a lactyl group in the linear peptide moiety of didemnin B [104]. Preclinical studies indicate that aplidine is active against several human tumor cell lines. Currently, aplidine has passed clinical phase I and II trials and is currently undergoing phase III trials for relapsed/refractory myeloma (NCT01102426) [105]. The exact mechanism of action of aplidine has not been fully elucidated. However, some researcher suggests that aplidine blocks the secretion of the angiogenic factor VEGF in human leukemia cells (MOLT-4) leading to the blockage of VEGF/VEGF-1 autocrine loop [106]. It has also been shown that aplidine induces a cell cycle perturbation with a block of MOLT-4 cells mainly in G1 phase of the cell cycle. Another mechanism of actions for the activity is aplidine induces cell apoptosis by inducing caspase-3 and -9 activation, cytochrome *c* and membrane dysfunction [107]. Aplidine also induces p53-independent apoptosis in different cancer cell lines in vitro. Similar to didemnin B, aplidine also has dose-limiting toxicities, including diarrhea, dermal toxicity, asthenia, and neuromuscular.

Tamandarins A and B are two naturally occurring cytotoxic cyclic depsipeptides which are closely related to didemnin; these peptides were isolated from a Brazilian ascidian of the family Didemnidae. The structures of are similar to that of didemnin B, the molecules were found to differ only by the presence of hydroxyisovaleric acid (Hiv2), instead of the hydroxyisovalerylpropionic acid (Hip2) unit which is present in didemnins [108]. Tamandarin A showed slightly more potent cytotoxicity against pancreatic carcinoma (BX-PC3) cells, prostate carcinoma (DU145) cells, and head and neck carcinoma (UMSCC10) cells as tested in vitro. The cytotoxic effect of tamandarins has been experimentally shown, but the precise molecular mechanism of action remains uncharacterized. Another cytotoxic peptides derived from ascidian with uncharacterized molecular mechanisms is mollamide. Mollamide is a cytotoxic cyclopeptide obtained from the ascidian *Didemnum molle* and it has shown cytotoxicity against P388, A549, HT29, and monkey kidney fibroblast (CV1) cells [102]. Trunkamide A is a cyclopeptide with a tiazoline ring and structurally analogs to mollamides [109]. Trunkamide A has already undergone preclinical trials with promising antitumor effects against cell lines derived from humans, including P-388, A-549, HT-29 and human melanoma (MEL-28) cells [110].

3.4. Anti-Cancer Peptides Derived from Mollusks

Mollusk is one of the most diverse groups of animals on the Earth. Apart from their important ecological role and commercial value for human food, their pharmacological roles are also of notable interest. Several anti-cancer peptides have been found in mollusks. Dolastatins, a group of cytotoxic peptides, have been isolated from marine mollusks *Dolabella auricularia*, with dolastatin 10 and dolastatin 15 the most prominent [102]. Dolastatin 10 is a pentapeptide containing several unique amino acid subunits. Cytotoxic activity of dolastatin 10 against mouse lymphocytic leukemia (L1210), human promyelocytic leukemia (HL-60), human acute myelomonocytic leukemia (ML-2), human monocytic (THP-1), multiple lymphoma, small cell lung cancer (NCI-H69, -H82, -H446, and -H510) and PC-3 cells have been reported [111,112]. It has been reported that anticancer activity of dolastatin involves microtubule assembly by interacting with tubulin and blocking tubulin-dependent GTP hydrolysis [113,114]. Dolastatin 10 also affects Bcl-2 level and an increase in p53 expression [115]. However, dolastatin 10 clinical trial result was unsatisfactory; hence, dolastatin 10 was withdrawn from further trials. Another cytotoxic peptide from marine mollusk is the Keenamide A isolated from *Pleurobranchus forskalii*. These hexapeptide exhibited significant activity against the P-388, A-549, MEL-20, and HT-29 tumor cell lines [58,102,115]. Liu et al. (2012) isolated a 15 kDa linier peptides (Mere15) derived from *Meretrix meretrix* [116,117]. Mere15 inhibited the growth of leukemia (K562) cells and the cytotoxicity was related to the apoptosis induction, cell cycle arrest and microtubule disassembly [116]. Further, in vivo analysis revealed that Mere15 inhibited the growth of A549 cells xenograft in nude mice by activating intrinsic pathway [117].

Kahalalides are cyclic depsipeptides that was originally isolated from the Hawaiian marine mollusks *Elysia rufescens*. Of the seven isolated Kahalalides (A–F), Kahalalide F showed significant cytotoxic activity against cell lines and tumor specimens derived from various human solid tumors, including prostate, breast, non-small-cell lung, ovarian, and colon carcinomas [102,118]. Gonzales et al. (2003) demonstrated that cancer cells treated with Kahalalide F underwent a series of profound alterations including severe cytoplasmic swelling and vacuolization, dilation and vesiculation of the endoplasmic reticulum, mitochondrial damage, and plasma membrane rupture, suggesting that Kahalalide F induces cell death via oncosis preferentially in tumor cells. Subsequently, it was reported that ErbB3 and the downstream PI3K-Akt pathway is an important determinants of the cytotoxic activity of Kahalalide F in vitro [118]. Kahalalide F was dropped from phase II clinical trials due to a lack of efficacy despite results indicating a limited number of patients achieved a positive response. Based on the pharmacokinetic studies, it was suggested that Kahalalide F has a short half-life, which may affect its efficacy [119].

Ziconotide (formerly SNX-111, Neurex Pharmaceuticals, Menlo Park, CA, USA) is the synthetic equivalent of ω-conopeptide MVIIA, a 25-amino-acid polybasic peptide originally isolated from the venom of *Conus magus*, a marine snail [120]. Ziconotide is an analgesic agent administered intrathecally and has been for almost one decade for the treatment of chronic cancer pain [121]. However, the use of ziconotide can induce several and sometimes serious adverse events. Hence, a low initial dosage followed by slow titration is recommended to reduce serious adverse events.

3.5. Anti-Cancer Peptides Derived from Cyanobacteria

Cyanobacteria (blue-green algae) are a very old and diverse group of photosynthetic, prokaryotic organisms that produce a variety of secondary metabolites with various biological activities, including phenols, peptides, alkaloids or terpenoids [122]. Cyclic depsipeptides, grassypeptolides D and E, have been isolated from the marine cyanobacterium *Leptolyngbya* sp. [123]. These peptides have shown cytotoxic effect against mouse neuroblastoma (N2A) and HeLa cell line, which was confirmed by MTT cell viability assay. *Lyngbya majuscula*, a benthic filamentous marine cyanobacterium, has been extensively studied and has produced more than 250 compounds with diverse structural features. This diversity is in part attributable to the fact that a major theme in *L. majuscula* biochemistry relies on the production of metabolites via polyketide synthases and nonribosomal peptide synthetases within specialized biosynthetic pathways. Malyngamide 4, somocystinamide A, and hectochlorins are potent anti-cancer lipopeptides isolated from *L. majuscula* [124–126]. Hectochlorins have been reported to be strong actin-disrupting agents. Hectochlorin showed great anti-proliferative activity against colon, melanoma, ovarian, and renal cancer cells [127]. Shaala et al. (2013) demonstrate that malyngamide A inhibited proliferations of A549, HT29, and breast adenocarcinoma (MDA-MB-231) cells cultured in vitro. Another lipopeptide isolated from *L. majuscula*, Somocystinamide A showed potent cytotoxicity against N2A cells. Further, Somocystinamide was found as potent apoptosis inductor in a number of tumor cell lines and angiogenic endothelial cells via intrinsic and extrinsic pathways, but the more effective mechanism is the activation of caspase 8 [126]. Apratoxin A is a cyclodepsipeptide isolated from a *L. majuscula*. This peptide showed anti-proliferative activity in KB and LoVo cancer cells. Apratoxin A mediates its anti-proliferative activity through the induction of G1 cell cycle arrest and an apoptotic cascade, which partially initiated through antagonism of FGF signaling via STAT3 [128].

The blue-green colored pigment-protein complex, c-phycocyanin, isolated from marine cyanobacteria *Agmenellum quadruplicatum*, *Mastigocladus laminosus*, *Oscillatoria tenuis* appeared to be a potent activator of pro-apoptotic gene and downregulator of anti-apoptotic gene expression [129]. Transduction of apoptosis signals resulting apoptosis of HeLa cells in vitro [130]. Further, apoptosis features such as cell shrinkage, membrane blebbing, nuclear condensation and DNA fragmentation were observed in A549 and HT29 treated with c-phycocyanin [131].

Cyanobacteria possess several advantages to be developed as nutraceuticals for the prevention and treatment of cancer and other NCD. The advantages of cyanobacteria include simple growth requirement, ease of genetic manipulation, and attractive platforms for carbon neutral production process [132]. However, it should be noted that some cyanobacteria produce cyanotoxins, therefore an appropriate regulatory framework should be developed for pharmaceutical and nutraceutical products from cyanobacteria to ensure that safety and quality standards are met.

4. Marine-Derived Antioxidant Peptides

In addition to the general risk factors in the development of NCD, free radicals are also known to play a significant role in NCD. Marine-derived protein, protein hydrolysates, peptides and amino acids have been shown to have significant antioxidant effects. Marine organisms are probably the most extensively studied as an important source of antioxidants. Antioxidant activity of marine organisms has been determined by various in vitro and in vivo methods, such as 2,2-diphenyl-1-picrylhydrazyl (DPPH), peroxide, hydroxyl and superoxide anion radical scavenging activities which have been detected by electronspin resonance spectroscopy method as well as intra cellular free radical scavenging assays, such as DNA oxidation, ROS scavenging, membrane protein oxidation and membrane lipid oxidation [133]. Many studies reported that proteins from marine organisms exhibit potent antioxidant activity; however, in many cases, peptide fractions or protein hydrolysates showed greater antioxidant activity. These suggest that peptides play a significant role in antioxidant actions of marine proteins. Therefore, many individual bioactive peptides responsible for antioxidant activity of marine protein or protein hydrolysates were then purified and identified. Marine-derived peptides have varied antioxidant activities depending on the structure. The peptide structure including the size and amino acid sequences were influenced by the protein sources and extraction conditions. As an example, clam peptides, isolated from body or viscera of clam (*Meretrix casta*) protein hydrolyse with three different enzymes such as trypsin, pepsin and papain resulted in different DPPH radical scavenging activities, ranging from 9.1% to 82.5% and reducing power ranging from 0.1 to 0.7, measured as the ability of the hydrolysate to reduce iron (III) [134]. Rajapakse et al. (2005) identified four different molecular weight peptides from giant squid mussel by employing ultrafiltration membrane with three different molecular weight cut off membranes (10, 5 and 3 kDa). Lower molecular weight peptide was found to possess stronger antioxidant activity compared to the higher molecular weight peptides. They assumed that lower molecular weight improves contact ability with membrane lipids and or permeability [135]. Further, it is believed that aromatic amino acid and histidine act positively as direct radical scavengers within peptide sequences. The presence of aromatic amino acids in the structure of a peptide is an advantage in this regard because they can donate protons easily to electron-deficient radicals and, at the same time, maintain their stability via resonance structures. Hence, it can be speculated that difference in scavenging activity could be due to the molecular weight or the specific arrangement of amino acid residues in the peptide sequence [13].

In addition to marine peptides, marine processing by-products have also been explored for production of proteins, peptides, and hydrolysates with antioxidant potentials [136]. Purification of antioxidant peptides derived from marine by-product using enzymatic hydrolysis has been in practice during recent years. Antioxidant peptides derived from marine processing by-product were found to possess strong antioxidant activity in linoleic acid model [137,138]. Himaya et al. (2012) demonstrated that peptide isolated from Japanese flounder skin gelatin could protect against cellular oxidative damage. Some peptides derived from marine processing by-product were found to possess strong activity to inhibit lipid peroxidation in linoleic acid models. This activity was attributed to the ability of peptide to interfere propagation cycle of lipid peroxidation and there by slowing radical mediated linoleic acid oxidation. Hydrophobic amino acids in peptide sequences may contribute to peroxidation inhibition by increasing the solubility of peptide in lipid and thereby facilitating better interaction with radical species [139]. Position of hydrophobic amino acid, Leu at the *N*-terminus of the peptide sequences has been shown to increase the interaction between peptides and fatty acids. More importantly, hydrophobic peptides can protect macromolecule oxidation by donating photons to reactive radicals [13,140]. Moreover, the activity of histidine containing peptides has also been reported to act against lipid peroxidation. In addition, Shahidi and Zhong (2008) reported that in the case of tripeptide, tripeptides containing 2 tyrosine units had higher capacity than those containing 2 histidine units in inhibiting linoleic acid oxidation. Later, it was reported that histidine-containing peptides can act as metal chelator, active oxygen quencher, and hydroxyl radical scavenger, thus contributing to the antioxidant activity of the protein hydrolysate and peptide.

Epidemiological studies show that a diet rich in antioxidants is associated with low prevalence of NCD, longevity and good health. Therefore, researchers are continually seeking for a good source of diet with potent antioxidant ability as an alternative for the dietary supplements and food. Bioactive peptides of marine origin have the potential to subside the biochemical imbalances induced by the formation of free radicals, and many of these peptides have been viewed as promising agents for the prevention and treatment of NCD. One of the commercially available products from marine organisms to reduce oxidative stress is Fortidium Liquamen, a hydolyzed skin of white fish (*Molva molva*) [141]. Based on those collective findings, it may be assumed that marine-derived bioactive peptides is a healthy choice to strengthen the body's fight against oxidative stress and other related NCD.

5. Anti-Diabetic and Hypocholesterolemic Effects of Marine-Derived Bioactive Peptides

Metabolic disorders comprise a collection of health disorders that increase the risk of morbidity and loss qualify of life, these includes diabetes and obesity. Marine-derived proteins and their peptides exert anti-diabetic effects. Zhu et al. (2010) have reported that treatment with oligopeptides from marine salmon skin modulated type 2 diabetes mellitus-related hyperglycemia and β-cell apoptosis in rats induced by high fat diet and low doses of streptozotocin. The anti-diabetic effect of salmon skin-derived oligopeptides was mediated by down-regulation of type 2 diabetes mellitus-related oxidative stresses and inflammation, which then protect the pancreatic β-cells from apoptosis [142].

A marine collagen peptide (MCP) isolated from wild marine fish caught from the East China Sea has shown anti-diabetic effects in patients with or without hypertension [143]. The levels of free fatty acid, hs-CRP, resistin and prostacyclin were decreased significantly following MCP treatment, indicating that MCP could offer protection against diabetes and hypertension by affecting levels of molecules involved in diabetic and hypertensive pathogenesis. Further, it was confirmed that MCP modulates glucose and lipid metabolism in patients with type 2 diabetes mellitus [144]. It was demonstrated that MCP is a peptide mixture containing two to six amino acid residues in length with molecular weight 100–800 Da. Unfortunately, the amino acid sequence of MCP is not elucidated yet. Peptide possess anti-metabolic disorder are generally low molecular weight (500–800 Da) [145]. Peptide sequence also plays an important role in anti-diabetic and anti-obesity effects. Generally, anti-diabetic and anti-obesity peptides are hydrophobic. Such a hydrophobic peptide is envisaged to be able to cross (biological) membranes. Vernaleken et al. (2007) described that specific functional tripeptide fragments (i.e., "Gln-Cys-Val" and "Gln-Cys-Pro") are potent inhibitors of monosaccharide-dependent exocytotic pathway of Na^+-D-Glucose co transporter SGLT1. The specific peptide sequence may influence negatively specific nutrient transporters/receptors in vivo which further lead to posttranscriptional down regulation of nutrient transporters and reduction of body weight [146]. It was also reported that high amounts of Gly amino acids in marine-derived proteins could contribute to an increase in fecal cholesterol and/or bile acid excretion, thus contributing to improvement in plasma lipid variables [147]. In addition, low molecular weight peptides derived from Salmon rich in Gly significantly alleviated obesity-linked inflammation. Many studies have shown that pro-inflammatory mediators including tumor necrosis factor-α (TNF-α), interleukin-1β (IL-1β) and interleukin-6 (IL-6) are increased during obesity and diabetes. The suppression of these pro-inflammatory mediators may decrease the risk of developing metabolic disorders-associated inflammation and insulin [148].

Hyperlipidemia, particularly hypercholesterolemia, is an obesity related condition common in diabetic patients, and also one of the most important risk factors contributing to the development of NCD. Natural extracts with cholesterol-lowering effect have been explored for their potential in prevention and treatment of hypercholesterolemia. In vivo study showed that protein derived from microalgae (*Spirulina platensis*) c-phycocyanin, plays a crucial role in the hypocholesterolemic activities [149]. In addition, Colla et al. (2008) demonstrated that *Spirulina platensis* when added in rabbit feed for 30–60 days reduced the levels of total cholesterol, high-density lipoprotein and triacylglycerols [150].

Current food environments are unhealthy which dominated by energy-dense, nutrient-poor processed food products which are widely available and relatively inexpensive [151]. These seem to create a supply-side "push" effect on unhealthy diets which is the prevailing driver of population unhealthy weight gain and NCD. To reduce hypercholesterolemic, diabetes, and other diet-related NCD, there needs to be a central focus on creating "healthy food environments" which shift population diets, especially those of socially disadvantaged populations, towards healthy diets. Marine-derived bioactive peptides have excellent potential as functional food ingredients to reduce NCD as they possess advantageous physiological effects, with medicinal characteristics and added health benefits such as anti-diabetic and hyocholesterolemic activities.

6. Future Perspectives of Marine-Derived Bioactive Peptides

Successful characterization of marine-derived bioactive peptides and investigations of their cardio protective, antihypertensive, anticancer, anti-diabetic, and anti-oxidative effects suggest their promising future for NCD. However, current marine peptides are still unable to meet the design parameters for drugs for NCD due to their low metabolic stability, low membrane permeability, and their high costs of manufacture [152]. Therefore, marine-derived bioactive peptides can be administered using different delivery vehicles such as functional food and or nutraceuticals. In order to be used as ingredients in food products, different studies should be carried out to determine if bioactivity of marine peptides is maintained after manufacturing and cooking processes. For example, wheat bread containing the hydrolysate from red algae retained renin inhibitory bioactivity after the baking process [19]. Furthermore, biological effect of marine-derived peptides is strongly influenced by their bioavailability, which is predominantly determined by their susceptibility to degradation into inactive fragments by digestive enzymes peptidase and intestinal absorption. Bioavailability should be taken into account when developing food and beverages products containing marine-derived bioactive peptides for the prevention and treatment of NCD.

Bioavailability of peptides can be defined as the quantity that passes through the cell membranes in the intestine and is available for action within the cells [153]. Bioavailability of peptides are generally affected by physicochemical properties of the peptides such as molecular size, charge, sequence, and solubility; smaller peptides are transported across the enterocytes through intestinal-expressed peptide transporters, whereas oligopeptides may be absorbed by passive transport through hydrophobic regions of membrane epithelia or tight junctions [154]. Many studies demonstrated that marine-derived peptides are mostly peptides of small molecular weights, especially tripeptides from marine algae and small oligopeptides. These small molecular weight peptides are too small for the substrates of digestive proteases, and therefore they have high resistance to gastrointestinal digestion and are easily to be absorbed. In addition, small molecular weight peptides are convenient and cheaper to be synthesized through chemical method. Thus, chemical synthesis can be used to produce large quantities of marine-derived bioactive peptides to be used in functional foods and pharmaceuticals to meet the needs for NCD remedy.

Several studies have demonstrated the bioavailability of marine-derived bioactive peptides for the treatment of NCD using both animal models and human volunteers. For example, long-term oral administration of peptides derived from jellyfish reduced systolic blood pressure and diastolic blood pressure of the renovascular hypertension rats [155]. Interestingly, these bioactive peptides affected the production of Angiotensin II only in kidney but not in plasma. In addition, Lee et al. (2010) demonstrated that oral administration of peptide-derived from tuna frame significantly reduced systolic blood pressure and diastolic blood pressure in spontaneously hypertensive rats. That information provides basic information that peptide-derived from tuna frame show stability against gastrointestinal proteases and original peptide sequences that displayed anti-hypertensive activity are delivered to the cellular sites of action. These provide evidence that marine-derived bioactive peptides can be used for the preparation of oral treatment for blood pressure homeostasis which further protects cardiovascular system.

Up to now, many marine peptides are unable to meet the requirements for food (e.g., taste, bioavailability, or stability). Bitterness of some marine-derived peptides is an undesirable property, which should be reduced during food, beverages and or pharmaceuticals production. Marine-derived bioactive peptides hosting residues with hydrophobic side chains have a distinct bitter taste. Therefore, further studies on controlling these properties are needed. These can be achieved by several methods including chemical or physical modifications of the peptides (i.e., microencapsulation, and quantitating the bitter taste relationship). Microencapsulation not only increases consumer's acceptance, but also ensures that the marine-derived peptide sequences that displayed bioactivity are conserved and delivered to the cellular sites of action in NCD. Further, microencapsulation will enhance their stability and absorption.

In order to develop food and beverages product containing marine-derived bioactive peptides, methods must be developed to enhance their availability and bioactivity. Bioactive peptides can be obtained from marine organisms by organic solvent extraction, fermentation and enzymatic hydrolysis by proteolytic enzymes. In food industries, the last methods are more preferred due to the lack of residual organic solvents or toxic chemicals in the products and or microbial residue. Notably, physico-chemical conditions of the reaction media, such as temperature and pH of the protein solution, must then be adjusted in order to optimize the activity of the enzyme used. Further, to obtain desired molecular weight and functional properties of marine-derived bioactive peptides, a suitable method is the use of an ultrafiltration membrane system. This system has the main advantage that the molecular weight distribution of the desired peptide can be controlled by adoption of an appropriate ultrafiltration membrane.

The number of marine organism's consumption is estimated to rise each year due to the increasing consumer knowledge about their health benefit effects. Marine organisms are viewed as "natural and healthy" by consumers, and this promotes a positive response in consumers, who often regard natural entities. Therefore, marine organisms may be considered a consumer friendly source of functional foods which may use to prevent and treat NCD. Last but not least, scientists should work out sustainable ways to refine bioactive peptides derived from marine organisms, and develop food and pharmaceuticals products to alleviate NCD.

7. Conclusions

Many studies have shown that marine-derived bioactive peptides possess remarkable activities relevant to the prevention and treatment of NCD. The possibilities of designing new functional foods, nutraceuticals, and pharmaceuticals derived from marine bioactive peptides for the prevention and treatment of NCD are promising. While much information is available on biological activities of marine-derived bioactive peptides, future studies should be directed towards evaluation of bioavailability in human subjects as well as clinical trials. In addition, safety and quality standards of marine-derived peptides-based products should be evaluated prior to commercialization.

Acknowledgments: This study is supported by Thematic Research Program Research Center for Oceanography, Indonesian Institute of Sciences (P2O-LIPI).

Author Contributions: R.P. prepared the review article under the supervision of S.-K.K.

Conflicts of Interest: The authors declare no conflict of interest.

Abbreviations

The following abbreviations are used in this manuscript:

ACE	Angiotensin converting enzymes
CVD	Cardiovascular disease
DPPH	2,2-diphenyl-1-picrylhydrazyl
FAO	Food and Agriculture Organization
FOSHU	foods for specified health uses
IC50	half maximal inhibitory concentration
IL-6	interleukin-6
IL-1β	interleukin-1β
MCP	marine collagen peptide
MW	Molecular weight
NCD	Non communicable diseases
PUFA	polyunsaturated fatty acids
RAS	renin-angiotensin system
ROS	reactive oxygen species
SHR	spontaneously hypertensive rat
SBP	systolic blood pressure
TNF-α	tumor necrosis factor-α
WHO	World Health Organization

References

1. World Health Organization. *Global Status Report on Noncommuniacble Diseases*; World Health Organization: Genève, Switzerland, 2014; p. 280.
2. World Health Organization. *Global Status Report on Noncommunicable Diseases*; World Health Organization: Genève, Switzerland, 2010; p. 162.
3. Bhandari, G.P.; Angdembe, M.R.; Dhimal, M.; Neupane, S.; Bhusal, C. State of non-communicable diseases in Nepal. *BMC Public Health* **2014**, *14*, 23. [CrossRef] [PubMed]

4. Collins, K.G.; Fitzgerald, G.F.; Stanton, C.; Ross, R.P. Looking Beyond the Terrestrial: The Potential of Seaweed Derived Bioactives to Treat Non-Communicable Diseases. *Mar. Drugs* **2016**, *14*, 60. [CrossRef] [PubMed]
5. Wagner, K.-H.; Brath, H. A global view on the development of non communicable diseases. *Prev. Med.* **2012**, *54*, S38–S41. [CrossRef] [PubMed]
6. Erdmann, K.; Cheung, B.W.Y.; Schröder, H. The possible roles of food-derived bioactive peptides in reducing the risk of cardiovascular disease. *J. Nutr. Biochem.* **2008**, *19*, 643–654. [CrossRef] [PubMed]
7. Kris-Etherton, P.M.; Hecker, K.D.; Bonanome, A.; Coval, S.M.; Binkoski, A.E.; Hilpert, K.F.; Griel, A.E.; Etherton, T.D. Bioactive compounds in foods: Their role in the prevention of cardiovascular disease and cancer. *Am. J. Med.* **2002**, *113* (Suppl. 9), 71S–88S. [CrossRef]
8. Wijesekara, I.; Pangestuti, R.; Kim, S.K. Biological activities and potential health benefits of sulfated polysaccharides derived from marine algae. *Carbohydr. Polym.* **2010**, *84*, 14–21. [CrossRef]
9. Pangestuti, R.; Kim, S.-K. Biological activities and health benefit effects of natural pigments derived from marine algae. *J. Funct. Foods* **2011**, *3*, 255–266. [CrossRef]
10. Yusuf, S.; Wood, D.; Ralston, J.; Reddy, K.S. The World Heart Federation's vision for worldwide cardiovascular disease prevention. *Lancet* **2015**, *386*, 399–402. [CrossRef]
11. Nichols, M.; Townsend, N.; Scarborough, P.; Rayner, M. Cardiovascular disease in Europe 2014: Epidemiological update. *Eur. Heart J.* **2014**, *35*, 2950–2959. [CrossRef] [PubMed]
12. Kearney, P.M.; Whelton, M.; Reynolds, K.; Muntner, P.; Whelton, P.K.; He, J. Global burden of hypertension: Analysis of worldwide data. *Lancet* **2005**, *365*, 217–223. [CrossRef]
13. Himaya, S.; Ngo, D.-H.; Ryu, B.; Kim, S.-K. An active peptide purified from gastrointestinal enzyme hydrolysate of Pacific cod skin gelatin attenuates angiotensin-1 converting enzyme (ACE) activity and cellular oxidative stress. *Food Chem.* **2012**, *132*, 1872–1882. [CrossRef]
14. Kher, V. Renin inhibition—Benefit beyond hypertension control. *J. Assoc. Phys. India* **2009**, *57*, 518–521.
15. Allikmets, K. Aliskiren—An orally active renin inhibitor. Review of pharmacology, pharmacodynamics, kinetics, and clinical potential in the treatment of hypertension. *Vasc. Health Risk Manag.* **2007**, *3*, 809. [PubMed]
16. Fitzgerald, C.N.; Mora-Soler, L.; Gallagher, E.; O'Connor, P.; Prieto, J.; Soler-Vila, A.; Hayes, M. Isolation and characterization of bioactive pro-peptides with in vitro renin inhibitory activities from the macroalga *Palmaria palmata*. *J. Agric. Food Chem.* **2012**, *60*, 7421–7427. [CrossRef] [PubMed]
17. Fitzgerald, C.; Aluko, R.E.; Hossain, M.; Rai, D.K.; Hayes, M. Potential of a renin inhibitory peptide from the red seaweed *Palmaria palmata* as a functional food ingredient following confirmation and characterization of a hypotensive effect in spontaneously hypertensive rats. *J. Agric. Food Chem.* **2014**, *62*, 8352–8356. [CrossRef] [PubMed]
18. Suetsuna, K.; Maekawa, K.; Chen, J.-R. Antihypertensive effects of *Undaria pinnatifida* (wakame) peptide on blood pressure in spontaneously hypertensive rats. *J. Nutr. Biochem.* **2004**, *15*, 267–272. [CrossRef] [PubMed]
19. Fitzgerald, C.; Gallagher, E.; Doran, L.; Auty, M.; Prieto, J.; Hayes, M. Increasing the health benefits of bread: Assessment of the physical and sensory qualities of bread formulated using a renin inhibitory *Palmaria palmata* protein hydrolysate. *LWT Food Sci. Technol.* **2014**, *56*, 398–405. [CrossRef]
20. Suetsuna, K.; Osajika, K. The inhibitroy activity of angiotensin-1 converting enzyme of basic peptides from sardine and hair tail meat. *Bull. Jpn. Soc. Sci. Fish.* **1986**, *52*, 1981–1984. [CrossRef]
21. Kim, S.; Wijesekara, I. Development and biological activities of marine-derived bioactive peptides: A review. *J. Funct. Foods* **2010**, *2*, 1–9. [CrossRef]
22. Wijesekara, I.; Kim, S.K. Angiotensin-I-converting enzyme (ACE) inhibitors from marine resources: Prospects in the pharmaceutical industry. *Mar. Drugs* **2010**, *8*, 1080–1093. [CrossRef] [PubMed]
23. Nakamura, Y.; Yamamoto, N.; Sakai, K.; Okubo, A.; Yamazaki, S.; Takano, T. Purification and Characterization of Angiotensin I-Converting Enzyme Inhibitors from Sour Milk. *J. Dairy Sci.* **1995**, *78*, 777–783. [CrossRef]
24. Ariyoshi, Y. Angiotensin-converting enzyme inhibitors derived from food proteins. *Trends Food Sci. Technol.* **1993**, *4*, 139–144. [CrossRef]
25. Suetsuna, K.; Nakano, T. Identification of an antihypertensive peptide from peptic digest of wakame (*Undaria pinnatifida*). *J. Nutr. Biochem.* **2000**, *11*, 450–454. [CrossRef]
26. Nomura, A.; Noda, N.; Maruyama, S. Purification of angiotensin I-converting enzyme inhibitors in pelagic thresher *Alopias pelagicus* muscle hydrolysate and viscera extracts. *Fish. Sci.* **2002**, *68*, 954–956. [CrossRef]

27. Ikeda, A.; Ichino, H.; Kiguchiya, S.; Chigwechokha, P.; Komatsu, M.; Shiozaki, K. Evaluation and Identification of Potent Angiotensin-I Converting Enzyme Inhibitory Peptide Derived from Dwarf Gulper Shark (*Centrophorus atromarginatus*). *J. Food Process. Preserv.* **2015**, *39*, 107–115. [CrossRef]

28. Wu, H.; He, H.-L.; Chen, X.-L.; Sun, C.-Y.; Zhang, Y.-Z.; Zhou, B.-C. Purification and identification of novel angiotensin-I-converting enzyme inhibitory peptides from shark meat hydrolysate. *Proc. Biochem.* **2008**, *43*, 457–461. [CrossRef]

29. Jian, P.W.; Aluko, R.E.; Nakai, S. Structural Requirements of Angiotensin I-Converting Enzyme Inhibitory Peptides: Quantitative Structure-Activity Relationship Study of Di- and Tripeptides. *J. Agric. Food Chem.* **2006**, *54*, 732–738.

30. Sato, M.; Hosokawa, T.; Yamaguchi, T.; Nakano, T.; Muramoto, K.; Kahara, T.; Funayama, K.; Kobayashi, A.; Nakano, T. Angiotensin I-converting enzyme inhibitory peptides derived from wakame (*Undaria pinnatifida*) and their antihypertensive effect in spontaneously hypertensive rats. *J. Agric. Food Chem.* **2002**, *50*, 6245–6252. [CrossRef] [PubMed]

31. Suetsuna, K. Purification and identification of angiotensin I-converting enzyme inhibitors from the red alga *Porphyra yezoensis*. *J. Mar. Biotechnol.* **1998**, *6*, 163–167. [PubMed]

32. Suetsuna, K.; Chen, J.-R. Identification of Antihypertensive Peptides from Peptic Digest of Two Microalgae, *Chlorella vulgaris* and *Spirulina platensis*. *Mar. Biotechnol.* **2001**, *3*, 305–309. [CrossRef] [PubMed]

33. Kohama, Y.; Matsumoto, S.; Oka, H.; Teramoto, T.; Okabe, M.; Mimura, T. Isolation of angiotensin-converting enzyme inhibitor from tuna muscle. *Biochem. Biophys. Res. Commun.* **1988**, *155*, 332–337. [CrossRef]

34. Matsumura, N.; Fujii, M.; Takeda, Y.; Sugita, K.; Shimizu, T. Angiotensin I-converting enzyme inhibitory peptides derived from bonito bowels autolysate. *Biosci. Biotechnol. Biochem.* **1993**, *57*, 695–697. [CrossRef] [PubMed]

35. Kim, S.-K.; Kim, Y.-T.; Byun, H.-G.; Nam, K.-S.; Joo, D.-S.; Shahidi, F. Isolation and characterization of antioxidative peptides from gelatin hydrolysate of Alaska pollack skin. *J. Agric. Food Chem.* **2001**, *49*, 1984–1989. [CrossRef] [PubMed]

36. Ono, S.; Hosokawa, M.; Miyashita, K.; Takahashi, K. Isolation of Peptides with Angiotensin I-converting Enzyme Inhibitory Effect Derived from Hydrolysate of Upstream Chum Salmon Muscle. *J. Food Sci.* **2003**, *68*, 1611–1614. [CrossRef]

37. Enari, H.; Takahashi, Y.; Kawarasaki, M.; Tada, M.; Tatsuta, K. Identification of angiotensin I-converting enzyme inhibitory peptides derived from salmon muscle and their antihypertensive effect. *Fish. Sci.* **2008**, *74*, 911–920. [CrossRef]

38. Ngo, D.-H.; Kang, K.-H.; Ryu, B.; Vo, T.-S.; Jung, W.-K.; Byun, H.-G.; Kim, S.-K. Angiotensin-I converting enzyme inhibitory peptides from antihypertensive skate (*Okamejei kenojei*) skin gelatin hydrolysate in spontaneously hypertensive rats. *Food Chem.* **2015**, *174*, 37–43. [CrossRef] [PubMed]

39. García-Moreno, P.J.; Espejo-Carpio, F.J.; Guadix, A.; Guadix, E.M. Production and identification of angiotensin I-converting enzyme (ACE) inhibitory peptides from Mediterranean fish discards. *J. Funct. Foods* **2015**, *18*, 95–105. [CrossRef]

40. He, H.L.; Chen, X.L.; Sun, C.Y.; Zhang, Y.Z.; Zhou, B.C. Analysis of novel angiotensin 1 converting enzyme inhibitory peptides from protease-hydrolyzed marine shrimp *Acetes chinensis*. *J. Pept. Sci.* **2006**, *12*, 726–733.

41. Wang, Y.-K.; He, H.-L.; Chen, X.-L.; Sun, C.-Y.; Zhang, Y.-Z.; Zhou, B.-C. Production of novel angiotensin I-converting enzyme inhibitory peptides by fermentation of marine shrimp *Acetes chinensis* with *Lactobacillus fermentum* SM 605. *Appl. Microbiol. Biotechnol.* **2008**, *79*, 785–791. [CrossRef] [PubMed]

42. Cao, W.; Zhang, C.; Hong, P.; Ji, H.; Hao, J. Purification and identification of an ACE inhibitory peptide from the peptic hydrolysate of Acetes chinensis and its antihypertensive effects in spontaneously hypertensive rats. *Int. J. Food Sci. Technol.* **2010**, *45*, 959–965. [CrossRef]

43. Nii, Y.; Fukuta, K.; Yoshimoto, R.; Sakai, K.; Ogawa, T. Determination of antihypertensive peptides from an izumi shrimp hydrolysate. *Biosci. Biotechnol. Biochem.* **2008**, *72*, 861–864. [CrossRef] [PubMed]

44. Liu, X.; Zhang, M.; Jia, A.; Zhang, Y.; Zhu, H.; Zhang, C.; Sun, Z.; Liu, C. Purification and characterization of angiotensin I converting enzyme inhibitory peptides from jellyfish *Rhopilema esculentum*. *Food Res. Int.* **2013**, *50*, 339–343. [CrossRef]

45. Du, L.; Fang, M.; Wu, H.; Xie, J.; Wu, Y.; Li, P.; Zhang, D.; Huang, Z.; Xia, Y.; Zhou, L. A novel angiotensin I-converting enzyme inhibitory peptide from *Phascolosoma esculenta* water-soluble protein hydrolysate. *J. Funct. Foods* **2013**, *5*, 475–483. [CrossRef]

46. Suetsuna, K. Identification of antihypertensive peptides from peptic digest of the short-necked clam *Tapes philippinarum* and the pearl oyster *Pinctada fucata martensii*. *Fish. Sci.* **2002**, *68*, 233–235. [CrossRef]

47. Fujita, H.; Yamagami, T.; Ohshima, K. Effects of an ACE-inhibitory agent, katsuobushi oligopeptide, in the spontaneously hypertensive rat and in borderline and mildly hypertensive subjects. *Nutr. Res.* **2001**, *21*, 1149–1158. [CrossRef]

48. Fujita, H.; Yoshikawa, M. LKPNM: A prodrug-type ACE-inhibitory peptide derived from fish protein. *Immunopharmacology* **1999**, *44*, 123–127. [CrossRef]

49. Kouno, K.; Hirano, S.-I.; Kuboki, H.; Kasai, M.; Hatae, K. Effects of Dried Bonito (*Katsuobushi*) and Captopril, an Angiotensin I-Converting Enzyme Inhibitor, on Rat Isolated Aorta: A Possible Mechanism of Antihypertensive Action. *Biosci. Biotechnol. Biochem.* **2005**, *69*, 911–915. [CrossRef] [PubMed]

50. Lee, J.K.; Jeon, J.-K.; Byun, H.-G. Antihypertensive effect of novel angiotensin I converting enzyme inhibitory peptide from chum salmon (*Oncorhynchus keta*) skin in spontaneously hypertensive rats. *J. Funct. Foods* **2014**, *7*, 381–389. [CrossRef]

51. Lee, S.-H.; Qian, Z.-J.; Kim, S.-K. A novel angiotensin I converting enzyme inhibitory peptide from tuna frame protein hydrolysate and its antihypertensive effect in spontaneously hypertensive rats. *Food Chem.* **2010**, *118*, 96–102. [CrossRef]

52. Ko, S.-C.; Kang, N.; Kim, E.-A.; Kang, M.C.; Lee, S.-H.; Kang, S.-M.; Lee, J.-B.; Jeon, B.-T.; Kim, S.-K.; Park, S.-J. A novel angiotensin I-converting enzyme (ACE) inhibitory peptide from a marine *Chlorella ellipsoidea* and its antihypertensive effect in spontaneously hypertensive rats. *Proc. Biochem.* **2012**, *47*, 2005–2011. [CrossRef]

53. Roy, M.; Mukherjee, A.; Sarkar, R.; Mukherjee, S.; Biswas, J. In search of natural remediation for cervical cancer. *Anti-Cancer Agents Med. Chem.* **2015**, *15*, 57–65. [CrossRef]

54. He, G.; Karin, M. NF-κB and STAT3—Key players in liver inflammation and cancer. *Cell Res.* **2011**, *21*, 159–168. [CrossRef] [PubMed]

55. Hall, E.; Cameron, D.; Waters, R.; Barrett-Lee, P.; Ellis, P.; Russell, S.; Bliss, J.; Hopwood, P.; Investigators, T.T. Comparison of patient reported quality of life and impact of treatment side effects experienced with a taxane-containing regimen and standard anthracycline based chemotherapy for early breast cancer: 6 year results from the UK TACT trial (CRUK/01/001). *Eur. J. Cancer* **2014**, *50*, 2375–2389. [CrossRef] [PubMed]

56. Kim, S.-K.; Kalimuthu, S. Introduction to Anticancer Drugs from Marine Origin. In *Handbook of Anticancer Drugs from Marine Origin*; Springer: Berlin, Germany, 2015; pp. 1–13.

57. Zheng, L.-H.; Wang, Y.-J.; Sheng, J.; Wang, F.; Zheng, Y.; Lin, X.-K.; Sun, M. Antitumor peptides from marine organisms. *Mar. Drugs* **2011**, *9*, 1840–1859. [CrossRef] [PubMed]

58. Cheung, R.C.F.; Ng, T.B.; Wong, J.H. Marine peptides: Bioactivities and applications. *Mar. Drugs* **2015**, *13*, 4006–4043. [CrossRef] [PubMed]

59. Crews, P.; Manes, L.V.; Boehler, M. Jasplakinolide, a cyclodepsipeptide from the marine sponge, *Jaspis* sp. *Tetrahedron Lett.* **1986**, *27*, 2797–2800. [CrossRef]

60. Takeuchi, H.; Ara, G.; Sausville, A.E.; Teicher, B. Jasplakinolide: Interaction with radiation and hyperthermia in human prostate carcinoma and Lewis lung carcinoma. *Cancer Chemother. Pharmacol.* **1998**, *42*, 491–496. [CrossRef] [PubMed]

61. Robinson, S.J.; Morinaka, B.I.; Amagata, T.; Tenney, K.; Bray, W.M.; Gassner, N.C.; Lokey, R.S.; Crews, P. New Structures and Bioactivity Properties of Jasplakinolide (Jaspamide) Analogues from Marine Sponges. *J. Med. Chem.* **2010**, *53*, 1651–1661. [CrossRef] [PubMed]

62. Rangel, M.; Prado, M.P.; Konno, K.; Naoki, H.; Freitas, J.C.; Machado-Santelli, G.M. Cytoskeleton alterations induced by *Geodia corticostylifera* depsipeptides in breast cancer cells. *Peptides* **2006**, *27*, 2047–2057. [CrossRef] [PubMed]

63. Freitas, V.M.; Rangel, M.; Bisson, L.F.; Jaeger, R.G.; Machado-Santelli, G.M. The geodiamolide H, derived from Brazilian sponge *Geodia corticostylifera*, regulates actin cytoskeleton, migration and invasion of breast cancer cells cultured in three-dimensional environment. *J. Cell. Physiol.* **2008**, *216*, 583–594. [CrossRef] [PubMed]

64. Chatterjee, J.; Rechenmacher, F.; Kessler, H. *N*-Methylation of Peptides and Proteins: An Important Element for Modulating Biological Functions. *Angew. Chem. Int. Ed.* **2013**, *52*, 254–269. [CrossRef] [PubMed]

65. Tran, T.D.; Pham, N.B.; Fechner, G.A.; Hooper, J.N.; Quinn, R.J. Potent cytotoxic peptides from the Australian marine sponge *Pipestela candelabra*. *Mar. Drugs* **2014**, *12*, 3399–3415. [CrossRef] [PubMed]

66. Kobayashi, M.; Aoki, S.; Ohyabu, N.; Kurosu, M.; Wang, W.; Kitagawa, I. Arenastatin A, a potent cytotoxic depsipeptide from the okinawan marine sponge *Dysidea arenaria*. *Tetrahedron Lett.* **1994**, *35*, 7969–7972. [CrossRef]

67. Silva, M.A.D.; Bierhalz, A.C.K.; Kieckbusch, T.G. Alginate and pectin composite films crosslinked with Ca^{2+} ions: Effect of the plasticizer concentration. *Carbohydr. Polym.* **2009**, *77*, 736–742. [CrossRef]

68. Murakami, N.; Tamura, S.; Koyama, K.; Sugimoto, M.; Maekawa, R.; Kobayashi, M. New analogue of arenastatin A, a potent cytotoxic spongean depsipeptide, with anti-tumor activity. *Bioorg. Med. Chem. Lett.* **2004**, *14*, 2597–2601. [CrossRef] [PubMed]

69. Pettit, G.R.; Tan, R. Isolation and Structure of Phakellistatin 14 from the Western Pacific Marine Sponge *Phakellia* sp. 1. *J. Nat. Prod.* **2005**, *68*, 60–63. [CrossRef] [PubMed]

70. Li, W.-L.; Yi, Y.-H.; Wu, H.-M.; Xu, Q.-Z.; Tang, H.-F.; Zhou, D.-Z.; Lin, H.-W.; Wang, Z.-H. Isolation and Structure of the Cytotoxic Cycloheptapeptide Phakellistatin 13. *J. Nat. Prod.* **2003**, *66*, 146–148. [CrossRef] [PubMed]

71. Pettit, G.R.; Tan, R.; Ichihara, Y.; Williams, M.D.; Doubek, D.L.; Tackett, L.P.; Schmidt, J.M.; Cerny, R.L.; Boyd, M.R.; Hooper, J.N. Antineoplastic agents, 325. Isolation and structure of the human cancer cell growth inhibitory cyclic octapeptides phakellistatin 10 and 11 from *Phakellia* sp. *J. Nat. Prod.* **1995**, *58*, 961–965. [CrossRef] [PubMed]

72. Pettit, G.R.; Xu, J.-P.; Dorsaz, A.-C.; Williams, M.D.; Boyd, M.R.; Cerny, R.L. Isolation and structure of the human cancer cell growth inhibitory cyclic decapeptides phakellistatins 7, 8 and 9 1, 2. *Bioorg. Med. Chem. Lett.* **1995**, *5*, 1339–1344. [CrossRef]

73. Pelay-Gimeno, M.; Meli, A.; Tulla-Puche, J.; Albericio, F. Rescuing biological activity from synthetic phakellistatin 19. *J. Med. Chem.* **2013**, *56*, 9780–9788. [CrossRef] [PubMed]

74. Zhan, K.-X.; Jiao, W.-H.; Yang, F.; Li, J.; Wang, S.-P.; Li, Y.-S.; Han, B.-N.; Lin, H.-W. Reniochalistatins A–E, cyclic peptides from the marine sponge Reniochalina stalagmitis. *J. Nat. Prod.* **2014**, *77*, 2678–2684. [CrossRef] [PubMed]

75. Edrada-Ebel, R.; Jaspars, M. The 9th European Conference on Marine Natural Products. *Mar. Drugs* **2015**, *13*, 7150–7249. [CrossRef] [PubMed]

76. Ernst, E.; Consortium, C.C. *Shark Cartilage: Concerted Action for Complementary and Alternative Medicine Assessment in the Cancer Field (CAM-Cancer)*; NAFKAM: Tromsø, Norway, 2013; pp. 1–53.

77. Ulbricht, C.; Hammerness, P.; Barrette, E.-P.; Boon, H.; Szapary, P.; Sollars, D.; Smith, M.; Tsouronis, C.; Bent, S. Shark cartilage monograph: A clinical decision support tool. *J. Herb. Pharmacother.* **2002**, *2*, 71–93. [CrossRef]

78. Dupont, E.; Falardeau, P.; Mousa, S.A.; Dimitriadou, V.; Pepin, M.-C.; Wang, T.; Alaoui-Jamali, M.A. Antiangiogenic and antimetastatic properties of Neovastat (AE-941), an orally active extract derived from cartilage tissue. *Clin. Exp. Metastasis* **2002**, *19*, 145–153. [CrossRef] [PubMed]

79. Gingras, D.; Boivin, D.; Deckers, C.; Gendron, S.; Barthomeuf, C.; Béliveau, R. Neovastat—A novel antiangiogenic drug for cancer therapy. *Anti-Cancer Drugs* **2003**, *14*, 91–96. [CrossRef] [PubMed]

80. Dredge, K. AE-941 (AEterna). *Curr. Opin. Investig. Drugs* **2004**, *5*, 668–677. [PubMed]

81. Zheng, L.; Ling, P.; Wang, Z.; Niu, R.; Hu, C.; Zhang, T.; Lin, X. A novel polypeptide from shark cartilage with potent anti-angiogenic activity. *Cancer Biol. Ther.* **2007**, *6*, 775–780. [CrossRef] [PubMed]

82. Chen, J.-Y.; Lin, W.-J.; Wu, J.-L.; Her, G.M.; Hui, C.-F. Epinecidin-1 peptide induces apoptosis which enhances antitumor effects in human leukemia U937 cells. *Peptides* **2009**, *30*, 2365–2373. [CrossRef] [PubMed]

83. Lin, W.-J.; Chien, Y.-L.; Pan, C.-Y.; Lin, T.-L.; Chen, J.-Y.; Chiu, S.-J.; Hui, C.-F. Epinecidin-1, an antimicrobial peptide from fish (*Epinephelus coioides*) which has an antitumor effect like lytic peptides in human fibrosarcoma cells. *Peptides* **2009**, *30*, 283–290. [CrossRef] [PubMed]

84. Hsu, K.-C.; Li-Chan, E.C.; Jao, C.-L. Antiproliferative activity of peptides prepared from enzymatic hydrolysates of tuna dark muscle on human breast cancer cell line MCF-7. *Food Chem.* **2011**, *126*, 617–622. [CrossRef]

85. Song, R.; Wei, R.-B.; Luo, H.-Y.; Yang, Z.-S. Isolation and identification of an antiproliferative peptide derived from heated products of peptic hydrolysates of half-fin anchovy (*Setipinna taty*). *J. Funct. Foods* **2014**, *10*, 104–111. [CrossRef]

86. Wang, M.; Nie, Y.; Peng, Y.; He, F.; Yang, J.; Wu, C.; Li, X. Purification, characterization and antitumor activities of a new protein from *Syngnathus acus*, an officinal marine fish. *Mar. Drugs* **2011**, *10*, 35–50. [CrossRef] [PubMed]

87. Freitas-Júnior, A.C.; Costa, H.M.; Icimoto, M.Y.; Hirata, I.Y.; Marcondes, M.; Carvalho, L.B.; Oliveira, V.; Bezerra, R.S. Giant Amazonian fish pirarucu (*Arapaima gigas*): Its viscera as a source of thermostable trypsin. *Food Chem.* **2012**, *133*, 1596–1602. [CrossRef]

88. Shai, Y.; Fox, J.; Caratsch, C.; Shih, Y.-L.; Edwards, C.; Lazarovici, P. Sequencing and synthesis of pardaxin, a polypeptide from the Red Sea Moses sole with ionophore activity. *FEBS Lett.* **1988**, *242*, 161–166. [CrossRef]

89. Huang, T.-C.; Lee, J.-F.; Chen, J.-Y. Pardaxin, an antimicrobial peptide, triggers caspase-dependent and ROS-mediated apoptosis in HT-1080 cells. *Mar. Drugs* **2011**, *9*, 1995–2009. [CrossRef] [PubMed]

90. Boivin, D.; Gendron, S.; Beaulieu, É.; Gingras, D.; Béliveau, R. The Antiangiogenic Agent Neovastat (Æ-941) Induces Endothelial Cell Apoptosis 1 Supported by Æterna Laboratories, Québec City, Québec, Canada. 1. *Mol. Cancer Ther.* **2002**, *1*, 795–802. [PubMed]

91. Oren, Z.; Shai, Y. A Class of Highly Potent Antibacterial Peptides Derived from Pardaxin, A Pore-Forming Peptide Isolated from Moses Sole Fish Pardachirus marmoratus. *Eur. J. Biochem.* **1996**, *237*, 303–310. [CrossRef] [PubMed]

92. Hsu, J.-C.; Lin, L.-C.; Tzen, J.T.; Chen, J.-Y. Pardaxin-induced apoptosis enhances antitumor activity in HeLa cells. *Peptides* **2011**, *32*, 1110–1116. [CrossRef] [PubMed]

93. Pan, C.-Y.; Lin, C.-N.; Chiou, M.-T.; Yu, C.Y.; Chen, J.-Y.; Chien, C.-H. The antimicrobial peptide pardaxin exerts potent anti-tumor activity against canine perianal gland adenoma. *Oncotarget* **2015**, *6*, 2290–2301. [CrossRef] [PubMed]

94. Han, Y.; Cui, Z.; Li, Y.-H.; Hsu, W.-H.; Lee, B.-H. In vitro and in vivo anticancer activity of pardaxin against proliferation and growth of oral squamous cell carcinoma. *Mar. Drugs* **2015**, *14*, 2. [CrossRef] [PubMed]

95. Mukherjee, A.K.; Saviola, A.J.; Burns, P.D.; Mackessy, S.P. Apoptosis induction in human breast cancer (MCF-7) cells by a novel venom L-amino acid oxidase (Rusvinoxidase) is independent of its enzymatic activity and is accompanied by caspase-7 activation and reactive oxygen species production. *Apoptosis* **2015**, *20*, 1358–1372. [CrossRef] [PubMed]

96. Food and Agriculture Organization (FAO). *The State of World Fisheries and Aquaculture 2016*; FAO: Roma, Italy, 2016; p. 200.

97. Boutayeb, A.; Boutayeb, S. The burden of non communicable diseases in developing countries. *Int. J. Equity Health* **2005**, *4*, 2. [CrossRef] [PubMed]

98. Tahergorabi, R.; Jaczynski, J. Isoelectric solubilization/precipitation as a means to recover protein and lipids from seafood by-products. In *Seafood Processing By-Products*; Springer: New York, NY, USA, 2014; pp. 101–123.

99. Schwartsmann, G.; da Rocha, A.B.; Berlinck, R.G.; Jimeno, J. Marine organisms as a source of new anticancer agents. *Lancet Oncol.* **2001**, *2*, 221–225. [CrossRef]

100. Cain, J.M.; Liu, P.; Alberts, D.E.; Gallion, H.H.; Laufman, L.; O'Sullivan, J.; Weiss, G.; Bickers, J.N. Phase II trial of didemnin-B in advanced epithelial ovarian cancer. *Investig. New Drugs* **1992**, *10*, 23–24. [CrossRef]

101. Urdiales, J.; Morata, P.; De Castro, I.N.; Sánchez-Jiménez, F. Antiproliferative effect of dehydrodidemnin B (DDB), a depsipeptide isolated from Mediterranean tunicates. *Cancer Lett.* **1996**, *102*, 31–37. [CrossRef]

102. Suarez-Jimenez, G.-M.; Burgos-Hernandez, A.; Ezquerra-Brauer, J.-M. Bioactive peptides and depsipeptides with anticancer potential: Sources from marine animals. *Mar. Drugs* **2012**, *10*, 963–986. [CrossRef] [PubMed]

103. Kucuk, O.; Young, M.L.; Habermann, T.M.; Wolf, B.C.; Jimeno, J.; Cassileth, P.A. Phase II Trial of Didemnin B in Previously Treated Non-Hodgkin's Lymphoma: An Eastern Cooperative Oncology Group (ECOG) Study. *Am. J. Clin. Oncol.* **2000**, *23*, 273–277. [CrossRef] [PubMed]

104. Da Rocha, A.B.; Lopes, R.M.; Schwartsmann, G. Natural products in anticancer therapy. *Curr. Opin. Pharmacol.* **2001**, *1*, 364–369. [CrossRef]

105. Cooper, E.L.; Albert, R. Tunicates: A vertebrate ancestral source of antitumor compounds. In *Handbook of Anticancer Drugs from Marine Origin*; Springer: Berlin, Germany, 2015; pp. 383–395.

106. Broggini, M.; Marchini, S.; Galliera, E.; Borsotti, P.; Taraboletti, G.; Erba, E.; Sironi, M.; Jimeno, J.; Faircloth, G.; Giavazzi, R. Aplidine, a new anticancer agent of marine origin, inhibits vascular endothelial growth factor (VEGF) secretion and blocks VEGF-VEGFR-1 (flt-1) autocrine loop in human leukemia cells MOLT-4. *Leukemia* **2003**, *17*, 52–59. [CrossRef] [PubMed]

107. Kitagaki, J.; Shi, G.; Miyauchi, S.; Murakami, S.; Yang, Y. Cyclic depsipeptides as potential cancer therapeutics. *Anti-Cancer Drugs* **2015**, *26*, 259–271. [CrossRef] [PubMed]

108. Vervoort, H.; Fenical, W.; Epifanio, R.D.A. Tamandarins A and B: New cytotoxic depsipeptides from a Brazilian ascidian of the family Didemnidae. *J. Org. Chem.* **2000**, *65*, 782–792. [CrossRef] [PubMed]

109. Blanco-Míguez, A.; Gutiérrez-Jácome, A.; Pérez-Pérez, M.; Pérez-Rodríguez, G.; Catalán-García, S.; Fdez-Riverola, F.; Lourenço, A.; Sánchez, B. From amino acid sequence to bioactivity: The biomedical potential of antitumor peptides. *Protein Sci.* **2016**, *25*, 1084–1095. [CrossRef] [PubMed]

110. Bowden, B.; Gravalos, D.G. Cyclic Hepta-Peptide Derivative from Colonial Ascidians, *Lissoclinum* sp. U.S. Patent No. US 20040033940 A1, 3 April 2004.

111. Kalemkerian, P.G.; Ou, X.; Adil, R.M.; Rosati, R.; Khoulani, M.M.; Madan, K.S.; Pettit, R.G. Activity of dolastatin 10 against small-cell lung cancer in vitro and in vivo: Induction of apoptosis and bcl-2 modification. *Cancer Chemother. Pharmacol.* **1999**, *43*, 507–515. [CrossRef] [PubMed]

112. Aneiros, A.; Garateix, A. Bioactive peptides from marine sources: Pharmacological properties and isolation procedures. *J. Chromatogr. B* **2004**, *803*, 41–53. [CrossRef] [PubMed]

113. Margolin, K.; Longmate, J.; Synold, T.W.; Gandara, D.R.; Weber, J.; Gonzalez, R.; Johansen, M.J.; Newman, R.; Baratta, T.; Doroshow, J.H. Dolastatin-10 in Metastatic Melanoma: A Phase II and Pharmokinetic Trial of the California Cancer Consortium. *Investig. New Drugs* **2001**, *19*, 335–340. [CrossRef]

114. Turner, T.; Jackson, W.H.; Pettit, G.R.; Wells, A.; Kraft, A.S. Treatment of human prostate cancer cells with dolastatin 10, a peptide isolated from a marine shell-less mollusc. *Prostate* **1998**, *34*, 175–181. [CrossRef]

115. Zheng, L.; Lin, X.; Wu, N.; Liu, M.; Zheng, Y.; Sheng, J.; Ji, X.; Sun, M. Targeting cellular apoptotic pathway with peptides from marine organisms. *BBA Rev. Cancer* **2013**, *1836*, 42–48. [CrossRef] [PubMed]

116. Liu, M.; Zhao, X.; Zhao, J.; Xiao, L.; Liu, H.; Wang, C.; Cheng, L.; Wu, N.; Lin, X. Induction of apoptosis, G_0/G_1 phase arrest and microtubule disassembly in K562 leukemia cells by Mere15, a novel polypeptide from *Meretrix meretrix* Linnaeus. *Mar. Drugs* **2012**, *10*, 2596–2607. [CrossRef] [PubMed]

117. Wang, H.; Wei, J.; Wu, N.; Liu, M.; Wang, C.; Zhang, Y.; Wang, F.; Liu, H.; Lin, X. Mere15, a novel polypeptide from Meretrix meretrix, inhibits adhesion, migration and invasion of human lung cancer A549 cells via down-regulating MMPs. *Pharm. Biol.* **2013**, *51*, 145–151. [CrossRef] [PubMed]

118. Janmaat, M.L.; Rodriguez, J.A.; Jimeno, J.; Kruyt, F.A.; Giaccone, G. Kahalalide F induces necrosis-like cell death that involves depletion of ErbB3 and inhibition of Akt signaling. *Mol. Pharmacol.* **2005**, *68*, 502–510. [CrossRef] [PubMed]

119. Wang, B.; Waters, A.L.; Valeriote, F.A.; Hamann, M.T. An efficient and cost-effective approach to kahalalide F *N*-terminal modifications using a nuisance algal bloom of *Bryopsis pennata*. *Biochim. Biophys. Acta* **2015**, *1850*, 1849–1854. [CrossRef] [PubMed]

120. Staats, P.S.; Yearwood, T.; Charapata, S.G.; Presley, R.W.; Wallace, M.S.; Byas-Smith, M.; Fisher, R.; Bryce, D.A.; Mangieri, E.A.; Luther, R.R. Intrathecal ziconotide in the treatment of refractory pain in patients with cancer or AIDS: A randomized controlled trial. *JAMA* **2004**, *291*, 63–70. [CrossRef] [PubMed]

121. Olivier Brenet, M.; Sabine de Bourmont, M.; Florence Dixmerias, M.; Nadia Buisset, M.; Nathalie Lebrec, M.; Dominique Monnin, M. Ziconotide adverse events in patients with cancer pain: A multicenter observational study of a slow titration, multidrug protocol. *Pain Phys.* **2012**, *15*, 395–403.

122. Mundt, S.; Kreitlow, S.; Nowotny, A.; Effmert, U. Biochemical and pharmacological investigations of selected cyanobacteria. *Int. J. Hyg. Environ. Health* **2001**, *203*, 327–334. [CrossRef] [PubMed]

123. Thornburg, C.C.; Thimmaiah, M.; Shaala, L.A.; Hau, A.M.; Malmo, J.M.; Ishmael, J.E.; Youssef, D.T.; McPhail, K.L. Cyclic depsipeptides, grassypeptolides D and E and Ibu-epidemethoxylyngbyastatin 3, from a Red Sea *Leptolyngbya* cyanobacterium. *J. Nat. Prod.* **2011**, *74*, 1677–1685. [CrossRef] [PubMed]

124. Do Rosário Martins, M.; Costa, M. Marine Cyanobacteria Compounds with Anticancer Properties: Implication of Apoptosis. In *Handbook of Anticancer Drugs from Marine Origin*; Kim, S.-K., Ed.; Springer: Cham, Germany, 2015; pp. 621–647.

125. Shaala, L.A.; Youssef, D.T.; McPhail, K.L.; Elbandy, M. Malyngamide 4, a new lipopeptide from the Red Sea marine cyanobacterium *Moorea producens* (formerly *Lyngbya majuscula*). *Phytochem. Lett.* **2013**, *6*, 183–188. [CrossRef]

126. Wrasidlo, W.; Mielgo, A.; Torres, V.A.; Barbero, S.; Stoletov, K.; Suyama, T.L.; Klemke, R.L.; Gerwick, W.H.; Carson, D.A.; Stupack, D.G. The marine lipopeptide somocystinamide A triggers apoptosis via caspase 8. *Proc. Natl. Acad. Sci. USA* **2008**, *105*, 2313–2318. [CrossRef] [PubMed]

127. Marquez, B.L.; Watts, K.S.; Yokochi, A.; Roberts, M.A.; Verdier-Pinard, P.; Jimenez, J.I.; Hamel, E.; Scheuer, P.J.; Gerwick, W.H. Structure and absolute stereochemistry of hectochlorin, a potent stimulator of actin assembly. *J. Nat. Prod.* **2002**, *65*, 866–871. [CrossRef] [PubMed]
128. Luesch, H.; Chanda, S.K.; Raya, R.M.; DeJesus, P.D.; Orth, A.P.; Walker, J.R.; Belmonte, J.C.I.; Schultz, P.G. A functional genomics approach to the mode of action of apratoxin A. *Nat. Chem. Biol.* **2006**, *2*, 158–167. [CrossRef] [PubMed]
129. Singh, R.K.; Tiwari, S.P.; Rai, A.K.; Mohapatra, T.M. Cyanobacteria: An emerging source for drug discovery. *J. Antibiot.* **2011**, *64*, 401–412. [CrossRef] [PubMed]
130. Li, B.; Gao, M.H.; Zhang, X.C.; Chu, X.M. Molecular immune mechanism of c-phycocyanin from *Spirulina platensis* induces apoptosis in HeLa cells in vitro. *Biotechnol. Appl. Biochem.* **2006**, *43*, 155–164. [PubMed]
131. Thangam, R.; Suresh, V.; Princy, W.A.; Rajkumar, M.; SenthilKumar, N.; Gunasekaran, P.; Rengasamy, R.; Anbazhagan, C.; Kaveri, K.; Kannan, S. C-Phycocyanin from *Oscillatoria tenuis* exhibited an antioxidant and in vitro antiproliferative activity through induction of apoptosis and G_0/G_1 cell cycle arrest. *Food Chem.* **2013**, *140*, 262–272. [CrossRef] [PubMed]
132. Lau, N.-S.; Matsui, M.; Abdullah, A.A.-A. Cyanobacteria: Photoautotrophic microbial factories for the sustainable synthesis of industrial products. *BioMed Res. Int.* **2015**, *2015*, 754934. [CrossRef] [PubMed]
133. Ngo, D.H.; Wijesekara, I.; Vo, T.S.; Van Ta, Q.; Kim, S.K. Marine food-derived functional ingredients as potential antioxidants in the food industry: An overview. *Food Res. Int.* **2011**, *44*, 523–529. [CrossRef]
134. Nazeer, R.; Prabha, K.D.; Kumar, N.S.; Ganesh, R.J. Isolation of antioxidant peptides from clam, *Meretrix casta* (Chemnitz). *J. Food Sci. Technol.* **2013**, *50*, 777–783. [CrossRef] [PubMed]
135. Rajapakse, N.; Mendis, E.; Byun, H.-G.; Kim, S.-K. Purification and in vitro antioxidative effects of giant squid muscle peptides on free radical-mediated oxidative systems. *J. Nutr. Biochem.* **2005**, *16*, 562–569. [CrossRef] [PubMed]
136. Shahidi, F.; Zhong, Y. Bioactive Peptides. *J. AOAC Int.* **2008**, *91*, 914–931. [PubMed]
137. Je, J.-Y.; Park, P.-J.; Kim, S.-K. Antioxidant activity of a peptide isolated from Alaska pollack (*Theragra chalcogramma*) frame protein hydrolysate. *Food Res. Int.* **2005**, *38*, 45–50. [CrossRef]
138. Jun, S.-Y.; Park, P.-J.; Jung, W.-K.; Kim, S.-K. Purification and characterization of an antioxidative peptide from enzymatic hydrolysate of yellowfin sole (*Limanda aspera*) frame protein. *Eur. Food Res. Technol.* **2004**, *219*, 20–26.
139. Mendis, E.; Rajapakse, N.; Kim, S.-K. Antioxidant properties of a radical-scavenging peptide purified from enzymatically prepared fish skin gelatin hydrolysate. *J. Agric. Food Chem.* **2005**, *53*, 581–587. [CrossRef] [PubMed]
140. Himaya, S.; Ryu, B.; Ngo, D.-H.; Kim, S.-K. Peptide Isolated From Japanese Flounder Skin Gelatin Protects Against Cellular Oxidative Damage. *J. Agric. Food Chem.* **2012**, *60*, 9112–9119. [CrossRef] [PubMed]
141. Guérard, F.; Decourcelle, N.; Sabourin, C.; Floch-Laizet, C.; Le Grel, L.; Le Floc'H, P.; Gourlay, F.; Le Delezir, R.; Jaouen, P.; Bourseau, P. Recent developments of marine ingredients for food and nutraceutical applications: A review. *J. Sci. Halieut. Aquat.* **2010**, *2*, 21–27.
142. Zhu, C.-F.; Peng, H.-B.; Liu, G.-Q.; Zhang, F.; Li, Y. Beneficial effects of oligopeptides from marine salmon skin in a rat model of type 2 diabetes. *Nutrition* **2010**, *26*, 1014–1020. [CrossRef] [PubMed]
143. Zhu, C.-F.; Li, G.-Z.; Peng, H.-B.; Zhang, F.; Chen, Y.; Li, Y. Effect of marine collagen peptides on markers of metabolic nuclear receptors in type 2 diabetic patients with/without hypertension. *Biomed. Environ. Sci.* **2010**, *23*, 113–120. [CrossRef]
144. Zhu, C.-F.; Li, G.-Z.; Peng, H.-B.; Zhang, F.; Chen, Y.; Li, Y. Treatment with marine collagen peptides modulates glucose and lipid metabolism in Chinese patients with type 2 diabetes mellitus. *Appl. Physiol. Nutr. Metab.* **2010**, *35*, 797–804. [CrossRef] [PubMed]
145. Chance, W.T.; Tao, Z.; Sheriff, S.; Balasubramaniam, A. WRYamide, A NPY-based tripeptide that antagonizes feeding in rats. *Brain Res.* **1998**, *803*, 39–43. [CrossRef]
146. Vernaleken, A.; Veyhl, M.; Gorboulev, V.; Kottra, G.; Palm, D.; Burckhardt, B.-C.; Burckhardt, G.; Pipkorn, R.; Beier, N.; van Amsterdam, C. Tripeptides of RS1 (RSC1A1) inhibit a monosaccharide-dependent exocytotic pathway of Na^+-D-glucose cotransporter SGLT1 with high affinity. *J. Biol. Chem.* **2007**, *282*, 28501–28513. [CrossRef] [PubMed]

147. Chevrier, G.; Mitchell, P.L.; Rioux, L.-E.; Hasan, F.; Jin, T.; Roblet, C.R.; Doyen, A.; Pilon, G.; St-Pierre, P.; Lavigne, C. Low-molecular-weight peptides from salmon protein prevent obesity-linked glucose intolerance, inflammation, and dyslipidemia in LDLR$^{-/-}$/ApoB$^{100/100}$ mice. *J. Nutr.* **2015**, *145*, 1415–1422. [CrossRef] [PubMed]

148. Dandona, P.; Aljada, A.; Bandyopadhyay, A. Inflammation: The link between insulin resistance, obesity and diabetes. *Trends Immunol.* **2004**, *25*, 4–7. [CrossRef] [PubMed]

149. Nagaoka, S.; Shimizu, K.; Kaneko, H.; Shibayama, F.; Morikawa, K.; Kanamaru, Y.; Otsuka, A.; Hirahashi, T.; Kato, T. A novel protein c-phycocyanin plays a crucial role in the hypocholesterolemic action of *Spirulina platensis* concentrate in rats. *J. Nutr.* **2005**, *135*, 2425–2430. [PubMed]

150. Colla, L.M.; Muccillo-Baisch, A.L.; Costa, J.A.V. *Spirulina platensis* effects on the levels of total cholesterol, HDL and triacylglycerols in rabbits fed with a hypercholesterolemic diet. *Braz. Arch. Biol. Technol.* **2008**, *51*, 405–411. [CrossRef]

151. Lordan, S.; Ross, R.P.; Stanton, C. Marine Bioactives as Functional Food Ingredients: Potential to Reduce the Incidence of Chronic Diseases. *Mar. Drugs* **2011**, *9*, 1056–1110. [CrossRef] [PubMed]

152. Craik, D.J.; Fairlie, D.P.; Liras, S.; Price, D. The future of peptide-based drugs. *Chem. Biol. Drug Des.* **2013**, *81*, 136–147. [CrossRef] [PubMed]

153. Shahidi, F.; Chandrasekara, A. Millet grain phenolics and their role in disease risk reduction and health promotion: A review. *J. Funct. Foods* **2013**, *5*, 570–581. [CrossRef]

154. Udenigwe, C.C.; Aluko, R.E. Food protein-derived bioactive peptides: Production, processing, and potential health benefits. *J. Food Sci.* **2012**, *77*, R11–R24. [CrossRef] [PubMed]

155. Zhuang, Y.; Sun, L.; Zhang, Y.; Liu, G. Antihypertensive Effect of Long-Term Oral Administration of Jellyfish (*Rhopilema esculentum*) Collagen Peptides on Renovascular Hypertension. *Mar. Drugs* **2012**, *10*, 417–426. [CrossRef] [PubMed]

marine drugs

MDPI

Review

Marine Peptides as Potential Agents for the Management of Type 2 Diabetes Mellitus—A Prospect

En-Qin Xia, Shan-Shan Zhu, Min-Jing He, Fei Luo, Cheng-Zhan Fu and Tang-Bin Zou *

Dongguan Key Laboratory of Environmental Medicine, School of Public Health, Guangdong Medical University, Dongguan 523808, China; enqinxia@163.com (E.-Q.X.); zss90y@163.com (S.-S.Z.); minjinghe0818@163.com (M.-J.H.); luofei_00@163.com (F.L.); fuchengzhan@126.com (C.-Z.F.)
* Correspondence: zoutb@163.com; Tel.: +86-769-2289-6572

Academic Editor: Se-Kwon Kim
Received: 30 January 2017; Accepted: 20 March 2017; Published: 23 March 2017

Abstract: An increasing prevalence of diabetes is known as a main risk for human health in the last future worldwide. There is limited evidence on the potential management of type 2 diabetes mellitus using bioactive peptides from marine organisms, besides from milk and beans. We summarized here recent advances in our understanding of the regulation of glucose metabolism using bioactive peptides from natural proteins, including regulation of insulin-regulated glucose metabolism, such as protection and reparation of pancreatic β-cells, enhancing glucose-stimulated insulin secretion and influencing the sensitivity of insulin and the signaling pathways, and inhibition of bioactive peptides to dipeptidyl peptidase IV, α-amylase and α-glucosidase activities. The present paper tried to understand the underlying mechanism involved and the structure characteristics of bioactive peptides responsible for its antidiabetic activities to prospect the utilization of rich marine organism proteins.

Keywords: marine protein; bioactive peptide; regulation; glucose metabolism; structure active relationship

1. Introduction

Diabetes mellitus is known as a seriously chronic metabolism disorder. According to the forecast, the prevalence of type 2 diabetes mellitus (T2DM) will increase from 350 million today to 592 million by 2035 [1]. The nutrient overload in prolonged periods was the key factor causing the bad situation due to impairing the pancreatic β cell function. In sequence, insulin resistance, the impaired secretary insulin and glucose tolerance become true, and the following result is to develop T2DM [2–4]. Fortunately, recent studies of Zhu et al. suggest that peptides and protein hydrolysates from wild *Chum Salmon* (*Oncorhynchus kern*) skin markedly decreased the level of fasting glucose level and the pancreatic apoptosis of islet cells [5]. Pandey et al. reported that the bacteria associated with marine sponge, *Aka coralliphaga*, produce many glucosidase inhibitory peptides [6]. Among marine microalgae, *Chlorella vulgaris* (*C. vulgaris*) is regarded as a complementary medicine, due to its supplements, and exhibited benefits for some health disorders, such as dyslipidemia, hyperglycemia, and hypertension as well as weight loss in several studies [7,8]. These results indicated that the natural marine bioactive peptides can improve the deleterious process of T2DM. However, few reports involved antidiabetic activities of marine natural peptides can be found in literature in our knowledge. Therefore, the present review will summarize all evidence on antidiabetic activities of natural peptides from milk, bean and marine organisms. Its primary coverage involved the response of insulin-regulated glucose metabolism and dipeptidyl peptidase IV (DPP-IV), α-amylase and α-glucosidase activities on the bioactive peptides. The mechanism underlying each antidiabetic activity and the structure characteristics of bioactive peptides responsible for its antidiabetic activities were also carefully discussed for the future novel marine peptides investigation.

2. Regulation of Bioactive Peptides on the Insulin-Regulated Glucose Metabolism

2.1. Protecting Pancreatic β-Cells of Bioactive Peptides

Adequate insulin secretion is necessary to maintain blood glucose levels within a physiological range, and competent pancreatic β-cells are responsible for that task. However, it becomes insufficient for the T2DM individuals due to their pancreatic β-cell failure [9]. The major factor causing the adverse effect on pancreatic β cells was reported as chronic nutrient overload, which causes a cell to increase its function and mass to match the increasing nutrient availability and insulin resistance [10]. As it can not adapt to maintaining glucose homeostasis at the high glucose challenges over prolonged periods [4], hyperglycemia was developed, which was verified to generate over-production of free radical species including reactive oxygen species (ROS) and nitric oxide (NO) radicals. These free radicals can cause defects in the mitochondrial respiratory chain by excessive requirements of oxidative enzymes activities (e.g., NOX), protein glycation and non-enzymatic oxidation and peroxidation of carbohydrates and lipids, and so on. Ultimately, increasing oxygen free radicals generate in various tissues including pancreas and kidneys [11–13]. Based on in vitro cell models and animal models of diabetes assays, researchers found that increased intracellular ROS production can significantly promote peripheral insulin resistance and induce endoplasmic reticulum (ER) stress, mitochondrial dysfunction, apoptosis and cell death. Subsequently, pancreatic β cell mass become deficient and its secretory function was impaired [14–20]. Using Min6 cells and pancreatic islets, Mailloux and coworkers found that an increase in mitochondrial matrix ROS can reverse the glutathionylation of uncoupling protein-2, which subsequently impedes glucose-stimulated insulin secretion from β cells [21]. In addition, it is reported that insulin-secreting β cells of the pancreatic islets contain gene expression and activity of the H_2O_2-reducing enzymes catalase and glutathione peroxidase (GPx) in islets accounting for only 1%–5% of the values in liver, and levels of cytosolic and mitochondrial superoxide dismutases (SOD) in islets only around 30% of those in the liver [22,23]. Obviously, as induced by chronic hyperglycemia and exposure to proinflammatory cytokines, the poor antioxidant defense capability of pancreatic β cells was sensitive towards oxidative stress [23–25]. Overexpression of these antioxidant enzymes has been observed to protect various β cell lines against oxidative damage [26–28]. Reactive oxygen species sensitized by metallothionein and catalase in nonobese diabetic mice was predicted to protect the pancreatic β cells from autoimmune destruction in male non-obese diabetic (NOD) [29]. Therefore, control of oxidative stress and inflammatory may be key approaches to reduce pancreatic β cell damage and the development of T2DM [30].

Recently, researchers reported that natural nutrient metabolites can exert a significant role to preserve β cell functions and mass, prolong the pre-diabetic phase and delay the progress to frank T2DM. For instance, consumption of dairy products was found linking with a decrease in the risk of type 2 diabetes [31,32]. Nasri et al. also reported that orally administered goby fish protein hydrolysates, not undigested goby fish protein, can significantly attenuate hyperglycemia and restored the antioxidant status under high-fat-high-fructose diet-induced oxidative stress in rats [33]. The same results reported that the effectiveness of the natural administration of fish protein hydrolysates, produced from *Sardinella aurita* and *Salaria basilisca*, in improving the oxidative status antioxidant for cholesterol-fed rats and alloxan-induced diabetic rats, respectively [34,35]. The results suggest that the presence of potent active peptides in fish protein hydrolysates was effective in enhancement of the antioxidant status.

Using in vitro assays, scavenging free radical capacity of bioactive peptides was observed by several researchers [14,36–38]. After being treated with pinto Durango bean (*P. vulgaris* L.) alcalase hydrolysates for 20 h at the concentration of 100 μg/mL, ROS production due to *tert*butyl hydroperoxide (t-BOOH) was almost eradicated in in vitro cell assay [36]. Fernández-Tomé [37] also found that lunasin, a soy peptide, exerted an effective scavenger as high as 190% of the intracellular

ROS generation in HepG2 cells due to exposed to t-BOOH, compared to the control. In addition, the receptor in HepG2 cells for advanced glycation end products (RAGE) showed the lowest expression treated with the complete protein hydrolysates. While RAGE was found in pancreatic islets acting as an inducer of pancreatic β-cell apoptosis and developing of chronic diabetic complications via nicotinamide adenine dinucleotide phosphate oxidase mediated ROS generation in vitro assays [38]. Treatment with β-casomorphin-7, a milk-derived bioactive peptide, a considerable reduction in H_2O_2 content ($p < 0.05$) and a remarkable increase in the activity of GSH-peroxidase, SOD and catalase of the anti-oxidation system were observed. Simultaneously, the abatement of free-radical-mediated oxidative stress in blood and myocardium and cardiac indexes were also observed [39]. Protective effect of peptides on pancreatic β-cells against intracellular ROS due to a high glucose exposure has also been observed [14].

Natural peptides were also reported to efficiently ameliorate the diabetes symptoms. The levels of blood glucose of streptozotocin-induced diabetic rats markedly decreased after treatment with β-casomorphin-7, compared with model control group ($p < 0.01$) [39]. Bioactive peptides were observed to reduce the expression of cytokines such as interleukin-1β and tumor necrosis factor-α in pancreatic β-cells, which both generate as the cells were exposed to high glucose in vitro [40]. A Chlorella-11 peptide was also able to suppress lipopolysaccharide-induced nitric oxide (NO), serum TNF-α and inflammation [41]. In addition, it was reported that the common bean peptides can upregulated the expression of insulinlike growth factor 2 (IGF-II), a kind of adipokines in pancreatic β-cells now being believed to play a negative role in the development of obesity-associated insulin resistance and anti-inflammation [42].

2.2. Enhancement of Glucose-Stimulated Insulin Secretion

It has been revealed that T2DM develops when the insulin secretory capacity is unable to compensate for the increase of insulin resistance. The incretins, gut-derived hormones released from small intestine enteroendocrine cells (EECs), i.e., glucagonlike peptide 1 (GLP-1) and glucose dependent insulinotropic peptide (GIP), exert the significant role in regulation of food digestion by stimulation of glucose-dependent insulin secretion, as well food intake by promoting satiety to decrease appetite [43–45]. However, studies showed that circulating GLP-1 levels increase after meal intake but rapidly decrease 80%–90% due to cleaved by dipeptidyl peptidase IV (DDP-IV) [46]. Therefore, the DPP-IV inhibitors have indirect effects on islet function via contributing to insulin secretion and lowering blood glucose by increasing incretin action [47]. As early as 1988, Liddle et al. found that protein digestion can stimulate gut hormone secretion and expression in rats [48]. According to Caron et al., intestinal digestion derived from bovine haemoglobin exhibited significant efficiency on gut hormone release and DPP-IV activity inhibition, and those hormones' gene expression was also up-expressed [49]. The DPP-IV inhibition capacity of some diet origin peptides above 200 µM of in literature is displayed in Table 1.

Table 1. The precursors, sequences, inhibition capacity (IC 50) of some natural origin peptides with dipeptidyl peptidase IV inhibitory activity in literature with IC 50 < 200 µM.

Food	Precursor Protein	Peptide Sequence	IC 50 (µM)	Reference
Plant Protein	Macroalga *Palmaria palmate* protein	ILAP LLAP MAGVDHI	43.40 53.67 159.37	[50]
Collagen	Halibut skin gelatin	SPGSSGPQGFTG GPVGPAGNPGANGLN PPGPTGPRGQPGNIGF	101.6 81.3 146.7	[51]
	Tilapia skin gelatin	IPGDPGPPGPPGP LPGERGRPGAPGP GPKGDRGLPGPPGRDG	65.4 76.8 89.6	
	Tuna cooking juice hydrolysates	PACGGFWISGRPG CAYQWQRPVDRIR PGVGGPLGPIGPCYE	96.4 78 116.1	[52]

Table 1. *Cont.*

Food	Precursor Protein	Peptide Sequence	IC 50 (μM)	Reference
Collagen	Deer skin protein	GPVGXAGPPGK GPVGPSGPXGK	83.3 93.7	[53]
Milk protein	α-Lactalbumin	LKPTPEGDL LAHKALCSEKL LCSEKLDQ TKCEVFRE	45 165 186 166	[54]
	β-Lactalbumin	VAGTWY IPAVF	174 44.7	[55] [56]
	Atlantic salmon collagen/gelatin	GPAE GPGA	49.6 41.9	[52]
	Gouda-type cheese	VPITPTL VPITPT LPQNIPPL VAGTWY LPQ	110 130 46 174 82	[57]
	Whey protein	LAHKALCSEKL WLAHKALCSEKLDQ LKPTPEGDL LKPTPEGDLEIL WLAHKALCSEKLDQ WR IPIQY WCKDDQNPHS TKCEVFRE IPA VA3, VL, WL, WI LKPTPEGDLE LKALPMH	165 141 45 9 57 141 31.4 28.2 75.0 166 49 <170 42 193	[58–63]
	Milk protein	WA WR WK LPYPY WQ WI WN YPYY	92.6 37.8 40.6 108.3 120.3 138.7 148.5 194.4	[46,64–66]
Milk protein	Milk protein	WN IP IPI IPIQY FLQP WV LPVPQ IPM HL VA WL WP	148.5 149.6 3.5 35.2 65.3 65.7 48.2 73.9 143.2 168.2 43.6 44.5	[46,64–66]

From Table 1, milk is the main source of peptides with efficient DPP-IV inhibitors in literature. Skin from halibut, tilapia and deer also showed significant DPP-IV inhibition capacity with IC 50 lower than 200 μM. Plant proteins digested in vitro or in vivo have been investigated the DPP-IV inhibitory peptides by some researchers, such as cowpea bean [67], Quinoa [68], rice bran [69], raw amaranth flour, soybean flour, and wheat flour [70]. However, except for Macroalga *Palmaria palmate*, DPP-IV inhibition capacity were exhibited with IC 50 far higher than 200 μM. The collected data showed that novel original peptides from natural proteins, especially from marine organisms, have been widely investigated for the management of T2DM.

2.3. Regulation of Glucose Uptake and Lipid Accumulation

Hyperglycemia has been identified as a key factor to induce to deficiency in insulin secretion and/or decreased reaction of the organs to insulin (World Health Organization, 1999). Besides diet and lifestyle modifications, control and prevention of hyperglycemia are primary approaches in the management of diabetes mellitus, which is involved in several physiological processes, such as increasing utilization of the glucose by the peripheral tissues and lowering hepatic glucose output and adipocyte fat-accumulation [71].

Some studies showed that natural peptides can correct high blood sugar, even without regulation of insulin secretion. For instance, the bean hydrolysates from pinto Durango had a dose-dependent insulin sensitizing effect ($p < 0.05$) comparing to the control. The most potent fraction was pinto Durango-alcalase < 1 kDa, which caused insulin resistant cells to increase (67 ± 3.2)% of glucose uptake compared to the non-insulin resistant cells [37]. The plasma glucose was also significantly decreased (25%–34%), after simultaneously intervening rats high-fat-high-fructose diet (HFFD) and goby fish protein hydrolysates, compared to the HFFD group [33]. β-casomorphin-7, a peptide from milk, was also found to restrain the elevation of blood glucose, and its effect is slightly inferior to insulin (11.18 ± 0.72 to 14.92 ± 0.66 mmol/L) [39]. The same results were found that the hypoglycemic effect of protein hydrolysates from muscle fish *Zebra blenny* in alloxan-induced diabetic rats [35] and from the vegetable *Momordica charantia* L. in alloxan-induced diabetic mice [33]. Yuh et al. found significantly enhanced hypoglycemic effects of *chlorella* consumption on streptozocin (STZ) induced diabetic mice [72]. By the similar assays, Jeong et al. observed significant improvement of insulin sensitivity in type 2 diabetic and normal Wistar rats., but the glucose-stimulated insulin secretion had not been influenced by chlorella consumption [73] Aglycin, a peptide from soy, exhibited effectively in preventing hyperglycemia in a diabetic animal model with impaired glucose tolerance and insulin resistance, which were induced in BALB/c mice (i.e., the laboratory bred strain of albino mice specially used for the study of cancer, neurological diseases) with a high-fat diet and received a single intraperitoneal injection of STZ [74]. Aglycin reduced blood glucose levels by 45.0% after long-term treatment (aglycin vs. model day 21 7.3 ± 0.5 vs. 11.3 ± 0.4 mmol/L; day 28 7.1 ± 0.2 vs. 12.4 ± 0.6 mmol/L, Pb.01, respectively) [74]. The notable efficiency has also been observed by Veloso et al. [75]. However, insulin secretion and body weight control in aglycin treated mice was not affected. Glucose levels were lowered after insulin loading in aglycin-treated mice in the insulin tolerance test. It indicated that glucose control induced by aglycin is largely mediated by enhancing glucose utilization and insulin sensitivity in peripheral insulin target sites [74]. Peptides from salmon hydrolysate, separated by electrodialysis with filtration membrane, also enhanced glucose uptake in L6 skeletal muscle cells by up to 40% without insulin increase [76].

Recently, researchers found that there was an inverse relationship between amount of secretory adiponectin, known as an insulin sensitizor, and the percentage of adipose tissue in the internal organs. It indicates that the decreased fat accumulation may improve glucose tolerance by the enhancement of insulin sensitivity [77]. According to results of Toledo et al., lipid accumulation was inhibited from 13% to 28% when adipocytes were treated with the bean hydrolysates throughout the differentiation process, and their <1 kDa peptide fraction showed slightly higher than whole hydrolysates [36]. Similarly, Martinez-Villaluenga et al. showed an inhibition of lipid accumulation from 27% to 46% in 3T3-L1 adipocytes when treated with alcalase soy hydrolysates at a concentration of 100 μM (1000 μg/mL) with an average of molecular mass of 10 kDa [78]. This study showed that common bean hydrolysates have the inhibitory effect on lipid accumulation. Lipid accumulation in adipose tissue can be reduced by different mechanisms, e.g., reducing lipid uptake through suppressing lipoprotein lipase or reducing lipid synthesis through inhibiting fatty acid synthase (FAS) [79]. To investigate the effect of peptides from natural protein on liquid accumulation, human white pre-adipocytes (HWP) received intervention from 11 peptides from fish, seaweed, shellfish, in order to investigate proliferation, differentiation and maturation. The results showed that Ala-Pro, Val-Ala-Pro and Ala-Lys-Lys greatly affected viability of HWP during the proliferation period, while Lys-Trp and Val-Trp reduced the number of viable cells during the differentiation stage. The decrease of their final lipid content and of the mRNA level of adipocyte markers (aP2, GLUT4, LPL and AGT) was also involved. Kim et al. showed that the peptide GAGVGY also downregulates lipid accumulation modulating of gene expression such as sterol regulatory element-binding proteins-1c, Peroxisome proliferator-activated receptor gamma and fatty acid synthase [80], which exhibited a dual effect regulating glucose uptake and lipid accumulation in a similar way as the present results [81–83]. In literature, some peptides from goat and soybean have been reported to inhibit growth of preadipocytes, decrease the differentiation process and decrease the final lipid content in human white preadipocytes and lipid accumulation [79,81]. These results suggest that natural peptides may be a potential compound on regulation of T2DM via inhibitory lipid accumulation.

2.4. Regulation of the Insulin-Signaling Pathways

Correction insulin resistance is an important therapeutic strategy for T2DM. Insulin resistance is a physiological condition, in which cells fail to respond to the normal actions of the hormone insulin, and reduction or impairment of insulin-stimulated glucose uptake. Insulin receptor substrate-1/phosphoinositide-3-kinase/protein kinase B (IRS-1/PI3K/Akt) signaling pathways was found as a main target to correct insulin regulating glucose uptake [84]. A defect in protein kinase B (PKB or Akt) signaling that reduces the translocation of the glucose transporter protein GLUT4 to the cellular membrane may be a main impairment of insulin-stimulated glucose uptake under insulin resistance conditions [85]. The action mechanism of IRS-1/PI3K/Akt signaling pathways on regulation of blood sugar is illustrated in Figure 1. In vivo assays showed that under insulin resistance of diabetes, phosphorylation level of Akt Ser473 decreases, and insulin signal transduction also significantly abates due to a decrease in insulin receptor concentration and kinase activity [86,87]. Therefore, insulin receptor (IR), insulin receptor substrate-1/2 (IRS-1/2), PI3K and Akt all might be efficient targets to regulate downstream signaling cascade to lower blood sugar level.

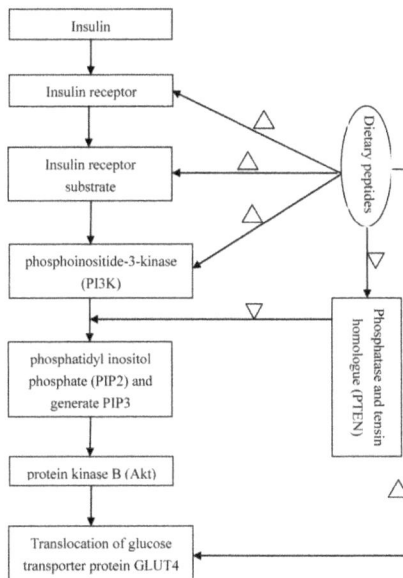

Figure 1. Some regulation evidence of natural peptides on the insulin-signaling pathways. Note: 'Δ' and '∇' mean the natural peptides display upregulation and downregulation on the corresponding bioprecessed, respectively.

Some evidence has reported on the upregulation of GLUT4 in T2DM individuals by natural peptides. The effect of aglycin, a peptide from soy, on insulin signaling in the mice skeletal muscle showed that a significant increase in the expression of IR and IRS1 genes, as well as total IR, IRS1, p-Akt protein and membrane GLUT4, was observed. An increase of 75% of basal glucose uptake was found in both normal and insulin-resistant C2C12 cells [74]. B-casomorphin-7 and insulin increased ($p < 0.05$) 1.37-fold and 1.62-fold of the expression of GLUT-4 in myocardium, respectively. In contrast with the model group, soybean peptides also have been reported to improve insulin action via increasing the expressions of GLUT4 and insulin regulatory genes in diabetic animals [88,89]. The peptide GAGVGY, a fibroin derivative, increased both basal and insulin stimulated glucose uptake through enhancement of GLUT1 expression and PI3K-dependent GLUT4 translocation [80]. Tripeptides such as GEY and GYG, derived from the peptide E5K6 from silk, stimulated glucose uptake in 3T3-L1 adipocytes by inducing the expression of glucose transporters GLUT1 and GLUT4 [90].

In adipose tissue, insulin is also responsible for the enhanced uptake of glucose by GLUT4. When adipocytes show insulin resistance, GLUT4 does not translocate to cell membrane in response to insulin release by the pancreas, and this leads to reduced glucose uptake. Common bean peptides modulating glucose transporters in insulin resistant adipocyte 3T3-L1 were evaluated by confocal microscopy. Results showed that there was a dose-dependence between the upregulation expression of glucose transporters GLUT4 and fraction peptides with <1 kDa derived from alcalase and bromelain [36]. In addition, a fermented soybean extract significantly increased the expression of GLUT4 and glucose uptake in 3T3-L1 adipocytes [91].

In vivo assays showed that fat accumulation was strongly associated with the inhibition of the PI3k signaling pathway, which was involved in the inhibition of insulin signaling [92]. Recovery of the activity of PI3K/Akt could protect the liver from non-alcoholic fatty liver disease induced injury [93]. These results might explain the fact that long-term feeding of soy peptide induced weight loss in obese mice both in healthy and diabetic animal models [74,94].

In addition, Akt is one of the major downstream targets of PI3K and responsible for the physiological function of insulin in adipocytes [95]. Phosphatase and tensin homologue (PTEN) is a lipid phosphatase that downregulates the action of PI3K decreasing insulin signaling, playing a role in regulating glucose metabolism [96]. A reduction of PTEN was presented with pinto Durango-bromelain bean hydrolysate and its <1 kDa peptide fraction treatments [36].

2.5. Clinical Trials

In very rare clinical human studies on anti-diabetes peptides, the efficacy of *Chlorella vulgaris* (*C. Vulgaris*) in prevention and treatment of dyslipidemia, hyperglycemia, hypertension as well as weight loss was found in literature [97–99]. *C. vulgaris*, a single-celled marine green algae, has been regarded as a complementary medicine [100]. Studies showed consumption of chlorella tablets for 16 weeks led to ameliorate insulin signaling pathways and noticeable reduction in serum glucose concentrations [97]. Panahi et al. also reported that *C. vulgaris* supplementation results in a marked decrease in insulin resistance and fasting serum glucose level in non-alcoholic fatty liver disease (NAFLD) patients [98]. Recently, Ebrahimi-Mameghani et al. recruited 70 obese patients with NAFLD aged 20–50 years to interfere with *C. vulgaris* supplementation. The results showed that 1200 mg *C. vulgaris* supplementation brought several potential beneficial effects, such as loss of weight, lowering serum glucose level, improvement of inflammatory biomarkers and liver function in NAFLD patients [99].

3. Inhibition of Bioactive Peptides to α-Amylase and α-Glucosidase Activities

Other approaches to decrease hyperglycemia are to control or delay glucose absorption by inhibition of α-glucosidase or α-amylase in the gastrointestinal tract. α-amylase is an enzyme that hydrolyses α-bonds polysaccharide such as glycogen and starch to oligosaccharides. α-Glucosidase is present in the epithelial mucosa of the small intestine and cleaves glycosidic bonds in oligosaccharides, releasing monosaccharides into blood sugar [30,101]. Thus, inhibition of α-amylase and α-glucosidase is an alternative pathway to management of the blood glucose levels and T2DM [102].

A significant increase in the serum α-amylase activity (by 86.08%) was found in HFFD-fed rats compared to control rats ($p < 0.05$). As oral administration of goby fish protein hydrolysates, the α-amylase activity in that of high-fat-high-fructose feed rats decreased by about 62% compared to the HFFD group [35]. Pinto bean peptides (<3 kDa fraction) and cumin seed-derived peptides also showed α-amylase inhibitory capacity at 62.1% and 24.54%, respectively [103,104]. Some inhibitory peptides and its IC50 and sequences has been detected or identified in literature and are illustrated in Table 2. To our knowledge, reports about inhibitory α-amylase peptides were limited, and less references could be obtained on the α-glucosidase inhibitory peptides in literature.

Table 2. The sequences, inhibition capacity (IC 50) and precursors of natural peptides with α-amylase and α-glucosidase inhibitory activity in literature.

Ingredient	Peptides Sequence	IC 50	Precursors	Reference
α-amylase	PPHMLP PLPWGAGF PPHMGGP PLPLHMLP LSSLEMGSLGALFVCM	1.97 (mg mL^{-1}) 8.96 (mg mL^{-1}) 14.63 (mg mL^{-1}) 18.45 (mg mL^{-1}) 20.56 (mg mL^{-1})	Pinto bean	[103]
	FFRSKLLSDGAAAAKGALLPQYW RCMAFLLSDGAAAAQQLLPQYW DPAQPNYPWTAVLVFRH	0.02 (μM) 0.04 (μM) 0.03 (μM)	Cumin seed protein	[104]
	RCMAFLLSDGAAAAQQLLPQYW DPAQPNYPW TAVLVFRH	0.04 (μM) 0.15 (μM)	Cumin seed protein	[105]
	WEVM AKSPLF	- -	Black bean protein	[106]
	<3 kDa fraction	-	Rice bran protein	[107]
	KLPGF NVLQPS	120.0 ± 4.0 (μM) 110.0 ± 6.2 (μM)	Albumin	[108]
α-glucosidase	-	36.3%–50.1% mg^{-1} DW	*Bean protein*	[109]
	TTGGKGGK	-	Black bean protein	[107]
	KLPGF NVLQPS	59.5 ± 5.7 (μM) 100.0 ± 5.7 (μM)	Albumin	[108]

4. The Structure Characteristics of Antidiabetes Peptides

Different approaches have been suggested potential for the treatment and management of Type 2 diabetes by natural peptides. A limited number of studies focus on the structural features that govern the properties of peptides in literature, including the structure of DPP-IV inhibitory, insulinotropic and α-amylase peptides.

For DPP-IV inhibitory peptides, they exert their effect by binding either at the active site and/or outside the catalytic center of the enzyme. In silico studies predicted that the active site of DPP-IV comprises a hydrophobic S1 (Tyr662 and Tyr666) pocket and a charged S2 (Phe357 and Arg125) pocket with an overall negative charge [110,111]. Hydrogen bonds and hydrophobic interactions were involved between N-terminal amino acids of DPP-IV inhibitory peptides and the catalytic active site of DPPIV. Thus, the structural features of DPP-IV inhibitors generally were inferred as a hydrophobic or aromatic amino acid at the N-terminus, such as Ile, Leu, Val, Phe, Trp or Tyr [52]. However, several non-inhibitory peptides possessing hydrophobic or aromatic amino acids at their N-terminus were also found (Table 1). Statistically, 77% of all hydrophobic peptides (with an hydrophobicity index >0) and 53% of the hydrophilic peptides were detected with DPP-IV inhibition [46]. It indicates that N-terminal hydrophobicity or aromaticity is a desirable characteristic, but not sufficient for inhibition.

To investigate the preferential amino acids involving in DPP-IV inhibition, the amount of each amino acid occurring in the DPP-IV inhibitory peptides was calculated based on Table 1. The preferential amino acids were also found as Pro, Leu, Gly, Ala, Trp in decreasing order, and Pro occurred most frequently in DDP-IV inhibitory peptides. Studies also showed that collagen from fish and mammals has also attracted notable attention as a potential source of DPP-IV inhibitory peptides partly due to its high content in Pro residue [51,52]. In addition, peptides with the presence of a Trp at the N-terminus was found as more potent DPP-IV inhibitors with an IC 50 value <200 mM [65]. A positive correlation between the presence of Trp-containing peptides within plant (hemp, pea, rice and soy) protein hydrolysates and their DPP-IV inhibitory properties have been observed [68].

However, containing the most preferring amino acid, i.e., Pro, Ile-Pro and Pro-Tyr were shown to be a DPP-IV inhibitor while Pro-Ile and Tyr-Pro were not [69,112]. The results indicated that the different stereochemistry between them may exert a role in the biological activity of the peptide. Indeed, the existence of exclusion volumes in the S1 pocket of DPP-IV might explain the results, which may restrict the access of bulky amino acids and allow access to smaller residues such as Pro, Ala and Gly [113].

In literature, hydrolysis fractions (<1 kDa) of hard-to-cook bean proteins and whey proteins hydrolysis both showed insulin secretagogue action and improved insulin signaling in adipocytes [36,114]. In vitro studies using pancreatic β-cell lines or primary islet cells displayed significant insulinotropic effects of different amino acid residues, including Ala, Leu, Arg and Gln [115,116]. Without carbohydrates, only ingestion of amino acids (Leu, Arg, Ile, Phe and Ala) or milk-derived peptides also exhibited an increase of insulin secretion [117,118]. Furthermore, studies verified that branched chain amino acids were closely associated with insulinotropic effects [115,116,119,120].

Investigation of whey protein hydrolysis showed that the most potent insulinotropic fractions obtained were hydrophilic, which might be responsible for the activity observed [114]. However, other peptides with hydrophobic characteristics may also contribute to the insulinotropic properties of the whey protein hydrolysate [63]. In addition, the levels of Arg and Phe were found to be associated with an insulinotropic activity [116]. According to the results above, it was inferred that free amino acids and dipeptides would be bioavailable and therefore may reach pancreatic β cells in vivo.

The complex mechanisms of these amino acids exerting their action involve mitochondrial metabolism. During fasting periods, glutamine and alanine are important factors to modulate glucagon release from pancreatic α-cells and subsequently influence insulin secretion from β-cells. On the other hand, high glucose levels raise ATP/ADP ratio in β-cells, and inhibit glutamate oxidation to amplify insulin signals [36].

The structural features of α-amylase peptide inhibitor were investigated by Yu et al. [108]. The amino acids, such as Leu, Pro, Gly and Phe, are frequently found in α-amylase inhibitory activity. Peptides with Pro at the N-terminal of Gly or Phe and C-terminal of Phe or Leu were found owning α-amylase inhibitory activity. The positioning of these amino acids at the N- or C-terminal were believed to be the contributors to α-amylase inhibitory activity of peptides extracted from Pinto bean [121]. Some reports showed that the high molecular weights of amino acids with aromatic ring, such as Arg, Phe, Trp, and Tyr, were crucial for interacting with the active site of human pancreatic α-amylases [104,122]. However, the results on the structural active relationship of α-amylase inhibitor peptides are still limited and not many studies have been conducted. Evenly, the data on the structure features for peptides such as potential α-glucosidase inhibitors have not been found in literature [108].

5. Conclusions

Nowadays, the discovery of novel ocean bioactive peptides is one of the most exciting new directions of pharmaceutical science due to their nutritional attributes, large output and uniqueness in terms of diversity, and structural and functional features with respect to peptides isolated from terrestrial plants. The diverse nature of T2DM means that food ingredients, such as natural peptides, will be more suitable for combating it and its associated complications than the synthetic and other drugs with significant side effects in the long term. This review concluded that natural origin peptides derived from several kinds of marine organisms, for instance, macro- and micro-algae, marine sponge, fish skin gelatin, and even tuna cooking juice hydrolysates, besides from milk and beans, showed great potential to regulate glucose metabolism for insulin resistance individuals. *Chlorella vulgaris*, one type of marine microalgae with large biomass and high quality protein accounting for over 60% (Wt), has been reported with significant antidiabetic activities in rare clinical trials. This evidence suggested that valuable antidiabetic activities associated with marine bioactive peptides, especially derived from marine microorganisms and their metabolites might be used in future potentialities in nutraceutical and pharmaceutical industries. Investigation of the structural features of peptides linked with anti-diabetic activity, using bioinformatics combined with molecular biological technology, will be a powerful tool to exploit new peptides from abundant marine proteins. In addition, the performances of research studies using human models or clinical trials are necessary in the future for their further application.

Acknowledgments: This work was supported by grants from the National Natural Science Foundation of China (81302416), the Guangdong Science and Technology Planning Project (2014A020212297), the Guangdong Training Plan for Outstanding Young Teachers (YQ201405), the Dongguan Science and Technology Planning Project (2014108101053), and a grant from the Dongguan Key Laboratory of Environmental Medicine.

Author Contributions: Tang-Bin Zou designed the review; En-Qin Xia and Shan-Shan Zhu analyzed and wrote drafts of the manuscript; Min-Jing He, Fei Luo and Cheng-Zhan Fu helped rewrite the revised manuscript; Tang-Bin Zou and En-Qin Xia revised the manuscript and approved it in its final form. All authors read and approved the final manuscript.

Conflicts of Interest: The author declares no conflict of interest.

References

1. Lee, J.E.; Min, S.H.; Lee, D.H.; Oh, T.J.; Kim, K.M.; Moon, J.H.; Choi, S.H.; Park, K.S.; Jang, H.C.; Lim, S. Comprehensive assessment of lipoprotein subfraction profiles according to glucose metabolism status, and association with insulin resistance in subjects with early-stage impaired glucose metabolism. *Int. J. Cardiol.* **2016**, *225*, 327–331. [CrossRef] [PubMed]

2. Leavens, K.F.; Birnbaum, M.J. Insulin signaling to hepatic lipid metabolism in health and disease. *Crit. Rev. Biochem. Mol. Biol.* **2011**, *46*, 200–215. [CrossRef] [PubMed]

3. So, W.Y.; Leung, P.S. Irisin ameliorates hepatic glucose/lipid metabolism and enhances cell survival in insulin-resistant human HepG2 cells through adenosine monophosphate-activated protein kinase signaling. *Int. J. Biochem. Cell Biol.* **2016**, *78*, 237–247. [PubMed]

4. Maulucci, G.; Daniel, B.; Cohen, O.; Avrahami, Y.; Sasson, S. Hormetic and regulatory effects of lipid peroxidation mediators in pancreatic β cells. *Mol. Aspects Med.* **2016**, *49*, 49–77. [CrossRef] [PubMed]

5. Zhu, C.F.; Peng, H.B.; Liu, G.Q.; Zhang, F.; Li, Y. Beneficial effects of oligopeptides from marine salmon skin in a rat model of type 2 diabetes. *Nutrition* **2010**, *26*, 1014–1020. [CrossRef] [PubMed]

6. Pandey, S.; Sree, A.; Dash, S.S.; Sethi, D.P.; Chowdhury, L. Diversity of marine bacteria producing β-glucosidase inhibitors. *Microb. Cell Fact.* **2013**, *12*, 35. [CrossRef] [PubMed]

7. Shibata, S.; Hayakawa, K.; Egashira, Y.; Sanada, H. Hypocholesterolemic mechanism of Chlorella: Chlorella and its indigestible fraction enhance hepatic cholesterol catabolism through upregulation of cholesterol 7α-hydroxylase in rats. *Biosci. Biotechnol. Biochem.* **2007**, *71*, 916–925. [CrossRef] [PubMed]

8. Mello-Sampayo, C.; Luisa-Corvo, M.; Mendes, R.; Duarte, D.; Lucas, J.; Pinto, R. Insights on the safety of carotenogenic Chlorella vulgaris in rodents. *Algal Res.* **2013**, *2*, 409–915.

9. Belgardt, B.F.; Ahmed, K.; Spranger, M.; Latreille, M.; Denzler, R.; Kondratiuk, N.; von, Meyenn, F.; Villena, F.N.; Herrmanns, K.; Bosco, D.; et al. The microRNA-200 family regulates pancreatic β cell survival in type 2 diabetes. *Nat. Med.* **2015**, *21*, 619–627. [CrossRef] [PubMed]

10. Kaiser, N.; Leibowitz, G. Failure of β-cell adaptation in type 2 diabetes Lessons from animal models. *Front. Biosci. (Landmark Ed).* **2009**, *14*, 1099–1115. [CrossRef] [PubMed]

11. Houstis, N.; Rosen, E.D.; Lander, E.S. Reactive oxygen species have a causal role in multiple forms of insulin resistance. *Nature* **2006**, *440*, 944–948. [CrossRef] [PubMed]

12. Khan, S.R. Is oxidative stress; a link between nephrolithiasis and obesity; hypertension; diabetes; chronic kidney disease; metabolic syndrome? *Urol. Res.* **2012**, *40*, 95–112. [CrossRef] [PubMed]

13. Roberts, C.K.; Sindhu, K.K. Oxidative stress and metabolic syndrome. *Life Sci.* **2009**, *84*, 705–712. [PubMed]

14. Fernandez-Millan, E.; Cordero-Herrera, I.; Ramos, S.; Escriva, F.; Alvarez, C.; Goya, L.; Martin, M.A. Cocoa-rich diet attenuates β cell mass loss and function in young Zucker diabetic fatty rats by preventing oxidative stress and β cell apoptosis. *Mol. Nutr. Food Res.* **2015**, *59*, 820–824. [CrossRef] [PubMed]

15. Bayod, S.; Del, Valle, J.; Lalanza, J.F.; Sanchez-Roige, S.; de Luxan-Delgado, B.; Coto-Montes, A.; Canudas, A.M.; Camins, A.; Escorihuela, R.M.; Pallas, M. Long-term physical exercise induces changes in sirtuin 1 pathway and oxidative parameters in adult rat tissues. *Exp. Gerontol.* **2012**, *47*, 925–935. [CrossRef] [PubMed]

16. Carnagarin, R.; Dharmarajan, A.M.; Dass, C.R. PEDF-induced alteration of metabolism leading to insulin resistance. *Mol. Cell Endocrinol.* **2015**, *40*, 98–104.

17. Cnop, M.; Igoillo-Esteve, M.; Cunha, D.A.; Ladriere, L.; Eizirik, D.L. An update on lipotoxic endoplasmic reticulum stress in pancreatic β-cells. *Biochem. Soc. Trans.* **2008**, *36*, 909–915.

18. Lenzen, S. Oxidative stress the vulnerable β-cell. *Biochem. Soc. Trans.* **2008**, *36*, 343–347. [CrossRef] [PubMed]

19. Newsholme, P.; Haber, E.P.; Hirabara, S.M.; Rebelato, E.L.; Procopio, J.; Morgan, D.; Oliveira-Emilio, H.C.; Carpinelli, A.R.; Curi, R. Diabetes associated cell stress and dysfunction: Role of mitochondrial and non-mitochondrial ROS production and activity. *J. Physiol.* **2007**, *583*, 9–24. [CrossRef] [PubMed]

20. Poitout, V.; Amyot, J.; Semache, M.; Zarrouki, B.; Hagman, D.; Fontes, G. Glucolipotoxicity of the pancreatic β cell. *Biochim. Biophys. Acta* **2010**, *1801*, 289–298. [CrossRef]

21. Mailloux, R.J.; Fu, A.; Robson-Doucette, C.; Allister, E.M.; Wheeler, M.B.; Screaton, R.; Harper, M.E. Glutathionylation state of uncoupling protein-2 and the control of glucose-stimulated insulin secretion. *J. Biol. Chem.* **2012**, *287*, 39673–39685. [PubMed]

22. Lenzen, S.; Drinkgern, J.; Tiedge, M. Low antioxidant enzyme gene expression in pancreatic islets compared with various other mouse tissues. *Free Radic. Biol. Med.* **1996**, *20*, 463–466. [PubMed]

23. Tiedge, M.; Lortz, S.; Drinkgern, J.; Lenzen, S. Relation between antioxidant enzyme gene expression and antioxidative defense status of insulin-producing cells. *Diabetes* **1997**, *46*, 1733–1742. [PubMed]

24. Corbett, J.A.; Wang, J.L.; Hughes, J.H.; Wolf, B.A.; Sweetland, M.A.; Lancaster, J.R.; McDaniel, M.L. Nitric oxide and cyclic GMP formation induced by interleukin 1 β in islets of Langerhans. Evidence for an effector role of nitric oxide in islet dysfunction. *Biochem. J.* **1992**, *287*, 229–235. [PubMed]

25. Morgan, D.; Oliveira-Emilio, H.R.; Keane, D.; Hirata, A.E.; Santos, da, Rocha, M.; Bordin, S.; Curi, R.; Newsholme, P.; Carpinelli, A.R. Glucose, palmitate and pro-inflammatory cytokines modulate production and activity of a phagocyte-like NADPH oxidase in rat pancreatic islets and a clonal β cell line. *Diabetologia* **2007**, *50*, 359–369. [PubMed]

26. Lei, X.G.; Vatamaniuk, M.Z. Two tales of antioxidant enzymes on β cells and diabetes. *Antioxid. Redox Signal.* **2011**, *14*, 489–503. [PubMed]

27. Tiedge, M.; Lortz, S.; Munday, R.; Lenzen, S. Protection against the co-operative toxicity of nitric oxide and oxygen free radicals by overexpression of antioxidant enzymes in bioengineered insulin-producing RINm5F cells. *Diabetologia* **1999**, *42*, 849–855.

28. Wolf, G.; Aumann, N.; Michalska, M.; Bast, A.; Sonnemann, J.; Beck, J.F.; Lendeckel, U.; Newsholme, P.; Walther, R. Peroxiredoxin III protects pancreatic ss cells from apoptosis. *J. Endocrinol.* **2010**, *207*, 163–175. [PubMed]

29. Li, X.; Chen, H.; Epstein, P.N. Metallothionein and catalase sensitize to diabetes in nonobese diabetic mice reactive oxygen species may have a protective role in pancreatic β-cells. *Diabetes* **2006**, *55*, 1592–1604. [PubMed]

30. Ibrahim, M.A.; Koorbanally, N.A.; Islam, M.S. Antioxidative activity and inhibition of key enzymes linked to type-2 diabetes (α-glucosidase and α-amylase) by Khaya senegalensis. *Acta Pharm.* **2014**, *64*, 311–324. [CrossRef]

31. Choi, H.K.; Willett, W.C.; Stampfer, M.J.; Rimm, E.; Hu, F.B. Dairy consumption and risk of type 2 diabetes mellitus in men a prospective study. *Arch. Intern. Med.* **2005**, *165*, 997–1003. [PubMed]

32. Tremblay, A.; Gilbert, J.A. Milk products; insulin resistance syndrome and type 2 diabetes. *J. Am. Coll. Nutr.* **2009**, *28* (Suppl. S1), 91S–102S. [PubMed]

33. Nasri, R.; Abdelhedi, O.; Jemil, I.; Daoued, I.; Hamden, K.; Kallel, C.; Elfeki, A.; Lamri-Senhadji, M.; Boualga, A.; Nasri, M.; et al. Ameliorating effects of goby fish protein hydrolysates on high-fat-high-fructose diet-induced hyperglycemia; oxidative stress and deterioration of kidney function in rats. *Chem-Biol. Interact.* **2015**, *24*, 271–280.

34. Ben, Khaled, H.; Ghlissi, Z.; Chtourou, Y.; Hakim, A.; Ktari, N.; Fatma, M.A.; Barkia, A.; Sahnoun, Z.; Nasri, M. Effect of protein hydrolysates from sardinelle (*Sardinella. aurita*) on the oxidative status and blood lipid profile of cholesterol-fed rats. *Food Res. Int.* **2012**, *45*, 60–68.

35. Ktari, N.; Nasri, R.; Mnafgui, K.; Hamden, K.; Belguith, O.; Boudaouara, T.; El, Feki, A.; Nasri, M. Antioxidative and ACE inhibitory activities of protein hydrolysates from zebra blenny (*Salaria. basilisca*) in alloxan-induced diabetic rats. *Process Biochem.* **2014**, *49*, 890–897.

36. Oseguera, Toledo, M.E.; Gonzalez de Mejia, E.; Sivaguru, M.; Amaya-Llano, S.L. Common bean (*Phaseolus. vulgaris* L.) protein-derived peptides increased insulin secretion; inhibited lipid accumulation; increased glucose uptake and reduced the phosphatase and tensin homologue activation in vitro. *J. Func. Foods* **2016**, *27*, 160–177.

37. Fernández-Tomé, S.; Ramos, S.; Cordero-Herrera, I.; Recio, I.; Goya, L.; Hernández-Ledesma, B. In vitro chemo-protective effect of bioactive peptide lunasin against oxidative stress in human HepG2 cells. *Food Res. Int.* **2014**, *62*, 793–800.

38. Lee, K.W.; Kim, S.J. Uptake of modified LDLs in HepG2 cells and cholesterol accumulation by modified LDLs in THP-1 macrophages. *Toxicol. Lett.* **2010**, *196*, s243. [CrossRef]

39. Han, D.N.; Zhang, D.H.; Wang, L.P.; Zhang, Y.S. Protective effect of β-casomorphin-7 on cardiomyopathy of streptozotocin-induced diabetic rats via inhibition of hyperglycemia and oxidative stress. *Peptides* **2013**, *44*, 120–126. [CrossRef]

40. Donath, M.Y.; Storling, J.; Maedler, K.; Mandrup-Poulsen, T. Inflammatory mediators and islet β-cell failure a link between type 1 and type 2 diabetes. *J. Mol. Med. (Berl.).* **2003**, *81*, 455–470. [CrossRef]

41. Cherng, J.Y.; Liu, C.C.; Shen, C.R.; Lin, H.H.; Shih, M.F. Beneficial effects of Chlorella-11 peptide on blocking LPS-induced macrophage activation and alleviating thermal injury-induced inflammation in rats. *Int. J. Immuno. Pathol. Pharmacol.* **2010**, *23*, 811–820. [CrossRef] [PubMed]

42. Polyzos, S.A.; Kountouras, J.; Mantzoros, C.S. Adipokines in nonalcoholic fatty liver disease. *Metabolism* **2016**, *65*, 1062–1079. [PubMed]

43. Moran, T.H.; Dailey, M.J. Minireview Gut peptides targets for antiobesity drug development? *Endocrinology* **2009**, *150*, 2526–2530. [PubMed]

44. Perry, B.; Wang, Y. Appetite regulation and weight control the role of gut hormones. *Nutr. Diabetes* **2012**, *2*, e26. [CrossRef] [PubMed]

45. Troke, R.C.; Tan, T.M.; Bloom, S.R. The future role of gut hormones in the treatment of obesity. *Ther. Adv. Chronic Dis.* **2014**, *5*, 4–14.

46. Nongonierma, A.B.; FitzGerald, R.J. Inhibition of dipeptidyl peptidase IV (DPP-IV) by proline containing casein-derived peptides. *J. Funct. Foods* **2013**, *5*, 1909–1917.

47. Omar, B.; Ahlkvist, L.; Yamada, Y.; Seino, Y.; Ahren, B. Incretin hormone receptors are required for normal β cell development and function in female mice. *Peptides* **2016**, *79*, 58–65. [CrossRef] [PubMed]

48. Liddle, R.A.; Carter, J.D.; McDonald, A.R. Natural regulation of rat intestinal cholecystokinin gene expression. *J. Clin. Invest.* **1988**, *81*, 2015–2019.

49. Caron, J.; Domenger, D.; Belguesmia, Y.; Kouach, M.; Lesage, J.; Goossens, J.F.; Dhulster, P.; Ravallec, R.; Cudennec, B. Protein digestion and energy homeostasis How generated peptides may impact intestinal hormones? *Food Res. Int.* **2016**, *88*, 310–318.

50. Harnedy, P.A.; O'Keeffe, M.B.; FitzGerald, R.J. Purification and identification of dipeptidyl peptidase (DPP) IV inhibitory peptides from the macroalga *Palmaria. palmata. Food Chem.* **2015**, *172*, 400–406. [PubMed]

51. Wang, T.Y.; Hsieh, C.H.; Hung, C.C.; Jao, C.L.; Chen, M.C.; Hsu, K.C. Fish skin gelatin hydrolysates as dipeptidyl peptidase IV inhibitors and glucagon-like peptide-1 stimulators improve glycaemic control in diabetic rats: A comparison between warm- and cold-water fish. *J. Func. Foods* **2015**, *18*, 330–340.

52. Huang, S.L.; Jao, C.L.; Ho, K.P.; Hsu, K.C. Dipeptidyl-peptidase IV inhibitory activity of peptides derived from tuna cooking juice hydrolysates. *Peptides* **2012**, *35*, 114–121. [CrossRef] [PubMed]

53. Jin, Y.; Yan, J.; Yu, Y.; Qi, Y. Screening and identification of DPP-IV inhibitory peptides from deer skin hydrolysates by an integrated approach of LC–MS/MS and in silico analysis. *J. Funct. Foods* **2015**, *18*, 344–357. [CrossRef]

54. Lacroix, I.M.; Li-Chan, E.C. Overview of food products and natural constituents with antidiabetic properties and their putative mechanisms of action a natural approach to complement pharmacotherapy in the management of diabetes. *Mol. Nutr. Food Res.* **2014**, *58*, 61–78. [CrossRef] [PubMed]

55. Uchida, M.; Ohshiba, Y.; Mogami, O. Novel dipeptidyl peptidase-4-inhibiting peptide derived from β-lactoglobulin. *J. Pharmacol. Sci.* **2011**, *117*, 63–66. [CrossRef] [PubMed]

56. Silveira, S.T.; Martinez-Maqueda, D.; Recio, I.; Hernandez-Ledesma, B. Dipeptidyl peptidase-IV inhibitory peptides generated by tryptic hydrolysis of a whey protein concentrate rich in β-lactoglobulin. *Food Chem.* **2013**, *141*, 1072–1077. [PubMed]

57. Uenishi, K. Diabetes mellitus and osteoporosis. Natural therapy of diabetes related osteoporosis. *Clin. Calcium.* **2012**, *22*, 1398–1402. [PubMed]

58. Lacroix, I.M.; Li-Chan, E.C. Inhibition of dipeptidyl peptidase (DPP)-IV and α-glucosidase activities by pepsin-treated whey proteins. *J. Agric. Food Chem.* **2013**, *61*, 7500–7506. [CrossRef]

59. Lacroix, I.M.; Li-Chan, E.C. Isolation and characterization of peptides with dipeptidyl peptidase-IV inhibitory activity from pepsin-treated bovine whey proteins. *Peptides* **2014**, *54*, 39–48. [CrossRef] [PubMed]

60. Lacroix, I.M.; Li-Chan, E.C. Comparison of the susceptibility of porcine and human dipeptidyl-peptidase IV to inhibition by protein-derived peptides. *Peptides* **2015**, *69*, 19–25. [PubMed]

61. Lacroix, I.M.E.; Meng, G.; Cheung, I.W.Y.; Li-Chan, E.C.Y. Do whey protein-derived peptides have dual dipeptidyl-peptidase IV and angiotensin I-converting enzyme inhibitory activities? *J. Funct. Foods* **2016**, *21*, 87–96. [CrossRef]

62. Tulipano, G.; Sibilia, V.; Caroli, A.M.; Cocchi, D. Whey proteins as source of dipeptidyl dipeptidase IV (dipeptidyl peptidase-4) inhibitors. *Peptides* **2011**, *32*, 835–838. [CrossRef] [PubMed]

63. Le Maux, S.; Nongonierma, A.B.; FitzGerald, R.J. Improved short peptide identification using HILIC-MS/MS retention time prediction model based on the impact of amino acid position in the peptide sequence. *Food Chem.* **2015**, *17*, 3847–3854. [CrossRef]

64. Nongonierma, A.B.; FitzGerald, R.J. Dipeptidyl peptidase IV inhibitory and antioxidative properties of milk protein-derived dipeptides and hydrolysates. *Peptides* **2013**, *39*, 157–163. [CrossRef] [PubMed]

65. Nongonierma, A.B.; Mooney, C.; Shields, D.C.; FitzGerald, R.J. In silico approaches to predict the potential of milk protein-derived peptides as dipeptidyl peptidase IV (DPP-IV) inhibitors. *Peptides* **2014**, *57*, 43–51. [CrossRef] [PubMed]

66. Nongonierma, A.B.; FitzGerald, R.J. Strategies for the discovery; identification and validation of milk protein-derived bioactive peptides. *Trends. Food Sci. Tec.* **2016**, *50*, 26–43.

67. De Souza Rocha, T.; Hernandez, L.M.R.; Chang, Y.K.; de Mejía, E.G. Impact of germination and enzymatic hydrolysis of cowpea bean (*Vigna. unguiculata*) on the generation of peptides capable of inhibiting dipeptidyl peptidase IV. *Food Res. Int.* **2014**, *64*, 799–809.

68. Nongonierma, A.B.; Le, Maux, S.; Dubrulle, C.; Barre, C.; FitzGerald, R.J. Quinoa (*Chenopodium. quinoa* Willd.) protein hydrolysates with In Vitro dipeptidyl peptidase IV (DPP-IV) inhibitory and antioxidant properties. *J. Cereal Sci.* **2015**, *65*, 112–118.

69. Hatanaka, T.; Inoue, Y.; Arima, J.; Kumagai, Y.; Usuki, H.; Kawakami, K.; Kimura, M.; Mukaihara, T. Production of dipeptidyl peptidase IV inhibitory peptides from defatted rice bran. *Food Chem.* **2012**, *134*, 797–802. [PubMed]

70. Velarde-Salcedo, A.J.; Barrera-Pacheco, A.; Lara-Gonzalez, S.; Montero-Morán, G.M.; Díaz-Gois, A.; de Mejia, E.G.; Barba, de la Rosa, A.P. In Vitro inhibition of dipeptidyl peptidase IV by peptides derived from the hydrolysis of amaranth (*Amaranthus hypochondriacus* L.) proteins. *Food Chem.* **2013**, *136*, 758–764. [CrossRef] [PubMed]

71. Zeng, Z.; Shuai, T.; Yi, L.J.; Wang, Y.; Song, G.M. Effect of case management on patients with type 2 diabetes mellitus: A meta-analysis. *Chinese Nursing Research* **2016**, *3*, 71–76.

72. Jong-Yuh, C.; Mei-Fen, S. Potential hypoglycemic effects of Chlorella in streptozotocin-induced diabetic mice. *Life Sci.* **2005**, *77*, 980–990. [CrossRef]

73. Jeong, H.; Kwon, H.J.; Kim, M.K. Hypoglycemic effect of Chlorella vulgaris intake in type 2 diabetic Goto-Kakizaki and normal Wistar rats. *Nutr. Res. Prac.* **2009**, *3*, 23–30.

74. Lu, J.; Zeng, Y.; Hou, W.; Zhang, S.; Li, L.; Luo, X.; Xi, W.; Chen, Z.; Xiang, M. The soybean peptide aglycin regulates glucose homeostasis in type 2 diabetic mice via IR/IRS1 pathway. *J. Nutr. Biochem.* **2012**, *23*, 1449–1457. [CrossRef] [PubMed]

75. Veloso, R.V.; Latorraca, M.Q.; Arantes, V.C.; Reis, M.A.; Ferreira, F.; Boschero, A.C.; Carneiro, E.M. Soybean diet improves insulin secretion through activation of cAMP/PKA pathway in rats. *J. Nutr. Biochem.* **2008**, *19*, 778–784. [PubMed]

76. Roblet, C.; Akhtar, M.J.; Mikhaylin, S.; Pilon, G.; Gill, T.; Marette, A.; Bazinet, L. Enhancement of glucose uptake in muscular cell by peptide fractions separated by electrodialysis with filtration membrane from salmon frame protein hydrolysate. *J. Funct. Foods* **2016**, *22*, 337–346.

77. Kwon, D.Y.; Daily, J.W.; Kim, H.J.; Park, S. Antidiabetic effects of fermented soybean products on type 2 diabetes. *Nutr. Res.* **2010**, *30*, 1–13. [CrossRef]

78. Martinez-Villaluenga, C.; Bringe, N.A.; Berhow, M.A.; Gonzalez, de Mejia, E. B-conglycinin embeds active peptides that inhibit lipid accumulation in 3T3-L1 adipocytes In Vitro. *J. Agric. Food Chem.* **2008**, *56*, 10533–10543. [CrossRef] [PubMed]

79. Martinez-Villaluenga, C.; Dia, V.P.; Berhow, M.; Bringe, N.A.; Gonzalez de Mejia, E. Protein hydrolysates from β-conglycinin enriched soybean genotypes inhibit lipid accumulation and inflammation In Vitro. *Mol. Nutr. Food Res.* **2009**, *53*, 1007–1018. [CrossRef] [PubMed]

80. Kim, E.D.; Kim, E.; Lee, J.H.; Hyun, C.K. Gly-Ala-Gly-Val-Gly-Tyr: A novel synthetic peptide; improves glucose transport and exerts beneficial lipid metabolic effects in 3T3-L1 adipoctyes. *Eur. J. Pharmacol.* **2011**, *650*, 479–485. [CrossRef]

81. Hammé, V.; Sannier, F.; Piot, J.M.; Bordenave-Juchereau, S. Effects of lactokinins from fermented acid goat whey on lipid content and adipogenesis of immortalised human adipocytes. *Int. Dairy J.* **2010**, *20*, 642–645. [CrossRef]

82. Yim, M.J.; Hosokawa, M.; Mizushina, Y.; Yoshida, H.; Saito, Y.; Miyashita, K. Suppressive effects of Amarouciaxanthin A on 3T3-L1 adipocyte differentiation through downregulation of PPARγ and C/EBPα mRNA expression. *J. Agric. Food Chem.* **2011**, *59*, 1646–1652. [PubMed]

83. Ben, Henda, Y.; Laamari, M.; Lanneluc, I.; Travers, M.A.; Agogué, H.; Arnaudin, I.; Bridiau, N.; Maugard, T.; Piot, J.M.; Sannier, F.; et al. Di and tripeptides from marine sources can target adipogenic process and contribute to decrease adipocyte number and functions. *J. Funct. Foods* **2015**, *17*, 1–10.

84. Wang, L.L.; Hao, S.; Zhang, S.; Guo, L.J.; Hu, C.Y.; Zhang, G.; Gao, B.; Zhao, J.J.; Jiang, Y.; Tian, W.G.; et al. PTEN/PI3K/AKT protein expression is related to clinicopathologic features and prognosis in breast cancer with axillary lymph node metastases. *Hum. Pathol.* **2017**, *61*, 49–57. [PubMed]

85. Govers, R. Molecular mechanisms of GLUT4 regulation in adipocytes. *Diabetes Metab.* **2014**, *40*, 400–410. [CrossRef]

86. Morino, K.; Neschen, S.; Bilz, S.; Sono, S.; Tsirigotis, D.; Reznick, R.M.; Moore, I.; Nagai, Y.; Samuel, V.; Sebastian, D.; et al. Muscle-specific IRS-1 Ser->Ala transgenic mice are protected from fat-induced insulin resistance in skeletal muscle. *Diabetes* **2008**, *57*, 2644–2651. [CrossRef] [PubMed]

87. Bozulic, L.; Hemmings, B.A. PIKKing on PKB regulation of PKB activity by phosphorylation. *Curr. Opin. Cell Biol.* **2009**, *21*, 256–261. [PubMed]

88. Davis, J.; Higginbotham, A.; O'Connor, T.; Moustaid-Moussa, N.; Tebbe, A.; Kim, Y.C.; Cho, K.W.; Shay, N.; Adler, S.; Peterson, R.; et al. Soy protein and isoflavones influence adiposity and development of metabolic syndrome in the obese male ZDF rat. *Ann. Nutr. Metab.* **2007**, *51*, 42–52. [PubMed]

89. Nordentoft, I.; Jeppesen, P.B.; Hong, J.; Abudula, R.; Hermansen, K. Increased insulin sensitivity and changes in the expression profile of key insulin regulatory genes and β cell transcription factors in diabetic KKAy-mice after feeding with a soy bean protein rich diet high in isoflavone content. *J. Agric. Food Chem.* **2008**, *56*, 4377–4385. [CrossRef] [PubMed]

90. Han, B.K.; Lee, H.J.; Lee, H.S.; Suh, H.J.; Park, Y. Hypoglycaemic effects of functional tri-peptides from silk in differentiated adipocytes and streptozotocin-induced diabetic mice. *J. Sci. Food Agric.* **2016**, *96*, 116–121.

91. Huang, K.C.; Huang, H.J.; Chen, C.C.; Chang, C.T.; Wang, T.Y.; Chen, R.H.; Chen, Y.C.; Tsai, F.J. Susceptible gene of stasis-stagnation constitution from genome-wide association study related to cardiovascular disturbance and possible regulated traditional Chinese medicine. *BMC Complement. Altern. Med.* **2015**, *15*, 229. [CrossRef] [PubMed]

92. McCurdy, C.E.; Klemm, D.J. Adipose tissue insulin sensitivity and macrophage recruitment Does PI3K pick the pathway? *Adipocyte* **2013**, *21*, 135–142.

93. Zhang, Y.; Hai, J.; Cao, M.; Zhang, Y.; Pei, S.; Wang, J.; Zhang, Q. Silibinin ameliorates steatosis and insulin resistance during non-alcoholic fatty liver disease development partly through targeting IRS-1/PI3K/Akt pathway. *Int. Immunopharmacol.* **2013**, *17*, 714–720. [PubMed]

94. Ishihara, K.; Oyaizu, S.; Fukuchi, Y.; Mizunoya, W.; Segawa, K.; Takahashi, M.; Mita, Y.; Fukuya, Y.; Fushiki, T.; Yasumoto, K. A soybean peptide isolate diet promotes postprandial carbohydrate oxidation and energy expenditure in type II diabetic mice. *J. Nutr.* **2003**, *133*, 752–757. [PubMed]

95. Zhu, K.N.; Jiang, C.H.; Tian, Y.S.; Xiao, N.; Wu, Z.F.; Ma, Y.L.; Lin, Z.; Fang, S.Z.; Shang, X.L.; Liu, K.; et al. Two triterpeniods from Cyclocarya paliurus (β1) Iljinsk (Juglandaceae) promote glucose uptake in 3T3-L1 adipocytes: The relationship to AMPK activation. *Phytomedicine* **2015**, *22*, 837–846.

96. Butler, A.E.; Janson, J.; Bonner-Weir, S.; Ritzel, R.; Rizza, R.A.; Butler, P.C. β-cell deficit and increased β-cell apoptosis in humans with type 2 diabetes. *Diabetes* **2003**, *52*, 102–110. [PubMed]

97. Mizoguchi, T.; Takehara, I.; Masuzawa, T.; Saito, T.; Naoki, Y. Nutrigenomic studies of effects of *Chlorella* on subjects with high-risk factors for lifestyle-related disease. *J. Med. Food* **2008**, *11*, 395–404. [PubMed]

98. Panahi, Y.; Ghamarchehreh, M.E.; Beiraghdar, F.; Zare, R.; Jalalian, H.R.; Sahebkar, A. Investigation of the effects of *Chlorella vulgaris* supplementation in patients with non-alcoholic fatty liver disease: A randomized clinical trial. *Hepatogastroenterology* **2012**, *59*, 2099–2103. [PubMed]

99. Ebrahimi-Mameghani, M.; Sadeghi, Z.; Farhangi, M.A.; Vaghef-Mehrabany, E.; Aliashrafi, S. Glucose homeostasis, insulin resistance and inflammatory biomarkers in patients with non-alcoholic fatty liver disease: Beneficial effects of supplementation with microalgae *Chlorella vulgaris*: A double-blind placebo-controlled randomized clinical trial. *Clin. Nutr.* **2016**. [CrossRef]

100. Jo, B.H.; Lee, C.S.; Song, H.R.; Lee, H.G.; Oh, H.M. Development of novel microsatellite markers for strain-specific identification of *Chlorella vulgaris*. *J. Microbiol. Biotechnol.* **2014**, *24*, 1189–1195.

101. Kim, M.; Kim, E.; Kwak, H.S.; Jeong, Y. The ingredients in Saengshik; a formulated health food; inhibited the activity of α-amylase and α-glucosidase as anti-diabetic function. *Nutr. Res. Pract.* **2014**, *8*, 602–606. [PubMed]

102. Ren, Y.; Liang, K.; Jin, Y.; Zhang, M.; Chen, Y.; Wu, H.; Lai, F. Identification and characterization of two novel α-glucosidase inhibitory oligopeptides from hemp (*Cannabis sativa* L.) seed protein. *J. Funct. Foods* **2016**, *26*, 439–450. [CrossRef]

103. Ngoh, Y.Y.; Gan, C.Y. Enzyme-assisted extraction and identification of antioxidative and α-amylase inhibitory peptides from Pinto beans (*Phaseolus. vulgaris* cv. Pinto). *Food Chem.* **2016**, *190*, 331–337. [PubMed]

104. Siow, H.L.; Gan, C.Y. Extraction; identification; and structure–activity relationship of antioxidative and α-amylase inhibitory peptides from cumin seeds (*Cuminum. cyminum*). *J. Funct. Foods* **2016**, *22*, 1–12. [CrossRef]

105. Siow, H.L.; Lim, T.S.; Gan, C.Y. Development of a workflow for screening and identification of α-amylase inhibitory peptides from food source using an integrated Bioinformatics-phage display approach Case study-Cumin seed. *Food Chem.* **2017**, *214*, 67–76. [CrossRef]

106. Mojica, L.; de Mejia, E.G. Optimization of enzymatic production of anti-diabetic peptides from black bean (*Phaseolus. vulgaris* L.) proteins; their characterization and biological potential. *Food. Funct.* **2016**, *7*, 713–727.

107. Uraipong, C.; Zhao, J. Rice bran protein hydrolysates exhibit strong in vitro α-amylase; β-glucosidase and ACE-inhibition activities. *J. Sci. Food Agric.* **2016**, *96*, 1101–1110. [CrossRef]

108. Yu, Z.; Yin, Y.; Zhao, W.; Liu, J.; Chen, F. Anti-diabetic activity peptides from albumin against α-glucosidase and α-amylase. *Food Chem.* **2012**, *135*, 2078–2085. [CrossRef]

109. Mojica, L.; Luna-Vital, D.A.; Gonzalez de Mejia, E. Characterization of peptides from common bean protein isolates and their potential to inhibit markers of type-2 diabetes; hypertension and oxidative stress. *J. Sci. Food Agric.* **2016**. [CrossRef]

110. Engel, M.; Hoffmann, T.; Wagner, L.; Wermann, M.; Heiser, U.; Kiefersauer, R.; Huber, R.; Bode, W.; Demuth, H.U.; Brandstetter, H. The crystal structure of dipeptidyl peptidase IV (CD26) reveals its functional regulation and enzymatic mechanism. *Proc. Natl. Acad. Sci. USA* **2003**, *100*, 5063–5068. [CrossRef]

111. Juillerat-Jeanneret, L. Dipeptidyl peptidase IV and its inhibitors therapeutics for type 2 diabetes and what else? *J. Med. Chem.* **2014**, *57*, 2197–2212.

112. Nongonierma, A.B.; FitzGerald, R.J. Susceptibility of milk protein-derived peptides to dipeptidyl peptidase IV (DPP-IV) hydrolysis. *Food Chem.* **2014**, *145*, 845–852.

113. Lu, I.L.; Tsai, K.C.; Chiang, Y.K.; Jiaang, W.T.; Wu, S.H.; Mahindroo, N.; Chien, C.H.; Lee, S.J.; Chen, X.; Chao, Y.S.; et al. A three-dimensional pharmacophore model for dipeptidyl peptidase IV inhibitors. *Eur. J. Med. Chem.* **2008**, *43*, 1603–1611. [CrossRef]

114. Nongonierma, A.B.; Gaudel, C.; Murray, B.A.; Flynn, S.; Kelly, P.M.; Newsholme, P.; FitzGerald, R.J. Insulinotropic properties of whey protein hydrolysates and impact of peptide fractionation on insulinotropic response. *Int. Dairy J.* **2013**, *32*, 163–168.

115. Dixon, G.; Nolan, J.; McClenaghan, N.; Flatt, P.R.; Newsholme, P. A comparative study of amino acid consumption by rat islet cells and the clonal β-cell line BRIN-BD11-the functional significance of L-alanine. *J. Endocrinol.* **2003**, *179*, 447–454.

116. Bender, K.; Newsholme, P.; Brennan, L.; Maechler, P. The importance of redox shuttles to pancreatic β-cell energy metabolism and function. *Biochem. Soc. Trans.* **2006**, *34*, 811–814. [CrossRef]

117. Power, O.; Hallihan, A.; Jakeman, P. Human insulinotropic response to oral ingestion of native and hydrolysed whey protein. *Amino Acids* **2009**, *37*, 333–339. [CrossRef]

118. Horner, K.; Drummond, E.; Brennan, L. Bioavailability of milk protein-derived bioactive peptides a glycaemic management perspective. *Nutr. Res. Rev.* **2016**, *29*, 91–101. [CrossRef]

119. Manders, R.J.; Koopman, R.; Sluijsmans, W.E.; van den Berg, R.; Verbeek, K.; Saris, W.H.; Wagenmakers, A.J.; van Loon, L.J. Co-ingestion of a protein hydrolysate with or without additional leucine effectively reduces postprandial blood glucose excursions in Type 2 diabetic men. *J. Nutr.* **2006**, *136*, 1294–1299.

120. Manders, R.J.; Praet, S.F.; Meex, R.C.; Koopman, R.; De Roos, A.L.; Wagenmakers, A.J.; Saris, W.H.; Van Loon, L.J. Protein hydrolysate/leucine co-ingestion reduces the prevalence of hyperglycemia in type 2 diabetic patients. *Diabetes Care* **2006**, *29*, 2721–2722.

121. Ngoh, Y.Y.; Lim, T.S.; Gan, C.Y. Screening and identification of five peptides from pinto bean with inhibitory activities against α-amylase using phage display technique. *Enzyme. Microb. Technol.* **2016**, *89*, 76–84.

122. Ochiai, T.; Sugita, T.; Kato, R.; Okochi, M.; Honda, H. Screening of an α-amylase inhibitor peptide by photolinker-peptide array. *Biosci. Biotechnol. Biochem.* **2012**, *76*, 819–824. [CrossRef]

marine drugs

MDPI

Article

In Vitro Antioxidant Activities of Enzymatic Hydrolysate from *Schizochytrium* sp. and Its Hepatoprotective Effects on Acute Alcohol-Induced Liver Injury In Vivo

Xixi Cai [1,2], Ana Yan [2], Nanyan Fu [1] and Shaoyun Wang [2,*]

[1] The Key Lab of Analysis and Detection Technology for Food Safety of the MOE, College of Chemistry, Fuzhou University, Fuzhou 350108, China; caixx_0123@163.com (X.C.); nanyan_fu@fzu.edu.cn (N.F.)
[2] College of Biological Science and Technology, Fuzhou University, Fuzhou 350108, China; m18144065085@163.com
* Correspondence: shywang@fzu.edu.cn; Tel.: +86-591-2286-6375

Academic Editors: Se-Kwon Kim and Anake Kijjoa
Received: 1 October 2016; Accepted: 7 April 2017; Published: 10 April 2017

Abstract: *Schizochytrium* protein hydrolysate (SPH) was prepared through stepwise enzymatic hydrolysis by alcalase and flavourzyme sequentially. The proportion of hydrophobic amino acids of SPH was 34.71%. The molecular weight (MW) of SPH was principally concentrated at 180–3000 Da (52.29%). SPH was divided into two fractions by ultrafiltration: SPH-I (MW < 3 kDa) and SPH-II (MW > 3 kDa). Besides showing lipid peroxidation inhibitory activity in vitro, SPH-I exhibited high DPPH and ABTS radicals scavenging activities with IC_{50} of 350 µg/mL and 17.5 µg/mL, respectively. In addition, the antioxidant activity of SPH-I was estimated in vivo using the model of acute alcohol-induced liver injury in mice. For the hepatoprotective effects, oral administration of SPH-I at different concentrations (100, 300 mg/kg BW) to the mice subjected to alcohol significantly decreased serum alanine aminotransferase (ALT) and aspartate aminotransferase (AST) activities and hepatic malondialdehyde (MDA) level compared to the untreated mice. Besides, SPH-I could effectively restore the hepatic superoxide dismutase (SOD), catalase (CAT), and glutathione peroxidase (GSH-Px) activities and glutathione (GSH) level. Results suggested that SPH was rich in biopeptides that could be exploited as antioxidant molecules against oxidative stress in human body.

Keywords: *Schizochytrium*; protein hydrolysate; antioxidant; hepatoprotective effects; alcohol-induced liver injury

1. Introduction

Schizochytrium sp., a kind of heterotrophic marine fungus, is well known for the production of Ω-3 fatty acids, pigments, proteins, polysaccharides, etc. [1,2]. A number of researchers have focused on the industrial production of docosahexaenoic acid for *Schizochytrium* sp. studies [3]. However, there is little information on the utilization of *Schizochytrium* sp. byproduct. In addition to a high content of fat, *Schizochytrium* sp. also contains a high amount of protein, which is about 40% (dry weight). Therefore, great efforts are needed to transform these biological wastes into value-added bioproducts. Thus, the utilization of protein recovered from the defatted byproduct presents an opportunity to develop pharmaceutical products and food ingredients.

Free radicals such as the superoxide anion radical ($O_2 \cdot^-$) and hydroxyl radical ($\cdot OH$) are highly reactive oxygen species (ROS) with single and unpaired electrons that are involved in biological oxidation process and can cause many adverse effects on food and biological systems [4]. In human organs, free radicals, which are inevitably produced through oxidative metabolism, can induce several diseases such as arteriosclerosis and cancer. Liver injury is a widespread disease that can be caused by an overload of xenobiotics, such as alcohol, CCl_4, and bromobenzene. Alcohol-induced liver injury has been one of the most frequent causes of liver diseases. The mechanism of liver dysfunction induced by alcohol is thought to involve the generation of free radicals, oxidative stress, and lipid peroxidation [5,6]. More attention has been paid to search for safe antioxidants for effective therapy of oxidative stress-induced diseases. Small molecules with strong antioxidant activities from plants [7,8] and algal [9] have been widely investigated. In addition, preparation of bioactive peptides from proteins through enzymatic hydrolysis has been a hot topic. Peptides from the hydrolysates of Alaska Pollock skin collagen [10], egg white protein [11], chickpea protein [12], and algae protein waste [13] have been prepared and shown to possess antioxidant activities in different oxidation systems.

In this study, *Schizochytrium* sp. byproduct protein hydrolysate was prepared by stepwise enzymatic hydrolysis. The in vitro antioxidant activities of the enzymatic hydrolysates and the hepatoprotective effects on acute alcohol-induced liver injury in vivo were evaluated. The present study suggests that *Schizochytrium* protein hydrolysates have the potential in increasing resistivity against oxidative stress in the human body.

2. Results and Discussion

2.1. Analyses of Amino Acid Composition and Molecular Weight Distribution of SPH

Schizochytrium sp. protein isolates (SP) were enzymatically hydrolyzed by alcalase and flavourzyme sequentially for the preparation of antioxidant peptides. It has been recognized that the amino acid composition of the peptides plays critical roles in their antioxidant activities. The amino acid composition of SPH was determined by amino acid automatic analyzer. Results showed that SPH was rich in Glx and Asx, which accounted for 17.66% and 15.89%, respectively (Table 1). In addition, the total hydrophobic amino acids content in SPH constituted 34.71%. Udenigwe et al. [14] indicated that acidic amino acids such as Glu and Asp contributed to the antioxidant activities of peptides due to the presence of excess electrons which could be donated during interaction with free radicals. For protein hydrolysates and peptides, an increase in hydrophobicity would increase their interaction with lipid targets or entry of the peptides into target organs through hydrophobic associations, which was good for enhancing their antioxidant effects [15–17]. In addition, SPH contained 5.79% Lys and 7.81% Arg. Reports have demonstrated that peptides containing amino acids with carboxyl or amino side chains, such as Glu, Gln, Lys, and Arg, could donate electrons or hydrogen atoms to interact with pro-oxidants and inactivate their activity [18–20]. Moreover, the amino acids that contained nucleophilic sulphur-containing side chains (Met and Cys), aromatic side chains (Phe and Tyr), or imidazole-containing side chains (His) could donate electron to convert radicals into stable molecules [21].

Table 1. Amino acid composition of *Schizochytrium* protein hydrolysate (SPH).

Amino Acids	Content (%)
Ile	3.52
Leu	9.96
Met	1.35
Phe	5.27
Thr	3.93
Val	5.17
Lys	5.79
Cys	0.62
Tyr	2.72
Asx [a]	15.89
Ser	5.42
Glx [b]	17.66

Table 1. *Cont.*

Amino Acids	Content (%)
Gly	3.88
Ala	5.28
Pro	4.16
Arg	7.81
His	1.57
THAA [c]	34.71
Total	100

[a] Asx: containing Asp and Asn; [b] Glx: containing Glu and Gln; [c] THAA: total hydrophobic amino acid.

Besides amino acid composition, the molecular weight of peptides is also a significant factor that reflects the antioxidant activities of peptides. MW distribution of SPH was determined using HPLC and the results are shown in Figure 1. The fraction of peptides with MW ranging from 180 to 3000 Da was abundant in SPH, accounting for 52.29%. There are several reports suggesting that peptides with low MW have stronger antioxidant activities than their high MW counterparts. In fact, peptides with low MW could cross the intestinal barrier and further exert their antioxidant effects [22,23].

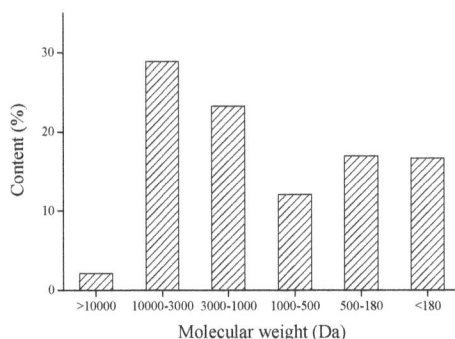

Figure 1. Molecular mass distribution of SPH.

2.2. In Vitro Antioxidant Activities of SPH and Its Fractions

The degrees of hydrolysis (DHs) and DPPH radical scavenging activity of hydrolysate were studied at different hydrolysis stages (Data not shown). The DPPH radical scavenging activity increased from 12.16% (SP) to 38.08% after the first step of hydrolysis by alcalase and the DH reached 8.37%. The activity was further enhanced to 58.06% at the second step of hydrolysis by flavourzyme with a DH of 21.48%.

In order to study the effect of MW on the antioxidant activities of the peptides, SPH was further fractionated by ultrafiltration to obtain SPH-I (MW < 3 kDa) and SPH-II (MW > 3 kDa). To evaluate the antioxidant activities of SPH and its fractions in vitro, different antioxidant parameters were obtained.

2.2.1. Free Radical Scavenging Activities

The ability of SPH and its fractions to scavenge DPPH and ABTS radicals is shown in Figure 2. DPPH and ABTS radicals scavenging activities of SPH and its fractions increased in a concentration dependent manner. SPH-I (MW < 3 kDa) had higher DPPH and ABTS radicals scavenging activities than SPH and SPH-II (MW > 3 kDa) at the same concentration. A lower IC_{50} value was indicative of higher scavenging activity, and the IC_{50} values of SPH-I against DPPH and ABTS radicals were 350 µg/mL and 17.5 µg/mL, respectively. These results indicated that free radical scavenging activities of peptides were related to their MW. Similar results were reported by Li et al. [12], who found that the fraction with low MW of chickpea protein hydrolysate had the highest DPPH radical scavenging activity compared to other fractions. In addition, the peptides with MW < 1 kDa from egg white protein hydrolysate and ethanol-soluble proteins hydrolysate of the *Sphyrna lewini* muscle were found to exhibit higher antioxidant activities than the high MW fractions [11,24].

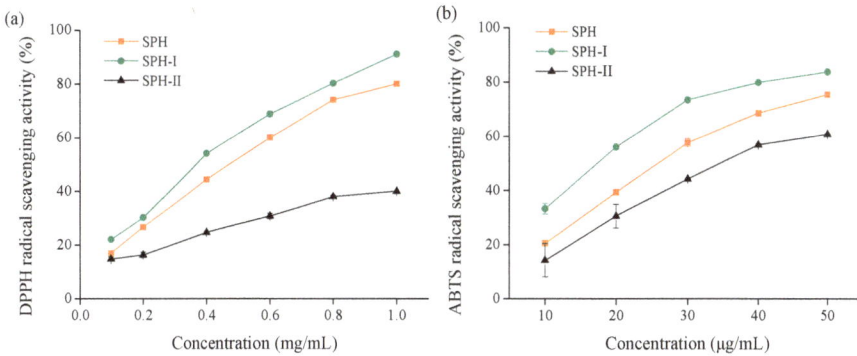

Figure 2. Free radical scavenging activities of SPH and its fractions. (**a**) 1,1-diphenyl-2-picrylhydrazyl (DPPH) radical scavenging activity; (**b**) 2,2'-azinobis-3-ethylbenzthiazoline-6-sulphonate (ABTS) radical scavenging activity.

2.2.2. Reducing Power

Reducing power was measured to evaluate the capacity of compounds to donate electrons or hydrogen atoms, and was related to their ability to inhibit the transformation of Fe^{3+} to Fe^{2+} [21,23]. The reducing power of SPH and its fractions was determined and the results are shown in Figure 3. SPH-I had the highest reducing power as compared with SPH and SPH-II in a concentration dependent manner. At the concentration of 1 mg/mL, the absorbance at 700 nm of SPH, SPH-I, and SPH-II was 0.43, 0.54, and 0.33, respectively. This result suggested that all three fractions have the potential to react with free radicals and block radical chain reactions.

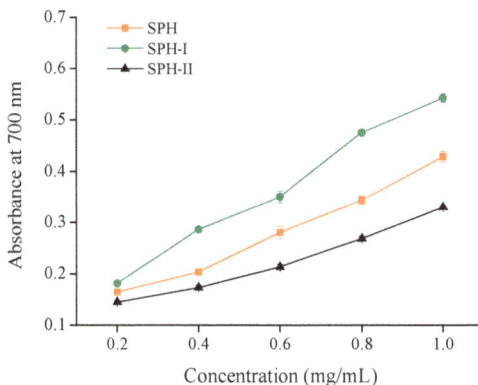

Figure 3. Reducing power of SPH and its fractions.

2.2.3. Inhibition of Linoleic Acid Peroxidation

Lipid peroxidation was thought to proceed via radical mediated abstraction of hydrogen atoms from methylene carbons in polyunsaturated fatty acids [16]. The process of lipid peroxidation generated a series of potentially toxic substances such as electrophilic aldehydes and ketones [25,26]. The inhibitory ability of SPH and its fractions on lipid peroxidation was determined in a linoleic acid system. As shown in Figure 4, the control had the highest absorbance at 500 nm, indicating the highest oxidation degree, while the samples with SPH and its fractions (1 mg/mL) could lower the absorbance. SPH-I exhibited the strongest lipid peroxidation inhibition activity, which was in accordance with the previous report [27] showing that low MW peptides were more effective against linoleic acid peroxidation. The lipid peroxidation inhibition activity of SPH and the ultrafiltration fractions may be related to the high content of hydrophobic amino acids, molecular size, and the amino acid residues at the terminal end of the peptides [28].

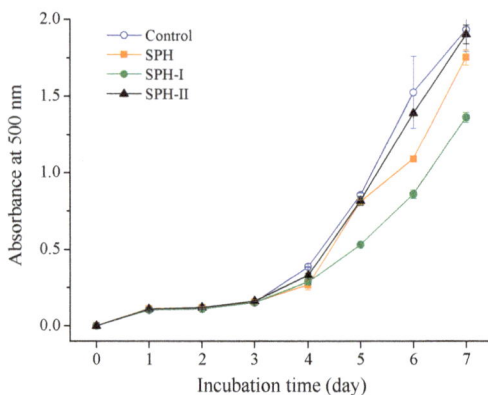

Figure 4. Inhibition activity of SPH and its fractions on linoleic acid peroxidation.

2.3. Effects of SPH-I on Acute Alcohol-Induced Liver Injury in Mice

Various pathways involving multiple types of enzymes and oxidative stress were thought to be associated with the pathological process of alcohol-induced liver injury [6,29]. Oxidative stress, caused by partially-reduced ROS such as superoxide anion ($O_2 \cdot^-$), hydroxyl free radical ($\cdot OH$), and hydrogen peroxide (H_2O_2), played a part in the pathogenesis of alcohol-induced liver injury [25,30]. To study the antioxidant effect of SPH-I in vivo, the well-described alcohol-induced mice hepatotoxicity was used. Alcohol administration was likely to enhance production of free radicals that would initiate lipid peroxidation and decreased activities of antioxidative enzymes, leading to cell membrane damage, intracellular enzyme leakage, and even cell necrosis [18,31].

In this study, forty male Kunming (KM) mice were randomly divided into four groups of ten mice each. Group I served as the normal control and group II was the alcohol model group. Group III and IV were mice treated with SPH-I at 100 and 300 mg/kg BW, respectively, for 24 consecutive days. At the end of the experiment, the mice were euthanized and related biochemical indices were measured.

2.3.1. Effects of SPH-I on Serum ALT and AST Activities

ALT is a cytosolic enzyme that mainly exists in the liver, while AST is primarily present in mitochondria and cytoplasm in the liver. Once hepatocytes are damaged, ALT and AST will leak through the liver cell membrane into circulation and the levels of these enzymes will increase in the serum [30].

The effects of SPH-I on the serum ALT and AST activities are shown in Figure 5. Mice with alcohol administration (groups II, III, and IV) showed a significant increase of serum ALT and AST activities compared with those of group I ($p < 0.05$) and the values of AST/ALT were less than 1, indicating that the alcohol-induced liver injury model in mice was well-established. Administration of SPH-I at 100 and 300 mg/kg BW revealed a significant protective effect on the alcohol-induced liver injury by attenuating the elevation of the activity of ALT by 38.9% and 41.4%, respectively (Figure 5a) and depressing the increase of the activity of AST by 23.8% and 25.8%, respectively, compared with the alcohol model group (Figure 5b).

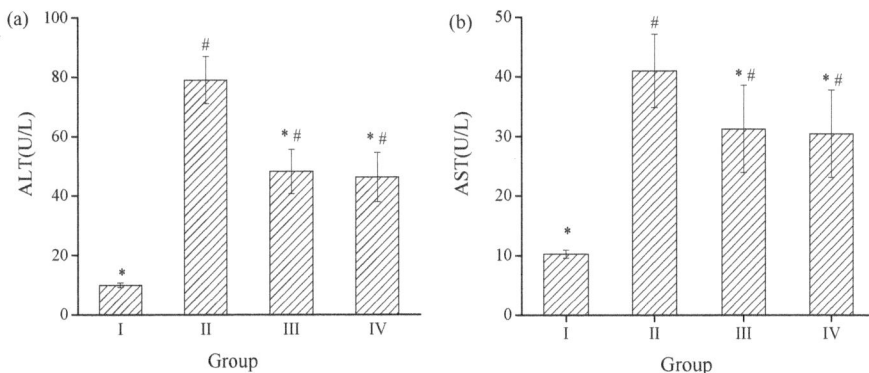

Figure 5. Effects of SPH-I on the activities of serum (**a**) alanine aminotransferase (ALT) and (**b**) aspartate aminotransferase (AST). Group I, normal control; Group II, alcohol model; Group III, SPH-I (100 mg/kg BW) + alcohol; Group IV, SPH-I (300 mg/kg BW) + alcohol; each group contained 10 KM mice.* Statistical significance $p < 0.05$, compared with alcohol-treated group. # Statistical significance $p < 0.05$, compared with control group.

2.3.2. Effect of SPH-I on Hepatic MDA Level

MDA is the end-product of lipid peroxidation, whose levels could reflect the extent of cellular damage, serving as a marker of free radical-mediated lipid peroxidation [32]. Results shown in Figure 6 manifested that the hepatic MDA level of group II was remarkably enhanced after exposure to alcohol by 93.3%, indicating oxidative damage to the liver. Treatment of mice with SPH-I at the doses of 100 and 300 mg/kg significantly reversed the elevation of MDA levels by 27.0% and 38.7%, respectively, compared to the alcohol model (group II), suggesting that SPH-I could inhibit alcohol induced lipid peroxidation in the liver.

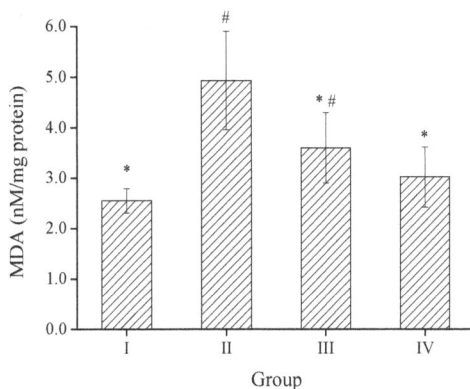

Figure 6. Effect of SPH-I on the hepatic malondialdehyde (MDA) level. Group I, normal control; Group II, alcohol model; Group III, SPH-I (100 mg/kg BW) + alcohol; Group IV, SPH-I (300 mg/kg BW) + alcohol; each group contained 10 Kunming (KM) mice. * Statistical significance $p < 0.05$, compared with alcohol-treated group. # Statistical significance $p < 0.05$, compared with control group.

2.3.3. Effects of SPH-I on Hepatic SOD, CAT, GSH-Px Activities, and GSH Level

Antioxidant enzymes play important roles in elimination of ROS derived from the redox reactions of xenobiotics in liver [33]. SOD is an efficient enzyme that catalyzes the conversion of superoxide into O_2 and H_2O_2, and H_2O_2 could be further decomposed into H_2O and O_2 by CAT, GSH-Px, and the participation of GSH [34]. As a main non-enzymatic antioxidant in cells, GSH plays a critical role in antioxidant defense to protect cells from oxidative damage of ROS such as hydroxyl radical, lipid peroxyl radical, and H_2O_2 [35]. The effects of SPH-I on hepatic SOD, CAT, GSH-Px activities and GSH level were shown in Figure 7. Compared to the control group, the GSH level and GSH-Px, CAT, SOD activities were significantly decreased after exposure to alcohol by 67.6%, 22.8%, 33.3%, and 11.5%, respectively. The levels of GSH were 45.7% and 114% higher than those of group II with administration of SPH-I at the doses of 100 and 300 mg/kg BW respectively. Pretreatment of mice with SPH-I could also remarkably increase the hepatic SOD, CAT, and GSH-Px activities at the same time ($p < 0.05$), indicating that the hepatoprotective effects of SPH-I against acute alcohol-induced liver injury were due to the stabilization of intracellular antioxidant defense systems.

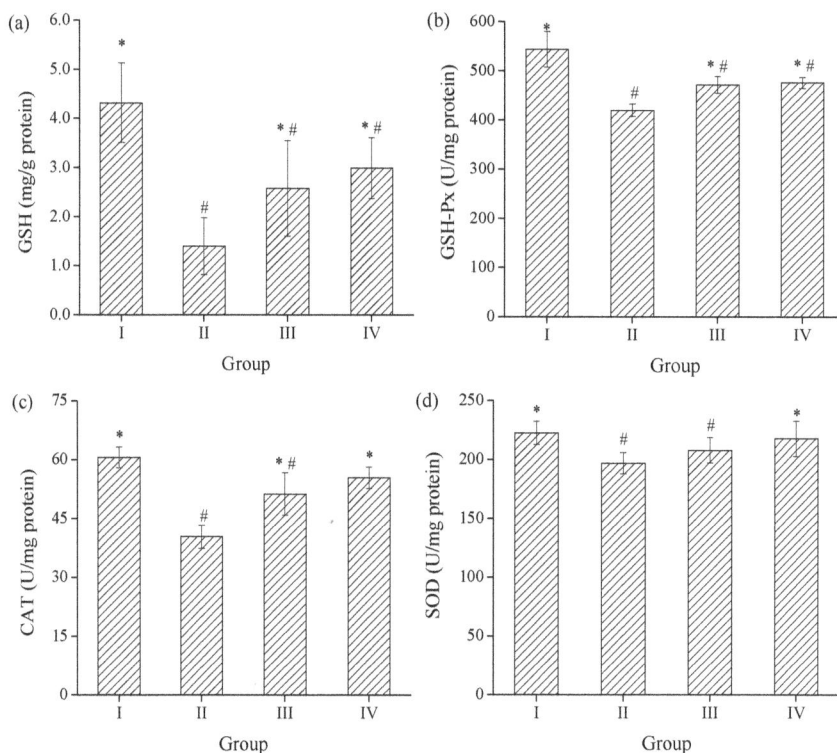

Figure 7. Effects of SPH-I on the level of hepatic (**a**) glutathione (GSH) level and (**b**) glutathione peroxidase (GSH-Px), (**c**) superoxide dismutase (SOD), (**d**) catalase (CAT) activities. Group I, normal control; Group II, alcohol model; Group III, SPH-I (100 mg/kg BW) + alcohol; Group IV, SPH-I (300 mg/kg BW) + alcohol; each group contained 10 KM mice.* Statistical significance $p < 0.05$, compared with alcohol-treated group. # Statistical significance $p < 0.05$, compared with control group.

Previous report showed that a peptide from duck skin byproducts hydrolysate with strong free radical scavenging activities could inhibit the production of ROS and cell death against alcohol-induced liver cell damage, and enhanced the antioxidative enzymes (SOD, CAT, GSH-Px) activities in response to alcohol-induced oxidative damage in rats [36]. The antioxidant activity of an antioxidant compound has been attributed to various mechanisms, among which are radical scavenging, binding of transition metal ion catalysts, reductive capacity, prevention of chain initiation, decomposition of peroxides, and prevention of continued hydrogen abstraction [37]. The results obtained from the present study clearly validated powerful antioxidant activity of SPH-I against various oxidation systems in vitro, which contributed to its hepatoprotective effects of SPH-I in alcohol-induced liver injury in mice.

3. Materials and Methods

3.1. Materials

Schizochytrium processing byproduct was kindly provided by Fisheries Research Institute of Fujian, China, and was stored at −20 °C before use. The commercial protease, alcalase (EC. 3.4.21.62, 2.2×10^5 U/g) and flavourzyme (EC. 3.4.11.1, 7.8×10^4 U/g) were purchased from Novozymes (Copenhagen, Denmark). 2,2′-azinobis-3-ethylbenzthiazoline-6-sulphonate (ABTS), 1,1-diphenyl-2-picrylhydrazyl (DPPH) were obtained from Sigma Chemical Co. (St. Louis, MO, USA). All the kits for biochemical analyses used in the animal experiment were the products of Nanjing Jiancheng Bioengineering Institute (Nanjing, China). All other chemicals and reagents were of analytical grade and commercially available.

3.2. Preparation of SP

SP was extracted by using alkaline extraction and acid precipitation as described previously [38]. The *Schizochytrium* byproduct was ground to powder (sieved through a 50 mesh sieve). One percent (*w/v*) *Schizochytrium* powder in 0.39 M NaOH solution was stirred at 90 °C for 30 min and then centrifuged at 11,000× *g*, 20 °C for 20 min. The supernatant was adjusted to pH 3.0 by 6 M HCl solution and kept for 30 min (pH 3.0 was confirmed to precipitate most of the protein from the alkaline extract in our preliminary experiments). The mixture was centrifuged at 11,000× *g*, 20 °C for 20 min. The precipitated SP was lyophilized for further enzymatic hydrolysis.

3.3. Preparation of SPH

SPH was prepared through stepwise enzymatic hydrolysis by alcalase and flavourzyme sequentially. Two percent (*w/v*) SP was first hydrolyzed by alcalase at a ratio of alcalase to SP of 10% (*w/w*), pH 9.0 at 50 °C for 6 h. Then the mixture was hydrolyzed for another 8 h at 50 °C, pH 6.7 by flavourzyme (the ratio of flavourzyme to SP was 12.5%, *w/w*). The hydrolysate was heated at 100 °C for 10 min to inactive the enzymes and then cooled to room temperature. The SPH in the supernatant was collected by centrifugation at 11,000× *g* for 20 min, and then lyophilized and stored at −20 °C for further analysis.

3.4. Analysis of Amino Acid Composition

The lyophilized hydrolysate was digested at 110 °C for 24 h with HCl (6 M) under nitrogen atmosphere. A High Speed Amino Acid Analyzer Model L-8900 (Hitachi High-Technologies Co., Tokyo, Japan) was used to analyze the amino acid composition of the hydrolysate.

3.5. Determination of MW Distribution of SPH

MW distribution of SPH was determined using HPLC. The sample was applied to a Waters 650E Advanced Protein Purification System (Waters Corporation, Milford, MA, USA) equipped with TSKgel2000 SWXL column (300 mm × 7.8 mm). The mobile phase was 45% acetonitrile and 55% deionized water containing 0.1% trifluoroacetic acid. Chromatographic analysis was carried out with a flow rate of 0.5 mL/min and a column temperature at 30 °C. The absorbance was monitored at 220 nm. A calibration curve was obtained with bovine carbonic anhydrase (29,000 Da), horse heart cytochrome C (12,500 Da), aprotinin (6500 Da), bacitracin (1450 Da), gly–gly–tyr–arg (451 Da) and gly–gly–gly (189 Da). With the help of elution time of calibration materials, the linear regression equation was obtained for the calculation of MW. The results were processed with Millennium32 version 3.05 (Waters Corporation, Milford, MA, USA).

3.6. Ultrafiltration of SPH

SPH obtained from alcalase and flavourzyme digestion was fractionated through ultrafiltration membrane with a MW cut-off of 3 kDa (Millipore, Billerica, MA, USA). All fractions recovered were collected as SPH-I (MW < 3 kDa) and SPH-II (MW > 3 kDa).

3.7. Detemination of Antioxidant Activity In Vitro

3.7.1. DPPH Radical Scavenging Activity

The scavenging activity of SPH and its fractions against DPPH radical was tested according to the method of Wu et al. [39] with slight modification. DPPH was dissolved in ethanol to a final concentration of 0.1 mM. 1 mL of sample was mixed with 1 mL of DPPH solution and then kept in the dark for 30 min at room temperature. Distilled water instead of the sample was used for control. The absorbance values of samples and control were measured at 517 nm. The scavenging rate of DPPH radical of the sample was evaluated with the following equation:

$$\text{DPPH radical scavenging activity } (\%) = \left(A_{\text{control}} - A_{\text{sample}} \right) / A_{\text{control}} \times 100, \tag{1}$$

where A_{sample} and A_{control} were the absorbances of sample and control group, respectively.

3.7.2. ABTS Radical Scavenging Activity

The ABTS radical scavenging assay was carried out according to the method of Wang et al. [24]. The ABTS radical was generated by mixing ABTS stock solution (7 mM) with equal volume of potassium persulfate (2.45 mM), and the mixture was incubated in the dark at room temperature for 12–16 h. The ABTS radical solution was diluted in phosphate buffer (5 mM, pH 7.4) to an absorbance of 0.70 ± 0.02 at 734 nm before used. 1 mL ABTS radical solution was added to 1 mL sample solution. The mixture was then incubated in the dark for 10 min and the absorbance was read at 734 nm. Distilled water instead of the sample was used for control. The ABTS radical scavenging activity of the samples was calculated by the following equation:

$$\text{ABTS radical scavenging activity } (\%) = \left(A_{\text{control}} - A_{\text{sample}} \right) / A_{\text{control}} \times 100, \tag{2}$$

where A_{sample} and A_{control} were the absorbances of sample and control group, respectively.

3.7.3. Reducing Power

The reducing power of SPH and its fractions was estimated according to Oyaizu [40] with some modification. 1 mL sample was mixed with 1 mL of phosphate buffer (0.2 M, pH 6.6) and 1 mL of 1% of potassium ferricyanide. The mixture was then incubated at 50 °C for 20 min followed by addition of 1 mL of 10% trichloroacetic acid. The mixture was then centrifuged at $2500 \times g$ for 10 min. 1 mL of the supernatant was mixed with 1 mL distilled water and 0.2 mL of 0.1% FeCl$_3$. After 10 min, the absorbance was recorded at 700 nm.

3.7.4. Inhibition of Linoleic Acid Peroxidation

The capacity of inhibiting linoleic acid peroxidation of SPH and its fractions was measured according to the method described by Osawa and Namiki [41] with some modification. Briefly, samples were dissolved in distilled water to a concentration of 1 mg/mL and then mixed with 2 mL of ethanol, 26 μL of linoleic acid and 2 mL of phosphate buffer (50 mM, pH 7.0). The mixture was incubated in a colorimetric tube with plug at 40 °C in the dark. The degree of oxidation was measured at 24 h intervals using the ferric thiocyanate (FTC) method of Mitsuda et al. [42]. 100 μL of the reaction mixture was added to a solution of 4.7 mL of 75% ethanol, 0.1 mL of 30% ammonium thiocyanate, and 0.1 mL of 20 mM $FeCl_2$ solution in 3.5% of HCl. After 3 min, the degree of color development that represented the linoleic acid oxidation was measured spectrophotometrically at 500 nm.

3.8. Evaluation of Hepatoprotective Effects of SPH-I in Mice

3.8.1. Animals and Treatments

Forty male KM mice with body weight (BW) of about 20 g were purchased from Slac Laboratory Animal Center (Shanghai, China). Throughout the experiments, mice were fed with standard pellet laboratory animal feed and had free access to food and water. The experiments were carried out in accordance with the guidelines issued by the Ethical Committee of Fujian Medical University (Fujian, China).

After a seven-day period of acclimatizing, forty mice were randomly divided into four groups with ten mice in each group. Group I served as normal control. Group II was alcohol model group in which mice were treated with alcohol alone. Group III and IV were mice treated with SPH-I at 100 and 300 mg/kg BW respectively for 24 consecutive days (the test dosages of SPH-I were decided by preliminary tests). Mice in group I and II were orally given the same volume of deionized water instead of SPH-I solution. One hour after substances administration at the 24th day, the mice except the normal control group were treated with single dose of 50% of alcohol (12 mL/kg BW), while group I was treated with the same volume of water. Whole blood was collected after a fasting period of 24 h. The mice were then euthanized and their livers were excised.

3.8.2. Analysis of Serum Biochemical Indices

Blood samples were collected immediately and the serum was separated by centrifugation at $1500 \times g$ for 10 min at 4 °C. The activities of serum ALT and AST were analyzed using ALT and AST assay kits according to the manufacturer's protocol.

3.8.3. Analysis of Hepatic Biochemical Indices

Liver tissues were excised and homogenized in 0.1 g/mL of cold normal saline. The supernatant of the homogenate was collected after centrifugation at $3000 \times g$ for 10 min at 4 °C. The activities of SOD, CAT, GSH-Px, and the level of GSH and MDA were determined with T-SOD assay kit (hydroxylamine method), CAT assay kit (visible light), GSH-PX assay kit (colorimetric method), GSH assay kit, and MDA assay kit (TBA method) according to the manufacturer's protocol, respectively. The total protein content of liver homogenate was determined according to the Bradford method [43].

3.9. Statistical Analysis

All results are presented as means ± standard deviation (SD). Statistical analysis was carried out with IBM SPSS 17.0 software (SPSS, Chicago, IL, USA). Statistical analysis was performed by one-way analysis of variance (ANOVA) with Duncan's test for post hoc analysis and $p < 0.05$ values were considered as statistically significant.

4. Conclusions

In this study, *Schizochytrium* sp. protein hydrolysate was prepared by alcalase and flavourzyme sequentially, and mainly composed of Glu (17.66%), Asp (15.89%), Leu (9.96%), and Arg (7.81%) along with small amounts of Phe (5.27%), Tyr (2.71%), and His (1.57%). After ultrafiltration of SPH with 3 kDa membrane, SPH-1, peptides with MW below 3 kDa, exhibited the highest DPPH and ABTS radicals scavenging activities, reducing power and lipid peroxidation inhibition potential. In addition, SPH-I could significantly alleviate alcohol-induced hepatotoxicity in mice. Results of the present study indicated that SPH-I could be developed as a potential functional antioxidant additive for effective therapy of alcohol-induced liver diseases.

Acknowledgments: This work was supported by the Natural Science Foundation of China (No. 31571779), the High & New project of Fujian Marine Fisheries Department (No. [2015]20), and the Fujian Production & Study project of Provincial Science & Technology Hall (No. 2016N5006), China.

Author Contributions: Xixi Cai, Ana Yan and Shaoyun Wang conceived and designed the experiments; Ana Yan and Xixi Cai carried out the experiments; Xixi Cai and Nanyan Fu analyzed the data; Xixi Cai drafted the paper; Shaoyun Wang and Nanyan Fu reviewed the manuscript and provided useful suggestion to improve the manuscript. All authors read and approved the final manuscript.

Conflicts of Interest: The authors declare no conflict of interest.

References

1. Wu, S.T.; Yu, S.T.; Lin, L.P. Effect of culture conditions on docosahexaenoic acid production by *Schizochytrium* sp. S31. *Process Biochem.* **2005**, *40*, 3103–3108. [CrossRef]
2. Morita, E.; Kumon, Y.; Nakahara, T.; Kagiwada, S.; Noguchi, T. Docosahexaenoic acid production and lipid-body formation in *Schizochytrium limacinum* SR21. *Mar. Biotechnol.* **2006**, *8*, 319–327. [CrossRef] [PubMed]
3. Yaguchi, T.; Tanaka, S.; Yokochi, T.; Nakahara, T.; Higashihara, T. Production of high yields of docosahexaenoic acid by *Schizochytrium* sp. strain SR21. *J. Am. Oil Chem. Soc.* **1997**, *74*, 1431–1434. [CrossRef]
4. Mendis, E.; Rajapakse, N.; Kim, S.K. Antioxidant properties of a radical-scavenging peptide purified from enzymatically prepared fish skin gelatin hydrolysate. *J. Agric. Food Chem.* **2005**, *53*, 581–587. [CrossRef] [PubMed]
5. Sindhu, E.R.; Preethi, K.C.; Kuttan, R. Antioxidant activity of carotenoid lutein in vitro and in vivo. *Indian J. Exp. Biol.* **2010**, *48*, 843–848. [PubMed]
6. Wang, M.C.; Zhu, P.L.; Jiang, C.X.; Ma, L.P.; Zhang, Z.J.; Zeng, X.X. Preliminary characterization, antioxidant activity In Vitro and hepatoprotective effect on acute alcohol-induced liver injury in mice of polysaccharides from the peduncles of *Hovenia dulcis*. *Food Chem. Toxicol.* **2012**, *50*, 2964–2970. [CrossRef] [PubMed]
7. Impei, S.; Gismondi, A.; Canuti, L.; Canini, A. Metabolic and biological profile of autochthonous *Vitis vinifera* L. ecotypes. *Food Funct.* **2015**, *6*, 1526–1538. [CrossRef] [PubMed]
8. Giovannini, D.; Gismondi, A.; Basso, A.; Canuti, L.; Braglia, R.; Canini, A.; Mariani, F.; Cappelli, G. *Lavandula angustifolia* Mill. essential oil exerts antibacterial and anti-Inflammatory effect in macrophage mediated immune response to *Staphylococcus aureus*. *Immunol. Investig.* **2016**, *45*, 11–28. [CrossRef] [PubMed]
9. Gammone, M.; Riccioni, G.; Orazio, N. Marine carotenoids against oxidative stress: Effects on human health. *Mar. Drugs* **2015**, *13*, 6226–6246. [CrossRef] [PubMed]
10. Sun, L.P.; Chang, W.D.; Ma, Q.Y.; Zhuang, Y.L. Purification of antioxidant peptides by high resolution mass spectrometry from simulated gastrointestinal digestion hydrolysates of Alaska Pollock (*Theragra chalcogramma*) skin collagen. *Mar. Drugs* **2016**, *14*, 186. [CrossRef] [PubMed]
11. Lin, S.Y.; Jin, Y.; Liu, M.Y.; Yang, Y.; Zhang, M.S.; Guo, Y.; Jones, G.; Liu, J.B.; Yin, Y.G. Research on the preparation of antioxidant peptides derived from egg white with assisting of high-intensity pulsed electric field. *Food Chem.* **2013**, *139*, 300–306. [CrossRef] [PubMed]
12. Li, Y.H.; Jiang, B.; Zhang, T.; Mu, W.M.; Liu, J. Antioxidant and free radical-scavenging activities of chickpea protein hydrolysate (CPH). *Food Chem.* **2008**, *106*, 444–450. [CrossRef]
13. Sheih, I.C.; Wu, T.K.; Fang, T.J. Antioxidant properties of a new antioxidative peptide from algae protein waste hydrolysate in different oxidation systems. *Bioresour. Technol.* **2009**, *100*, 3419–3425. [CrossRef] [PubMed]
14. Udenigwe, C.C.; Aluko, R.E. Chemometric analysis of the amino acid requirements of antioxidant food protein hydrolysates. *Int. J. Mol. Sci.* **2011**, *12*, 3148–3161. [CrossRef] [PubMed]
15. Sarmadi, B.H.; Ismail, A. Antioxidative peptides from food proteins: A review. *Peptides* **2010**, *31*, 1949–1956. [CrossRef] [PubMed]
16. Rajapakse, N.; Mendis, E.; Jung, W.K.; Je, J.Y.; Kim, S.K. Purification of a radical scavenging peptide from fermented mussel sauce and its antioxidant properties. *Food Res. Int.* **2005**, *38*, 175–182. [CrossRef]

17. Saiga, A.; Tanabe, S.; Nishimura, T. Antioxidant activity of peptides obtained from porcine myofibrillar proteins by protease treatment. *J. Agric. Food Chem.* **2003**, *51*, 3661–3667. [CrossRef] [PubMed]

18. Giménez, B.; Alemán, A.; Montero, P.; Gómez-Guillén, M.C. Antioxidant and functional properties of gelatin hydrolysates obtained from skin of sole and squid. *Food Chem.* **2009**, *114*, 976–983. [CrossRef]

19. Xia, Y.C.; Bamdad, F.; Gänzle, M.; Chen, L.Y. Fractionation and characterization of antioxidant peptides derived from barley glutelin by enzymatic hydrolysis. *Food Chem.* **2012**, *134*, 1509–1518. [CrossRef] [PubMed]

20. Hong, J.; Chen, T.T.; Hu, P.; Yang, J.; Wang, S.Y. Purification and characterization of an antioxidant peptide (GSQ) from Chinese leek (*Allium tuberosum* Rottler) seeds. *J. Funct. Foods* **2014**, *10*, 144–153. [CrossRef]

21. Shi, Y.N.; Kovacs-Nolan, J.; Jiang, B.; Tsao, R.; Mine, Y. Antioxidant activity of enzymatic hydrolysates from eggshell membrane proteins and its protective capacity in human intestinal epithelial Caco-2 cells. *J. Funct. Foods* **2014**, *10*, 35–45. [CrossRef]

22. Chen, M.; Li, B. The effect of molecular weights on the survivability of casein-derived antioxidant peptides after the simulated gastrointestinal digestion. *Innov. Food Sci. Emerg. Technol.* **2012**, *16*, 341–348. [CrossRef]

23. Chen, N.; Yang, H.M.; Sun, Y.; Niu, J.; Liu, S.Y. Purification and identification of antioxidant peptides from walnut (*Juglans regia* L.) protein hydrolysates. *Peptides* **2012**, *38*, 344–349. [CrossRef] [PubMed]

24. Wang, B.; Li, Z.R.; Chi, C.F.; Zhang, Q.H.; Luo, H.Y. Preparation and evaluation of antioxidant peptides from ethanol-soluble proteins hydrolysate of *Sphyrna lewini* muscle. *Peptides* **2012**, *36*, 240–250. [CrossRef] [PubMed]

25. Niki, E. Assessment of antioxidant capacity in vitro and in vivo. *Free Radic. Biol. Med.* **2010**, *49*, 503–515. [CrossRef] [PubMed]

26. Winczura, A.; Zdżalik, D.; Tudek, B. Damage of DNA and proteins by major lipid peroxidation products in genome stability. *Free Radic. Res.* **2012**, *46*, 442–459. [CrossRef] [PubMed]

27. Je, J.Y.; Park, P.J.; Kim, S.K. Antioxidant activity of a peptide isolated from Alaska pollack (*Theragra chalcogramma*) frame protein hydrolysate. *Food Res. Int.* **2005**, *38*, 45–50. [CrossRef]

28. Cheung, I.W.Y.; Cheung, L.K.Y.; Tan, N.Y.; Li-Chan, E.C.Y. The role of molecular size in antioxidant activity of peptide fractions from Pacific hake (*Merluccius productus*) hydrolysates. *Food Chem.* **2012**, *134*, 1297–1306. [CrossRef] [PubMed]

29. Cederbaum, A.I.; Lu, Y.K.; Wu, D.F. Role of oxidative stress in alcohol-induced liver injury. *Arch. Toxicol.* **2009**, *83*, 519–548. [CrossRef] [PubMed]

30. Obogwu, M.B.; Akindele, A.J.; Adeyemi, O.O. Hepatoprotective and in vivo antioxidant activities of the hydroethanolic leaf extract of *Mucuna pruriens* (Fabaceae) in antitubercular drugs and alcohol models. *Chin. J. Nat. Med.* **2014**, *12*, 273–283. [CrossRef]

31. Yazdanparast, R.; Bahramikia, S.; Ardestani, A. *Nasturtium officinale* reduces oxidative stress and enhances antioxidant capacity in hypercholesterolaemic rats. *Chem. Biol. Interact.* **2008**, *172*, 176–184. [CrossRef] [PubMed]

32. Cheng, N.; Du, B.; Wang, Y.; Gao, H.; Cao, W.; Zheng, J.B.; Feng, F. Antioxidant properties of jujube honey and its protective effects against chronic alcohol-induced liver damage in mice. *Food Funct.* **2014**, *5*, 900–908. [CrossRef] [PubMed]

33. Choi, D.W.; Lee, J.H.; Chun, H.H.; Song, K.B. Isolation of a calcium-binding peptide from bovine serum protein hydrolysates. *Food Sci. Biotechnol.* **2012**, *21*, 1663–1667. [CrossRef]

34. Cai, X.X.; Yang, Q.; Wang, S.Y. Antioxidant and hepatoprotective effects of a pigment-protein complex from *Chlorella vulgaris* on carbon tetrachloride-induced liver damage in vivo. *RSC Adv.* **2015**, *5*, 96097–96104. [CrossRef]

35. Wu, G.Y.; Fang, Y.Z.; Yang, S.; Lupton, J.R.; Turner, N.D. Glutathione metabolism and its implications for health. *J. Nutr.* **2004**, *134*, 489–492. [PubMed]

36. Lee, S.J.; Kim, Y.S.; Hwang, J.W.; Kim, E.K.; Moon, S.H.; Jeon, B.T.; Jeon, Y.J.; Kim, J.M.; Park, P.J. Purification and characterization of a novel antioxidative peptide from duck skin byproducts that protects liver against oxidative damage. *Food Res. Int.* **2012**, *49*, 285–295. [CrossRef]

37. GüLCIN, I.; Alici, H.A.; Cesur, M. Determination of in vitro antioxidant and radical scavenging activities of propofol. *Chem. Pharm. Bull.* **2005**, *53*, 281–285. [CrossRef] [PubMed]

38. Lin, J.P.; Cai, X.X.; Tang, M.R.; Wang, S.Y. Preparation and evaluation of the chelating nanocomposite fabricated with marine algae *Schizochytrium* sp. protein hydrolysate and calcium. *J. Agric. Food Chem.* **2015**, *63*, 9704–9714. [CrossRef] [PubMed]

39. Wu, H.C.; Chen, H.M.; Shiau, C.Y. Free amino acids and peptides as related to antioxidant properties in protein hydrolysates of mackerel (*Scomber austriasicus*). *Food Res. Int.* **2003**, *36*, 949–957. [CrossRef]

40. Oyaizu, M. Antioxidative activities of browning products of glucosamine fractionated by organic solvent and thin-layer chromatography. *J. Jpn. Soc. Food Sci.* **1988**, *35*, 771–775. [CrossRef]

41. Osawa, T.; Namiki, M. A novel type of antioxidant isolated from leaf wax of *Eucalyptus* leaves. *Agric. Biol. Chem.* **1981**, *45*, 735–739. [CrossRef]

42. Mitsuda, H.; Yuasumoto, K.; Iwami, K. Antioxidation action of indole compounds during the autoxidation of linoleic acid. *Eiyo Shokuryo* **1966**, *19*, 210–214. [CrossRef]

43. Bradford, M.M. A rapid and sensitive method for the quantitation of microgram quantities of protein utilizing the principle of protein-dye binding. *Anal. Biochem.* **1976**, *72*, 248–254. [CrossRef]

marine drugs

MDPI

Article

Hydrolysates of Fish Skin Collagen: An Opportunity for Valorizing Fish Industry Byproducts

María Blanco *, José Antonio Vázquez, Ricardo I. Pérez-Martín and Carmen G. Sotelo

Instituto de Investigaciones Marinas (IIM-CSIC), Eduardo Cabello, 6, Vigo, Galicia 36208, Spain; jvazquez@iim.csic.es (J.A.V.); ricardo@iim.csic.es (R.I.P.-M.); carmen@iim.csic.es (C.G.S.)
* Correspondence: mblanco@iim.csic.es; Tel.: +34-986-231-930; Fax: +34-986-292-762

Academic Editors: Se-Kwon Kim and Peer B. Jacobson
Received: 21 March 2017; Accepted: 2 May 2017; Published: 5 May 2017

Abstract: During fish processing operations, such as skinning and filleting, the removal of collagen-containing materials can account for up to 30% of the total fish byproducts. Collagen is the main structural protein in skin, representing up to 70% of dry weight depending on the species, age and season. It has a wide range of applications including cosmetic, pharmaceutical, food industry, and medical. In the present work, collagen was obtained by pepsin extraction from the skin of two species of teleost and two species of chondrychtyes with yields varying between 14.16% and 61.17%. The storage conditions of the skins appear to influence these collagen extractions yields. Pepsin soluble collagen (PSC) was enzymatically hydrolyzed and the resultant hydrolysates were ultrafiltrated and characterized. Electrophoretic patterns showed the typical composition of type I collagen, with denaturation temperatures ranged between 23 °C and 33 °C. In terms of antioxidant capacity, results revealed significant intraspecific differences between hydrolysates, retentate, and permeate fractions when using β-Carotene and DPPH methods and also showed interspecies differences between those fractions when using DPPH and ABTS methods. Under controlled conditions, PSC hydrolysates from *Prionace glauca*, *Scyliorhinus canicula*, *Xiphias gladius*, and *Thunnus albacares* provide a valuable source of peptides with antioxidant capacities constituting a feasible way to efficiently upgrade fish skin biomass.

Keywords: collagen; enzymatic hydrolysis; antioxidant activity; β-carotene; DPPH; ABTS

1. Introduction

As the human population is growing and their consumption behavior changing, the worldwide demand for fishery products is increasing as is the demand for ready to cook meals in the form of loins or steaks. These kinds of processed products generate a large amount of by-products in the form of skin, bones, viscera, heads, scales, etc. Those organic materials are considered postharvest fish losses (by-products) and are a main concern for current fishery management policies because they represent a significant source of valuable compounds as proteins, fat, minerals, etc. Although part of these by-products are already being used, either for fish meal or oil production (35% of world fishmeal production was obtained from fish byproducts) [1]; this kind of utilization is considered to produce very little added-value, but due to present technological developments, a more valuable and profitable use is possible [2].

Fishing activity in Galicia (North-West Spain) constitutes a key sector for the economy of the region, with a high concentration of small, medium, and big businesses dedicated to fish processing activities that render a wide variety of by-products susceptible to valorization. During fish processing operations the removal of collagen-containing materials (mainly skin, bones and scales) could account for as much as 30% of the total by-products generated after filleting (75% of the total catch weight) [3,4].

Although collagen is the main protein component of fish skin and its particular heterotrimeric structure $[\alpha_1(I)]_2 \; \alpha_2(I)$ has been previously described, there have been only a few publications describing the properties of fish skin collagen hydrolysates [5–7], and even less research has been conducted on the characterization of hydrolysates obtained from pepsin soluble collagen of marine origin [7]. As acid solubilisation of collagen has been shown to render low yields, enzymatic proteolysis has been studied as an alternative to enhance the yield and at the same time obtaining hydrolysates with good nutritional composition, increased solubility and better emulsifying, foaming, and gelating properties, as well as biologically active peptides [8–10].

Two sharks, blue shark (*Prionace glauca*; PGLA) and small-spotted catshark (*Scyliorhinus canicula*; SCAN), and two bonny fishes, yellowfin tuna (*Thunnus albacares*; TALB) and swordfish (*Xiphias gladius*; XGLA) were selected since a significant amount of these are industrially processed generating significant amounts of skin [11–13]. The objective of this study was to evaluate the potential use of skins which are obtained as a by-product of the fish processing industry to obtain fish skin collagen hydrolysates and to test the influence of some biochemical properties, as the amino acid content or molecular weight, on antioxidant capacity of hydrolysates. This is the first time, as far as we know, that the extraction, characterization and comparison of collagen hydrolysates from these species, is described.

2. Results and Discussion

Fish skin can be an important by-product for some fishery industries, for example some companies produce pieces of skinned and deboned fish which render important amounts of skins and bones as by-products. One of the problems associated with these by-products is the heterogeneity of them: they are originated from different species, previous frozen storage conditions can be different (frozen storage in brine), they can be mixed with bones or other by-products, etc. Appropriate management of these by-products should take into account these problems, and one important and initial step is to estimate the value associated with each type of product. Therefore, the initial chemical characterization and the estimations of collagen content are important data in evaluating the potential value of these by-products. Low yields of collagen extraction can be expected in industrial conditions because of the previous treatment and storage history of the raw materials. Hydrolysis would help to overcome some of the problems associated with these previous treatments, increasing the yield of a valuable product, collagen hydrolysates, which has many interesting properties, such as antioxidant activity [14,15].

2.1. Chemical Composition of Skin By-Products

2.1.1. Proximate Composition

Table 1 shows the chemical composition of the skins of the four species analysed, these were similar to the skins of other fish species. Skin of the two elasmobranch contained similar amounts of protein, while swordfish skin presented the lowest protein content of all species, while those from tuna were the highest. In the case of swordfish, it is remarkably the highest lipid content (30.53%), which may also be the target of valorisation for this type of by-product. The higher ash content in the skin of the small-spotted catshark is remarkable and it could be attributed to its particular skin structure; a thinner skin with a higher proportion of scales compared to the skin of the blue shark. The skin of the blue shark is thicker and presents two different layers with scales only present in the upper layer.

2.1.2. Hydroxyproline (HPro) Content

Hydroxyproline has been used as a method to quantify the amount of collagen in a particular tissue [16]. This analytical approach was used to estimate the collagen content in the skin of all the species analyzed, assuming that all HPro content of skin is due to collagen and taking into account that the ratio of HPro in collagen is 12.5 g of HPro/100 g of collagen [17]. Table 2 shows that the collagen content was higher in the skin of TALB, followed by the two species of elasmobranch which showed similar values (SCAN and PGLA), and finally the lowest value corresponded to the skin of XGLA, these results are in coherence with the protein content found in the skin of these species (Table 1). Collagen content reported previously for other fish species was similar with slight variations depending on the species [18].

Sotelo et al. [19] have reported a low collagen content in the skin of SCAN (11.6% in a wet basis), which may be explained by differences in the previous treatment of skins for this species (used fresh in this study).

Table 1. Chemical composition of fish skins from the four species used for the study. Values, expressed in a wet basis, are means of 3 determinations ± standard deviation (Protein = $N \times 5.4$).

Species	Composition (%)			
	Moisture	Protein	Lipids	Ash
PGLA	76.03 ± 0.83	20.14 ± 0.97	0.24 ± 0.03	4.24 ± 0.24
SCAN	61.5 ± 0.79	22.09 ± 0.96	0.36 ± 0.01	14.01 ± 0.5
XGLA	42.87 ± 0.54	16.28 ± 2.21	30.53 ± 1.99	2.49 ± 0.21
TALB	62.57 ± 2.4	26.96 ± 2.04	3.22 ± 0.72	0.67 ± 0.14

Table 2. Hydroxyproline (OHPro) content in skin (g OHPro/100 g skin), collagen content calculated from the hydroxiproline values, and yield of PSC_1 (g collagen/100 g skin), and PSC_2 (g collagen/100 g collagen of the skin). The average values (±SD) expressed in a wet weight basis are means of three replicates.

	Hydroxyproline Content in Skin (%)	Collagen Content (%)	PSC_1 Yield (%)	PSC_2 Yield (%)
PGLA	1.23 ± 0.11	9.84 ± 0.88	5.87 ± 0.49	61.17 ± 5.15
SCAN	1.85 ± 0.14	14.8 ± 1.14	4.89 ± 0.85	33.00 ± 5.25
XGLA	1.08 ± 0.16	8.64 ± 1.28	2.59 ± 0.22	31.33 ± 5.55
TALB	2.69 ± 0.26	21.53 ± 2.09	2.97 ± 0.98	14.16 ± 6.14

2.2. Extraction of Collagen

2.2.1. Yield of PSC

Previous reports have shown that pepsin enhances the extraction efficiency in collagen because it is able to cleave specifically telopeptide regions of collagen [20,21]. Besides, by hydrolysing the non-triple helice domain, non-collagen proteins are more easily removed, and thus collagen becomes readily solubilized in acid solution and the antigenicity caused by telopeptides is reduced, obtaining a collagen with higher purity with the possibility of using it in different applications [22–24].

Table 2 shows PSC yields obtained for PGLA, SCAN, TALB, and XGLA. Extraction yields obtained for PGLA and SCAN were similar to other PSC extracted from different fish species, such as bigeye snapper skin [25], brownstripe red snapper skin [26], or largefin longbarbel catfish [27]. However, the yields obtained for TALB and XGLA are lower than those values. While TALB showed the highest collagen content values (determined by means of hydroxyproline analysis in skin), it also showed (together with XGLA skins) lower extraction yields (PSC_1 and PSC_2). These results could be attributed to several factors such as differences in the structure of the collagen fibers or the storage conditions; processing of tuna usually involves freezing and frozen storage, most of the times in brine. This treatment may cause protein denaturation, a higher degree of crosslinking and therefore lower collagen solubility and extraction yields [27–30].

2.2.2. Characterization of PSC

Polyacrylamide Gel Electrophoresis (SDS-PAGE)

Figure 1 shows the PSC electrophoretic patterns of the analysed species. The PSC SDS-PAGE pattern from PGLA and TALB were more similar to the type I collagen pattern where two identical α_1-chains (120 kDa), one α_2-chain (110 kDa), and one β dimer band of about 200 kDa can be observed [16,31]. The molecular weight data obtained for α and β chains of PSC from TALB are similar to those previously published for the same species [23,32]. The cross-linking rate of collagen has been reported to be low; which might explain why highly cross-linked components (γ-component) in PGLA, TALB, and XGLA are shown only as a faint bands in Figure 1 [33,34]. This result indicates that pepsin was able to hydrolyse the cross-links in the telopeptide region without damaging the integrity of the triple-helix.

PSC from SCAN was characterized by a high susceptibility to pepsin hydrolysis, as revealed by the fact that neither dimer nor trimer could be observed in SDS-PAGE, and also by the presence of several weak α subunits lower than 110 kDa, which could be products of enzymatic hydrolysis of collagen components (Figure 1). In fact, previous publications have shown that β and γ-components were present in acid soluble collagen from SCAN skin [19].

In the electrophoretic pattern of XGLA, one intermediate band was observed between the β and α component with an approximate molecular weight of about 150 kDa. The presence of similar components have also been reported for PSC from different species, suggesting either an incomplete hydrolysis of β dimers, or the presence of a mixture of different collagens [35,36].

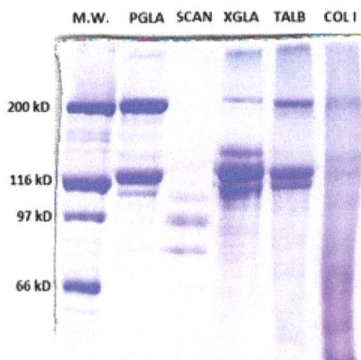

Figure 1. 7% Sodium Dodecyl Sulfate-Polyacrylamide Gel Electrophoresis (SDS-PAGE) showing Pepsin soluble collagen (PSC) from *Prionace glauca* (PGLA), *Scyliorhinus canicula* (SCAN), *Thunnus albacares* (TALB) and *Xiphias gladius* (XGLA). M.W: Molecular Weight Standards. Col I: standard collagen type I from mammal.

Amino Acid Content

Table 3 shows the amino acid composition of the PSC of the four studied species and also that from calf skin (data obtained from Zhang et al. [21]). To our knowledge, amino acid composition has never previously been reported for PSC collagen of these species except for TALB [32]. Although, Glycine was the most abundant amino acid in all the species studied, yet did not represent one third of the total amino acid residues as expected [19,20]. Similar results have been previously reported in PSC obtained from yellowfin tuna skin [32] and squid skin collagen [7]. This result might be explained due to the presence of telopeptide fractions in which the repetitive occurrence of glycine every three amino acid is absent [30].

The lower imino acid content found in SCAN PSC, contributes to the low stability of the triple helix structure [35], which is a result that is in agreement with the SDS profiles shown above, indicating the higher susceptibility of this species to the action of pepsin.

Table 3. Amino acid composition of PSC of PGLA, SCAN, TALB and XGLA (residues/1000). Data from calf skin collagen is also included [21]. Imino acids includes proline and hydroxyproline.

Amino Acid	PSC				CALF
	PGLA	SCAN	TALB	XGLA	
Hydroxyproline	84.62 ± 0.98	88.28 ± 0.62	87.38 ± 0.60	76.55 ± 0.87	94
Aspartic acid	46.58 ± 0.42	52.16 ± 0.43	55.40 ± 0.54	61.32 ± 0.46	45
Serine	35.98 ± 0.42	54.02 ± 0.14	35.53 ± 0.25	39.89 ± 0.74	33
Gultamic acid	92.02 ± 1.00	92.10 ± 0.47	97.89 ± 0.43	94.64 ± 0.96	75
Glycine	214.80 ± 2.92	234.69 ± 1.36	217.22 ± 1.32	210.20 ± 3.22	330
Histidine	15.80 ± 0.20	17.35 ± 0.10	12.70 ± 0.05	15.67 ± 0.34	5
Arginine	111.50 ± 1.09	91.26 ± 1.08	92.16 ± 2.97	89.54 ± 2.26	50
Threonine	33.59 ± 0.16	33.41 ± 0.44	40.00 ± 1.81	42.89 ± 1.60	18
Alanine	108.57 ±0.87	89.79 ± 0.97	111.78 ± 2.58	105.20 ± 2.39	119
Proline	107.68 ± 0.76	95.22 ± 0.29	114.86 ± 0.45	121.89 ± 1.30	121
Cystine	0.88 ± 0.01	0.31 ± 0.00	0.07 ± 0.00	0.61 ± 0.01	0
Tyrosine	3.39 ± 0.05	1.36 ± 0.00	4.42 ± 0.07	6.45 ± 0.15	3
Valine	27.77 ± 0.39	34.13 ± 0.12	25.64 ± 0.15	26.95 ± 0.40	21
Methionine	13.51 ± 0.33	14.06 ± 0.20	6.29 ± 0.13	3.53 ± 0.15	6
Lysine	33.48 ± 0.36	37.78 ± 0.13	35.37 ± 0.23	31.52 ± 0.43	26
Isoleucine	24.62 ± 0.30	18.29 ± 0.02	14.26 ± 0.15	20.47 ± 0.38	11
Leucine	25.97 ± 0.36	27.30 ± 0.07	28.28 ± 0.21	31.19 ± 0.68	23
Phenylalanine	19.25 ± 0.22	18.49 ± 0.01	20.75 ± 0.15	21.50 ± 0.47	3
Iminoacids	192.3	183.5	202.24	198.44	215
% hydroxylation of proline	44.00	48.10	43.20	38.57	44

Determination of Denaturation Temperature

DSC analyses of lyophilized PSC were performed. Calf skin type I collagen was used for comparison purposes. Denaturation temperatures for PGLA, SCAN, TALB, and XGLA PSCs were 33 °C, 23.6 °C, 30.6 °C, and 31.4 °C respectively, which are similar to those found in literature for other PSC in different marine organisms: paper nautilus [37], striped catfish [38], bighead carp [35], or blueshark [39]. Denaturation temperatures of PSC in all species were lower than that of collagen type I of calf skin (T_d = 40 °C). Among the four species studied, the lower denaturation temperature was found in SCAN PSC. These results agree with the lower imino acid content (hydroxyproline and proline) found in the collagen obtained from this species. Thermal stability of collagen is related to the restriction of the secondary structure imposed by the pyrrolidine rings of proline and hydroxyproline, contributing to the strength of the triple helix [20,40]. Sotelo et al. [19] have found a higher denaturation temperature for ASC obtained from small-spotted catshark skin, suggesting the influence of pepsin cross-link cleavage on lower thermal stability found in PSC. Similar results were obtained for ASC and PSC from the skin of brownbanded bamboo shark [32].

2.3. Enzymatic Hydrolysis of PSC

2.3.1. Degree of Hydrolysis

Hydrolysis curves were similar to others previously reported for different marine skin proteins [41,42]. The hydrolysis degree (DH) (average values ±SD) calculated using the pH-STAT method were 16.52 ± 3.74%, 15.80 ± 0.99%, 11.49 ± 1.5%, and 12.56 ± 1.79% for PGLA, SCAN, TALB, and XGLA, respectively. Enzymatic proteolysis and the resulting degree of hydrolysis are key parameters influencing peptide length and other related characteristics such as solubility, nutritional, functional, or sensory properties [7,9].

2.3.2. Antioxidant Activities in Hydrolysates

Table 4 shows data of antioxidant analysis in collagen unfractionated hydrolysates (H) and 3kDa ultrafiltration fractions: retentates (R) and permeates (P). The antioxidant capacities were evaluated using 3 methods, including two based on free radical scavenging capacity, that is, DPPH and ABTS, and one based on the inhibition of lipid peroxidation, determined by the β-carotene assay.

The precise mechanism explaining the antioxidant activity of peptides has not been entirely elucidated, however several authors suggested the influence of hydrolysis degree [14,15]. As it was expected, hydrolysate (H) fractions, determined with DPPH and ABTS exhibited lower values of antioxidant activity in the hydrolysate with the highest hydrolysis degree (PGLA). However, the highest values of antioxidant activity were found in XGLA which showed a higher hydrolysis degree than TALB, suggesting the influence not only of the hydrolysis degree but also to the presence of some amino acids such as cysteine which may interact with free radicals by their SH groups [14,43–45]. Thus, while XGLA hydrolysate presented the highest values of cysteine content (53.03/1000 residues), PGLA hydrolysate showed a low cysteine content (8.93/1000 residues) (Table 5). On the other hand, the β-carotene method showed highest antioxidant capacity with those hydrolysates with the highest DH (SCAN and PGLA), while those with the lowest DH showed also the lowest antioxidant capacity (Table 4).

Table 4. Antioxidant activities (Mean ± SD) of collagen unfractionated hydrolysates (H), retentates (R) and permeates (P) quantified by means of three methods (DPPH, ABTS, and β-carotene) and calculated as equivalents (in µg) of BHT per mL of hydrolysate.

Species	Fraction	DPPH (mg BHT Eq/mL)	ABTS (mg BHT q/mL)	β-Carotene (mg BHT Eq/mL)
XGLA	H	677.20 ± 114.42	253.77 ± 1.85	7.59 ± 1.93
TALB	H	578.87 ± 57.81	199.57 ± 37.54	5.67 ± 0.61
SCAN	H	494.17 ± 210.3	159.17 ± 30.78	20.86 ± 3.53
PGLA	H	405.30 ± 9.89	151.20 ± 43.49	15.26 ± 5.02
XGLA	R	465.63 ± 30.47	247.27 ± 10.70	5.91 ± 1.04
TALB	R	435.97 ± 85.54	174.10 ± 70.05	11.94 ± 3.86
SCAN	R	603.40 ± 30.88	143.57 ± 29.80	7.38 ± 11.69
PGLA	R	422.97 ± 41.32	124.90 ± 35.76	19.18 ± 1.92
XGLA	P	448.0 ± 66.45	264.87 ± 18.86	8.08 ± 0.33
TALB	P	457.67 ± 95.61	192.83 ± 56.66	15.26 ± 2.91
SCAN	P	601.70 ± 175.33	209.70 ± 53.71	12.40 ± 9.14
PGLA	P	416.03 ± 18.88	134.87 ± 26.76	17.03 ± 2.64

To test the influence of molecular size reduction of peptides on the functional properties of collagen hydrolysates [10,14,46], the antioxidant capacity of unfractionated hydrolysates (H), retentates (R) and permeates (P) were statistically analyzed. One-way ANOVA analysis of data revealed some significant intraspecific differences between H, R, and P when using β-Carotene and DPPH methods (Figure 2) and also showed interspecies differences between H, R, and P when using DPPH and ABTS methods (Figure 3). The unfractionated hydrolysate (H) of XGLA showed significant higher value ($p \leq 0.05$) of antioxidant activity determined with DPPH compared to retentate or permeate fractions (Figure 2). Significant differences were also observed in TALB, when data from the β-Carotene method were analyzed, between unfractionated hydrolysate and the other two fractions (R and P). Interspecies significant differences of hydrolysates, retentates, and permeates are presented in Figure 3 ($p \leq 0.05$). Figure 3A shows the differences found for unfractionated hydrolysates with ABTS; XGLA showed the highest antioxidant activity whereas SCAN and PGLA were the lowest. However, unfractionated hydrolysates did not show significant differences between species when the antioxidant activity was determined with DPPH or the β-Carotene method (data not shown). In Figure 3B, it can be also observed that the retentate fraction of SCAN presented the highest activity compared to other three species when DPPH was used, while ABTS data (Figure 3C) showed significant differences in retentate fractions only between XGLA and PGLA (lowest). Regarding permeate fractions (Figure 3D), significant differences were observed only between XGLA and PGLA when ABTS data were analyzed.

Table 5. Amino acid composition of collagen hydrolysates of four species (residues/1000). Imino acids includes proline and hydroxyproline.

Amino Acid	HYDROLYSATES			
	PGLA	SCAN	TALB	XGLA
Hydroxyproline	84.65 ± 0.80	87.50 ± 1.22	86.97 ± 0.54	75.15 ± 0.36
Aspartic acid	48.56 ± 0.45	53.33 ± 0.77	53.08 ± 0.24	59.39 ± 0.34
Serine	36.39 ± 0.34	52.45 ± 0.65	34.81 ± 0.20	38.83 ± 0.19
Gultamic acid	92.49 ± 0.89	90.97 ± 1.27	90.69 ± 0.42	92.02 ± 0.43
Glycine	230.71 ± 2.10	227.17 ± 2.96	215.82 ± 0.66	211.01 ± 1.06
Histidine	16.53 ± 0.13	16.49 ± 0.18	11. 18 ± 0.12	14.91 ± 0.03
Arginine	93.64 ± 0.98	93.00 ± 1.08	90.92 ± 0.65	76.46 ± 0.16
Threonine	27.99 ± 0.32	36.62 ± 0.59	40.00 ± .035	39.00 ± 0.26
Alanine	105.81 ±1.11	93.50 ± 1.27	108.72 ± 0.74	97.97 ± 0.62
Proline	106.47 ± 1.14	89.31 ± 1.26	100.22 ± 0.77	99.87 ± 0.61
Cystine	8.93 ±0.16	8.29 ± 0.33	31.91 ± 0.33	53.03 ± 0.16
Tyrosine	2.17 ± 0.01	1.68 ± 0.02	1.84 ± 0.02	2.24 ± 0.00
Valine	27.84 ± 0.28	34.12 ± 0.42	26.17 ± 0.17	27.61 ± 0.12
Methionine	13.68 ± 0.15	17.06 ± 0.26	15.19 ± 0.24	12.39 ± 0.09
Lysine	34.16 ± 0.32	37.55 ± 0.48	33.88 ± 0.14	32.70 ± 0.17
Isoleucine	24.65 ± 0.26	17.45 ± 0.20	13.05 ± 0.10	19.15 ± 0.09
Leucine	26.11 ± 0.25	25.95 ± 0.27	26.20 ± 0.13	28.58 ± 0.07
Phenylalanine	19.23 ± 0.19	17.56 ± 0.17	19.34 ± 0.10	19.67 ± 0.04
Iminoacids	191.12	176.81	187.19	175.02
% hydroxylation of prol	44.29	49.48	46.45	42.93

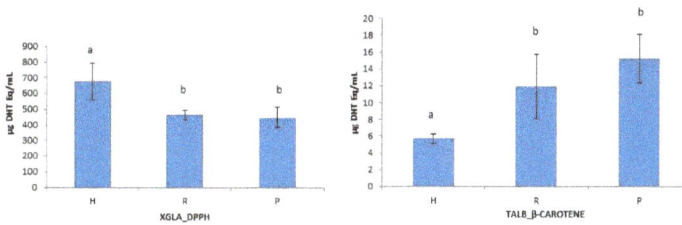

Figure 2. Intraspecific differences between hydrolysate (H), retentate (R) and permeate (P) in XGLA analyzed by DPPH method and in TALB analyzed by β-Carotene method. Different letters indicate significant differences among means ($p \leq 0.05$).

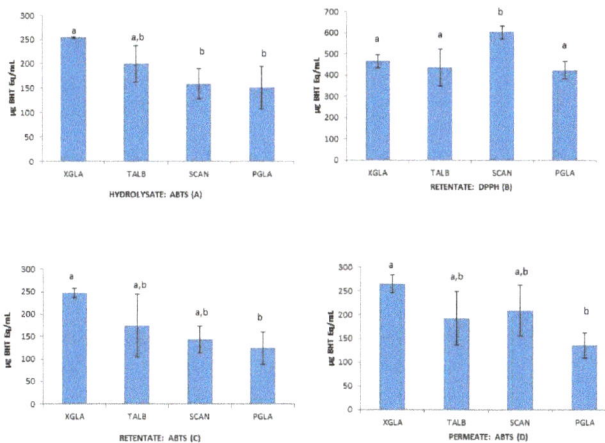

Figure 3. Interspecies differences in hydrolysate fraction using ABTS (**A**); in retentate fraction using DPPH (**B**) and ABTS (**C**); in permeate fraction using ABTS (**D**). Different letters indicate significant differences among means ($p \leq 0.05$).

Significant differences ($p \leq 0.05$) were observed between the antioxidant capacity of unfractionated hydrolysates of teleost (XGLA and TALB) and chondrychtyes (PGLA and SCAN) with the β-carotene assay. Thus, the two teleost species XGLA and TALB showed lower antioxidant capacity than chondrychtyes, results that might be in relation with the higher content of hydrophilic amino acids (Asp, Ser, Gly, His, Arg, Thr, and Cys) in chondrychtyes hydrolysates compared to teleost (Table 5). This result agree with other studies suggesting differences on the antioxidant defense system between elasmobranchs and teleosts, due to different evolutionary rates and also due to different physical activity, nutrient intake and environment in which each species develops [47].

In summary, antioxidant capacity results suggest that there is not a unique factor responsible for this antioxidant capacity of hydrolysates, which seems to be influenced by the species which is being studied, the type and length of the peptides present in the sample and the methodology employed to determine the antioxidant activity.

2.3.3. Amino Acid Content

Table 5 shows the amino acid content of unfractionated collagen hydrolysates. Besides the influence of amino acid composition and other factors on antioxidant activity (discussed above), it is also of importance to highlight the increase in Cystine content in hydrolysates, in comparison to non-hydrolyzed collagen (PSC). These variations might be explained because the alkaline pH achieved during hydrolysis promotes reoxidation of cysteine residues to generate the original disulfide bond [48]. The higher Cystine content found in TALB and XGLA hydrolysates is therefore related to the low collagen yield obtained for those skins (Section 2.2.1). As it was previously reported, the positive correlation between high disulfide bond content and low extraction yields is because of a higher stabilization of supramolecular assemblies [49]. The higher content of methionine in SCAN hydrolysates compared to the other species is also noteworthy.

3. Experimental Section

3.1. Raw Material

Fresh skin of the small-spotted catshark was obtained by a local fishing fleet, while frozen skin of blue shark, swordfish, and yellowfin tuna was provided by a Lumar S.L industry (Galicia, Spain) and stored at $-20\,°C$ until used. Fins, fat, and muscle residues were removed from skins, then skin was cut into small pieces ($0.5\,cm \times 0.5\,cm$) and mixed thoroughly. The skin pieces of each species were divided into three batches which were kept frozen at $-20\,°C$ until collagen extraction.

Identification of fish species was performed by DNA analysis, following the methodology of Blanco et al. [50].

3.1.1. Proximate Composition

Skin was analyzed for crude protein content by Kjeldhal method [51] in a DigiPREP HT digestor (SCP Science, Quebec, QC, Canada) and a TitroLine easy titration unit (SCHOTT, Mainz, Germany). Lipid content was determined by Bligh and Dyer [52]. Moisture was determined after heating the sample overnight at $105\,°C$ and ash content was determined after heating the sample overnight at $600\,°C$. The conversion factor used for calculating the protein content from Kjeldahl nitrogen data was 5.4 as collagen, the main protein present in skin, contains approximately 18.7% nitrogen [53,54].

3.1.2. Hydroxyproline Content

30 mg of dried grinded skin was introduced in hydrolysis microwaves tubes and 4 mL of 6 M HCl were added. Hydrolysis was performed in a microwave (speed wave MWS-2) (Berghof GmbH, Eningen, Germany) at a $150\,°C$ for 90 min at 70% power. Once the hydrolysis step finished, samples were allowed to cool down to room temperature and were made up to a known volume with 6 M HCl. 400 µL of this solution were transferred to glass vials and left to dry in a vacuum desiccator

at 60 °C in the presence of solid NaOH, after drawing air for 3 days. The resulting dry matter was suspended in 8 mL of buffer (0.13 M citric acid, 0.75% glacial acetic acid, 0.6 M sodium acetate, 0.15 M sodium hydroxide and 20.13% *n*-propanol, pH was adjusted to 6.5 with 0.2 M NaOH and volume was brought to 660 mL with distilled water).

Hydroxyproline primary standard was prepared by dissolving 50 mg of hydroxyproline (Sigma-Aldrich, St. Louis, MO, USA) in 100 mL of buffer. From this primary standard a calibration curve of hydroxyproline, ranging from 0.5 µg/mL up to 10 µg/mL, was prepared. Chloramine-T reagent was freshly prepared just before using it (0.05 M Chloramine in distilled water). 3 mL of either samples or standards were placed in a tube and 1.5 mL of Chloramine-T reagent was added, the mixture was allowed to react for 25 min. Upon completion of that time, chromogenic reagent (15 g of *p*-dimethyl-amino-benzaldehyde, 60 mL of *n*-propanol, 26 mL of 70% perchloric acid were made up to a volume of 100 mL with distilled water) was added and tubes introduced in a water bath at 60 °C for 15 min. Samples were left to cool to room temperature and after, absorbance was read at 550 nm in a Beckman UV-VIS spectrophotometer (Beckman-Coulter, Brea, CA, USA).

3.2. Extraction of Pepsin Soluble Collagen (PSC) from Skin

Collagen from skin was extracted according to the methodology of Liu et al. [35] with minor modifications (Figure 4). All procedures were performed at 4 °C. Skin pieces of blue shark and small-spotted-catshark were first treated with 0.1 N NaOH (1:15, *w/v*) and stirred for 24 h. Then, skins were washed with cold distilled water until a neutral pH was found, and skin residues were extracted with 0.5 M acetic acid containing 0.1% (*w/v*) pepsin (0.5 U/mg; Acros Organics, Janssen Pharmaceuticalaan 3a, Geel, Belgium), at a sample solution ratio of 1:40 (*w/v*) for 24 h. Suspension was centrifuged at 6000× *g* for 20 min, the residue discarded and the supernatant was salted-out by adding NaCl (final concentration of 2 M). The precipitate was dissolved in 0.5 M acetic acid and dialyzed against water using 12,000 Da cut-off membranes for 3 days. Aliquots were obtained and freeze-dried for analysis of Kjeldahl nitrogen, amino acid content, denaturation temperature, and electrophoresis. The remaining liquid volume of dialyzed PSC was stored frozen at −20 °C until used for hydrolysis.

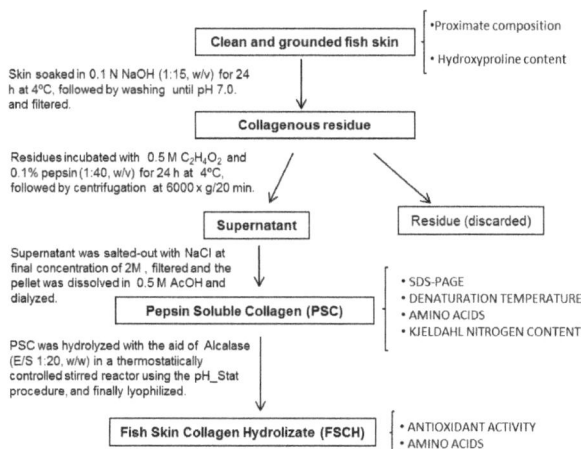

Figure 4. Scheme for the recovery of pepsin soluble collagen (PSC), preparation of the hydrolysate and analytical determinations.

The procedure used for swordfish and yellowfin tuna skin was slightly different than the one employed with sharks. Higher fat content in both swordfish and tuna skin required that after alkaline treatment and before the acid pepsin extraction, samples were soaked in 10% butyl alcohol for 24 h to remove any remaining fat at a sample/solid ratio of 1:10 (w/v), and then washed until neutral pH. Also thre time for pepsin extraction of these skins was increased up to 3 days.

PSC yields were calculated using Kjeldahl nitrogen values (data not shown) in the collagen solution considering that collagen contains approximately 18.7% of nitrogen [53,54].

3.3. Characterization of Pepsin Soluble Collagen (PSC) from Skin

3.3.1. Polyacrylamide Gel Electrophoresis

PSC samples for Sodium Dodecyl Sulfate-Polyacrylamide Gel Electrophoresis (SDS-PAGE) were prepared according to methodology reported by Sotelo et al. [19]. Molecular weights of PSC subunits were estimated using high range molecular weight standards (BIO-RAD): Myosin (200 kDa); β-Galactosidase (116 kDa); phosphorylase B (97 kDa) and analyzing the gel with the software Quantity One (BIO-RAD).

3.3.2. Differential Scanning Calorimetry

Freeze-dried PSC samples were solubilized in 50 mM acetic acid (1 mg of freeze-dried sample/mL). Thermostability of PSC solutions was measured in a DSC III microcalorimeter (Setaram, France) by differential scanning calorimetry (DSC). The samples were weighed accurately in a Mettler AE-240 balance, introduced into the calorimeter at 283.15°K and left for one hour to stabilize. Afterwards, temperature increase was set to 1°K/min up to 343°K. The denaturation temperature was calculated by difference with the apparent specific heat of ultrapure water.

3.3.3. Nitrogen Content

PSC was analyzed in terms of nitrogen content by Kjeldahl method described in Section 3.1.1 considering a 5.4 factor to obtain the collagen content.

3.3.4. Amino Acid Composition

100 mg of lyophilized PSC samples were hydrolyzed using 6 N hydrochloric acid under vacuum pressure at 110 °C for 24 h. HPLC-fluorescence determination of amino acids, using AccQ-Tag Amino acid analysis column (Waters Co., Milford, MA, USA), was carried out after derivatization using the AccQ-Tag Chemistry kit (Waters-WAT052875).

3.4. Enzymatic Hydrolysis of Pepsin Soluble Collagen

Enzymatic hydrolysis was carried out according to the methodology of Liu et al. [35] with minor modifications. Prior to the hydrolysis process, the selected volume of each PSC collagen batch was thawed. Hydrolysates were prepared in a stirred and thermostated reactor connected to a pH electrode and a temperature probe, using the pH-Stat procedure, as described by Adler-Nissen [55]. Temperature and pH were recorded by a visual display at all time. Food-grade Alcalase (2.4 AU-A/g) provided by Novo Nordisk (Bagsvaerd, Denmark) was used for the hydrolysis. The 2 L of thawed PSC were introduced in the reactor and heated up to 55 °C (Alcalase optimum temperature), pH was adjusted to pH 8.0 with 1 N NaOH and maintained constant during the hydrolysis reaction by automatically adding 1 N NaOH. Hydrolysis started with the addition of enzyme (enzyme/protein ratio of 1:20 w/w). The hydrolysis reaction was allowed to continue for 3 h under constant stirring. At the end of hydrolysis, the enzyme was inactivated by heating at 90 °C for 5 min. The resulting hydrolysates were freeze-dried and kept frozen at −20 °C until characterization analysis.

Degree of Hydrolysis

Degree of hydrolysis (DH) was obtained according to the following expression [55,56] where DH is the percent ratio between the total number of peptide bonds cleaved and the total number of peptide bonds in the initial protein.

$$DH(\%) = \frac{B \times N_b}{\alpha \times M_p \times h_{tot}} \tag{1}$$

where B is the volume (mL) of 1 M NaOH consumed during hydrolysis; N_b is the normality of NaOH; M_p is the mass (g) of initial protein (nitrogen $\times$ 5.4); h_{tot} is the total number of peptide bonds available for proteolytic hydrolysis, and α is the average degree of dissociation of the amino groups in the protein substrate and was calculated as follows:

$$\alpha = \frac{10^{pH-pK}}{1 + 10^{pH-pK}} \tag{2}$$

The pK value dependent on the temperature of hydrolysis was calculated according to the following expression, where T is the temperature (K):

$$pK = \left[7.8 + \frac{298 - T}{298T}\right] \times 2400 \tag{3}$$

h_{tot} was calculated considering a mean molecular weight of amino acids around 125 g/mol [57], and total content of amino acid in each PSC obtained from different species (PGLA: 78.4 g/100 g; SCAN: 96.02 g/100 g; TALB: 92.75 g/100 g; XGLA: 80.84 g/100 g). h_{tot} of PSC collagen were 6.8 meq/g protein, 8.3 meq/g protein, 8.06 meq/g protein and 7.02 meq/g protein for PGLA, SCAN, TALB, and XGLA respectively.

3.5. Antioxidant Capacity of Pepsin Soluble Collagen Hydrolysates

3.5.1. Ultrafiltration

To test the influence of molecular weight on antioxidant capacity, four grams of freeze-dried hydrolysates were dissolved in distilled water (1%) and ultrafiltrated in two steps using ultrafiltration centrifugal devices (Amicon Ultra-15 Unit) (Merck Millipore, Billerica, MA, USA) with molecular weight cut-off of 10 kDa and 3 kDa. After this process, fractions containing peptides with molecular weight between 10,000 Da and 3000 Da (retentate fraction) and fractions containing peptides below 3000 Da (permeate fraction) were then freeze-dried and stored at −20 °C until subjected to antioxidant capacity analysis.

3.5.2. Antioxidant Activity Determinations

β-Carotene Bleaching Method

The β-carotene bleaching assay was performed according to Prieto et al. 2012 [58] with a microplate spectrophotometer. Reactions were performed by combining in each well of a 96-well microplate, 25 μL of antioxidant (butyl hydroxytoluene (BHT) at 0–22.7 μM or hydrolysate samples) with 125 μL of the β-carotene/linoleic emulsion. The microplate spectrophotometer (Multiskan Spectrum Microplate Spectrophotometer) (Thermo Fisher Scientific, Waltham, MA, USA) was programmed to record the absorbance at 470 nm and 45 °C every three minutes during a period of 200 min with agitation at 660 cycles/min (1 mm amplitude).

1,1-Diphenyl-2-Picryhydrazyl (DPPH) Radical-Scavenging Capacity

The antioxidant activity as radical-scavenging capacity was assessed with DPPH as a free radical, using an adaptation to the microplate of the method described by Brand-Williams et al. [59,60].

The decrease in the absorbance of hydrolysates and the BHT control (0–108 μM) was followed at 515 nm every 3 min during 200 min at 30 °C.

ABTS Bleaching Method

The ABTS (2,2′-azinobis-(3-ethyl-benzothiazoline-6-sulphonic acid) radical scavenging activities were assessed according the protocol developed by Prieto et al. [60]. The absorbance at 414 nm and 30 °C (maintaining continuous agitation) of samples and BHT (0–9.1 μM) were measured each 3 min in the microplate reader.

In all methods, the kinetics of reaction were performed in triplicate following the methodology of Amado et al. (2016) [61].

3.5.3. Amino Acid Composition

Hydrolysates were analyzed for amino acid content following the methodology described in Section 3.3.3.

3.5.4. Statistical Analysis

Interspecific and intraspecific differences regarding antioxidant capacity between unfractionated hydrolysates (H) and 3 kDa MWCO ultrafiltrated fractions: permeates (P) and retentates (R) were tested by one-way analysis of variance (ANOVA). It was applied to a Post hoc comparison test. Significance levels were set at $p \leq 0.05$. Statistical tests were performed with IBM SPSS 23 (IBM Corporation, Armonk, NY, USA).

Acknowledgments: The authors would like to acknowledge the financial support through the projects MARMED (ref: Atlantic Area Programme 2011-1/164) and NOVOMAR (ref: FEDER POCTEP_0687-POCTEP Programme). Authors are also grateful to Marta P. Testa, Araceli Menduiña and Ana Durán for their professional work and dedication.

Author Contributions: Carmen G. Sotelo, José Antonio Vázquez, Ricardo I. Pérez-Martín and María Blanco conceived and designed the experiments; María Blanco performed the experiments; Carmen G. Sotelo, José Antonio Vázquez, Ricardo I. Pérez-Martín and María Blanco analyzed the data; María Blanco wrote the paper. Carmen G. Sotelo participated in the redaction of the manuscript. Carmen G. Sotelo and Ricardo I. Pérez-Martín critically revised the manuscript.

Conflicts of Interest: The authors declare no conflict of interest.

References

1. Food and Agriculture Organization. *El Estado Mundial de la Pesca y la Acuicultura*; FAO: Roma, Italy, 2016.
2. Blanco, M.; Fraguas, J.; Sotelo, C.G.; Pérez-Martín, R.I.; Vázquez, J.A. Production of Chondroitin sulphate from head, skeleton and fins of *Scyliorhinus canicula* by-products by combination of enzymatic, chemical precipitation and ultrafiltration methodologies. *Mar. Drugs* **2015**, *13*, 3287–3308. [CrossRef] [PubMed]
3. Gómez-Guillén, M.C.; Turnay, J.; Férnandez-Díaz, M.D.; Ulmo, N.; Lizarbe, M.A.; Montero, P. Structural and physical properties of gelatin extracted from different marine species: A comparative study. *Food Hydrocoll.* **2002**, *16*, 25–34. [CrossRef]
4. Karayannakidis, P.D.; Chatziantoniou, S.E.; Zotos, A. Effects of selected process parameters on physical and sensorial properties of yellowfin tuna (*Thunnus albacares*) skin gelatin. *J. Food Process Eng.* **2014**, *37*, 461–473. [CrossRef]
5. Chi, C.F.; Cao, Z.H.; Wang, B.; Hu, F.Y.; Li, Z.R.; Zhang, B. Antioxidant and functional properties of collagen hydrolysates from Spanish mackerel skin as influenced by average molecular weight. *Molecules* **2014**, *19*, 11211–11230. [CrossRef] [PubMed]
6. Halim, N.R.A.; Yusof, H.M.; Sarbon, N.M. Functional and bioactive properties of fish protein hydrolysates and peptides: A comprehensive review. *Trends Food Sci. Technol.* **2016**, *51*, 24–33. [CrossRef]

7. Nam, K.A.; You, S.G.; Kim, S.M. Molecular and physical characteristics of squid (*Todarodes pacificus*) skin collagens and biological properties of their enzymatic hydrolysates. *J. Food Sci.* **2008**, *73*, 249–255. [CrossRef] [PubMed]

8. Byun, H.G.; Kim, S.K. Purification and characterization of angiotensis I converting enzyme (ACE) inhibitory peptides from Alaska Pollack (*Theragra chalcogramma*) skin. *Process Biochem.* **2001**, *36*, 1155–1162. [CrossRef]

9. Chalamaiah, M.; Kumar, B.D.; Hemalatha, R.; Jyothirmayi, T. Fish protein hydrolysate: Proximate composition, amino acid composition, antioxidant activities and applications: A review. *Food Chem.* **2012**, *135*, 3020–3038. [CrossRef] [PubMed]

10. Jia, J.; Zhou, Y.; Lu, J.; Chen, A.; Li, Y.; Zheng, G. Enzymatic hydrolysis of Alaska Pollack (*Theragra chalcogramma*) skin and antioxidant activity of the resulting hydrolysate. *J. Sci. Food Agric.* **2010**, *90*, 635–640. [CrossRef] [PubMed]

11. Vázquez, J.A.; Blanco, M.; Fraguas, J.; Pastrana, L.; Pérez-Martín, R.I. Optimisation of the extraction and purification of chondroitin sulphate from head by-products of *Prionace glauca* by environmental friendly process. *Food Chem.* **2016**, *198*, 28–35. [CrossRef] [PubMed]

12. Autoridad Portuaria de Vigo. *Memoria Anual 2015*; Autoridad Portuaria de Vigo: Pontevedra, Spain, 2015.

13. Blanco, M. Valorización de Descartes y Subproductos de Pintarroja (*Scyliorhinus canicula*). Ph.D. Thesis, Universidad de Vigo, Pontevedra, Spain, December 2015.

14. Klompong, V.; Benjakul, S.; Kantachote, D.; Shahidi, F. Antioxidative activity and functional properties of protein hydrolyste of yellow stripe trevally (*Selaroides leptolepis*) as influenced by the degree of hydrolysis and enzyme type. *Food Chem.* **2007**, *102*, 1317–1327. [CrossRef]

15. Theodore, A.E.; Raghavan, S.; Kristinsson, H.G. Antioxidant activity of protein hydrolysates prepared from alkaline-aided channel catfish protein isolates. *J. Agric. Food Chem.* **2008**, *56*, 7459–7466. [CrossRef] [PubMed]

16. Kittiphattanabawon, P.; Benjakul, S.; Visessanguan, W.; Nagai, T.; Tanaka, M. Characterisation of acid-soluble collagen from skin and bone of bigeye snapper (*Pricanthus tayenus*). *Food Chem.* **2005**, *89*, 363–372. [CrossRef]

17. Edwards, C.A.; O'Brien, W.D., Jr. Modified assay for determination of hydroxyproline in a tissue hydrolyzate. *Clin. Chim. Acta* **1980**, *104*, 161–167. [CrossRef]

18. Ahmad, M.; Benjakul, S.; Nalinanon, S. Compositional and physicochemical characteristics of acid solubilized collagen extracted from the skin of unicorn leatherjacket (*Aluterus monoceros*). *Food Hydrocoll.* **2010**, *24*, 588–594. [CrossRef]

19. Sotelo, C.G.; Blanco, M.; Ramos-Ariza, P.; Pérez-Martín, R.I. Characterization of collagen from different discarded fish species of the West coast of the Iberian Peninsula. *J. Aquat. Food Prod. Technol.* **2015**, *25*, 388–399. [CrossRef]

20. Benjakul, S.; Thiansilakul, Y.; Visessanguan, W.; Roytrakul, S.; Kishimura, H.; Prodpran, T. Extraction and characterisation of pepsin-solubilised collagens from the skin of bigeye snapper (*Priacanthus tayenus and Prianthus macracanthus*). *J. Sci. Food Agric.* **2010**, *90*, 132–138. [CrossRef] [PubMed]

21. Zhang, M.; Liu, W.; Li, G. Isolation and characterisation of collagens from the skin of largefin longbarbel catfish (*Mystus macropterus*). *Food Chem.* **2009**, *115*, 826–831. [CrossRef]

22. Lynn, A.K.; Yannas, I.V.; Bonfield, W. Antigenicity and immunogenicity of collagen. *J. Biomed. Mater. Res. B* **2004**, *71*, 343–354. [CrossRef] [PubMed]

23. Morimoto, K.; Kunii, S.; Hamano, K.; Tonomura, B. Preparation and structural analysis of actinidain-processed atelocollagen of yellowfin tuna (*Thunnus albacares*). *Biosci. Biotechnol. Biochem.* **2004**, *68*, 861–867. [CrossRef] [PubMed]

24. Nagai, T.; Araki, Y.; Suzuki, N. Collagen of the skin of ocellate puffer fish (*Takifugu rubripes*). *Food Chem.* **2002**, *78*, 173–177. [CrossRef]

25. Jongjareonrak, A.; Benjakul, S.; Visessanguan, W.; Tanaka, M. Isolation and characterization of collagen from bigeye snapper (*Priacanthus macracanthus*) skin. *J. Sci. Food Agric.* **2005**, *85*, 1203–1210. [CrossRef]

26. Jongjareonrak, A.; Benjakul, S.; Visessanguan, W.; Tanaka, M. Isolation and characterization of acid and pepsin-solubilised collagens from the skin of browntripe red snapper (*Lutjanus vitta*). *Food Chem.* **2005**, *93*, 475–484. [CrossRef]

27. Zelechowska, E.; Sadowska, M.; Turk, M. Isolation and some properties of collagen from the backbone of Baltic cod (*Gadus morhua*). *Food Hydrocoll.* **2010**, *24*, 325–329. [CrossRef]

28. Borderías, A.J.; Montero, P. Changes in fish muscle collagen during frozen storage. In *Storage Lives of Chilled and Frozen Fish and Fish Products*; International Institute of Refrigeration: Hong Kong, China, 1985; pp. 85–91.

29. Badij, F.; Howell, N. Elucidation of the effect of formaldehyde and lipids on frozen stored cod collagen by FT-Raman spectroscopy and differential scanning calorimetry. *J. Agric. Food Chem.* **2003**, *51*, 1440–1446.

30. Foegeding, E.A.; Lanier, T.C.; Hultin, H.O. Characteristics of edible muscle tissues. In *Food Chemistry*; Fennema, O.R., Ed.; Marcel Dekker, Inc.: New York, NY, USA, 1996; pp. 902–906.

31. Kittiphattanabawon, P.; Benjakul, S.; Visessanguan, W.; Kishimura, H.; Shahidi, F. Isolation and characteridsation of collagen from the skin of brownbanded bamboo shark (*Chiloscyllium punctatum*). *Food Chem.* **2010**, *119*, 1519–1526. [CrossRef]

32. Woo, J.; Yu, S.; Cho, S.; Lee, Y.; Kim, S. Extraction optimization and properties of collagen from yellowfin tuna (*Thunnus albacares*) dorsal skin. *Food Hydrocoll.* **2008**, *22*, 879–887. [CrossRef]

33. Love, R.M.; Yamaguchi, K.; Créach, Y.; Lavéty, J. The connective tissues and collagens of cod during starvation. *Comp. Biochem. Physiol. B* **1976**, *55*, 487–492. [CrossRef]

34. Sikorski, Z.E.; Kolakowska, A.; Pan, B.S. The nutritive composition of the major groups of marine food organisms. In *Seafood: Resources, Nutritional Composition and Preservation*; Sikorski, Z.E., Ed.; CRC Press: Boca Raton, FL, USA, 1990; pp. 29–54.

35. Liu, D.; Liang, L.; Regenstein, J.M.; Zhow, P. Extraction and characterisation of pepsin-solubilised collagen from fins, scales, skins, bones and swim bladders of bighead carp (*Hypophthalmichthys nobilis*). *Food Chem.* **2012**, *133*, 1441–1448. [CrossRef]

36. Nishimoto, M.; Sakamoto, R.; Mizuta, S.; Yoshinaka, R. Identification and characterization of molecular species of collagen in ordinary muscle and skin of the Japanese flounder *Paralichthys olivaceus*. *Food Chem.* **2005**, *90*, 151–156. [CrossRef]

37. Nagai, T.; Suzuki, N. Preparation and partial characterization of collagen from paper nautilus (*Argonauta argo*, Linnaeus) uter skin. *Food Chem.* **2002**, *76*, 149–153. [CrossRef]

38. Singh, P.; Benjakul, S.; Maqsood, S.; Kishimura, H. Isolation and characterization of collagen extracted from the skin of the striped catfish (*Pangasianodon hypophtalmus*). *Food Chem.* **2011**, *124*, 97–105. [CrossRef]

39. Nomura, Y.; Toki, S.; Ishii, Y.; Shirai, K. The physicochemical property of shark type I collagen gel and membrane. *J. Agric. Food Chem.* **2000**, *48*, 2028–2032. [CrossRef] [PubMed]

40. Wong, D.S. *Mechanism and Theory in Food Chemistry*; Van Nostrand Reinhold: New York, NY, USA, 1989.

41. Bougatef, A.; Nedjar-Arroume, N.; Manni, L.; Ravallec, R.; Barkia, A.; Guillochon, D.; Nasri, M. Purification and identification of novel antioxidant peptides from enzymatic hydrolysates of sardinelle (*Sardinella aurita*) by-products proteins. *Food Chem.* **2010**, *118*, 559–565. [CrossRef]

42. Kristinsson, H.G.; Rasco, B.A. Fish protein hydrolysates: Production, biochemical and functional properties. *Crit. Rev. Food Sci. Nutr.* **2000**, *40*, 43–81. [CrossRef] [PubMed]

43. Quian, Z.-J.; Jung, W.K.; Kim, S.K. Free radical scavenging activity of a novel antioxidative peptide purified from hydrolysate of bullfrog skin, *Rana catesbeiana Shaw*. *Bioresour. Technol.* **2008**, *99*, 1690–1698. [CrossRef] [PubMed]

44. Harman, L.S.; Mottley, C.; Mason, R. Free radical metabolites of L-cysteine oxidation. *J. Boil. Chem.* **1984**, *259*, 5606–5611.

45. Sarmadi, B.H.; Ismail, A. Antioxidative peptides from food proteins: A review. *Peptides* **2010**, *31*, 1949–1956. [CrossRef] [PubMed]

46. Chi, C.-F.; Wang, B.; Deng, Y.Y.; Wang, Y.M.; Deng, S.G.; Ma, J.Y. Isolation and characterization of three antioxidant pentapeptides from protein hydrolsates of monkfish (*Lophius litulon*) muscle. *Food Res. Int.* **2014**, *55*, 222–228. [CrossRef]

47. Vélez-Alavez, M. Evaluación de los Indicadores de Estrés Oxidativo Asociados a las Características de Nado en Elasmobranquios y Teleósteos. Ph.D. Thesis, Centro de Investigaciones Biológicas del Noroeste, S.C., La Paz, Mexico, 2015.

48. Lundblad, R. *Techniques in Protein Modification*; CRC Press: Boca Raton, FL, USA, 1994.

49. Barth, D.; Kyrieleis, O.; Frank, S.; Renner, C.; Moroder, L. The role of cystine knots in collagen folding and stability, part II. Conformational properties of (Pro-Hyp-Gly)n model trimers with N- and C-terminal collagen type III cystine knots. *Chemistry* **2003**, *9*, 3703–3714. [CrossRef] [PubMed]

50. Blanco, M.; Perez-Martin, R.I.; Sotelo, C.G. Identification of Shark Species in Seafood Products by Forensically Informative Nucleotide Sequencing (FINS). *J. Agric. Food Chem.* **2008**, *56*, 9868–9874. [CrossRef] [PubMed]

51. Association of Official Analytical Chemistry. *Methods of Analysis*, 15th ed.; Association of Official Analytical Chemistry: Washington, DC, USA, 1997.

52. Bligh, E.G.; Dyer, W.J. A rapid method of total lipid extraction and purification. *Can. J. Biochem. Phsiol.* **1959**, *37*, 911–917. [CrossRef] [PubMed]

53. Muyonga, J.H.; Cole, C.G.B.; Duodu, K.G. Characterisation of acid soluble collagen from skins of Young and adult Nile perch (*Lates niloticus*). *Food Chem.* **2004**, *85*, 81–89. [CrossRef]

54. Eastoe, J.; Eastoe, B. A method for the determination of total nitrogen in proteins. *Br. Gel. Glue Res. Assoc. Res. Rep.* **1952**, *5*, 1–17.

55. Adler-Nissen, J. Control of the proteolytic reaction and of the level of bitterness in protein hydrolysis processes. *J. Chem. Technol. Biotechnol.* **1984**, *34*, 215–222. [CrossRef]

56. Camacho, F.; González-Tello, P.; Páez-Dueñas, M.P.; Guadix, E.M.; Guadix, A. Correlation of base consumption with the degree of hydrolysis in enzymic protein hydrolysis. *J. Dairy Res.* **2001**, *68*, 251–265. [CrossRef] [PubMed]

57. Nielsen, P.M.; Petersen, D.; Dambmann, C. Imporoved method for determining food protein degree of hydrolysis. *J. Food Sci.* **2001**, *66*, 642–646. [CrossRef]

58. Prieto, M.A.; Rodríguez-Amado, I.; Vázquez, J.A.; Murado, M.A. β-Carotene assay revisited. Application to characterize and quantify antioxidant and prooxidant activities in a microplate. *J. Agric. Food Chem.* **2012**, *60*, 8983–8993. [CrossRef] [PubMed]

59. Brand-Williams, W.; Cuvelier, M.E.; Berset, C. Use of a free radical method to evaluate antioxidant activity. *LWT-Food Sci. Technol.* **1995**, *28*, 25–30. [CrossRef]

60. Prieto, M.A.; Curran, T.P.; Gowen, A.; Vázquez, J.A. An efficient methodology for quantification of synergy and antagonismin single electron transfer antioxidant assays. *Food Res. Int.* **2015**, *67*, 284–298. [CrossRef]

61. Amado, I.R.; González, M.P.; Murado, M.A.; Vázquez, J.A. Shrimp wastewater as a source of astaxanthin and bioactive peptides. *J. Chem. Technol. Biotechnol.* **2016**, *91*, 793–805. [CrossRef]

![marine drugs logo](marine drugs)

MDPI

Review

Marine Fish Proteins and Peptides for Cosmeceuticals: A Review

Jayachandran Venkatesan [1], Sukumaran Anil [2], Se-Kwon Kim [3,*] and Min Suk Shim [1,*]

[1] Division of Bioengineering, Incheon National University, Incheon 406-772, Korea; venkatjchem@gmail.com
[2] Department of Preventive Dental Sciences, College of Dentistry, Prince Sattam Bin Abdulaziz University, Riyadh, Post Box 153, AlKharj 11942, Saudi Arabia; drsanil@gmail.com
[3] Department of Marine Life Sciences, Korean Maritime and Ocean University, 727 Taejong-ro, Yeongdo-Gu, Busan 49112, Korea
* Correspondence: sknkim@pknu.ac.kr (S.-K.K.); msshim@inu.ac.kr (M.S.S.);
  Tel.: +82-51-629-7550 (S.-K.K.); +82-32-835-8268 (M.S.S.)

Academic Editor: Orazio Taglialatela-Scafati
Received: 28 February 2017; Accepted: 11 May 2017; Published: 18 May 2017

Abstract: Marine fish provide a rich source of bioactive compounds such as proteins and peptides. The bioactive proteins and peptides derived from marine fish have gained enormous interest in nutraceutical, pharmaceutical, and cosmeceutical industries due to their broad spectrum of bioactivities, including antioxidant, antimicrobial, and anti-aging activities. Recently, the development of cosmeceuticals using marine fish-derived proteins and peptides obtained from chemical or enzymatical hydrolysis of fish processing by-products has increased rapidly owing to their activities in antioxidation and tissue regeneration. Marine fish-derived collagen has been utilized for the development of cosmeceutical products due to its abilities in skin repair and tissue regeneration. Marine fish-derived peptides have also been utilized for various cosmeceutical applications due to their antioxidant, antimicrobial, and matrix metalloproteinase inhibitory activities. In addition, marine fish-derived proteins and hydrolysates demonstrated efficient anti-photoaging activity. The present review highlights and presents an overview of the current status of the isolation and applications of marine fish-derived proteins and peptides. This review also demonstrates that marine fish-derived proteins and peptides have high potential for biocompatible and effective cosmeceuticals.

Keywords: marine fish; cosmeceuticals; proteins; peptides; hydrolysates; collagen; antioxidant; anti-photoaging

1. Introduction

Oceans cover about 70% of the earth's surface and are inhabited by a large variety of living organisms. The marine environment serves as an enormous resource that provides abundant bioactive substances in the form of food, cosmeceuticals, and pharmaceutical products. Recently, much attention has been paid to obtaining bioactive proteins and peptides from various marine organisms, including fish, algae, crustaceans, and sponges, for cosmeceutical and pharmaceutical applications [1,2]. Marine bioactive proteins and peptides, depending on their structures and amino acid sequences, exhibit a wide range of biological activities including antioxidant, antimicrobial, anticancer, immunomodulatory, antihypertensive, anticoagulant, and anti-diabetic effects [3,4].

Marine fish is mostly used as a source of food for human consumption, which has resulted in several fish processing industries producing fish meat. However, these industries discard huge amounts of waste containing fish skin and bones, which in turn aggravate the problem of environmental pollution. To avoid such issues, by-products generated by seafood processing industries are utilized to isolate bioactive compounds beneficial for human health. This process not only assists in decreasing the pollution but also increases the value of the by-products from fish processing [5–7]. Fish processing waste contains significant amounts of useful proteins, which represent a source for bioactive peptide mining. For example, collagen is one of the most abundant proteins that can be extracted from the skin, bones, and scales of fish. Collagen has been extensively utilized for various applications, including cosmeceuticals [8], functional foods [9], tissue engineering [10,11], and anti-diabetic medications [12].

In addition to bioactive proteins, various bioactive peptides can be produced from marine fish via chemical or enzymatical hydrolysis. The peptides, which are present in the inactive form within the protein chains, are activated after their hydrolysis using enzymes, including trypsin, proteinases, chymotrypsin, alcalase, and pepsin [3,13–15]. Marine fish waste-derived bioactive peptides have gained tremendous interest in nutraceutical and cosmeceutical industries due to their broad spectrum of bioactivities, including antioxidant, antimicrobial, antihypertensive, calcium-binding, and obesity control properties [3,16]. This review describes various bioactive proteins and peptides, which were identified in marine processing waste with emphasis on their potential bioactivities for cosmeceutical applications. Moreover, it outlines current technologies used in the production and purification of the marine fish-derived proteins and peptides.

2. Marine Fish Proteins and Peptides

Figure 1 depicts the increasing number of the studies on marine fish-derived proteins and peptides in the last two decades. Marine fish proteins mainly consist of collagen, which has been widely utilized in cosmeceutical areas owing to its moisturizing properties. In addition, it has been extensively studied in pharmaceuticals, nutraceuticals, and food applications. Collagen can be isolated from by-products of fish processing, such as fish bones and fish skin [17–19].

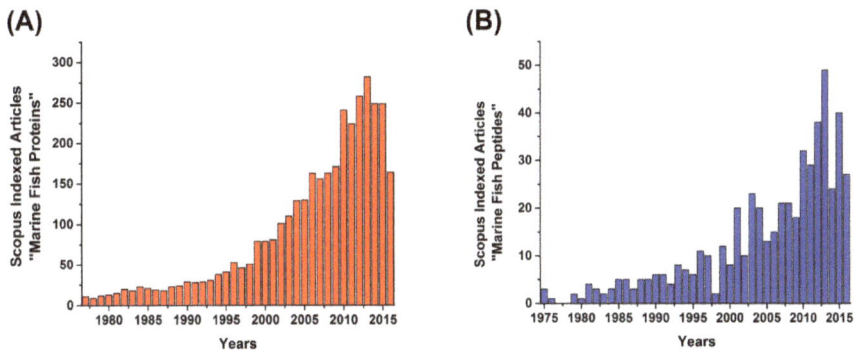

Figure 1. Articles indexed in Scopus with the keywords (**A**) marine fish proteins and (**B**) marine fish peptides. Graph shows the continuous research growth on marine fish proteins and peptides. The bar graph highlights the number of articles indexed in Scopus on "marine fish proteins", which is greater than that of "marine fish peptides".

3. Marine Fish-Derived Collagen

Collagen is a main structural protein in connective tissues of skin and bone. It is commonly obtained from bovine and porcine skin. The bovine and porcine collagens have been extensively used for pharmaceutical, cosmeceutical, and nutraceutical purposes. However, the outbreak of certain transmissible diseases such as bovine spongiform encephalopathy and some religious issues associated with the use of bovine proteins hamper their use. Hence, there has been a need to find a suitable alternative to solve these issues, which has led several researchers to turn toward marine sources for the production of collagen. Marine-derived collagen has an ability to scavenge free radicals, and thus can be utilized for skin care products [17,20,21]. Marine-derived collagen has also been widely used as a scaffold for tissue engineering due to its excellent bioactive properties, including biocompatibility, low antigenicity, high biodegradability, and cell growth potential [22–24]. There are two types of collagen: fibrillar and nonfibrillar. Marine fish often contain Type I fibrillar collagen in skin and bones [25].

3.1. Isolation of Marine Fish-Derived Collagen

Although around 75% of the fish weight consists of skin, bones, head, and scales, they are often discarded as by-products by the seafood processing industries [26]. These by-products are a rich source of collagen with a variety of bioactivities. Figure 2 shows the common procedures for isolating collagen from the skin and bones of marine fish [19,26,27]. Acid solubilization and pepsin solubilization are major methods for isolating collagen from various parts of fish species (e.g., skin, bones, and scales). For the acid-soluble collagen (ASC) method, 0.5 M acetic acid is used to digest the fish skin in sufficient time, whereas 10% w/v pepsin is used for the pepsin-soluble collagen (PSC) method. Table 1 shows a list of some important marine fish species used for collagen isolation. It is observed that the PSC method leads to higher amounts of collagen as compared to the ASC method [19,27,28]. This implies that pepsin in the PSC method is more efficient in digesting skin or bone tissues as compared to acid solution in the ASC method.

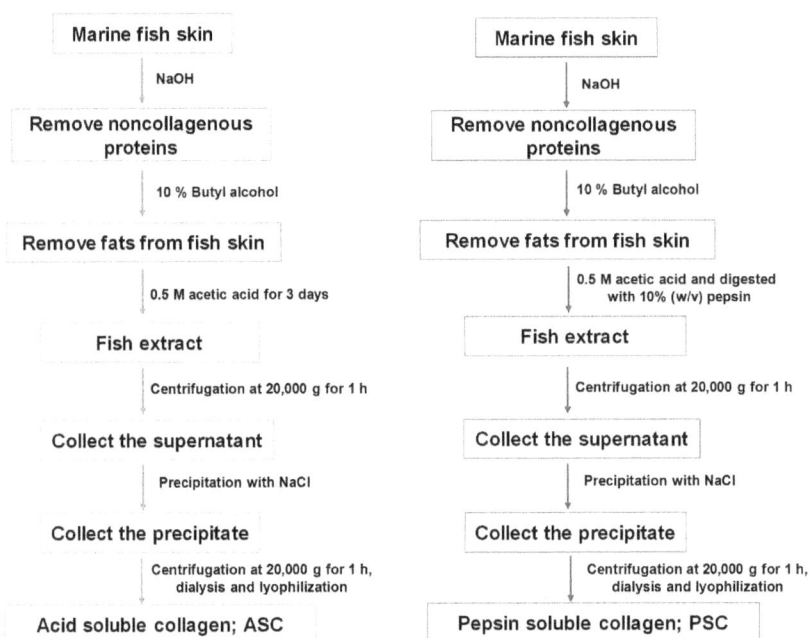

Figure 2. A flowchart for the isolation of collagen from marine fish skin. (**A**) acid-soluble collagen (ASC) method and (**B**) pepsin-soluble collagen (PSC) method.

Table 1. Important marine fish species used to isolate collagen.

Fish Species Name	Parts	Method	Yield (%)	Reference
Lagocephalus gloveri	Skin	PSC	54.3	[17]
Thunnus obesus	Bone	ASC and PSC	–	[19]
Paralichthys olivaceus, Sebastes schlegeli, Lateolabrax maculatus, Pagrus major	Skin	ASC	–	[22]
Takifugu rubripes	Skin	ASC and PSC	10.7 and 44.7	[27]
Sepiella inermis	Skin	ASC and PSC	0.58 and 16.23	[28]
Lutjanus vitta	Skin	ASC and PSC	9.0 and 4.7	[29]
Magalaspis cordyla	Bone	ASC and PSC	30.5 and 27.6	[30]
Otolithes ruber	Bone	ASC and PSC	45.1 and 48.6	[30]
Evenchelys macrura	Skin	ASC and PSC	80 and 7.1	[31,32]
Saurida spp., *Trachurus japonicus, Mugil cephalis, Cypselurus melanurus, Dentex tumifrons*	Scales	ASC	0.13–1.5%	[33]
Cyanea nozakii Kishinouye	All parts	ASC and PSC	13.0 and 5.5	[34]
Sardinella longiceps	Scales	ASC and PSC	1.25 and 3	[35]
Priacanthus tayenus	Skin and bone	ASC	10.94 and 1.59 (Skin and bone)	[36]
Priacanthus tayenus	Skin	PSC	–	[37]
Priacanthus tayenus and Priacanthus macracanthus	Skin	PSC	–	[38]
Parupeneus heptacanthus	Scale	ASC and PSC	0.46 and 1.2	[39]
Mystus macropterus	Skin	ASC and PSC	16.8 and 28	[40]
Syngnathus schlegeli	All parts	ASC and PSC	5.5 and 33.2	[41]
Jellyfish	All parts	PSC	46.4	[42]
Chrysaora sp.	All parts	PSC	9–19	[43]

3.2. Marine Fish-Derived Collagen in Cosmeceuticals

Marine fish-derived collagen is extensively employed in the development of cosmeceutical products due to its excellent bioactivity toward skin repair and regeneration. The marine fish-derived collagen possesses a higher absorbing capacity than the collagen from animal sources [44]. In addition, marine fish-derived collagen has low odor and improved mechanical strength, prerequisites for cosmetic products [8]. Skin-hydrating and skin-firming effects of cosmetic formulations (cream or serum formulations) using collagen derived from fish were evaluated [45]. The result suggested that serum formulations displayed a better moisturizing effect within a short duration [44,45]. The cream formulations appeared to become more active later, particularly following the repetitive applications. However, a sustained tensor (firming) effect was observed during the treatment using both the lotion and the cream [45].

3.3. Marine Fish-Derived Collagen in Wound Healing and Tissue Engineering

Tissue-engineered skin substitutes serve as a promising therapeutic agent in replacing the skin lost in wounds such as burns by providing cells, bioactive compounds, bioactive polymers, and proper microenvironments, thereby initiating the wound healing process [46–48]. Currently, a main source of collagen is bovine skin and tendons as well as porcine skin, which suffer from drawbacks such as transmission of prions [49]. Therefore, marine organism-derived materials have become initiators or co-initiators of hundreds of promising pharmaceutical and tissue-engineered skin substitutes [50]. Many studies based on marine organism-derived collagen scaffolds for skin tissue regeneration have demonstrated a high potential in clinical applications [51]. In this regard, a composite film comprising salmon milt DNA and salmon collagen showed a remarkable efficacy in wound regeneration [52]. The implantation of the film into a full-thickness wound in the rat dorsal region resulted in tissue regeneration with a morphological appearance similar to that of native rat dermis tissues. In addition, it significantly enhanced the formation of blood capillaries [52].

The abundant presence of type I collagen in fish bone tissues has widely increased the applications of collagen-based scaffolds for bone tissue engineering [53–55]. Collagen plays an important role in stimulating the differentiation of bone progenitor cells into osteoblasts through interaction with transmembrane α2β1 integrin receptors, and subsequently eliciting cell growth and mineral production [56,57]. The incorporation of glycosaminoglycans (GAGs) into collagen has shown to enhance osteoblastic differentiation of mesenchymal stem cells (MSCs) both in vivo and in vitro [58,59].

4. Marine Fish-Derived Peptides

Marine fish proteins consist of small peptides, which are often present in the inactive form with a full protein sequence. Enzymatic hydrolysis is frequently used to isolate short and bioactive peptides from marine organisms and seafood waste products. A large amount of histidine-containing dipeptides, carnosine (β-alanylhistidine), and anserine (β-alanyl-1-methylhistidine) are present in tuna, salmon, and eels [60]. Peptides serve as important active ingredients for several pharmaceutical and cosmeceutical applications [4,61,62]. The bioactive peptides are usually made up of 3–20 amino acid residues. Marine fish-derived peptides exhibit various biological activities such as antioxidant, antimicrobial, and angiotensin-I-converting inhibitory activity, as well as cancer metastasis inhibition, and immunostimulant activity [63–65]. The most commonly used proteinases for the hydrolysis of fish proteins include alcalase, chymotrypsin, and pepsin [66–68].

4.1. Isolation of Marine Fish Peptides

Enzymatic hydrolysis is one of the commonly used methods to obtain bioactive peptides. The mechanistic study of the enzymatic hydrolysis of fish proteins is described elsewhere [69]. The general procedures to produce collagen peptides from marine fish skin and bone are shown in Figure 3. Various antioxidant marine fish-derived peptides were obtained through enzymatic hydrolysis methods [70,71]. Different kinds of enzymes (e.g., alcalase, α-chymotrypsin, neutrase, papain, pepsin, and trypsin) were used for the optimized conditional buffer system (Table 2) [70]. The peptides are commonly separated using chromatographic techniques and ultrafiltration membranes. The same group also reported the use of a series of ultrafiltration membranes to separate the peptides [72]. Fast protein liquid chromatography (FPLC) and reverse phase high-pressure liquid chromatography (RP-HPLC) were widely utilized to purify the peptides.

Procedures for the isolation of marine fish peptides

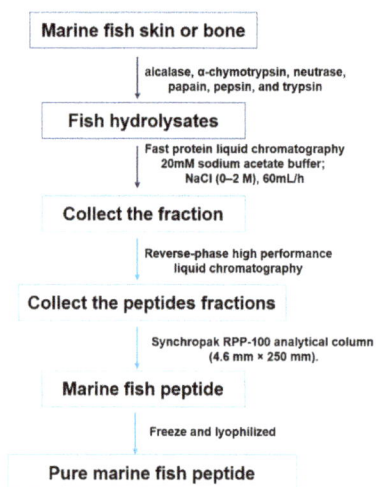

Marine fish skin or bone

alcalase, α-chymotrypsin, neutrase, papain, pepsin, and trypsin

Fish hydrolysates

Fast protein liquid chromatography
20mM sodium acetate buffer;
NaCl (0–2 M), 60mL/h

Collect the fraction

Reverse-phase high performance
liquid chromatography

Collect the peptides fractions

Synchropak RPP-100 analytical column
(4.6 mm × 250 mm).

Marine fish peptide

Freeze and lyophilized

Pure marine fish peptide

Figure 3. The flowchart showing the common procedures for the isolation and identification of the marine fish-derived peptides through enzymatic hydrolysis methods [70].

Table 2. Conditions for the enzymatic hydrolysis of tuna backbone proteins.

Enzymes for Hydrolysis	Buffer	pH	Temperature (°C)
alcalase	0.1 M Na_2HPO_4–NaH_2PO_4	7	50
α-chymotrypsin	0.1 M Na_2HPO_4–NaH_2PO_4	8	37
papain	0.1 M Na_2HPO_4–NaH_2PO_4	6	37
pepsin	0.1 M Glycine–HCl	2	37
neutrase	0.1 M Na_2HPO_4–NaH_2PO_4	8	50
trypsin	0.1 M Na_2HPO_4–NaH_2PO_4	8	37

4.2. Biological Activities of Marine Fish Peptides as Cosmeceuticals

An increasing interest in health, well-being, and physical appearance has resulted in high demand for various cosmetics. Recently, a combination of cosmetics with pharmaceuticals and marine-derived biologically active ingredients has become the hallmark of cosmetic industries [15,73–76]. Antioxidant, anti-inflammatory, reduction of melanin synthesis, tyrosinase inhibition, and matrix metalloproteinase (MMP) inhibitor tests are important in the development of cosmeceuticals against aging and wrinkling of the skin (Table 3).

Table 3. Biological activities for cosmeceutical applications. MMP: matrix metalloproteinase.

Activity	Cosmeceutical Applications	Reference
Antioxidant	Anti-aging, photo-protective effects	[15]
Tyrosinase inhibitor	Whitening	[75]
MMP inhibitor	Anti-wrinkle	[76]
Anti-inflammatory	Skin soothing	[77]

4.3. Antioxidant Fish Peptides

Antioxidants play an important role in providing protection against oxidative stress. The generation of oxidative stress is attributed to the formation of several reactive oxygen species, including alkyl radicals, hydroxyl radicals, superoxide radicals, peroxide radicals, and singlet oxygen species. In the human body, an imbalance between the free radicals and antioxidants leads to skin damage, inflammation, cancer, and neuron-related diseases [78]. The highly reactive free radicals can easily damage cellular membranes, DNA, proteins, and lipids, and are widely accepted as the primary reason for skin aging [79]. The human body possesses various antioxidant enzymes (e.g., catalase, superoxide dismutase, and glutathione peroxidase) and biomolecules (e.g., vitamin C, vitamin glutathione, and ubiquinone) to control the free radicals inside [79]. In addition, several synthetic products are often used to inhibit free radical activity (e.g., butylated hydroxyanisole (BHA), butylated hydroxytoluene (BHT), tert-butylhydroquinone (TBHQ), and propyl gallate [80]). However, the major drawback of using these antioxidants is the safety concern. Therefore, considerable attention has been diverted to the use of naturally-derived antioxidants [81–85]. Recently, a number of studies have demonstrated that various peptides derived from marine fish serve as effective antioxidants (Table 4) [71,86–98]. Enzymes for the isolation of antioxidant peptides from the marine fish are also described in Table 4.

Various types of methods have been used to evaluate the antioxidant activity of fish-derived peptides, including the 2,2-diphenyl-1-picrylhydrazyl (DPPH) radical scavenging assay, the 2,2′-azino-bis(3-ethylbenzothiazoline-6-sulphonic acid) (ABTS) radical scavenging assay, hydroxyl radical scavenging activity, Cu^{2+} chelating activity, and Fe^{2+} chelating activity [99–105].

Table 4. Potential bioactive antioxidant peptides from marine fish resources.

Fish Species Name	Enzymes for Hydrolysis	Peptides (Amino Acid Sequence)	Reference
Scomber austriasicus	protease N	–	[60]
Thunnus obesus	alcalase, α-chymotrypsin, neutrase, papain, pepsin, and trypsin	H-Leu-Asn-Leu-Pro-Thr-Ala-Val-Tyr-Met-Val-Thr-OH	[71]
Salmon	alcalase, flavourzyme, neutrase, pepsin, protamex, and trypsin	Peptides (unknown sequence, 1000–2000 Da)	[86]
Decapterus maruadsi	alcalase, neutral protease, papain, pepsin, and trypsin	His-Asp-His-Pro-Val-Cys and His-Glu-Lys-Val-Cys	[87]
Johnius belengerii	pepsin, trypsin, papain, α-chymotrypsin, alcalase, and neutrase	Glu-Ser-Thr-Val-Pro-Glu-Arg-Thr-His-Pro-Ala-Cys-Pro-Asp-Phe-Asn	[88]
Paralichthys olivaceus	papain, pepsin, trypsin, neutrase, alcalase, kojizyme, protamex, and α-chymotrypsin	Val-Cys-Ser-Val and Cys-Ala-Ala-Pro	[89]
Magalaspis cordyla	pepsin, trypsin, and α-chymotrypsin	Ala-Cys–Phe–Leu (518.5 Da),	[90]
Magalaspis cordyla	pepsin/trypsin, and α-chymotrypsin	Asn-His-Arg-Tyr-Asp-Arg (856 Da)	[91]
Otolithes ruber	pepsin/trypsin and α-chymotrypsin	Gly-Asn-Arg-Gly-Phe-Ala-Cys-Arg-His-Ala (1101.5 Da)	[91]
Johnius belengerii	trypsin, R-chymotrypsin, and pepsin	His-Gly-Pro-Leu-Gly-Pro-Leu	[92]
Otolithes ruber	pepsin, trypsin, and α-chymotrypsin	Lys-Thr-Phe-Cys-Gly-Arg-His	[93]
Oreochromis niloticus	alcalase, pronase E, pepsin, and trypsin	Asp-Pro-Ala-Leu-Ala-Thr-Glu-Pro-Asp-Pro-Met-Pro-Phe	[94]
Merluccius productus	Validase® BNP (V) and Flavourzyme®	–	[95]
Oreochromis niloticus	properase E and multifect neutral	Glu-Gly-Leu (317.33 Da) and Tyr-Gly-Asp-Glu-Tyr	[96]
Hypoptychus dybowskii	alcalase, neutrase, α-chymotrypsin, papain, pepsin, and trypsin	Ile–Val–Gly–Gly–Phe–Pro–His–Tyr–Leu	[97]

4.4. Antimicrobial Fish Peptides

Antimicrobial peptides possess cationic moieties, which facilitate their interaction with membranes of microbial pathogens [106]. Antimicrobial peptides from marine organisms constitute a new generation of antibiotics. They are currently extensively studied in the development of cosmeceutical products, including lotions, shampoos, and moisture creams. Numerous studies have reported that marine fish-derived peptides can be used as antimicrobial agents, as shown in Table 5 [103–106]. The enzymes used for the isolation of antimicrobial fish peptides and the microorganisms susceptible to these antimicrobial peptides were listed in Table 5.

Table 5. Marine fish species and enzymes used in the isolation of antimicrobial peptides. Targeted microorganisms used to check the marine fish-derived antimicrobial peptides are shown.

Name of Fish Species	Enzymes for Hydrolysis	Microorganisms	Reference
Setipinna taty	pepsin	*Escherichia coli*	[107]
Setipinna taty	papain, pepsin, trypsin, alkaline protease, acidic protease, and flavoring protease	*Escherichia coli, Pseudomonas fluorescens Proteus vulgaris, Bacillus megaterium Staphylococcus aureus, Bacillus subtilis, Bacillus megaterium, Sarcina lutea*	[108]
Scomber scombrus	–	*Listeria innocua, Escherichia coli*	[109]
Scomber scombrus	protamex, neutrase, papain, and flavourzyme.	*Listeria innocua HPB13 and Escherichia coli*	[110]

4.5. Matrix Metalloproteinases Inhibiting Fish Peptides

MMPs are endopeptidases containing zinc metal ion with an ability to degrade extracellular components. MMPs are produced by a variety of cells, including fibroblasts, keratinocytes, mast cells, macrophages, and neutrophils. Six different kinds of MMPs are available, which consist of collagenases, gelatinases, stromelysins, matrilysins, membrane-type MMPs, and other MMPs. The MMPs are categorized into three major functional groups. They include interstitial collagenases with affinities toward collagen types I, II, and III, (MMP-1, -8, and -13, respectively), stromelysins with specificity for laminin, fibronectin, and proteoglycans (MMP-3, -10, and -11, respectively), and gelatinases that effectively cleave type IV and V collagens (MMP-2 and -9) [67].

Wrinkles are a typical symptom of skin aging, and are associated with the reduction in the amount of collagen that dominates the elasticity of the skin dermal tissues. Since collagen fibers and other extracellular matrix are readily degraded by MMPs, formation of wrinkles is closely associated with increased expression of MMPs throughout the skin aging. Therefore, a variety of MMP inhibitors have been utilized to prevent the formation of wrinkles. However, studies on the use of fish-derived hydrolysates, proteins, and peptides as MMP inhibitors and their applications for cosmeceuticals are limited. Only a few studies focus on the MMP inhibitory activity of marine fish-derived peptides. Ryu et al. reported the isolation of novel peptides from seahorses that effectively increased collagen release through the suppression of collagenases 1 and 3 [111]. The same group isolated a protein from seahorse with an ability to inhibit MMP-1, MMP-3, and MMP-13 [112]. Shen et al. reported the hydrolysis of fish muscle from *Collichthys niveatus* using four commercial enzymes, namely alcalase, neutrase, protamex, and flavourzyme to isolate the peptides [113]. The major amino acids observed in the hydrolysate were threonine, glutamic acid, phenylalanine, tryptophan, and lysine. The total content of essential amino acids was calculated to be 970.7 ng/mL. The study was performed to check the effects of enzymatic hydrolysis conditions on the composition and properties of the peptides obtained from the hydrolysate, which could be utilized as health supplements [113]. A proteinase inhibitor (21 kDa) with similar properties to human tissue inhibitor of MMP-2 (TIMP-2) was obtained from Atlantic cod muscle and then identified by using gelatin affinity chromatography, real-time reverse zymography, and mass spectroscopy [114]. The amino acid sequences of the two peptides obtained from the inhibitor showed a high similarity to those of the human TIMP-2. The inhibitor was found to inhibit the gelatin-degrading enzymes.

5. Photo-Protective and Anti-Photoaging Activity of Fish Peptides and Fish Protein Hydrolysates

Skin is made up of three different layers, namely epidermis, dermis, and hypodermis. It acts as a chemical and physical barrier to protect the body against harmful foreign pollutants [115]. Skin can be damaged by various external environmental attacks, including harmful chemicals, ultraviolet (UV) light exposure, and temperature changes [116]. Photoaging and inflammation are often caused by UV radiation. Photoaging, also known as dermatoheliosis, is characterized by changes in the skin due to exposure of UV-A (400 to 320-nm wavelength) and UV-B (320 to 290-nm wavelength) light, which is main light source for photoaging [117]. The UV-A can permeate more deeply into the dermal matrix than UV-B, whereas UV-B is more carcinogenic compared to UV-A [118]. Considerable attention has been given to the utilization of marine fish-derived peptides for skin protection due to an excellent bioactivity, biocompatibility, penetration ability, and skin-repairing ability. Various fish-derived proteins and peptides have been investigated for their usage in the protection of skin from UV exposure [119–121].

Fish skin collagen and hydrolysates demonstrated a high biocompatibility with an ability to provide protection against the detrimental effects of UV radiation (Table 6). Zhuang et al. reported that jellyfish (*Rhopilema esculentum*) collagen (JC) and jellyfish collagen hydrolysate (JCH) alleviated UV-induced abnormal changes of antioxidant defense systems such as superoxide dismutase and glutathione peroxidase [116]. Both JC and JCH significantly protected the skin lipid and collagen from UV radiation. In addition, the UV-induced changes in the total ceramide and glycosaminoglycans in the skin were recovered, thus maintaining the balance of lipid compositions in the skin. The mechanism is mainly based on the antioxidative properties of the both JC and JCH along with stimulation of skin collagen synthesis. The study indicated that JCH that has lower molecular weights as compared to JC provides a much stronger protection against UV-induced photoaging [116]. The importance of jellyfish collagen on the antioxidant activities is further strengthened by another study that reports jellyfish as an abundant source of collagen with a high potential for nutraceutical applications [122]. The effects of JC and JCH on UV-induced skin damage of mice were evaluated by the analysis of skin moisture as well as microscopic analyses of skin and immunity indexes [123]. It was observed that the moisture retention ability of UV-induced mice skin increased upon treatment with JC and JCH. Further histological analysis demonstrated that JC and JCH could repair the endogenous collagen and elastin protein fibers, thus maintaining the natural ratio of type I to type III collagen. The immunity indexes showed that JC and JCH played a pivotal role in enhancing the immunity of photoaging mice in vivo. Again, as mentioned above, JCH exhibited a much higher protective ability than JC [123].

Hou et al. evaluated the effects of collagen polypeptides isolated from cod skin on UV-induced damage to mouse skin [124]. Collagen polypeptide fractions (CP1 (2 kDa < Mr < 6 kDa) and CP2 (Mr < 2 kDa)) were obtained through pepsin digestion and alkaline protease hydrolysis methods. Collagen polypeptides provided good moisture absorption and retention properties, and CP2 was more efficient than CP1. In vivo studies demonstrated that both of the peptides provided protective effects against UV-induced wrinkle formation and destruction of skin structures (Figure 4). The action mechanisms of the collagen polypeptides mainly involve increasing immunity, decreasing the loss of moisture and lipid, and repairing endogenous collagen and elastin protein fibers [124].

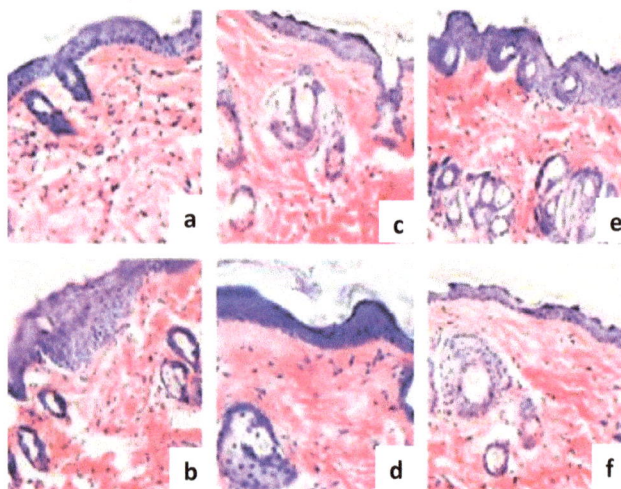

Figure 4. Effects of collagen polypeptide 1 and collagen polypeptide 2 on the morphology of photoaging skin (magnification 200×). (**a**) normal; (**b**) model; (**c**) collagen polypeptide 1 (50 mg/kg); (**d**) collagen polypeptide 1 (200 mg/kg); (**e**) collagen polypeptide 2 (50 mg/kg); and (**f**) collagen polypeptide 2 (200 mg/kg). Adapted with permission from [124].

Chen et al. studied the effects of gelatin hydrolysate extracted from the Pacific cod (*Gadus macrocephalus*) skin on UV radiation-induced inflammation and collagen reduction in photoaging mouse skin. Oral administration of gelatin hydrolysate suppressed UV radiation-induced damage to the skin by inhibiting the depletion of endogenous antioxidant enzyme activity, and by suppressing the expression of nuclear factor-κB (NF-κB) as well as NF-κB-mediated expression of pro-inflammatory cytokines. Furthermore, gelatin hydrolysate inhibited type I procollagen synthesis by up-regulating the type II transforming growth factor β (TGFβ) receptor (TβRII) level and down-regulating Smad7 levels, which demonstrates that gelatin hydrolysate is involved in matrix collagen synthesis by activating the TGF-β/Smad pathway in the photoaging skin [125].

Age-related skin thinning is involved in a decrease in the content of collagen in the skin. Co-treatment with collagen peptide and vitamin C upregulates the type I collagen in vivo. Shibuya et al.

demonstrated that the collagen peptides supplemented with vitamin C reduced the superoxide dismutase 1 (Sod-1) [126]. In vitro studies further revealed that collagen oligopeptide, a digestive product of ingested collagen peptide, significantly enhanced the bioactivity of the vitamin C derivative with respect to the migration and proliferation of fibroblasts [126]. The collagen peptide and the vitamin C derivative additively increased the skin thickness of hairless Sod1-deficient mice.

Recently, gelatin and its hydrolysates from salmon skin were used to protect the skin from photoaging [127]. The average molecular weights of the gelatin and gelatin hydrolysates were found to be 65 kDa and 873 kDa, respectively [127]. In another study, dose effects of orally administered collagen hydrolysates on the UV-B-irradiated skin damage were investigated using UV-B-irradiated hairless mice [128]. The low dose of collagen hydrolysates increased the skin hydration and reduced the transepidermal water loss in the damaged skin [128]. In addition to this, tilapia gelatin peptides were investigated against UV-induced damage to mouse skin [129]. The results suggested that tilapia gelatin had an ability to avoid the UV damage by protecting the collagen and lipid in the skin. The antioxidant peptide, Leu-Ser-Gly-Tyr-Gly-Pro (592.26 Da), was identified from the tilapia gelatin peptides, and the peptide has an ability to scavenge the hydroxyl radicals with the IC_{50} value of 22.47 µg/mL [129,130].

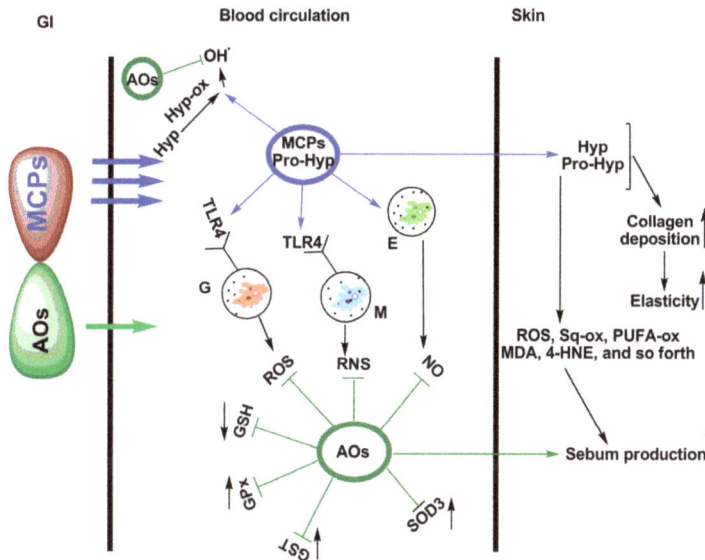

Figure 5. Scheme of the hypothesized redox-dependent mechanisms of physiological effects after co-treatment of marine collagen peptides (MCPs) and skin-targeting antioxidants (AOs). Redrawn with permission from [131]. In the figure, the three arrows (blue) indicate that MCPs easily penetrate the gastrointestinal wall (GI) through blood circulation and are mainly deposited in the skin. The single arrow (green) indicates that AOs are partially metabolized. However, AOs can reach the different layers of skin. While circulating in the blood, MCPs activate blood phagocytes (i.e., granulocytes (G) and monocytes (M)) and endotheliocytes (E) to generate reactive oxygen species (ROS) and reactive nitrogen species (RNS) by provoking Toll-like receptor-4 (TLR4)-mediated signals. Co-administered antioxidants can prevent systemic oxidative stress by blocking glutathione (GSH) oxidation, and activation of glutathione peroxidase (GPx), glutathione-S-transferase (GST), and superoxide dismutase 3 (SOD3).

The major concern regarding the safety and clinical feasibility of administration of marine collagen peptides (MCPs) has been raised because MCPs from different origin can activate innate immune response through Toll-like receptor 4 (TLR4)-mediated NADPH-oxidase (NOX4) activation and over-production of reactive oxygen species (ROS) [131]. Figure 5 represents the hypothesized redox-dependent mechanisms behind the physiological effects of fish skin MCPs combined with plant-derived skin-targeting antioxidants (coenzyme Q10 + grape-skin extract + luteolin + selenium) [131]. The MCPs were derived from the skin of deep sea fish (e.g., *Pollachius virens*, *Hippoglossus hippoglossus, and Pleuronectes platessa*). MCPs easily penetrate the gastrointestinal (GI) wall (three arrows) through blood circulation and are mainly deposited in the skin [131]. The clinical study demonstrated that combination treatment of MCPs with skin-targeting antioxidants can remarkably improve skin elasticity and sebum production while lowering the oxidative damage [131]. These results clearly indicate that skin-targeting antioxidants are essential components of MCPs-containing cosmeceuticals for more effective and safe treatment.

Table 6. Photo-protective and anti-photoaging proteins and peptides from marine fish.

Name of Fish Species and Parts	Fish-Derived Proteins and Peptides	Enzymes for Hydrolysis	Reference
Jellyfish	Collagen	properase E	[123]
Cod skin	Collagen polypeptides	alkaline protease and pepsin	[124]
Cod skin	Gelatin hydrolysate	alkaline protease and trypsin	[125]
Salmon skin	Gelatin	alkaline protease and trypsin	[127]
Tilapia	Gelatin peptides	properase E	[129]
Pollachius virens, Hippoglossus hippoglossus, and Pleuronectes platessa	Marine collagen peptides	complex proteases	[131]

6. Conclusions

Marine fish-derived proteins and peptides are becoming the important resource for cosmetic industries. Several bioactive proteins and peptides were produced from marine fish via chemical or enzymatical hydrolysis and regarded as a safer option for the development of cosmeceutical products. The use of marine fish-derived proteins and peptides contribute to alleviating the environmental pollution caused by the waste generated by fish processing industries. Much attention has been paid to marine fish collagen for cosmeceutical applications owing its properties for skin hydration, with low odor and improved mechanical strength. In addition, marine fish-derived peptides have been extensively explored for cosmeceutical applications due to their various biological properties including antioxidant, antimicrobial, MMP inhibitory, photo-protective, and anti-photoaging activities. These biological activities of the marine fish peptides have led to the development of several types of anti-aging, skin care, and anti-wrinkle products. Despite the great potential of marine fish-derived proteins and peptides for cosmeceutical applications, most of them are still in the experimental stage and need to be further investigated with regard to their formulations and long-term safety for successful commercialization. Moreover, development of supplements that can further increase the bioavailability and tissue regeneration efficacy of marine fish-derived proteins and peptides is also required to increase their potential for cosmeceuticals.

Acknowledgments: This work was supported by the Post-Doctor Research Program (2016) through Incheon National University (INU), Incheon, Republic of Korea. This work was also supported by Basic Science Research Program through the National Research Foundation of Korea (NRF) funded by the Ministry of Education (NRF-2016R1D1A1B03933136 to M.S.S.).

Author Contributions: Jayachandran Venkatesan and Sukumaran Anil developed the concept for the review and wrote the manuscript. Se-Kwon Kim and Min Suk Shim wrote and edited the manuscript.

Conflicts of Interest: The authors declare no conflict of interest.

References

1. Malve, H. Exploring the ocean for new drug developments: Marine pharmacology. *J. Pharm. Bioallied Sci.* **2016**, *8*, 83–91. [CrossRef] [PubMed]
2. Kim, S.-K.; Venkatesan, J. Introduction to marine biotechnology. In *Springer Handbook of Marine Biotechnology*; Kim, S.-K., Ed.; Springer: Berlin/Heidelberg, Germany, 2015; pp. 1–10.

3. Najafian, L.; Babji, A.S. A review of fish-derived antioxidant and antimicrobial peptides: Their production, assessment, and applications. *Peptides* **2012**, *33*, 178–185. [CrossRef] [PubMed]

4. Cheung, R.C.F.; Ng, T.B.; Wong, J.H. Marine peptides: Bioactivities and applications. *Mar. Drugs* **2015**, *13*, 4006–4043. [CrossRef] [PubMed]

5. Senevirathne, M.; Kim, S.-K. Utilization of seafood processing by-products: Medicinal applications. In *Advances in Food and Nutrition Research*; Kim, S.-K., Ed.; Academic Press: Waltham, MA, USA, 2012; Volume 65, pp. 495–512.

6. Rustad, T. Physical and chemical properties of protein seafood by-products. In *Maximising the Value of Marine By-Products*; Shahidi, F., Ed.; Woodhead Publishing Limited: Cambridge, UK, 2007; pp. 3–21.

7. Nilsang, S.; Lertsiri, S.; Suphantharika, M.; Assavanig, A. Optimization of enzymatic hydrolysis of fish soluble concentrate by commercial proteases. *J. Food Eng.* **2005**, *70*, 571–578. [CrossRef]

8. Allard, R.; Malak, N.A.; Huc, A. Collagen Product Containing Collagen of Marine Origin with a Low Odor and Preferably with Improved Mechanical Properties, and Its Use in the Form of Cosmetic or Pharmaceutical Compositions or Products. U.S. Patent 6,660,280, 9 December 2003.

9. Shahidi, F.; Kamil, Y.J. Enzymes from fish and aquatic invertebrates and their application in the food industry. *Trends Food Sci. Technol.* **2001**, *12*, 435–464. [CrossRef]

10. Hoyer, B.; Bernhardt, A.; Heinemann, S.; Stachel, I.; Meyer, M.; Gelinsky, M. Biomimetically mineralized salmon collagen scaffolds for application in bone tissue engineering. *Biomacromolecules* **2012**, *13*, 1059–1066. [CrossRef] [PubMed]

11. Hayashi, Y.; Yamada, S.; Guchi, K.Y.; Koyama, Z.; Ikeda, T. Chitosan and fish collagen as biomaterials for regenerative medicine. In *Advances in Food and Nutrition Research*; Kim, S.K., Ed.; Academic Press: Waltham, MA, USA, 2012; Volume 65, pp. 107–120.

12. Lauritano, C.; Ianora, A. Marine organisms with anti-diabetes properties. *Mar. Drugs* **2016**, *14*, 220. [CrossRef] [PubMed]

13. Venugopal, V. Cosmeceuticals from marine fish and shellfish. In *Marine Cosmeceuticals: Trends and Prospects*; Kim, S.-K., Ed.; CRC Press: Boca Raton, FL, USA, 2011; pp. 211–232.

14. Senevirathne, M.; Kim, S.-K. Development of bioactive peptides from fish proteins and their health promoting ability. In *Advances in Food and Nutrition Research*; Kim, S.-K., Ed.; Academic Press: Waltham, MA, USA, 2012; Volume 65, pp. 235–248.

15. Ngo, D.H.; Vo, T.S.; Ngo, D.N.; Wijesekara, I.; Kim, S.K. Biological activities and potential health benefits of bioactive peptides derived from marine organisms. *Int. J. Biol. Macromol.* **2012**, *51*, 378–383. [CrossRef] [PubMed]

16. Pangestuti, R.; Kim, S.-K. Bioactive peptide of marine origin for the prevention and treatment of non-communicable diseases. *Mar. Drugs* **2017**, *15*, 67. [CrossRef] [PubMed]

17. Senaratne, L.; Park, P.-J.; Kim, S.-K. Isolation and characterization of collagen from brown backed toadfish (*Lagocephalus gloveri*) skin. *Bioresour. Technol.* **2006**, *97*, 191–197. [CrossRef] [PubMed]

18. Pati, F.; Adhikari, B.; Dhara, S. Isolation and characterization of fish scale collagen of higher thermal stability. *Bioresour. Technol.* **2010**, *101*, 3737–3742. [CrossRef] [PubMed]

19. Jeong, H.-S.; Venkatesan, J.; Kim, S.-K. Isolation and characterization of collagen from marine fish (*Thunnus obesus*). *Biotechnol. Bioprocess Eng.* **2013**, *18*, 1185–1191. [CrossRef]

20. Xu, Y.; Han, X.; Li, Y. Effect of marine collagen peptides on long bone development in growing rats. *J. Sci. Food Agric.* **2010**, *90*, 1485–1491. [CrossRef] [PubMed]

21. Swatschek, D.; Schatton, W.; Kellermann, J.; Müller, W.E.; Kreuter, J. Marine sponge collagen: Isolation, characterization and effects on the skin parameters surface-pH, moisture and sebum. *Eur. J. Pharm. Biopharm.* **2002**, *53*, 107–113. [CrossRef]

22. Cho, J.K.; Jin, Y.G.; Rha, S.J.; Kim, S.J.; Hwang, J.H. Biochemical characteristics of four marine fish skins in Korea. *Food Chem.* **2014**, *159*, 200–207. [CrossRef] [PubMed]

23. Haug, I.J.; Draget, K.I.; Smidsrød, O. Physical and rheological properties of fish gelatin compared to mammalian gelatin. *Food Hydrocoll.* **2004**, *18*, 203–213. [CrossRef]

24. Subhan, F.; Ikram, M.; Shehzad, A.; Ghafoor, A. Marine collagen: An emerging player in biomedical applications. *J. Food Sci. Technol.* **2015**, *52*, 4703–4707. [CrossRef] [PubMed]

25. Muralidharan, N.; Jeya Shakila, R.; Sukumar, D.; Jeyasekaran, G. A Skin, bone and muscle collagen extraction from the trash fish, leather jacket (*Odonus niger*) and their characterization. *J. Food Sci. Technol.* **2013**, *50*, 1106–1113. [CrossRef] [PubMed]

26. Silva, T.H.; Moreira-Silva, J.; Marques, A.L.; Domingues, A.; Bayon, Y.; Reis, R.L. Marine origin collagens and its potential applications. *Mar. Drugs* **2014**, *12*, 5881–5901. [CrossRef] [PubMed]

27. Nagai, T.; Araki, Y.; Suzuki, N. Collagen of the skin of ocellate puffer fish (*Takifugu rubripes*). *Food Chem.* **2002**, *78*, 173–177. [CrossRef]

28. Shanmugam, V.; Ramasamy, P.; Subhapradha, N.; Sudharsan, S.; Seedevi, P.; Moovendhan, M.; Krishnamoorthy, J.; Shanmugam, A.; Srinivasan, A. Extraction, structural and physical characterization of type I collagen from the outer skin of *Sepiella inermis* (Orbigny, 1848). *Afr. J. Biotechnol.* **2012**, *11*, 14326–14337. [CrossRef]

29. Jongjareonrak, A.; Benjakul, S.; Visessanguan, W.; Nagai, T.; Tanaka, M. Isolation and characterisation of acid and pepsin-solubilised collagens from the skin of Brownstripe red snapper (*Lutjanus vitta*). *Food Chem.* **2005**, *93*, 475–484. [CrossRef]

30. Kumar, N.S.S.; Nazeer, R.A. Wound healing properties of collagen from the bone of two marine fishes. *Int. J. Pept. Res. Ther.* **2012**, *18*, 185–192. [CrossRef]

31. Veeruraj, A.; Arumugam, M.; Balasubramanian, T. Isolation and characterization of thermostable collagen from the marine eel-fish (*Evenchelys macrura*). *Process Biochem.* **2013**, *48*, 1592–1602. [CrossRef]

32. Veeruraj, A.; Arumugam, M.; Ajithkumar, T.; Balasubramanian, T. Isolation and characterization of drug delivering potential of type-I collagen from eel fish *Evenchelys macrura*. *J. Mater. Sci. Mater. Med.* **2012**, *23*, 1729–1738. [CrossRef] [PubMed]

33. Minh Thuy, L.T.; Okazaki, E.; Osako, K. Isolation and characterization of acid-soluble collagen from the scales of marine fishes from Japan and Vietnam. *Food Chem.* **2014**, *149*, 264–270. [CrossRef] [PubMed]

34. Zhang, J.; Duan, R.; Huang, L.; Song, Y.; Regenstein, J.M. Characterisation of acid-soluble and pepsin-solubilised collagen from jellyfish (*Cyanea nozakii* Kishinouye). *Food Chem.* **2014**, *150*, 22–26. [CrossRef] [PubMed]

35. Muthumari, K.; Anand, M.; Maruthupandy, M. Collagen extract from marine finfish scales as a potential mosquito larvicide. *Protein J.* **2016**, *35*, 391–400. [CrossRef] [PubMed]

36. Kittiphattanabawon, P.; Benjakul, S.; Visessanguan, W.; Nagai, T.; Tanaka, M. Characterisation of acid-soluble collagen from skin and bone of bigeye snapper (*Priacanthus tayenus*). *Food Chem.* **2005**, *89*, 363–372. [CrossRef]

37. Nalinanon, S.; Benjakul, S.; Visessanguan, W.; Kishimura, H. Use of pepsin for collagen extraction from the skin of bigeye snapper (*Priacanthus tayenus*). *Food Chem.* **2007**, *104*, 593–601. [CrossRef]

38. Benjakul, S.; Thiansilakul, Y.; Visessanguan, W.; Roytrakul, S.; Kishimura, H.; Prodpran, T.; Meesane, J. Extraction and characterisation of pepsin-solubilised collagens from the skin of bigeye snapper (*Priacanthus tayenus* and *Priacanthus macracanthus*). *J. Sci. Food Agric.* **2010**, *90*, 132–138. [CrossRef] [PubMed]

39. Matmaroh, K.; Benjakul, S.; Prodpran, T.; Encarnacion, A.B.; Kishimura, H. Characteristics of acid soluble collagen and pepsin soluble collagen from scale of spotted golden goatfish (*Parupeneus heptacanthus*). *Food Chem.* **2011**, *129*, 1179–1186. [CrossRef] [PubMed]

40. Zhang, M.; Liu, W.; Li, G. Isolation and characterisation of collagens from the skin of largefin longbarbel catfish (*Mystus macropterus*). *Food Chem.* **2009**, *115*, 826–831. [CrossRef]

41. Khan, S.B.; Qian, Z.-J.; Ryu, B.; Kim, S.-K. Isolation and biochemical characterization of collagens from seaweed pipefish, *Syngnathus schlegeli*. *Biotechnol. Bioprocess Eng.* **2009**, *14*, 436–442. [CrossRef]

42. Nagai, T.; Ogawa, T.; Nakamura, T.; Ito, T.; Nakagawa, H.; Fujiki, K.; Nakao, M.; Yano, T. Collagen of edible jellyfish exumbrella. *J. Sci. Food Agric.* **1999**, *79*, 855–858. [CrossRef]

43. Barzideh, Z.; Latiff, A.A.; Gan, C.Y.; Benjakul, S.; Karim, A.A. Isolation and characterisation of collagen from the ribbon jellyfish (*Chrysaora* sp.). *Int. J. Food Sci. Technol.* **2014**, *49*, 1490–1499. [CrossRef]

44. Available online: https://www.justvitamins.co.uk/blog/bovine-collagen-vs-marine-collagen/ (accessed on 11 May 2017).

45. Xhauflaire-Uhoda, E.; Fontaine, K.; Pierard, G. Kinetics of moisturizing and firming effects of cosmetic formulations. *Int. J. Cosmet. Sci.* **2008**, *30*, 131–138. [CrossRef] [PubMed]

46. Chandika, P.; Ko, S.-C.; Oh, G.-W.; Heo, S.-Y.; Nguyen, V.-T.; Jeon, Y.-J.; Lee, B.; Jang, C.H.; Kim, G.; Park, W.S. Fish collagen/alginate/chitooligosaccharides integrated scaffold for skin tissue regeneration application. *Int. J. Biol. Macromol.* **2015**, *81*, 504–513. [CrossRef] [PubMed]

47. Gautam, M.; Purohit, V.; Agarwal, M.; Singh, A.; Goel, R. In vivo healing potential of *Aegle marmelos* in excision, incision, and dead space wound models. *Sci. World J.* **2014**, *2014*, 740197. [CrossRef] [PubMed]

48. Nithya, M.; Suguna, L.; Rose, C. The effect of nerve growth factor on the early responses during the process of wound healing. *Biochim. Biophys. Acta Gen. Subj.* **2003**, *1620*, 25–31. [CrossRef]

49. Song, E.; Yeon Kim, S.; Chun, T.; Byun, H.J.; Lee, Y.M. Collagen scaffolds derived from a marine source and their biocompatibility. *Biomaterials* **2006**, *27*, 2951–2961. [CrossRef] [PubMed]

50. Yeo, M.; Jung, W.-K.; Kim, G. Fabrication, characterisation and biological activity of phlorotannin-conjugated PCL/β-TCP composite scaffolds for bone tissue regeneration. *J. Mater. Chem.* **2012**, *22*, 3568–3577. [CrossRef]

51. Hoyer, B.; Bernhardt, A.; Lode, A.; Heinemann, S.; Sewing, J.; Klinger, M.; Notbohm, H.; Gelinsky, M. Jellyfish collagen scaffolds for cartilage tissue engineering. *Acta Biomater.* **2014**, *10*, 883–892. [CrossRef] [PubMed]

52. Shen, X.; Nagai, N.; Murata, M.; Nishimura, D.; Sugi, M.; Munekata, M. Development of salmon milt DNA/salmon collagen composite for wound dressing. *J. Mater. Sci. Mater. Med.* **2008**, *19*, 3473–3479. [CrossRef] [PubMed]

53. Yamada, S.; Yamamoto, K.; Ikeda, T.; Yanagiguchi, K.; Hayashi, Y. Potency of fish collagen as a scaffold for regenerative medicine. *Biomed Res. Int.* **2014**, *2014*, 302932. [CrossRef] [PubMed]

54. Elango, J.; Zhang, J.; Bao, B.; Palaniyandi, K.; Wang, S.; Wenhui, W.; Robinson, J.S. Rheological, biocompatibility and osteogenesis assessment of fish collagen scaffold for bone tissue engineering. *Int. J. Biol. Macromol.* **2016**, *91*, 51–59. [CrossRef] [PubMed]

55. Jeong, S.I.; Kim, S.Y.; Cho, S.K.; Chong, M.S.; Kim, K.S.; Kim, H.; Lee, S.B.; Lee, Y.M. Tissue-engineered vascular grafts composed of marine collagen and PLGA fibers using pulsatile perfusion bioreactors. *Biomaterials* **2007**, *28*, 1115–1122. [CrossRef] [PubMed]

56. Chen, D.C.; Lai, Y.L.; Lee, S.Y.; Hung, S.L.; Chen, H.L. Osteoblastic response to collagen scaffolds varied in freezing temperature and glutaraldehyde crosslinking. *J. Biomed. Mater. Res.* **2007**, *80*, 399–409. [CrossRef] [PubMed]

57. Mullen, C.; Haugh, M.; Schaffler, M.; Majeska, R.; McNamara, L. Osteocyte differentiation is regulated by extracellular matrix stiffness and intercellular separation. *J. Mech. Behav. Biomed.* **2013**, *28*, 183–194. [CrossRef] [PubMed]

58. Byrne, E.M.; Farrell, E.; McMahon, L.A.; Haugh, M.G.; O'Brien, F.J.; Campbell, V.A.; Prendergast, P.J.; O'Connell, B.C. Gene expression by marrow stromal cells in a porous collagen–glycosaminoglycan scaffold is affected by pore size and mechanical stimulation. *J. Mater. Sci. Mater. Med.* **2008**, *19*, 3455–3463. [CrossRef] [PubMed]

59. Keogh, M.B.; O'Brien, F.J.; Daly, J.S. A novel collagen scaffold supports human osteogenesis—Applications for bone tissue engineering. *Cell Tissue Res.* **2010**, *340*, 169–177. [CrossRef] [PubMed]

60. Wu, H.-C.; Chen, H.-M.; Shiau, C.-Y. Free amino acids and peptides as related to antioxidant properties in protein hydrolysates of mackerel (*Scomber austriasicus*). *Food Res. Int.* **2003**, *36*, 949–957. [CrossRef]

61. Kim, S.-K. *Marine Proteins and Peptides: Biological Activities and Applications*; John Wiley & Sons: Hoboken, NJ, USA, 2013.

62. Lintner, K.; Peschard, O. Biologically active peptides: From a laboratory bench curiosity to a functional skin care product. *Int. J. Cosmet. Sci.* **2000**, *22*, 207–218. [CrossRef] [PubMed]

63. Khora, S.S. Marine fish-derived bioactive peptides and proteins for human therapeutics. *Int. J. Pharm. Pharm. Sci.* **2013**, *5*, 31–37.

64. Halim, N.; Yusof, H.; Sarbon, N. Functional and bioactive properties of fish protein hydolysates and peptides: A comprehensive review. *Trends Food Sci. Technol.* **2016**, *51*, 24–33. [CrossRef]

65. Sila, A.; Hedhili, K.; Przybylski, R.; Ellouz-Chaabouni, S.; Dhulster, P.; Bougatef, A.; Nedjar-Arroume, N. Antibacterial activity of new peptides from barbel protein hydrolysates and mode of action via a membrane damage mechanism against *Listeria monocytogenes*. *J. Funct. Foods* **2014**, *11*, 322–329. [CrossRef]

66. Kim, S.K. Marine cosmeceuticals. *J. Cosmet. Dermatol.* **2014**, *13*, 56–67. [CrossRef] [PubMed]

67. Thomas, N.V.; Kim, S.-K. Beneficial effects of marine algal compounds in cosmeceuticals. *Mar. Drugs* **2013**, *11*, 146–164. [CrossRef] [PubMed]

68. Kim, S.-K.; Ravichandran, Y.D.; Khan, S.B.; Kim, Y.T. Prospective of the cosmeceuticals derived from marine organisms. *Biotechnol. Bioprocess Eng.* **2008**, *13*, 511–523. [CrossRef]

69. Kristinsson, H.G.; Rasco, B.A. Fish protein hydrolysates: Production, biochemical, and functional properties. *Crit. Rev. Food Sci. Nutr.* **2000**, *40*, 43–81. [CrossRef] [PubMed]

70. Je, J.-Y.; Qian, Z.-J.; Byun, H.-G.; Kim, S.-K. Purification and characterization of an antioxidant peptide obtained from tuna backbone protein by enzymatic hydrolysis. *Process Biochem.* **2007**, *42*, 840–846. [CrossRef]

71. Je, J.-Y.; Qian, Z.-J.; Lee, S.-H.; Byun, H.-G.; Kim, S.-K. Purification and antioxidant properties of bigeye tuna (*Thunnus obesus*) dark muscle peptide on free radical-mediated oxidative systems. *J. Med. Food* **2008**, *11*, 629–637. [CrossRef] [PubMed]

72. Jeon, Y.-J.; Byun, H.-G.; Kim, S.-K. Improvement of functional properties of cod frame protein hydrolysates using ultrafiltration membranes. *Process Biochem.* **1999**, *35*, 471–478. [CrossRef]

73. Martins, A.; Vieira, H.; Gaspar, H.; Santos, S. Marketed marine natural products in the pharmaceutical and cosmeceutical industries: Tips for success. *Mar. Drugs* **2014**, *12*, 1066–1101. [CrossRef] [PubMed]

74. Yoon, N.Y.; Eom, T.-K.; Kim, M.-M.; Kim, S.-K. Inhibitory effect of phlorotannins isolated from Ecklonia cava on mushroom tyrosinase activity and melanin formation in mouse B16F10 melanoma cells. *J. Agric. Food Chem.* **2009**, *57*, 4124–4129. [CrossRef] [PubMed]

75. Schurink, M.; van Berkel, W.J.; Wichers, H.J.; Boeriu, C.G. Novel peptides with tyrosinase inhibitory activity. *Peptides* **2007**, *28*, 485–495. [CrossRef] [PubMed]

76. Kim, M.-M.; Van Ta, Q.; Mendis, E.; Rajapakse, N.; Jung, W.-K.; Byun, H.-G.; Jeon, Y.-J.; Kim, S.-K. Phlorotannins in *Ecklonia cava* extract inhibit matrix metalloproteinase activity. *Life Sci.* **2006**, *79*, 1436–1443. [CrossRef] [PubMed]

77. Kim, J.-A.; Kim, S.-K. Bioactive peptides from marine sources as potential anti-inflammatory therapeutics. *Curr. Protein Pept. Sci.* **2013**, *14*, 177–182. [CrossRef] [PubMed]

78. Birben, E.; Sahiner, U.M.; Sackesen, C.; Erzurum, S.; Kalayci, O. Oxidative stress and antioxidant defense. *World Allergy Organ. J.* **2012**, *5*, 9–19. [CrossRef] [PubMed]

79. Rinnerthaler, M.; Bischof, J.; Streubel, M.K.; Trost, A.; Richter, K. Oxidative stress in aging human skin. *Biomolecules* **2015**, *5*, 545–589. [CrossRef] [PubMed]

80. Gülçin, İ.; Huyut, Z.; Elmastaş, M.; Aboul-Enein, H.Y. Radical scavenging and antioxidant activity of tannic acid. *Arab. J. Chem.* **2010**, *3*, 43–53. [CrossRef]

81. Winata, A.; Lorenz, K. Antioxidant potential of 5-n-pentadecylresorcinol. *J. Food Process. Preserv.* **1996**, *20*, 417–429. [CrossRef]

82. Becker, G. Preserving food and health: Antioxidants make functional, nutritious preservatives. *Food Process.* **1993**, *12*, 54–56.

83. Osawa, T.; Namiki, M. Natural antioxidants isolated from *Eucalyptus* leaf waxes. *J. Agric. Food Chem.* **1985**, *33*, 777–780. [CrossRef]

84. Byun, H.-G.; Lee, J.K.; Park, H.G.; Jeon, J.-K.; Kim, S.-K. Antioxidant peptides isolated from the marine rotifer, *Brachionus rotundiformis*. *Process Biochem.* **2009**, *44*, 842–846. [CrossRef]

85. Kim, S.-K.; Wijesekara, I. Development and biological activities of marine-derived bioactive peptides: A review. *J. Funct. Foods* **2010**, *2*, 1–9. [CrossRef]

86. Ahn, C.-B.; Je, J.-Y.; Cho, Y.-S. Antioxidant and anti-inflammatory peptide fraction from salmon byproduct protein hydrolysates by peptic hydrolysis. *Food Res. Int.* **2012**, *49*, 92–98. [CrossRef]

87. Jiang, H.; Tong, T.; Sun, J.; Xu, Y.; Zhao, Z.; Liao, D. Purification and characterization of antioxidative peptides from round scad (*Decapterus maruadsi*) muscle protein hydrolysate. *Food Chem.* **2014**, *154*, 158–163. [CrossRef] [PubMed]

88. Kim, S.-Y.; Je, J.-Y.; Kim, S.-K. Purification and characterization of antioxidant peptide from hoki (*Johnius belengerii*) frame protein by gastrointestinal digestion. *J. Nutr. Biochem.* **2007**, *18*, 31–38. [CrossRef] [PubMed]

89. Ko, J.-Y.; Lee, J.-H.; Samarakoon, K.; Kim, J.-S.; Jeon, Y.-J. Purification and determination of two novel antioxidant peptides from flounder fish (*Paralichthys olivaceus*) using digestive proteases. *Food Chem. Toxicol.* **2013**, *52*, 113–120. [CrossRef] [PubMed]

90. Kumar, N.S.; Nazeer, R.; Jaiganesh, R. Purification and biochemical characterization of antioxidant peptide from horse mackerel (*Magalaspis cordyla*) viscera protein. *Peptides* **2011**, *32*, 1496–1501. [CrossRef] [PubMed]

91. Kumar, N.S.; Nazeer, R.; Jaiganesh, R. Purification and identification of antioxidant peptides from the skin protein hydrolysate of two marine fishes, horse mackerel (*Magalaspis cordyla*) and croaker (*Otolithes ruber*). *Amino Acids* **2012**, *42*, 1641–1649. [CrossRef] [PubMed]

92. Mendis, E.; Rajapakse, N.; Kim, S.-K. Antioxidant properties of a radical-scavenging peptide purified from enzymatically prepared fish skin gelatin hydrolysate. *J. Agric. Food Chem.* **2005**, *53*, 581–587. [CrossRef] [PubMed]

93. Nazeer, R.; Kumar, N.S.; Ganesh, R.J. In vitro and in vivo studies on the antioxidant activity of fish peptide isolated from the croaker (*Otolithes ruber*) muscle protein hydrolysate. *Peptides* **2012**, *35*, 261–268. [CrossRef] [PubMed]

94. Ngo, D.-H.; Qian, Z.-J.; Ryu, B.; Park, J.W.; Kim, S.-K. In vitro antioxidant activity of a peptide isolated from Nile tilapia (*Oreochromis niloticus*) scale gelatin in free radical-mediated oxidative systems. *J. Funct. Foods* **2010**, *2*, 107–117. [CrossRef]

95. Samaranayaka, A.G.; Li-Chan, E.C. Autolysis-assisted production of fish protein hydrolysates with antioxidant properties from Pacific hake (*Merluccius productus*). *Food Chem.* **2008**, *107*, 768–776. [CrossRef]

96. Zhang, Y.; Duan, X.; Zhuang, Y. Purification and characterization of novel antioxidant peptides from enzymatic hydrolysates of tilapia (*Oreochromis niloticus*) skin gelatin. *Peptides* **2012**, *38*, 13–21. [CrossRef] [PubMed]

97. Lee, W.-S.; Jeon, J.-K.; Byun, H.-G. Characterization of a novel antioxidative peptide from the sand eel *Hypoptychus dybowskii*. *Process Biochem.* **2011**, *46*, 1207–1211. [CrossRef]

98. Kong, Y.-Y.; Chen, S.-S.; Wei, J.-Q.; Chen, Y.-P.; Lan, W.-T.; Yang, Q.-W.; Huang, G.-R. Preparation of antioxidative peptides from spanish mackerel (*Scomberomorus niphonius*) processing byproducts by enzymatic hydrolysis. *Biotechnology* **2015**, *14*, 188–193.

99. Jeevithan, E.; Bao, B.; Zhang, J.; Hong, S.; Wu, W. Purification, characterization and antioxidant properties of low molecular weight collagenous polypeptide (37 kDa) prepared from whale shark cartilage (*Rhincodon typus*). *J. Food Sci. Technol.* **2015**, *52*, 6312–6322. [CrossRef] [PubMed]

100. Gajanan, P.G.; Elavarasan, K.; Shamasundar, B.A. Bioactive and functional properties of protein hydrolysates from fish frame processing waste using plant proteases. *Environ. Sci. Pollut. Res.* **2016**, *23*, 24901–24911. [CrossRef] [PubMed]

101. Li, J.; Li, Q.; Li, J.; Zhou, B. Peptides derived from *Rhopilema esculentum* hydrolysate exhibit angiotensin converting enzyme (ACE) inhibitory and antioxidant abilities. *Molecules* **2014**, *19*, 13587–13602. [CrossRef] [PubMed]

102. Harada, K.; Maeda, T.; Hasegawa, Y.; Tokunaga, T.; Ogawa, S.; Fukuda, K.; Nagatsuka, N.; Nagao, K.; Ueno, S. Antioxidant activity of the giant jellyfish *Nemopilema nomurai* measured by the oxygen radical absorbance capacity and hydroxyl radical averting capacity methods. *Mol. Med. Rep.* **2011**, *4*, 919–922. [PubMed]

103. Samanta, J.K.M.P.K.; Khora, S. Antioxidant activity of fish protein hydrolysates from *Sardinella longiceps*. *Int. J. Drug Dev. Res.* **2014**, *6*, 137–145.

104. Chi, C.-F.; Cao, Z.-H.; Wang, B.; Hu, F.-Y.; Li, Z.-R.; Zhang, B. Antioxidant and functional properties of collagen hydrolysates from spanish mackerel skin as influenced by average molecular weight. *Molecules* **2014**, *19*, 11211–11230. [CrossRef] [PubMed]

105. Kangsanant, S.; Thongraung, C.; Jansakul, C.; Murkovic, M.; Seechamnanturakit, V. Purification and characterisation of antioxidant and nitric oxide inhibitory peptides from Tilapia (*Oreochromis niloticus*) protein hydrolysate. *Int. J. Food Sci. Technol.* **2015**, *50*, 660–665. [CrossRef]

106. Bardan, A.; Nizet, V.; Gallo, R.L. Antimicrobial peptides and the skin. *Expert Opin. Biol. Ther.* **2004**, *4*, 543–549. [CrossRef] [PubMed]

107. Song, R.; Wei, R.-B.; Luo, H.-Y.; Wang, D.-F. Isolation and characterization of an antibacterial peptide fraction from the pepsin hydrolysate of half-fin anchovy (*Setipinna taty*). *Molecules* **2012**, *17*, 2980–2991. [CrossRef] [PubMed]

108. Song, R.; Wei, R.; Zhang, B.; Wang, D. Optimization of the antibacterial activity of half-fin anchovy (*Setipinna taty*) hydrolysates. *Food Bioprocess Technol.* **2012**, *5*, 1979–1989. [CrossRef]

109. Ennaas, N.; Hammami, R.; Beaulieu, L.; Fliss, I. Purification and characterization of four antibacterial peptides from protamex hydrolysate of Atlantic mackerel (*Scomber scombrus*) by-products. *Biochem. Biophys. Res. Commun.* **2015**, *462*, 195–200. [CrossRef] [PubMed]

110. Ennaas, N.; Hammami, R.; Beaulieu, L.; Fliss, I. Production of antibacterial fraction from Atlantic mackerel (*Scomber scombrus*) and its processing by-products using commercial enzymes. *Food Bioprod. Process.* **2015**, *96*, 145–153. [CrossRef]

111. Ryu, B.; Qian, Z.J.; Kim, S.K. SHP-1, A novel peptide isolated from seahorse inhibits collagen release through the suppression of collagenases 1 and 3, nitric oxide products regulated by NF-κB/p38 kinase. *Peptides* **2010**, *31*, 79–87. [CrossRef] [PubMed]

112. Ryu, B.; Qian, Z.J.; Kim, S.K. Purification of a peptide from seahorse, that inhibits TPA-induced MMP, iNOS and COX-2 expression through MAPK and NF-κB activation, and induces human osteoblastic and chondrocytic differentiation. *Chem. Biol. Interact.* **2010**, *184*, 413–422. [CrossRef] [PubMed]

113. Shen, Q.; Guo, R.; Dai, Z.; Zhang, Y. Investigation of enzymatic hydrolysis conditions on the properties of protein hydrolysate from fish muscle (*Collichthys niveatus*) and evaluation of its functional properties. *J. Agric. Food Chem.* **2012**, *60*, 5192–5198. [CrossRef] [PubMed]

114. Lødemel, J.B.; Egge-Jacobsen, W.; Olsen, R.L. Detection of TIMP-2-like protein in Atlantic cod (*Gadus morhua*) muscle using two-dimensional real-time reverse zymography. *Biosci. Biotechnol. Biochem.* **2004**, *139*, 253–259. [CrossRef] [PubMed]

115. Katiyar, S.; Elmets, C.A.; Katiyar, S.K. Green tea and skin cancer: Photoimmunology, angiogenesis and DNA repair. *J. Nutr. Biochem.* **2007**, *18*, 287–296. [CrossRef] [PubMed]

116. Zhuang, Y.; Hou, H.; Zhao, X.; Zhang, Z.; Li, B. Effects of collagen and collagen hydrolysate from jellyfish (*Rhopilema esculentum*) on mice skin photoaging induced by UV irradiation. *J. Food Sci.* **2009**, *74*, H183–H188. [CrossRef] [PubMed]

117. Diffey, B.L. Solar ultraviolet radiation effects on biological systems. *Phys. Med. Biol.* **1991**, *36*, 299–328. [CrossRef] [PubMed]

118. Tanino, Y.; Budiyanto, A.; Ueda, M.; Nakada, A.; Nyou, W.T.; Yanagisawa, M.; Ichihashi, M.; Yamamoto, Y. Decrease of antioxidants and the formation of oxidized diacylglycerol in mouse skin caused by UV irradiation. *J. Dermatol. Sci. Suppl.* **2005**, *1*, S21–S28. [CrossRef]

119. Han, Y.-T.; Han, Z.-W.; Yu, G.-Y.; Wang, Y.-J.; Cui, R.-Y.; Wang, C.-B. Inhibitory effect of polypeptide from *Chlamys farreri* on ultraviolet A-induced oxidative damage on human skin fibroblasts in vitro. *Pharmacol. Res.* **2004**, *49*, 265–274. [CrossRef] [PubMed]

120. Yu, Y.; Li, Z.; Liu, X.; Wang, Y. Effects of polypeptides from *Chlamys farreri* on the structure of skin and the content of antioxidants in hairless mice irradiated by ultraviolet B. *China J. Lepr. Skin Dis.* **2004**, *20*, 20–23.

121. Wang, C.-B.; Ding, B.-X.; Guo, S.-B.; Wang, Y.-Z.; Han, Y.-T.; Wang, Y.-J. Protective effect of polypeptide from *Chlamys farreri* on mitochondria in human dermal fibroblasts irradiated by ultraviolet B. *Acta Pharmacol. Sin.* **2003**, *24*, 692–696. [PubMed]

122. Leone, A.; Lecci, R.M.; Durante, M.; Meli, F.; Piraino, S. The bright side of gelatinous blooms: Nutraceutical value and antioxidant properties of three Mediterranean jellyfish (Scyphozoa). *Mar. Drugs* **2015**, *13*, 4654–4681. [CrossRef] [PubMed]

123. Fan, J.; Zhuang, Y.; Li, B. Effects of collagen and collagen hydrolysate from jellyfish umbrella on histological and immunity changes of mice photoaging. *Nutrients* **2013**, *5*, 223–233. [CrossRef] [PubMed]

124. Hou, H.; Li, B.; Zhang, Z.; Xue, C.; Yu, G.; Wang, J.; Bao, Y.; Bu, L.; Sun, J.; Peng, Z.; et al. Moisture absorption and retention properties, and activity in alleviating skin photodamage of collagen polypeptide from marine fish skin. *Food Chem.* **2012**, *135*, 1432–1439. [CrossRef] [PubMed]

125. Chen, T.; Hou, H. Protective effect of gelatin polypeptides from Pacific cod (*Gadus macrocephalus*) against UV irradiation-induced damages by inhibiting inflammation and improving transforming growth factor-β/Smad signaling pathway. *J. Photochem. Photobiol. B Biol.* **2016**, *162*, 633–640. [CrossRef] [PubMed]

126. Shibuya, S.; Ozawa, Y.; Toda, T.; Watanabe, K.; Tometsuka, C.; Ogura, T.; Koyama, Y.I.; Shimizu, T. Collagen peptide and vitamin C additively attenuate age-related skin atrophy in Sod1-deficient mice. *Biosci. Biotechnol. Biochem.* **2014**, *78*, 1212–1220. [CrossRef] [PubMed]

127. Chen, T.; Hou, H.; Lu, J.; Zhang, K.; Li, B. Protective effect of gelatin and gelatin hydrolysate from salmon skin on UV irradiation-induced photoaging of mice skin. *J. Ocean Univ. China* **2016**, *15*, 711–718. [CrossRef]

128. Jimbo, N.; Kawada, C.; Nomura, Y. Optimization of dose of collagen hydrolysate to prevent UVB-irradiated skin damage. *Biosci. Biotechnol. Biochem.* **2016**, *80*, 356–359. [CrossRef] [PubMed]

129. Sun, L.; Zhang, Y.; Zhuang, Y. Antiphotoaging effect and purification of an antioxidant peptide from tilapia (*Oreochromis niloticus*) gelatin peptides. *J. Funct. Foods* **2013**, *5*, 154–162. [CrossRef]

130. Zhuang, Y.; Sun, L. Preparation of reactive oxygen scavenging peptides from tilapia (*Oreochromis niloticus*) skin gelatin: Optimization using response surface methodology. *J. Food Sci.* **2011**, *76*, C483–C489. [CrossRef] [PubMed]

131. De Luca, C.; Mikhal'Chik, E.V.; Suprun, M.V.; Papacharalambous, M.; Truhanov, A.I.; Korkina, L.G. Skin antiageing and systemic Redox effects of supplementation with marine collagen peptides and plant-derived antioxidants: A single-blind case-control clinical study. *Oxid. Med. Cell. Longev.* **2016**, *2016*, 4389410. [CrossRef] [PubMed]

marine drugs

MDPI

Review

Microbial Diseases of Bivalve Mollusks: Infections, Immunology and Antimicrobial Defense

Carla Zannella [1], Francesco Mosca [2], Francesca Mariani [2], Gianluigi Franci [1], Veronica Folliero [1], Marilena Galdiero [1], Pietro Giorgio Tiscar [2] and Massimiliano Galdiero [1,*]

[1] Department of Experimental Medicine—University of Campania "Luigi Vanvitelli", Via Costantinopoli 16, 80138 Napoli, Italy; carlazannella88@gmail.com (C.Z.); gianluigi.franci@unina2.it (G.F.); veronica.folliero@unina2.it (V.F.); marilena.galdiero@unina2.it (M.G.)
[2] Faculty of Veterinary Medicine, University of Teramo, Piano d'Accio, 64100 Teramo, Italy; fmosca@unite.it (F.M.); fmariani@unite.it (F.M.); pgtiscar@unite.it (P.G.T.)
* Correspondence: massimiliano.galdiero@cantab.net

Academic Editors: Se-Kwon Kim and Peer B. Jacobson
Received: 16 January 2017; Accepted: 8 June 2017; Published: 17 June 2017

Abstract: A variety of bivalve mollusks (phylum Mollusca, class Bivalvia) constitute a prominent commodity in fisheries and aquacultures, but are also crucial in order to preserve our ecosystem's complexity and function. Bivalve mollusks, such as clams, mussels, oysters and scallops, are relevant bred species, and their global farming maintains a high incremental annual growth rate, representing a considerable proportion of the overall fishery activities. Bivalve mollusks are filter feeders; therefore by filtering a great quantity of water, they may bioaccumulate in their tissues a high number of microorganisms that can be considered infectious for humans and higher vertebrates. Moreover, since some pathogens are also able to infect bivalve mollusks, they are a threat for the entire mollusk farming industry. In consideration of the leading role in aquaculture and the growing financial importance of bivalve farming, much interest has been recently devoted to investigate the pathogenesis of infectious diseases of these mollusks in order to be prepared for public health emergencies and to avoid dreadful income losses. Several bacterial and viral pathogens will be described herein. Despite the minor complexity of the organization of the immune system of bivalves, compared to mammalian immune systems, a precise description of the different mechanisms that induce its activation and functioning is still missing. In the present review, a substantial consideration will be devoted in outlining the immune responses of bivalves and their repertoire of immune cells. Finally, we will focus on the description of antimicrobial peptides that have been identified and characterized in bivalve mollusks. Their structural and antimicrobial features are also of great interest for the biotechnology sector as antimicrobial templates to combat the increasing antibiotic-resistance of different pathogenic bacteria that plague the human population all over the world.

Keywords: marine bivalve mollusks; antimicrobial peptides; bivalve immune system

1. Introduction

Marine bivalve mollusks may be affected by numerous infectious diseases. In this review, we will consider the most important diseases caused by viruses, bacteria and protistans, which are responsible for mortality outbreaks and have a substantial commercial impact. To a lesser extent, other diseases are also caused by fungi (*Aspergillus*, *Penicillium* and *Fusarium*) [1], Porifera (*Cliona* spp.) [2], and helminth parasites, such as trematodes, cestodes and nematodes [3].

Currently, the main infectious diseases of marine bivalve mollusks, such as herpes virus infection and bonamiasis, have taken on a worldwide distribution due to trade globalization. Since the transmission among bivalve mollusks is direct and horizontal, high-density production systems and environmental changes might have contributed to increasing the spread of diseases [4]; therefore, to avoid the current risk of further spreading of illnesses throughout the world, the World Organisation for Animal Health (OIE) *Aquatic Code* has set out standards and recommendations to improve the safety of international trade in aquatic animals, including marine bivalve mollusks. Nowadays, the aim is to prevent the pathogen's introduction into an importing country, so as to avoid the onset of disease outbreak rather than to eradicate the pathogen, which would be more difficult and expensive because of the high-density production systems used in commercial hatcheries and nurseries and the continuous stock movements around the world. Due to the absence of effective and specific chemotherapy and anti-viral treatments or vaccines available to prevent illnesses [5], the surveillance and control plan of these diseases based on their prevention have a key role. For this reason, it is important to implement high levels of on-farm and live-holding facility biosecurity and to restrict stock movements. Avoiding stressors, such as exposition to intense temperatures, a high or low level of salinity, handling, substantial co-infection with other parasites, as well as decreasing density, should help to reduce the impact of diseases [4]. The present review will describe bacterial and viral pathogens that affect bivalve mollusks and will illustrate the immune responses generated in bivalves and the repertoire of immune cells and their activation upon infection. An aspect of importance in the immune defense mechanisms operated by bivalve mollusks is the expression of antimicrobial peptides; therefore, the second part of the review will be dedicated to providing an outline of the antimicrobial peptides that have been identified and characterized in bivalve mollusks.

2. Infectious Diseases of Marine Bivalve Mollusks

Taking into account all of these considerations, the development of reliable and useful diagnostic tools is of paramount importance for the prevention and control of diseases. Considering that the lack of bivalve molluscan cell lines has greatly limited the possibility for viral isolation and the study of the experimental transmission of these pathogenic microorganisms, currently, the detection of the causative agent of microbial disease is mainly based on direct diagnostic methods. Moreover, classic serological methods are not suitable for diagnostic purposes since mollusks do not produce antibodies; in fact, histology is considered as the standard screening diagnostic method because it supplies a wider amount of information, but macroscopic examination usually gives no pathognomonic signs or indicative information. It is difficult to diagnose an infection based exclusively on morphological differences between species that have similar morphological characteristics, such as for example *Bonamia ostreae* and *Bonamia exitiosa*, when observed under the microscope [6]. To overcome these problems, nowadays, more specific and sensitive molecular diagnostic techniques are used in addition to electron microscopy for specific identification of the pathogen. The OIE *Manual of Diagnostic Tests for Aquatic Animals* describes specific protocols designed to detect a certain pathogen agent, to be employed to confirm histological examination results and provides a species-specific diagnosis. Methods used in targeted surveillance programs should not be time consuming; in fact, in most cases, PCR and subsequently the sequencing of 16S or other candidate loci are recommended for the identification of the isolate. For a presumptive diagnosis of a disease, the standard method is histopathology, but for a confirmatory diagnosis, the standard methods are polymerase chain reaction (PCR) and in situ hybridization (ISH); however, sequencing and transmission electron microscope (TEM) are recommended.

Viral infectious diseases: the most important viruses associated with disease outbreaks and the major cause of mortality in bivalves are currently members of the families of *Herpesviridae* and *Iridoviridae*; nevertheless there are also other viruses that can infect bivalves belonging to the families of *Picornaviridae*, *Papovaviridae*, *Birnaviridae*, *Retroviridae* and *Reoviridae* [7]. Viral pathogens are often highly infectious and easily transmissible. High-density production systems and environmental changes might have contributed to increasing the spread of the disease [4]. Molecular tools such as PCR, ISH and immunochemistry are used to detect viral pathogens in mollusks [8–10].

Herpes-like viruses: The first description of herpes-like viral infection was reported by Farley et al. (1972) [11] in *Crassostrea virginica* from the east coast of the USA. Afterwards, disease outbreaks, associated with high mortality rates, particularly in larvae and spawn during the summer period, have been reported from Pacific oysters (*Crassostrea gigas*) in France, where higher mortalities were observed in 2008 and increased in 2009 and 2010 [12]. Oyster herpesvirus type 1 variant μvar (OsHV-1 μvar) was also found associated with Pacific oyster mass mortalities in Ireland, Italy, The Netherlands, Spain, the U.K. and in Australia, New Zealand and Korea, but is known to be detected elsewhere in the absence of oyster mortalities (e.g., Japan) [13]. Recently, OsHV-1 was detected in *C. gigas* in Japan and South Korea associated with mass mortality rate [14,15] and in Sweden and Norway [16]. In the Thau Lagoon (France), OsHV-1 and, secondarily *Vibrio splendidus* are responsible for mass mortality of *C. gigas* [17]. During the summer of 2012 and 2013, OsHV-1 caused high mortality of *Scapharca broughtonii* in China [10]. TEM analysis showed that larvae exhibited generalized infections, whereas focal infections usually occurred in juveniles. Adult stages were less sensitive than younger stages. Infected larvae showed a reduction in feeding and swimming activities, and mortality can reach 100% in a few days. The effects of the disease on the hosts manifested in velar and mantle lesions, and the attitude of larvae to swim weakly in circles. Histologically, fibroblastic-like cells exhibited abnormal cytoplasmic basophilia and enlarged nuclei with marginated chromatin; other cell types including hemocytes and myocytes showed extensive chromatin condensation [18].

Irido-like viruses: Maladie des branchies or gill disease is a disease caused by gill necrosis virus (GNV), responsible for recurrent mass mortalities in adult Portuguese oysters (*Crassostrea angulata*), from 1966 until the early 1970s along French coasts. To a lesser extent, gill disease also affected the Pacific oyster, *C. gigas*, imported in France, but with a negligible mortality. Because of its natural resistance to infection, *C. gigas* is currently the main species bred in Europe. Another Irido-like virus, hemocyte infection virus (HIV), caused mass mortality of Portuguese adult oysters (*C. angulata*) in France between 1970 and 1973. This virus is similar to GNV, and viral particles can be observed in the cytoplasm of atypical infected hemocytes in connective tissues. A third type of Irido-like virus, the oyster velar virus (OVV), caused a high mortality rate of Pacific larval oysters on the west coast of North America (Washington State, USA) from 1976–1984 [19]. Currently, Irido-like virus infections are uncommon in Europe.

Bacterial infectious diseases: Due to the marine bivalve filter-feeding habit, they concentrate a rich and diverse bacterial commensal microbiota, composed of various species belonging to different genera like *Vibrio*, *Pseudomonas*, *Acinetobacter*, *Photobacterium*, *Moraxella*, *Aeromonas*, *Micrococcus* and *Bacillus* [20]; some of them may be pathogenic in larval rearing systems and not in the wild; in fact, pathogenicity depends on the host species, their life stage, amount of bacteria and on environmental factors. The rate of larval mortality can reach 100%, especially if larvae are reared in static systems at high temperature and density. Diagnosis of bacterial diseases is based on macroscopic inspection of shell valves in combination with PCR, which is the most sensitive and rapid method [21].

Gram-negative bacteria: Most bacterial diseases of bivalves are caused by a large range of *Vibrio* species (*Vibrio alginolyticus*, *V. splendidus*, *Vibrio anguillarum*, *Vibrio tubiashi*, *Vibrio tapetis*, *Vibrio aestuarianus*, *Vibrio neptunius* and other *Vibrio* spp.), *Pseudomonas* and *Aeromonas* [22–25]. These bacteria are responsible for bacillary necrosis in a wide range of species of bivalve larvae [26]. *Vibrio* spp. produce exotoxins (ciliostatic factors and hemolysins), which cause deciliation, loss of velar epithelial and abnormal swimming behavior [27]. Necrosis has been well described with histological, immunofluorescent and ultrastructure techniques [28]. *V. tapetis* is the activating agent of an epizootic infection described in adult clams called brown ring disease (BRD); the most sensitive species is *Ruditapes philippinarum* [21,29]. The disease has been detected in France, Spain, Portugal, Italy, the United Kingdom, Ireland, Norway [30] and, less frequently, in the Mediterranean and Adriatic seas [31]. In 2006, *V. tapetis* was reported from Manila clams on the west coast of Korea [32]. Transmission passes through direct contact between infected clams [33]. *V. tapetis* adheres to and colonizes the surface of the periostracal lamina at the mantle edge of the shell, causing anomalous deposition of periostracum, an abnormal calcification process and the accumulation of brown organic material, which is a striking sign of the disease. From the extrapallial space, the bacteria can penetrate the mantle epithelium and the soft tissues, where they reproduces themselves and cause severe damage and subsequent death [34]. Juvenile oyster disease (JOD) is a similar syndrome to BRD that appeared in 1988 in juvenile *C. virginica* [35] caused by *Roseovarius crassostreae*; nowadays, the disease is known as *Roseovarius* oyster disease [36]. Major mortal outbreaks have been reported in cultured oysters from New York to Maine (USA). Symptoms, such as reduced growth rates, fragile shell development, cupping on the left valve, anomalous conchiolin deposit around the periphery of the mantle on the inner valves [37], occur when water temperature exceeds about 21–25 °C, in high salinity and high-density culture conditions. Mortality reaches up to 90% in animals <25 mm; instead, larger juveniles forms show lower rates of mortality [38].

Gram-positive bacteria: Few Gram-positive bacteria cause diseases in bivalves; the main pathogenic agent is represented by *Nocardia crassostreae*, the etiological agent of Pacific oyster nocardiosis (PON) infecting *C. gigas* and *Ostrea edulis* cultivated near infected *C. gigas* along the west coast of North America from the Strait of Georgia, British Columbia to California and Japan (Matsushima Bay) [39]. Carella et al. (2013) [40] have notified nocardiosis in Mediterranean bivalves. Mortality rate reaches up to 35%; infected animals show yellow-green pustules in the mantle, gills, adductor and cardiac muscle associated with intense hemocyte infiltration around the colonies of *Nocardia* [41].

Protozoan infectious diseases: The most important protozoan pathogens belong to the genera *Bonamia*, *Perkinsus*, *Haplosporidium* and *Marteilia*; they can mainly infect oyster and clam species, causing enormous damage to commercial productions. In particular, in this review, we will discuss in detail the diseases caused by *Perkinsus marinus*, *Perkinsus olseni*, *Marteilia refringens*, *B. ostreae* and *B. exitiosa* that are currently under surveillance and require mandatory notification by the World Organization for Animal Health. Prevalence and intensity of infections tend to increase during the warm season, depending on temperature and high salinity rates [42]. For correct identification of the pathogen, histological examination is needed, as well as ISH or PCR. However, for unknown susceptible species and unknown geographical range, confirmation by sequencing and description by TEM are recommended. Bonamiosis represents 63% of protozoan diseases in Europe [43] and is caused by a group of protists in the genus Bonamia. *B. ostreae* has spread in Europe (France, Ireland, Netherlands, Portugal, Spain and the U.K.), but also on the west coast of Canada and both coasts of the USA [44]. *B. ostreae* was found in *O. edulis* imported to China [45]; recently, this species was detected in New Zealand infecting the flat oyster *Ostrea chilensis* [46]. *B. exitiosa* was found in *O. chilensis* in South Island, New Zealand [47], and in *Ostrea angasi* in southeastern Australia [48]. Since 2003, the parasite has been observed in both the Atlantic and Pacific coasts of the USA [49], including California [50]; *B. exitiosa* was also detected in *O. edulis* from the Galician coast (Spain) and the Manfredonia Gulf, Italy (Adriatic Sea), including concurrent infections with *B. ostreae* [51] and in *Ostrea stentina* in Tunisia [52]. *B. exitiosa* was found, as well, on the Spanish Mediterranean coast [53], in southwestern England [54] and in southern Portugal [55]. Others species, *Bonamia perspora* and *Bonamia roughleyi*, have been found on the east coast of the USA [56] and in southeastern Australia [57]. With the exception of *B. perspora*, all of the other species can be normally observed within hemocytes of the host [58]; *B. perspora* occurs within connective tissues [56]. The pathogen infects the hemocytes, multiplies within blood cells and spreads to all tissues. In highly infected adult oysters, we can find a yellow discoloration of the tissue, extensive lesions on the gill and mantle, breakdown of connective tissue and significant mortality (>90%). Larvae can be infected and contribute to the spread of the parasite. A lack of resistance to infection and, therefore, high densities of oysters in closely-spaced beds favor the development of epizootics.

In the late 1960s, two protozoans, both in the genus *Marteilia*, *Marteilia refringens* and *Marteilia sydneyi*, were identified as the causative agents of disease and heavy mortalities in the flat oyster: *O. edulis*, in France, and in *Saccostrea glomerata*, in Australia, respectively [59];

afterward, the parasite become widespread: *M. refringens*, infecting *O. edulis*, has, to date, mainly been found in Europe (Albania, Croatia, France, Greece, Italy, Morocco, Portugal, Spain, Sweden, Tunisia and the United Kingdom); *Mytilus galloprovincialis*, in the Gulf of Thermaikos, northern Greece [60], on the north coast of the Adriatic Sea [61] and along the Campanian coast (Tyrrhenian Sea, South of Italy) [62]. Marteilia was also previously found in *M. galloprovincialis* bred in Puglia [63]. *M. sydneyi*, infecting *Saccostrea glomerata* and possibly other *Saccostrea* spp. [64], has been reported in New South Wales, Queensland and Western Australia [65]. Marteiliosis is also known as Aber disease (*M. refringens*, two types: M and O, as defined by Peruzzi et al. [66], and Queensland Unknown (QX) disease (*M. sydneyi*). Both infect the digestive system and sporulate in epithelial cells of the digestive gland causing paleness of the digestive glands, emaciation of the oyster, dissipation of its reserves of energy, tissue necrosis, cessation of growth and mortality up to 90% in summer. Juveniles and older life stages are susceptible to infection, but prevalence and infection intensity are generally higher in individuals of two years old or more [67]. By TEM analysis, *M. sydneyi* can be differentiated from *M. refringens* by a paucity of striated inclusions within the plasmodia, the formation of eight to sixteen sporangial primordia in each plasmodium (instead of eight for *M. refringens*), the occurrence of two or three spores in each sporangium (rather than four in *M. refringens*) and the presence of a thick coat of concentric biological membranes surrounding mature *M. sydneyi* spores.

Perkinsosis: Perkinsosis is caused by *Perkinsus marinus*, responsible for dermo disease in *C. virginica* and, to a lesser extent, in *C. gigas*, *Crassostrea rhizophorae* and *Crassostrea corteziensis* [68]. The pathogen, uncommon in Europe, was first described in the Gulf of Mexico [69]. It was found along the southeast coast of the USA from Maine to Florida [70], along the Pacific coast of Mexico [71], in the Gulf of California (northwest Mexico) [72] and in Brazil [73]. *Perkinsus olseni* causes perkinsosis in many clam species with distribution in Australia, Korea, China, Japan and Europe [32]. Both parasites were also found associated with ten oyster species collected from both Panamanian coasts, including the Panama Canal and Bocas del Toro [74]. Transmission is direct from oyster to oyster; viable cells are released in host feces or on the death of the host [32] and are acquired through host feeding mechanisms. Every life stage is susceptible to disease [75]. The pathogen infects and proliferates in the digestive epithelium, connective tissue of all organs and hemocytes causing hemocytosis and tissue lysis with a consequent severe emaciation; mortality arrives up to 80% based on environmental factors [76]. In order to simplify all microbiological diseases, we generated tables with details of the infections (Tables 1–3).

Table 1. Main viral infectious diseases of marine bivalve mollusks.

Disease (Pathogenic Agent)	Host Species	Effects on Host	Geographical Distribution	References
VIRUSES				
Herpes virus infection (oyster herpes virus)	Mainly hatchery-reared larvae of *Crassostrea gigas* and *Ostrea* spp.	Velar and mantle lesions; deterioration; swim in circles	Europe (France, Ireland, Italy, The Netherlands, Spain); U.K.; Australia; New Zealand, Mexico, USA, Japan, South Korea, China	[10,13,14]
Gill necrosis virus (GNV) Hemocyte infection virus (HIV)	*Crassostrea angulata* and *C. gigas*	Destruction of gill filaments Virus infected hemocytes	France, Portugal, Spain, U.K. France, Spain	[77]
Oyster velar virus disease (OVV)	*C. gigas* larvae	Larval movement affected through loss of infected epithelial cells from velum	Washington State, USA	

Table 2. Main bacterial infectious diseases of marine bivalve mollusks.

Disease (Pathogenic Agent)	Host Species	Effects on Host	Geographical Distribution	References
BACTERIA				
Larval and juvenile vibriosis (*Vibrio anguillarum, V. tubiashi, V. alginolyticus, V. splendidus, V. aestuarianus, V. neptunius*)	Wide range of hatchery-reared species	Tissue necrosis (due to production of exotoxin by the bacteria), up to 100% larval mortality	In all marine waters where bivalve hatchery culture is practiced	[78,79]
Brown ring disease (*Vibrio tapetis*)	*Ruditapes philippinarum*	Brown deposit on shell; degeneration of digestive gland followed by metabolic disorder and death	Entire European Atlantic coast to North Africa, including coasts of France, Portugal, Spain, Italy, U.K., Ireland and Norway, west coast of Korea	[30,32]
Roseovarius oyster disease (*Roseovarius crassostreae*)	*Crassostrea virginica* juveniles <25 mm shell length	Reduced growth rates, fragile shell development, cupping on the left valve, mantle lesions, up to 90% mortalities	USA	[37]
Pacific oyster nocardiosis (*Nocardia crassostreae*)	*Crassostrea gigas, Ostrea edulis* cultivated near infected *C. gigas*	Yellow-green pustules in the mantle, gills, adductor and cardiac muscle, up to 35% mortalities	West coast of North America from the Strait of Georgia, British Columbia to California, and Japan (Matsushima Bay), Mediterranean Sea	[39,40]

Table 3. Main protozoan infectious diseases of marine bivalve mollusks.

Disease (Pathogenic Agent)	Host Species	Effects on Host	Geographical Distribution	References
		PROTISTS		
Bonamiasis (*Bonamia ostreae, B. exitiosa, B. perspora, B. roughleyi*)	Wide range of oyster species	Yellow discoloration of tissue, extensive lesions on gill and mantle, breakdown of connective tissue, significant mortality (up to 90%)	Europe, U.K., west coast Canada, east and west coasts of USA, New Zealand and SE Australia	[44–46,48,50,51]
Digestive gland (or Aber) disease (*Marteilia refringens*)	*Ostrea edulis* and *Mytilus galloprovincialis*	Pale digestive gland, severe emaciation, tissue necrosis, cessation of growth, mortalities up to 90% in summer	In *O. edulis*: Europe (Albania, Croatia, France, Greece, Italy, Morocco, Portugal, Spain, Sweden, Tunisia, U.K.); in *M. galloprovincialis*: northern Greece, in Italy, along the Adriatic Sea and the Campanian coast (Tyrrhenian Sea)	[56–59]
QX disease (*Marteilia sydneyi*)	*Saccostrea glomerata* and *Saccostrea* spp.	Necrosis of digestive gland, loss of condition, gonad absorption, mortalities up to 90% in summer	New South Wales, Queensland and Western Australia.	[60,61]
Dermo disease (*Perkinsus marinus*)	*Crassostrea virginica*	Severe emaciation, loss of condition, high mortality rate depending on temperature and salinity	Gulf of Mexico, southeast coast of USA, Pacific coast of Mexico, Gulf of California, Brazil	[65–69]

In conclusion, we must as well remember that for a disease to occur, the synergy among three factors is required. This synergy is commonly named the epidemiological triangle, which is composed by a host, a pathogenic agent and the environment. In fact, some of the infectious agents may be pathogenic or nonpathogenic based on the host species, its life stage (larval, juvenile or adult form) and the immune system, on which the environmental factor plays a key role. We can, therefore, assert that, in general, the exposition to extreme temperatures, a too high or too low level of salinity, human handling, an increasing density in rearing systems and co-infection with other parasites may reduce the immune defenses of the host, as well as increase the pathogenic agent rate of growth and, hence, its pathogenicity, making the host more susceptible to illness.

Based on the knowledge we have today on how a disease can spread and acknowledging the key role of the environment, it is clear that an improvement of the surveillance on the environment of rearing systems is essential.

3. Defense Mechanisms in Marine Bivalve Mollusks

During the last few decades, the immunology of marine bivalve mollusks (MBM) has been investigated with great interest, leading to the development of different branches for basic and applied research. The hemocyte phagocytosis constitutes the major immune response in MBM, and the study of this highly conserved process has contributed to better understanding not only the pathogenetic mechanisms of infectious diseases in MBM [81], but also the role of the filter-feeding organisms as passive carriers of pathogens to humans, considering the involvement of the hemocytes and hemolymph factors on the microbial clearance from mollusk's tissues [82].

The study of the hemocyte properties as biomarkers for monitoring the biological effects of anthropogenic stressors in polluted sites [83], as well as measuring the economic impact of environmental stressors on shellfish productions [84] represents another interesting immunological field of investigation for MBM. Moreover, the basic investigation of the hemocyte/hemolymph system represents a simplified phylogenetic model for understanding the ancestral interactions and integrations that occur between immunity and neuroendocrine response [85] and, more generally, between defense mechanisms and host homeostasis.

In the present section, we have mainly described the most important phases that characterize the hemocyte phagocytosis in MBM, giving, in parallel, importance to the humoral factors that participate in the recognition and opsonization of foreign particles, thus focusing on the common features of the innate immunity that are shared by invertebrate hemocytes and phagocytic cells of higher vertebrates.

Hemocyte-mediated immunity: The cell-mediated immunity represents the main internal defense response of marine bivalve mollusks. The hemocyte phagocytosis constitutes the key activity, leading to the recognition, engulfment and demolition of biotic and abiotic foreign particles [86]. The innate immune properties of the hemocytes rely on their ancestral role in food digestion and nutrient transport [87]. Indeed, the interplay between phagocytosis and nutrition in invertebrates has been ascribed to the primary function of the hemocytes phagosome as the digestive organelle, where microorganisms are degraded as nutrients source, then evolving into a more specialized compartment to kill pathogens (Figure 1) [88].

Marine bivalve mollusks possess an open circulatory system, and the hemocytes are found either in hemolymph or in tissues, respectively as circulating or infiltrating cells [89]. From an ontogenetic point of view, some authors suggested that hemocytes are derived from connective tissue cells [90]; however, different models have been proposed about the types of progenitor cell lines [91]. The nomenclature of mollusks hemocytes still represents a subject of debate, and the efforts at developing a uniform classification have resulted in the recognition of two main types of cells, such as granulocytes and hyalinocytes, based on morphological appearance and granularity under microscopic examinations. However, flow cytometry [92], electron microscopy [93,94] and monoclonal-antibodies based assays [95] have suggested the presence of various hemocyte sub-types. Such diversity may reveal a broad array of activities, and in particular, it is widely accepted that granulocytes play the most active role in the phagocytosis response [96]. Nevertheless, all of the hemocyte populations contribute to the overall immune response, operating in a differential fashion on the basis of the different stimulations [97]. The relative concentration of the various circulating hemocyte types can be exposed to reversible and selective modifications through physiological and molecular mechanisms comparable to the margination/demargination processes, which take place in humans and other mammals [98]. Therefore, the quantitative diversity of the cellular hemolymph configuration represents an important factor for the modulation of the immune response [99].

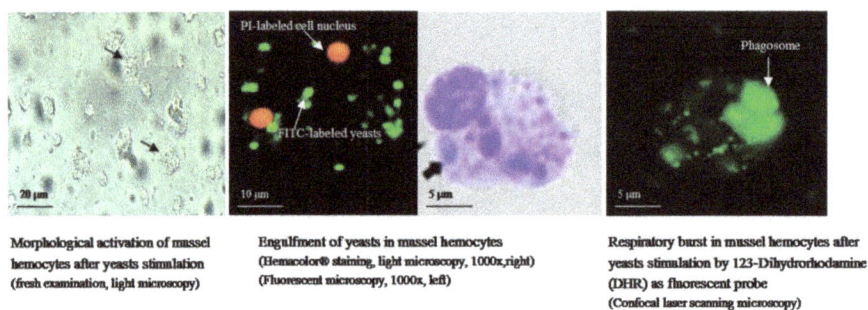

| Morphological activation of mussel hemocytes after yeasts stimulation (fresh examination, light microscopy) | Engulfment of yeasts in mussel hemocytes (Hemacolor® staining, light microscopy, 1000x, right) (Fluorescent microscopy, 1000x, left) | Respiratory burst in mussel hemocytes after yeasts stimulation by 123-Dihydrorhodamine (DHR) as fluorescent probe (Confocal laser scanning microscopy) |

Figure 1. The figure reports the main phases of the hemocyte phagocytosis, as well as previously investigated in mussel (author's unpublished figures) that are particularly described in the present section, in combination with the main humoral opsonizing and degradative factors.

The interest for the study of the hemocyte immunity mainly derives from the role of marine bivalve mollusks as sentinel organisms in environmental monitoring [100]. Indeed, many hemocyte parameters have been investigated as biomarkers in field or laboratory studies [101–103], and detailed data are available in the literature about the strong influence of natural and anthropic stressors on the hemocyte activity. In particular, the disruption of their morpho-functional properties has been described after exposure to low [104] and high [105] temperatures, pH acidification [106], mechanical stress [107], salinity changes [108], exposure to air [109], harmful algal bloom [110], organic and inorganic contaminants [111]. Nevertheless, some endogenous factors seem to have also an important influence on the modulation of the hemocyte activity, such as age [112], gender [113] and reproductive stage [114].

The hemocyte phagocytosis, mechanisms and kinetics: Bivalve hemocytes resemble the vertebrate monocyte/macrophage lineage, both in structure and function [115]. The chemotactic ability of hemocytes to migrate toward foreign particles and to incorporate them inside phagosomes is closely dependent on morphological activation through the projections of membrane ruffles or pseudopodia [116]. The cytoskeleton re-arrangement following proper stimulation represents the pivotal mechanism that allows the hemocytes to acquire a morphological spreading, from roundish to irregular shape [117]. Previous studies demonstrated that bacterial products, such as lipopolysaccharides and formylated tripeptide (N-FMLP), were able to elicit chemotactic and/or chemokinetic reactions in hemocytes [118], and the type of cell movement appeared as dependent on the nature of chemoattractant, thus hypothesizing a receptor-dependent mechanism [119]. Indeed, a differential migration activity was detected in hemocytes on the basis of the bacteria types that were encountered [120]. Both chemotaxis and chemokinesis augment the probability of physical association between hemocytes and foreign particles, but to date, the knowledge about pattern recognition receptors (PRRs) is still limited in bivalve hemocytes. PRRs recognize the conserved highly repeated microbial structures, termed pathogen-associated molecular patterns (PAMPs), and the Toll-like receptors (TRLs) have a prominent role within PRRs group, being traced to the most ancestral multicellular invertebrates [121]. TLRs belong to type I membrane receptors and contain an extracellular leucine-rich repeat (LRR) domain mediating the recognition of PAMPs [122]. Unlike mammal TLRs, invertebrate TLRs could not directly recognize PAMPs, but they seem to require the cytokine-like molecule Spatzle as an assistant [123]. Indeed, some authors suggested a hybrid function in pattern recognition for the primitive mollusk TLR, being characterized by broader ligands affinity and involving the assistance of some serum components [124].

In the Pacific oyster *C. gigas*, a putative TLR was cloned and named CgToll-1, showing upregulation in the hemolymph after challenge with *V. anguillarum* [125]. Similarly, in the scallop *Chlamys farreri*, a Toll homologue was detected and named CfToll-1, revealing transcripts modulation in the hemocytes after exposure to LPS [126]. In the mussel *Mytilus edulis*, the transcriptome analysis indicated a wide repertoire of innate recognition receptors, including transcripts for 27 TLR, particularly expressed in hemocytes [127]. Following binding of the ligand to the extracellular segment of TLR, signal transduction takes place by the intracellular toll-interleukin domain (TIR) containing adaptor molecules [128]. Each TLR recognizes distinct microbial components and activates different signaling pathways using selected adaptor molecules, then leading to the engagement of the signaling cascade of protein kinases that ultimately activate transcription factors and the expression of genes involved in the immune response [129,130]. In contrast to the large amount of data on TLR signaling systems from higher vertebrates, relatively little is known in bivalve mollusks. The existence of genes/transcripts mediating the Toll signaling pathway in hemocytes was reported in *M. galloprovincialis*, showing upregulation after bacterial challenge, particularly by Gram-negative, whereas a marginal response was detected following stimulation with purified PAMPs (LPS, β-glucans) [131]. Intermediate transcripts of the Toll signaling pathway were also detected in scallop [132], clams [133] and oysters [134]. Moreover, the intensity and duration of intermediate components activation, such as kinase-mediated cascade, appeared as dependent on the type of extracellular stimuli [135]. This kind of evidence contributes to support the existence of a differential hemocyte response depending on the bacteria types that are used for challenge [136]. Although most of the studies focused on the presence of Toll pathways in marine bivalve mollusks, other types of receptors have been investigated. Recently, the mRNA transcripts of a new putative phagocytic receptor (CgNimc), belonging to the Nimrod superfamily, were identified in hemocytes of *C. gigas*, revealing upregulation after bacterial challenge, whereas the recombinant protein showed higher binding affinity toward LPS rather than peptidoglycan [137]. In the scallop *Argopecten irradians*, a peptidoglycan recognition protein (PGRP) was cloned, sharing high identity with PGRPs of higher organisms and showing upregulation in the hemocytes exposed to peptidoglycan, but not to LPS [138]. Two short PGRPs were also detected in the bivalve *Solen grandis*, and they were particularly induced by peptidoglycan and β-1,3-glucan [139]. Following an engulfment of foreign particles within hemocyte phagosome, the activation of lysosomes granules leads to the formation of phago-lysosomes vacuoles where intracellular digestion takes place [140]. The lysosomal enzymes strongly participate in the degradation of ingested, material and the hydrolytic activity of β-glucuronidase, phosphatases, esterases and sulfatases has been detected in mussels [141], clams [142], cockles [143] and oysters [58]. Oxidative enzymes, such as peroxidase and phenoloxidase, are also involved in degradative mechanisms, but their presence is not a common feature in any marine bivalve mollusks [144,145]. The respiratory burst represents another heavy microbicidal mechanism, and the generation of the highly oxidant reactive oxygen species (ROS) inside phago-lysosomal vacuoles of stimulated hemocytes was suggested as an NADPH-oxidase-dependent mechanism [146–148]. The ROS synthesis has been widely reported in mussels [149], oysters [150], scallops and clams [151], although some authors indicated the lack of the NADPH-oxidase activity in the family Veneridae. The detection of ROS has been mostly investigated in terms of defense mechanism; however, the role of these molecules has been also considered in cellular and tissue homeostasis [152]. Indeed, previous reports indicated the ability of mussel hemocytes to generate ROS in the absence of phagocytic stimulation [153], and more recent evidence has suggested that mitochondria represent the main source of ROS in the unstimulated hemocytes, rather than the activity of lysosomal NADPH-oxidase [154].

Humoral defense factors: The hemocyte degranulation and the extracellular release of lysosomal enzymes represent the first humoral defense mechanism that was investigated in marine bivalve mollusks [155], a strategy commonly described as a response to pathogens [156]. However, marine bivalve mollusks possess more selective extracellular tools to contrast invaders, including recognition and effector proteins, such as lectins, complement-like molecules, lipopolysaccharide- (LBP) and β-1,3-glucan-binding proteins (β-GBP), fibrinogen-related proteins (FREPs) and antimicrobial peptides (AMPs) [156–158]. Lectins represent carbohydrate-recognition proteins, and their agglutinating and opsonizing activities have been previously described in marine bivalve mollusks, revealing heterogeneous binding specificity towards microbial surface sugars [159,160]. In particular, C-type lectins can recognize and bind terminal sugars on glycoproteins and glycolipids in a calcium-dependent manner. Recent studies in different marine bivalve mollusks have demonstrated both their gene upregulation following bacterial challenge and the binding activity of the recombinant proteins towards purified PAMPs [161–163]. Galectins, formerly known as S-type lectins, represent another conserved and ubiquitous family of carbohydrates-binding proteins, particularly characterized by their affinity for β-galactosides [164–167]. In the clam *R. philippinarum* [168] and scallop *A. irradians* [169], galectins have been cloned and characterized, showing gene upregulation and agglutination activity following bacterial challenge. Galectins seem to possess also an opsonizing role by promoting the hemocyte phagocytosis through cross-linking between extracellular glycocalyx and hemocyte surface, as observed for the oyster galectin CvGal1 [170]. Homologues of the vertebrate complement cascade have been investigated in marine bivalve mollusks for their immune role against pathogens. C1q represents the first sub-component of the classical complement pathway, and to date, C1q domain-containing proteins have been characterized at molecular level in oysters [171], mussels [172], scallops [173] and clams [174], revealing high molecular diversification of this family [175]. From a functional point of view, the recombinant proteins showed binding activity towards whole bacterial cells, as well as isolated PAMPs [176]. In addition, some authors have identified in oyster a complement component C3-like gene, particularly expressed in the hemocytes [177]. Although AMPs represent the most examined group of antimicrobial proteins, further discussed in the present review, other bactericidal compounds have been identified in marine bivalve molluscan integrative components of their humoral defense system. Member homologues of the bactericidal/permeability-increasing protein (BPI) family were isolated, showing binding activity toward LPS and bactericidal properties against Gram-negative bacteria [178–180]. Members of the lysozyme families (N-acetylmuramide glycanhydrolase) have been characterized in mucosal tissues and secretions of several bivalve species, as described in mussels [181], clams [182], scallops [183] and oysters [184], displaying a broad spectrum of antimicrobial activity and playing a dual role both in nutrition and immunity [185]. Moreover, the presence of plasma proteases' activity was previously described in marine bivalve mollusks as a microbicidal mechanism [186,187], and such evidence has accounted for the isolation and characterization of genes encoding proteases, such as cathepsins [188].

In conclusion, the great part of the studies on bivalve immunity has been directed to investigating the morpho-functional properties of circulating hemocytes and the humoral defense factors, providing limited information about the spatial and temporal heterogeneity of the immune response. In the future, a better understanding of microbe-bivalve interactions at mucosal interfaces is required, considering the interplay between mutualistic, commensal and pathogenic microbes at the initial encounter/colonization sites [189].

4. AMPs and Their Mechanism of Action

Marine ecosystems constitute more than 70% of the Earth's surface, are associated with astonishing species diversity and, therefore, represent an enormous resource of pharmacologically-active molecules. Marine living beings can be reconsidered as a potentially unlimited reservoir of bioactive molecules, either derived by complex metabolic reactions or gene-encoded peptides [190,191]. For each milliliter of seawater, approximately 10^6 bacteria and 10^9 viruses are generally present; therefore, seawater is to be considered an abundant source of pathogens. Most marine organisms reside in intimate coexistence with pathogenic microbes, and their survival in such a hostile surrounding is directly dependent on the development of a vigorous and successful immune system. In fact, living marine organisms are continuously exposed to microbial hazards, and to maintain their safeguard in such a harsh environment, they need a strong defensive mechanism to control all microbial pathogens that are inglobated with nutrients [192,193]. In fact, microbes are accumulated in bivalves, and microbial densities in their tissues are generally greater than in seawater. If the filtered microbes are pathogenic, their concentration in bivalve tissues can be deleterious. In order to defend themselves against such detrimental pathogens, bivalves depend on cellular defense mechanisms, as earlier described, and humoral defense factors, among which AMPs play an important role.

As a matter of fact, bivalves having evolved in the constant proximity of microorganisms must rely on their innate immune system effector molecules to contrast microbial pathogens. An ancient mechanism of innate immunity is represented by the production of anti-microbial substances, primarily peptides or polypeptides, which are produced by different types of cells and secretions and are either constitutively synthesized or induced at the time of infection [191]. In this regard, marine bivalves represent a valuable and scarcely delved source for novel antimicrobials [194].

Since the innate immunity system is supposed to represent the primary line of host defense against invading pathogens, it is of paramount importance to maintain host-microbe homeostasis and AMPs as ancient evolutionary molecules universally distributed in most of the multicellular organisms, which perform a broad-spectrum antimicrobial activity and, often, also present an immunomodulatory capacity. Therefore, AMPs play a crucial role in host defense against a wide range of microorganisms including Gram-positive and Gram-negative bacteria, viruses, fungi and parasites. Since their first discovery, with the isolation of a peptide named cecropin [195–197], from the insect *Hyalophora cecropia*, almost 2000 sequences encoding putative AMPs have been described and included in "The Antimicrobial Peptide Database" (http://aps.unmc.edu/AP/main.php).

AMPs are relatively small peptides (<60 amino acids) and may play polyvalent roles, which expand beyond their ability to serve as antibiotics [198]. Several of these peptides have been proven to possess anticancer activity, to be able to stimulate the immune system by promoting cytokine release, promote chemotaxis, antigen presentation, angiogenesis, inflammatory responses and adaptive immune induction [199–201]. During the last few decades, they have been purified from plants, invertebrates and vertebrates and are consequently considered to be part of the immune process probably in all Metazoa, representing innate immunity actors conserved along evolution in all biological kingdoms [202–204]. Despite their ample variation in biophysical characteristics, such as mass, composition and primary structure, several functional correlates have been identified [205]. In fact, the embracing features of most AMPs include rapid killing mechanisms, broad spectra of action, a clear net cationic charge and a strong propensity to give rise to amphipathic surfaces able to promote peptide: membrane interactions [206]. Although the primary structures of AMPs are diverse, based on genomic and protein sequences analysis coupled with structural and functional studies, AMPs have been sorted into several groups, including: (i) linear peptides able to adopt an α-helical conformation in a membrane-mimetic environment, (ii) peptides stabilized by one or several pairs of cysteine residues able to form disulfide bridges that have structures predominantly composed of β-sheets, (iii) peptides with a high content of specific amino acids, such as prolines, arginines, tryptophans, histidines and glycines, but with no uniform secondary structures, and (iv) peptides derived by partial hydrolysis of bulky precursor proteins with unknown or limited antimicrobial activities before the enzymatic degradation. AMPs' antimicrobial activity derives from membrane disruption and osmotic lysis of bacteria as opposed to the usual mechanism of action of most antibiotics where specific sites during bacterial growth and replication are targeted. Moreover, some AMPs proved to be also efficient in inhibiting viral infections. The putative mechanism for exerting an antiviral activity seems to be: (i) blocking early steps of viral entry by surface carbohydrate interaction, (ii) blocking viral attachment or penetration into the host cells by interactions with specific cellular receptors,

(iii) interaction and inactivation of viral envelope glycoproteins, (iv) modulation of host cell antiviral responses, (v) blocking intracellular expression of viral genes and/or production of viral proteins. However, no unequivocal correlation between AMP structures and microbial inhibition or killing mechanisms has so far become obvious; in fact, striking differences from peptide to peptide and specificity for particular AMP-microbe combinations are generally observed. The forthright antibiotic action of AMPs is considered to hinge on their cationic and amphiphilic nature, which empowers these molecules with the ability of interact with negatively-charged bacterial surfaces and membranes, therefore leading to membrane disruption or alteration [207]. In fact, AMPs essentially take advantage of the broad differences found in the organization of bacterial against eukaryotic membranes in order to promote damage of the membrane. Several differences, such as the absence of cholesterol, the abundance of anionic lipids and an electric field with a strong inward direction, are decisive for the correlation of specificity in favor of AMPs action against bacterial pathogens and a lower toxicity toward host cells [208,209]. Moreover, it is unlikely that bacteria can spoil these features and develop resistance since it would necessitate a profound modification of the bacterial membranes and their functions. Even though the exact AMP mechanisms of action remain a subject of discussion, the majority of them share similar biophysical characteristics that allow peptides to interact with microbes. Importantly, AMPs' antibacterial power is directly attributable to some of these features, such as a net positive charge, enabling AMP-bacterial membrane interactions via electrostatic forces and the propensity to form amphipathic structures in hydrophobic environments allowing penetration into the bacterial phospholipid bilayer. Therefore, regardless of any variations in size and structure, AMPs are often portrayed with an intrinsic cationic and hydrophobic nature that is the key for their first interaction with target bacterial cells. In agreement with literature data, AMPs' mode of action seems to proceed similarly to a pore-forming action or to a detergent effect. Several models have been put forth to explain such mechanisms, namely the barrel-stave model, the toroidal model and the carpet-like model (Figure 2) [210]. The three models were all elaborated following the assumption that AMPs have the tendency of being attracted by the bacterial membrane in virtue of electrostatic bonding forming between the peptide cationic feature and the low electric charge conferred to outer bacterial membranes by their surface components, such as phosphate groups within the lipopolysaccharide (LPS) of Gram-negative bacteria or lipoteichoic acids abundant on the exterior of Gram-positive bacteria [211].

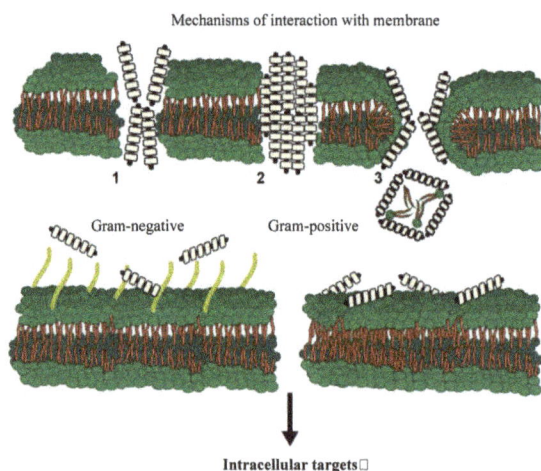

Figure 2. Mechanisms of interaction of AMPs with membranes. Top: The main proposed modes of action are: carpet model (1), barrel stave model (2) and toroidal-pore model (3). Bottom: Interactions of AMPs with Gram-negative and Gram-positive bacteria.

These models can be described in brief as follows: in the barrel-stave model, peptides (more often α-helical peptides with marked hydrophobic-hydrophilic domains), once attached to the phospholipids, aggregate and enter inside the membrane bilayer with the hydrophilic peptide domains forming the inside of the pore, and the hydrophobic peptide parts lined up facing the lipidic region of membrane phospholipids [212]. As a result, transmembrane pores, made up of a bundle of amphipathic helices, are created in a perpendicular topology within the membrane. A toroidal model with lipids intercalating between helices can be envisioned when peptides remain linked with lipid head groups, also when peptides are located upright within the highly-curved lipid bilayer. In this case, diffusion of lipids between the outer and inner membrane layers is granted by the continuous surface formed by both membrane leaflets [203,213]. A carpet-like model revealed instrumental for describing AMPs mechanism devoid of the straight insertion within the hydrophobic core of the membrane, but with peptides accumulation in an oriented array on the membrane surface forming a real carpet-like structure. When a threshold concentration is attained, the formation of transitory pores with disruption of lipid assembly and a detergent-like cell lysis ensues. This also leads to the micellization of the bilayer. Hence, all models considered, membrane perturbation is driven by cationic and hydrophobic residues shaping the interactions between peptides and phospholipids [214]. Nevertheless, an oversimplification of the hypothetical pore models is generally applied to the description of the mechanisms of action of AMPs [215]. More likely, a disordered or chaotic pore can be envisaged where peptides may shift their conformation, mutating the charges and relative interactions with lipids and, therefore, allowing the flickering of pores [216].

In conclusion, the interfacial activity is a leading determinant of the permeabilizing activity of several peptides [217], and it also provides a useful means to differentiate between peptides with antibacterial power that have a detrimental effect on membrane bilayers and cell-penetrating peptides, which seem to move past the bilayer without producing serious damages [218,219]. A further possibility for the antibacterial mechanisms is the action AMPs can exert on microbial intracellular targets where peptides can block cell-wall and/or nucleic acid synthesis, protein production and enzymatic activities [203].

5. Marine Bivalve Antimicrobial Peptides

Marine AMPs have been discussed thoroughly in other reviews [220–223]. However, in the present review, we describe AMPs derived from marine mollusks and their application in fighting infectious diseases. One unanswered enigma remains: the understanding of how mollusk survive in the absence of an acquired immune system since they are in close contact with a magnitude of putative pathogens such as viruses, bacteria, fungi and parasites, as a consequence of their filtering activities. Fortunately, mollusks encode for several antimicrobial molecules, and several AMPs have been isolated from marine mollusks, such as mussels (major species analyzed: *M. galloprovincialis*, *M. edulis*), clams (major species analyzed: *Venerupis philippinarum*), scallops (major species analyzed: *A. irradians*, *Argopecten purpuratus* and *Chlamys nobilis*) and oysters (major species analyzed: *C. virginica* and *C. gigas*).

Initial identification of AMPs in bivalves dates almost 20 years ago with the pioneering studies that led to the characterization of the first AMPs in mussels [224–227]. To date, several AMPs have been described from mussels and other bivalves. The majority of them rank along to the group of cysteine-containing peptides, which include a huge variety of defensins and defensin-like peptide and larger proteins. The discovery of the first molluscan AMP was performed by Hubert et al., 1996, from the *M. galloprovincialis*. Subsequently, several AMPs have been identified and extensively studied in two main mussels species, *M. galloprovincialis* and *M. edulis*. These molecules with antimicrobial properties have been classified into four groups following a primary structure classification parameter: defensins, mytilins, myticins and mytimycin. Overall, mussels' AMPs are characterized by possessing strong hydrophobic and cationic properties and a signature amphipathic structure (α-helix, β-hairpin-like β-sheet, β-sheet or α-helix/β-sheet mixed structures), all considered fundamental for the antimicrobial activity displayed. The principal defensin from mussels is a 39-long peptide present in two isoforms MGD1 (PDB 1FJN) and MGD2 sharing significant sequence homology. Furthermore, the three-dimensional structure (Figure 3 shows the 3D structure of several mollusk-derived AMPs) has been solved using NMR analysis [228] and has shown the presence of a side helical part (spanning from Asn7 to Ser16) and two anti-parallel β-strands (spanning fromArg20 to Cys25 and from Cys33-Arg37), which constitute the common cysteine-stabilized motif.

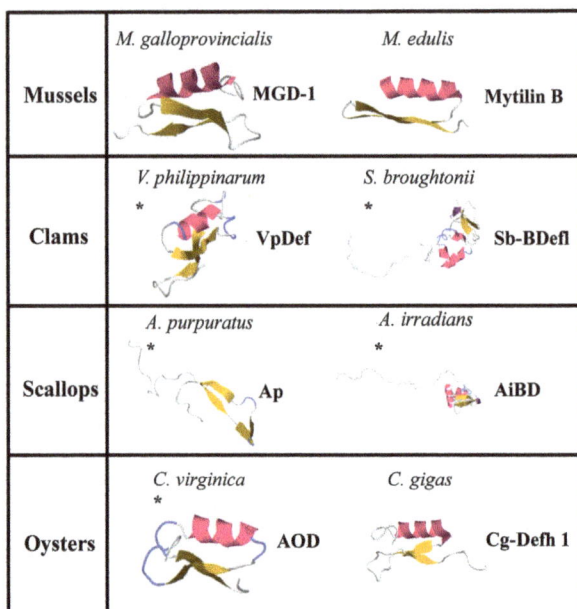

Figure 3. Examples of three-dimensional structures of bivalve AMPs. * The predicted structures were designed using AIDA (ab initio domain assembly) server, a tool for the prediction of protein tridimensional structures (http://ffas.burnham.org/AIDA).

A similar structural icon has been recently described in the attempt to unify all known classes of Cys-stabilized antimicrobial peptides. Yount and Yeaman identified this common structural signature and named the "γ-core motif" [229,230]. Conservation of the γ-core motif across all living organisms suggests it may represent an antimicrobial peptide archetype; in fact, several structural topologies resembling a γ-core motif can be described in a wide range of organisms, from unicellular organisms to humans. The γ-core present in many AMPs is not the only structural determinant to confer an antimicrobial activity, but in many instances, it can be used as a scaffold, to which further antimicrobial determinants (e.g., α-helices or β-sheets) can be attached in a modular fashion to yield various configurations. Higher organisms showed the most diversified range of γ-core polypeptides. This is, though, expected considering the necessity to provide protection to diverse tissues and to cooperate with other useful immune-system components. For example, several studies have been recently conducted in the analysis of such determinant in human β-defensins (HBDs), and the γ-core alone has been proven sufficient for retaining substantial antimicrobial activity [231–234]. Nevertheless, there is a significant difference with most of the know AMPs bearing the γ-core signature; that is, the fact that mussels defensins are characterized from being stabilized by four disulfide bonds (Cys4-Cys25, Cys10-Cys33, Cys14-Cys35 and Cys21-Cys38 in MGD-1), instead of the three disulfide bonds generally described in most other molecules, including arthropod defensins. MGD1 [224], a member of the arthropod defensin family from edible Mediterranean mussels

(*M. galloprovincialis*), and MGD2 [227] share the same size and sequence showing 80% identity with amino acids. Both contain an ORF encoding 81 amino acids including a 21-residue N-terminal sequence with a highly hydrophobic core representing a signal domain, followed by a 39-amino acid sequence corresponding to the active defensin and a 21-residue C-terminal extension. MGDs are principally active against Gram-positive bacteria, but MGD2 showed increased activity also against Gram-negative bacteria. Structural features of MGD1, cardinal for the supply of antimicrobial activity, were analyzed by Romestand et al., 2003 [235], by producing a set of synthetic peptides analogous to the described secondary structures of the molecule. The nonapeptide from residue 25 to residue 33 (CGGWHRLRC) displayed a consistent bacteriostatic activity, especially when cyclized by a disulfide bridge between Cys25 and Cys33.

The second group of mussels-derived AMPs is represented by the mytilins family, comprising five isoforms (A, B, C, D and G1). Isoforms A and B were found in *M. edulis* plasma [236], while isoforms B, C, D and G1 were isolated from *M. galloprovincialis* hemocytes [237]. Mytilin B (PDB 2EEM) is produced from a precursor molecule, which contains an initial region (22-amino acid residues) signal peptide region, a mature peptide 34 amino acids long, followed by a C-terminal domain of 48 residues rich in acidic amino acids [236]. The assorted mytilin isoforms have been shown to possess distinctive antimicrobial activities. In fact, mytilins A, B, C, and D showed a considerable activity against both Gram-positive and Gram-negative bacteria while mytilin G1 was revealed to be functioning only against Gram-positive bacteria. Mytilins B and D have also shown potency against the filamentous fungus *Fusarium oxysporum*. Moreover, experiments for describing the kinetics of bactericidal effects showed that, at high concentrations, several hours of incubation were needed for mytilins D and G1 to kill all bacteria in contrast to the few minutes necessary in the presence of mytilin B. A potent antiviral activity was also observed for mytilin B. Therefore, the different mytilin isoforms are endowed with complementary properties, which altogether contribute to the defense mechanisms, increasing the antimicrobial potential of mussels living in the context with a high diversity of pathogens.

A cysteine-rich peptide has also been isolated from mussels (*M. galloprovincialis*) and named myticin [226]. The mature molecule is 40 residues long and shows four intra-molecular disulfide bridges. Three different isoforms of myticin have been described (A, B and C) with the isoform C being the most abundantly expressed transcript in adult mollusks. All isoforms are highly active against Gram-positive bacteria and sometimes against Gram-negative bacteria, but myticin C is also a potent antiviral compound [238,239]. Constitutively-expressed myticin C-peptides in naive mussels render oysters resistant to ostreid herpesvirus 1 (OsHV-1) infections when oyster hemocytes are incubated with mussel hemolymph. Moreover, myticin C molecules retain antiviral activity in vitro against human herpes simplex viruses 1 (HSV-1) and 2 (HSV-2), showing a high potential for biotechnological applications [240].

A strictly antifungal peptide named mytimycin (MytM) containing 12 cysteines with a molecular weight of 6.2 KDa was derived [225] from the plasma of *M. edulis*. A novel cysteine-rich peptide with noteworthy antibacterial activity was recently isolated from *Mytilus coruscus* and was named myticusin-1 [241]. This is a 104-amino acid long polypeptide including 10 cysteine residues. Antimicrobial studies showed that myticusin-1 presented a more pronounced anti-microbial activity against Gram-positive bacteria compared to Gram-negative bacteria and fungus. From the same mussel (*M. coruscus*), a novel antimicrobial peptide with 55 amino acid residues was also identified [242]. This new antimicrobial peptide is endowed by predominant activity against fungi and Gram-positive bacteria and is characterized by possessing a chitin-biding domain and by six Cys residues forming three intra-molecular disulfide bridges. The recent advent of genome sequencing technologies has also allowed the identification of two previously uncharacterized mussel AMP families, big defensins and macins. A recent analysis [243] brought to light the existence of eight novel big defensins (MgBDs) and five novel macins (mytimacins) in the transcriptome of the Mediterranean mussel *M. galloprovincialis*, therefore further extending the vast antimicrobial peptides range present in this marine bivalve organism.

The Manila clam, *V. philippinarum*, is a meaningful marine bivalve for commercial purposes, and an amino acid sequence has been identified (named VpBD) that shares common features with other AMPs, such as an α-helical structure, a net positive charge and a high hydrophobic residue ratio. The display and spacing of cysteine residues and their flanking amino acid residues indicated that VpBD represents a member of the big defensin family. The structure of big defensins, generally, comprises a highly hydrophobic region located at the N-terminal, one C-terminal cysteine-rich and positively-charged region, as well as six cysteine residues arranged to form 1–5, 2–4, 3–6 disulfide bonds in the mature peptide, in a similar pattern to mammalian β-defensins. The microbicidal activities of VpBD (expressed in *Escherichia coli*) in vitro have been investigated and demonstrated a strong antibacterial activity towards various bacterial species, namely Gram-negative (*Pseudomonas putida*) and Gram-positive (*Staphylococcus aureus*) [244]. A further big defensin has been isolated from the ark shell, *Scapharca broughtonii* [245] and clam myticin isoforms 1, 2 and 3, and clam mytilin, (similar to myticins and mytilins from mussels) have been identified and characterized in *Ruditapes decussatus* [246].

Extracts from acidified gills of the American oyster *C. virginica* [247] delivered the first defensin molecules to be purified by oysters. The peptide (named *Cv-Def*) was 38 amino acids long with six cysteines, and the molecular mass was 4.2 KDa. The antimicrobial spectrum covered by *Cv-Def* included both Gram-positive bacteria and Gram-negative bacteria [247]. Successively, three more defensins were found and characterized from the mantle, denoted as *Cg-Def*, or hemocytes, designated as *Cg-defh*1 (PDB 2B68) and *Cg-defh*2, of the Pacific oyster *C. gigas* [248,249], which shared the cystine-stabilized alpha-beta motif (CS-αβ) [250]. Other AMPs are produced in *C. gigas*, such as Cg-Prp, which belong to the family of proline-rich peptides and has been identified from hemocytes [251,252]. A 5.5-kDa antimicrobial peptide 55 amino acids long, and named *cgMolluscidin*, has been recently purified from the acidified gill extract of *C. gigas*. This sequence has no homology with any known AMPs and showed a strong antimicrobial effect against both Gram-positive bacteria (*Bacillus subtilis*, *Micrococcus luteus* and *S. aureus*) and Gram-negative bacteria (*E. coli*, *Salmonella enterica* and *Vibrio parahaemolyticus*). Finally, the last group of AMPs, described in the present review, has been identified in scallops, mainly from *A. irradians* (AiBD) and *A. purpuratus* (Ap). The scallop AiBD consisted of 531 nucleotides and produced a peptide of 122 amino acids. Recombinant AiBD was able to block the growth of both Gram-positive and Gram-negative bacteria and also presented a strong fungicidal power [253]. AP was isolated from *A. purpuratus* hemocytes, consists of 47 residues and shares partial homology with reported effective AMPs. A modified version of 30 residues designed to increase hydrophobicity and cationicity was used in antimicrobial experiments and showed an excellent activity against *Saprolegnia* sp., a parasitic pathogen fungus that attacks the culture of fish in different stages of their life, from the egg stage to grown-up animal [254].

In order to summarize all those peptides in an intelligible overview, we generated a table with their detailed characteristics (Table 4).

Table 4. AMPs isolated from marine bivalves and main characteristics.

Name	Source	Sequence	Length	Net Charge	% Hydrophobic Residues	Structure	Antimicrobial Activity	Reference
Defensin MGD-1	*Mytilus galloprovincialis*	GFGCPNNYQCHRHCKSIPCGRCGGY CGGWHRLPCTCYRCG	39	5	30	Combined helix and β-sheet	Gram+	[224]
Defensin MGD-2	*Mytilus galloprovincialis*	GFGCPNNYACHQHCKSIRGCYGGY CAGWFRLRCTCYRCG	39	5	38	* Combined helix and β-sheet	Gram+ and Gram−	[227]
Mytilin A	*Mytilus edulis*	GCASRCKAKCAGRRCKGWASASFR GRCYCKCFRC	34	10	47	* Combined helix and β-sheet	Gram+ and Gram−	[225]
Mytilin B	*Mytilus edulis*	SCASRCKGHCRARRCGYYVSVLYRG RCYCKCLRC	34	9	41	Combined helix and β-sheet	Gram+ and Gram−, antiviral	[225]
Myticin A	*Mytilus galloprovincialis*	HSHACTSYWCGKFCGTASCTHYLC RVLHPGKMCACVHCSR	40	4	45	* Combined helix and β-sheet	Gram+ and Gram−; antifungal	[226]
Myticin B	*Mytilus galloprovincialis*	HPHVCTSYYCSKFCGTAGCTRYGCR NLHRGKLCFCLHCSR	40	6	37	* Combined helix and β-sheet	Gram+ and Gram−; antifungal	[226]
Myticin C	*Mytilus galloprovincialis*	QSVACTSYYCSKFCGSAGCSLYGCY LLHPGKICYCLHCSR	40	3	35	* Combined helix and β-sheet	Gram+ and Gram−; antifungal	[239]
Mytimycin	*Mytilus galloprovincialis*	MSLVLRMTLLFVVCCVVIGMSNAA CCHKPFWKHCWDCTAGTPYCGYRS CNIFGCGCTCRTEPYGKSCYERGNR CRCYTDKRKRRSLSFEDISPNIKFAGL DINSDGLIEQFEFIKALEQMDIIDNTT MFHHWSIMDEDKDGTITILEEFDK	150	−2	41	* Combined helix and β-sheet	Antifungal	[225]
Mytimacin	*Mytilus galloprovincialis*	MGYIGLCGVLSLSLLMLLQIPTSDA NVLGDCWEDWSRCTRQTNWFTNI AWQSCPNRCKCQGHAGGNCIQVR SNCFLWRNKRWMCNCYGRRSGPK PGWCGF	101	7	43	* Combined helix and β-sheet	Gram+ and Gram−	[243]
Big-Defensin	*Mytilus galloprovincialis*	MNRKAILCVLYATLLIIPAPILGRVV AKKKEEKRYAAVYPIAAYAGMTVS LPVFLAIVAAYGAWTVARYHIRSRS RSSHNSHNCANNRGWCRPNCFRR EYHDWYHSDTCCSYKCCRYR	119	14	42	* Combined helix and β-sheet	Gram+ and Gram−	[243]
Myticusin-1	*Mytilus coruscus*	TDHQMAQSACIGVSQDNAYASAIP RDCHGGKTCEGICADATATMDRYS DTGGPLSIARCVNAFFHFYKRRGEEN VSYKPFVVSWKYGVAGCFYTHCGP NFCCCIS	104	0	39	* Combined helix and β-sheet	Gram+ and Gram−, antifungal	[241]

Table 4. *Cont.*

Name	Source	Sequence	Length	Net Charge	% Hydrophobic Residues	Structure	Antimicrobial Activity	Reference
VpBD	*Venerupis philippinarum*	LCLDQKPEMEPFRKDAQQALEPSRQ RRWLHRRCLSGRGFHCRAICSIFEEPV RGNIDCYFGYNCCRRMFSHYRTS	74	5	36	* Helix	Gram+ and Gram−	[244]
MCdef	*Ruditapes philippinarum*	GFGCPNDYSCSNHCRDSIGCRGGYC KYQLICTCYGCKKRRSIQE	44	4	29	* Combined helix and β-sheet	Gram+ and Gram−	[133]
VpDef	*Venerupis philippinarum*	GFGCPEDEYECHNHCKNSVGCRGG YCDAGTLRQRCTCYGCNQKGRSIQE	49	0	26	* Combined helix and β-sheet	Gram+ and Gram−	[162]
Sb-BDef1	*Scapharca broughtonii*	MTHKIVLCCIYLLLSTSFILSKHLPEE RKQKKQVLLAAGAGVALSELLGPV LVGAGTLAGAALLNQAVSSNRWVI PCANNRGWCRTDCHRGEHIDDYHS DICHSGYKCCRY	111	3	45	* Combined helix and β-sheet	Gram−	[245]
Ap	*Argopecten purpuratus*	TYMPVEEGEYTVNISYADQPKKNSPF TAKKQPGPKVDLSGVKAYGPG	47	1	25	* Polyproline rich β-sheet	Gram+, antifungal	[254]
AiBD	*Argopecten irradians*	MTRPSLVRCYSLFFTALIVMAIICPA WSEIIPKSRKKRAIPIAYVGMAVAP QVFRWLVRAYGAAAVTAAGVTLRR VINRSRSNDNHSCYGNRGWCRSSCR SYEREYRGGNLGVCCSYKCCVT	122	14	44	* Combined helix and β-sheet	Gram+ and Gram−, antifungal	[253]
AOD	*Crassostrea virginica*	GFGCPWNRYQCHSHCRSIGCRLGGY CAGSLRLTCTCYRS	38	5	34	* Combined helix and β-sheet	Gram+ and Gram−	[247]
Cg-Prp	*Crassostrea gigas*	ILENLLARSTNEDREGSIFDTGPIRRP KPRPRPPEG	37	2	21	* Proline-rich peptide	Synergistic antimicrobial activity with Cg-Def	[248]
cgMolluscidin	*Crassostrea gigas*	AATAKKGAKKADAPAKPKKATKP KSPKKAAKKAGAKKGVKRAGKKG AKKTTKAKK	55	23	29	* Helix	Gram+ and Gram−	[247]
Cg-Defh1	*Crassostrea gigas*	GFGCPRDQYKCNSHCQSIGCRAGY CDAVTLWLRCTCTDCNGKK	43	3	37	Combined helix and β-sheet	Gram+ and Gram−	[249]
Cg-Defh2	*Crassostrea gigas*	GFGCPGDQYECNRHCRSIGCRAGY CDAVTLWLRCTCTGCSGKK	43	3	37	* Combined helix and β-sheet	Gram+ and Gram−	[229]

* The predicted structures were designed using AIDA (http://ffas.burnham.org/AIDA).

6. Applications of AMPs in Medicine and in Preventing Diseases of Aquatic Animals

In the present review, we first outlined the infectious diseases of marine bivalve and their defense mechanisms including the classification and mode of action of AMPs. In the remaining and concluding part of the review, we will focus on the potential utilization of AMPs as substitutes for antibiotics in aquaculture and in the medical field.

The economic development of several countries relies on the aquaculture of mollusks since a considerable amount of shellfish (mainly mollusks) for human consumption is produced by aquaculture. Clam, scallop, oyster and mussel farming represents a noticeable share of the aquaculture market worldwide, accounting for more than 50% of the shellfish present on the global market. The increasing demand for seafood, including bivalves, will surely extend in the forthcoming years, and higher capability productions will need to be accomplished to meet this demand. The major obstacle to the thriving of the aquaculture industry is posed by the emergence and spread of many infectious diseases, which are exacerbated by the densely-populated culture conditions in limited space. Future research needs to include the development of novel methods to control diseases in hatcheries in order to minimize the occurrences of mass mortalities caused by either obligate or opportunistic pathogens. Research efforts are mandatory to explore inexpensive and effective treatments when diseases occur in a hatchery situation to avoid production losses. Moreover, safer compounds are highly desired considering the environmental risks and that bivalve production is mainly devoted to human consumption. The possibilities to contain the spread of diseases in mollusks is not facilitated by the fact that there is limited information to fully understand the physiology of marine bivalves, in particular concerning their immune defense system. A detailed description of the immune defense mechanisms of mollusks has been provided in the first part of the review with the intent to foster the characterization of immune effectors to provide new understanding into healthiness and disease management in mollusk aquaculture. Since the introduction of antibiotics for the treatment of infectious diseases, they have been widely used in medicine and animal breeding, as well as in aquaculture with successful regimens. However, due to the overuse and misuse of antibiotics, the resistance of bacteria to antibiotics has dramatically increased, posing new challenges to human health and to the sustainability of the aquaculture industry. Moreover, the worldwide threat of a rapid increase in pathogenic multidrug-resistant (MDR) bacteria is paralleled by the environmental risks with antibiotic compounds being distributed throughout the environment. The quest for novel compounds with antibacterial activity that could overcome resistance emergence events is, therefore, an urgent societal challenge. Nature-derived AMPs are regarded as convenient templates for the development of substitutes to traditional antibiotic [255], and bivalve AMPs have been shown to be structurally different from their analogue peptides derived from the terrestrial habitat and usually present novel and unexploited structures [223,256]. AMPs' antimicrobial activity relies on the early electrostatic interactions with the negatively-charged surface of the bacteria, therefore, free ions produced by the high salt concentrations in the surrounding medium, typical of some illnesses, could efficiently decrease interaction and antimicrobial activity. In general, marine AMPs have evolved to easily adapt to the high salt concentration of seawater, and probably, this has been obtained by the substitution of lysines with arginines. Therefore, marine bivalves are a hopeful reservoir of novel bioactive molecules for the development of alternative antimicrobials. In fact, AMPs can be regarded as encouraging candidates for selecting new and more environmentally-friendly antimicrobials. Notwithstanding the abundance of scientific knowledge on their activities in vitro, major challenges need to be outflanked to allow for their clinical application. The major obstacles to be considered are: (1) rapid degradation by proteases; (2) uncertainty of the antimicrobial activities under physiological salt, pH and serum conditions; (3) poor oral availability; (4) laborious routes of administration; (5) burdensome transportation across cell membranes; (6) non-selective receptor binding; and (7) costs associated with their production [223]. Excessive amounts of common antibiotics are used in aquaculture in some countries for both therapeutic and prophylactic purposes [257–259]. Moreover, veterinary antimicrobials also include compounds used clinically in human medicine [260]. Current knowledge regarding the genetic aspects of antimicrobial resistance in aquatic bacteria is highly suggestive of the possibility that antibiotics used in fish farming are likely to select antimicrobial-resistant bacteria in aquacultural environments and of their subsequent diffusion to terrestrial counterpart [261–263]. Therefore, in order to safeguard public health, it is of paramount importance to adopt novel methodologies used in aquaculture. The use of AMPs is one of the main strategies currently under deep investigation. It is, though, imperative to devote future efforts to design improved antimicrobial molecules with a broad spectrum of activity against a wide array of pathogenic microorganisms through the modification of native AMPs in order to achieve more selective and efficient drugs that could substitute conventional antibiotics both in aquaculture and in clinical human medicine.

One of the most awaited goals is the use of rational design to produce AMPs with improved characteristics, such as: (1) stronger antibacterial activity, (2) lower cytotoxicity and (3) ease of production on an industrial scale for obtaining a marketable drug. Native AMPs can be used as templates for the design of new antibacterial agents through peptidomimetics, where only structural key elements of the native peptide are conserved to provide a scaffold able to preserve the AMP characteristic of easy interaction with the biological target and produce an enhanced antimicrobial biological response [264]. Domains responsible for activity are, therefore, analyzed in detail and modified to obtain functionally-improved AMPs. Several authors have described the modification of primary AMP sequences to enhance their effectiveness and stability in order to obtain promising lead compounds for the development of therapeutic agents [265]. Attempt to increase AMP activities have generally involved the methodical change of amino acid residues or alternative chemical alterations, which permit the achievement of improved activities, such as: chemical modification of terminal ends of peptides [266], development of analogues containing unnatural amino acids [267], β-peptides, shortening of the native sequence, modifications of their amphipathic character, cyclization [234], hybrid peptide-peptidomimetic structures, lipidation, etc. [268]. One of the reasons for the high interest in AMPs derived from marine peptides is their ability to sustain physiological salt concentration and protease activity. In fact, a foremost role for the improvement of AMPs therapeutic impact is represented by the suitable harmonization of their hydrophobicity, amphipathicity and positive charge [269,270]. Peptides with sufficient positive charge can be modified to decrease interactions with mammalian cells, while favoring the preferential binding to bacterial cell membranes by reducing the overall hydrophobicity of the molecule [271,272]. Exploiting the variety of post-translational modifications displayed by marine AMPs could be instrumental in the design of AMPs with enhanced stability and efficacy, for therapeutic utilization in human medicine [273]. For example, the high salinity (up to 600 mM) of the marine habitat may have forged marine AMPs to be naturally endowed with a sharpened salt resistance, allowing them to conserve strong antibacterial effectiveness in relatively high-salt environments, such as in saliva, gastrointestinal fluid, serum or other body fluids [231,232]. Understanding the chemical propriety backing this salt independent activity could subside in the construction of novel AMPs that could be better endowed for facing pathogens notwithstanding the wide range of salt concentrations that could be encountered. For example, human beta defensin-1 (hBD-1) being unable to inhibit *P. aeruginosa* due to a 120 mM concentration of NaCl in the lungs of cystic fibrosis patients [274] has been engineered by constructing a chimera with β-defensin-3 (hBD-3), which shows antibacterial activity also at high salt concentrations [231,232]. The C-terminal domain of hBD-3 presents an abundance of arginine residues considered to be involved in the activity at high ionic conditions. Therefore, marine AMPs, dictating many of the universal rules featuring the ability of an AMP to inhibit microorganisms under physiological salt concentrations (120–150 mM), are of paramount importance for indicating novel strategies for the improvement of native AMP sequences [275].

7. Conclusions

Bivalve farming has recently reached a large portion of the fish market worldwide; therefore, from both a human health perspective and for reducing economic losses, huge efforts are devoted to the improvement of the microbiological quality of the product. We are awaiting a comprehensive understanding and knowledge of all of the infectious diseases that could affect bivalve mollusks, and insights on their defense mechanisms are only recently being deeply investigated. Several studies on bivalve immunity have expanded our understanding of the morpho-functional properties of circulating hemocytes and the humoral defense factors, but most of the patterns for microbe recognition and downhill immune pathways activation are still to be explored. At the same time, in the last decade, a large amount of data on mollusk-derived AMPs has been gathered, to allow the beginning of the initial studies to attempt to modify these lead compounds in search of a wider applicability of their antimicrobial properties in both aquaculture and human medicine.

Acknowledgments: We thank Albert Bonaminio (M. Sci.) for linguistic editing. This work was supported by: Regional funds "Pepmar".

Author Contributions: All Authors contributed equally.

Conflicts of Interest: The authors declare no conflict of interest.

References

1. Santos, A.; Hauser-Davis, R.A.; Santos, M.J.; De Simone, S.G. Potentially toxic filamentous fungi associated to the economically important *Nodipecten nodosus* (Linnaeus, 1758) scallop farmed in southeastern Rio De janeiro, Brazil. *Mar. Pollut. Bull.* **2016**, *115*, 75–79. [CrossRef] [PubMed]

2. Carver, C.E.; Thériault, I.; Mallet, A.L. Infection of cultured eastern oysters *Crassostrea virginica* by the boring sponge *Cliona celata*, with emphasis on sponge life history and mitigation strategies. *J. Shellfish Res.* **2010**, *29*, 905–915. [CrossRef]

3. Gagne, N.; Cochennec, N.; Stephenson, M.; McGladdery, S.; Meyer, G.R.; Bower, S.M. First report of a Mikrocytos-like parasite in European oysters *Ostrea edulis* from Canada after transport and quarantine in France. *Dis. Aquat. Organ.* **2008**, *80*, 27–35. [CrossRef] [PubMed]

4. Guo, X.; Ford, S.E. Infectious diseases of marine molluscs and host responses as revealed by genomic tools. *Philos. Trans. R. Soc. Lond. B Biol. Sci.* **2016**, *371*. [CrossRef] [PubMed]

5. Sirisinha, S. Evolutionary insights into the origin of innate and adaptive immune systems: Different shades of grey. *Asian Pac. J. Allergy Immunol.* **2014**, *32*, 3–15. [PubMed]

6. Ramilo, A.; Gonzalez, M.; Carballal, M.J.; Darriba, S.; Abollo, E.; Villalba, A. Oyster parasites *Bonamia Ostreae* and *B. exitiosa* co-occur in Galicia (NW Spain): Spatial distribution and infection dynamics. *Dis. Aquat. Organ.* **2014**, *110*, 123–133. [CrossRef] [PubMed]

7. Suffredini, E.; Lanni, L.; Arcangeli, G.; Pepe, T.; Mazzette, R.; Ciccaglioni, G.; Croci, L. Qualitative and quantitative assessment of viral contamination in bivalve molluscs harvested in Italy. *Int. J. Food Microbiol.* **2014**, *184*, 21–26. [CrossRef] [PubMed]

8. Arzul, I.; Langlade, A.; Chollet, B.; Robert, M.; Ferrand, S.; Omnes, E.; Lerond, S.; Couraleau, Y.; Joly, J.P.; Francois, C.; et al. Can the protozoan parasite *Bonamia ostreae* infect larvae of flat oysters *Ostrea edulis*? *Vet. Parasitol.* **2011**, *179*, 69–76. [CrossRef] [PubMed]

9. Batista, F.M.; Arzul, I.; Pepin, J.F.; Ruano, F.; Friedman, C.S.; Boudry, P.; Renault, T. Detection of ostreid herpesvirus 1 DNA by PCR in bivalve molluscs: A critical review. *J. Virol. Methods* **2007**, *139*, 1–11. [CrossRef] [PubMed]

10. Bai, C.; Gao, W.; Wang, C.; Yu, T.; Zhang, T.; Qiu, Z.; Wang, Q.; Huang, J. Identification and characterization of ostreid herpesvirus 1 associated with massive mortalities of *Scapharca broughtonii* broodstocks in China. *Dis. Aquat. Organ.* **2016**, *118*, 65–75. [CrossRef] [PubMed]

11. Farley, C.A.; Banfield, W.G.; Kasnic, G., Jr.; Foster, W.S. Oyster herpes-type virus. *Science* **1972**, *178*, 759–760. [CrossRef] [PubMed]

12. Segarra, A.; Pepin, J.F.; Arzul, I.; Morga, B.; Faury, N.; Renault, T. Detection and description of a particular *Ostreid herpesvirus 1* genotype associated with massive mortality outbreaks of Pacific oysters, *Crassostrea gigas*, in France in 2008. *Virus Res.* **2010**, *153*, 92–99. [CrossRef] [PubMed]

13. Lynch, S.A.; Carlsson, J.; Reilly, A.O.; Cotter, E.; Culloty, S.C. A previously undescribed Ostreid Herpes Virus 1 (OsHV-1) genotype detected in the Pacific oyster, *Crassostrea gigas*, in Ireland. *Parasitology* **2012**, *139*, 1526–1532. [CrossRef] [PubMed]

14. Bai, C.; Wang, C.; Xia, J.; Sun, H.; Zhang, S.; Huang, J. Emerging and endemic types of *Ostreid herpesvirus 1* were detected in bivalves in China. *J. Invertebr. Pathol.* **2015**, *124*, 98–106. [CrossRef] [PubMed]

15. Hwang, J.Y.; Park, J.J.; Yu, H.J.; Hur, Y.B.; Arzul, I.; Couraleau, Y.; Park, M.A. Ostreid herpesvirus 1 infection in farmed Pacific oyster larvae *Crassostrea gigas* (Thunberg) in Korea. *J. Fish Dis.* **2013**, *36*, 969–972. [PubMed]

16. Mortensen, S.; Strand, A.; Bodvin, T.; Alfjorden, A.; Skar, C.K.; Jelmert, A.; Aspan, A.; Saelemyr, L.; Naustvoll, L.J.; Albretsen, J. Summer mortalities and detection of ostreid herpesvirus microvariant in Pacific oyster *Crassostrea gigas* in Sweden and Norway. *Dis. Aquat. Organ.* **2016**, *117*, 171–176. [CrossRef] [PubMed]

17. Pernet, F.; Barret, J.; le Gall, P.; Corporeau, C.; Dégremont, L.; Lagarde, F.; Pépin, J.F.; Keck, N. Mass mortalities of Pacific oysters *Crassostrea gigas* reflect infectious diseases and vary with farming practices in the Mediterranean Thau lagoon, France. *Aquacult. Environ. Interact.* **2012**, *2*, 215–237. [CrossRef]

18. Da Silva, P.M.; Renault, T.; Fuentes, J.; Villalba, A. Herpesvirus infection in European flat oysters *Ostrea edulis* obtained from brood stocks of various geographic origins and grown in Galicia (NW Spain). *Dis. Aquat. Organ.* **2008**, *78*, 181–188. [CrossRef] [PubMed]

19. Leibovitz, L.; Elston, R.; Lipovsky, V.P.; Donaldson, J. A new disease of larval Pacific oysters (*Crassostrea gigas*). *J. World Aquacult. Soc.* **1978**, *9*, 603–615. [CrossRef]

20. Kueh, C.S.; Chan, K.Y. Bacteria in bivalve shellfish with special reference to the oyster. *J. Appl. Bacteriol.* **1985**, *59*, 41–47. [CrossRef] [PubMed]

21. Bidault, A.; Richard, G.G.; le Bris, C.; Paillard, C. Development of a Taqman real-time PCR assay for rapid detection and quantification of *Vibrio tapetis* in extrapallial fluids of clams. *PeerJ* **2015**, *3*, e1484. [CrossRef] [PubMed]

22. Garnier, M.; Labreuche, Y.; Nicolas, J.L. Molecular and phenotypic characterization of *Vibrio aestuarianus* subsp. *francensis* subsp. nov., a pathogen of the oyster *Crassostrea gigas*. *Syst. Appl. Microbiol.* **2008**, *31*, 358–365. [PubMed]

23. Prado, S.; Dubert, J.; da Costa, F.; Martinez-Patino, D.; Barja, J.L. Vibrios in hatchery cultures of the razor clam, *Solen marginatus* (Pulteney). *J. Fish Dis.* **2014**, *37*, 209–217. [CrossRef] [PubMed]

24. Biel, F.M.; Allen, F.A.; Hase, C.C. Autolysis in *Vibrio tubiashii* and *Vibrio coralliilyticus*. *Can. J. Microbiol.* **2014**, *60*, 57–63. [CrossRef] [PubMed]

25. Kwan, T.N.; Bolch, C.J. Genetic diversity of culturable *Vibrio* in an Australian blue mussel *Mytilus galloprovincialis* hatchery. *Dis. Aquat. Organ.* **2015**, *116*, 37–46. [CrossRef] [PubMed]

26. Sugumar, G.; Nakai, T.; Hirata, Y.; Matsubara, D.; Muroga, K. *Vibrio splendidus* biovar II as the causative agent of bacillary necrosis of Japanese oyster *Crassostrea gigas* larvae. *Dis. Aquat. Organ.* **1998**, *33*, 111–118. [CrossRef] [PubMed]

27. Valerio, E.; Chaves, S.; Tenreiro, R. Diversity and impact of prokaryotic toxins on aquatic environments: A review. *Toxins (Basel)* **2010**, *2*, 2359–2410. [CrossRef] [PubMed]

28. Gomez-Leon, J.; Villamill, L.; Salger, S.A.; Sallum, R.H.; Remacha-Trivino, A.; Leavitt, D.F.; Gomez-Chiarri, M. Survival of eastern oysters *Crassostrea virginica* from three lines following experimental challenge with bacterial pathogens. *Dis. Aquat. Organ.* **2008**, *79*, 95–105. [CrossRef] [PubMed]

29. Balboa, S.; Romalde, J.L. Multilocus sequence analysis of *Vibrio tapetis*, the causative agent of Brown Ring Disease: Description of *Vibrio tapetis* subsp. *britannicus* subsp. Nov. *Syst. Appl. Microbiol.* **2013**, *36*, 183–187. [CrossRef] [PubMed]

30. Paillard, C.; Korsnes, K.; Le Chevalier, P.; Le Boulay, C.; Harkestad, L.; Eriksen, A.G.; Willassen, E.; Bergh, O.; Bovo, C.; Skar, C.; et al. *Vibrio tapetis*-like strain isolated from introduced Manila clams *Ruditapes philippinarum* showing symptoms of brown ring disease in Norway. *Dis. Aquat. Organ.* **2008**, *81*, 153–161. [CrossRef] [PubMed]

31. Paillard, C. A short-review of brown ring disease, a vibriosis affecting clams, *Ruditapes philippinarum* and *Ruditapes decussatus*. *Aquat. Living Resour.* **2004**, *17*, 467–475. [CrossRef]

32. Park, K.I.; Yang, H.S.; Kang, H.S.; Cho, M.; Park, K.J.; Choi, K.S. Isolation and identification of *Perkinsus olseni* from feces and marine sediment using immunological and molecular techniques. *J. Invertebr. Pathol.* **2010**, *105*, 261–269. [CrossRef] [PubMed]

33. Ramilo, A.; Iglesias, D.; Abollo, E.; Gonzalez, M.; Darriba, S.; Villalba, A. Infection of Manila clams *Ruditapes philippinarum* from Galicia (NW Spain) with a *Mikrocytos*-like parasite. *Dis. Aquat. Organ.* **2014**, *110*, 71–79. [CrossRef] [PubMed]

34. Allam, B.; Paillard, C.; Ford, S.E. Pathogenicity of *Vibrio tapetis*, the etiological agent of brown ring disease in clams. *Dis. Aquat. Organ.* **2002**, *48*, 221–231. [CrossRef] [PubMed]

35. Boettcher, K.J.; Barber, B.J.; Singer, J.T. Additional evidence that juvenile oyster disease is caused by a member of the *Roseobacter* group and colonization of nonaffected animals by *Stappia stellulata*-like strains. *Appl. Environ. Microbiol.* **2000**, *66*, 3924–3930. [CrossRef] [PubMed]

36. Kessner, L.; Spinard, E.; Gomez-Chiarri, M.; Rowley, D.C.; Nelson, D.R. Draft genome sequence of *Aliiroseovarius crassostreae* CV919–312, the causative agent of *Roseovarius* oyster disease (formerly juvenile oyster disease). *Genome Announc.* **2016**, *4*. [CrossRef] [PubMed]

37. Ford, S.E.; Borrero, F.J. Epizootiology and pathology of juvenile oyster disease in the Eastern oyster, *Crassostrea virginica*. *J. Invertebr. Pathol.* **2001**, *78*, 141–154. [CrossRef] [PubMed]

38. Ford, S.E. *Roseovarius Oyster Disease (ROD) Caused by Roseovarius crassostreae*; ICES Identification Leaflets for Diseases and Parasites of Fish and Shellfish; ICES: Copenhagen, Denmark, 2011.

39. Friedman, C.S.; Cloney, D.F.; Manzer, D.; Hedrick, R.P. Haplosporidiosis of the Pacific oyster, *Crassostrea gigas*. *J. Invertebr. Pathol.* **1991**, *58*, 367–372. [CrossRef]

40. Carella, F.; Carrasco, N.; Andree, K.B.; Lacuesta, B.; Furones, D.; De Vico, G. Nocardiosis in Mediterranean bivalves: First detection of *Nocardia crassostreae* in a new host *Mytilus galloprovincialis* and in *Ostrea edulis* from the Gulf of Naples (Italy). *J. Invertebr. Pathol.* **2013**, *114*, 324–328. [CrossRef] [PubMed]

41. Friedman, C.S.; Beattie, J.H.; Elston, R.A.; Hedrick, R.P. Investigation of the relationship between the presence of a Gram-positive bacterial infection and summer mortality of the Pacific oyster, *Crassostrea gigas* Thunberg. *Aquaculture* **1991**, *94*, 1–15. [CrossRef]

42. Queiroga, F.R.; Marques-Santos, L.F.; De Medeiros, I.A.; Da Silva, P.M. Effects of salinity and temperature on in vitro cell cycle and proliferation of *Perkinsus marinus* from Brazil. *Parasitology* **2016**, *143*, 475–487. [CrossRef] [PubMed]

43. Fernandez Robledo, J.A.; Vasta, G.R.; Record, N.R. Protozoan parasites of bivalve molluscs: Literature follows culture. *PLoS ONE* **2014**, *9*, e100872. [CrossRef] [PubMed]

44. Carnegie, R.B.; Meyer, G.R.; Blackbourn, J.; Cochennec-Laureau, N.; Berthe, F.C.; Bower, S.M. Molecular detection of the oyster parasite *Mikrocytos mackini*, and a preliminary phylogenetic analysis. *Dis. Aquat. Organ.* **2003**, *54*, 219–227. [CrossRef] [PubMed]

45. Feng, C.; Lin, X.; Wang, F.; Zhang, Y.; Lv, J.; Wang, C.; Deng, J.; Mei, L.; Wu, S.; Li, H. Detection and characterization of *Bonamia ostreae* in *Ostrea edulis* imported to China. *Dis. Aquat. Organ.* **2013**, *106*, 85–91. [CrossRef] [PubMed]

46. Lane, H.S.; Webb, S.C.; Duncan, J. *Bonamia ostreae* in the New Zealand oyster *Ostrea chilensis*: A new host and geographic record for this haplosporidian parasite. *Dis. Aquat. Organ.* **2016**, *118*, 55–63. [CrossRef] [PubMed]

47. Hine, P.M.; Cochennec-Laureau, N.; Berthe, F.C. *Bonamia exitiosus* n. sp. (Haplosporidia) infecting flat oysters *Ostrea chilensis* in New Zealand. *Dis. Aquat. Organ.* **2001**, *47*, 63–72. [CrossRef] [PubMed]

48. Corbeil, S.; Arzul, I.; Robert, M.; Berthe, F.C.; Besnard-Cochennec, N.; Crane, M.S. Molecular characterisation of an Australian isolate of *Bonamia exitiosa*. *Dis. Aquat. Organ.* **2006**, *71*, 81–85. [CrossRef] [PubMed]

49. Dungan, C.F.; Carnegie, R.B.; Hill, K.M.; McCollough, C.B.; Laramore, S.E.; Kelly, C.J.; Stokes, N.A.; Scarpa, J. Diseases of oysters *Crassostrea ariakensis* and *C. virginica* reared in ambient waters from the Choptank River, Maryland and the Indian River Lagoon, Florida. *Dis. Aquat. Organ.* **2012**, *101*, 173–183. [CrossRef] [PubMed]

50. Hill, K.M.; Stokes, N.A.; Webb, S.C.; Hine, P.M.; Kroeck, M.A.; Moore, J.D.; Morley, M.S.; Reece, K.S.; Burreson, E.M.; Carnegie, R.B. Phylogenetics of *Bonamia* parasites based on small subunit and internal transcribed spacer region ribosomal DNA sequence data. *Dis. Aquat. Organ.* **2014**, *110*, 33–54. [CrossRef] [PubMed]

51. Narcisi, V.; Arzul, I.; Cargini, D.; Mosca, F.; Calzetta, A.; Traversa, D.; Robert, M.; Joly, J.P.; Chollet, B.; Renault, T.; et al. Detection of *Bonamia ostreae* and *B. exitiosa* (Haplosporidia) in *Ostrea edulis* from the Adriatic Sea (Italy). *Dis. Aquat. Organ.* **2010**, *89*, 79–85. [CrossRef] [PubMed]

52. Hill, K.M.; Carnegie, R.B.; Aloui-Bejaoui, N.; Gharsalli, R.E.; White, D.M.; Stokes, N.A.; Burreson, E.M. Observation of a *Bonamia* sp. infecting the oyster *Ostrea stentina* in Tunisia, and a consideration of its phylogenetic affinities. *J. Invertebr. Pathol.* **2010**, *103*, 179–185. [CrossRef] [PubMed]

53. Carrasco, N.; Villalba, A.; Andree, K.B.; Engelsma, M.Y.; Lacuesta, B.; Ramilo, A.; Gairin, I.; Furones, M.D. *Bonamia exitiosa* (Haplosporidia) observed infecting the European flat oyster *Ostrea edulis* cultured on the Spanish Mediterranean coast. *J. Invertebr. Pathol.* **2012**, *110*, 307–313. [CrossRef] [PubMed]

54. Longshaw, M.; Stone, D.M.; Wood, G.; Green, M.J.; White, P. Detection of *Bonamia exitiosa* (Haplosporidia) in European flat oysters *Ostrea edulis* cultivated in mainland Britain. *Dis. Aquat. Organ.* **2013**, *106*, 173–179. [CrossRef] [PubMed]

55. Batista, F.M.; Lopez-Sanmartin, M.; Grade, A.; Navas, J.I.; Ruano, F. Detection of *Bonamia exitiosa* in the European flat oyster *Ostrea edulis* in Southern Portugal. *J. Fish Dis.* **2016**, *39*, 607–611. [CrossRef] [PubMed]

56. Carnegie, R.B.; Burreson, E.M.; Hine, P.M.; Stokes, N.A.; Audemard, C.; Bishop, M.J.; Peterson, C.H. *Bonamia perspora* n. sp. (Haplosporidia), a parasite of the oyster *Ostreola equestris*, is the first *Bonamia* species known to produce spores. *J. Eukaryot. Microbiol.* **2006**, *53*, 232–245. [CrossRef] [PubMed]

57. Lynch, S.A.; Armitage, D.V.; Coughlan, J.; Mulcahy, M.F.; Culloty, S.C. Investigating the possible role of benthic macroinvertebrates and zooplankton in the life cycle of the haplosporidian *Bonamia ostreae*. *Exp. Parasitol.* **2007**, *115*, 359–368. [CrossRef] [PubMed]

58. Cochennec-Laureau, N.; Auffret, M.; Renault, T.; Langlade, A. Changes in circulating and tissue-infiltrating hemocyte parameters of European flat oysters, *Ostrea edulis*, naturally infected with *Bonamia ostreae*. *J. Invertebr. Pathol.* **2003**, *83*, 23–30. [CrossRef]

59. Kleeman, S.N.; Adlard, R.D. Molecular detection of *Marteilia sydneyi*, pathogen of Sydney rock oysters. *Dis. Aquat. Organ.* **2000**, *40*, 137–146. [CrossRef] [PubMed]

60. Schneider, O.; Sereti, V.; Machiels, M.A.; Eding, E.H.; Verreth, J.A. The potential of producing heterotrophic bacteria biomass on aquaculture waste. *Water Res.* **2006**, *40*, 2684–2694. [CrossRef] [PubMed]

61. Gombac, M.; Kusar, D.; Ocepek, M.; Pogacnik, M.; Arzul, I.; Couraleau, Y.; Jencic, V. Marteiliosis in mussels: A rare disease? *J. Fish Dis.* **2014**, *37*, 805–814. [CrossRef] [PubMed]

62. Carella, F.; Aceto, S.; Marrone, R.; Maiolino, P.; De Vico, G. *Marteilia refringens* infection in cultured and natural beds of mussels (*Mytilus galloprovincialis*) along the Campanian coast (Tirrenian sea, South of Italy). *Bull. Eur. Ass. Fish Pathol.* **2010**, *30*, 189–196.

63. Tiscar, P.G.; Chagot, D.; Tempesta, M.; Marsilio, F.; Buonavoglia, D. Presenza di *Marteilia* sp. in mitili (*Mytilus galloprovincialis*, Lmk) allevati in Puglia. *Boll. Soc. Ital. Patol. Ittica* 1993, *12*, 40–45.

64. Roubal, F.R.; Masel, J.; Lester, R.J.G. Studies on *Marteilia sydneyi*, agent of QX disease in the Sydney rock oyster, *Saccostrea commercialis*, with implications for its life cycle. *Aust. J. Mar. Freshw. Res.* **1989**, *40*, 155–167. [CrossRef]

65. Rubio, A.; Frances, J.; Coad, P.; Stubbs, J.; Guise, K. The onset and termination of the Qx disease window of infection in Sydney rock oyster (*Saccostrea glomerata*) cultivated in the Hawkesbury River, NSW, Australia. *J. Shellfish Res.* **2013**, *32*, 483–496. [CrossRef]

66. Peruzzi, L.; Gianoglio, B.; Porcellini, G.; Conti, G.; Amore, A.; Coppo, R. Neonatal chronic kidney failure associated with cyclo-oxygenase-2 inhibitors administered during pregnancy. *Minerva Urol. Nefrol.* **2001**, *53*, 113–116. [PubMed]

67. Audemard, C.; Barnaud, A.; Collins, C.M.; Le Roux, F.; Sauriau, P.; Coustau, C.; Blachier, P.; Berthe, F.C. Claire ponds as an experimental model for *Marteilia refringens* life-cycle studies: New perspectives. *J. Exp. Mar. Biol. Ecol.* **2001**, *257*, 87–108. [CrossRef]

68. Ford, S.E. *Dermo Disease of Oysters Caused by Perkinsus marinus*; Ford, S.E., Ed.; ICES Identification Leaflets for Diseases and Parasites of Fish and Shellfish; ICES: Copenhagen, Denmark, 2011.

69. Mackin, J.G.; Owen, H.M.; Collier, A. Preliminary note on the occurrence of a new protistan parasite, *Dermocystidium marinum* n. sp. in *Crassostrea virginica* (Gmelin). *Science* 1950, *111*, 328–329. [CrossRef] [PubMed]

70. Remacha-Trivino, A.; Borsay-Horowitz, D.; Dungan, C.; Gual-Arnau, X.; Gomez-Leon, J.; Villamil, L.; Gomez-Chiarri, M. Numerical quantification of *Perkinsus marinus* in the American oyster *Crassostrea virginica* (Gmelin, 1791) (Mollusca: Bivalvia) by modern stereology. *J. Parasitol.* **2008**, *94*, 125–136. [CrossRef] [PubMed]

71. Caceres-Martinez, J.; Madero-Lopez, L.H.; Padilla-Lardizabal, G.; Vasquez-Yeomans, R. Epizootiology of *Perkinsus marinus*, parasite of the pleasure oyster *Crassostrea corteziensis*, in the Pacific coast of Mexico. *J. Invertebr. Pathol.* **2016**, *139*, 12–18. [CrossRef] [PubMed]

72. Enriquez-Espinoza, T.L.; Grijalva-Chon, J.M.; Castro-Longoria, R.; Ramos-Paredes, J. *Perkinsus marinus* in *Crassostrea gigas* in the Gulf of California. *Dis. Aquat. Organ.* **2010**, *89*, 269–273. [CrossRef] [PubMed]

73. Queiroga, F.R.; Vianna, R.T.; Vieira, C.B.; Farias, N.D.; Da Silva, P.M. Parasites infecting the cultured oyster *Crassostrea gasar* (Adanson, 1757) in Northeast Brazil. *Parasitology* **2015**, *142*, 756–766. [CrossRef] [PubMed]

74. Pagenkopp Lohan, K.M.; Hill-Spanik, K.M.; Torchin, M.E.; Aguirre-Macedo, L.; Fleischer, R.C.; Ruiz, G.M. Richness and distribution of tropical oyster parasites in two oceans. *Parasitology* **2016**, *143*, 1119–1132. [CrossRef] [PubMed]

75. Paynter, K.T.; Politano, V.; Lane, H.A.; Allen, S.M.; Meritt, D. Growth rates and prevalence of *Perkinsus marinus* in restored oyster populations in Maryland. *J. Shellfish Res.* **2010**, *29*, 309–317. [CrossRef]

76. Smolowitz, R. A review of current state of knowledge concerning *Perkinsus Marinus* effects on *Crassostrea virginica* (Gmelin) (the eastern oyster). *Vet. Pathol.* **2013**, *50*, 404–411. [CrossRef] [PubMed]

77. Arzul, I.; Corbeil, S.; Morga, B.; Renault, T. Viruses infecting marine molluscs. *J. Invertebr. Pathol.* **2017**. [CrossRef] [PubMed]

78. Gomez-Leon, J.; Villamil, L.; Lemos, M.L.; Novoa, B.; Figueras, A. Isolation of *Vibrio alginolyticus* and *Vibrio splendidus* from aquacultured carpet shell clam (*Ruditapes decussatus*) larvae associated with mass mortalities. *Appl. Environ. Microbiol.* **2005**, *71*, 98–104. [CrossRef] [PubMed]

79. Froelich, B.A.; Noble, R.T. *Vibrio* bacteria in raw oysters: Managing risks to human health. *Philos. Trans. R. Soc. Lond. B Biol. Sci.* **2016**, *371*. [CrossRef] [PubMed]

80. Carnegie, R.B.; Hill, K.M.; Stokes, N.A.; Burreson, E.M. The haplosporidian *Bonamia exitiosa* is present in Australia, but the identity of the parasite described as *Bonamia* (formerly *Mikrocytos*) *roughleyi* is uncertain. *J. Invertebr. Pathol.* **2014**, *115*, 33–40. [CrossRef] [PubMed]

81. Engelsma, M.Y.; Culloty, S.C.; Lynch, S.A.; Arzul, I.; Carnegie, R.B. *Bonamia* parasites: A rapidly changing perspective on a genus of important mollusc pathogens. *Dis. Aquat. Organ.* **2014**, *110*, 5–23. [CrossRef] [PubMed]

82. Canesi, L.; Pezzati, E.; Stauder, M.; Grande, C.; Bavestrello, M.; Papetti, A.; Vezzulli, L.; Pruzzo, C. *Vibrio cholerae* interactions with *Mytilus galloprovincialis* hemocytes mediated by serum components. *Front. Microbiol.* **2013**, *4*, 371. [CrossRef] [PubMed]

83. Galloway, T.S.; Depledge, M.H. Immunotoxicity in invertebrates: Measurement and ecotoxicological relevance. *Ecotoxicology* **2001**, *10*, 5–23. [CrossRef] [PubMed]

84. Perez, D.G.; Fontanetti, C.S. Hemocitical responses to environmental stress in invertebrates: A review. *Environ. Monit. Assess.* **2011**, *177*, 437–447. [CrossRef] [PubMed]

85. Malagoli, D.; Ottaviani, E. Cross-talk among immune and neuroendocrine systems in molluscs and other invertebrate models. *Horm. Behav.* **2017**, *88*, 41–44. [CrossRef] [PubMed]

86. Bachère, E.; Mialhe, E.; Noël, D.; Boulo, V.; Morvan, A.; Rodriguez, J. Knowledge and research prospects in marine mollusc and crustacean immunology. *Aquaculture* **1995**, *132*, 17–32. [CrossRef]

87. Evariste, L.; Auffret, M.; Audonnet, S.; Geffard, A.; David, E.; Brousseau, P.; Fournier, M.; Betoulle, S. Functional features of hemocyte subpopulations of the invasive mollusk species *Dreissena polymorpha*. *Fish Shellfish Immunol.* **2016**, *56*, 144–154. [CrossRef] [PubMed]

88. Boulais, J.; Trost, M.; Landry, C.R.; Dieckmann, R.; Levy, E.D.; Soldati, T.; Michnick, S.W.; Thibault, P.; Desjardins, M. Molecular characterization of the evolution of phagosomes. *Mol. Syst. Biol.* **2010**, *6*, 423. [CrossRef] [PubMed]

89. Song, L.; Wang, L.; Qiu, L.; Zhang, H. Bivalve immunity. *Adv. Exp. Med. Biol.* **2010**, *708*, 44–65. [PubMed]

90. Smolowitz, R.M.; Miosky, D.; Reinisch, C.L. Ontogeny of leukemic cells of the soft shell clam. *J. Invertebr. Pathol.* **1989**, *53*, 41–51. [CrossRef]

91. Moore, M.N.; Lowe, D.M. The cytology and cytochemistry of the hemocytes of *Mytilus edulis* and their responses to experimentally injected carbon particles. *J. Invertebr. Pathol.* **1977**, *29*, 18–30. [CrossRef]

92. Grandiosa, R.; Merien, F.; Pillay, K.; Alfaro, A. Innovative application of classic and newer techniques for the characterization of haemocytes in the New Zealand black-footed abalone (*Haliotis iris*). *Fish Shellfish Immunol.* **2016**, *48*, 175–184. [CrossRef] [PubMed]

93. Rebelo Mde, F.; Figueiredo Ede, S.; Mariante, R.M.; Nobrega, A.; de Barros, C.M.; Allodi, S. New insights from the oyster *Crassostrea rhizophorae* on bivalve circulating hemocytes. *PLoS ONE* **2013**, *8*, e57384. [CrossRef] [PubMed]

94. Wang, Y.; Hu, M.; Chiang, M.W.; Shin, P.K.; Cheung, S.G. Characterization of subpopulations and immune-related parameters of hemocytes in the green-lipped mussel *Perna viridis*. *Fish Shellfish Immunol.* **2012**, *32*, 381–390. [CrossRef] [PubMed]

95. Xue, Q.; Renault, T. Monoclonal antibodies to European flat oyster *Ostrea edulis* hemocytes: Characterization and tissue distribution of granulocytes in adult and developing animals. *Dev. Comp. Immunol.* **2001**, *25*, 187–194. [CrossRef]

96. Lambert, C.; Soudant, P.; Choquet, G.; Paillard, C. Measurement of *Crassostrea gigas* hemocyte oxidative metabolism by flow cytometry and the inhibiting capacity of pathogenic vibrios. *Fish Shellfish Immunol.* **2003**, *15*, 225–240. [CrossRef]

97. Parisi, M.G.; Li, H.; Jouvet, L.B.; Dyrynda, E.A.; Parrinello, N.; Cammarata, M.; Roch, P. Differential involvement of mussel hemocyte sub-populations in the clearance of bacteria. *Fish Shellfish Immunol.* **2008**, *25*, 834–840. [CrossRef] [PubMed]

98. Renwrantz, L.; Siegmund, E.; Woldmann, M. Variations in hemocyte counts in the mussel, *Mytilus edulis*: Similar reaction patterns occur in disappearance and return of molluscan hemocytes and vertebrate leukocytes. *Comp. Biochem.Physiol. A Mol. Integr. Physiol.* **2013**, *164*, 629–637. [CrossRef] [PubMed]

99. Anderson, R.S.; Ozbay, G.; Kingsley, D.H.; Strauss, M.A. Oyster hemocyte mobilization and increased adhesion activity after β-glucan administration. *J. Shellfish Res.* **2011**, *30*, 635–641. [CrossRef]

100. Taylor, A.M.; Edge, K.J.; Ubrihien, R.P.; Maher, W.A. The freshwater bivalve *Corbicula australis* as a sentinel species for metal toxicity assessment: An in situ case study integrating chemical and biomarker analyses. *Environ. Toxicol. Chem.* **2016**, *36*, 709–719. [CrossRef] [PubMed]

101. Mosca, F.; Lanni, L.; Cargini, D.; Narcisi, V.; Bianco, I.; Tiscar, P.G. Variability of the hemocyte parameters of cultivated mussel *Mytilus galloprovincialis* (Lmk 1819) in Sabaudia (Latina, Italy) coastal lagoon. *Mar. Environ. Res.* **2013**, *92*, 215–223. [CrossRef] [PubMed]

102. Farcy, E.; Burgeot, T.; Haberkorn, H.; Auffret, M.; Lagadic, L.; Allenou, J.P.; Budzinski, H.; Mazzella, N.; Pete, R.; Heydorff, M.; et al. An integrated environmental approach to investigate biomarker fluctuations in the blue mussel *Mytilus edulis* L. in the Vilaine estuary, France. *Environ. Sci. Pollut. Res. Int.* **2013**, *20*, 630–650. [CrossRef] [PubMed]

103. Hannam, M.L.; Bamber, S.D.; Sundt, R.C.; Galloway, T.S. Immune modulation in the blue mussel *Mytilus edulis* exposed to north sea produced water. *Environ. Pollut.* **2009**, *157*, 1939–1944. [CrossRef] [PubMed]

104. Camus, L.; Grosvik, B.E.; Borseth, J.F.; Jones, M.B.; Depledge, M.H. Stability of lysosomal and cell membranes in haemocytes of the common mussel (*Mytilus edulis*): Effect of low temperatures. *Mar. Environ. Res.* **2000**, *50*, 325–329. [CrossRef]

105. Dimitriadis, V.K.; Gougoula, C.; Anestis, A.; Portner, H.O.; Michaelidis, B. Monitoring the biochemical and cellular responses of marine bivalves during thermal stress by using biomarkers. *Mar. Environ. Res.* **2012**, *73*, 70–77. [CrossRef] [PubMed]

106. Matozzo, V.; Chinellato, A.; Munari, M.; Finos, L.; Bressan, M.; Marin, M.G. First evidence of immunomodulation in bivalves under seawater acidification and increased temperature. *PLoS ONE* **2012**, *7*, e33820. [CrossRef] [PubMed]

107. Lacoste, A.; Malham, S.K.; Gelebart, F.; Cueff, A.; Poulet, S.A. Stress-induced immune changes in the oyster *Crassostrea gigas. Dev. Comp. Immunol.* **2002**, *26*, 1–9. [CrossRef]

108. Gagnaire, B.; Frouin, H.; Moreau, K.; Thomas-Guyon, H.; Renault, T. Effects of temperature and salinity on haemocyte activities of the Pacific oyster, *Crassostrea gigas* (Thunberg). *Fish Shellfish Immunol.* **2006**, *20*, 536–547. [CrossRef] [PubMed]

109. Boyd, J.N.; Burnett, L.E. Reactive oxygen intermediate production by oyster hemocytes exposed to hypoxia. *J. Exp. Biol.* **1999**, *202*, 3135–3143. [PubMed]

110. Buratti, S.; Franzellitti, S.; Poletti, R.; Ceredi, A.; Montanari, G.; Capuzzo, A.; Fabbri, E. Bioaccumulation of algal toxins and changes in physiological parameters in Mediterranean mussels from the North Adriatic Sea (Italy). *Environ. Toxicol.* **2013**, *28*, 451–470. [CrossRef] [PubMed]

111. Hoher, N.; Kohler, A.; Strand, J.; Broeg, K. Effects of various pollutant mixtures on immune responses of the blue mussel (*Mytilus edulis*) collected at a salinity gradient in danish coastal waters. *Mar. Environ. Res.* **2012**, *75*, 35–44. [CrossRef] [PubMed]

112. Mosca, F.; Narcisi, V.; Cargini, D.; Calzetta, A.; Tiscar, P.G. Age related properties of the Adriatic clam *Chamelea gallina* (L. 1758) hemocytes. *Fish Shellfish Immunol.* **2011**, *31*, 1106–1112. [CrossRef] [PubMed]

113. Dang, C.; Tan, T.; Moffit, D.; Deboutteville, J.D.; Barnes, A.C. Gender differences in hemocyte immune parameters of bivalves: The Sydney rock oyster *Saccostrea glomerata* and the pearl oyster *Pinctada fucata. Fish Shellfish Immunol.* **2012**, *33*, 138–142. [CrossRef] [PubMed]

114. Li, Y.; Qin, J.G.; Abbott, C.A.; Li, X.; Benkendorff, K. Synergistic impacts of heat shock and spawning on the physiology and immune health of *Crassostrea gigas*: An explanation for summer mortality in Pacific oysters. *Am. J. Physiol. Regul. Integr. Comp. Physiol.* **2007**, *293*, R2353–R2362. [CrossRef] [PubMed]

115. Hughes, T.K., Jr.; Smith, E.M.; Barnett, J.A.; Charles, R.; Stefano, G.B. LPS stimulated invertebrate hemocytes: A role for immunoreactive TNF and IL-1. *Dev. Comp. Immunol.* **1991**, *15*, 117–122. [CrossRef]

116. Ottaviani, E.; Franchini, A.; Malagoli, D.; Genedani, S. Immunomodulation by recombinant human interleukin-8 and its signal transduction pathways in invertebrate hemocytes. *Cell. Mol. Life Sci.* **2000**, *57*, 506–513. [CrossRef] [PubMed]

117. Panara, F.; Di Rosa, I.; Fagotti, A.; Simoncelli, F.; Mangiabene, C.; Pipe, R.K.; Pascolini, R. Characterization and immunocytochemical localization of actin and fibronectin in haemocytes of the mussel *Mytilus galloprovincialis. Histochem. J.* **1996**, *28*, 123–131. [CrossRef] [PubMed]

118. Schneeweiss, H.; Renwrantz, L. Analysis of the attraction of haemocytes from *Mytilus edulis* by molecules of bacterial origin. *Dev. Comp. Immunol.* **1993**, *17*, 377–387. [CrossRef]

119. Fawcett, L.B.; Tripp, M.R. Chemotaxis of *Mercenaria mercenaria* hemocytes to bacteria in vitro. *J. Invertebr. Pathol.* **1994**, *63*, 275–284. [CrossRef] [PubMed]

120. Kumazawa, N.H.; Morimoto, N. Chemotactic activity of hemocytes derived from a brackish-water clam, *Corbicula japonica*, to *Vibrio parahaemolyticus* and *Escherichia coli* strains. *J. Vet. Med. Sci.* **1992**, *54*, 851–855. [CrossRef] [PubMed]

121. Buchmann, K. Evolution of innate immunity: Clues from invertebrates via fish to mammals. *Front. Immunol.* **2014**, *5*, 459. [CrossRef] [PubMed]

122. Kawai, T.; Akira, S. The role of pattern-recognition receptors in innate immunity: Update on toll-like receptors. *Nat. Immunol.* **2010**, *11*, 373–384. [CrossRef] [PubMed]

123. Lemaitre, B.; Hoffmann, J. The host defense of *Drosophila melanogaster. Annu. Rev. Immunol.* **2007**, *25*, 697–743. [CrossRef] [PubMed]

124. Wang, M.; Wang, L.; Guo, Y.; Sun, R.; Yue, F.; Yi, Q.; Song, L. The broad pattern recognition spectrum of the toll-like receptor in mollusk Zhikong scallop *Chlamys farreri. Dev. Comp. Immunol.* **2015**, *52*, 192–201. [CrossRef] [PubMed]

125. Zhang, L.; Li, L.; Zhang, G. A *Crassostrea gigas* toll-like receptor and comparative analysis of TLR pathway in invertebrates. *Fish Shellfish Immunol.* **2011**, *30*, 653–660. [CrossRef] [PubMed]

126. Qiu, L.; Song, L.; Xu, W.; Ni, D.; Yu, Y. Molecular cloning and expression of a toll receptor gene homologue from Zhikong scallop, *Chlamys farreri. Fish Shellfish Immunol.* **2007**, *22*, 451–466. [CrossRef] [PubMed]

127. Philipp, E.E.; Kraemer, L.; Melzner, F.; Poustka, A.J.; Thieme, S.; Findeisen, U.; Schreiber, S.; Rosenstiel, P. Massively parallel RNA sequencing identifies a complex immune gene repertoire in the lophotrochozoan *Mytilus edulis. PLoS ONE* **2012**, *7*, e33091. [CrossRef] [PubMed]

128. Watters, T.M.; Kenny, E.F.; O'Neill, L.A. Structure, function and regulation of the Toll/IL-1 receptor adaptor proteins. *Immunol. Cell Biol.* **2007**, *85*, 411–419. [CrossRef] [PubMed]

129. Kawasaki, T.; Kawai, T. Toll-like receptor signaling pathways. *Front. Immunol.* **2014**, *5*, 461. [CrossRef] [PubMed]

130. O'Neill, L.A. How Toll-like receptors signal: What we know and what we don't know. *Curr. Opin. Immunol.* **2006**, *18*, 3–9. [CrossRef] [PubMed]

131. Toubiana, M.; Rosani, U.; Giambelluca, S.; Cammarata, M.; Gerdol, M.; Pallavicini, A.; Venier, P.; Roch, P. Toll signal transduction pathway in bivalves: Complete CDS of intermediate elements and related gene transcription levels in hemocytes of immune stimulated *Mytilus galloprovincialis. Dev. Comp. Immunol.* **2014**, *45*, 300–312. [CrossRef] [PubMed]

132. Wang, M.; Yang, J.; Zhou, Z.; Qiu, L.; Wang, L.; Zhang, H.; Gao, Y.; Wang, X.; Zhang, L.; Zhao, J.; et al. A primitive Toll-like receptor signaling pathway in mollusk Zhikong scallop *Chlamys farreri. Dev. Comp. Immunol.* **2011**, *35*, 511–520. [CrossRef] [PubMed]

133. Moreira, R.; Balseiro, P.; Planas, J.V.; Fuste, B.; Beltran, S.; Novoa, B.; Figueras, A. Transcriptomics of in vitro immune-stimulated hemocytes from the Manila Clam *Ruditapes philippinarum* using high-throughput sequencing. *PLoS ONE* **2012**, *7*, e35009. [CrossRef] [PubMed]

134. Qu, F.; Xiang, Z.; Wang, F.; Zhang, Y.; Li, J.; Zhang, Y.; Xiao, S.; Yu, Z. Identification and function of an evolutionarily conserved signaling intermediate in toll pathways (ECSIT) from *Crassostrea hongkongensis. Dev. Comp. Immunol.* **2015**, *53*, 244–252. [CrossRef] [PubMed]

135. Canesi, L.; Betti, M.; Ciacci, C.; Lorusso, L.C.; Gallo, G.; Pruzzo, C. Interactions between *Mytilus* haemocytes and different strains of *Escherichia coli* and *Vibrio cholerae* O1 El Tor: Role of kinase-mediated signalling. *Cell. Microbiol.* **2005**, *7*, 667–674. [CrossRef] [PubMed]
136. Ciacci, C.; Citterio, B.; Betti, M.; Canonico, B.; Roch, P.; Canesi, L. Functional differential immune responses of *Mytilus galloprovincialis* to bacterial challenge. *Comp. Biochem. Physiol. B Biochem. Mol. Biol.* **2009**, *153*, 365–371. [CrossRef] [PubMed]
137. Wang, W.; Liu, R.; Zhang, T.; Zhang, R.; Song, X.; Wang, L.; Song, L. A novel phagocytic receptor (CgNimC) from Pacific oyster *Crassostrea gigas* with lipopolysaccharide and Gram-negative bacteria binding activity. *Fish Shellfish Immunol.* **2015**, *43*, 103–110. [CrossRef] [PubMed]
138. Ni, D.; Song, L.; Wu, L.; Chang, Y.; Yu, Y.; Qiu, L.; Wang, L. Molecular cloning and mRNA expression of peptidoglycan recognition protein (PGRP) gene in bay scallop (*Argopecten irradians*, Lamarck 1819). *Dev. Comp. Immunol.* **2007**, *31*, 548–558. [CrossRef] [PubMed]
139. Wei, X.; Yang, J.; Liu, X.; Yang, D.; Xu, J.; Fang, J.; Wang, W.; Yang, J. Identification and transcriptional analysis of two types of lectins (SgCTL-1 and SgGal-1) from mollusk *Solen grandis*. *Fish Shellfish Immunol.* **2012**, *33*, 204–212. [CrossRef] [PubMed]
140. Canesi, L.; Gallo, G.; Gavioli, M.; Pruzzo, C. Bacteria-hemocyte interactions and phagocytosis in marine bivalves. *Microsc. Res. Tech.* **2002**, *57*, 469–476. [CrossRef] [PubMed]
141. Carballal, M.J.; Lopez, C.; Azevedo, C.; Villalba, A. Enzymes involved in defense functions of hemocytes of mussel *Mytilus galloprovincialis*. *J. Invertebr. Pathol.* **1997**, *70*, 96–105. [CrossRef] [PubMed]
142. Cima, F.; Matozzo, V.; Marin, M.G.; Ballarin, L. Haemocytes of the clam *Tapes philippinarum* (Adams & Reeve, 1850): Morphofunctional characterisation. *Fish Shellfish Immunol.* **2000**, *10*, 677–693. [PubMed]
143. Matozzo, V.; Rova, G.; Marin, M.G. Haemocytes of the cockle *Cerastoderma glaucum*: Morphological characterisation and involvement in immune responses. *Fish Shellfish Immunol.* **2007**, *23*, 732–746. [CrossRef] [PubMed]
144. Wootton, E.C.; Dyrynda, E.A.; Ratcliffe, N.A. Bivalve immunity: Comparisons between the marine mussel (*Mytilus edulis*), the edible cockle (*Cerastoderma edule*) and the razor-shell (*Ensis siliqua*). *Fish Shellfish Immunol.* **2003**, *15*, 195–210. [CrossRef]
145. Pampanin, D.M.; Marin, M.G.; Ballarin, L. Morphological and cytoenzymatic characterization of haemocytes of the Venus Clam *Chamelea gallina*. *Dis. Aquat. Organ.* **2002**, *49*, 227–234. [CrossRef] [PubMed]
146. Anderson, R.S. Reactive oxygen species and antimicrobial defenses of invertebrates: A bivalve model. *Adv. Exp. Med. Biol.* **2001**, *484*, 131–139. [PubMed]
147. Adema, C.M.; van Deutekom-Mulder, E.C.; van der Knaap, W.P.; Sminia, T. NADPH-oxidase activity: The probable source of reactive oxygen intermediate generation in hemocytes of the gastropod *Lymnaea stagnalis*. *J. Leukoc. Biol.* **1993**, *54*, 379–383. [PubMed]
148. Connors, V.A.; Lodes, M.J.; Yoshino, T.P. Identification of a *Schistosoma mansoni* sporocyst excretory-secretory antioxidant molecule and its effect on superoxide production by *Biomphalaria glabrata* hemocytes. *J. Invertebr. Pathol.* **1991**, *58*, 387–395. [CrossRef]
149. Ordas, M.C.; Novoa, B.; Figueras, A. Modulation of the chemiluminescence response of Mediterranean mussel (*Mytilus galloprovincialis*) haemocytes. *Fish Shellfish Immunol.* **2000**, *10*, 611–622. [CrossRef] [PubMed]
150. Goedken, M.; De Guise, S. Flow cytometry as a tool to quantify oyster defense mechanisms. *Fish Shellfish Immunol.* **2004**, *16*, 539–552. [CrossRef] [PubMed]
151. Bugge, D.M.; Hegaret, H.; Wikfors, G.H.; Allam, B. Oxidative burst in hard clam (*Mercenaria mercenaria*) haemocytes. *Fish Shellfish Immunol.* **2007**, *23*, 188–196. [CrossRef] [PubMed]
152. Bartosz, G. Reactive oxygen species: Destroyers or messengers? *Biochem. Pharmacol.* **2009**, *77*, 1303–1315. [CrossRef] [PubMed]
153. Winston, G.W.; Moore, M.N.; Kirchin, M.A.; Soverchia, C. Production of reactive oxygen species by hemocytes from the marine mussel, *Mytilus edulis*: Ysosomal localization and effect of xenobiotics. *Comp. Biochem. Physiol. C Pharmacol. Toxicol. Endocrinol.* **1996**, *113*, 221–229. [CrossRef]
154. Donaghy, L.; Kraffe, E.; Le Goic, N.; Lambert, C.; Volety, A.K.; Soudant, P. Reactive oxygen species in unstimulated hemocytes of the Pacific oyster *Crassostrea gigas*: A mitochondrial involvement. *PLoS ONE* **2012**, *7*, e46594. [CrossRef] [PubMed]
155. Ge, W.; Li, D.; Gao, Y.; Cao, X. The roles of lysosomes in inflammation and autoimmune diseases. *Int. Rev. Immunol.* **2015**, *34*, 415–431. [CrossRef] [PubMed]

156. Mateo, D.R.; Spurmanis, A.; Siah, A.; Araya, M.T.; Kulka, M.; Berthe, F.C.; Johnson, G.R.; Greenwood, S.J. Changes induced by two strains of *Vibrio splendidus* in haemocyte subpopulations of *Mya Arenaria*, detected by flow cytometry with LysoTracker. *Dis. Aquat. Organ.* **2009**, *86*, 253–262. [CrossRef] [PubMed]

157. Allam, B.; Ashton-Alcox, K.A.; Ford, S.E. Flow cytometric comparison of haemocytes from three species of bivalve molluscs. *Fish Shellfish Immunol.* **2002**, *13*, 141–158. [CrossRef] [PubMed]

158. Garcia-Garcia, E.; Prado-Alvarez, M.; Novoa, B.; Figueras, A.; Rosales, C. Immune responses of mussel hemocyte subpopulations are differentially regulated by enzymes of the PI 3-K, PKC, and ERK kinase families. *Dev. Comp. Immunol.* **2008**, *32*, 637–653. [CrossRef] [PubMed]

159. Tunkijjanukij, S.; Giaever, H.; Chin, C.C.; Olafsen, J.A. Sialic acid in hemolymph and affinity purified lectins from two marine bivalves. *Comp. Biochem. Physiol. B Biochem. Mol. Biol.* **1998**, *119*, 705–713. [CrossRef]

160. Olafsen, J.A. Bacterial antigen priming of marine fish larvae. *Adv. Exp. Med. Biol.* **1995**, *371A*, 349–352. [PubMed]

161. Yang, J.; Huang, M.; Zhang, H.; Wang, L.; Wang, H.; Wang, L.; Qiu, L.; Song, L. CfLec-3 from scallop: An entrance to non-self recognition mechanism of invertebrate C-type lectin. *Sci. Rep.* **2015**, *5*, 10068. [CrossRef] [PubMed]

162. Mu, C.; Chen, L.; Zhao, J.; Wang, C. Molecular cloning and expression of a C-type lectin gene from *Venerupis philippinarum*. *Mol. Biol. Rep.* **2014**, *41*, 139–144. [CrossRef] [PubMed]

163. Huang, M.; Wang, L.; Yang, J.; Zhang, H.; Wang, L.; Song, L. A four-CRD C-type lectin from *Chlamys farreri* mediating nonself-recognition with broader spectrum and opsonization. *Dev. Comp. Immunol.* **2013**, *39*, 363–369. [CrossRef] [PubMed]

164. Vasta, G.R. Galectins as pattern recognition receptors: Structure, function, and evolution. *Adv. Exp. Med. Biol.* **2012**, *946*, 21–36. [PubMed]

165. Ahmed, H.; Vasta, G.R. Galectins: Conservation of functionally and structurally relevant amino acid residues defines two types of carbohydrate recognition domains. *Glycobiology* **1994**, *4*, 545–548. [CrossRef] [PubMed]

166. Ahmed, H.; Bianchet, M.A.; Amzel, L.M.; Hirabayashi, J.; Kasai, K.; Giga-Hama, Y.; Tohda, H.; Vasta, G.R. Novel carbohydrate specificity of the 16-kDa galectin from *Caenorhabditis elegans*: Binding to blood group precursor oligosaccharides (type 1, type 2, Talpha, and Tbeta) and gangliosides. *Glycobiology* **2002**, *12*, 451–461. [CrossRef] [PubMed]

167. Bianchet, M.A.; Ahmed, H.; Vasta, G.R.; Amzel, L.M. Soluble beta-galactosyl-binding lectin (galectin) from toad ovary: Crystallographic studies of two protein-sugar complexes. *Proteins* **2000**, *40*, 378–388. [CrossRef]

168. Kim, J.Y.; Kim, Y.M.; Cho, S.K.; Choi, K.S.; Cho, M. Noble tandem-repeat galectin of Manila Clam *Ruditapes philippinarum* is induced upon infection with the protozoan parasite *Perkinsus olseni*. *Dev. Comp. Immunol.* **2008**, *32*, 1131–1141. [CrossRef] [PubMed]

169. Song, X.; Zhang, H.; Wang, L.; Zhao, J.; Mu, C.; Song, L.; Qiu, L.; Liu, X. A galectin with quadruple-domain from bay scallop *Argopecten irradians* is involved in innate immune response. *Dev. Comp. Immunol.* **2011**, *35*, 592–602. [CrossRef] [PubMed]

170. Feng, C.; Ghosh, A.; Amin, M.N.; Giomarelli, B.; Shridhar, S.; Banerjee, A.; Fernandez-Robledo, J.A.; Bianchet, M.A.; Wang, L.X.; Wilson, I.B.; et al. The galectin CvGal1 from the eastern oyster (*Crassostrea virginica*) binds to blood group a oligosaccharides on the hemocyte surface. *J. Biol. Chem.* **2013**, *288*, 24394–24409. [CrossRef] [PubMed]

171. Jiang, S.; Li, H.; Zhang, D.; Zhang, H.; Wang, L.; Sun, J.; Song, L. A C1q domain containing protein from *Crassostrea gigas* serves as pattern recognition receptor and opsonin with high binding affinity to LPS. *Fish Shellfish Immunol.* **2015**, *45*, 583–591. [CrossRef] [PubMed]

172. Gerdol, M.; Manfrin, C.; De Moro, G.; Figueras, A.; Novoa, B.; Venier, P.; Pallavicini, A. The C1q domain containing proteins of the Mediterranean mussel *Mytilus galloprovincialis*: A widespread and diverse family of immune-related molecules. *Dev. Comp. Immunol.* **2011**, *35*, 635–643. [CrossRef] [PubMed]

173. Kong, P.; Zhang, H.; Wang, L.; Zhou, Z.; Yang, J.; Zhang, Y.; Qiu, L.; Wang, L.; Song, L. AiC1qDC-1, a novel gC1q-domain-containing protein from bay scallop *Argopecten irradians* with fungi agglutinating activity. *Dev. Comp. Immunol.* **2010**, *34*, 837–846. [CrossRef] [PubMed]

174. Li, C.; Yu, S.; Zhao, J.; Su, X.; Li, T. Cloning and characterization of a sialic acid binding lectins (SABL) from Manila Clam *Venerupis Philippinarum*. *Fish Shellfish Immunol.* **2011**, *30*, 1202–1206. [CrossRef] [PubMed]

175. Gerdol, M.; Venier, P.; Pallavicini, A. The genome of the Pacific oyster *Crassostrea gigas* brings new insights on the massive expansion of the C1q gene family in Bivalvia. *Dev. Comp. Immunol.* **2015**, *49*, 59–71. [CrossRef] [PubMed]

176. Wang, L.; Wang, L.; Zhang, D.; Jiang, Q.; Sun, R.; Wang, H.; Zhang, H.; Song, L. A novel multi-domain C1qDC protein from Zhikong scallop *Chlamys farreri* provides new insights into the function of invertebrate C1qDC proteins. *Dev. Comp. Immunol.* **2015**, *52*, 202–214. [CrossRef] [PubMed]

177. Xu, T.; Xie, J.; Li, J.; Luo, M.; Ye, S.; Wu, X. Identification of expressed genes in cDNA library of hemocytes from the RLO-challenged oyster, *Crassostrea ariakensis* gould with special functional implication of three complement-related fragments (CaC1q1, CaC1q2 and CaC3). *Fish Shellfish Immunol.* **2012**, *32*, 1106–1116. [CrossRef] [PubMed]

178. Mao, Y.; Zhou, C.; Zhu, L.; Huang, Y.; Yan, T.; Fang, J.; Zhu, W. Identification and expression analysis on bactericidal permeability-increasing protein (BPI)/lipopolysaccharide-binding protein (LBP) of ark shell, *Scapharca broughtonii*. *Fish Shellfish Immunol.* **2013**, *35*, 642–652. [CrossRef] [PubMed]

179. Zhang, Y.; He, X.; Li, X.; Fu, D.; Chen, J.; Yu, Z. The second bactericidal permeability increasing protein (BPI) and its revelation of the gene duplication in the Pacific oyster, *Crassostrea gigas*. *Fish Shellfish Immunol.* **2011**, *30*, 954–963. [CrossRef] [PubMed]

180. Gonzalez, M.; Gueguen, Y.; Destoumieux-Garzon, D.; Romestand, B.; Fievet, J.; Pugniere, M.; Roquet, F.; Escoubas, J.M.; Vandenbulcke, F.; Levy, O.; et al. Evidence of a bactericidal permeability increasing protein in an invertebrate, the *Crassostrea gigas* Cg-BPI. *Proc. Natl. Acad. Sci. USA* **2007**, *104*, 17759–17764. [CrossRef] [PubMed]

181. Balbi, T.; Fabbri, R.; Cortese, K.; Smerilli, A.; Ciacci, C.; Grande, C.; Vezzulli, L.; Pruzzo, C.; Canesi, L. Interactions between *Mytilus galloprovincialis* hemocytes and the bivalve pathogens *Vibrio aestuarianus* 01/032 and *Vibrio splendidus* LGP32. *Fish Shellfish Immunol.* **2013**, *35*, 1906–1915. [CrossRef] [PubMed]

182. Liu, X.; Zhao, J.; Wu, H.; Wang, Q. Metabolomic analysis revealed the differential responses in two pedigrees of clam *Ruditapes philippinarum* towards *Vibrio harveyi* challenge. *Fish Shellfish Immunol.* **2013**, *35*, 1969–1975. [CrossRef] [PubMed]

183. Zhou, Z.; Wang, L.; Shi, X.; Zhang, H.; Gao, Y.; Wang, M.; Kong, P.; Qiu, L.; Song, L. The modulation of catecholamines to the immune response against bacteria *Vibrio anguillarum* challenge in scallop *Chlamys farreri*. *Fish Shellfish Immunol.* **2011**, *31*, 1065–1071. [CrossRef] [PubMed]

184. Morga, B.; Renault, T.; Faury, N.; Chollet, B.; Arzul, I. Cellular and molecular responses of haemocytes from *Ostrea edulis* during in vitro infection by the parasite *Bonamia ostreae*. *Int. J. Parasitol.* **2011**, *41*, 755–764. [CrossRef] [PubMed]

185. Xue, Q.G.; Schey, K.L.; Volety, A.K.; Chu, F.L.; La Peyre, J.F. Purification and characterization of lysozyme from plasma of the eastern oyster (*Crassostrea virginica*). *Comp. Biochem. Physiol. B Biochem. Mol. Biol.* **2004**, *139*, 11–25. [CrossRef] [PubMed]

186. Allam, B.; Pales Espinosa, E.; Tanguy, A.; Jeffroy, F.; Le Bris, C.; Paillard, C. Transcriptional changes in Manila clam (*Ruditapes philippinarum*) in response to Brown Ring Disease. *Fish Shellfish Immunol.* **2014**, *41*, 2–11. [CrossRef] [PubMed]

187. Munoz, P.; Vance, K.; Gomez-Chiarri, M. Protease activity in the plasma of American oysters, *Crassostrea virginica*, experimentally infected with the protozoan parasite *Perkinsus marinus*. *J. Parasitol.* **2003**, *89*, 941–951. [CrossRef] [PubMed]

188. Niu, D.; Jin, K.; Wang, L.; Feng, B.; Li, J. Molecular characterization and expression analysis of four cathepsin L genes in the razor clam, *Sinonovacula constricta*. *Fish Shellfish Immunol.* **2013**, *35*, 581–588. [CrossRef] [PubMed]

189. Allam, B.; Pales Espinosa, E. Bivalve immunity and response to infections: Are we looking at the right place? *Fish Shellfish Immunol.* **2016**, *53*, 4–12. [CrossRef] [PubMed]

190. Mora, C.; Tittensor, D.P.; Adl, S.; Simpson, A.G.; Worm, B. How many species are there on earth and in the ocean? *PLoS Biol.* **2011**, *9*, e1001127. [CrossRef] [PubMed]

191. Mayer, A.M.; Rodriguez, A.D.; Berlinck, R.G.; Fusetani, N. Marine pharmacology in 2007–8: Marine compounds with antibacterial, anticoagulant, antifungal, anti-inflammatory, antimalarial, antiprotozoal, antituberculosis, and antiviral activities; affecting the immune and nervous system, and other miscellaneous mechanisms of action. *Comp. Biochem. Physiol. C Toxicol. Pharmacol.* **2011**, *153*, 191–222. [PubMed]

192. Charles, F.; Grémare, A.; Amouroux, J.-M.; Cahet, G. Filtration of the enteric bacteria *Escherichia coli* by two filter-feeding bivalves, *Venus verrucosa* and *Mytilus galloprovincialis*. *Mar. Biol.* **1992**, *113*, 125–131. [CrossRef]

193. McHenery, J.G. Uptake and processing of cultured microorganisms by bivalves. *J. Exp. Mar. Biol. Ecol.* **1985**, *90*, 145–163. [CrossRef]

194. Otero-Gonzalez, A.J.; Magalhaes, B.S.; Garcia-Villarino, M.; Lopez-Abarrategui, C.; Sousa, D.A.; Dias, S.C.; Franco, O.L. Antimicrobial peptides from marine invertebrates as a new frontier for microbial infection control. *FASEB J.* **2010**, *24*, 1320–1334. [CrossRef] [PubMed]

195. Hultmark, D.; Steiner, H.; Rasmuson, T.; Boman, H.G. Insect immunity. Purification and properties of three inducible bactericidal proteins from hemolymph of immunized pupae of *Hyalophora cecropia*. *Eur. J. Biochem.* **1980**, *106*, 7–16. [CrossRef] [PubMed]

196. Steiner, H.; Hultmark, D.; Engstrom, A.; Bennich, H.; Boman, H.G. Sequence and specificity of two antibacterial proteins involved in insect immunity. *Nature* **1981**, *292*, 246–248. [CrossRef] [PubMed]

197. Hancock, R.E.; Lehrer, R. Cationic peptides: A new source of antibiotics. *Trends Biotechnol.* **1998**, *16*, 82–88. [CrossRef]

198. Brown, K.L.; Hancock, R.E. Cationic host defense (antimicrobial) peptides. *Curr. Opin. Immunol.* **2006**, *18*, 24–30. [CrossRef] [PubMed]

199. Hancock, R.E.; Diamond, G. The role of cationic antimicrobial peptides in innate host defenses. *Trends Microbiol.* **2000**, *8*, 402–410. [CrossRef]

200. Gaspar, D.; Veiga, A.S.; Castanho, M.A. From antimicrobial to anticancer peptides. A review. *Front. Microbiol.* **2013**, *4*, 294. [CrossRef] [PubMed]

201. Radek, K.; Gallo, R. Antimicrobial peptides: Natural effectors of the innate immune system. *Semin. Immunopathol.* **2007**, *29*, 27–43. [CrossRef] [PubMed]

202. Zasloff, M. Antimicrobial peptides of multicellular organisms. *Nature* **2002**, *415*, 389–395. [CrossRef] [PubMed]

203. Brogden, K.A. Antimicrobial peptides: Pore formers or metabolic inhibitors in bacteria? *Nat. Rev. Microbiol.* **2005**, *3*, 238–250. [CrossRef] [PubMed]

204. Hancock, R.E.; Scott, M.G. The role of antimicrobial peptides in animal defenses. *Proc. Natl. Acad. Sci. USA* **2000**, *97*, 8856–8861. [CrossRef] [PubMed]

205. Jenssen, H.; Hamill, P.; Hancock, R.E. Peptide antimicrobial agents. *Clin. Microbiol. Rev.* **2006**, *19*, 491–511. [CrossRef] [PubMed]

206. Boman, H.G. Peptide antibiotics and their role in innate immunity. *Annu. Rev. Immunol.* **1995**, *13*, 61–92. [CrossRef] [PubMed]

207. Hancock, R.E.; Rozek, A. Role of membranes in the activities of antimicrobial cationic peptides. *FEMS Microbiol. Lett.* **2002**, *206*, 143–149. [CrossRef] [PubMed]

208. Matsuzaki, K. Control of cell selectivity of antimicrobial peptides. *Biochim. Biophys. Acta* **2009**, *1788*, 1687–1692. [CrossRef] [PubMed]

209. Zhang, L.; Rozek, A.; Hancock, R.E. Interaction of cationic antimicrobial peptides with model membranes. *J. Biol. Chem.* **2001**, *276*, 35714–35722. [CrossRef] [PubMed]

210. Oren, Z.; Shai, Y. Mode of action of linear amphipathic alpha-helical antimicrobial peptides. *Biopolymers* **1998**, *47*, 451–463. [CrossRef]

211. Shai, Y. Mode of action of membrane active antimicrobial peptides. *Biopolymers* **2002**, *66*, 236–248. [CrossRef] [PubMed]

212. Ehrenstein, G.; Lecar, H. Electrically gated ionic channels in lipid bilayers. *Q. Rev. Biophys.* **1977**, *10*, 1–34. [CrossRef] [PubMed]

213. Yang, L.; Harroun, T.A.; Weiss, T.M.; Ding, L.; Huang, H.W. Barrel-stave model or toroidal model? A case study on melittin pores. *Biophys. J.* **2001**, *81*, 1475–1485. [CrossRef]

214. Yeaman, M.R.; Yount, N.Y. Mechanisms of antimicrobial peptide action and resistance. *Pharmacol. Rev.* **2003**, *55*, 27–55. [CrossRef] [PubMed]

215. Lee, T.H.; Hall, K.N.; Aguilar, M.I. Antimicrobial peptide structure and mechanism of action: A focus on the role of membrane structure. *Curr. Top. Med. Chem.* **2016**, *16*, 25–39. [CrossRef] [PubMed]

216. Marrink, S.J.; de Vries, A.H.; Tieleman, D.P. Lipids on the move: Simulations of membrane pores, domains, stalks and curves. *Biochim. Biophys. Acta* **2009**, *1788*, 149–168. [CrossRef] [PubMed]

217. Wimley, W.C. Describing the mechanism of antimicrobial peptide action with the interfacial activity model. *ACS Chem. Biol.* **2010**, *5*, 905–917. [CrossRef] [PubMed]
218. Falanga, A.; Galdiero, M.; Galdiero, S. Membranotropic cell penetrating peptides: The outstanding journey. *Int. J. Mol. Sci.* **2015**, *16*, 25323–25337. [CrossRef] [PubMed]
219. Galdiero, S.; Vitiello, M.; Falanga, A.; Cantisani, M.; Incoronato, N.; Galdiero, M. Intracellular delivery: Exploiting viral membranotropic peptides. *Curr. Drug Metab.* **2012**, *13*, 93–104. [CrossRef] [PubMed]
220. Ponnappan, N.; Budagavi, D.P.; Yadav, B.K.; Chugh, A. Membrane-active peptides from marine organisms—Antimicrobials, cell-penetrating peptides and peptide toxins: Applications and prospects. *Probiotics Antimicrob. Proteins* **2015**, *7*, 75–89. [CrossRef] [PubMed]
221. Masso-Silva, J.A.; Diamond, G. Antimicrobial peptides from fish. *Pharmaceuticals (Basel)* **2014**, *7*, 265–310. [CrossRef] [PubMed]
222. Kang, H.K.; Seo, C.H.; Park, Y. Marine peptides and their anti-infective activities. *Mar. Drugs* **2015**, *13*, 618–654. [CrossRef] [PubMed]
223. Falanga, A.; Lombardi, L.; Franci, G.; Vitiello, M.; Iovene, M.R.; Morelli, G.; Galdiero, M.; Galdiero, S. Marine antimicrobial peptides: Nature provides templates for the design of novel compounds against pathogenic bacteria. *Int. J. Mol. Sci.* **2016**, *17*, 785. [CrossRef] [PubMed]
224. Hubert, F.; Noel, T.; Roch, P. A member of the arthropod defensin family from edible Mediterranean mussels (*Mytilus galloprovincialis*). *Eur. J. Biochem.* **1996**, *240*, 302–306. [CrossRef] [PubMed]
225. Charlet, M.; Chernysh, S.; Philippe, H.; Hetru, C.; Hoffmann, J.A.; Bulet, P. Innate immunity. Isolation of several cysteine-rich antimicrobial peptides from the blood of a mollusc, *Mytilus edulis*. *J. Biol. Chem.* **1996**, *271*, 21808–21813. [PubMed]
226. Mitta, G.; Hubert, F.; Noel, T.; Roch, P. Myticin, a novel cysteine-rich antimicrobial peptide isolated from haemocytes and plasma of the mussel *Mytilus galloprovincialis*. *Eur. J. Biochem.* **1999**, *265*, 71–78. [CrossRef] [PubMed]
227. Mitta, G.; Vandenbulcke, F.; Hubert, F.; Roch, P. Mussel defensins are synthesised and processed in granulocytes then released into the plasma after bacterial challenge. *J. Cell Sci.* **1999**, *112*, 4233–4242. [PubMed]
228. Yang, Y.S.; Mitta, G.; Chavanieu, A.; Calas, B.; Sanchez, J.F.; Roch, P.; Aumelas, A. Solution structure and activity of the synthetic four-disulfide bond Mediterranean mussel defensin (MGD-1). *Biochemistry* **2000**, *39*, 14436–14447. [CrossRef] [PubMed]
229. Yeaman, M.R.; Yount, N.Y. Unifying themes in host defense effector polypeptides. *Nat. Rev. Microbiol.* **2007**, *5*, 727–740. [CrossRef] [PubMed]
230. Yount, N.Y.; Yeaman, M.R. Multidimensional signatures in antimicrobial peptides. *Proc. Natl. Acad. Sci. USA* **2004**, *101*, 7363–7368. [CrossRef] [PubMed]
231. Scudiero, O.; Galdiero, S.; Cantisani, M.; Di Noto, R.; Vitiello, M.; Galdiero, M.; Naclerio, G.; Cassiman, J.J.; Pedone, C.; Castaldo, G.; et al. Novel synthetic, salt-resistant analogs of human beta-defensins 1 and 3 endowed with enhanced antimicrobial activity. *Antimicrob. Agents Chemother.* **2010**, *54*, 2312–2322. [CrossRef] [PubMed]
232. Scudiero, O.; Galdiero, S.; Nigro, E.; Del Vecchio, L.; Di Noto, R.; Cantisani, M.; Colavita, I.; Galdiero, M.; Cassiman, J.J.; Daniele, A.; et al. Chimeric beta-defensin analogs, including the novel 3NI analog, display salt-resistant antimicrobial activity and lack toxicity in human epithelial cell lines. *Antimicrob. Agents Chemother.* **2013**, *57*, 1701–1708. [CrossRef] [PubMed]
233. Nigro, E.; Colavita, I.; Sarnataro, D.; Scudiero, O.; Zambrano, G.; Granata, V.; Daniele, A.; Carotenuto, A.; Galdiero, S.; Folliero, V.; et al. An ancestral host defense peptide within human beta-defensin 3 recapitulates the antibacterial and antiviral activity of the full-length molecule. *Sci. Rep.* **2015**, *5*, 18450. [CrossRef] [PubMed]
234. Scudiero, O.; Nigro, E.; Cantisani, M.; Colavita, I.; Leone, M.; Mercurio, F.A.; Galdiero, M.; Pessi, A.; Daniele, A.; Salvatore, F.; et al. Design and activity of a cyclic mini-beta-defensin analog: A novel antimicrobial tool. *Int. J. Nanomed.* **2015**, *10*, 6523–6539. [PubMed]
235. Romestand, B.; Molina, F.; Richard, V.; Roch, P.; Granier, C. Key role of the loop connecting the two beta strands of mussel defensin in its antimicrobial activity. *Eur. J. Biochem.* **2003**, *270*, 2805–2813. [CrossRef] [PubMed]

236. Mitta, G.; Vandenbulcke, F.; Hubert, F.; Salzet, M.; Roch, P. Involvement of mytilins in mussel antimicrobial defense. *J. Biol. Chem.* **2000**, *275*, 12954–12962. [CrossRef] [PubMed]
237. Mitta, G.; Hubert, F.; Dyrynda, E.A.; Boudry, P.; Roch, P. Mytilin Band MGD2, two antimicrobial peptides of marine mussels: Gene structure and expression analysis. *Dev. Comp. Immunol.* **2000**, *24*, 381–393. [CrossRef]
238. Balseiro, P.; Falco, A.; Romero, A.; Dios, S.; Martinez-Lopez, A.; Figueras, A.; Estepa, A.; Novoa, B. *Mytilus galloprovincialis* Myticin C: A chemotactic molecule with antiviral activity and immunoregulatory properties. *PLoS ONE* **2011**, *6*, e23140. [CrossRef] [PubMed]
239. Domeneghetti, S.; Franzoi, M.; Damiano, N.; Norante, R.; El Halfawy, N.M.; Mammi, S.; Marin, O.; Bellanda, M.; Venier, P. Structural and antimicrobial features of peptides related to Myticin C, a special defense molecule from the Mediterranean mussel *Mytilus galloprovincialis*. *J. Agric. Food Chem.* **2015**, *63*, 9251–9259. [CrossRef] [PubMed]
240. Novoa, B.; Romero, A.; Alvarez, A.L.; Moreira, R.; Pereiro, P.; Costa, M.M.; Dios, S.; Estepa, A.; Parra, F.; Figueras, A. Antiviral activity of Myticin C peptide from mussel: An ancient defense against herpesviruses. *J. Virol.* **2016**, *90*, 7692–7702. [CrossRef] [PubMed]
241. Liao, Z.; Wang, X.C.; Liu, H.H.; Fan, M.H.; Sun, J.J.; Shen, W. Molecular characterization of a novel antimicrobial peptide from *Mytilus coruscus*. *Fish Shellfish Immunol.* **2013**, *34*, 610–616. [CrossRef] [PubMed]
242. Qin, C.L.; Huang, W.; Zhou, S.Q.; Wang, X.C.; Liu, H.H.; Fan, M.H.; Wang, R.X.; Gao, P.; Liao, Z. Characterization of a novel antimicrobial peptide with chitin-biding domain from *Mytilus coruscus*. *Fish Shellfish Immunol.* **2014**, *41*, 362–370. [CrossRef] [PubMed]
243. Gerdol, M.; De Moro, G.; Manfrin, C.; Venier, P.; Pallavicini, A. Big defensins and mytimacins, new AMP families of the Mediterranean mussel *Mytilus galloprovincialis*. *Dev. Comp. Immunol.* **2012**, *36*, 390–399. [CrossRef] [PubMed]
244. Zhao, J.; Li, C.; Chen, A.; Li, L.; Su, X.; Li, T. Molecular characterization of a novel big defensin from clam *Venerupis philippinarum*. *PLoS ONE* **2010**, *5*, e13480. [CrossRef] [PubMed]
245. Cheng-Hua, L.; Jian-Min, Z.; Lin-Sheng, S. Molecular characterization and expression of a novel big defensin (Sb-BDef1) from ark shell, *Scapharca broughtonii*. *Fish Shellfish Immunol.* **2012**, *33*, 1167–1173.
246. Gestal, C.; Costa, M.; Figueras, A.; Novoa, B. Analysis of differentially expressed genes in response to bacterial stimulation in hemocytes of the carpet-shell clam *Ruditapes decussatus*: Identification of new antimicrobial peptides. *Gene* **2007**, *406*, 134–143. [CrossRef] [PubMed]
247. Seo, J.K.; Crawford, J.M.; Stone, K.L.; Noga, E.J. Purification of a novel arthropod defensin from the American oyster, *Crassostrea virginica*. *Biochem. Biophys. Res. Commun.* **2005**, *338*, 1998–2004. [CrossRef] [PubMed]
248. Gueguen, Y.; Herpin, A.; Aumelas, A.; Garnier, J.; Fievet, J.; Escoubas, J.M.; Bulet, P.; Gonzalez, M.; Lelong, C.; Favrel, P.; et al. Characterization of a defensin from the oyster *Crassostrea gigas*. Recombinant production, folding, solution structure, antimicrobial activities, and gene expression. *J. Biol. Chem.* **2006**, *281*, 313–323. [CrossRef] [PubMed]
249. Gonzalez, M.; Gueguen, Y.; Desserre, G.; de Lorgeril, J.; Romestand, B.; Bachere, E. Molecular characterization of two isoforms of defensin from hemocytes of the oyster *Crassostrea gigas*. *Dev. Comp. Immunol.* **2007**, *31*, 332–339. [CrossRef] [PubMed]
250. Li, C.; Zhao, J.-M.; Song, L.-S. A review of advances in research on marine molluscan antimicrobial peptides and their potential application in aquaculture. *Molluscan Res.* **2009**, *29*, 17–26.
251. Gueguen, Y.; Bernard, R.; Julie, F.; Paulina, S.; Delphine, D.G.; Franck, V.; Philippe, B.; Evelyne, B. Oyster hemocytes express a proline-rich peptide displaying synergistic antimicrobial activity with a defensin. *Mol. Immunol.* **2009**, *46*, 516–522. [CrossRef] [PubMed]
252. Schmitt, P.; de Lorgeril, J.; Gueguen, Y.; Destoumieux-Garzon, D.; Bachere, E. Expression, tissue localization and synergy of antimicrobial peptides and proteins in the immune response of the oyster *Crassostrea gigas*. *Dev. Comp. Immunol.* **2012**, *37*, 363–370. [CrossRef] [PubMed]
253. Zhao, J.; Song, L.; Li, C.; Ni, D.; Wu, L.; Zhu, L.; Wang, H.; Xu, W. Molecular cloning, expression of a big defensin gene from bay scallop *Argopecten irradians* and the antimicrobial activity of its recombinant protein. *Mol. Immunol.* **2007**, *44*, 360–368. [CrossRef] [PubMed]
254. Arenas, G.; Guzman, F.; Cardenas, C.; Mercado, L.; Marshall, S.H. A novel antifungal peptide designed from the primary structure of a natural antimicrobial peptide purified from *Argopecten purpuratus* hemocytes. *Peptides* **2009**, *30*, 1405–1411. [CrossRef] [PubMed]

255. Galdiero, S.; Falanga, A.; Berisio, R.; Grieco, P.; Morelli, G.; Galdiero, M. Antimicrobial peptides as an opportunity against bacterial diseases. *Curr. Med. Chem.* **2015**, *22*, 1665–1677. [CrossRef] [PubMed]

256. Cheung, R.C.; Ng, T.B.; Wong, J.H. Marine peptides: Bioactivities and applications. *Mar. Drugs* **2015**, *13*, 4006–4043. [CrossRef] [PubMed]

257. Sapkota, A.; Sapkota, A.R.; Kucharski, M.; Burke, J.; McKenzie, S.; Walker, P.; Lawrence, R. Aquaculture practices and potential human health risks: Current knowledge and future priorities. *Environ. Int.* **2008**, *34*, 1215–1226. [CrossRef] [PubMed]

258. Wain, L.V.; Shrine, N.; Miller, S.; Jackson, V.E.; Ntalla, I.; Soler Artigas, M.; Billington, C.K.; Kheirallah, A.K.; Allen, R.; Cook, J.P.; et al. Novel insights into the genetics of smoking behaviour, lung function, and chronic obstructive pulmonary disease (UK BiLEVE): A genetic association study in UK Biobank. *Lancet Respir. Med.* **2015**, *3*, 769–781. [CrossRef]

259. Barton, M.D.; Ndi, O.L. Can we feel it in our waters? Antimicrobials in aquaculture. *Med. J. Aust.* **2012**, *197*, 487. [CrossRef] [PubMed]

260. Cabello, F.C.; Godfrey, H.P.; Tomova, A.; Ivanova, L.; Dolz, H.; Millanao, A.; Buschmann, A.H. Antimicrobial use in aquaculture re-examined: Its relevance to antimicrobial resistance and to animal and human health. *Environ. Microbiol.* **2013**, *15*, 1917–1942. [CrossRef] [PubMed]

261. Aarestrup, F.M.; Agerso, Y.; Gerner-Smidt, P.; Madsen, M.; Jensen, L.B. Comparison of antimicrobial resistance phenotypes and resistance genes in *Enterococcus faecalis* and *Enterococcus faecium* from humans in the community, broilers, and pigs in denmark. *Diagn. Microbiol. Infect. Dis.* **2000**, *37*, 127–137. [CrossRef]

262. Nikaido, H. Multidrug resistance in bacteria. *Annu. Rev. Biochem.* **2009**, *78*, 119–146. [CrossRef] [PubMed]

263. Buschmann, A.H.; Tomova, A.; Lopez, A.; Maldonado, M.A.; Henriquez, L.A.; Ivanova, L.; Moy, F.; Godfrey, H.P.; Cabello, F.C. Salmon aquaculture and antimicrobial resistance in the marine environment. *PLoS ONE* **2012**, *7*, e42724. [CrossRef] [PubMed]

264. Avan, I.; Hall, C.D.; Katritzky, A.R. Peptidomimetics via modifications of amino acids and peptide bonds. *Chem. Soc. Rev.* **2014**, *43*, 3575–3594. [CrossRef] [PubMed]

265. Qvit, N.; Rubin, S.J.; Urban, T.J.; Mochly-Rosen, D.; Gross, E.R. Peptidomimetic therapeutics: Scientific approaches and opportunities. *Drug Discov. Today* **2017**, *22*, 454–462. [CrossRef] [PubMed]

266. Danial, M.; van Dulmen, T.H.; Aleksandrowicz, J.; Potgens, A.J.; Klok, H.A. Site-specific PEGylation of HR2 peptides: Effects of PEG conjugation position and chain length on HIV-1 membrane fusion inhibition and proteolytic degradation. *Bioconjug. Chem.* **2012**, *23*, 1648–1660. [CrossRef] [PubMed]

267. Papo, N.; Oren, Z.; Pag, U.; Sahl, H.G.; Shai, Y. The consequence of sequence alteration of an amphipathic alpha-helical antimicrobial peptide and its diastereomers. *J. Biol. Chem.* **2002**, *277*, 33913–33921. [CrossRef] [PubMed]

268. Mojsoska, B.; Jenssen, H. Peptides and peptidomimetics for antimicrobial drug design. *Pharmaceuticals (Basel)* **2015**, *8*, 366–415. [CrossRef] [PubMed]

269. Cantisani, M.; Finamore, E.; Mignogna, E.; Falanga, A.; Nicoletti, G.F.; Pedone, C.; Morelli, G.; Leone, M.; Galdiero, M.; Galdiero, S. Structural insights into and activity analysis of the antimicrobial peptide myxinidin. *Antimicrob. Agents Chemother.* **2014**, *58*, 5280–5290. [CrossRef] [PubMed]

270. Cantisani, M.; Leone, M.; Mignogna, E.; Kampanaraki, K.; Falanga, A.; Morelli, G.; Galdiero, M.; Galdiero, S. Structure-activity relations of myxinidin, an antibacterial peptide derived from the epidermal mucus of hagfish. *Antimicrob. Agents Chemother.* **2013**, *57*, 5665–5673. [CrossRef] [PubMed]

271. Kustanovich, I.; Shalev, D.E.; Mikhlin, M.; Gaidukov, L.; Mor, A. Structural requirements for potent versus selective cytotoxicity for antimicrobial dermaseptin S4 derivatives. *J. Biol. Chem.* **2002**, *277*, 16941–16951. [CrossRef] [PubMed]

272. Zelezetsky, I.; Pag, U.; Sahl, H.G.; Tossi, A. Tuning the biological properties of amphipathic alpha-helical antimicrobial peptides: Rational use of minimal amino acid substitutions. *Peptides* **2005**, *26*, 2368–2376. [CrossRef] [PubMed]

273. Wang, G. Post-translational modifications of natural antimicrobial peptides and strategies for peptide engineering. *Curr. Biotechnol.* **2012**, *1*, 72–79. [CrossRef] [PubMed]

274. Goldman, M.J.; Anderson, G.M.; Stolzenberg, E.D.; Kari, U.P.; Zasloff, M.; Wilson, J.M. Human beta-defensin-1 is a salt-sensitive antibiotic in lung that is inactivated in cystic fibrosis. *Cell* **1997**, *88*, 553–560. [CrossRef]

275. Fedders, H.; Michalek, M.; Grotzinger, J.; Leippe, M. An exceptional salt-tolerant antimicrobial peptide derived from a novel gene family of haemocytes of the marine invertebrate *Ciona intestinalis*. *Biochem. J.* **2008**, *416*, 65–75. [CrossRef] [PubMed]

MDPI AG

St. Alban-Anlage 66

4052 Basel, Switzerland

Tel. +41 61 683 77 34

Fax +41 61 302 89 18

http://www.mdpi.com

Marine Drugs Editorial Office

E-mail: marinedrugs@mdpi.com

http://www.mdpi.com/journal/marinedrugs

www.ingramcontent.com/pod-product-compliance
Lightning Source LLC
Chambersburg PA
CBHW051702210326
41597CB00032B/5345